Powertrain

More information about this series at http://www.springer.com/series/7569

Richard van Basshuysen
Editor

Natural Gas and Renewable Methane for Powertrains

Future Strategies for a Climate-Neutral Mobility

Michael Bargende and 55 Co-authors

Springer

Editor
Richard van Basshuysen
Bad Wimpfen, Germany

Translation from the German language edition:
Erdgas und erneuerbares Methan für den Fahrzeugantrieb - Wege zur klimaneutralen Mobilität by Richard van Basshuysen

ISSN 1613-6349
Powertrain
ISBN 978-3-319-23224-9 ISBN 978-3-319-23225-6 (eBook)
DOI 10.1007/978-3-319-23225-6

Library of Congress Control Number: 2016931709

Springer Cham Heidelberg New York Dordrecht London

Printed on acid-free paper

Springer International Publishing AG Switzerland is part of Springer Science+Business Media (www.springer.com)

Preface

One day they will come to grief
(Unknown origin)

Humanity's energy needs are continuing to grow. However, fossil fuel resources are limited. Economy and ecology are engaged in a clinch. The search for alternatives has therefore begun.

Although power is available from solar energy, wind and water, biological raw materials and geothermal energy, it will take decades of research and development together with considerable social and political changes to become independent on fossil fuels. To help us bridge the gap, natural gas, which consists of a high percentage of methane, is available.

Natural gas is second only to coal as the most prevalent fossil energy carrier and has the best H/C ratio for energy exploitation and thus has the lowest CO_2 burden for the atmosphere when combusted. It is also relatively low-cost on the global market. Above all, it can be produced from a variety of biological and synthetic means from sustainable sources, something that is being increasingly practiced today, and together with the utilization of existing infrastructure facilitates the previously noted transition to a non-fossil energy system.

If in the future more and more regenerative energy carriers in the form of chemically stored energy will be available—for example as gaseous or liquid hydrocarbons, there will be no necessity to replace the combustion engine with the electric motor, since the combustion is then CO_2 neutral—which is not the case with the electric motor—and its energy consumption (well-to-wheel) is identical to or lower than that of an electric motor, depending on the electricity mix.

In contrast to the electric motor and with an appropriate design and exhaust aftertreatment, it is even able to reduce the pollution in megacities (sub-zero emission vehicle). This secures "eternal life" for the piston engine. Policy-makers can be criticized for the one-sided sponsoring of electro-mobility despite this insight. Instead of this, lawmakers should only determine limits. Only scientific and engineering (R&D) solutions can show the technological way forwards due to the complex interdependencies.

I was inspired by the automotive manufacturer Audi to develop this book. Audi, who not only produce natural gas vehicles, are also the first automotive manufacturer worldwide to offer their customers artificially produced methane from their own "power-to-gas" plant.

More than 50 authors and co-authors from the scientific community, from industry and from politics have participated in this work. Redundancy was thus not always avoidable, but has been preserved to render the individual chapters easier to understand for the readers, since there is no necessity to cross-reference with other chapters. There are also several places where inconsistencies occur, for example the question of what percentage of hydrogen is acceptable in the natural gas pipeline network. Such cases require further research and development work in order to obtain definitive answers.

The book is aimed primarily at product developers for natural gas and methane applications, and at production management in the automotive and supplier industry. It is intended to be an important source of advice for science, research, teaching and politics. It is designed to fill a gap in the book market.

I would like to thank the highly competent authors from the automotive industry, universities, scientific institutes, associations and politics. I am grateful to Professor Michael Bargende from the University of Stuttgart, Dr Michael Specht from the Centre for Solar Energy and Hydrogen Research (ZSW), Reiner Mangold and Reinhard Otten from Audi and Professor Helmut Eichsleder from the Technical University of Graz, who contributed the lion's share of the content of this book. Almost 300 diagrams and graphics illustrate the content and 572 references to external literature provide the opportunity for further study.

Thank you also to AVL, who helped with the design of the book's content and who made the publication of this book possible in the series "The Vehicle Powertrain" whose editor is Helmut List. Thanks also to Gisela Großmann from the University of Stuttgart for the comprehensive support.

Finally, I wish this work its deserved attention.

Bad Wimpfen, Germany
February 2016

Richard van Basshuysen

Contents

About the Editor

Dr. Richard van Basshuysen has been considered the father of the passenger car diesel engine with direct fuel injection since his time at Audi. As head of development for luxury class vehicles and engine and transmission development, he was responsible for the first detoxified exhaust, turbocharged diesel engine with direct fuel injection for passenger cars. He thus revolutionized the diesel engine throughout the world in the direction of drastically improved fuel consumption with simultaneous increases in power and torque.

His many years of expert scientific work as an author and publisher resulted in a reputation as a visionary thinker in the domain of powertrain technology with combustion engines.

His work over many years up to the present day has linked him with publishing houses Franckh-Kosmos, Vieweg, Springer, SAE International Text Books und China Machine Press. He received the VDI "Benz-Daimler-Maybach-Honorary Medal 2001" for the series implementation of the passenger car diesel engine with direct fuel injection and the highly prestigious Ernst-Blickle-Preis 2000. He was also awarded an honorary doctorate by the University of Magdeburg for his life's work.

This book project demonstrates once again his instinct for climate-neutral mobility and proves that low emission and low fuel consumption mobility is already possible with today's series technology and can provide a significant contribution to the sustainability of mobility.

Forewords

1

Rainer Bomba, Ulrich Eichhorn, and Jakob Seiler

1.1 Foreword by the German Federal Ministry of Transport and Digital Infrastructure

Rainer Bomba

As fuels, natural gas (CNG) and renewable methane (biomethane) can make a major contribution towards achieving the Federal Government's energy and climate change objectives. In a first step, we want to reduce final energy consumption in the transport sector by 10 % by 2020, and we want to do this with the aid of natural gas, among others.

Natural gas mobility has an important role to play here. It can help to increase energy efficiency, enhance the security of supply and reduce noise emissions. This is one of the key messages of the Federal Government's Mobility and Fuel Strategy.

For this reason, the Federal Government supports the commercialization of natural gas and biomethane in Germany. We are promoting natural gas as an alternative fuel by imposing a lower rate of fuel tax, as set out in the Energy Tax Act. This makes natural gas competitive with conventional fuels (diesel/gasoline) in terms of total costs. The Coalition Agreement states that this tax privilege will be continued beyond 2018 in order to give consumers the planning certainty they need.

However, whole-life value for money is only one aspect of consumer acceptance, which ultimately results in them deciding to purchase a gas-powered vehicle. The

R. Bomba (✉)
German Federal Ministry of Transport and Digital Infrastructure, Berlin, Germany
e-mail: al-g@bmvi.bund.de

U. Eichhorn (✉) • J. Seiler (✉)
VDA Berlin, Germany
e-mail: seiler@vda.de; eichhorn@vda.de

R. van Basshuysen (ed.), *Natural Gas and Renewable Methane for Powertrains*, Powertrain, DOI 10.1007/978-3-319-23225-6_1

Mobility and Fuel Strategy has developed further recommendations for action: improvement of consumer information and user friendliness, expansion of the vehicle market, EU-wide harmonization and the targeted installation of the refuelling infrastructure.

As the Mobility and Fuel Strategy evolves, it will assume the character of a "learning strategy", and we will conduct further studies into the potential inherent in gaseous alternative fuels as we progress along the path towards a more sustainable supply of transport energy. Thus, for instance, hydrogen produced from renewable electricity could be processed into methane and used in vehicles. This "Power-to-Gas" alternative is another interesting fuel option.

We are linking this initiative to the Clean Power for Transport Directive, which has been adopted by the European Union. This directive establishes the framework for the activities of the years ahead at a European level. One of its main objectives is transparent price labelling, which is designed to enable consumers to identify at a glance the economic advantages of both alternative and conventional fuels.

The directive also contains important requirements regarding the infrastructure for clean fuels and common EU-wide standards for the necessary equipment. Germany has already met the objective set out in the directive of providing an adequate number of natural gas filling stations on the core European transport network by 2025 at the latest.

But this does not mean that there is no room for further improvement. The road immediately ahead is paved with further challenges, and the Mobility and Fuel Strategy urges us to tackle them. All the stakeholders—from government, industry and society—have to stand shoulder to shoulder. The main aim must be to use natural gas and biomethane, among others, to diversify the supply of energy, thereby reducing our dependence on imports.

Prudent policymakers establish the framework for this but do not replace market forces. This must continue to be the responsibility of vehicle manufacturers and petroleum companies. However, the Federal Government is not just a spectator, but sets the future course by establishing a suitable regulatory framework.

This also includes not putting all our eggs in one basket. There is no silver bullet. We must consider all options when it comes to diversifying the energy supply and introducing alternative drivetrain technologies. Intensifying the use of gaseous fuels in the transport sector is definitely part of this.

I have no doubt that the present manual will provide us with some very interesting ideas.

1.2 Foreword by the German Association of the Automotive Industry (VDA)

Ulrich Eichhorn and Jakob Seiler

Today, mobility is a fundamental requirement for growth, prosperity and quality of life in Europe, with road passenger transportation and road freight transportation forming the key pillars of individual mobility. This places a special responsibility on road transport and on our industry, which indeed we take seriously.

Our vehicles are becoming more efficient all the time and fuel consumption continues to fall. In the first half of 2014, CO_2 emissions from all newly registered passenger cars in Germany decreased to an average of only 132.4 g/km. That is the result of many efficiency-enhancing measures implemented in our vehicles, and not least the continual optimisation of our drivetrains and engines. German car manufacturers currently have 112 models on offer with CO_2 emissions below 100 g/km.

Following far-reaching improvements in classical internal combustion engines and the extensive exploitation of classical potential, attention is now turning to alternative CO_2 reduction technologies. The most significant of these are vehicle electrification and—increasingly—alternative, low-CO_2 fuels. Natural gas plays a major part in the German automotive industry's CO_2 reduction strategy. Gas propulsion scores more than 20 % better than gasoline on CO_2 output, and can reduce CO_2 by up to 10 % compared with diesel fuel. This gives natural gas a hugely valuable CO_2 advantage in view of Europe's extremely ambitious fleet target of 95 g CO_2.

Most of this CO_2 saving is due to the high hydrogen-to-carbon ratio. But natural gas has the added benefit of being a very homogeneous, clean and knock-resistant fuel. This means that an engine optimized for this fuel has energy-saving potentials of several per cent, resulting mainly from higher possible compression ratios. This advantage is apparent in the products from the German automotive industry. A gas-powered vehicle with 110 hp in the compact class has a CO_2 output of 88 g/km, and a family estate car with 150 hp has a CO_2 output of 117 g/km. Both of these are in efficiency class A!

Natural gas also offers advantages when it comes to pollutant emissions. Although Euro 6 already marks the point where today's passenger cars—from gasoline and gas-powered cars all the way to diesels—are generally classified as clean, natural gas engines need far less complex exhaust aftertreatment than diesels. With stoichiometric combustion, a gas engine at operating temperature practically cleans the air. Furthermore, gas engines will continue to satisfy exhaust legislation in the future.

The global reserves of natural gas and the global availability of natural gas significantly exceed the liquid fossil reserves. Therefore natural gas is also an important part of the ever-broader diversification in energy supplying the transport sector. In Europe, natural gas is obtained from a wide range of sources. For example, we have reserves on the European mainland and in the North Sea. In addition to the Eastern European pipelines, a large proportion of European natural gas comes from the Mediterranean region. However, a growing role is played by major European ports where natural gas arrives on large LNG ships. LNG provides a firm element in supply, alongside the established gas pipelines.

As LNG becomes more widely available, it will present an increasingly important alternative to diesel power in classical, heavy long-distance freight transport. Heavy road freight suffers from a lack of alternatives to diesel fuel, but LNG is an exception: the LNG engine is ready for series production, and LNG propulsion is both efficient and suitable for long-distance application. The engergy density of LNG is sufficiently high for long-haul transport. Filling up is simple and fast, and therefore does not represent a technical hurdle. The expansion of the LNG infrastructure, as in the EU's "Blue Corridors" project for instance, brings LNG-based mobility within reach. Indeed, gas-powered freight transport is already a reality close to the coast and the large European ports where there is already an LNG supply.

Liquefied natural gas is competitive not least on price. The world-wide extraction of natural gas, the world-wide expansion of the LNG supply, and the large production and liquefaction capacities have left the price of LNG less and less closely tied to the price of crude oil. So also in economic terms LNG is an appealing alternative to fossil fuels. The attractive prices provide the best basis for developing mobility with natural gas, because the investment by individual customers in LNG-based mobility must pay off, in particular while the system is being established.

At present, LNG enjoys reduced rates of taxation in many parts of the world. This is the case in Germany, too—at least until 2018. In its coalition agreement, the German Federal Government committed to extending tax relief for LNG; however, this has yet to be implemented.

Moreover, it is relatively easy to find sustainable substitute components for LNG. "Biomethane", for example, has a very small CO_2 footprint because of its biogenic production. It can be manufactured via classical biogas production with qualitative after treatment of the biogas generated. In terms of CO_2 and efficient land use, biogas is the best biofuel currently available in large quantities. The German Energy Agency (dena) states that biomethane can potentially reduce CO_2 emissions by over 90 %. Today it makes up around 20 % of natural gas fuel sold at the pumps. Irrespective of the production pathway—fermentation or synthesis—methane (CH_4) is always the product, and it can be added to the CNG network in any proportion. Therefore renewable methane is the ideal drop-in fuel.

A new, pioneering method is to manufacture biomethane using electrical power. Electricity generated from renewable sources—preferably excess electricity—is used to produce "green hydrogen" which is then methanized in a synthesis process. A production plant belonging to the firm Audi demonstrates that this technology can be operated efficiently on a large scale.

Natural gas is a new fuel for vehicle propulsion, yet the automotive manufacturers already have a wide range of models on offer, from small cars to the upper medium segment, and from family vans to heavy-duty commercial vehicles. Despite this relatively broad and attractive portfolio, gas-powered vehicles today still only account for a small proportion of the total vehicle fleet, although the figures are rising. In the first quarter of 2014 a total of 2174 natural gas vehicles were newly registered in Germany (up by 20 %). Natural gas is now available at more than 920 filling stations across the country.

CNG is a topic that will grow in the future. Here, too, developments are moving towards direct-injection and dual-fuel operation. New biofuel production processes should continue to be developed and expanded, which can be integrated into a new renewable energy supply system. This means that methane will also gain importance in the future as a stable source and carrier of energy. In our highly innovative country we want to utilise and build on this opportunity.

We hope you enjoy reading this exciting book.

Climate Neutral Mobility: Natural Gas and Methane as Part of the Solution

2

Kirsten Adlunger, Katrin Dziekan, Martin Lange, and Lars Mönch

2.1 Introduction

Mobility is an important prerequisite for economic exchange and individual prosperity and good transportation systems fulfil the mobility needs of the population. Not only has the passenger car assumed a dominant role within the domain of private transport, road transport is the mainstay of freight traffic. Motorized traffic, as it exists today, has many negative effects on the environment and also on people: greenhouse gas emissions, air pollution, noise, land consumption and the use of resources for the creation of infrastructure, the vehicles themselves and the fuels (see also [1]). The focus of this contribution lies on the greenhouse gases caused by motorized traffic.

This book considers natural gas as a current and future source of energy for the transportation sector. This prompts us to describe additional options and scenarios in an introductory section that are necessary if one is to reform transportation—and in particular motorized traffic—to be more climate neutral. After the introductory section and the problem statement from the point of view of climate protection in Sect. 2.2, Sect. 2.3 provides a short overview of possible non-technical measures, followed by the technical measures for energy supply in the transportation sector in Sect. 2.4. An investigation into the existing infrastructure in Sect. 2.5 will reveal that natural gas can be one step on the road towards greenhouse gas neutral transportation and that it may be worthwhile to reflect on technologies particularly in conjunction with renewable methane generated from electricity.

K. Adlunger • K. Dziekan • M. Lange • L. Mönch (✉)
Dessau/Berlin, Germany
e-mail: lars.moench@uba.de

R. van Basshuysen (ed.), *Natural Gas and Renewable Methane for Powertrains*, Powertrain, DOI 10.1007/978-3-319-23225-6_2

What Is Driving Growth in the Transportation Sector and the Problem of Greenhouse Gas Emission?
Over the last century, transport performance (measured in person-km or ton-km) has risen drastically, whereby the increase over the last 20 years has been over-proportional to economic growth. The most important reasons for this were and still are on the one hand the relatively favorable price development of fuel in all transport domains and the associated widespread increasing amount of traffic per manufactured unit of the gross domestic product (GDP), and on the other the integration of Germany into the growing European Union (EU). Diagram 2.1 shows a comparison of the development of GDP and traffic volume over time in Germany. All highly developed countries experienced identical development trends.

There has been a consensus in the scientific world for the past 20 years that current climatic changes are mainly linked to human activities and the associated emission of greenhouse gases (GHG) [5]. This awareness has also reached international politics and is concentrated in, for example, the Kyoto process and other measures in order to limit the growth of or to obtain a reduction of climate gas emissions. There are fundamental initiatives that primarily affect areas other than transportation, for example emissions trading systems in the stationary industrial sector. In parallel, there is an almost shocking development in the transportation sector on a global scale. Diagram 2.2 shows scenarios of the development of the transport sector in Germany and the associated climate gas emissions. The amount of traffic is increasing: there is a simultaneous albeit dampened growth in climate gas emissions that decreases in the long-term. Compared to other sources of GHG emissions however, traffic itself has shown substantial growth since 1990, although there has been an increase in technical efficiency and the energy required

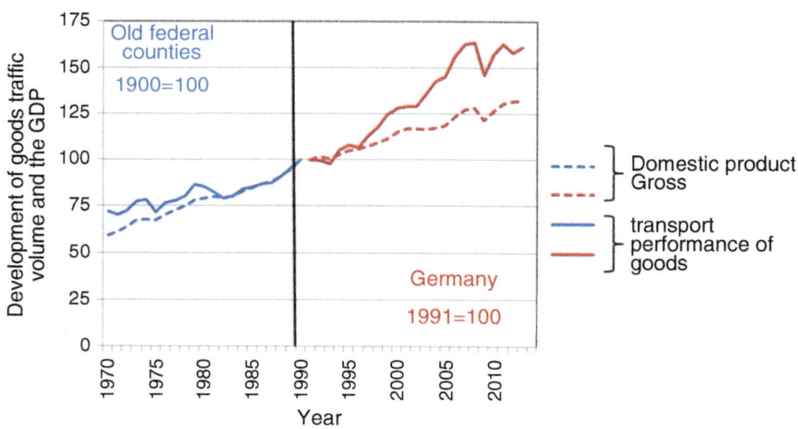

Diagram 2.1 Development of the transport performance of goods and the gross domestic product (adjusted for seasonal and calendar effects) in Germany. Data from economic statistics [2] and traffic in numbers [3, 4]

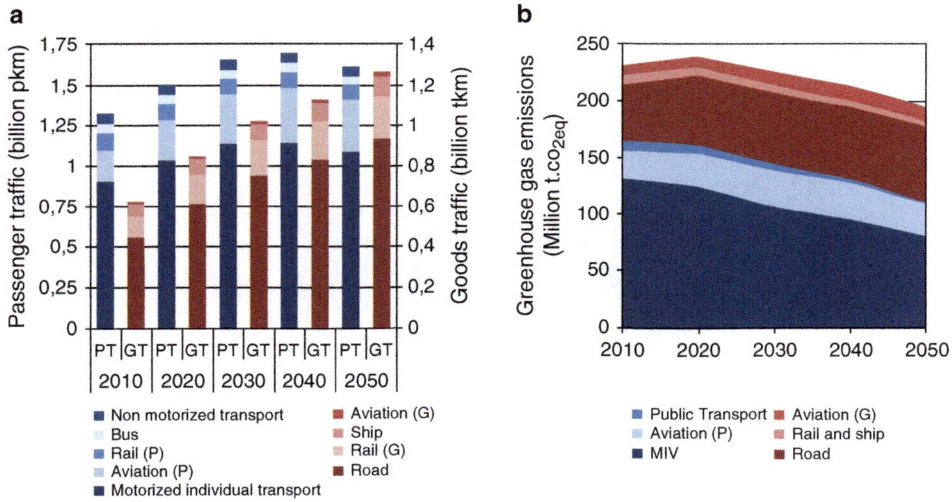

Diagram 2.2 (**a**) Development of person and goods transport performance in Germany and (**b**) the corresponding transport-related GHG emissions according to [6]

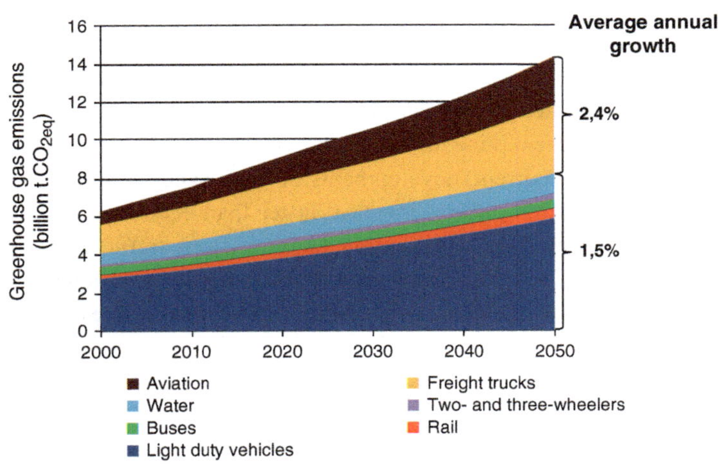

Diagram 2.3 Development of greenhouse gases (incl. upstream emissions) caused by global transport, split according to different means of transport with growth rates [7]

to fulfil one unit of the transport performance has fallen. Diagram 2.3 and the underlying study forecasts a further increase of the transport performance on a global scale.

There are two main drivers that can be differenciated: On the one hand, there is a global increase in long-haul goods transportation and air traffic; on the other hand, the Asian states are in the throes of rapid economic growth (see Diagram 2.4).

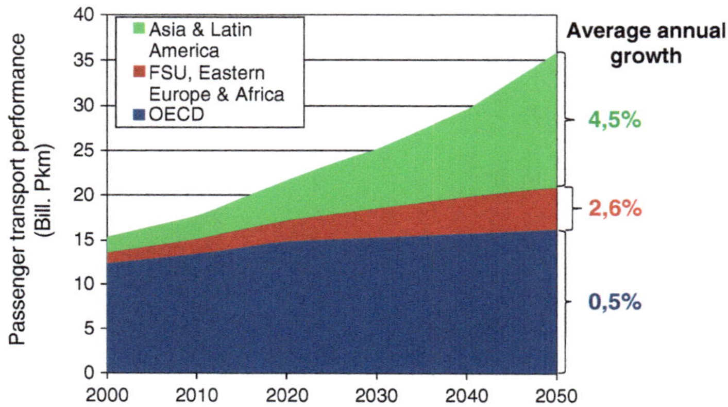

Diagram 2.4 Development of passenger transport performance by light duty vehicles are passenger cars and light duty commercial vehicles light-duty vehicles in various regions with average growth rate. Data from [7]

Even if the focus is on climate gas emissions, it is important to point out the other consequences of this development. A major deterioration of air quality is associated with the development of the Asian region, which generally speaking is mainly due to the burning of fossil primary fuels, a large part of which originates from traffic.

The diagrams show that considerable effort is required in the transport sector such that transportation can contribute anything at all to climate protection. All signs indicate that there will be no major decoupling of GDP and the growth in transport, not even a dampening effect in the current development. A first success would be if energy consumption in transportation did not increase faster than economic growth. It is also clear that the climate protection contribution required from traffic is not possible by concentrating on a single task. It is inevitable that the focus must actively shift to the avoidance of traffic, to modal shifts on more environmental compatible means of transport and to the environmental compatible handling of the remaining transport. Both of the first two points are necessary on a global scale, but are probably very difficult to implement, even in highly developed industrial nations. Even the technical issue of increasing efficiency alone will not be sufficient to make the necessary contribution.

Additional fundamental and globally compatible solutions must be found. The use renewable energy in transportation seems obvious, although closer inspection of various options quickly reveals that the possibilities are limited. The use of cultivated biomass for example has been closely examined by the German Environment Agency [8]. The conclusion was that it does not represent a long-term solution. The reasons lie in the domain of associated social issues, for example "tank versus plate", environmental damage such as a drastic increase in water consumption, intensive fertilization and the negative consequences for the soil, and the insufficiently positive climatic impact. For these reasons, current discussions are focussed on the direct or indirect use of regenerative and practically unlimited electrical energy [6, 9]. A global approach is required since the subject of energy supply for transportation should not be evaluated on a limited regional basis.

2.2 Contribution of Transport to Climate Change or Alternatively, the Non-contribution of Transport to Climate Protection

Currently, transport contributes significantly to global greenhouse gas emissions and hence climate change due to the burning of fossil fuels by the various modes of transportation. Since 1990, first attempts have been made to limit climatic change to a value of 2° or below. Over the same period of time, the contribution of emissions by traffic has increased disproportionately from 12.9 to 14.5 % [10].

Climate change is an environmental problem with non-local effects, i.e. emissions from the whole world contribute to a global effect, without a correlation between local effects and local greenhouse gas emission. For this reason, climate protection must be driven on an international level, whereby nation states and groups of nations have not been able to agree to date on how to distribute the burdens of climate protection. The UN's Intergovernmental Panel of Climate Change (IPCC) recommended a greenhouse gas reduction target corridor for the year 2050 for differently developed states [11] in its fourth status report (2007). The fifth status report (2014) no longer contains this suggestion—it now derives only globally acting reduction targets for the 2-degree target [10]. IPCC scenarios in which the 2-degree target is complied with in the twenty-first century are characterized by a reduction in emissions of 40–70 % by 2050 compared to 2010 and emissions being practically zero in 2100. If transport were to contribute to this reduction in accordance with the size of its emissions share, then greenhouse emissions in 2050 would only lie between 2 and 4 billion $tCO_{2,eq}$. This range is contrasted in Diagram 2.4 with the predicted emissions according to WBCSD (World Business Council for Sustainable Development). The WBCSD scenarios only contain, however, moderate improvements in efficiency, low avoidance of transport and transport modal shift measures.

The discrepancy between the forecast emissions—as shown here with the WBCSD example—and the "permissible" emissions from the point of view of climate protection show how important it is to make substantial reductions in greenhouse gas emissions. This can be achieved through increased avoidance of traffic and modal shifts, increases in efficiency of transportation means and the use of fuels with low specific CO_2 emissions per useful energy. The short-term replacement of gasoline by natural gas is conceivable for the second aspect, whereas in the medium and long term, it appears expedient to employ electro-mobility and fuels generated using electricity from an environmental point of view. Low volume potential could also be covered with alternative fuels based on wood and straw residues (Diagram 2.5).

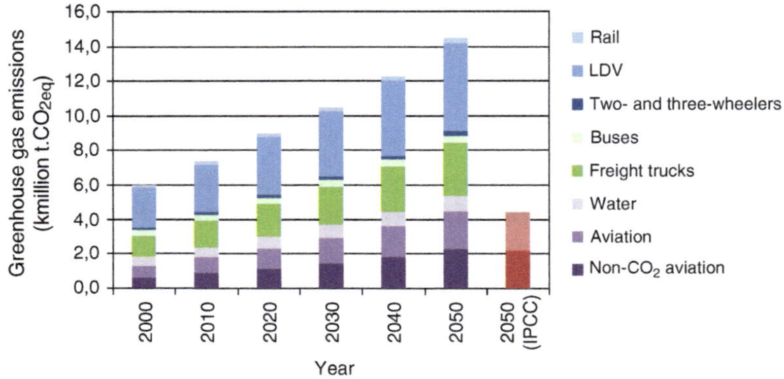

Diagram 2.5 Comparison of direct greenhouse gas emissions for a reduction target of between 40 and 70 % for the year 2050 (based on [10]). The modes of transport are clustered and colored according to their potential for electrification. The non-CO_2 effects of aviation have been taken into account. Data from [7]

It is conceivable that a part of the transportat could be largely or completely electrified by the year 2050 (for example passenger cars and two-wheelers) and hence no longer directly emit greenhouse gases and, if supplied with electricity from renewable sources, only have very low upstream emissions. For others, for example long-distance buses or long-haul trucks, this is principally technically possible, however it is not clear to what degree electrification is really possible. It does not appear possible that international shipping and aviation will have a significant proportion of electrification by 2050. Further discussion on the options for energy supply for transportation will be covered in Sect. 2.4.

Aviation is proving to be the problem child due to the direct CO_2 climatic effects, the additional non-CO_2 effects that cannot be avoided even with greenhouse gas-neutral fuels, the strongly limites potential of electrification and forecasts of huge growth rates of 5 % per year. In 2050, the non-CO_2 effects of aviation emissions will already correspond to the "permissible" greenhouse gas emissions based on a reduction of 70 % compared to 2010 in order to keep global warming below 2°C.

Another mode of transport with strongly increasing greenhouse emissions is road-based long-haul goods transport, which could however be electrified in various ways, e.g. overhead lines, power rails or inductive systems.

Currently, it is not clear whether sufficiently high traffic densities can be achieved on a global level that would make widespread electrification economical (for example in central Europe or North America, densities are possibly higher). Such an undertaking would also require a huge effort in standardization in order to develop international systems. Alternative, low-greenhouse gas fuels such as methane or hydrogen produced from renewable electricity together with liquid fuels produced from renewable electricity can contribute to a low-GHG supply of such modes of transportation. The emission of nitrogen oxides and particulates/soot by long haul transport can be reduced via the currently adopted technical solutions of exhaust aftertreatment and filter systems, even though these systems are sometimes very complex. The target for the future must be to guarantee the endurance and performance of these systems their service in the field to a high degree.

It may seem appropriate at this point to enter into deeper discussions concerning the technical issues concerning fuels. However, since there are numerous non-technical measures that can be employed to significantly reduce greenhouse gas emissions (and also to reduce many other negative impacts of transport), these will be discussed first.

2.3 Non-technical Measures for Climate-Friendly Transport

A catalogue of measures is needed to achieve the ambitious greenhouse gas reduction targets, since single measures do not have the required effect and only an integrated approach can be effective, particularly in the transport sector. This section will deal with so-called non-technical measures as examples of what can contribute to climate-friendly transport. In this section the examples describe the situation in Germany, if not stated differently.

Economic Measures

Economic incentives can be created with appropriately structured energy taxes, amongst other things. Currently, different energy taxes lead to a distortion of competition. For example, there is no taxation of kerosene in aviation, whereas railway companies are burdened with taxes on their electricity by the German Renewable Energy Act (EEG). A further hidden subsidy is the tax on diesel fuel. The energy tax advantage of approximately 18 cents per litre of diesel over gasoline is not justified from a climate-political point of view. The German Environment Agency therefore recommends a gradual alignment and the levying of energy taxes on the basis of fuel carbon content [12]. A further economic measure would be the elimination of the company car privilege and a differentiation of tax according to CO_2 emissions. Company cars represent a large part of the fleet in Germany.

"Company cars are vehicles that can be used by the employee for private use. This private use is subject to a monthly income tax of 1 % of the vehicle list price upon first registration as monetary value. [...] Company cars are normally larger vehicles with above-average fuel consumption. The company car privilege also pushes the passenger car as a mode of transport and contributes to the environmental pollution caused by road traffic. [...]"[13].

Allocating the true costs of traffic is a further point for discussion. For example, since 2012, external costs for air pollution and noise can be added to road tolls, according to the EU guideline 1999/62/EG, however only up to a relatively low-capped value. As stated by the current road cost report from the Federal Ministry of Transport and Digital Infrastructure, the true external costs are much higher [14]. Further external costs caused by traffic, for example land fragmentation or traffic accidents, are not allowed to be considered so far.

The increase and extension of the truck toll scheme could be an appropriate step to burden goods transportation with a part of the external costs caused by this sector and to participate them in infrastructure maintenance. According to the study "Renewability II", truck tolls should be extended to all truck categories upwards from a total permitted weight of 3.5 tons for all road categories and gradually increased [15].

The instrument of car tolls could also be implemented to transfer costs to the passenger car sector and simultaneously make citycenters more accessible and liveable. London and Stockholm have good experience with this model and passenger car traffic has been reduced by approximately a fifth. In general, from an environmental point of view, only kilometre-related passenger car tolls are recommendable, as they can have a steering effect [16].

Estate Structure

The structure of an estate creates transport performance: planning mistakes create traffic. Estate, production and infrastructure in the past few years have developed in such a manner that distances between the start and end of journeys have increased. At the same time, there was a focus on mono-functional estates, for example shopping centers on greenfields or suburbs on the edge of large cities, while inner city sites and brownfields remain unused.

"The elimination of the commuting allowance would be an important step against the trend to suburbanization and to promote more traffic-free estate structures. Employees can write off the commuting allowance for journey to the workplace against income tax, currently 30 cents per kilometre distance between place of abode and the workplace. This reduces the tax burden as soon as the annual professional expenses allowance is exceeded (currently 1000 €; 2010: 920 €). Such a comparable tax relief does not exist in the majority of EU countries. The commuting allowance supports the growth in transport performance, the trend towards long journeys to work and the urban sprawl in the countryside. Above all, it favors passenger car traffic, since the availability of public transport particularly in places with low population density is very limited and is therefore not an alternative for many employees. The commuting allowance works against climate protection and contributes to air pollution and noise. The use of land as a result of urban sprawl is an important reason for the loss of bio-diversity and has further negative environmental effects. [...]" [13] The elimination of the commuting allowance could reduce CO_2 emissions by 1.8 million tons by 2015 [17]. The recommendation is thus to eliminate the commuting allowance and to permit the journey costs to be recognized as tax-relief in the form of exceptional burden to support those in dire need. According to the latest calculations by the Ministry of Finance, the state could increase tax income by approximately five billion euros.

The implementation of the planning concept "City of Short Distances" is a further measure that can help reduce greenhouse gas emissions originating from traffic. Federal states, counties and municipalities should follow a twofold strategy: (1) Prioritized use of existing space in estate inventory instead of previously undeveloped areas in exterior areas or city borders (inner development), (2) Simultaneous urban developmental by the upgrading of residential areas concerning the quality of the environment, so that residing and living in a city is attractive. This includes new forms of residential layout (for example modern townhouses, car-free estates, communal housing) that can compete with detached single-family houses in suburban countryside. The main areas for action concerning the control of residential areas are city planning, building and planning regulations, and instruments of estate-related financial politics. Regional planning must contain and limit the growth of estates on the outskirts of cities. In regions where further estate development is necessary due to continuing population expansion, development must be concentrated in focal points of estate development [12]. Such a trial in space trading is currently taking place for German communities (www.flaechenhandel.de).

Regulatory and Planning Measures

Lower speed equals less energy consumption—this is a well-known physical law. Reducing the speed therefore means directly saving energy. According to the German Environment Agency, a speed limit of 120 km/h on German motorways would save 2.9 $MtCO_2$ in 2020 and 3.2 $MtCO_2$ in 2030 [12]. Even though these calculations are based on numbers from the 1990s and require updating, it can be said that a speed limit would be a step towards effective energy saving. Lower permissible top speeds would also mean that vehicles could be constructed lighter, since the safety requirements and design of the corresponding vehicle components also depend on the expected collision speeds. The project Renewability II calculated that a speed limit of 120 km/h on German motorways would reduce the average fuel consumption by 6.5 % [15]. Apart from better quality of life and a reduced number of accidents in cities, a blanket speed limit of 30 km/h in inner cities would also offer the potential to reduce CO_2 [18].

The challenge to reduce greenhouse gases in transport requires support from all political levels and also cities and communities. The setting of concrete targets (such as Stockholm's goal of being greenhouse gas neutral by 2030) by those responsible gives them a development path and hence steps to take. A suitable instrument is a traffic development plan [19], or also "Sustainable Urban Mobility Plans" [20], as they are called in a European context. The participative creation of such traffic development plans with the goal of reducing greenhouse gases can unfold a long-term saving effect, since it is a consensual and integrated plan with a high probability of implementation in the communities.

Measures Targeted on Behavioral Change

Behavioral change begins in the mind and is therefore only achievable through a paradigm shift. Nevertheless, steps for behavioral change offer the possibility of being sustainably and permanently effective with relatively low input of financial means. Behavioral changes are desirable so that people avoid unnecessary car journeys for example, by walking short distances or using a bicycle if possible, or public transport.

The promotion of active mobility (walking, bicycling, public transport) is therefore recommendable, since multi-modality is the key to sustainable mobility: The best combination of transport media is used to satisfy the mobility needs. On one day, a person could cycle to work, on another day travel by train and walk to the station, for example if it is raining. Car sharing and bicycle sharing systems offer good opportunities to do justice to the slogan "Use instead of Own" [21].

Good communication work and campaigns such as "Mind On, Engine Off", new resident marketing, residential location advice or campaigns such as "On Your Bike" or "City Biking" can help to promote behavioral change.

Ecodriving, i.e. driving in manner that saves fuel, can reduce fuel consumption in passenger cars by 7 % and in trucks by up to 8 % [15].

Finally, there is huge potential offered by mobility management to change traffic behavior. Although the effects of CO_2 saving are difficult to verify since it is a bundle of initiatives, it can be assumed that the effect of job tickets, mobility advice and communication campaigns is positive—as demonstrated by projects such as the German promotion *effizient mobil* (efficiently mobile) or the Austrian *klima.aktiv* (active climate).

Infrastructure Measures

It is urgently recommended to prioritise the preservation of existing infrastructure over new road construction—on the one hand because the existing budgets do not permit new construction, and on the other because it is well-known that new roads inexorably lead to more traffic and thus directly to an increase in CO_2 emissions [12]. The support of combined traffic and optimization of logistics promises positive effects [15]. The desired shift towards the environmentally friendlier rail transport can only be achieved with a focussed expansion and upgrading of the German rail network [22].

A further appropriate step is the expansion of public transport and making it more attractive.

The economical viability of so-called non-technical measures in the transport sector has been proved for individual bundles of activities [23].

The complexity of the total system must always be considered. A key aspect is integrated traffic planning. One study [20] showed, as an example, how the promotion of bicycle routes as a system could save 40 $MtCO_2$ per year.

2.4 Transport's Energy Supply: Sustainable, Climate-Friendly Fuels for Transport

Up to now, traffic is practically completely based upon mineral oil products: In 2010 the share of fossil oil in the energy required for traffic in the EU was 94 %. It is important to promote low CO_2 alternatives to de-carbonise transport. In this context, the EU underlined gaseous fossil fuels as a part of the proposal package "Clean Energy in Transport" in 2013. The use of Liquefied Petroleum Gas, or LPG is already relatively widespread in the EU. It has a share of 3 % of fuels and is used by nine million vehicles. Liquefied Natural Gas, LNG and Compressed Natural Gas, CNG play an increasingly important role in the gas market for vehicles.

The advantages over oil are the lower greenhouse gas emissions per heat unit. The German energy agency (dena) states that the life-cycle GHG emissions for natural gas are 124 $gCO_{2,eq}$/km, whereas it is 164 $gCO_{2,eq}$/km for gasoline and 156 $gCO_{2,eq}$/km for diesel fuel [24]. Compared to gasoline, the emissions are almost a quarter less. The mobility and fuel strategy devised by the Federal Government also reflects the high importance of gas use and suggests steps for the intensification of its use, for example via the expansion of highly frequented filling stations to include natural gas pumps, or an increase in the number of natural gas filling stations.

A much larger greenhouse gas reduction can be achieved on the basis of electricity; either directly (as for example already practiced by the railways) or via gaseous or liquid fuels produced by electricity on the basis of regenerative energies. However, compared to natural gas, the production processes for the latter fuels are still in the early stages of development and test. Electricity is used to split water into hydrogen (PtG-hydrogen) and oxygen using electrolysis. Further chemical reactions convert this into methane (PtG-methane). Fischer-Tropsch synthesis is able to combine carbon-based gases with hydrogen and form hydrocarbon-based liquid fuels (PtL). To ensure that the result is climate-friendly fuels, the power to gas (PtG) and power to liquid (PtL) process steps need to use energy from renewable sources. An extension of current natural gas infrastructure can be used for electricity-generated methane, since this can be transported in pipelines and, in contrast to hydrogen, can be mixed in large volumes, stored and implemented in current natural gas tanks. The existing distribution network can also be used for liquid fuels. Which fuel and powertrain options are best suited for which mode of transport and which changes to infrastructure will be necessary, will be discussed in the following chapters.

The different modes of transport pose very different technological requirements on the energy supply independently of their effect on the environment and eliminate certain fuels or fuel storage systems because of them. The volumetric and gravimetric energy density of the fuels in particular, together with the energy storage itself, and the mass and volume of the fuel store are decisive.

This is particularly prominent in aviation, where alternatives to kerosene are conceivable at best in the long-term and then only for short-haul flights. These could be future hybrid planes with batteries and hydrogen fuel cells, or even jet engines powered by hydrogen.

The gravimetric energy density is mainly influenced by the choice of fuel, whereby the volumetric energy density can be increased to a certain degree, for example by the compression or liquefaction of fuels that are gaseous under normal conditions. The compression of natural gas, methane or hydrogen should suffice in the medium term to satisfy the range requirements of passenger cars, whereby the liquefaction of gaseous fuels for commercial long-haul trucks cannot avoid large amounts of cooling in order to guarantee the required range. Similar requirements also eliminate other energy supply options for transport such as battery-driven long-haul trucking at least until the middle of the twenty-first century, unless an unexpected technological breakthrough occurs.[1]

If the additional requirements are considered, i.e. that the fuels and energy stores must be manufactured in a sustainable and environmentally friendly manner and enable at least a practically greenhouse gas neutral transport system, then the number of energy distribution options is limited further.

The strict requirements for a particularly large reduction in greenhouse gas emissions from fuels up to greenhouse gas neutrality result from the wide-ranging necessities for climate protection that are required to limit global warming to a value below 2°C. This also requires huge reductions compared to the forecast in greenhouse gases in the transport sector—both in developed countries and in developing countries. Both criteria named result in fossil energy carriers not being an option for supply in 2050. Only the following post-fossil energy supply options remain for transport:

- direct use of electricity: battery-driven and cable-dependent forms of electro-mobility
- indirect use of electricity: gaseous and liquefied fuels on the basis of renewable energies [for example Power-to-Gas (PtG: hydrogen or methane) and power-to-liquid (PtL)]
- biogenically based fuels from residual waste.

Biofuels of the first generation produced from cultivated biomass cannot fulfil the requirements for fuel sustainability, since they are associated with social and ecological problems such as land use and competitive use of land, polluted water and over-fertilized land [8]. Only first generation biofuels based on biological waste and second generation biofuels from that use wood or straw have a high probability of fulfilling the necessary large reductions in GHGs and have total ecological advantages [25]. The potential amount for these fuels is, however, very limited and can only cover a share (approx. 10 %) of the global energy supply for transportation in 2050. In the long-term, biofuels of the third generation on the basis of algae are conceivable, but will not be discussed further here, since sound predictions concerning costs and efficiency are not possible. Thus large disruptions in energy distribution are necessary for sustainable and climate-friendly supply of the transport sector, since practically all fuels must be replaced. This is the reason the talk is of energy transition in transport. A transition towards a shift and an avoidance of traffic is unavoidable in order to reduce the energy requirements of the transport sector on a global basis and thus to enable a first successful energy transition in the transport sector.

---

[1] A significant increase in the volumetric and gravimetric energy density has been considered in this timescale within the scope of assessable technological development.

Contribution to the Discussion for Efficiency

A fundamental element of future action is to increase the technical efficiency of each and every mode of transport. This means that the energy consumption per relative value must be reduced. To achieve this, a variety of steps are necessary in the area of the total vehicle. The current discussion around improving efficiency continually calls for further tightening of the CO_2 fleet emissions for passenger cars, or that they should be introduced for commercial vehicles. It is also important that such targets and limits be introduced for shipping and aviation. This is to be welcomed in principle.

In the future though, this CO_2 limitation is no longer systematically useful, since this approach originates from a time when mainly gasoline, diesel, heavy oil and kerosene were used in solely combustion engine driven vehicles. This will change in the future. Other energy carriers (for example hydrogen, compressed natural gas, liquefied natural gas, synthetic fuels, direct use of electricity) and hybrid systems will prevail. Because of this development, CO_2 will not be the unique indicator of energy consumption (for example, see [26]). From the point of view of the environment, it will also be necessary in the future to have not only climate-friendly energy distribution, but also resource-friendly energy distribution. In face of this, it is wise to use energetic consumption as an indicator of the efficiency of a vehicle [26]. This can provide orientation concerning the depletion of resources. It can also be the basis to provide information on CO_2 emissions under constantly changing boundary conditions (for example the degree of expansion of renewable energies).

Diagram 2.6 shows which energy distribution options will be principally suited for which mode of transport in 2050. This evaluation is based in results from a study by the Federal Environment Agency "Post-fossil energy supply options for a greenhouse-neutral transport sector in the year 2050: A cross-transport mode evaluation" [27].

However, in order to gain a holistic view of energy supply, not only are sustainability requirements placed on fuel production and greenhouse gas reductions decisive, but ecological, economical, technical, infrastructural and systemic aspects must also be

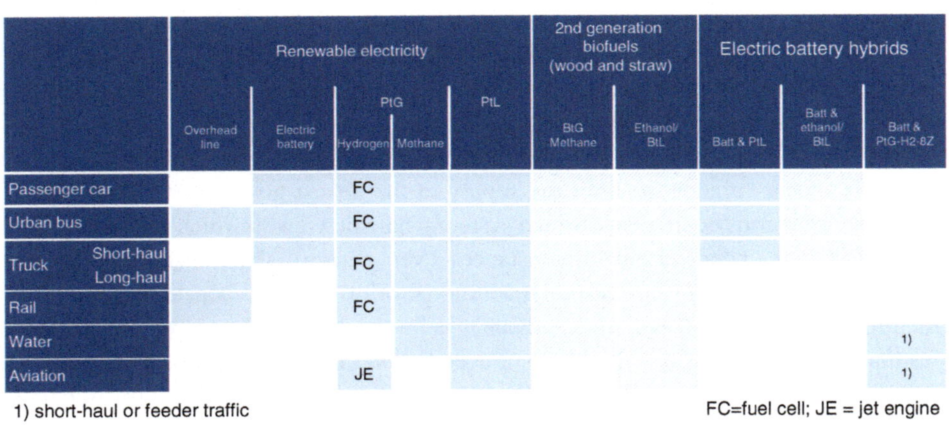

Diagram 2.6 Post fossil energy supply options for transport in 2050 for different modes of transport. Options with limited volume potential are shaded, see [27]

evaluated. The direct use of renewable electricity normally represents the most climate-friendly and the most economic variant. For this reason, wherever technically possible, battery-driven or plug-in hybrid vehicles should be an important pillar of transportation. This is true for passenger cars, light-duty trucks and short-haul trucks, buses and to a certain degree long-haul trucks. Fuels produced on the basis of renewable energy are of interest to these means of transport, with the exception of niche applications, that cannot directly use electricity such as shipping and aviation, since they offer large volume potential. Considerable amounts of regenerative electricity are required for their production that generally require additional renewable energy plants such that it needs the erection of PtL and PtG plants on internationally suitable renewable energy sites.

2.5 Infrastructure for Transportation Energy Supply

The supply of transportation with energy, currently in the form of fossil fuels mainly requires, apart from a few exceptions, an infrastructure exclusively used by vehicles. This infrastructure of transport is mainly connected to the overall energy supply infrastructure, as are other consumers. The construction and maintenance of the transportation supply infrastructure is associated with huge economical effort and negative ecological consequences. For these reasons, when regarding various energy supply options for transportation, it is worth checking how worthwhile and sustainable it is to have different supply infrastructure (at least for a time of transition or even long-term) for the same means of transport, or at least the same transportation platform (for example road transport). First, road transport will be considered, where natural gas is already offered as a fuel, followed by a short overview of other means of transport.

When considering road transport, it is conceivable that a large proportion of the vehicle fleet could be battery-driven for climate-protection reasons and that this would make a corresponding charging infrastructure for passenger cars and light duty trucks unavoidable. Hybrid vehicles would require at least one other supply infrastructure. It needs to be critically examined whether other fuels with different physical states or different types of fossil or renewable source should be made available.

Regarding only the effort required for a supply infrastructure for the entire transport volume, it appears prudent to limit the supply to the fewest possible and most similar fuels. Other scenarios could result from a systemic holistic view in which different means of transport and under certain circumstances even parts of the transport means are optimally provided for. In this case, the additional effort required concerning the infrastructure must be weighed against the advantages for the means of transport. It is however questionable, how far the state is able to or even has an interest to control the market to a large degree to avoid too many parallel transportation energy supply infrastructures. A diversification of transportation energy supply and the different transportation platforms may even be advantageous in reducing the dependency on certain countries and regions.

Generally speaking, different fuels require separate supply lines; particularly if they have different physical properties under standard ambient temperature and pressure. The effort required by the road traffic supply infrastructure for various fuels is different. For example, it is somewhat higher for natural gas than for liquid fuels, since compression to pipeline pressure is required when first connecting to a pipeline infrastructure and at the filling station to filling pressure. Whereas liquid fuels can normally be delivered by truck to the filling station, which requires little effort, is relatively simple and permits the delivery of different fuels in liquid form, natural gas typically requires a pipeline. The filling station is connected to the general natural gas network with its additional function as a storage. There are non-negligible losses in the transportation of natural gas through the pipeline to the vehicle's tank, but they are significantly lower for example than for hydrogen [28]. Nowadays there is already a sufficiently widespread supply infrastructure for natural gas and liquid fuels for road transport.[2] There is also a very well developed general infrastructure for both fuel types to supply transport as consumer of these fuels.

The statements concerning the availability of infrastructure are also valid for alternative fuels based on renewable energy. Methane and liquid fuels produced from electricity already have a well-developed general infrastructure and also a well-developed traffic supply infrastructure: for fossil natural gas and fossil liquid fuels. It is probable that PtG-methane can be mixed with fossil natural gas, since there is only one natural gas network, so that the user only fills up with a share of the fuel.[3] In the case of PtL-fuels, depending on how the system establishes itself, both balanced or physical delivery are thinkable. There is currently no sufficiently developed supply infrastructure for PtG-hydrogen, neither for transport purposes or otherwise. However, there are initiatives in Germany and Europe to develop a hydrogen supply network.[4] What this application of PtG-hydrogen will look like regarding road transport depends on the general development trend.

Similar basic demands on the infrastructure also exist for alternative fuels based on biomass. In the case of biogenic synthetic natural gas, it is possible to physically deliver it or mix it with fossil natural gas. Biogenic liquid fuels can also be mixed with gasoline or diesel and can than be used by vehicles, with possible slight modification, can be refueled with the fuels in a pure state. In both cases, slight modifications to the supply infrastructure may be needed.

[2] In Germany there were about 900 natural gas filling stations and a total of approximately 14,000 filling stations for liquid fuels, mainly based oil-based.

[3] Similar to renewable electricity that is not physically consumed, but whereby the renewable electricity is fed into a general production pool and the customer consumes power from an unknown source: the renewable power is then charged by share. This system is also used for example by Audi for Audi-e-gas, whereby customer consumption of e-Gas is registered via a filling card and the corresponding amount of PtG-methane is fed in to the natural gas network.

[4] In Germany there are approximately 40 hydrogen filling stations, seven of them of operated publicly.

Other means of transport, for example inland water and maritime traffic, are also currently intensively being checked to see whether liquid natural gas (LNG) could be used on a large scale instead of heavy oil, due to increasingly strict pollution emission requirements. From a supply infrastructure point of view, these fuels will require their own supply network. However, the possibility of deploying LNG in ports is easily conceivable in the medium and long term, since natural gas is already transported in liquid form by seafaring vessels to selected ports and it is probable that in the future appropriate infrastructure for landing and storing such fuels will be build up in more and more ports.

2.6 Summary

An element in the action plan to comply with the global 2-degree limit is that transportation at least massively limits the growth in its greenhouse gas emissions or rather contributes significantly to a reduction in emissions. This is necessary on both a national and a global level. The whole world is experiencing a massive surge in the growth of the transport performance: all forecasts point to further growth. It is therefore a problem of global proportions for which an exclusively national view is insufficient. Apart from the different boundary conditions in the various regions, it must be noted that the different transportation platforms have different growth and have different potential and options for the reduction of greenhouse gases. Forecasts assume that the GHG emissions due to transportation will more than double by 2050, from 6 $GtCO_{2,eq}$ to 14 $GtCO_{2,eq}$.

Since economic growth and traffic growth are coupled in many countries, steps to drastically reduce the volume of traffic or to reduce its growth are politically difficult to impose. The conclusion is that it is highly probable that only minor successes can be achieved in this area. It appears that the key to the transport-based climate issue lies in the domain of transport's energy supply.

Global transport must reduce its forecasted greenhouse gas emissions in 2050 by about 80 %, if it is to comply with the 2-degree goal set out in climate protection. To do this, there is a range of indicators that even deem it necessary to reduce the levels by 100 % after 2050.

A drastic reduction of greenhouse gas emissions in global transportation cannot be achieved alone through steps to avoid transport shifting transport to more environmental modes of transport and reducing the specific energy consumption by the required amount. The targeted emissions reduction in the transportation sector is only possible on a global scale if energy supply options are available that contribute to substantial reductions.

There are a variety of options available for the different transportation platforms: The direct use of electricity based on renewable energy is practicable. The current level of knowledge states that there are no short or medium term globally available options for the direct use of electricity for heavy-duty traffic, shipping or aviation. However it must be mentioned that the direct use of electricity in commercial vehicles is in its infancy. Since the greenhouse gas emissions from the transport sector from a global perspective are currently still increasing, the pressure to act and solve issues in traffic is growing rapidly.

It is therefore high time to develop strategies to make technologies and energy carriers available for the transportation sector and if possible for other sectors too that have no negative impact on the climate.

An important element in such a strategy is, according the wisdom of today, the production of synthetic methane utilizing electricity. This affects the purely technological availability of production and its future development. It can be the basis for a climate-neutral energy supply for transportation, where for example direct use of electricity (rail, passenger car, light-duty trucks) is not possible. Without a massive increase in the development of renewable electricity supply up to the time corridor 2050–2070, this technology, as a contribution towards climate-protection, is however not very practicable. In parallel to this important initiative, the basis of every action in the transport sector is to significantly reduce the specific energy consumption of each individual traffic platform to dampen the growth of transportation effort.

The use of synthetic methane is practicable in view of the already widely available infrastructure.

Natural gas has the potential of reducing the CO_2 emissions in transport due to its physical properties (C/H ratio): It increases the efficiency of Otto engines since higher compression ratios are possible; methane can also be produced synthetically. Natural gas represents a step towards greenhouse gas neutral transportation. The next step would be the fastest possible formulation of a globally effective strategy for the industrial production and use of electricity-based fuels from renewable energy sources (for example PtG-methane) and the direct use of electrical energy in the transportation sector. At the same time, directed further development of the technical processes needs to be initiated, unless other more realistic methods can be found how to comply with the 2-degree limit without having to use electricity-based fuels.

References

1. SRU (2005) Umwelt und Straßenverkehr, Hohe Mobilität—Umweltverträglicher Verkehr. Sachverständigenrat für Umweltfragen, Berlin, http://www.umweltrat.de/SharedDocs/Downloads/DE/02_Sondergutachten/2005_SG_Umwelt_und_Strassenverkehr.html
2. Statistisches Bundesamt (2014) Volkswirtschaftliche Gesamtrechnungen, Bruttoinlandsprodukt ab 1970, Vierteljahres- und Jahresergebnisse. Statistisches Bundesamt, Wiesbaden, https://www.destatis.de/DE/ZahlenFakten/GesamtwirtschaftUmwelt/VGR/Inlandsprodukt/Tabellen/Bruttoinland Viertel- jahresdaten.xlsx
3. DIW (1991) Verkehr in Zahlen. Deutsches Institut für Wirtschaftsforschung e.V., Berlin
4. DIW (2012) Verkehr in Zahlen. Deutsches Institut für Wirtschaftsforschung e.V., Berlin
5. IPCC (1990) Climate change: the IPCC scientific assessment (1990). Report prepared for Intergovernmental Panel on climate change by Working Group I. Cambridge, Great Britain, New York, NY, and Melbourne, Australia

6. Öko-Institut (2013) Treibhausgasneutraler Verkehr 2050: Ein Szenario zur zunehmenden Elektrifizierung und dem Einsatz stromerzeugter Kraftstoffe im Verkehr. Sachverständigengutachten im Auftrag des Umweltbundesamtes. http://www.oeko.de/oekodoc/1829/2013-499-de.pdf
7. WBCSD (2014) Mobility 2030: meeting the challenges to sustainability, the sustainable mobility project. World Business Council for Sustainable Development, Hertfordshire, http://www.wbcsd.org/web/mobilitypubs.htm
8. Jering A, Klatt A, Seven J, Ehlers K, Günther J, Ostermeier A, Mönch L et al (2012) Globale Landflächen und Biomasse, nachhaltig und ressourcenschonend nutzen. Umweltbundesamt, Dessau Roßlau, http://www.umweltbundesamt.de/sites/default/files/medien/479/publikationen/globale_landflaechen_biomasse_bf_klein.pdf
9. Purr K et al (2014) Treibhausgasneutrales Deutschland im Jahr 2050. Climate change 07/2014. Umweltbundesamt, Dessau-Roßlau. http://www.umweltbundesamt.de/ sites/default/files/medien/376/publikationen/climate-change_07_2014_treibhausgasneutra- les_deutschland_2050_korr_18.6.2014.pdf
10. IPCC (2014) Climate change 2014: mitigation of climate change, contribution of Working Group III to the fifth assessment report of the intergovernmental panel on climate change. Cambridge and New York, NY
11. IPCC (2007) Contribution of working group i to the fourth assessment report of the intergovernmental panel on climate change. Cambridge und New York, NY
12. Rodt S et al (2010) CO_2-Emissionsminderung im Verkehr in Deutschland, Mögliche Maßnahmen und ihre Minderungspotenziale, Ein Sachstandsbericht des Umweltbundesamtes. UBA Texte 5/2010. Umweltbundesamt, Dessau-Roßlau. http://www.umweltbundesamt.de/publikationen/co2-emissionsminderung-im-verkehr-in-deutschland
13. Köder L et al (2014) Umweltschädliche Subventionen. Umweltbundesamt, Dessau-Roßlau
14. Alfen Consult—Aviso—IVM (2014) Berechnung der Wegekosten für das Bundesfernstraßennetz sowie der externen Kosten nach Maßgabe der Richtlinie 1999/62/EG für die Jahre 2013 bis 2017. Endbericht. Bundesministerium für Verkehr und digitale Infrastruktur, Berlin. http://www.bmvi.de/SharedDocs/DE/Anlage/VerkehrUndMobilitaet/Strasse/wegekostengut- achten-2013-2017-endbericht.pdf
15. Öko-Institut (2013) Weiterentwicklung des Analyseinstruments Renewability, Renewability II—Szenario für einen anspruchsvollen Klimaschutzbeitrag des Verkehrs. UBA Texte 84/2013. Umweltbundesamt, Dessau-Roßlau. http://www.umweltbundesamt.de/publikationen/weiter entwicklung-des-analyseinstruments
16. Erdmenger C et al (2010) Pkw-Maut in Deutschland? Eine umwelt- und verkehrspolitische Bewertung. Hintergrund. Umweltbundesamt, Dessau-Roßlau. http://www.umweltbundesamt.de/en/publikationen/pkw-maut-in-deutschland
17. Matthes FC et al (2008) Politikszenarien für den Klimaschutz IV, Szenarien bis 2030. Climate change 1/2008. Öko-Institut/Forschungszentrum Jülich IEF-STE/DIW Berlin/FhG-ISI im Auftrag des Umweltbundesamtes, Dessau-Roßlau. http://www.umweltbundesamt.de/sites/default/files/medien/publikation/long/3361.pdf
18. Wissenschaftlicher Beirat beim Bundesminister für Verkehr, Bau und Stadtentwicklung (2010) Sicherheit zuerst—Möglichkeiten zur Erhöhung der Straßenverkehrssicherheit in Deutschland. Zeitschrift für Verkehrssicherheit 56:S 171–194. http://vplno1.vkw.tu-dresden.de/ psycho/download/wissBeiratGutachten_Schlag_2010.pdf
19. FGSV (2013) Hinweise zur Verkehrsentwicklungsplanung, FGSV-Nr. 162. Forschungsgesellschaft für Straßen- und Verkehrswesen e.V., FGSV Verlag, Köln
20. Rupprecht Consult (2013) Guidelines developing and implementing a sustainable urban mobility plan. Rupprecht Consult—Forschung und Beratung GmbH, Köln, http://www.mobility-plans.eu/docs/SUMP_guidelines_web.pdf

21. Ahrens G-A, Klotzsch J, Wittwer R (2014) Autos nutzen, statt besitzen—Treiber des multimodalen Verkehrsverbundes. Zeitschrift für die gesamte Wertschöpfungskette Automobilwirtschaft (ZfAW) 2:S 6–20
22. KCW (2010) Schienennetz 2025/2030—Ausbaukonzeption für einen leistungsfähigen Schienengüterverkehr in Deutschland. Holzhey, Michael, UBA Texte 42 (2010). Umweltbundesamt, Dessau-Roßlau. http://www.umweltbundesamt.de/publikationen/schienennetz-2025-2030
23. Doll C et al (2013) Wirtschaftliche Aspekte nichttechnischer Maßnahmen zur Emissionsminderung im Verkehr. UBA Texte 11 (2013). Fraunhofer ISI/INFRAS/IFEU im Auftrag des Umweltbundesamtes, Dessau-Roßlau. http://www.umweltbundesamt.de/publikationen/ wirtschaftliche-aspekte-nichttechnischer-massnahmen
24. Dena (2011) Erdgas und Biomethan im künftigen Kraftstoffmix, Handlungsbedarf und Lösungen für eine beschleunigte Etablierung im Verkehr. Deutsche Energie-Agentur GmbH, Berlin, http://www.dena.de/fileadmin/user_upload/Publikationen/Verkehr/Dokumente/Erdgas_und_Biomethan_im_kA1_4nftigen_Kraftstoffmix.pdf
25. BFE (2012) Harmonization and extension of the bioenergy inventories and assessment. Bundesamt für Energie (Hrsg.), Bern. http://www.empa.ch/plugin/template/empa/*/125527
26. Ifeu (2013) Konzept zur zukünftigen Beurteilung der Effizienz von Kraftfahrzeugen. UBA Texte 95. Umweltbundesamt, Dessau-Roßlau. http://www.umweltbundesamt.de/sites/ default/files/medien/378/publikationen/texte_95_2013_konzept_zur_zukuenftigen_beurteilung_der_effizienz_von_kraftfahrzeugen.pdf
27. INFRAS—Quantis (2014) Postfossile Energieversorgungsoptionen für einen treibhausgasneutralen Verkehr im Jahr 2050: Eine verkehrsträgerübergreifende Bewertung. Umweltbundesamt, Dessau-Roßlau
28. Öko-Institut (2013) Strombasierte Kraftstoffe im Vergleich—Stand heute und die Langfristperspektive (Aktualisierte Fassung). Working paper, Öko-Institut e.V., Freiburg. http://www.oeko.de/uploads/oeko/oekodoc/1826/2013-496-de.pdf

A Historical Overview

3

Reinhold Bauer

3.1 History of Natural Gas

The Industrial Revolution that took place in Great Britain in the latter third of the eighteenth century was largely based on an "Energy Revolution" that has resulted in a roughly 15-fold increase in the average per-person energy consumption worldwide over the past 200 years. To be sure, this is an average figure for the entire global population, which in the same time frame has increased approximately sevenfold. This Energy Revolution was made possible by the transition from renewable energy sources such as wood, manual labor, wind and hydro-power based on the agrarian energy systems in pre-industrial times to the fossil fuel energy system of the industrial age. The now widely used fossil fuels coal and, increasingly in the twentieth century, petroleum appeared to have unlimited and above all inexpensive availability (Diagram 3.1).

Natural gas, another fossil fuel which is a combustible gas existing in underground deposits and primarily consisting of methane, has been known since the days of ancient China and pre-modern Europe, but has been rarely used. It began to see use only on a modest scale starting in the 1880s (particularly in the US state of Pennsylvania) in the production of process heat for industrial needs and for illumination, but failed to achieve broader application due to transportation challenges. It was not until the development of pressure-resistant seamless pipelines and new welding techniques led to a true pipeline boom in the United States starting from the end of the 1920s that it became possible to transport gas economically over long distances. This also signaled the beginning of the harvest of the vast natural gas fields which had in the meantime been discovered in the southern United States.

R. Bauer (✉)
Department of History, University of Stuttgart, Keplerstr. 17, 70174 Stuttgart, Germany
e-mail: Reinhold.bauer@hi.uni-stuttgart.de

R. van Basshuysen (ed.), *Natural Gas and Renewable Methane for Powertrains*,
Powertrain, DOI 10.1007/978-3-319-23225-6_3

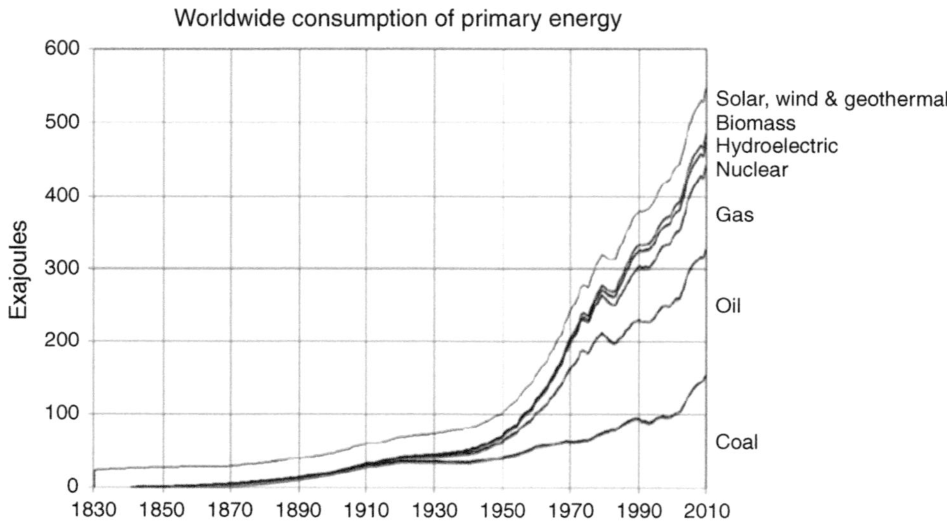

Diagram 3.1 Global primary energy consumption. *Source*: International Energy Agency (ed.): World Energy Outlook 2011 (http://www.iea.org/publications/freepublications/publication/weo2011_web.pdf, 05 August 2014). Smil, Vaclav: Energy Transitions: History, Requirements, Prospects, Santa Barbara (Ca) 2010, p. 62. Fernandes, Suneeta D.; Trautmann, Nina M.; Streets, David G.; Roden, Christoph A.; Bond, Tami C.: Global Biofuel Use, 1850–2000. In: Global Biogeochemical Cycles, 21(2007) (http://onlinelibrary.wiley.com/doi/10.1029/2006GB002836/pdf, 05 August 2014). Podonik, Bruce: Toward a Sustainable Energy Regime: A Long-Wave Interpretation of Global Energy Shifts. In: Technological Forecasting and Social Change, 62(1999), pp. 155–172. The BP Statistical Review of World Energy (http://www.bp.com/en/global/corporate/about-bp/energy-economics/statistical-review-of-world-energy/statistical-review-1951-2011.html, 05 August 2014).

At that time, a market for gas for lighting, heating, cooking and power had certainly already existed for decades both in the United States and in Europe. The history of the use of natural gas in the industrial age can be traced back to the gas works emerging starting in the early nineteenth century that began building gas generation and delivery systems in a rapidly growing number of cities. On the eve of World War I in the German *Reich* alone, there were nearly 1400 gas works producing and distributing gas to cities. This so-called city gas, or coal gas, is a synthetic gas that is created through dry distillation of coal, that is, by heating coal in the absence of air. Depending on the type of coal used and the precise manufacturing process, it consists of many different agents, but mainly hydrogen, methane, nitrogen and carbon monoxide plus several other substances. It was initially produced almost exclusively for lighting purposes, with the resulting gas lighting facilities leading to the achievement of a new quality of artificial lighting. This made it possible for the first time ever to decouple the work and life routines of larger segments of the population from the natural day/night cycle. Beginning in the 1880s, a diversification of use began to take place, particularly in the face of the newly emerging competition in the form of electric light. It was the gas works themselves who, with their twin objectives of tapping new markets and reducing the fluctuations in their daily and seasonal demand, propagated the use of gas for heating and cooking purposes and, for a time, also for the propulsion of stationary internal combustion engines, and promoted it with, among other things, special rates.

While the use of city gas for lighting and combustion became increasingly less important in the 1920s and 1930s with the expansion of the electricity supply and the concurrent drop in energy prices, the success of cooking gas and particularly of gas for heating more than compensated for this development. City gas consumption thus continued to grow in the industrialized nations primarily in Europe, including through the interruption of World War II. It was not until the 1950s that the commercial success of the gas industry was threatened by a reversal in the trend of private and especially commercial use of city gas appeared due to the influence and development of the "fully electric household" accompanied by the triumph of inexpensive mineral oil.

However, roughly parallel to this unfolding shift, large natural gas fields were being discovered and tapped in Europe, in around the 1950s in the Netherlands, Russia, northern Italy and then in North Africa as well, and in the 1960s most notably in the North Sea. In comparison to city gas, natural gas offered significant advantages, particularly a calorific value nearly twice that of city gas and cleaner burning due to its more favorable chemical composition. In addition, natural gas is not poisonous due to the absence of carbon monoxide and—certainly a decisive factor—it is fundamentally more favorable to obtain and distribute than synthetic city gas.

Lastly, a development had been taking place since the 1950s similar to the one that had been observed in the United States as early as the end of the 1920s: In the face of the newly discovered deposits and in light of the economic and technical advantages of natural gas, a pipeline infrastructure for the long distance transport of the gas was constructed, followed in the 1960s by the growing conversion of the gas supply from city gas to natural gas. This process was for the most part complete in western Germany in the early 1980s, and the conversion was made in the new eastern federal states of Germany in the 1990s. The switch to natural gas significantly enhanced the competitiveness of the gas industry in its competition with oil, such that in truth natural gas can most certainly be called "the savior of the gas industry."

However, the conversion to natural gas was not trouble-free, as it required a complex reconstruction of the system and new technical equipment for both the gas industry and the consumers. The expansion of the pipeline infrastructure including the compression facilities and the conversion or replacement of all gas-powered appliances due to the significantly different properties of natural gas as opposed to city gas initially caused high one-time costs. On the other hand, it was possible to use the existing city gas grids for the distribution and sale of natural gas. The previously decentralized power supply network was integrated into a Europe-wide comprehensive natural gas network via the resulting high-pressure long distance pipelines.

Typical of the German gas industry is its heterogeneous and above all decentralized structure. It includes a whole array of regionally operating utility companies existing alongside some 700 gas distributors commonly acting only locally, that is, usually public utilities overseeing the distribution to the end users. The long distance transport of natural gas is provided by special grid operators. By the way: this heterogeneous structure of the gas industry does not make it any easier to promote and establish new uses such as natural gas-powered vehicles or new mobility concepts based on natural gas, as we will discuss in greater detail below.

The consumption of natural gas has risen considerably especially since the 1960s in Europe and particularly in the Federal Republic of Germany where, as of the mid-1960s, it was still at about two billion cubic meters but by the end of the 1970s had already reached 60 billion cubic meters per year. In the middle of the 1990s, the then all-German consumption of natural gas reached a volume of about 100 billion cubic meters, with consumption in the West stagnating since the 1990s, due in particular to the increasing energy efficiency of the appliances and equipment in use. It's worth noting here that this stagnation in natural gas sales played a significant role in the development of natural gas vehicles in Germany.

Today, in Germany as in many other European countries, natural gas is employed primarily for the generation of process heat in industry and trades or for heating space and water in private homes. Contrary to the situation in some of the important producing countries, with a share of a good 10 % of the total German power generation, natural gas plays a significant but not a pre-eminent role. The share of natural gas in covering the entire primary energy needs in Germany is a solid 21 %, thus making it the second most important primary energy source after oil. Nearly 90 % of the natural gas used in Germany must be imported, with the most important supplier countries as follows: Russia in first place with a share of not quite 40 %, Norway in second place with a share of about 35 % and Holland in third place with a share of just about 5 %. Depending on the respective foreign political situation this high import quota is not without problems, which has become repeatedly very clear both in past and recent times.

Not least the current (global) political implications of the natural gas supply have refocused a great deal of public attention on a technology that has been intensively debated in recent years: hydraulic fracturing, recently more commonly called "fracking." Fracking for the most part involves tapping unconventional natural gas reserves by producing fractures in the reservoir rock in order to more effectively exploit them, as otherwise low extraction rates would result, due to the inadequate permeability of the reservoir rock. Breaking open the reservoir rock is achieved by pumping water under high pressure into the rock through bore holes, whereby the water is mixed with chemical additives and proppants. The technology itself is anything but new; in fact it had already been developed and was being employed in the United States by the latter half of the 1940s and has been in use in Western Europe and the former Soviet Union since the 1950s. It did not begin to attract greater attention until about 15 years ago when large quantities of natural gas began to be extracted through fracking. Not only did the yield increase significantly, but a long-running and heated public discussion on the potential dangers of fracking also began that has in the meantime also entered the European public domain. The fear is essentially that ground water will be contaminated by the drilling water or the to some extent poisonous and carcinogenic chemical additives, that surface waters will be contaminated by the so-called backwash that after some time leaks from the wellbore, and moreover that fracking may trigger larger earthquakes and, lastly, that fracking can lead to an uncontrolled leakage of methane which is harmful to the environment. The proponents of fracking see in it an effective way to substantially expand prospective natural gas reserves. On the other hand, there is still a considerable need for clarification in regard to the potential ecological and tectonic impacts. The future of the practice, particularly for Western Europe, thus appears uncertain, and its social and political implementability seems especially doubtful for Germany.

In connection with unconventional or alternative methods of natural gas extraction or production, a technology that has been discussed intensively in the past few years should also be mentioned here, that is to say the so-called "Power-to-Gas" technology. This technology concerns the conversion of (excess) electrical power generated in wind, water or solar power stations into chemical energy, which is then stored and distributed in the existing gas grid. In the first phase of the process, hydrogen is produced through water electrolysis, which then can be converted in a second phase using carbon dioxide into a synthetic methane, or natural gas. Special synergetic effects can result from the combination of biogas and Power-to-Gas plants regarding the gas input and the use of carbon dioxide with raw biogas. It is precisely here that the promising perspectives in connection with the sought-after energy turnaround in Germany for a climate-neutral system for supplying natural gas and ultimately natural gas mobility come into sharp relief.

In conclusion, we have seen that the rise of combustible gases as relevant energy sources began as early as in the "First Industrial Revolution." At that time it was synthetic city gas that served as the basis for the development of the market in various areas of application and for which the construction of the initially purely city-oriented supply infrastructure was completed. With the discovery of rich natural gas reserves in first half of the twentieth century and the development of a suitable long-distance transport technology, most notably in the form of natural gas pipelines, the rise of natural gas as a high-value fossil fuel began, particularly from the 1930s on. Today it is employed by commercial and private users most notably for heating purposes as well as in gas power plants. Natural gas also plays an important role as a fuel for motor vehicles in a few countries, which we will return to shortly. An expanded, internationally integrated pipeline system ensures the safe and relatively cost-effective transport of natural gas, with liquid natural gas tankers (LNG tankers) being increasingly employed for decidedly long-distance transport since the 1970s, of which approximately 200 are in service today. According to the latest estimates, the currently known world natural gas reserves that are technologically and commercially recoverable have a statistical range of approximately 60 years. To what extent this statistical range can be extended by the use of new extraction methods and the exploitation of unconventional reserves or by increasing production of "synthetic natural gas" will remain to be seen and is therefore a question that exceeds the scope of the historian's responsibility.

3.2 History of the Natural Gas Vehicle

The story of (natural) gas-powered vehicles has almost completely escaped the focus of historical researchers to date. Gas-driven vehicles have been glossed over not just by automotive historians, but also by business, municipal, environmental, and infrastructure historians in their discussions of the history of gas supply or the gas industry. Regarding automotive history in particular, historical hindsight to date continues to be heavily influenced by the dominance of gasoline and diesel engines running on liquid petroleum derivatives. Although more recent research has certainly addressed the competition between gasoline, steam, and electric vehicles in particular, especially in light of current issues, natural gas vehicles figure only rarely in the discussion. The situation is almost identical for the history of gas generation and use since the nineteenth century where very little discussion has been devoted to the use of gas as a vehicle fuel, despite the fact that there has been quite thorough study of the buildout of first city-wide, then regional and state-wide supply networks, the transition in use from gas for lighting to power generation and heating and the progression from city to natural gas (see above).

On the other hand, this neglect concerning gas-powered vehicles is hardly surprising, since it has so far been relegated to the status of a niche phenomenon which, at first glance at least, up to now has had very little relevance for transport, environmental, and energy history. In the case of vehicular engines, after all, liquid fuels have occupied the dominant position almost from the outset for both practical and pragmatic reasons. On the other hand, the multiple revivals of the gas vehicle "niche technology" in the twentieth and early twenty-first centuries are giving us a clear look at the changing discourses surrounding energy and energy-related issues over the past 100 years. And my first thesis is that precisely these discourses make natural gas mobility a historically relevant topic.

The second thesis is that the evolution of gas-powered vehicles can only be understood against the backdrop of changing perceptions of problems, which in turn have been influenced by changing political, economical, social, and especially cultural conditions. The gas-powered vehicle always has been and is still seen as an *alternative* to the established vehicle types or to other "alternative" vehicles, particularly as an alternative to the electric car. Depending on the issue at hand in which the evolution and use of gas-powered vehicles was (and is) embedded, the importance of the technology was seen in quite different ways, i.e. it has been *interpreted* in many different ways by its developers, manufacturers, and users, and by policy-makers and the media. The gas-powered vehicle has been characterized in a variety of different ways over the years, among which the following have predominated for a time in various combinations and degrees of overlap: a self-reliance machine, an oil-conservation machine, a fuel cost reduction machine, a gas market expansion machine, an environmental protection machine and, lastly, a climate protection machine.

Which brings us to my third thesis: This brief and certainly incomplete list shows clearly that the gas-powered vehicle is not just a *technical* artifact, but rather a *sociotechnical* one whose evolution can only be understood against the background of social change and one which furthermore was always meant to cause social change. It is precisely with this in mind that we will be tracing the history of the gas-powered vehicle, and in particular the natural gas-powered vehicle, or NGV, as we move through the important stages of its evolution and its respective interpretations. It is not, nor should it be, the goal of this exposition to provide an in-depth history of the *technology* behind the gas-powered vehicle, nor will we attempt to tell the complete story of the gas-powered vehicle at the global level. Our goal will rather be to choose examples of specific characteristic stages of development, with a special focus on its evolution in Germany.

Let us first review the early history of motorization: The period prior to the First World War when the motorized vehicle was becoming established was marked among other things by competition among vehicles powered by gasoline, electricity, and steam. Obviously, internal combustion engines running on liquid fuels emerged from this competition to become the long-term dominant engine technology. The first major wave of motorization that occurred in the inter-war era was thus already dominated by vehicles featuring essentially the same basic configuration as the ones of today. Moreover, during the period after the First World War the interaction between the automotive industry, the petroleum industry, the car-repair business, skilled labor, users, media, and politics led to the solid establishment of a stable automobile-industrial complex that alternative vehicles are able to make inroads into only with great difficulty even today. Since that time, the existence of special political and/or economic conditions has been an indispensable prerequisite to "reopen" this automobile-industrial complex for alternative vehicles.

The first "reopening shock" of this type occurred in the 1930s in Nazi Germany where, for reasons associated with the rearmament-related goal of conserving convertible currency and the goal of preparing for and waging war, a policy of self-reliance was pursued which was inter alia directed toward reducing petroleum imports for civilian use. The use of so-called "domestic fuels" suggested itself as a possible approach in this effort, with various types of gas coming into consideration as combustible fuels in this context.

Starting in 1934, municipal public transport organizations and vehicle fleets in particular experimented with gas-powered engines. Typically, both buses and waste disposal or street-cleaning vehicles were converted to operation using compressed gas (storage pressure: around 200 bar), with experimentation being done in some cases with city gas (see above), and in others with sewage gas (primarily methane, which we would call biogas today). After an initial period where vehicles were operated primarily using interchangeable gas cylinders, construction of public gas fueling stations began expanding in 1935 in response to the failure of the quite tedious process of switching cylinders during daily operation to prove viable. By 1938, a network of more than 50 fueling stations had come into being across nearly the entire country, of which ten supplied sewage gas and the others city gas (Diagram 3.2).

The partial conversion of the municipal vehicle fleets to gas-powered operation was accompanied by a spate of corresponding research efforts at technical universities, with the Berlin Technical University's Motorized Vehicle Research Institute in particular playing a leading role. Both test bench experiments and street trials were used to improve the adaptation of the engines and above all the load control to the new fuels.

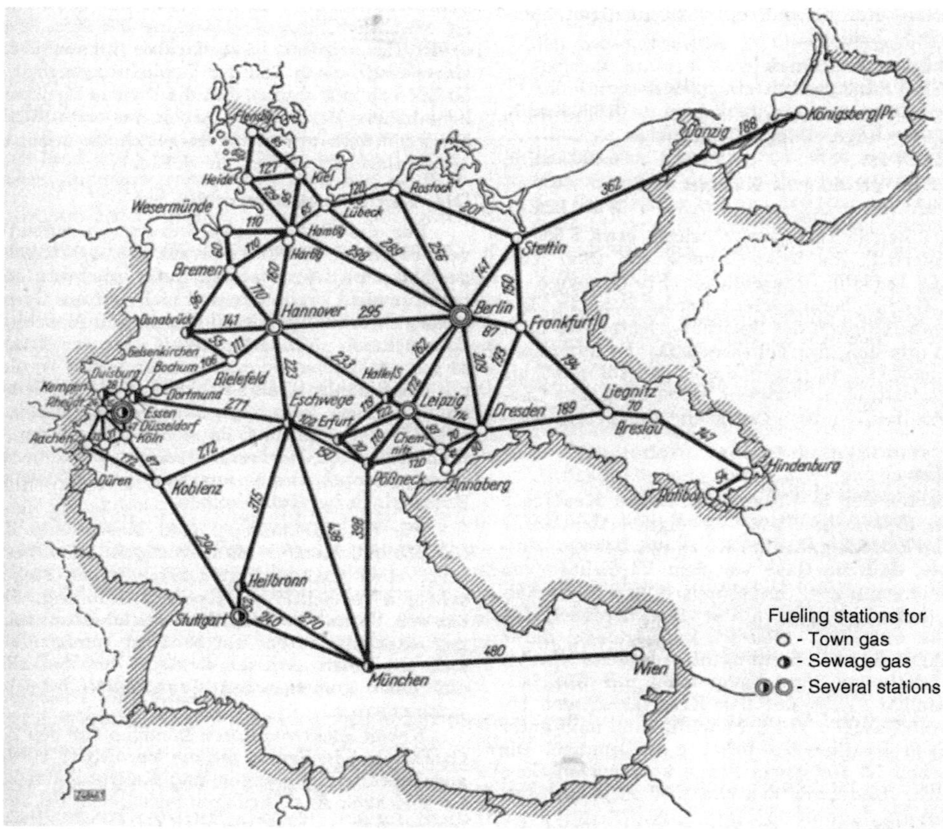

Diagram 3.2 Gas fueling stations in the German Reich in 1938 (The numbers indicate the distances between the fueling stations in kilometers). *Source*: Schumacher, Walter: Bemessung und Bau von Gastankanlagen [Dimensioning and Construction of Gas Fueling Stations]. In: ZVDI, 82(1938), pp. 585–590, here: p. 585

The great weight and considerable space requirements of the poorly built pressure storage tanks ruled out any wider use of gas-powered drive in passenger cars; this fact restricted it to fleet operation of commercial vehicles. That was also just as much due to the rather reluctant promotion of gas-powered operation by the Nazi leadership. The National Socialist government overall showed little interest in gas-powered vehicles, as it had placed its bets on the construction and expansion of hydrogenation plants for liquefaction of coal, in which it had invested billions since the middle of the 1930s.

A second form of city gas use adopted in response to crisis conditions did not arise until the war years, that is, running city buses primarily using gas supplied from low-pressure tanks (Diagram 3.3). From the end of 1941 on, airtight solid rubber sacks capable of holding 20 cubic meters of city gas were employed for this purpose; they were initially mounted on the vehicle roof under a metal frame with a covering pulled tightly over them.

Diagram 3.3 Magirus public transit bus for the city of Ulm in 1942 equipped with a low-pressure city gas storage tank on the roof. *Source*: Informationsblatt der Magdeburger Straßenbahnfreunde e.V., No. 2, 2005. http://www.msf-ev.de/Info_Archiv-Dateien/Info_02_2005.htm. 05 08 2014

Tanks pulled on trailers were added starting in 1942, making it possible to increase the range of the buses. According to contemporary reports, the maximum travel distance with a roof-mounted tank alone was 13 km, which could be improved to around 25 km with the use of trailers. In Germany, the first experiments with such low-pressure gas-powered buses were conducted in Wiesbaden, and the technology spread rapidly during 1942 and 1943 to a number of German cities, from Hamburg to Munich and from Constance to Berlin. Without this crisis-borne technology, it would have been simply impossible to maintain bus operation given the increasing shortage of fuel. Moreover, these low-pressure gas-powered vehicles were by no means restricted to use in Germany: Similar buses were also placed in service in the occupied countries, as well as in a number of cities in the British Isles. The shortage of resources therefore plainly gave a temporary boost to broader use of gas technology.

The same applies for the wood gas vehicle variant, which also began seeing increased use from the end of 1941 on. Work had begun on procedures for processing wood into combustible gas as early as during the First World War blockades and then, as might be expected, was taken up again from the mid-1930s on. Rapid spread of wood gas generators for operating vehicles did not occur until 1940, when war-related rationing forced a conversion to substitute fuels. The market in this sector was dominated almost completely by the Imbert company in Cologne, which built around 500,000 wood gas generators in the 1940s. Both passenger cars and commercial vehicles continued to be equipped with the quaint-looking wood gas boilers well into the initial postwar years.

The upshot of this is that gas-powered vehicles experienced a temporary boom brought about under the conditions of the policy of self-reliance and the war-related scarcity of fuel in the Third Reich. They can therefore be characterized in this phase as "self-reliance" or "oil conservation machines." In the initial postwar years as well, which continued to be characterized by resource scarcity, gas-powered vehicles were able to hold their own, but then rapidly faded in importance in the early 1950s (in West Germany, at least) with the advent of unfettered importation of petroleum products, available at ever lower prices, that accompanied the political and economic integration of the young Federal Republic of Germany into the West. Gas-powered vehicles disappeared from the streets or, to put it another way, the temporarily destabilized automobile-industrial complex was able to regain its footing, with all of the relevant actors, that is, manufacturers and users, university researchers and the energy industry, policy and the media, losing interest in gas-powered vehicles for a few decades.

This was also the situation in most of the (western) industrial nations, with one significant exception: Italy. In the Po valley of northern Italy and the adjacent piedmont to the south, natural gas had been being extracted since the 1930s, although it did not begin to grow by leaps and bounds until the "miracolo economico," northern Italy's economic miracle, starting around 1950. Massive increases in demands for energy, by the Italian economy, rapid expansion of the Italian natural gas pipeline network, tentative growth in living standards and with it the tentative beginning of motorization resulted in the idea of using domestic natural gas as a comparatively cost-effective vehicle fuel. This gave rise to a relatively dense infrastructure of nationally-subsidized natural gas fueling stations, especially in northern Italy, with conversion kits also being developed by the Italian parts industry with state support for use in converting conventional passenger cars to natural gas operation. Mass production of natural gas vehicles, on the other hand, remained a distant dream.

These special developments in Italy are relevant in the fact that the decades-long natural gas tradition may explain why, of all the Western European industrialized nations, Italy is today the one that has by far the most natural gas vehicles (around 750,000 in 2012). What may be more significant from a historical perspective is the fact that Italian natural gas know-how played an important role in the comeback of the NGV in the late 1970s. Lastly, in connection with the developments in Italy it must be noted that gas-powered vehicles received economic and economic policy support as "oil conservation machines," and remained relevant to users for decades thereafter as "fuel cost saving machines." Nevertheless, even in Italy, natural gas vehicles never occupied more than a small niche, which furthermore was regionally limited.

If we set the example of Italy aside, then the slow "rediscovery" of gas- and natural gas-powered vehicles did not begin until the 1970s, after the traditional oil derivative-based automotive system received a severe shock. The reason for the destabilization this time was a politically-induced exogenic shock, namely the first oil crisis beginning in 1973, which could be traced back to the OPEC states choking off oil production in reaction to the Yom Kippur War. As a consequence of this oil crisis, the dependence of the western industrial nations on oil as a source of energy and raw material came into stark relief. Moreover: With the first oil crisis, the illusion of the unlimited and permanent availability of cheap oil disappeared from public perception. The effect of this first "oil shock" was intensified by the second oil crisis in 1979/80, whereby this second crisis could be attributed to drops in supply and a general sense of insecurity after the Islamic Revolution in Iran.

In the medium term, a second significant social and cultural transformation influenced the further development of gas-powered vehicles, a change that also began taking shape primarily in the 1970s. We're speaking here of the environmental movement which was arising at this time, which must be characterized as part of a profound change in values. On the one hand, user and consumer preferences regarding automobiles changed as a consequence of this change in values, whereby these changes were neither nationally nor internationally uniform, nor truly consistent over the long term. On the other hand, what proved to have a more powerful effect was the change in national regulation policy emanating from the USA with respect to automobiles, in which nations (at least in the industrialized world) began increasingly to take on the role of "product regulators". This effectively brought the era of largely autonomous automobile design by manufacturers to an end.

All in all, the 1970s marked the beginning of an era in energy and environmental history whose consequences could also not fail to affect the evolution of gas-powered vehicles. From this moment on, it is possible to speak in terms of a more or less continuous, albeit still nationally fragmented, development of (natural) gas-powered vehicles.

In reaction to the first oil crisis, several countries (initially primarily the English-speaking nations) initiated government-sponsored research and development programs addressing natural gas mobility as well as state-subsidized market entry trials. NGVs garnered special attention during this time as a potential means of reducing dependence on oil as a vehicle fuel, whereas environmental aspects initially occupied a completely subordinate position. This is particularly true for countries with significant natural gas reserves, as the NGV offered them the possibility of reducing their own dependence on oil imports. To a certain extent, then, aspects of self-reliance are playing a role here once again, albeit one which must be viewed against the special backdrop of the energy market development of the 1970s as outlined above.

More comprehensive natural gas programs were established to some degree in the USA, Canada, and in New Zealand, whereby the New Zealand program was the one that was initially most successful and is, in my opinion, the most interesting one to date regarding its development. We shall therefore take a closer look at this program.

In response to the first oil crisis and the country's high degree of dependency on oil imports from the Middle East, the government of New Zealand embarked on a comprehensive energy program in the 1970s that was primarily designed to lead to a higher rate of domestic production of fuel for motorized transportation. The core of this program was the "Compressed Natural Gas (CNG) Vehicle Program," which was designed to aid in the development and establishment of natural gas vehicles. The energy base underlying the program consisted of New Zealand's natural gas reserves and a

Diagram 3.4 Early "natural gas fueling station" in New Zealand, 1979. *Source*: Energy Supply and Use. In: Tearra. The Encyclopedia of New Zealand. http://www.teara.govt.nz/en/photograph/21464/cng-service-station. 05 08 2014

pipeline infrastructure that was largely able to comprehensively supply at least the more densely populated northern island by the beginning of the 1970s.

The method of choice in New Zealand was converting conventional gasoline-powered vehicles to natural gas, whereby Italian conversion kits were used initially, which then of course had to be appropriately adapted to local conditions and came increasingly to be manufactured by domestic suppliers. By the mid-1980s, a network of shops certified to convert automobiles and a total of roughly 370 natural gas fueling stations had been established, so that a sufficiently dense fueling station infrastructure was in place, at least on the northern island. The New Zealand government subsidized the construction of these fueling stations as well as the vehicle conversions; the gas suppliers themselves further promoted the conversion through the use of fuel coupons (Diagram 3.4).

Under pressure from high gasoline and low natural gas prices, as well as under the effects of state support, the New Zealand CNG program quickly proved successful. By the end of the 1980s, around 120,000 natural gas vehicles were on the road in New Zealand, corresponding to a share of 11 % of all motor vehicles. A percentage, moreover, that had not been achieved anywhere else in the world. However, the first cracks in the New Zealand natural gas success story began to show in the mid-1980s: After a change in government from the National Party to the Labour Party in 1984, the latter abolished most of the financial incentives for converting to natural gas in 1986. In addition, the well-known drop in oil prices in the 1980s made conversion to natural gas operation less financially attractive in any case. Lastly, although a sufficiently dense conversion infrastructure had been built in New Zealand, it was not accompanied with corresponding service and repair capabilities, so that even minor technical defects in the vehicles led to problems that negatively affected customer satisfaction with natural gas-powered vehicles in a significant way. The combination of all these factors led to a situation where, after an NGV boom up to 1985, the annual number of conversions simply collapsed in the second half of the decade. In the end, only vehicle fleet operators such as taxi companies remained invested in natural gas, until finally these also lost interest in the alternative fuel. The most recent statistics for New Zealand for 2012 show a total of 65 (!) natural gas vehicles, which means that there is literally nothing left of the "CNG Vehicle Program" of the 1980s. While the New Zealand example may on the one hand serve to document the resurrected interest in gas- and especially natural gas-powered vehicles as "self-reliance" and "conservation machines" as a consequence of the oil crises, it also clearly shows how dependent any lasting entrenchment of NGVs is on a host of different background conditions. Moreover, New Zealand's example also serves to document how quickly a vacillating or spasmodic state-supported policy can undo achieved successes.

Let us return to the observation that the environmental movement taking shape in the 1970s, that the social and cultural alteration as a consequence of this, and that the corresponding change in national environmental policy brought about new conditions for the development, acceptance, and distribution of gas- and particularly natural gas-powered vehicles. This transformation was manifested in the old industrial nations primarily from the 1990s onwards, and as part of it new attributes began to gain significance, now adding the role as "environmental protection machine" to the NGV's role as "(oil) conservation machine."

However, this simple diagnosis is not enough to explain the significantly more intensive development, public perception, and use of natural gas vehicles on an international scale. Rather, it is also a matter of an increasingly comprehensive and more complex international system of players. A critical factor for the accelerating spread of NGVs was initially the development in a few quite different 'emerging nations' for whom the NGV was of particular interest in view of domestic or simply more accessible natural gas reserves primarily as a "self-reliance" or at least as an "(oil) conservation machine." Foremost among these are Iran, Pakistan, and India, as well as the South American nations of Argentina and Brazil.

If we look at the evolution in the Federal Republic of Germany, then the defining players here are initially a few natural gas suppliers, whose attention was drawn in the 1990s to the NGV as a potential "market expansion machine." The attribute of "environmental protection machine" became relevant only secondarily, while at the same time its nature as a "conservation machine" remained of primary importance for NGV marketing. We will take a closer look at these quite different developments below.

Let us begin with the emerging nations, which we might characterize as NGV "intensive user-nations": In 2012, the largest number of NGVs were operated in Iran, i.e. roughly three million, followed closely by Pakistan with some 2.9 million and Argentina with around 2.14 million vehicles. Around 1.73 million NGVs were on the road in Brazil in 2012, and around 1.25 million in India. Pakistan occupies a special position in that around two-thirds of all permitted automobiles were NGVs, thus reaching (albeit at an overall low level of motorization) a percentage of gas-powered vehicles that leaves the other intensive user-nations still far behind.

In spite of all the differences between the various countries where the integration of NGVs into the existing automobile system seems to have succeeded, what they have in common is that they all established comprehensive government support programs in the 1990s. The comparatively rapid spread of NGVs would be inconceivable without these support programs. While there were great differences in the details, the actions taken basically corresponded to the case study described for New Zealand: State support for vehicle conversion or procurement of new NGVs, tax credits or complete tax exemption for NGVs along with government-subsidized construction of a natural gas fueling station network, with Pakistan having built the most dense supply infrastructure with around 3000 natural gas stations. In all of the cases, the natural gas supply itself is being done almost exclusively by pipeline. Another thing all of the countries have in common is a low-price policy regarding natural gas, which further enhanced the relative benefits for the user of converting to natural gas over gasoline or diesel.

The motives and goals of the respective natural gas programs are vastly different. In the case of Iran, it was above all the lack of domestic refinery capacity that induced the government to initiate a natural gas program in the early 1990s, despite its adequate oil reserves. The goal was primarily to achieve greater independence from gasoline and diesel imports in the face of sanctions and threatened sanctions. In Pakistan, the national natural gas program began in 1992, with the most important development goals in this case being energy security and security against rising oil prices. In Brazil and Argentina as well, the focus was on the urgency of converting at least a portion of the vehicular traffic to domestic sources of energy. In Argentina, this conversion took place in parallel to the privatization of previously state-owned gas supply companies. In Brazil, the conversion to natural gas vehicles received an additional push when an expanded natural gas supply made possible by the startup of the GASBOL pipeline to the Bolivian natural gas fields at the end of the 1990s reduced prices even further. In both Argentina and Brazil, moreover, special emphasis was also placed from the official side on the environmental aspects associated with converting to natural gas. The goals of environmental protection and/or clean air were truly decisive for the natural gas programs in the metropolitan areas in India, where the incentive programs were primarily initiated at the municipal level. The explicit goal of these programs was to improve the emissions situation in major Indian cities, beginning in the 1990s in Delhi and Mumbai and then taken up over the course of the decade by some 30 Indian cities.

What all of the intensive user-nations have in common again is that it was predominantly conversion kits that formed the automotive technology basis for the use of natural gas. In the meantime, domestic vehicle producers in India (Tata Motors and Maruti Suzuki) and Iran (Khodro and Saipa) are now offering mass-produced natural gas vehicles. Bivalent vehicles, i.e., vehicles that can use both gasoline and natural gas, are used almost without exception in all of the countries mentioned above. This somewhat limits the adaptation of the engines to the specific properties of natural gas fuel—above all its antiknock characteristics—which in turn also limits the (environmental protection) potential of NGVs. Moreover, in all of the countries, it was the vehicle fleet operators, especially taxi and motorized rickshaw businesses, who were the first to convert to natural gas, and who therefore played an important role as pioneering users. Lastly, it should be noted that in all of the intensive user-nations, natural gas vehicles have been deployed for short-distance, and primarily local municipal transportation, rather than long-distance transportation. The fueling station infrastructure is accordingly concentrated in the city centers.

All of these examples serve to document the fact that rapid implementation of even large numbers of natural gas vehicles is possible under the conditions of proactive state promotion with significant economic benefit. In the Asian intensive user-nations at least, we are talking about countries where personal ownership of motorized transportation was still at a relatively low level at the time of the first natural gas vehicle incentive programs. It is hardly possible in these cases to speak of an established "car culture" that one had to "break into." In Brazil, on the other hand, the natural gas incentive program was appended to the already successful "Proàlcool" [pro-alcohol] program of the 1980s, which successfully propagated the use of ethanol fuels. A "flexible fuel culture" had thus already been established there, into which natural gas also fit as a new alternative fuel, so to speak. Lastly, we should note that, in Pakistan at least, a kind of "sorcerer's apprentice" problem has arisen in the meantime, in that the conversion to natural gas vehicles has been so successful that the growing demand can no longer be met with domestic reserves, and there is no infrastructure for importing natural gas. This has meanwhile made reversion to gasoline operation and abandoned natural gas fueling stations a widespread phenomenon in Pakistan.

Now let us turn our attention to the gas-powered vehicle renaissance in Germany: We should begin by noting that the developments surrounding oil prices in the 1970s discussed above also brought about some modest successes in Germany for another gas-powered vehicle, namely cars fueled with liquefied petroleum gas (LPG). LPG, a butane-propane mixture of varying composition, is produced as a "wet gas" during natural gas and oil production, as well as in the form of a byproduct of the petroleum refining process, and can be stored as a liquid at comparatively low pressures of five to ten bar. In the Federal Republic of Germany, vehicles converted to run on LPG enjoyed an initial wave of success in the late 1970s, saw their importance recede starting in the mid-1980s, only to experience a comeback in the last 10 years. A critical factor in this new upsurge has been the price advantage LPG has over gasoline and diesel fuel thanks to the corresponding tax policy, which we will come back to later. According to the Kraftfahr-Bundesamt [German Federal Motor Transport Authority], there were around 456,000 authorized LPG vehicles and a network of over 6000 LPG fueling stations in Germany in 2011.

Regarding natural gas vehicles, we have already briefly mentioned the fact that the gas industry was the initial driving force for more intensive deployment of NGVs in the 1990s. Thus it was the importers and sellers of natural gas who recognized in vehicular transportation a potentially receptive market for their product. This had been preceded by some three decades of expanding natural gas sales to German homes, but in the early 1990s the market began to show signs of stagnation (see above). In addition, the already well-developed gas supply infrastructure in Germany called for opening a new and potentially very absorptive market.

But opening up this market for NGVs was difficult in many respects for the gas suppliers. Although, as we have seen at least in the case of bivalent converted vehicles, the technology in the 1990s was already completely feasible for everyday use, the proper preconditions for rapid market introduction did not yet exist in Germany. Not even a rudimentary fueling station infrastructure existed, the gas suppliers had no experience whatsoever in the fuel market, there were institutionalized contacts with either the fuel industry or automobile manufacturers, and political support still had to be organized, while the NGV had up to that date received almost no public or media attention in Germany. Lastly, the attempts to break into the market also suffered from the aforementioned decentralized structure of the German gas industry and from the fact that within the sector itself there was absolutely no unity on how promising the fuel market actually was.

The 1990s thus were marked by a combination of different strategies. Individual gas suppliers initially converted their own vehicle fleets to natural gas and made further commitments to try and create fleets of natural gas buses in a series of German cities. Supported by the former Bundesministerium für Umwelt, Naturschutz und Reaktorsicherheit [Federal Ministry for the Environment, Nature Conservation, and Nuclear Safety] (BMU), a total of around 5000 vehicles had been permanently or temporarily converted to natural gas by the end of the 1990s, and some 160 fueling stations had sprung up in Germany, creating the basis for building a rudimentary network.

Another significant result of efforts to establish natural gas vehicles was the successful forging of new strategic alliances, which was manifested in particular in the founding of the "Trägerkreis Erdgasfahrzeuge" [NGV Support Group] in 2000, which still exists today. In addition to the energy and gas industry enterprises, founding members of the support group also included the oil company ARAL, the automobile club ADAC, the carmakers Fiat, Opel, Volkswagen, and Volvo, along with the BMU. This not only created the foundation for nationwide uniform information campaigns, the board also established a commitment to mass production of NGVs and for the expansion of the natural gas fueling station network, which was incentivized through investment subsidies of around 50 % per station.

The initiatives of the gas industry were accompanied by more proactive public relations activities, with the goal of establishing the perception of the NGV among potential institutional and private customers primarily as a "conservation machine" and as an "environmental protection machine" (the campaign's motto was "Natural Gas Vehicles: Thrifty, Safe, Cleaner"). Of central importance for both characterizations was the fact that NGVs had been receiving state support since 1994 in the form of a considerable reduction in fuel taxes for natural gas and biogas—by the way also for LPG (see above)—as well as exemption of monovalent natural gas vehicles from the vehicle tax. To be sure, NGVs were not a key ecological technology in German environmental policy; nevertheless, the fleet-based attempts alone discussed above show how thoroughly a series of projects was promoted by not just the Environmental Ministry, but also the Umweltbundesamt [Federal Environment Agency] and individual German states. The oil tax increase enacted in several stages in 1999 by the SPD-Green Party national coalition government ("eco-tax") also increased the relative economic advantage of (natural) gas-powered vehicles.

It seems worthwhile in this connection to insert a few remarks on the nature of the NGV as "environmental protection machine." The NGV can be described as a "conformist environmental technology." This means that inherent in the NGV is a promise to the user that it will behave in a (relatively) environmentally friendly manner, without requiring the user to give up his usual mode of transportation and its associated comfort level. This differentiates the NGV from other alternative vehicles, particularly the electric car which, due to its system-specific characteristics, forces its users to change their former habits. This applies as well to the newest generation of electric vehicles despite the progress in battery capacity and charging times. The attractiveness of the NGV, consists not least in the fact that it promises users "enjoyment without regret," as a 2005 market analysis literally expressed it. In other words: It is possible to integrate the conformist NGV environmental technology into the existing sociotechnical system of automobile transportation. It makes the system "greener" without destroying its familiar features. And that is precisely what makes this technology so attractive for its promoters and its users alike.

If the 1990s were a sort of dawn of the NGV in Germany, then a stage of development that with all due caution might be characterized as an early deployment phase began around the turn of the twenty-first century. One important result of this early deployment phase is the growing number of natural gas fueling stations in Germany. By 2013, a total of some 920 natural gas fueling stations had been built, whereby it should not be overlooked that a quite rapid period of expansion up to 2010 has been followed by some stagnation in the last few years. In spite of some successes, however, the goal of building 1000 fueling stations by 2004 as originally articulated by the NGV Support Group clearly proved to be unattainable.

A second significant result of the deployment phase is that the number of NGVs rose from the aforementioned number of around 5000 in 2000 to a solid 95,000 in 2012, albeit still comprising scarcely 0.2 % of the total German motorized vehicle inventory. What was unquestionably successful, however, is the fact that the NGV was able to move beyond only experimental and fleet operations to becoming established among at least a few private early-adopters. Nevertheless, in Germany on average only some 100 natural gas vehicles are operated per natural gas fueling station, which is well short of what is needed for the stations to be able to operate profitably. This determination underscores a fundamental and persistent dilemma arising from the close interaction between costly expansion of the fueling infrastructure and the true everyday feasibility of NGVs that is associated with it. Despite the progress achieved, a "chicken-and-egg" problem still exists today whose resolution appears to be of critical importance to the chances for success in the future of natural gas transportation (Diagram 3.5).

One important difference between the deployment and the early phases of natural gas transportation in Germany is the fact that in the meantime manufacturers have begun offering a series of mass-produced NGVs. In the mid-1990s, only Ford and Volvo had

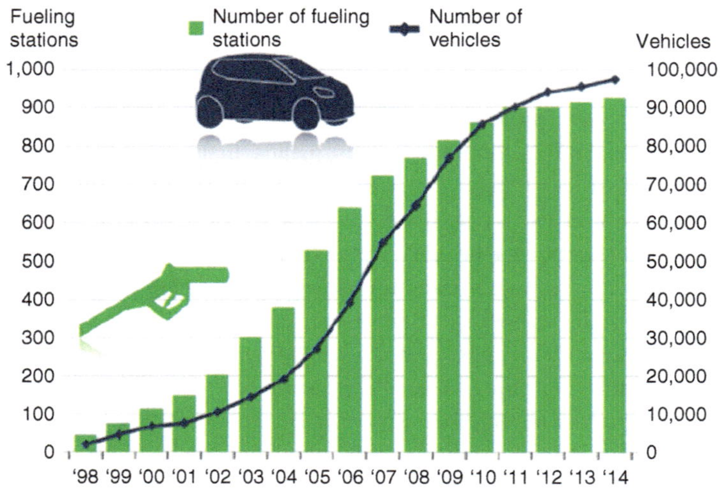

Diagram 3.5 Fueling station and vehicle inventory numbers in the Federal Republic of Germany from 1998 to 2014. *Source*: German Federal Motor Transport Agency, data as of March 1914. Illustration from: Erdgas-Mobil, press releases. http://www.erdgas-mobil.de/presse-aktuelles/bilder. 05 08 2014

NGV models available on the German market, while natural gas transportation remained primarily based on conversion of conventional vehicles, with all of its associated technical disadvantages, as discussed earlier. Since 1999, mass-produced NGVs have been available from Fiat, and in 2000 the GM subsidiary Opel, then Volkswagen and Mercedes finally introduced natural gas models. In the meantime, one Audi model and two models each from the VW subsidiaries Skoda and Seat have been added to the mix. The main advantage of mass-produced NGV models is derived on the one hand from the lower accessory costs in comparison to conversion, primarily with monovalent models—i.e. vehicles designed exclusively for use with natural gas—but on the other hand from the fact that the engines have been optimized to use natural gas, and are therefore able to take full advantage of the antiknock properties of natural gas through higher compression. This improves the efficiency of the engines, leading directly to reduced greenhouse gas emissions. It is also worth mentioning in this connection that at the end of the first decade of the twenty-first century, a new generation of NGV engines began appearing on the market. These NGVs are equipped with specially adapted turbochargers or compressors, thus making it possible to increase engine efficiency even more. These turbocharged engines now give NGVs a much greater range that is beginning to attain an order of magnitude approaching that of conventional gasoline and diesel vehicles. Despite these developments, however, it remains a fact that natural gas-powered engines still have considerable developmental potential that can be expected to be accompanied by additional increases in efficiency and reduced emissions. It likewise remains true that the number of mass-produced NGV models on offer remains quite small, making the switch to a gas-powered vehicle difficult for potential users.

All in all, more recent market analyses are showing that the coupling of customary transportation behavior with economical and ecological benefits is what users and potential users especially value with respect to conformist NGV environmental technology. On the other hand, the lack of suitable models and the as yet still widely scattered network of fueling stations are seen as strong negatives. It is worth noting that the questionnaire results naturally stem from relatively well-informed (potential) customers, whereas the overall perception of NGVs is still inadequate.

In my opinion, another very important development in the deployment phase of natural gas transportation in Germany was again closely linked with a changing attribution or interpretation of the technology. The term "greenhouse-gas emissions" used above clearly shows that the NGV began to play a role in addressing yet another now central issue in the twenty-first century, namely that of climate change or climate-compatible transportation. For the same consumption, NGVs are known to emit almost one-fourth less climate-related carbon dioxide as conventional vehicles due to methane's more favorable carbon-hydrogen ratio in comparison to petroleum-based fuels. This purely fuel-related benefit is further enhanced in combination with the greater efficiencies delivered by the newest generation of natural gas engines. In connection with the climate issue, the NGV has developed beyond its general role as an "environmental protection machine" to a "climate protection machine," considerably raising its ecological relevance. This is all the more true in connection with the "Power-to-Gas" technology described above, where in this case, to repeat ourselves once more, the promising prospects for a climate-neutral natural gas transportation system are becoming clear.

In conclusion, a cursory review of the selected chapters of the history of gas- and especially natural gas-powered vehicles presented above fully confirms the theses formulated at the outset. The only chance the (natural) gas-powered vehicle "niche technology" would ever have to penetrate the conventional automobile-industrial complex that has become so firmly established since the inter-war era was if the latter became destabilized by changing economic, political, societal, and cultural conditions. Even during these phases of the destabilization, persistent government or policy support would be needed to implement the (natural) gas vehicle and to provoke meaningful efforts at developing it further. The primary goal of the ongoing development in the Third Reich, no different from that in Italy of the 1950s, New Zealand in the 1970s and 1980s, or in the intensive user-nations since the 1990s, was to achieve relative independence from oil or fuel imports or, in other words, an at least partial degree of energy independence.

Thus, if the (natural) gas-powered vehicle began its career as a "self-reliance" or "oil conservation machine," it took on a second interpretation after the 1970s as an "environmental protection machine." This latter interpretation has been of increasing importance for the ongoing promotion, development, and implementation of gas-powered vehicles, especially in regard to securing the necessary political support for the NGV in the industrialized nations of Europe and North America. The recently added interpretation as a "climate protection machine" can be described as a preliminary culmination of this second career. In Germany in particular, further development of the NGV will doubtless be dependent on its positioning as an instrument for climate protection and as one of the spearheads of the energy turnaround announced by the federal government.

Moreover, the history of the (natural) gas-powered vehicle clearly shows that it will only be adopted by users if its relative advantage over conventional vehicles is sufficiently large. In other words: The implementation of the NGV can only succeed among larger user groups if it can meet the expectations associated with a "conservation machine" that is suitable for everyday use (!). In this regard as well it will require government commitment, since without this commitment, any appropriate fueling and maintenance infrastructure is as unimaginable as adequately ensuring low fuel costs through tax-preferred treatment of natural gas, for example. If this commitment is terminated prematurely, then the conventional automobile-industrial complex will regain its stability rapidly at the expense of the NGV, as seen in the example of New Zealand.

Only long-term government commitment will succeed in creating a sufficiently stable "development space," within which it will be possible to pursue a continuous further development of the technology and its transition to vehicle mass production. As we have shown, it has been only in the last 15 years that this has proven possible to achieve in the western industrialized nations, whereby it has only recently become possible to take practical advantage of the theoretical benefits arising from the good antiknock properties of natural gas. The potential for further development here continues to appear considerable.

From the historical perspective, we can say the following with respect to the future potential for development of the NGV, at least regarding the situation in Germany: If there is to be any chance of granting the NGV a future as a "climate protection machine" compatible with the reigning automobile-industrial complex, then the tax preference for natural gas as an automobile fuel must without question be extended beyond its current scheduled expiration date of 2018. Moreover, it should be kept in mind that the current strong political inclination towards seeing electric vehicles as the answer to the challenges of the energy turnaround and climate change carries with it the risk of making a premature decision in favor of only one of several possible technological paths. The potential of the natural gas vehicle as an important element of any future environmental and climate-compatible transportation culture should not in any case be abandoned unnecessarily.

Further Reading

1. Blümel H, Müller R (2005) 1000 Erdgastankstellen—und wer tankt? GWF. Gas, Erdgas 146:S 362–374
2. Blümel H, Müller R (2005) Kann das Wasserstoffauto vom Erdgasauto lernen? Erdgasauto-Pioniere und ihre Kundensicht—eine erste Längsschnittanalyse. In: Rammler S, Weider M (Hg). Wasserstoffauto. Zwischen Markt und Mythos, Münster, S 133–150
3. Bohn K, Cramer U, Wackertapp H (1994) Erdgaseinsatz im Verkehrssektor. Praxiserfahrung mit Erdgasfahrzeugen unter besonderer Berücksichtigung des Demonstrationsvorhabens Mainz/ Wiesbaden. GWF. Gas, Erdgas 135:S 545–553

4. Brunckhorst H-D (1978) Kommunalisierung im 19. Jahrhundert. Dargestellt am Beispiel der Gaswirtschaft in Deutschland. Tuduv-Verlags-gesellschaft, München
5. Bundesumweltministerium (Hg) (1998) Gasfahrzeuge von A bis Z. ohne Ort, Duisburg
6. Calabrese G (ed.) (2012) The greening of the automotive industry. Palgrave Macmillan, London
7. Cannon JS (1989) Drive for clean air: natural gas and methanol vehicles. INFORM, New York, NY
8. Cannon JS (1993) Paving the way to natural gas vehicles. INFORM, New York, NY
9. Carle G (2004) Erdgasfahrzeuge im Wettbewerb (internal working paper of the Federal Institute of Technology [ETH] in Zürich). ETH Zürich
10. Carle G (2006) Erdgasfahrzeuge und ihr Beitrag zu einer CO_2-Reduktion im motorisierten Personenverkehr in der Schweiz. Dissertation, ETH Zürich, Zürich
11. Castaneda CJ (1999) Invisible fuel: manufactured and natural gas in America, 1800–2000. Twayne, New York, NY
12. Di Pascoli S, Femia A, Luzzati T (2001) Natural gas, cars and the environment: a (relatively) 'clean' and cheap fuel looking for users. Ecol Econ 38:S 179–189
13. Eckermann E (2010) Alternative Antriebe. Die Zeit ist reif. Kultur Technik 03:S 6–10
14. Eckermann E (2008) Fahren mit Holz. Geschichte und Technik der Holzgasgeneratoren und Ersatzantriebe. Delius Klasing, Bielefeld
15. Eckermann E (2002) Vom Dampfwagen zum Auto. Die Motorisierung des Verkehrs. Delius Klasing, Bielefeld
16. Eckert F (1936) Leuchtgas-Zapfstelle für Kraftwagen. ZVDI 80:S 30
17. Erdgas Mobil: Erdgas: Natürlich Mobil (http://www.erdgas-mobil.de. 05.08.2014)
18. Ermo Vd' (1991) A statistical research on the role of natural gas in the world energy picture during the last 60 Years, Groningen
19. Fell C (2011) Kraftstoff CNG—Fahren mit Erdgas. 20 Jahre Erdgas-Fahrzeuge in Österreich, Forum Special, Sonderheft des Forum Gas, Wasser, Wärme, 5
20. Finkbeiner H (1935) Gaserzeuger für Kraftwagen. ZVDI 79:S 665–667
21. Flade J (2012) Gas, die treibende Kraft. Geschichte und Zukunft einer Antriebsenergie. Marketing-+-Wirtschaft-Verl.-Ges. Flade & Partner, München
22. Gibgas—Consulting/Medien: Gibgas. Mobil mit CNG. http://www.gibgas.de. 05 08 2014
23. Halblützel E (1940) Gas als Treibstoff für Automobile. Schweizerische Bauzeitung 116:S 179–181
24. Hamm M (2006) "Gib Gas?" Ökologische Ambivalenzen und Akzeptanzprobleme konformistischer Umwelttechnologien am Beispiel von Erdgasfahrzeugen. Technikgeschichte 73:S 207–225
25. Hayes MH, Jaffe AM, Victor DG (Hg) (2006) Natural gas and geopolitics: from 1970 to 2040. Cambridge University Press, Cambridge
26. Heymann M (2006) Ingenieure, Märkte und Visionen Die wechselvolle Geschichte der Erdgasverflüssigung. Piper Verlag, München
27. Högselius P (2013) Red Gas: Russia and the Origins of European Energy Dependence. Palgrave Macmillan, New York, NY
28. Holbein O (1943) Das Tanken der BVG-Omnibusse mit Flüssiggas. ZVDI 87:S 192f
29. International Association of Natural Gas Vehicles (IANGV) NGV global. Natural gas vehicle knowledge base. http://www.iangv.org. 05 08 2014
30. Illing F (2012) Energiepolitik in Deutschland. Die energiepolitischen Maßnahmen der Bundesregierung 1949–2013. Nomos, Baden-Baden
31. Ingersoll JG (1996) Natural gas vehicles. Fairmont Press, London
32. Kalischer P (1994) Erdgaseinsatz im Verkehrssektor. Ein Beitrag für eine ökologisch verträgliche Mobilität. GWF. Gas, Erdgas 135:S 536–544

33. Kirsch DA (2000) The electric vehicle and the burden of history. Rutgers University Press, New Brunswick, NJ
34. Kolb RW (2014) The natural gas revolution: at the pivot of the world's energy future. Pearson, Upper Saddle River, NJ
35. Körting J (1963) Geschichte der deutschen Gasindustrie. Mit Vorgeschichte und bestimmenden Einflüssen des Auslandes. Vulkan-Verlag Classen, Essen
36. Leunig G (1938) Motor und Kraftstoff. Fortschritte, Aufgaben, Erfahrungen. Ergebnisse der Wissenschaftlichen Herbsttagung des VDI in Augsburg 1938. ZVDI 82:S 1401–1409
37. Lübke A (1943) Das Deutsche Rohstoffwunder. 2 Bände, Stuttgart
38. Marcetti C (1984) On the long-term history of energy markets and the chances for natural gas. IIASA, Laxenburg
39. Mom GP (2013) The electric vehicle: technology and expectations in the automobile age. JHU Press, Baltimore, MD
40. Möser K (2002) Geschichte des Autos. Campus Verlag, Frankfurt/Main
41. Nijboer M (2010) The contribution of natural gas vehicles to sustainable transport. International Energy Agency: working paper, Paris
42. Verfasser O (1936) Leuchtgas als Treibstoff für Omnibusse. Schweizerische Bauzeitung 107/108:S 281
43. Verfasser O (1942) Omnibusbetrieb mit unverdichtetem Stadtgas. ZVDI 86: S 638
44. Verfasser O (1943) Omnibuszüge mit Stadtgas. ZVDI 87:S 29
45. Pastonesi G (1955) Produktion und Verwendung von Erdgas in Italien. Chemie—Ingenieur—Technik 27:S 465–469
46. Podonik B (1999) Toward a sustainable energy regime: a long-wave interpretation of global energy shifts. Technol Forecast Soc Chan 62:S 155–172
47. Riedel E (1935) Betrieb von Kraftfahrzeugen mit verflüssigten Gasen. ZVDI 79:S 579–585
48. Rixmann W (1936) Leuchtgasbetrieb von Fahrzeugmotoren. ZVDI 80:S 627–632
49. Ryssel W (1936) Klärgas als Kraftstoff für die Fahrzeuge der Gemeinden. ZVDI 80:S 1290
50. Schumacher W (1938) Bemessung und Bau von Gastankanlagen. ZVDI 82:S 585–590
51. Smil V (2010) Energy transitions: history, requirements, prospects. Praeger, Santa Barbara, CA
52. Stangier M (2008) Gasmotoren—Entwicklung und Geschichte sowie die exemplarische Behandlung des PKW Betriebes mit Autogas. GRIN-Verlag, München
53. Stocchetti A, Valpato G (2012) CNG cars: sustainable mobility is within reach. In: Calabrese G (Hg) The greening of the automotive industry. Palgrave Macmillan, London, S 140–163
54. Tomory L (2012) Progressive enlightenment: the origins of the gaslight industry, 1780–1820. MIT Press, Cambridge
55. Troesch M (1940) Ersatztreibstoffe für Automobile. Schweizerische Bauzeitung 116:S 235–239
56. Victor DG (2007) Natural gas and geopolitics. From 1970 to 2040. Cambridge University Press, Cambridge
57. Wedler M (2008) Alternative Kraftstoffe und Antriebssysteme für PKW-Fahrzeuge: Vergleich zwischen alternativen Kraftstoffen (Biodiesel/Pflanzenöle, Wasserstoff und Erdgas) und Antriebssystemen (Brennstoffzelle, Hybrid- und Elektromotoren). GRIN-Verlag, München
58. Wiborg K (1982) Erdöl und Erdgas: fünf Jahrzehnte Exploration und Produktion in Deutschland. Mobil-Oil-Aktiengesellschaft Deutschland, Hamburg
59. Wiesenberg R (2000) Erdgas als Treibstoff für den Straßenverkehr als Chance für Energieversorgungsunternehmen. Eine Szenario-Analyse für die swb AG, Hamburg

Natural Gas and Renewable Methane 4

Siegfried Bajohr, Michael Bargende, Tobias Block, Jochen Brellochs, Helmut Eichlseder, Volkmar Frick, Hendrik Gosda, Frank Graf, Elias Hammer, Janet Hochi, Oliver Jochum, Marco Klemm, Joachim Krassowski, Franziska Müller-Langer, Reinhard Otten, Michael Schlüter, Reinhard Schultz, Michael Specht, Bernd Stürmer, Adalbert Wolany, Nantje T. Zimmermann, and Ulrich Zuberbühler

S. Bajohr (✉)
KIT, Engler-Bunte-Institut, Bereich Chem. Energieträger – Brennstofftechnologie (EBI-ceb), Engler-Bunte-Ring 1, 76131 Karlsruhe, Germany
e-mail: Siegfried.bajohr@kit.edu

M. Bargende • A. Wolany (✉)
Forschungsinstitut für Kraftfahrwesen und Fahrzeugmotoren Stuttgart (FKFS), Pfaffenwaldring 12, 70569 Stuttgart, Germany
e-mail: Adalbert.wolany@ivk.uni-stuttgart.de

T. Block • H. Gosda • E. Hammer • R. Otten (✉)
Audi AG, I/EG-X, 85045 Ingolstadt, Germany
e-mail: Reinhard.otten@audi.de

J. Brellochs • V. Frick • M. Specht (✉) • B. Stürmer • U. Zuberbühler
Zentrum für Sonnenenergie- und Wasserstoff-Forschung Baden-Württemberg (ZSW), Industriestraße 6, 70565 Stuttgart, Germany
e-mail: michael.specht@zsw-bw.de

H. Eichlseder (✉)
Neue Welt Höhe 45c, 8042 Graz, Austria
e-mail: eichlseder@ivt.tugraz.at

F. Graf (✉)
DVGW-Forschungsstelle am Engler-Bunte-Institut des KIT, Engler-Bunte-Ring 1, 76131 Karlsruhe, Germany
e-mail: graf@dvgw-ebi.de

J. Hochi (✉) • R. Schultz, M.A. • N.T. Zimmermann, M.A.
Biogasrat, Mittelstraße 55, 10117 Berlin, Germany
e-mail: janet.hochi@biogasrat.de

O. Jochum • J. Krassowski (✉)
Fraunhofer-Institut für Umwelt-, Sicherheits- und Energietechnik UMSICHT, Osterfelder Str. 3, 46047 Oberhausen, Germany
e-mail: Joachim.krassowski@umsicht.fraunhofer.de

R. van Basshuysen (ed.), *Natural Gas and Renewable Methane for Powertrains*, Powertrain, DOI 10.1007/978-3-319-23225-6_4

4.1 The Basics of Natural Gas

Adalbert Wolany and Michael Bargende

According to the current hypothesis, natural gas is a fossil energy carrier that, just like oil, underwent organic transformation under high pressure and in the absence of oxygen in the interior of the earth and therefore its composition as a naturally occurring product varies according to its geological source. The main component is always methane, which varies between 75 and 98 %. Other components include nitrogen, ethane and carbon dioxide.

During natural gas recovery, sulfur and other unwanted components are removed, then water and long-chain hydrocarbons are eliminated in a drying process. The role of natural gas in the extension of alternative and renewable energies is of growing importance. Currently in Germany, 80 % of natural gas is used in the heating market, meaning that only a small amount of natural gas is used for electricity generation or in the area of mobility [1]. However, its use in the transportation sector is continually increasing.

Energy forecasts initiated by the Federal Ministry for Economic Affair and Technology (BMWi) [2] assume the establishment of natural gas in the transportation sector. By 2030, natural gas will make a contribution of approximately 4 % to the energy mix in transportation and 11 % by 2050. Hybrid and natural gas vehicles will probably enjoy the largest market share amongst alternative powertrains without further political interventions due to their comparably lower sales prices. By 2030, market shares of 9 % have been forecast for these passenger car powertrains, growing to 20 % in 2050 [2]. The main reason for the increase in importance of methane-based fuels—including natural gas, biogas, substitute natural gas, HCNG—as alternative energy sources is primarily the more favorable CO_2 emissions and the long-term availability of the resources. Based on data from Germany's Federal Institute for Geosciences and Natural Resources (BGR) [3], a static lifetime[1] has been calculated for conventional and unconventional natural gas resources[2] of 58 years and a further 294 years of resources.

M. Klemm • F. Müller-Langer (✉) • M. Schlüter
DBFZ Deutsches Biomasseforschungszentrum gemeinnützige GmbH, Bereich Bioraffinerien, Torgauer Straße 116, 04347 Leipzig, Germany
e-mail: Franziska.mueller-langer@dbfz.de

[1] The static lifetime of a raw material is the ratio of the reserves or the total resources (sum of reserves and resources) of a raw material to its annual output and corresponds to the number of years that a non-renewable raw material is globally available assuming constant consumption.

[2] A distinction is made in gas extraction between conventional and unconventional gas reserves. In conventional gas extraction, the gas is under pressure and escapes from the rock predominantly by itself. In unconventional gas reserves however, a fracking process is used, whereby the pressure required to extract the gas is created, for example using a mixture of water, sand and chemicals.

Other sources of literature [4] calculate static lifetimes for conventionally extracted gas reserves including resources of 165 years and a further 246 years for unconventional gas sources. The distribution of natural gas reserves is an energy-political advantage, since over two thirds lie outside OPEC states that already control approximately three quarters of the world's oil reserves. Thus a better security of supply can be achieved in an overall energy-political concept through much larger diversification. Additionally, the currently discussed production of biogas from renewable raw materials and its feeding in to the existing natural gas grid opens up new and sustainable perspectives. An increase of 95 % by 2030 to a share of 10 % is forecast for energy consumption via biogenic fuel and biogas in the transportation sector [2]. This increase in the use of biogas in transportation also has the potential of decarbonization. Even conventional natural gas of medium quality has a CO_2 formation potential of under 60 g CO_2/MJ for complete combustion. This reduces levels by approximately 20 % compared to gasoline engines due to the more favorable carbon/hydrogen ratio and is under the values of other alternative fuels such as ethanol and methanol. In addition to this, the emission advantages for CO, NO_x and higher hydrocarbons are up to 80 % below those for liquid fuels [5].

Further reductions in CO_2 can be achieved through increases in inner engine efficiency. This is due, amongst other things, to the very high knock resistance of methane (main component of natural gas) compared to gasoline and the associated possibility of higher compression ratios [6]. This is particularly true when using boost concepts [7]. Furthermore, the relatively wide ignition limits permit operation with air excess or high levels of EGR, whereby the gas exchange losses, amongst other things, can be reduced. A disadvantage for mobile use is primarily the low volume-specific energy content of natural gas, which results in the storage of the gas under pressure in heavy pressurized gas cylinders with the current level of technology, whereby a compromise must be found between the space requirements in the vehicle and the attainable vehicle range. The usual storage volume today is between 30 and 100 l, corresponding to ranges between 100 and 330 km. A further huge advantage for natural gas is the possible use in conventional gasoline engine concepts with manageable modifications. This permits so-called bivalent operation, whereby many natural gas vehicles have additional gasoline tanks of up to 15 l as emergency reserves to extend the vehicle's range. This advantage was often used in the development of natural gas powertrains because of the initially low density of natural gas filling stations for bivalent engine concepts.

The future however definitely belongs to monovalent natural gas engines. Monovalence ensures that the specific characteristics of natural gas as a fuel can be fully exploited using specifically design combustion processes and modified engine mechanics [5].

4.1.1 Physical Properties

Throughout production, transport and consumption, natural gas is used in all aggregate states according to requirements—solid, liquid, gaseous—. However, for engineering purposes using natural gas as a fuel, there is a distinction between CNG (*Compressed Natural Gas*) and LNG (*Liquefied Natural Gas*).

Gaseous Natural Gas

Under standard conditions at a room temperature of 25 °C and normal pressure of 1013.25 hPa, natural gas exists in a gaseous aggregate state. The gas is flammable, colorless and normally has no smell. In order to make natural gas detectable by human senses to protect the consumer, scents are added (low concentrations of thioether, for example tetrahydrothiophene), which give the gas its typical smell. Depending on the composition of natural gas, a distinction is made between natural gas type L (*low*) and H (*high*) that mainly differ in their composition in the amount of methane. L-gas has a maximum methane content of approximately 85 %, whereas H-gas has at least 89 % (North Sea) or up to 98 % (GUS states) [8], which affects the physical properties accordingly. The following chapter on chemical properties will provide more details in the composition and the resulting properties.

The density of natural gas is considerably lower than air ($\rho = 1.2$ kg/m^3 at 25 °C) and is $\rho = 0.66$ kg/m^3 at 25 °C for pure methane due to the high proportion of hydrogen in the molecule. The high level of hydrogen also results in the mass related calorific value H_i (previously called the lower heating value) at approximately 50 MJ/kg being considerably higher than the value for liquid hydrocarbons, as Diagram 4.1 shows for different fuels. The H_i is used for fuel comparison purposes. It is the energy content related to the amount of material that is determined for complete combustion by bringing all the products of combustion (oxidation) back to the original pre-combustion temperature.

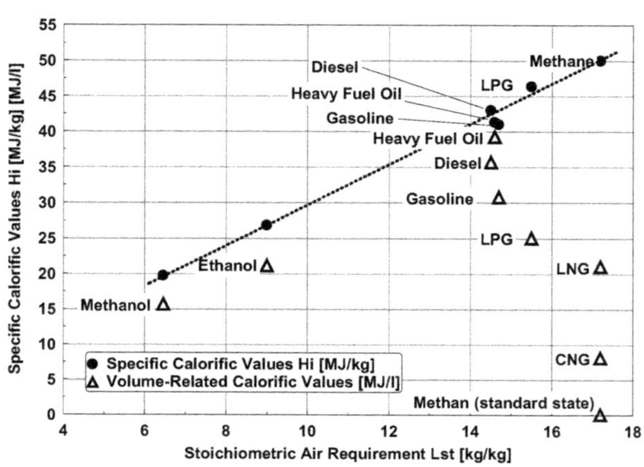

Diagram 4.1 Comparison of specific calorific values of various fuels

It does not contain the vaporization enthalpy of water, in contrast to the calorific value H_s [9]. The dotted line through the mass-related calorific values of the fuels shows that there is an almost constant relationship between calorific value and air requirement. However, this perspective ignores the considerable difference in density between gaseous and liquid fuels. To compensate for this, it is useful to relate the specific calorific values to a volume under normalized conditions. Methane then only has an infinitesimally small proportion (approximately 1/856) of the volume-related calorific value of super gasoline. Acceptable volume-related energy densities are only reached for methane if it is compressed or liquefied, giving CNG 26 % and LNG 70 % of the volume-related calorific value of super gasoline.

These circumstances already indicate the expected charging losses for mixture aspirating natural gas engines. The available calorific value of the mixture is decisive for the maximum attainable power of naturally aspirated engines with external mixture formation. The comparison will therefore be for a stoichiometrically operated engine that introduces fuel via the inlet manifold for operation with gasoline and natural gas. A completely evaporated liquid fuel would then have a density of 4.0 kg/m^3 natural gas would have 0.8 kg/m^3 with inlet manifold conditions of 1 bar, 300 °K. This results in filling losses of approximately 8 % when using natural gas in comparison to gasoline for the identical efficiency [5]. The associated performance loss at full-load due to the difference in density between the fuels and the more favorable knock resistance of natural gas leads to the necessity of turbocharging [10]. At partial load however, the lower density of natural gas brings with it a dethrottling effect [11] and hence the potential to reduce the gas exchange effort, since achieving the same load must be achievable with higher air mass flow, whereby the engine can be considerably dethrottled.

Due to the low density of natural gas, the volume-specific energy content must be increased for mobile use by storing the gas under pressure (CNG) or in liquid form (LNG) in order to achieve acceptable vehicle ranges. CNG use is most prevalent in vehicle applications. CNG is compressed to approximately 200 bar and stored in gas bottles made of steel or in lighter albeit more expensive aramid fibre reinforced aluminium bottles. The limitation of maximum bottle pressure to 200 bar is economically sound, since this value results in a maximum for the ratio pressure to fuel mass. The reason for this is the deviation of the behavior or real gases from the ideal gas law, which is based on the non-interaction of gas molecules. This deviation is described by the so-called real gas factor Z, shown for methane and hydrogen at 293 K in Diagram 4.2. A minimum of the real gas factor of methane is reached just under 200 bar and is thus a maximum value for density gain as opposed to the ideal gas law. This means that although the storage density is increased with increasing pressure, disproportionately more compression work must be done for pressures higher than 200 bar.

Over and above this, higher densities are limited because increasing pressure also increases the bottle mass and its cost, together with more complex compressors [5].

Methane as a representative of natural gas has a density of approximately 162 kg/m^3 at 200 bar and 20 °C. Diagram 4.3 compares storage volumes for different fuels with an

Diagram 4.2 Real gas factor of CH_4 and H_2 [5]

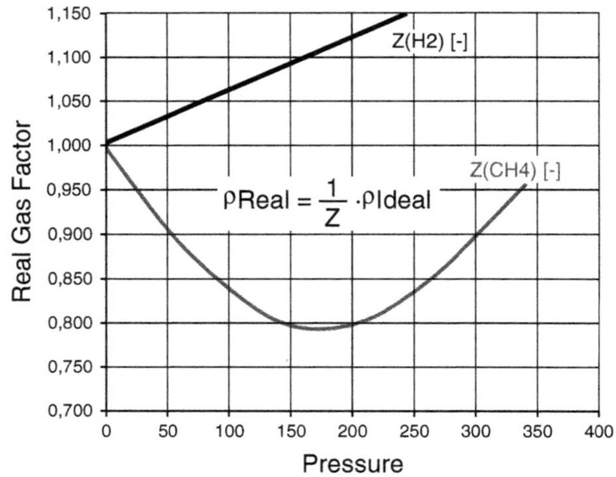

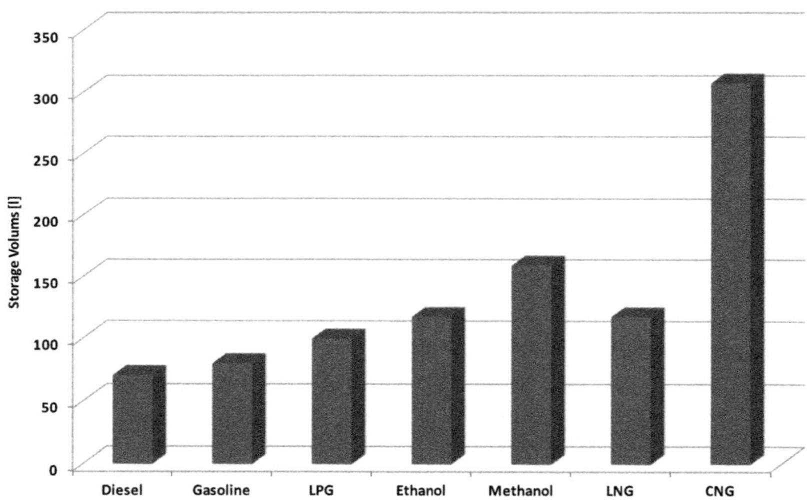

Diagram 4.3 Storage volumes of different fuels for an energy content of 250 MJ (80 l gasoline)

energy content of 250 MJ for vehicle applications, which corresponds to a gasoline tank capacity of 80 l. The material properties used for the calculation of the storage volumes are listed in Table 4.1 at the end of this chapter. It can be clearly seen from the results that only liquid natural gas (LNG) reaches the order of magnitude for storage volumes of conventional fuels.

Table 4.1 Properties of various fuels according to [8, 18]

	Density (kg/m^3)	Calorific value Hs[a] (kJ/kg)	Specific calorific value Hi[a] (kJ/kg)	Volume-related calorific values (MJ/m^3)	Lst (kg/kg)	Mixture calorific value HG[b] (MJ/m^3)	Ignition limits λ (–)	Boiling point[c] (°C)	Evaporation heat[c] (kJ/kg)
Methane (CH_4)	0.7175	55,515	50,028	36	17.2	3.75	0.7/2.1	−162	510
LNG (methane)	421			21,062					
CNG (methane)	162			8105					
Methane hydrate	900		6709	6038					
Gasoline	730..780	43,350	41,000	30,750	14.5	3.75	0.4/1.4	30...190	420
Diesel	815...855	45,400	43,000	35,690	14.5	3.87	0.48/1.35	170...350	300
Heavy fuel oil	950	–	41,300	39,235	14.6	3.66	0.5/1.35	175...450	–
LPG	540	–	46,330	25,018	–	–	–	–	–
Propane	2.01	–	46,300	93	15.6	3.35	0.4/2.2	−42	425
Butane	2.7	–	45,600	123	15.4	3.39	0.4/2.1	0	385
Methanol	795	23,847	21,099	16,774	6.46	3.44	0.34/2.0	65	1110
Ethanol	789	30,574	27,708	21,862	9.0	3.48	0.3/2.1	78	845

Note: Values at 0 °C and 1013 bar except LNG (−162 °C, 1 bar), CNG (25 °C, 200 bar), Methane Hydrate (<4 °C, >20 bar) LPG (25 °C, 5–10 bar)
[a]Reference temperature for the combustion 25 °C; [b]at $\lambda = 2$; [c]at 1013 bar

CNG however requires almost four times the volume compared to gasoline. A further storage possibility that avoids the technically complex low temperatures of LNG and the high pressures of CNG is Adsorbed Natural Gas (ANG). This is a technology currently under development where the natural gas molecules are adsorbed via the pore structure by surface-gas interactions on a micro-porous solid.

Physically speaking, storage volumes are possible for natural gas that lie between the values for CNG and LNG, but at room temperature and at pressures between 40 and 70 bar [12].

The biggest difficulty is however the search for a suitable adsorption material. Active carbon is presented in research investigations as a relatively low priced possibility, but it does not yet fulfill the defined targets for storage density [5], which is why particularly synthetically manufactured adsorbing materials, that have a very high surface-volume ratio, must be considered [13].

The viscosity as an effect of the inner friction between molecules increases in alkanes along with the chain length. Correspondingly, gaseous methane as the shortest alkane has the lowest dynamic viscosity ($\eta = 11 \times 10^{-6}$ kg/ms at 20 °C [8]), which is only 20 % of the value for gasoline fuel. The resulting absent lubrication and vaporization properties lead to increased requirements on the design of mechanical components that are in direct contact with natural gas. An example of this is the seating geometry and closing mechanism of a gas injector for direct injection (compare [14]). The addition of additives to gaseous fuels, as done conventionally to protect components under high duress with liquid fuels (for example valves or valve seats), is practically impossible. This results in the necessity to harden these elements or use other materials such as ceramics.

Liquefied Natural Gas

In order to transform methane from the gaseous to the liquid state under normal pressure, the temperature must fall below the evaporation point of −161.5 °C [15]. In this state, natural gas is described as liquid natural gas, or more commonly LNG (*Liquid Natural Gas*). It is important to note that liquid gas (LPG *Liquefied Petroleum Gas*; also called autogas) is not liquefied natural gas. LPG is a mixture of long chain C_3 and C_4 hydrocarbons (propane, butane) that can be liquefied at room temperature and at pressures from 5 to 10 bar [8].

Since LNG only has one 600th of the volume of gaseous natural gas (related to methane, compare Table 4.1), this is increasingly taken advantage of in transportation and storage, since natural gas could be sourced from areas that are not connected via pipeline for economic reasons (for example distance) or technical reasons (for example too high depth of water). LNG tankers are used for sea transport, however the effort economic viability of the process required for liquefaction, transport and re-evaporation must be taken into account, since this requires up to 15 % of the energy content of the gas [8].

Storing LNG in vehicles places very high demands on the insulation of the tank system. Special costly cryogenic storage processes are used in conjunction with the required control technology. The superinsulation of just a few centimetres thickness employed today corresponds to polystyrene walls of several metre thickness, but still cannot prevent the evaporation of approximately 10 % of the fuels after 72 h (average value). Moreover, these systems require effective liquefaction processes in filling stations and corresponding evaporation and compression devices, as appropriate, in the vehicle. Still, cryogenic storage technology offers the current optimum in terms of storage density and system weight [5], which has not yet asserted itself due to the mentioned technical challenges and is still in the development stage along with the ANG—(*Adsorbed Natural Gas*)—storage process for vehicle applications.

Solid Natural Gas

Methane occurs in nature in sea sediments and in permafrost ground in polar regions in the form of gas hydrates and also in solid aggregate form.

Under certain thermodynamic conditions with pressures >20 bar and temperatures <4 °C, naturally occurring gas hydrates, in particular methane, carbon dioxide and hydrogen sulphide are captured in crystalline form in water and form a clathrate. In methane hydrate water molecules completely surround the methane molecule. Compared to gaseous natural gas, the methane is bound in a very high density under normal conditions, which is why this form of storage is being investigated particularly for the transport of natural gas [16].

The density of methane hydrate under normal conditions is 900 kg/m^3 however with a mole ratio of 5.75 water per mole methane. This means for transportation that for every 1 kg of methane (mole mass 16,043 kg/mol), 6.5 kg of water (mole mass 18,015 kg/mol: density of ice: 918 kg/m^3) must also be transported, which results in a very low specific calorific value for methane hydrate, as shown in Table 4.1. However, this must be compared to LNG sea transport, as the technically difficult and economically costly effort to liquefy at -162 °C is eliminated.

The greatest resources in fossil energy carriers are assumed to be in the form of methane hydrate, however the complex production process is still under development and contains climatic dangers. Methane hydrate is only stable at high pressures and relatively low temperatures [17]. If methane is released into the atmosphere, it has a similar greenhouse gas effect to carbon dioxide. An increase in global warming enables methane hydrate to melt, which increases the greenhouse gas effect, creating an accelerating process. In addition, methane hydrate fulfills an important function similar to cement regarding submarine slope stability. If this is reduced by decomposition or mining, then this could cause submarine landslides that create huge tsunamis.

> Production at sea may seem more attractive than production from permafrost, due to the larger deposits, but economically rather improbable in the short and medium term due to the far more difficult production environments, the risk and the unknown environmental effects. However, gas production from gas hydrates in permafrost areas, as already done in small amounts in Siberia, could increase in importance in the near future. [17].

4.1.2 Chemical Properties

According to the chemical structure in organic chemistry, methane as the main component of natural gas belongs to the group of alkanes. Alkanes consist of chains of hydrocarbons with exclusively single bonds, which is why the carbon atoms bond with the highest possible number of hydrogen atoms. The simplest alkane is methane with the highest hydrogen to carbon ratio of 4 to 1, which is the reason for methane having approximately 20 % lower CO_2 emissions compared to gasoline.

CH_4 Molecular Structure and Properties of Alkanes

Structural analysis of CH_4 molecules has shown that the carbon atom is surrounded by four hydrogen atoms in the form of a tetrahedron. The bond angle H-C-H is 109°28′ (tetrahedron angle). The distances from the C atom to the H atoms are equidistant. Bonds, such as those in methane, are rotationally symmetrical around the atomic bond axes

Diagram 4.4 CH_4 tetrahedron [19]

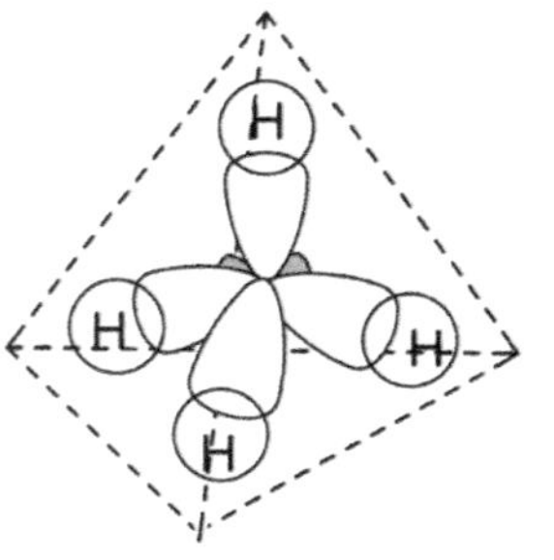

Table 4.2 Homologous series of alkanes [19]

Molecular formula	Name	Melting point (°C)	Boiling point (°C)
CH_4	Methane	−184	−162
C_2H_6	Ethane	−171	−93
C_3H_8	Propane	−190	−45
C_4H_{10}	Butane	−135	0
C_5H_{12}	Pentane	−130	36
C_6H_{14}	Hexane	−94	69
C_7H_{16}	Heptane	−90	98
⋮			
$C_{17}H_{36}$	Heptadecan	23	303
$C_{20}H_{42}$	Eicosane	37	–

between those atoms that are bonded. They are called σ-bonds and are shown in Diagram 4.4. Substances such as methane that form the largest possible number of σ-bonds and cannot accept any more hydrogen atoms are described as being saturated [19].

There is often a differentiation between alkanes with open chains and cyclical (in the form of rings) configurations with saturated hydrocarbons. There is a further distinction in open chain alkanes between unbranched (n-alkanes) and branched hydrocarbons (iso-alkanes), whereby the position of the branch significantly alters the substance properties and the description. Methane is the simplest open-chain alkane that becomes a homologous series of alkanes by the continuous addition of CH_2 molecules with the formula C_nH_{2n+2}. The chemical properties do not change much with larger chain lengths, but the physical properties such as melting and boiling point do change as a function of chain length (see Table 4.2). Under normal conditions (25 °C, 1 bar) the alkanes with a chain length C_1 to C_4 are gaseous, C_5 to C_{16} are liquids, and higher in solid aggregate form. The reason for the increase in melting and boiling points with chain length are the molecular van-der-Waals forces that increase with every additional CH_2 group. The molecules are nonpolar and therefore easily dissolved in other hydrocarbons, but not dissolvable in polar solutions, for example water (*hydrophobic*). Alkanes do not react chemically under most conditions, but do enable reactions that occur via radicals as an intermediate stage.

Reactive Properties

Alkanes are often called paraffins (*parum affinis* = lacking affinity or lacking reactivity), since they have an unreactive nature. Reactions occur via radicals as intermediates. A radical is characterized by an open electron shell, i.e. it has unpaired valence electrons that are created through photochemical or thermal breakup of a neutral molecule [19]. Typical reactions of alkanes are oxidation and radical substitution that occur under various radical chain mechanisms [20]:

- *Chain initiation*: Reactive species (radicals) are formed from stable species;
- *Chain propagation*: Reactive particles react with stable species and form a different radical;
- *Chain branching*: A radical reacts with a stable molecule and forms two new radicals;
- *Chain termination*: Radicals react to form stable molecules, for example on the container walls or via recombination of radicals.

Combustion is an exothermic chemical reaction in which a fuel is oxidized by oxidizers releasing light and heat. The individual processes in combustion such as ignition, flame propagation and fuel depletion all depend on how quickly radicals can be formed. Ignition occurs in areas where the chain branching reactions dominate the chain termination reactions [21]. Diagram 4.5 shows the reaction flow analysis of stoichiometric methane combustion in air, in which the degree of individual elementary reactions contribute to the depletion or formation of a specific substance. The main path of CH_4 oxidation to CO occurs via C_1 connections. 25 % of methyl (CH_3) reacts to CH_2O and finally via HCO also to CO. 25 % also occurs via the addition of OH to CH_3OH. Only a small proportion of 7 % of CH_3 reacts via combination to C_2H_6 and leads to the branch of C_2 chemistry [22].

Reaction flow analysis enables the identification of which reaction paths dominate during the oxidation of hydrocarbons, which especially contributes to the analysis of pollutant formation such as the emissions of unburned hydrocarbons or soot.

The main component of HC emissions in natural gas is methane. Since methane is very inert and has significantly higher activation energy compared to other hydrocarbons, the coating of the catalytic converter must be adapted. When operating an engine with natural gas, the exhaust temperatures are also lower than those of Otto gasoline engines, which result in a longer period before catalytic light-off temperature is reached [23]. Methane is not a radical as opposed to partially combusted hydrocarbons and is thus not a precursor to ozone. The ozone formation potential of methane is very low and is one to two degrees of

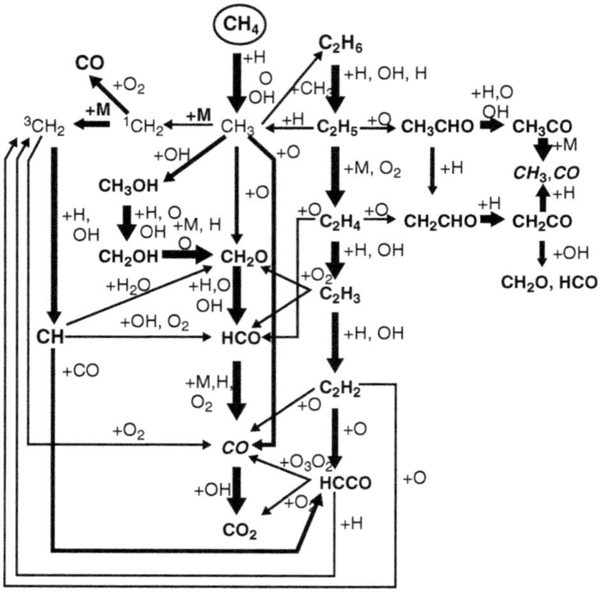

Diagram 4.5 Integral reaction flow analysis for a laminar stoichiometric CH_4/air flame. $Tu = 298$ K, $p = 1$ bar. The four arrow thicknesses correspond to the share of the path to the depletion of the species; *very thick*: 75–100 %, *thick*: 50–75 %; *medium*: 25–50 %, *narrow*: 5–25 % [22]

magnitude lower than the other emissions components of higher grade alkanes, so that the fuel natural gas has the lowest potential to form ground-level ozone of all organic fuels [5].

Much more worthy of consideration is the greenhouse effect of methane. If methane is compared with the most important greenhouse gas carbon dioxide, then the global warming potential (GWP) of methane is 75 times higher than the value for CO_2. In addition, the residence time of approximately 10 years (assumption from literature) must be considered, which is only one twenty-third of that of CO_2. This means that methane still has an approximately 7 times higher value than CO_2. However, even assuming that all the passenger cars in the world ran on natural gas, then an investigation into the worldwide methane emissions would conclude that the methane emissions of natural gas vehicles would only be a very small proportion compared to other large emitters (for example the oceans, insects, rice culture, livestock farming) [5].

Ignition delay times for higher hydrocarbons are much shorter than the compact and stable methane molecule, since the long-chain molecules tend to form radicals earlier [20]. Due to the low chemical reactivity of methane, the realization of pure self-ignition is limited by the high final compression temperatures, which are required. Between 1150 and 1200 K are necessary for the commencement of natural gas self-ignition. For this reason, only natural gas combustion processes can be considered in which an external addition of energy initializes ignition. These include conventional positively ignited Otto engines or systems with auxiliary ignition energy. In order to realize homogeneous charge compression ignition (HCCI) for example, or an operating process similar to diesel, auxiliary ignition energy in the form of hot surfaces (for example glow plug) or liquid fuel as pilot injection for ignition (ignition jet process) is required [5]. Dual-fuel technology offers a further alternative combustion process in which natural gas is injected into the inlet manifold and substitutes diesel fuel between 5 and 95 %. The ignition of the homogeneous natural gas-air mixture is done via compression and the resulting self-ignition of the small amount of diesel fuel injected into the combustion chamber. In contrast to the ignition jet procedure used in the domain of stationary large engines, the functionality of the diesel combustion process is retained.

Laminar Flame Speed

The flame speed is of critical importance for energy release and thus for the efficiency of the combustion process and also for the modeling and calculation of technical combustion processes. The laminar flame speed is defined as the flame propagation speed with which the flame moves perpendicular to the surface of the flame front relative to the unburned fresh gas.

One possibility of determining the speed by experiment is the cone angle method. Diagram 4.6 shows a laminar conical flame that is stabilized using suitable jets. An electronic camera captures the flame and the cone angle is calculated from digital evaluation of the pictures. The laminar flame speed S_L can be calculated from the flow speed v_u of the unburned fuel-air mixture and the flame angle α using a simple angular relationship [24]:

$$S_L = v_u \cdot \sin\alpha \tag{4.1}$$

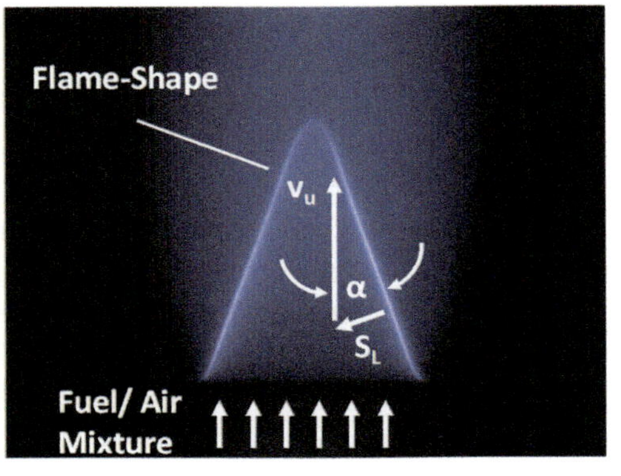

Diagram 4.6 Calculating the laminar flame speed S_L [24]

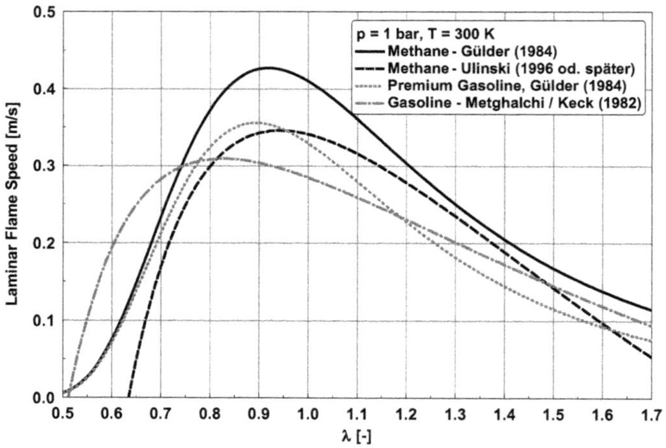

Diagram 4.7 Laminar flame speeds for methane and gasoline [14]

Diagram 4.7 portrays data from literature [25–27] showing that the maximum laminar flame speed of methane compared to normal and super gasoline lies significantly closer at the stoichiometric ratio of $\lambda = 1$. This situation creates the advantage that mixture enrichment at full-load to increase mean pressure, as it is usual with gasoline engines, can be omitted and thus the operation of the three-way catalytic converter is possible in its optimum range. However, due to the low chemical reactivity of methane, the converter coating must be adapted to be able to produce the significantly higher activation energy required by methane.

The measurement shows that even at higher λ values the laminar flame speed of methane is significantly higher than the speeds for liquid Otto fuels, which demonstrates the potential of natural gas for lean concepts.

As for the majority of carbon-based air flames, the laminar flame velocity is dependent on the mixture composition, pressure, form and in particular the curvature of the flame front, the fresh gas temperature and the inert gas additives. Diagram 4.8 clearly shows on the left the pressure dependence with a significant decrease in laminar burn velocity with increasing pressure. On the other hand, the right-hand side shows an exponential increase over the temperature up to 800 K.

In a combustion engine, there is both an increase in pressure and in temperature in the combustion chamber, whereby the opposing effect is imposed on the laminar flame velocity. In order to extend the so-far limited measurement data basis for natural gas, the project "Natural Gas Engine with Inner Mixture Formation" (EDI) [24] carried out wide-ranging investigations using a reference gas (92 % methane, 8 % ethane). The results showed a maximum S_L at approximately $\lambda = 1.1$ with a clear dependency on temperature

Diagram 4.8 Laminar flame velocity SL ($= \Lambda L$) dependency on pressure and fresh gas temperature for a stoichiometric CH_4—air flame for $Tu = 298$ K (*left*) und $p = 1$ bar (*right*) [28]

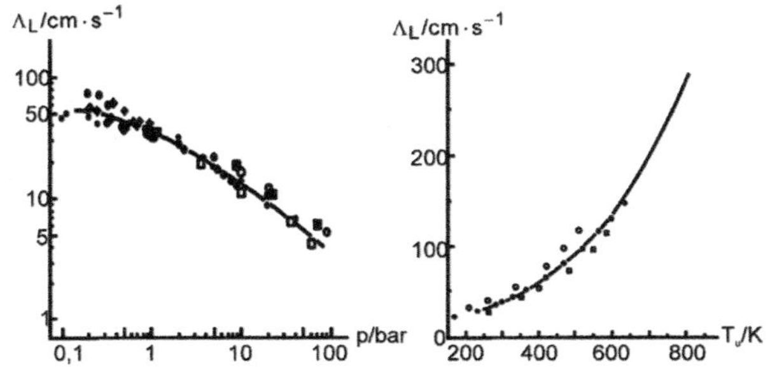

and pressure, however keeping the qualitative trend (maximum value at around $\lambda = 1.1$) over all variations.

Methane Number and Chemical Composition of Natural Gas

Gaseous fuels must fulfill requirements concerning energy content (heating value) and knock resistance for use in combustion engines. The methane number is used with combustion gases to evaluate the knock resistance, which is comparable with the research octane number (RON) for liquid fuels. The methane number gives the percentage methane volume ratio of a mixture of knock-resistant methane (MZ = 100) and knock-susceptible hydrogen (MZ = 0) that under defined test conditions shows the identical knock resistance in a test engine as the gas mixture to be tested [29]. Pure methane is characterized by very high knock resistance. The methane number of 100 corresponds to a RON of approximately 130. The methane number of natural gas is very heavily dependent on the gas composition and in particular on the proportion of methane gas, the proportion of higher value alkanes and the proportion of inert gas. Table 4.3 shows the average composition of different natural gas qualities from different geographical sources.

Table 4.3 clearly shows the dependency of the methane content on the geographical source of gas and its strong influence on the methane number. GUS gas has the highest methane content with a corresponding methane number of 95. Compared to North Sea natural gas H and composite gas H, the influence of low amounts of higher value hydrocarbons (butane, propane, ethane) on the methane number is clearly recognizable. The methane number of both gas mixtures can be classified in 9 classes, whereby North Sea gas H with its higher proportion of long-chain alkanes is significantly more knock-susceptible. In contrast to this, the high amount of inert gas (nitrogen) in composite gas H not only ensures a lower level of knock-susceptibility, but also a lower calorific value.

Diagram 4.9 uses the example of GUS natural gas H to show that the composition of the gas also varies within a geographical area. The high sensibility of the methane number

Table 4.3 Composition and properties of different qualities of natural gas, average values in Germany [29]

		Unit	North sea gas H	Compound gas H	CIS-gas HH	Holland-erdgas L	Compound gas L
Analysis values							
Carbon dioxide	CO_2	Vol.%	1.53	1.18	0.08	1.27	1.22
Nitrogen	N_2	Vol.%	1.10	4.48	0.81	10.64	10.01
Oxygen	O_2	Vol.%	0.00	<0.01	0.00	0.00	<0.01
Hydrogen	H_2	Vol.%	0.00	0.06	0.00	0.00	0.31
Carbon monoxide	CO	Vol.%	0.00	0.01	0.00	0.00	0.03
Methane	**CH_4**	**Vol.%**	**86.54**	**87.74**	**98.31**	**83.35**	**84.40**
Ethane	C_2H_6	Vol.%	8.02	4.86	0.50	3.71	3.14
Ethene	C_2H_4	Vol.%	0.00	<0.01	0.00	0.00	0.01
Propane	C_3H_8	Vol.%	2.06	1.21	0.19	0.70	0.61
Butane	C_4H_{10}	Vol.%	0.60	0.35	0.08	0.22	0.19
Pentane	C_5H_{12}	Vol.%	0.10	0.06	0.02	0.06	0.04
Hexane	C_6H_{14}	Vol.%	0.05	0.04	0.01	0.05	0.04
Specific values							
Calorific value		kWh/m^3	11.99	11.16	11.07	10.26	10.23
Spec. calorific value		kWh/m^3	10.85	10.09	9.98	9.27	9.24
Density		kg/m^3	0.835	0.812	0.732	0.83	0.818
Air requirement		m^3/m^3	10.34	9.63	9.54	8.85	8.82
Methane number		–	72	81	95	86	88

to just slight changes in volume of long-chain alkanes propane and ethane in natural gas is clearly shown by the fall in MN = 95 to MN = 91.

Further, it should be noted, that the addition of higher proportions of inert gases lowers the knock tendency and therefore values of methane numbers higher than 100 can be achieved. So for example, biogas, sewage gas and landfill gas can reach methane numbers of 130 due to the high proportion of carbon dioxide. However, in conjunction with the higher proportion of inert gases, the calorific value of the gas mixture is reduced considerably. When using biogas—desirable in the spirit of further reduction of CO_2 emissions—the composition of the gas must be significantly altered for use with combustion engines [5].

The methane number is very important when considering the compression ratio of a combustion engine. The thermal efficiency η_{th} of the process increases with the

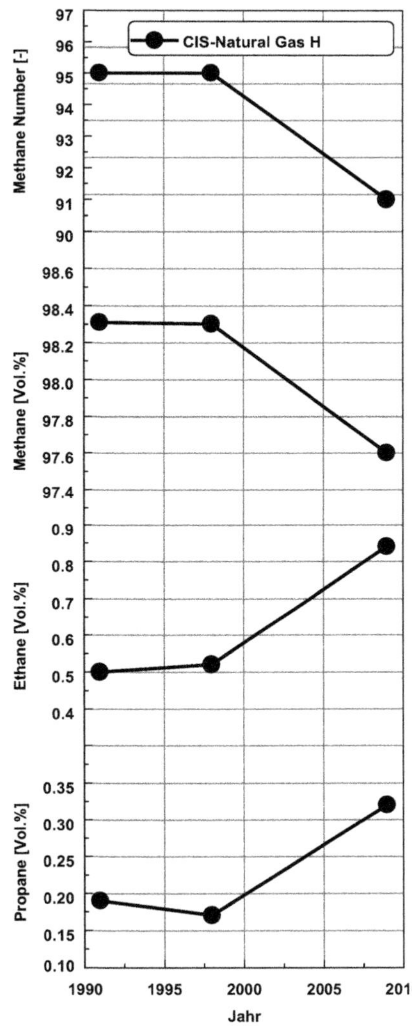

Diagram 4.9 Development of methane number from GUS natural gas H, data sourced from [29, 32, 33]

compression ratio ε. The higher knock resistance of methane proves to be a major advantage for higher efficiencies. Although the compression ratio is limited by the increasing mechanical loading of the crankshaft and unfavorable combustion chamber forms, naturally aspirated Otto engines using natural gas as a fuel still have efficiencies 5 % higher when using gasoline under the same boundary conditions [5]. The use of turbocharging offers a particularly good method of compensating for lower losses due to lower calorific values with external mixture formation. As a result of the possibly higher compression ratio for natural gas compared to gasoline fuel, a natural gas powered engine has a thermally superior efficiency of approximately 9 % [5]. The link between methane number and thermal efficiency and the sensibility of the methane number with regard to gas composition underlines the requirement for a minimum methane number.

The earlier weak standardization of natural gas from an engine perspective that mainly defined the minimum calorific value and not the knock resistance led to major variations in gas composition. Regional suppliers add higher quality alkanes, mainly propane and butane, to the natural gas because of their good storage properties in order to comply with legislative minimum calorific values. However, these additives severely reduce knock resistance. A 10 % addition of propane lowers the methane number below MN = 80, for 10 % butane even to MN = 50 for GUS H gas [5]. In order to fulfill requirements for a minimum methane number to permit monovalent engine design, a new fuel standard for natural gas was introduced in 2009, DIN 51624 [30]. The norm "Natural Gas as Fuel" is valid for natural gas exiting the nozzle at a natural gas filling station, whereas the requirements stated in DVGW-Arbeitsblattes G 260 [31] are valid for the connection of the natural gas filling station to the gas grid. The operator of the natural gas filling station must apply appropriate measures, for example a separator, to ensure that the values defined in Table 4.4 are complied with.

Table 4.4 Limit and reference values * (*see table) for the use of natural gas in motor vehicles (DIN 51624) [30]

Property	Limit
Density (abs.)	0.72 – 0.91 kg/m^3
Methane number	≥70
CH_4	≥80 %
Σ C_2-KW	≤12 %
Σ >C_2-KW*	≤8.5 %
C_3H_8	≤6 %
C_4H_{10}*	≤2 %
C_5H_{12}*	≤1 %
C_6H_{14} + höhere KW*	≤0.5 %
O_2	≤3 %
H_2	≤2 %
Σ $N_2 + CO_2 \leq 15$ %	≤15 %
H_2S	≤7 mg/kg
RSH	≤8 mg/kg
Σ $S_2 \leq 10$ mg/kg	≤10 mg/kg
H_2O	≤40 mg/kg

4.2 Natural Gas Sources and Extraction

Siegfried Bajohr

As shown in Diagram 4.10, in 2013 about 24 % of the primary world energy consumption of approximately 12,730.4 Mtoe (= 533 EJ) was met using natural gas [34]. Thus, natural gas is the third most important energy source in the world and in Germany, which can be taken as an example for the global natural gas situation. The German natural gas demand is about 90 billion m^3/year and makes up nearly 3 % of global natural gas consumption, which further emphasizes the interdependence of the German energy mix with the global natural gas industry [35]. The main sources of supply of natural gas in Germany were Russia and Norway [36]. The German domestic production of conventional natural gas in 2013 was about 10 %, which will be discussed in more detail in Sect. 4.3.

Although it is predicted by most committees of experts that global consumption will continue to rise sharply, the availability of natural gas, in contrast to that of crude oil, is not limited by bottlenecks in the short and medium term. The following sections will show that the existing natural gas reserves and resources for energy and as a fuel will, and must, make an important contribution to the energy supply of our society. In addition to the inherent limitations due to the physical and chemical properties, there are several binding "rules" to be observed, which are essential for the international gas market and all technical gas applications. Therefore, this should be done considering the global natural gas sources, as well as the essential technical rules and necessary process engineering for natural gas.

4.2.1 Types of Fuel Gases

As stated in the previous book chapters, gaseous fuels are used in many different ways in modern society. Gaseous fuels have been historically significant since the beginning of the

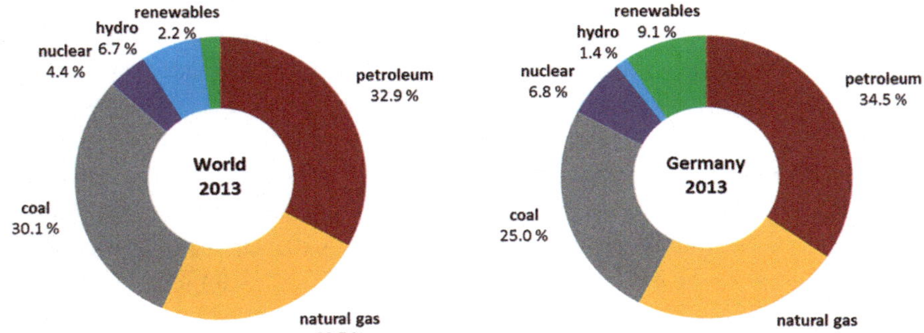

Diagram 4.10 Primary energy consumption of the World/Germany by energy source in 2013 (total: 12,730.4 Mtoe/533 EJ (world), 325.0 Mtoe/14 EJ (Germany), data from [34])

Table 4.5 Gaseous fuels in the public gas supply

Name		Definition
NG	Natural gas	Fossil based methane rich gas
SNG	Synthetic natural gas Substitute natural gas Biomethane	Gas produced from the anaerobic fermentation of biomass or Gas produced from the catalytic or biological methanation of synthesis gas (produced from biomass or fossil fuels)
LNG	Liquefied natural gas	Natural gas that is liquefied by cooling (liquid)
LBG	Liquefied biogas	Biogas that is liquefied by cooling (liquid)
CNG	Compressed natural gas	Highly compressed natural gas (gaseous)
LPG	Liquefied petroleum gas	Under pressure liquefied propane/butane mixtures, mostly from the refining of crude oil or Natural gas production

nineteenth century, particularly due to the rapid growth of cities. By the beginning of the twentieth century, the primary domestic applications such as cooking and heating were further promoted and later supplemented by other applications in the second half of the century. During this time an increasing changeover from gases produced from coal such as town gas to natural gas occurred, which was distributed as a raw material for the chemical industry, as a feed material for power production, and as a fuel itself.

The selection of the most suitable fuel gas depends on the intended use. Thus, it is of interest to make a quick overview of the different types of fuel gases that are currently available. Excluding non-carbon gases such as hydrogen, the technically relevant and publicly distributed gaseous fuels can be divided into the categories shown in Table 4.5.

Some of the gases listed in the above table are considered in further detail in the following chapters and are specifically evaluated and investigated for use in vehicle engines. However, in the following text, attention will be directed to the methane-rich gases in the public gas supply as well as a brief discussion of their origin. Here, a nomenclature has been established, which will be used in the following discussion. Table 4.6 provides an overview, identifying the various "gas sources" and their origin.

The sources of fossil natural gas are depicted graphically in Diagram 4.11. The type of natural gas source can be divided into two subcategories. First, **conventional sources** from conventional and associated gas, which in turn can be both stranded gases. Second, **unconventional sources** which include coal bed methane, tight gas, shale gas, and also aquifer gas and gas from gas hydrates that is not yet commercially used.

Globally, the piped public gas supply is now almost exclusively natural gas, which can be from one or a combination of several of the sources listed in Table 4.6. Since the composition and the resulting physical and chemical properties are heavily dependent on the nature and origin of the gas (see also the later chapter 4.2.4 "Natural Gas Upgrading"), binding limits and parameters for the composition as well as combustion specifications have been laid out since the early days of public gas utilization. In Germany, the German Technical and Scientific Association for Gas and Water (DVGW) uses their technical codes of practice to regulate all aspects of the public gas supply. These codes of practice

Table 4.6 Origin of methane rich gaseous fuels

Name	Origin
Conventional natural gas	Gas in "natural gas traps" under impermeable layers in suitable geological structures
Associated gas	Natural gas which is found together with oil and thereby separated from during the recovery process
Coalbed methane/ mine gas	Natural gas which is present in the pores within a coal seam
Stranded gas	Known but either physically or economically unusable gas sources
Aquifer gas	Natural gas that is dissolved in deep ground water deposits (aquifers)
Tight gas	Natural gas which is present in the pores of very dense rock formations
Shale gas	Natural gas which is present in hydrocarbon rich rock layers (i.e. oil shale)
Gas hydrates	Methane molecules which are trapped in ice crystals (deposits in the permafrost and deep ocean layers)
Biogas/Biomethane	Gas produced by the anaerobic fermentation of biomass

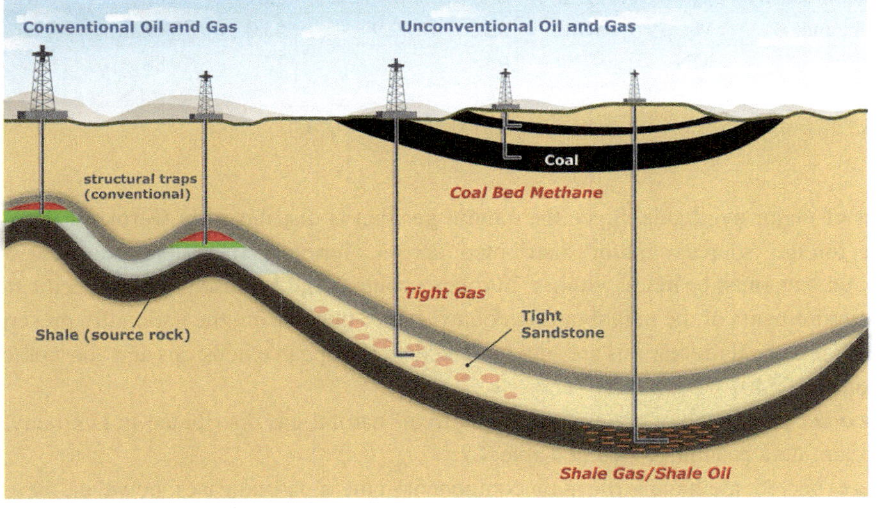

Diagram 4.11 Geological origins of methane rich fuel gases [35]

are binding for all partners participating in the market [37]. The DVGW Code of Practice G 260 "Gas Quality" is of central importance for all aspects of the natural gas supply [38]. This document specifies, in the form of a technical regulation, the requirements for the quality of fuel gases in the public gas supply and establishes conditions for the gas supply, gas transport, storage, operation of gas equipment and gas appliances (such as industrial gas applications), as well as providing the basis for the development, standardization, and testing. The first binding specification for natural gas was made in 1939 [39], which in subsequent years and with increasing use of natural gas experienced a

Table 4.7 Typical composition and other important characteristics of natural gases distributed in Germany [38]

			Natural gas H			Natural gas L	
Name	Symbol	Unit	Russia	North sea	Denmark	Holland	Germany
Methane	CH_4	mol-%	96.96	88.71	90.07	84.64	86.46
Nitrogen	N_2	mol-%	0.86	0.82	0.28	10.21	10.24
Carbon dioxide	CO_2	mol-%	0.18	1.94	0.60	1.68	2.08
Ethane	C_2H_6	mol-%	1.37	6.93	5.68	3.56	1.06
Propane	C_3H_8	mol-%	0.45	1.25	2.19	0.61	0.11
Butane	C_4H_{10}	mol-%	0.15	0.28	0.90	0.19	0.03
Pentane	C_5H_{12}	mol-%	0.02	0.05	0.22	0.04	0.01
Hexane and higher hydrocarbons	C_{6+}	mol-%	0.01	0.02	0.06	0.07	0.01
Total sulfur	S	mg/m^3	<3	<5	<3	<3	<3
Calorific value	$H_{S,n}$	MJ/m^3	40.3	41.9	43.7	36.8	35.4
Standard density	ρ_n	kg/m^3	0.74	0.81	0.81	0.83	0.81
Wobbe index	$W_{S,n}$	MJ/m^3	53.1	52.9	55.0	46.0	44.7
Methane number	MN	–	90	79	72	88	97

Index "n" = standard state for gases ($p_n = 1013$ mbar, $T_n = 0$ °C)

series of major overhauls. Since the natural gas that is distributed in Germany is largely from foreign sources and/or distributed across Germany to other countries, these requirements must be made within a European context and be based on the specifications and requirements of the neighbouring countries. For this reason, the nationally prescribed DVGW technical regulations are supplemented by European extensions and "best practice approaches" [40].

In order to give an overview of the quality of natural gas distributed in Germany, the important data is summarized in Table 4.7.

As expected, methane is the main component in these natural gases; however, they also contain many other components that have a corresponding effect on the chemical and physical properties (see Chap. 3).

4.2.2 Natural Gas Sources: Reserves and Resources

The question of the origins of the natural gas that is distributed throughout Germany as well the resulting dependencies is not only the result of modern political developments, but also stems from the beginnings of the use of natural gas. In order to assess the global commodity situation and obtain an understanding of the current and future resulting dependencies, natural gas sources—as is customary for all raw materials—are split up into reserves and resources.

▶ **Reserves** are sources that can be utilized economically with state-of-the-art technology.

▶ **Resources** are sources which are known to exist, but cannot be economically utilized.

Current data from [35] shows that conventional natural gases make up the lion's share of the global annual gas consumption of 129 EJ (2012) and the global reserves of 7455 EJ. Based on this data, a reserves-to-production ratio of 58 years can be calculated for natural gas (crude oil: 53 years). Taking into account all resources, the reserves-to-production ratio of natural gas at the present time can be expanded to 144 years (crude oil: 132 years). However, a look at the possible future reveals that these resources are mainly unconventional natural gas. In addition to these considerations, gases generated from renewable sources also need to be included (biogas/biomethane in Table 4.6), even though they only account for 1 % of the German natural gas consumption [41], the current political drive could increase their importance. These gases and their potential are considered in depth in Chap. 7 and thus will not be given further consideration in this chapter.

Diagram 4.12 clearly shows that over half of the world's natural gas reserves are concentrated in three countries: Russia, Iran, and Qatar. Together with the member countries of OPEC, Russia can access about 80 % of the global natural gas reserves.

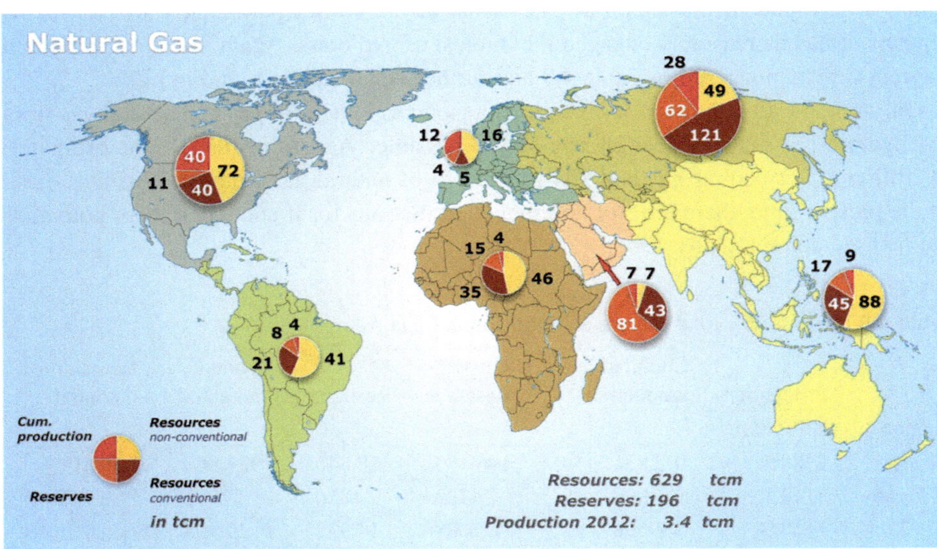

Diagram 4.12 Regional distribution of the total natural gas potentials (excluding aquifer gas and gas hydrates) [35]

4.2.2.1 Conventional Natural Gas Resources

In the following observations, conventional natural gas resources are the sum of conventional natural gas, associated gas, and stranded gas according to Table 4.6 at the beginning of this chapter. All other unconventional gases, including biogas, are not considered here. The natural gas world map in Diagram 4.12 shows that Russia has the largest natural gas resources by far, followed by China, the US, Canada, and Australia. Additionally, Russia has about a third of the conventional natural gas resources in the world, ahead of the United States, China, Saudi Arabia, and Turkmenistan. At the time of the publication of this book, the world's natural gas sources of both conventional and unconventional sources totalled about 837 trillion m^3. A list of the most important data is summarized in Table 4.8. Significantly more comprehensive data can be obtained from the 2013 BGR Energy Study [35].

4.2.2.2 Unconventional Natural Gas Resources

The unconventional natural gas resources are made up of the gases defined in Table 4.6 as coalbed methane, shale gas, tight gas, aquifer gas, and the recoverable natural gas from gas hydrates. The worldwide shale gas resources, with about 200 trillion m^3, make up the lion's share of unconventional natural gas, followed by tight gas und coalbed methane. Especially for Europe, the expected potential for shale gas is very attractive. Diagram 4.13 gives the predicted numbers and Diagram 4.14 shows a graphical depiction of the major deposits in Europe.

Since the previous estimates of tight gas potentials are incomplete, it can be assumed that the estimated 63 trillion m^3 will significantly increase in the future. Especially since tight gas is found in most natural gas basins in the world, particularly in Palaeozoic deposits, it has increasingly engaged the interest of explorers. Again, the most important diagrams are summarized in Table 4.9 and further detail can be found in [35].

Only rough estimates exist for the global resources of aquifer gas and natural gas from gas hydrates; a final evaluation is not yet possible. According to current estimates, 24 trillion m^3 of aquifer gas and 184 trillion m^3 of natural gas from gas hydrates exist. At the present time, there are still no reliable predictions for if and when these potentials can be used commercially.

Table 4.8 Data for the total sources and production (data from [35])

	Production	Cumulative production	Reserves	Resources	Total potential	Remaining potential
Region	Billion m^3					
World	3388.5	102,831	196,173	628,846	927,850	825,019
Germany	12.1	999	123	1870	2992	1993
EU-27	168.9	9898	2088	17,922	29,908	20,010

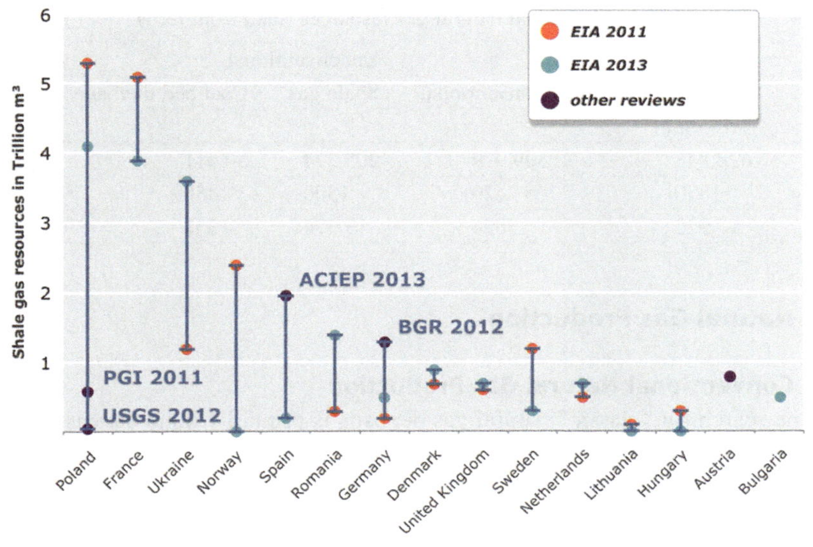

Diagram 4.13 Assessment of the technically recoverable shale gas resources in Europe (Diagram from [35]; cited sources: EIA 2011 = [42], EIA 2013 = [43], ACIEP 2013 = [44], BGR 2012 = [45], PGI 2011 = [46], USGS 2012 = [47])

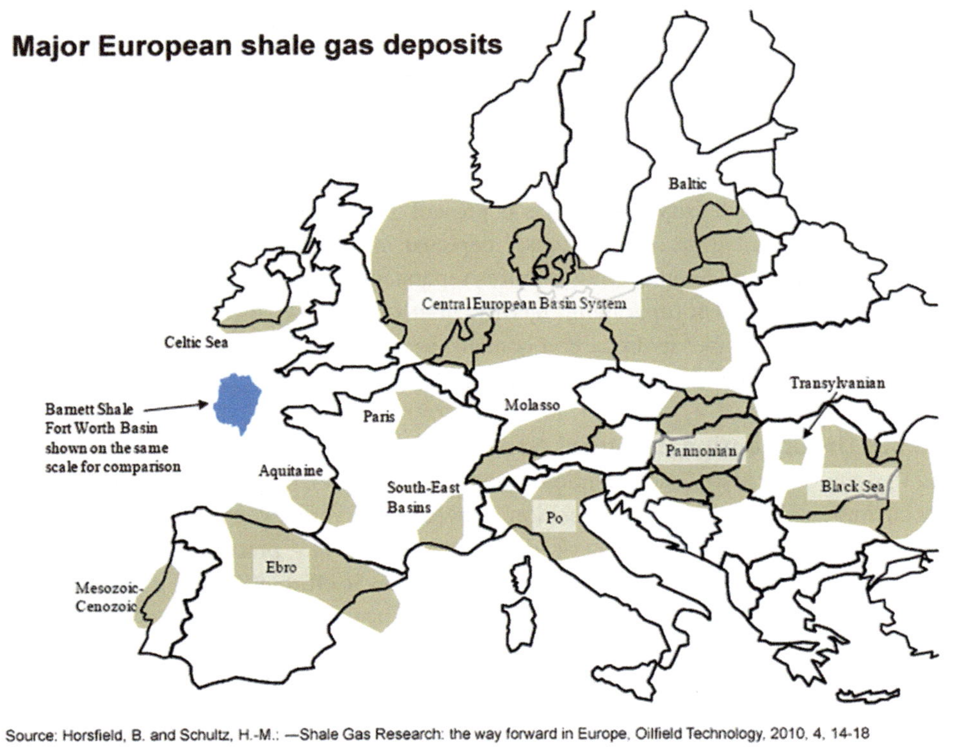

Diagram 4.14 Distribution of European shale gas deposits [48]

Table 4.9 Detailed data for the global natural gas resources (data from [35])

			Unconventional		
	Resources (total)	Conventional	Shale gas	Coal bed methane	Tight gas
Region	Trillion m^3				
World	628,846	309,979	205,374	50,411	63,082
Germany	1870	20	1300	450	100
EU-27	17,922	3094	13,374	1332	122

4.2.3 Natural Gas Production

4.2.3.1 Conventional Natural Gas Production

Today, the search for "classic" natural gas deposits is primarily done by seismic surveys of suitable rock formations. After evaluating the resulting data, an exploratory drilling (up to 10,000 m) is done in areas with promising natural gas sources. If the results are promising and gas reserves can be expected, then actual production wells are drilled. These wells are usually drilled with a drilling rig, as shown in Diagram 4.15, which introduces a natural gas probe into the relevant geological formation.

The natural gas probe is closed at the surface by a "Christmas tree". This consists of two main valves, one of which is an automatic safety device that locks the probe when a malfunction occurs. The pressurized gas leaving the well is transported by pipeline to a collection point, to a gas treatment facility and ultimately to distribution.

Further details on deposits, exploration as well as actual production can be found in [50]. An overview of the relevant natural gas production technical rules and standards can be found on the very informative website of the German Industrial Association of Oil and Gas Producers [51].

Due to its lower density, natural gas is present in the uppermost regions of crude oil deposits as associated gas, as previously depicted in Diagram 4.11. The natural gas is extracted with the crude oil, from which it is separated using suitable separators, before it is sent to a gas treatment (upgrading) facility. In most cases, these gases are characterized by a significantly higher hydrocarbon content (see Table 4.7; H-gases "North Sea" and "Denmark"), which is the result of the incomplete separation of gas and oil.

4.2.3.2 Unconventional Natural Gas Production

As the name suggests, unconventional natural gas production is only possible using unconventional methods. The transition between conventional and unconventional is thus a question of a momentary dissemination of the process. Here, the term unconventional does not refer to the nature of the gas itself, but instead to the nature of the sources from which the natural gas is extracted. The unconventional sources according to the classification made in Table 4.6 include coalbed methane, aquifer gas, tight gas, shale gas, and natural gas from gas hydrates. It has already been shown that shale gas (matrix of clay) and tight gas (matrix of sandstone or carbonates) are the most significant

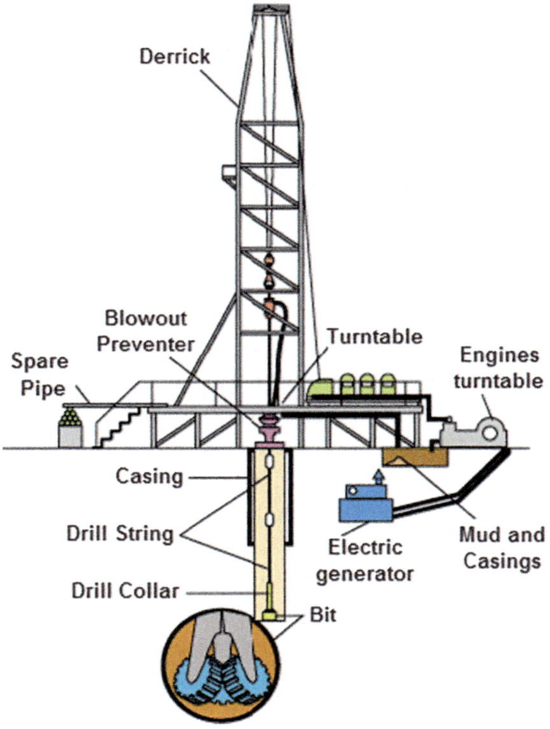

Diagram 4.15 Schematic of a natural gas drilling rig. Photo courtesy of the California Division of Oil, Gas and Geothermal Resources [49]

unconventional resources worldwide. For these reasons, the use of these natural gas resources is under discussion globally and is being intensively pursued. So far, large-scale commercial production is only done in North America using the controversial method of hydraulic fracturing or fracking.

Fracking is the shortened version of the term "hydraulic fracturing", which is defined as the hydraulic fracturing of rock layers in order to be able to extract entrapped natural gas or oil. In the case of natural gas production, a mixture consisting primarily of water, sand, and chemicals is forced at high pressure through a reinforced drilled hole into the gas-bearing rock strata. This is done in order to break the layer, to expand and to hold open by pressed sand in order to allow the enclosed gases to reach the production site. Thus, the freed gas and/or oil can escape through the cracks that have emerged and, just like in conventional natural gas extraction, pass through corresponding drill holes to the surface.

In recent years, the use of chemicals required for fracking is problematic, especially in Germany and other countries of the EU. However, in the heated debate it is often overlooked that since 1961 several hundred "fracs" were carried out in Germany for natural gas production. So far, none of these "fracs" have led to any complications or environmental pollution. Environmental protection agencies as well as internet and movies [52]—even through sensational false reports in the German media—have ignited a fear of groundwater contamination through the use of chemicals required for fracking. In general, the risk for groundwater contamination cannot be excluded and opens up the question of the acceptable residual risk that exists with all potentially dangerous technologies. The most common chemicals include not only different surfactants, but also various salts, alcohols, aldehydes, and amides. These compounds are needed to protect the fracking plant from corrosion and kill bacteria; they are used as lubricants and additives. The composition of the chemical mixture used varies greatly for each case and is highly dependent on the conditions of the local reservoir [53, 54]. The potential hazards of fracking have been repeatedly assessed by independent experts and discussed in relevant studies that are freely available on the website of the German Federal Environment Agency (UBA) [55, 56]. Irrespective of the studies listed, work will continue to improve the fracking process in order to further reduce the potential risk. Due the current very low acceptance of fracking by German society and therefore by politicians, scientists are consequently working very hard on the replacement of potentially hazardous chemicals with harmless substances.

Natural gas can also be present in coalbeds and extracted as coalbed methane (see Diagram 4.11). Usually, the methane is adsorbed on the inner surface of the coal and desorbs during depressurization (e.g. when drilling into the bed). The term coalbed methane refers to sources in beds not developed by mining and to those extracted from coal mines (mine/pit gas). Coalbed methane has been economically extracted around the world for decades, especially in the US, where it accounts for about 10 % of the national production of natural gas. Previously reported Diagrams show that coalbed methane represents a non-negligible source of natural gas [35]).

Methane can also exist in very deep underground aquifers, where it is dissolved due to the high pressure. In this case, the methane is referred to as aquifer gas. When extracting the water to the surface, the gas is released by the pressure relief and can then be used. Under certain conditions, but also at high pressure and low temperatures methane combines with water to form an ice-like solid that is commonly called a methane hydrate or **gas hydrate**. Gas hydrates are primarily found in permafrost and in sediments in deep water on the continental margins of the oceans (see Diagram 4.16). An economical use of aquifer gas or gas from methane hydrates has been extensively researched, but is not technically and economically feasible at the present time

4.2.4 Natural Gas Upgrading

4.2.4.1 Composition and Specifications

The composition of the various (raw-) natural gases varies widely and is highly dependent on the origin and the method of extraction. Table 4.10 shows representative gas compositions of gas at the top of the well, i.e. after separation of the crude gas from the crude oil and before further processing.

It is clear, that a "standard raw gas composition" does not exist and different challenges in gas upgrading are present depending on the production location. **Methane**, the actual

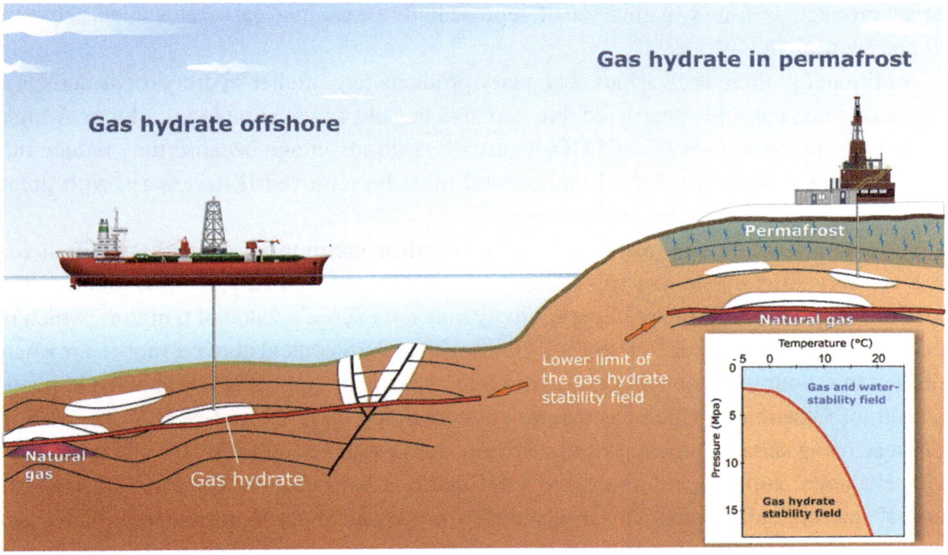

Diagram 4.16 Reserves and possible extraction methods for methane from gas hydrates [57]

Table 4.10 Typical (raw-) natural gas compositions prior to upgrading (data from [58])

	Alberta (Canada)	Colorado (USA)	Kansas (USA)	Bach Ho (Vietnam)	Misker (Tunisia)	Rio Arriba (USA)	Amarillo (USA)
Components	mol-%[a]						
He	0.0	0.0	0.45	0.0	0.0	0.0	1.8
N_2	3.2	26.10	14.65	0.21	16.903	0.68	25.6
CO_2	1.7	42.66	0.0	0.06	13.588	0.82	0.0
H_2S[b]	3.3	0.0	0.0	0.0	0.092	0.0	0.0
CH_4 (Methane)	**77.1**	**29.98**	**72.89**	**70.85**	**63.901**	**96.91**	**65.8**
C_2H_6 (Ethane)	6.6	0.55	6.27	13.41	3.349	1.33	3.8
C_3H_8 (Propane)	3.1	0.28	3.74	7.5	0.960	0.19	1.7
C_4H_{10} (Butane)	2.0	0.21	1.38	4.02	0.544	0.05	0.8
Pentane and high hydrocarbons	3.0	0.25	0.62	2.64	0.630	0.02	0.5

[a]Dry gas composition. Natural gas is often wet prior to upgrading (= saturated with water)
[b]Also at a value of "0" mol-% H_2S, the maximum limits are usually not reached without further sulfur removal

target product, is found in this set of representative example gases at values between 30 and 97 mol-%.

Additionally, there are various secondary products (e.g. higher hydrocarbons and He), that need to be carefully separated, but can also be sold as value-added products. A high content of the inert gases N_2 and CO_2 is usually a disadvantage because they reduce the calorific value/energy content of the gas and must be removed if necessary, with great effort, and have virtually no market value.

Particular attention must also be given to the **sulfur compounds**. Usually the limits for **H_2S** are very strict due to its toxic and corrosive nature. In Germany, the maximum sulfur content of natural gas in the public gas supply may not exceed a value of 6 mg/m^3, which is about 4 ppm or 0.0004 mol-% H_2S [38]. Gases with an H_2S content above 1 mol-% are often referred to as sour gases or acid gases. The largest natural gas production region in Germany is south of Oldenburg, which has natural gas containing up to 35 mol-% H_2S. The H_2S is removed using suitable processes and converted into elemental sulfur by the Claus process, with elemental sulfur being a storable solid that can be traded. In Germany, the largest natural gas upgrading plant in Großenkneten processes about 6 billion m^3 of raw gas annually, yielding approximately 800,000 t of sulfur. Together with a sister plant in Voigtei that is roughly half that size, the entire sulfur demand of the German industry can be met [59].

In order to inject natural gas into the transport grid of the public gas supply, other limits and requirements must be met, some of which have already been mentioned. Apart from the composition, combustion characteristics in accordance with the applicable rules and regulations must be complied with. The most important binding requirements (according to [38]) for Germany are summarized in Table 4.11.

Based on these requirements, the actual gas compositions from the natural gas grids that were shown in Table 4.6 differ significantly from the values of the raw gases shown in Table 4.10. Hence the need for gas upgrading, which will be discussed briefly in the following section.

Table 4.11 Combustion characteristics and limits for gas impurities according to [38]

Name	Symbol	Unit	L Gas	H Gas
Wobbe index	$W_{S,n}$	MJ/m^3	44.6 (+2.2/−5.0)	54.0 (+2.5/−5.0)
Heating value	$H_{S,n}$	MJ/m^3	30.2–47.2	
Relative density (in relation to air)	d_n	–	0.55–075	
Water content (Dew Point)	–	°C	−2 (for 1–70 bar)	
Total sulfur	–	mg/m^3	6	

This table contains only the most important specifications and limits. Details, exceptions, and applicable special exceptions are not included in this table and must be determined by consulting the technical rules of the DVGW

4.2.4.2 Upgrading Processes

The required process steps for upgrading the raw gas prior to injection into the gas grid or liquefaction to LNG (liquefied natural gas) are shown in Diagram 4.17. It should be noted that not all process steps are necessary or available at all natural gas upgrading facilities. For example, a He-separation and recovery only makes sense where a helium-containing raw gas is processed. The same is true for the hydrocarbon separation, which is only necessary if the high-calorific alkanes are not needed to increase the calorific value or are present in excess. Also, LNG liquefaction is only considered to be an option when pipeline transport is, for various reasons, out of the question. Furthermore, the order of the upgrading processes in a real gas processing plant may be different from the "example case" shown in Diagram 4.17. Moreover, several process steps may be combined into a single step. A classic example of this is the combination of gas cleaning and drying to a single step.

A more detailed description of the industrial processes shown in Diagram 4.17 can be found in [58], for example. Thus, the individual process stages are therefore only briefly described here.

Once the natural gas has been extracted from the reservoir or has been separated from the crude oil, a raw gas with a composition like those in Table 4.10 is sent to a **liquid separator**. These are usually gas-liquid separators, e.g. impact separators, in which the liquid hydrocarbons and water precipitate out together. Subsequently, the two substances are separated and processed further or disposed of accordingly.

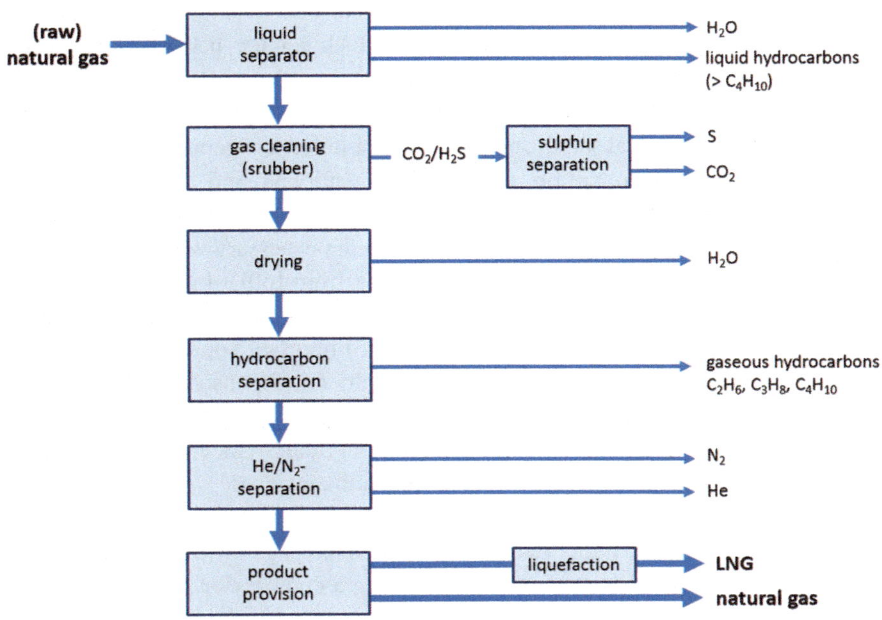

Diagram 4.17 Process chain for gas upgrading between natural gas extraction and distribution as a marketable product

After this first stage, the next step is **gas cleaning** or acid gas removal. Here, depending on the raw gas, gas purity requirements, and the other process steps, just CO_2 or CO_2 and/or H_2S are removed. Usually a chemical or physical scrubbing method is chosen for this particular process step and adjusted accordingly and CO_2 or H_2S is obtained as a product. The CO_2 is either released to the atmosphere or injected back into the geological strata from which the natural gas was extracted in order to maintain the reservoir pressure. Using the previously mentioned Claus process, the H_2S can be converted into elemental sulfur. From there it can either be stored or transported for use as an intermediate in other applications (metallurgy, fertilizer production, organic chemistry. . .).

Following the acid gas removal, or in combination with it, the natural gas is **dried**. In case of insufficient drying, a subsequent formation of undesirable methane hydrates may occur. The deposition of solid methane hydrates in pipelines and fittings may impair their ability to function and even cause damage. For these reasons, drying is generally carried out by scrubbing with suitable absorbents (i.e. glycol), by deep cooling through expansion, or by adsorption onto solid desiccants such as silica gel or zeolites. In order to ensure safe operation under all possible operating conditions, the actual water dew point (Table 4.11) is usually significantly lower than the required dew point.

The next stage shown is **hydrocarbon separation (it** is often beneficial to combine this with other process steps, i.e. natural gas liquefaction) or it can be excluded if the content of these compounds is low or their removal is undesirable, since otherwise the specifications of the gas grid cannot be reached. In natural gases that are extracted with oil (associated gas), the content of higher hydrocarbons is usually so large that their separation is necessary and useful, particularly since they are a valuable feedstock for the petrochemical industry. Usually the hydrocarbon recovery is carried out by cooling the gas to temperatures below the condensation temperature of the gas components to be removed. This requires a significant amount of electrical energy, mostly for the required refrigerators and compressors.

The **He- and/or N_2-separation** was briefly mentioned and should be regarded as a special case, which is only of concern for certain natural gases. Here, cooling methods are also used that are very expensive due to the necessary very low temperature, ($T_{Boiling,\ He} = -269\ °C$, $T_{Boiling,\ N2} = -196\ °C$; data from [60]). In the case of helium, which is a high-priced product amongst gases, this is less relevant. The helium-containing natural gases are the largest and economically most important sources of helium in the world. In the case of nitrogen, there is often another inexpensive alternative for its separation. At points where this is technically feasible, the N_2-rich gases are mixed with correspondingly high calorific gases that are N_2 poor in order to meet the specifications of the corresponding regulations before the mixture can be injected.

Following the described cleaning and upgrading process, is the product **provision** of the final natural gas product. If this gas is to be injected into a gas grid or pipeline, then it is usually compressed to the appropriate pressure using a compressor. As an alternative to long distance transport or in the absence of the necessary infrastructure for pipeline transport, then the natural gas is liquefied to LNG. This process is described in further detail in Chap. 5.

4.3 Transport and Storage of Natural Gas

Frank Graf

4.3.1 Natural Gas Transport via LNG

Natural gas can be liquefied by cooling to a temperature between −164 and −161 °C. As a result, the density drops to about 1/600 of the normal density of the gaseous state. Transport of LNG is a viable alternative to the grid-bound transport of natural gas, when distance or geological conditions prevent the construction of pipelines. The liquefaction of natural gas occurs in large plants with a liquefaction capacity of sometimes higher than 8 million t/a, which corresponds to about 1.2 million m^3/h natural gas (NTP) or a fuel capacity of 13.3 GW. Diagram 4.18 shows the example of the LNG plant Snøhvit in Norway, which can liquefy 4.2 million t/a. A simplified version of the LNG process chain is shown in Diagram 4.19. Following extraction, the raw gas contains impurities such as water, CO_2, N_2, and sulfur compounds, which prior to the actual liquefaction have to be mostly removed to reach residual contents of less than 1 ppmv (in the case of H_2O) in order to avoid fixed deposits (H_2O/CO_2), corrosion (e.g. Hg) or segregation in storage tanks (N_2), for example. Numerous methods are used for gas purification including gas

Diagram 4.18 LNG-Plant Snøhvit (Source: Linde AG)

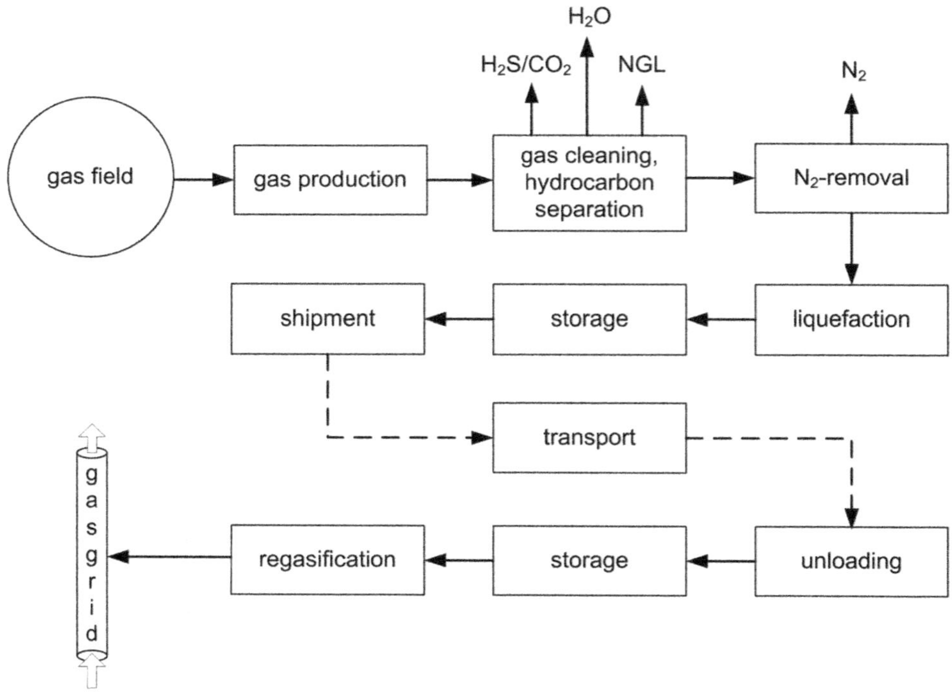

Diagram 4.19 LNG process chain

scrubbers, adsorption and membrane processes, as well as distillation. For the actual liquefaction, numerous industrial scale liquefaction processes are available. These processes differ in the conceptual design of the cooling circuits (stages, coolant) and in the design of apparatus. These include, for example, the C3MR® process from Air Products [61] or the MFC® process from Linde [62] (Mixed Fluid Cascade). All processes require a compression of the refrigerant and heat transfer processes suitable for the cooling of natural gas. As an example, the basic principle of the MFC® process is illustrated in Diagram 4.20. This process contains three cooling cycles that precool, liquefy, and subcool gaseous natural gas. To reduce the energy requirement, increase flexibility, and minimize the system demand (in particular, the design of the heat exchanger) different refrigerant mixtures can be used for the various cooling circuits. These refrigerant mixtures are composed of methane, ethane, propane, and nitrogen. Detailed information on the design of the individual process steps can be found in topic specific literature [63, 64].

Depending on the process and site-specific conditions (e.g. ambient/chilled water temperatures), the energy demand for LNG production is between 250 and 600 kWh/t. Depending on the feed gas, the energy requirement of large plants is between 6 and 11 %

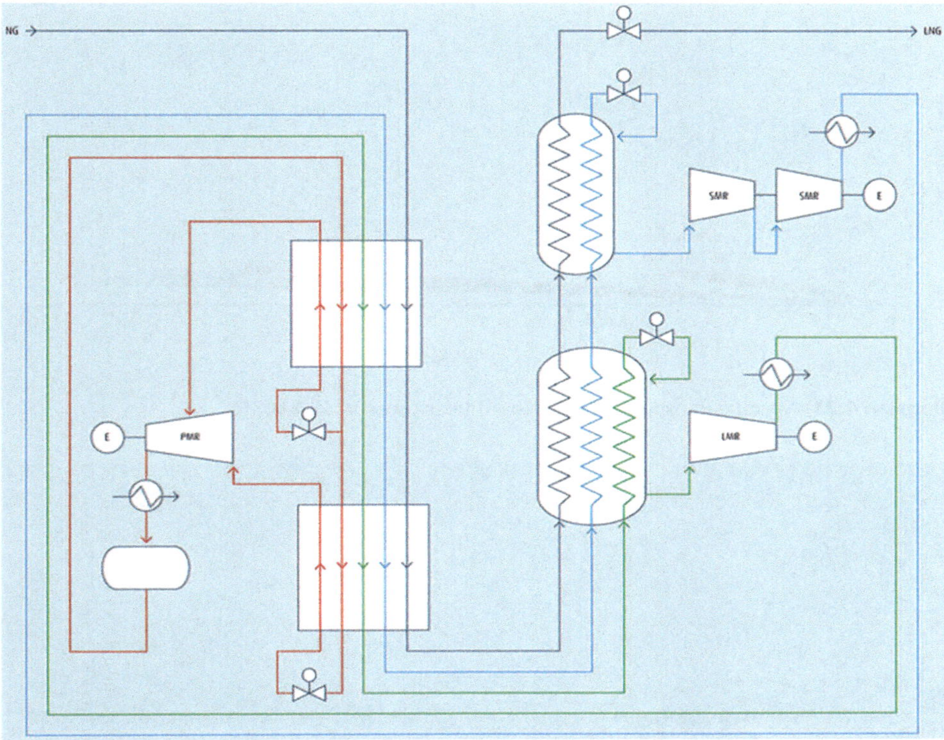

Diagram 4.20 Flow Diagram of MFC®-Process (*PMR* precooling mixed refrigerant, *LMR* liquefaction mixed refrigerant, *SMR* subcooling mixed refrigerant, Source: Linde AG)

of the energy content of the feed gas [65, 66]. Additionally, the distance-dependent energy consumption for transportation and the energy consumption for the regasification also needs to be considered. For transport over 12,000 km, the energy demand is approximately as high as the energy demand for the liquefaction [67]. Typically the transportation costs of LNG transport for distances between approximately 5000 km (Diagram 4.21) and 8000 km [67] are lower than for grid-based transport.

For the entire process chain from extraction to regasification, the average specific CO_2 emissions (as cited in various literature sources) are about 40 g/kWh [67, 69–71]. For the regasification step, the LNG is pumped to the required grid pressure and then vaporized in a heat exchanger. Additionally, an odorization is necessary. The overseas shipment of LNG is done via special tankers that can accommodate up to 120,000 tonnes of LNG (Diagram 4.22). The LNG is distributed through LNG terminals for storage and regasification. The biggest LNG terminals have a capacity of more than 30 million t/a. In addition to supplying LNG to the gas grid, it can also be distributed in the liquid

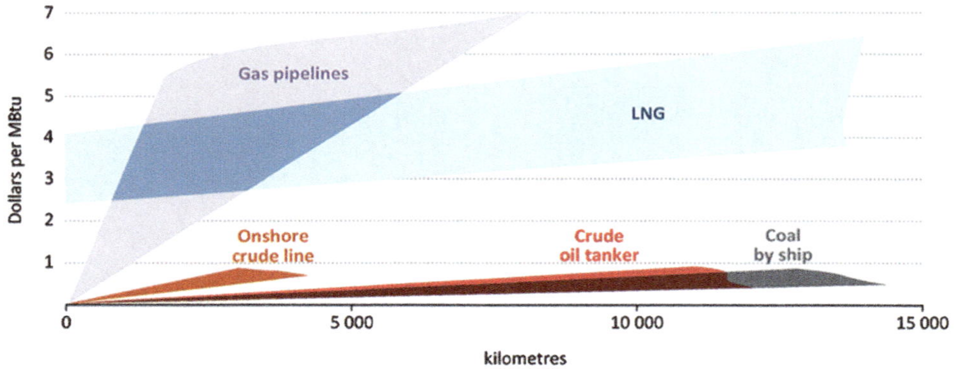

Diagram 4.21 Specific transport costs for fossil fuels (Source: IEA [68])

Diagram 4.22 LNG-Tanker (Source: Linde AG)

state via truck, rail, or inland waterway vessels in order to supply LNG fuelling stations, satellite plants, as well as peak load plants.

LNG plays an increasingly important role worldwide, even if global trade has stagnated in recent years with a volume of about 240 million t/a, which corresponds to about 10 % of the global natural gas market. In the last 20 years, the number of LNG importers has tripled to about 30. At the end of 2013, global liquefaction capacity was 290.7 million t/a, and the regasification capacity was 688 million t/a [72]. In Europe, there are currently 22 operating LNG terminals with a combined capacity of more than 150 million t/a. Six additional terminals will be built [73]. In particular, the use of LNG as fuel for ships and heavy vehicles is expected to be a growing market in the future.

4.3.2 Pipeline Based Natural Gas Transport

The natural gas pipeline network in Europe provides an excellent infrastructure to supply energy to all sectors of the economy. In Europe, there are 2,210,677 km of natural gas pipelines, which supply natural gas from 113 import and border crossing points to the consumer [73, 74]. A simplified depiction is shown in Diagram 4.23 With a total pipeline length of 508,128 km, Germany is fully supplied with natural gas. In Germany, a total of 470,433 km of pipeline is available for distribution purposes. Of these, there are 154,505 km installed at the low pressure level (<0.1 bar), 223,075 km are installed in the medium pressure range (0.1–1 bar), and 92,853 km are installed at the high pressure level (>1 bar). Additionally, there is a long distance gas grid with a pipeline length of 37,695 km (Diagram 4.24 that is divided into two market areas (Gaspool, NetConnect Germany). As illustrated in Diagram 4.25, the pipeline network in Germany has been steadily expanded in recent years. The further expansion and adaptation of the gas system are defined in the

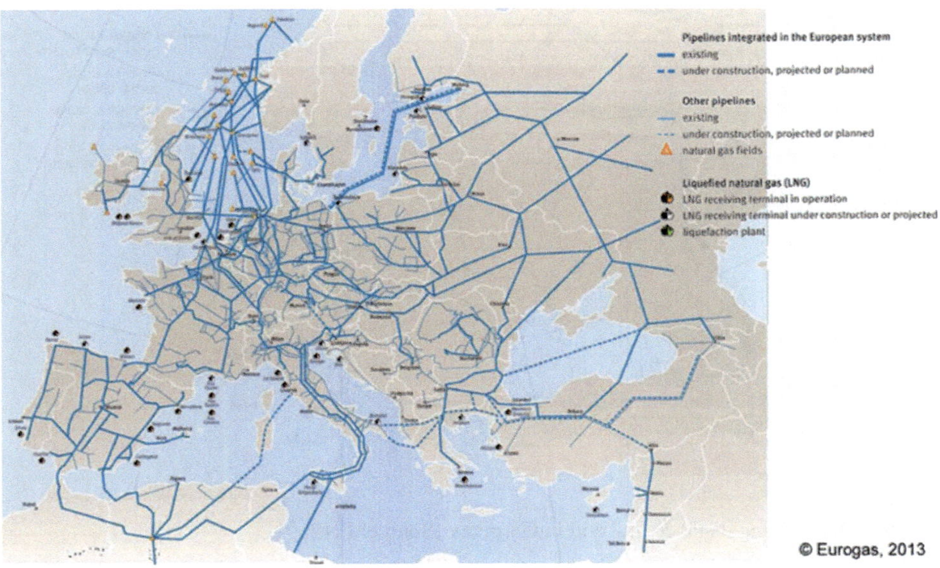

Diagram 4.23 The European natural gas grid in 2014 (Source: Eurogas [74])

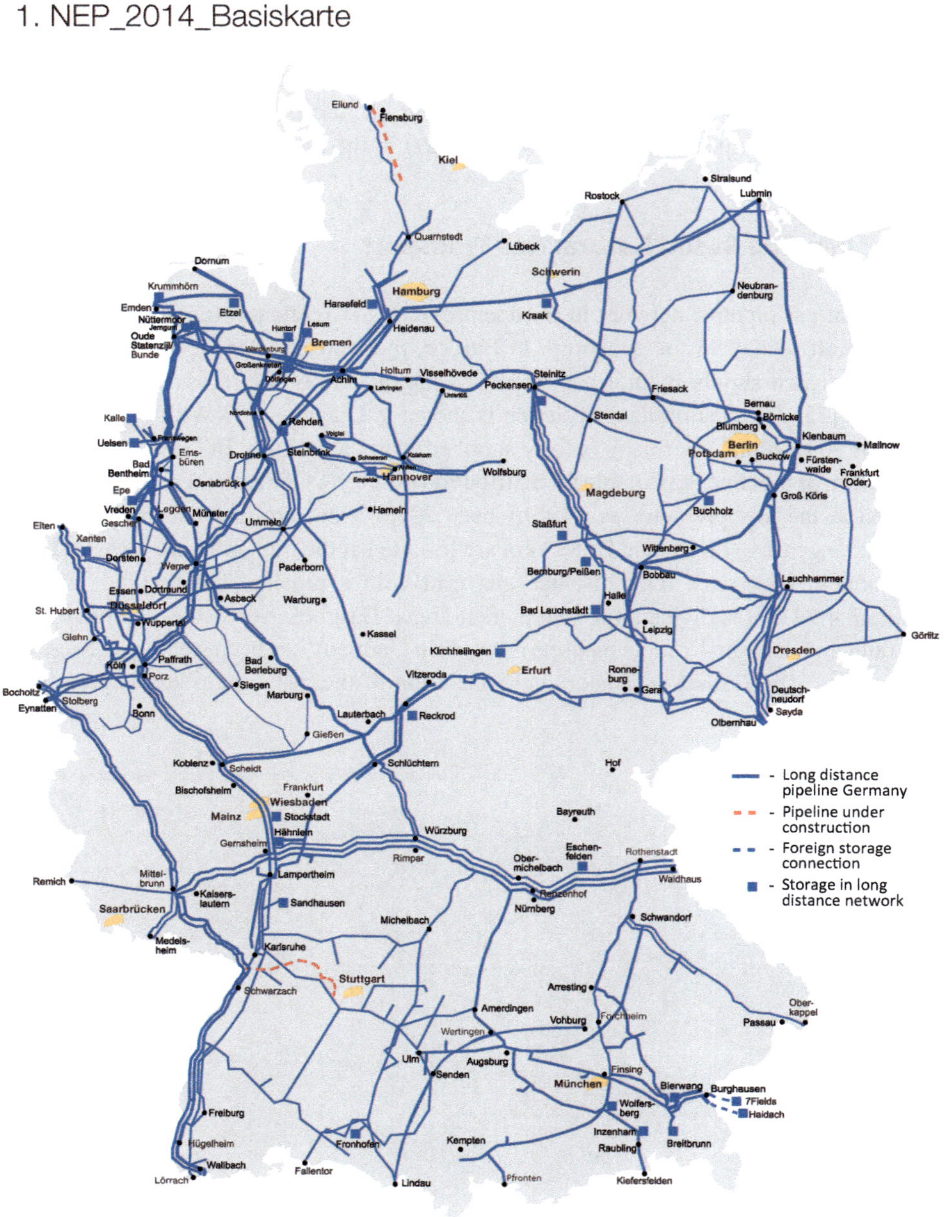

Diagram 4.24 Long distance gas grid in Germany (Source: FNB)

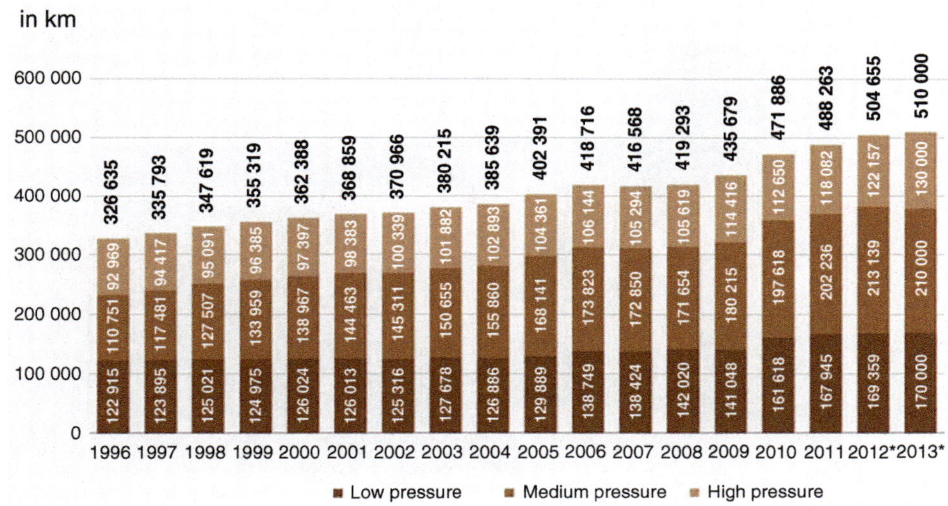

Diagram 4.25 Development of the natural gas network in Germany (Source: BDEW)

Network Development Plan Gas [75]. According to the Energy Industry Act (EnWG § 15a), an updated Network Development Plan Gas (NEP) is created by the gas transmission system operator (TSO) each year for the next decade and commented by the Federal Network Agency (FNA). The Network Development Plan includes a hypothetical scenario framework for future gas demand and requirements of technical adaptation measures, such as for L/H gas adjustment or injection of hydrogen into the natural gas infrastructure (Keyword: Power-to-Gas). The particular investment requirements for the infrastructure are determined from detailed analysis. Due to the annual update, new developments are continually considered.

According to European Gas Regulation (EC 715/2009) the European Network of Transmission System Operators for Gas (entsog) is in charge of the development of the European gas transmission system with respect to the European Union energy goals. For this purpose the ENTSOG Ten-year network development plan and Gas Regional Investment Plans are updated continuously [76].

Natural gas is largely transported by pipeline from the natural gas fields to consumers. A distinction is made between transport networks that transport large quantities of natural gas over long distances and distribution networks that handle the distribution to the final consumer. Transport pipelines in Germany (Diagram 4.26) are usually operated at pressures up to 100 bar.

The largest pipelines have an inner diameter of 1.4 m and reach transport capacity with a chemical energy of up to 60 GW, which is about one order of magnitude higher than for

Diagram 4.26 Pipes for natural gas pipelines (Source: OGE)

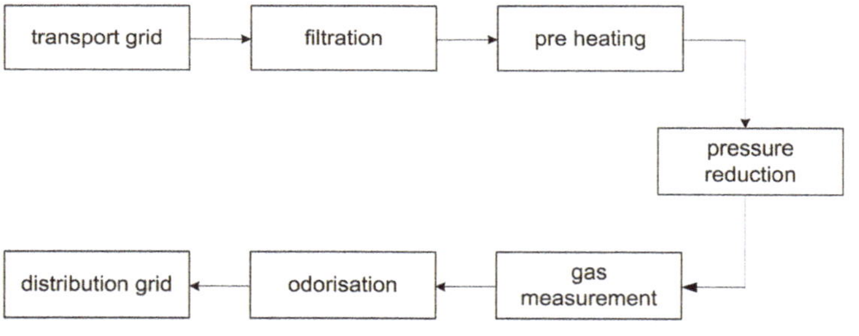

Diagram 4.27 Process steps for the distribution of natural gas

high voltage power transmission lines. To compensate for the pressure loss caused by pipe friction, the natural gas must be recompressed in compressor stations along the transportation path. Compressor systems consist of a drive unit (electric motor, gas turbine) and the actual compressor.

When passing from the long distance network system to regional transport networks and ultimately to the distribution network, various process steps are necessary (Diagram 4.27). These steps occur in so-called gas pressure regulating and metering stations. First, the supplied gas is filtered in order to separate out any impurities, especially impurities from abrasion in compressor stations or pipe dust. In order to avoid strong cooling and condensation of natural gas constituents (hydrocarbons, residual moisture) during pressure reduction due to the Joule-Thompson effect (approximately 0.45 K/bar), the gas is first preheated. Then the pressure is reduced to reach the grid pressure conditions which prevail in the downstream grid. During expansion, additional electric power can be obtained with the aid of expansion engines. Another important element is the gas measurement. This usually consists of a flow measurement and gas quality analysis and is typically used for official billing purposes. It should be noted that natural gas has real gas behavior and consists of numerous components (see Chap. 4). Another important issue for the distribution of natural gas is the odorization. In order for the mainly odourless gas to be perceived by the population in unwanted discharge such as leakages in buildings, it is odorized before distribution to the customer, i.e. provided with odours. In Germany, various odorants are used which consist of sulfur and sulfur-free components. A detailed description of these operations can be found in specialized literature [77–81].

Technical requirements for the construction, operation, and maintenance of gas handling equipment for public gas supply are defined in the DVGW standards. Europe-wide standards and guidelines are being increasingly developed. The possible contribution of the pipeline gas supply to the energy transition is currently under intense discussion in the gas industry. However, natural gas may be increasingly substituted by gases from renewable sources. With about 150 biogas plants, encompassing approximately 1 % of German gas consumption, Germany is the world leader for the injection of biogas into the natural gas grid. Many technical aspects of the production, processing, and injection have been studied in recent years by many market participants. Further details on the subject can be found in Chap. 4.4.1. Power-to-Gas (see Chap. 4.4.3) is another way to transport renewable energy using the natural gas grid in order to store and manage demand from the consumer. Power-to-Gas is the production of hydrogen or synthetic natural gas (SNG) from electrical energy from renewable sources. There are now pilot and demonstration plants in Germany. Together with flexible CHP technologies, natural gas grids can provide important support for the power supply. PtG fuels can also make an important contribution to reducing CO_2 emissions from the mobility sector. The DVGW, its member companies, and numerous research institutions have intensely investigated these aspects within the framework of the DVGW innovation campaign "Gas Technology" and provided important insights into the topic [82].

4.3.3 Gas Storage

Different types of storage are available for the storage of natural gas. In order to balance seasonal fluctuations, pore storage is used for supply optimization and for the provision of strategic gas reserves during crises, while cavern storage is used to cover daily peak demand. Natural gas storage facilities are also used to balance energy in natural gas transportation. At the regional and local level, there are also low pressure gas tanks, spherical gas tanks, and pipe storage. In recent years, the first two storage methods have significantly declined in importance and have been dismantled in many places. In the past, so-called peak-shaving LNG plants have been built to cover peak loads. Pore storage is an underground storage method. Pore storage has a suitable gas permeable rock structure

Diagram 4.28 Cavern storage (Source: KBB Underground Technologies GmbH)

which is covered by gas impermeable layers. The permeability and free pore volume are important parameters. Former natural gas and oil reservoirs or aquifers can be used as pore storage.

Storage caverns are cavities in salt domes with a gas-tight cover. These cavities are created artificially by solution mining. For this purpose, fresh water is introduced through a hole in the salt dome. This fresh water becomes saturated with salt, and the resulting brine is discharged, processed, or discharged into the sea. Examples of pore and cavern storage are shown in Diagrams 4.28 and 4.29.

In underground storage, the total storage volume is the volume of the working gas plus the volume of the cushion gas. The working gas volume is the actual usable volume for the

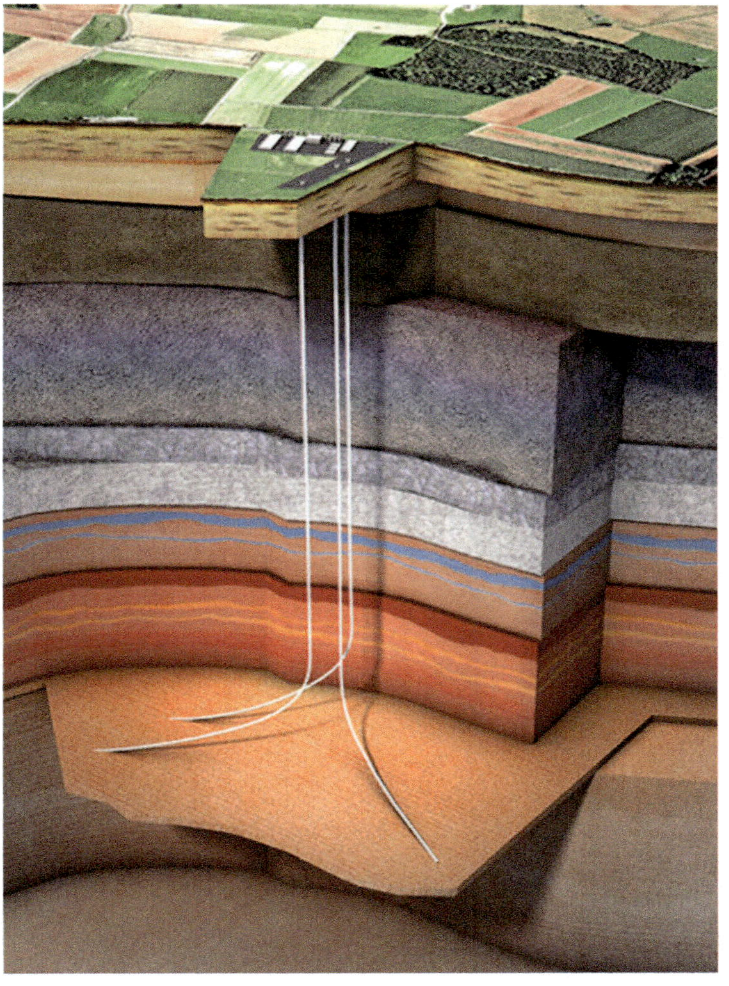

Diagram 4.29 Pore storage (Source: KBB Underground Technologies GmbH)

injection and withdrawal of the gas. The cushion gas volume is the non-usable volume that is required for pressure build-up.

In Europe, 139 storage facilities exist with a total storage capacity of about 96.4 billion m^3 [73].

There were 21 pore storages and 30 cavern storages with a total working gas volume of 23.8 billion m^3 operating in Germany at the end of 2013 [83]. Germany now has the world's fourth largest storage infrastructure, behind the USA, Russia, and the Ukraine. 15 more storages with a total working volume of 6.1 billion m^3 are being planned or under construction. An overview is given in Diagram 4.30. The working gas volume of pore storages in Germany is within a large range, between 18 and 4.400 billion m^3 (NTP)

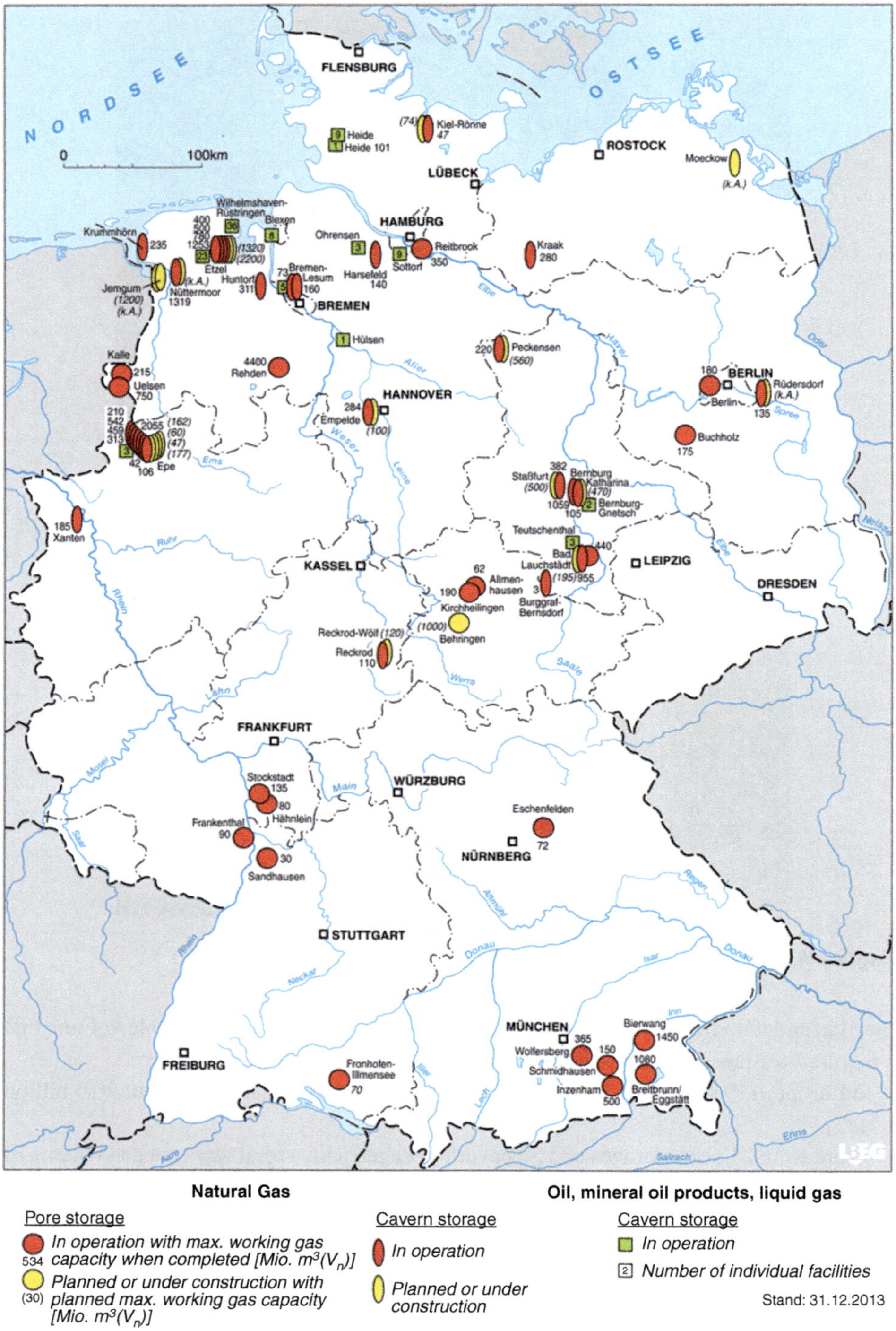

Diagram 4.30 Underground natural gas storage facilities in Germany (Source: LBEG)

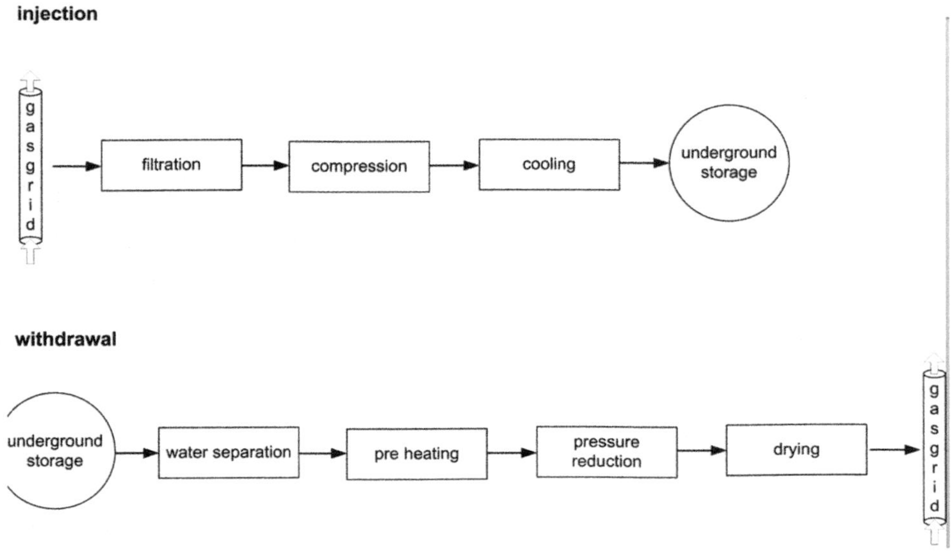

Diagram 4.31 Process steps for storage injection and withdrawal

adding up to a total working gas volume of 10.604 billion m^3 (NTP) and a total volume of 19.557 billion m^3 (NTP). A single cavern has a working gas volume between 5 and approximately 150 million m^3 (NTP), which is usually several individual caverns that are combined into a cavern storage facility (up to 39 individual caverns). In Germany, a total working gas volume of 12.068 billion m^3 (NTP) in a total storage volume of 16.176 billion m^3 (NTP) is available.

Furthermore, the degree of filling is dependent on the maximum withdrawal rate, which is storage specific (Diagram 4.31). This is typically higher for cavern storage than for pore storage. In the existing pore storages in Germany, the maximum withdrawal rate is between 45,000 and 2,400,000 m^3/h (NTP) and between 77,000 and 2,900,000 m^3/h (NTP) for cavern storages. Additionally, specific characteristics are defined for each storage. These characteristics include the start-up and changeover times as well as the minimum injection flow.

The injection and withdrawal of natural gas requires various process steps (Diagrams 4.32 and 4.33). For storage, the incoming natural gas from the transmission pipeline is first freed of particulate impurities before it is compressed from the pipeline pressure (<100 bar) to the storage pressure (<250 bar) using a compressor unit. The gas is heated during compression and must be cooled before it is injected into underground storage.

For the withdrawal, the moist underground gas is first freed of the liquid water contents. Prior to expansion of the natural gas to the desired pipeline pressure, the natural gas is preheated to avoid the formation of gas hydrates during the temperature drop caused

Diagram 4.32 Above ground natural gas storages Uelzen (Source: Storengy Deutschland GmbH)

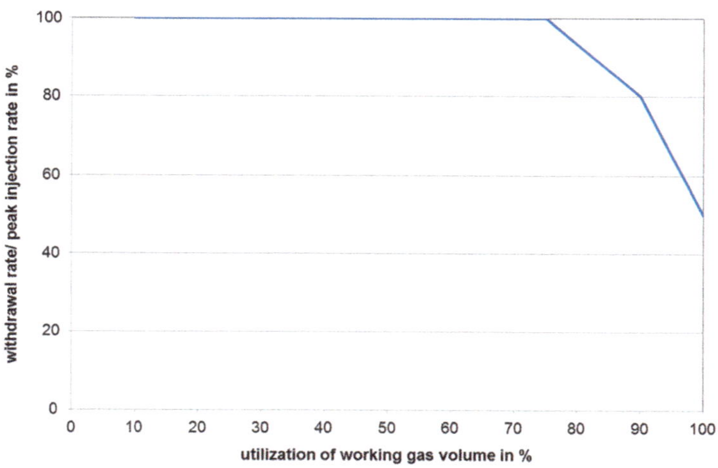

Diagram 4.33 Example specific withdrawal rate for an underground storage

by expansion (ca. 0.5 K/bar). Before being fed into the grid, there is a drying step. Glycol dehydration is typically used for this purpose. For more details, please refer to the technical literature [80].

Storage levels vary seasonally. During the winter months, which have high natural gas consumption, the stored gas is consumed. In the summer, the storages are filled.

The use of existing natural gas storage infrastructure is being considered for the chemical storage of surplus electric energy from renewable sources in the form of hydrogen. While storage caverns are already being used worldwide to store hydrogen, there are outstanding questions that need to be addressed for hydrogen compatibility, especially for pore storage [84].

4.3.4 Gas Market

Internationally, natural gas is on the rise and gas markets are facing major changes [85]. The extensive extraction of unconventional natural gas in the United States has caused the local gas prices to fall too much and for coal-based applications to be replaced by natural gas, which has increased the supply of coal in the world market. In 2014, the prices of natural gas in the US were half of those in Europe and a quarter of those in Asia (Diagram 4.34).

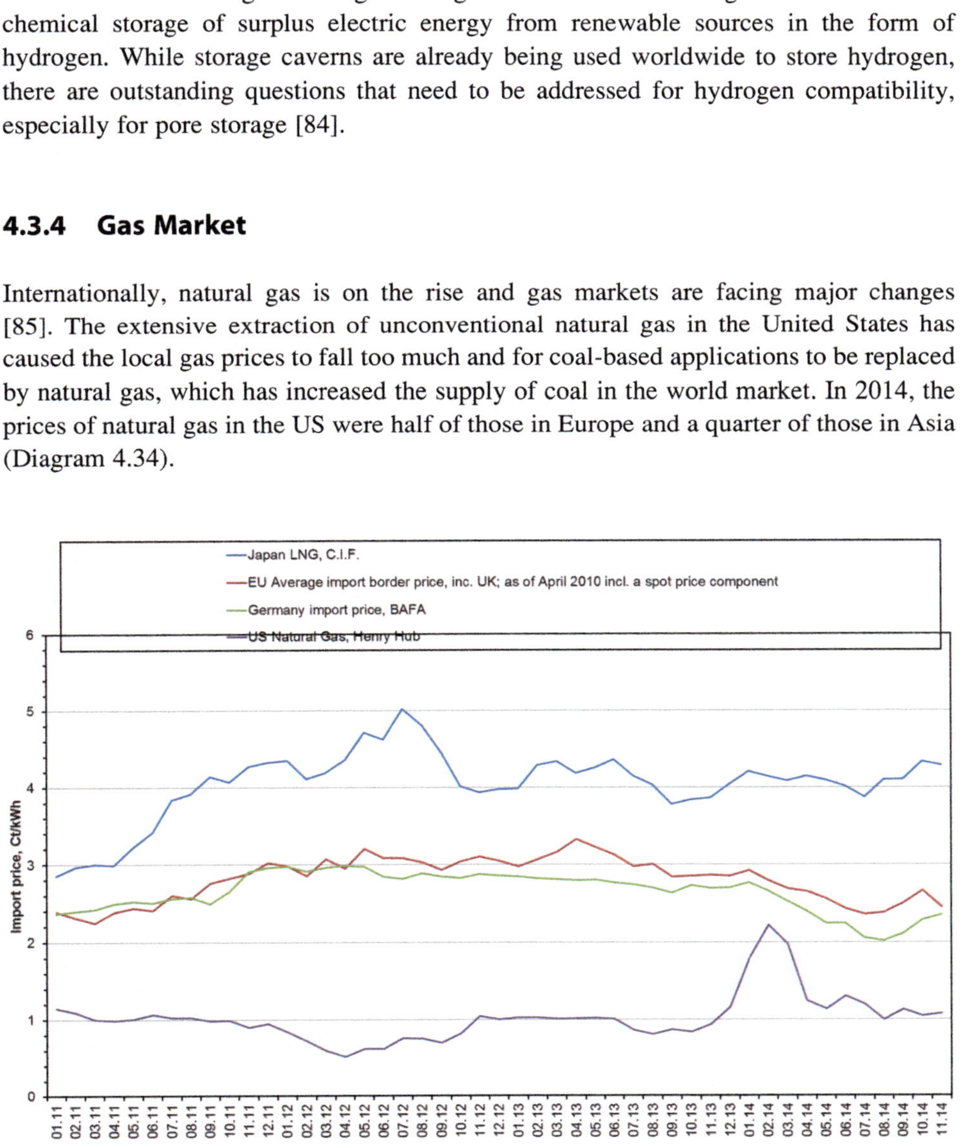

Diagram 4.34 Natural gas price development in USA, Japan and Europe

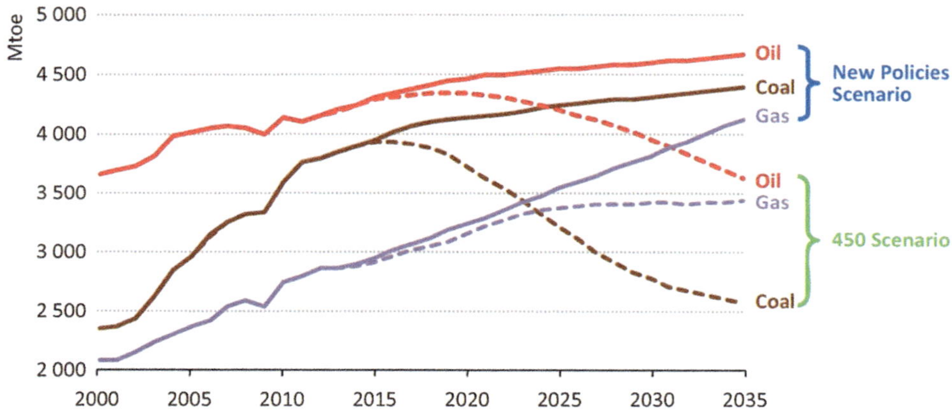

Diagram 4.35 Global expansion of the consumption of fossil fuels for various IEA Scenarios [69] (Source: IEA)

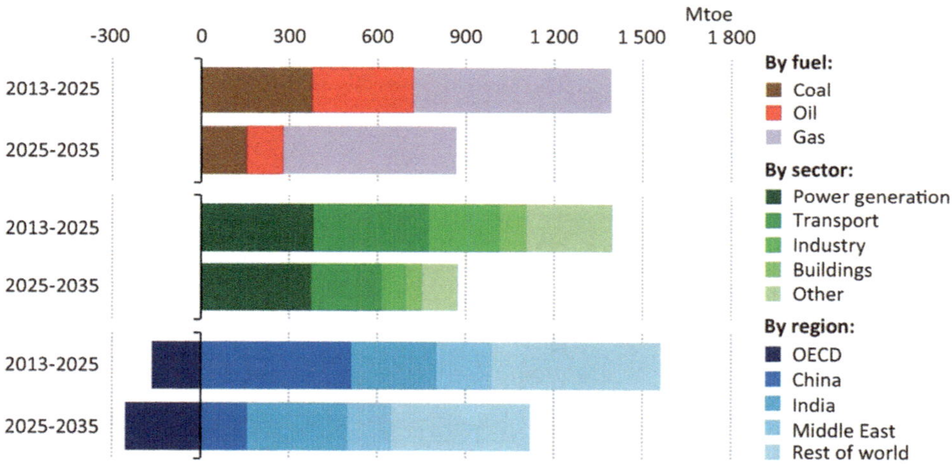

Diagram 4.36 Projected global increase in the consumption of fossil fuels from 2013 to 2035 [48] according to the IEA New Policies Scenario (Source: IEA)

According to studies from the IEA [86, 87], IGU [88], the EU Commission [89, 90] and BP [91] the use of natural gas is expected to increase worldwide. Depending on the scenario, the IEA expects an increase in the worldwide gas production of between 25 and 55 % by 2035 in comparison to 2010 (Diagrams 4.35 and 4.36). In particular, strong growth is expected in Asia. Due to the rapidly increasing global energy consumption, the share of fossil fuels in the total volume is only expected to moderately decline from the current 82 % to 75 %, despite a strong expansion of renewable energies by 2035.

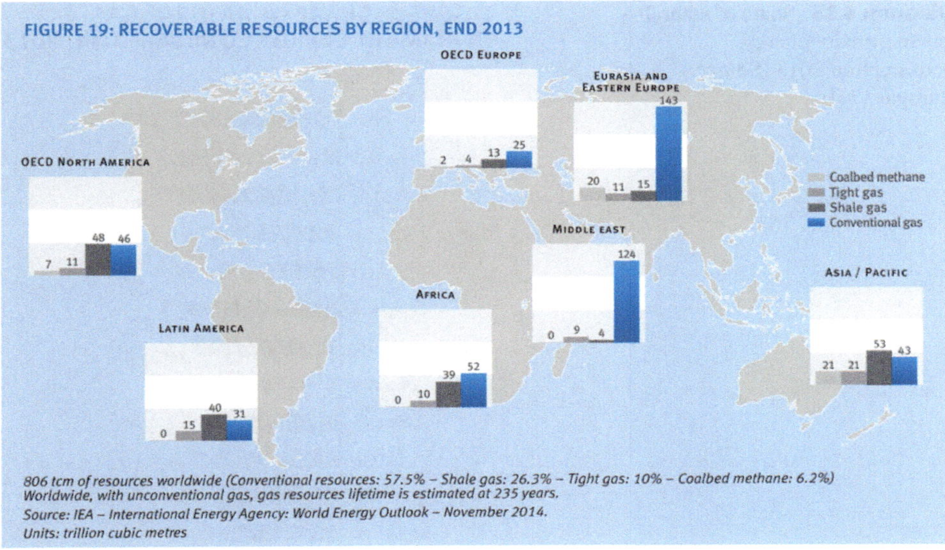

Diagram 4.37 Recoverable natural gas resources by region 2013 (Source: Eurogas [72])

About 65 % of this increase is due to unconventional natural gas, whose share will amount to about 32 % of total extraction in 2035. The reserves-to-production ratio for natural gas is currently 55 years [92]. Furthermore, relevant reserves from unconventional resources are expected (Diagram 4.37). According to the IEA, the largest producers of unconventional natural gas will be the US, China, Canada and Australia. Due to the increasing promotion of unconventional natural gas, the distribution of LNG as well as the liquefaction capacity will continue to increase in the next few years [73].

In Europe, the share of natural gas in the primary energy consumptions averaged out at 23.1 % in 2013 with a wide range between 0 and 43 % in the different member countries (Diagram 4.38). The total natural gas consumption was 4996 TWh. The residential and commercial market is still the most important sector in gas sales with 43 %, followed by industry (30 %) and power plants (23 %). The EU has a diverse supplier structure. The indigenous production (plus Norway) accounts for 55 %. Russia is the most important supplier with a share of 27 % (Diagram 4.39). The LNG imports decreased significantly by 29.1 % in 2013 to 447 TWh (Diagram 4.40). Quatar delivered about 51 % of the European LNG demand.

Diagram 4.38 Share of natural gas in primary energy consumption 2013 (Source: Eurogas [74])

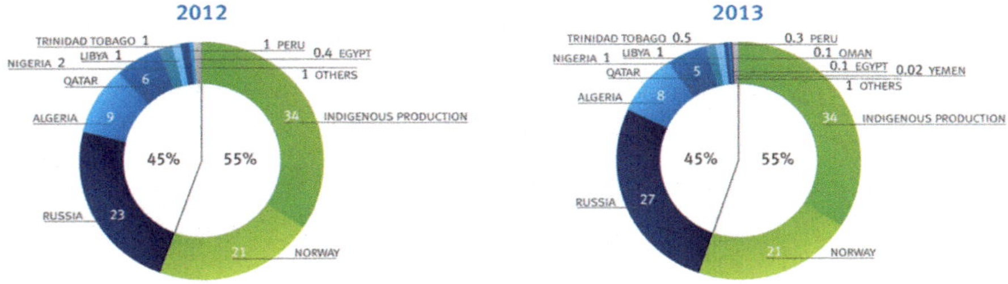

Diagram 4.39 EU-28 gas suppliers 2012 and 2013 (Source: Eurogas [74])

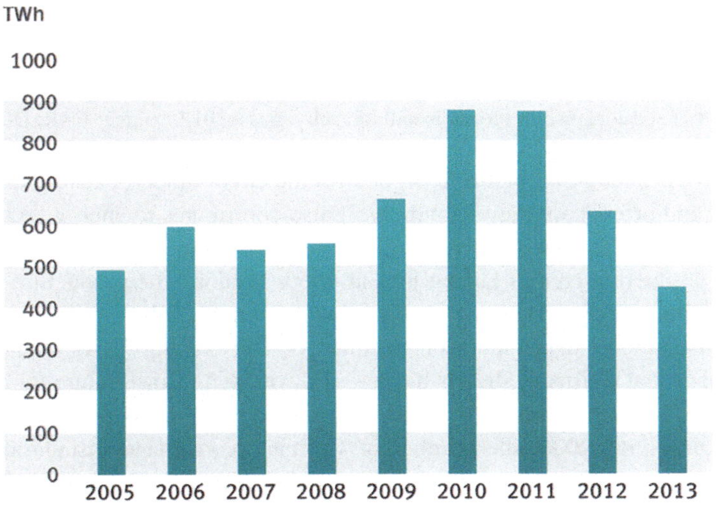

Diagram 4.40 LNG imports to EU-28 and Turkey 2005–2013 (Source: Eurogas [74])

4.4 Sustainable Natural Gas/Methane Production

4.4.1 Biomethane from Fermentation of Organic Waste and Renewable Raw Materials

Janet Hochi, Nantje T. Zimmermann, and Reinhard Schultz

4.4.1.1 Introduction and Summary

Biomethane is a fuel of many talents—a real all-rounder, the joker in the energy transition process. After upgrading to biomethane, biogas produced from organic waste or renewable raw materials is chemically identical to natural gas and can be used as an alternative green fuel for electricity and heat generation, or as a motor fuel. Because it is

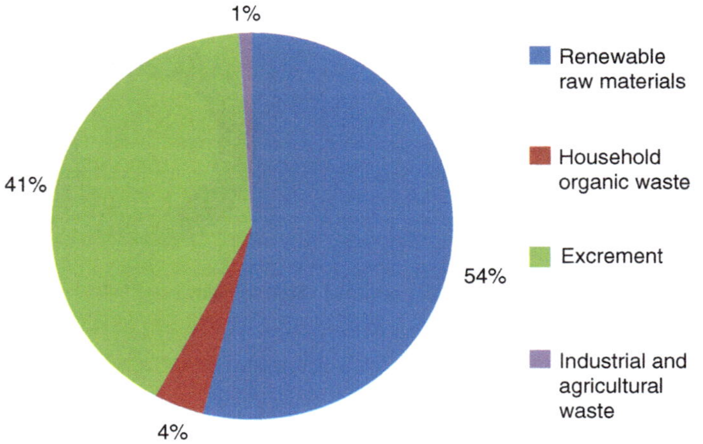

Diagram 4.41 Feedstock use in biogas installations by mass, 2012; source: FNR, DBFZ

more or less CO_2 neutral, biomethane is good for the environment; because it is location-independent and offers continuity of supply, it also contributes to energy security. Local generation concentrated in often economically disadvantaged rural areas means that biomethane production creates lasting jobs all the way along the production and supply chain—from feedstock production and collection, to delivery, plant management and gas sales—and also reduces dependency on gas imports. The existing 510,000 km natural gas grid [97] means that Germany already has a well-developed infrastructure for distributing and storing biomethane produced at small local sites (Diagram 4.41).

In the years since 2006, the number of German biomethane installations feeding renewable gas into the natural gas network has grown to 151. A further 46 plants are currently under construction or planned. With a feed-in capacity of 93,650 Nm^3/h (2014 Diagram), this makes Germany the leading producer in Europe [103].

Also in 2006, the Biofuels Quota Act (*Biokraftstoffquotengesetz*) set the wheels in motion for greater use of biofuels by requiring a specific percentage of biofuels to be added to conventional commercial motor fuels. Biomethane from biogenic waste has been being sold as a motor fuel since 2009, and the market share of biomethane—a second generation biofuel—has risen steadily since then. The EU and many of its member states have recognized the potential of biomethane used as a motor fuel to help reduce greenhouse gas emissions and have encouraged its use by creating an appropriate legislative framework. Car makers ranging from the VW Group, Fiat, Mercedes and Opel to MAN, Scania and Volvo Trucks are also increasingly offering models able to run on natural gas as a fuel. The natural gas and biomethane filling station infrastructure is also becoming less patchy, offering the potential for climate-friendly biomethane/natural gas-based mobility Germany-wide and in the future even Europe-wide.

Until 2012, biogas installations were largely run using renewable raw materials such as wholecrop maize, grass or cereal silage. Due primarily to sustainability concerns, the 2012 Renewable Energy Act (*Erneuerbare-Energien-Gesetz*) shifted the focus to the utilization of waste materials. Both feedstock groups offer plenty of potential, enabling biomethane to extend its reach as a secure, environmentally-friendly component of the electricity, heat and fuels market.

Market Developments, Markets and Trends

Germany's first biomethane installation was connected to the gas grid in December 2006. The installation fed 480 Nm^3/h of biomethane from renewable raw materials into the Stadtwerke München gas grid [93]. In July 2014, there were a total of 151 installations in Germany, upgrading biogas to produce 665 million m^3 of biomethane. A further 46 installations are under construction or planned. The state of Saxony-Anhalt leads the field in terms of capacity, producing 15,940 m^3/h of biomethane. In terms of number of installations, Lower Saxony leads the way with 27 projects across the state.

The restrictions imposed in the 2014 Renewable Energy Act mean there is likely to be little or no further expansion in the near future (Diagram 4.42).

Markets

Because it has the same chemical composition as natural gas, biomethane is suitable for used in a wide range of applications. The German government has promoted the use of biomethane in combined heat and power (CHP) plants, i.e. for co-generation of electricity and heat. Payments are regulated by the Renewable Energy Act, with payments being made for 20 years plus the year in which the installation comes online. This is currently the largest market by sales volume for biomethane. At the same time, biomethane has also gained in attractiveness as a fuel for the (pure) heating market, as the Renewable Energies Heat Act (*Erneuerbare-Energien-Wärmegesetz*) provides for the use of renewable energy sources for heating in new and public buildings. From 5 to 100 % biomethane can be mixed with or used as a substitute for fossil natural gas, making a major contribution to reducing a building's primary energy factor and to meeting the requirements of the Renewable Energies Heat Act. Either mixed with or as a substitute for natural gas, biomethane is also used in the transport sector. The advantages of using this highly efficient fuel were given a major boost (as were transport sector sales) when it became eligible for inclusion in the biofuels quota in 2009. With both EU and German strategies

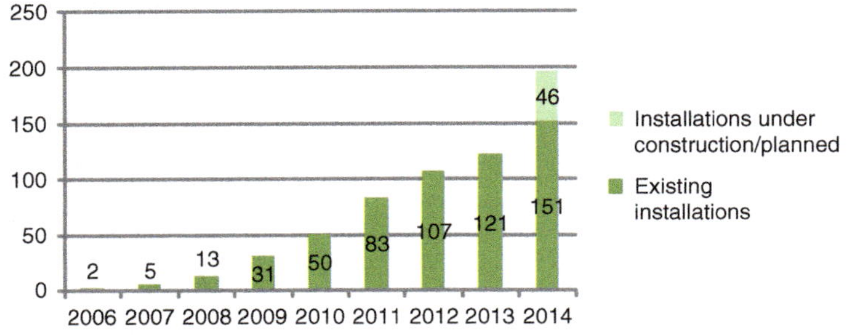

Diagram 4.42 Growth in number of installations 2006–2014; source: dena

Table 4.12 Feed-in volume and number of installations by state in 2013; source: dena

State	Feed-in volume	Installations
Baden-Württemberg	4470 m^3/h	12
Bavaria	10,355 m^3/h	17
Berlin	490 m^3/h	1
Brandenburg	12,670 m^3/h	17
Bremen	0 m^3/h	0
Hamburg	300 m^3/h	1
Hesse	5410 m^3/h	12
Mecklenburg-Western Pomerania	10,540 m^3/h	9
Lower Saxony	11,240 m^3/h	27
North Rhine-Westphalia	5665 m^3/h	10
Rhineland-Palatinate	1030 m^3/h	2
Saarland	550 m^3/h	1
Saxony	5520 m^3/h	8
Saxony-Anhalt	15,940 m^3/h	18
Schleswig-Holstein	2490 m^3/h	5
Thuringia	3420 m^3/h	6

for the mobility and fuels sector anticipating a major role for biomethane, car makers and filling station operators have a large part to play in widening availability and use of this green gas as a motor fuel (Table 4.12).

Trends

An important factor in the expansion of the biomethane economy is the increased utilization of (biogenic) waste materials. The 2012 Renewable Energy Act and the inclusion of biomethane produced from waste materials in biofuels quotas have given added impetus to this process. The industry is also working with the scientific and research community to explore new, higher yield energy crops able to meet more stringent sustainability criteria.

As well as optimizing feedstock utilization, the biogas industry is also working to improve performance at a number of different points along the production and supply chain. Better monitoring of fermenter biology, more efficient engine cooling systems and heating system pumps, improved gas upgrading processes and technologies, power generation from waste heat and more efficient auxiliary systems (mixture cooler) all result in better utilization of the process energy and of the energy produced. Selected measures to upgrade an installation can improve installation operation and reduce the cost of electricity generation.

4.4.1.2 Organic Waste

As a feedstock for biogas installations, organic waste—waste of animal, vegetable or fungal origin—which is able to be broken down by micro-organisms, soil organisms or enzymes can make a major contribution to a sustainable energy supply. The passage of the

2014 Renewable Energy Act meant that waste materials would in future have to make up the primary feedstock for biogas production. This has significantly boosted the value of this feedstock and has made finding new sources of organic waste a key factor for achieving further expansion of the biogas sector. Companies involved in building new installations and installation operators are having to adapt to a new feedstock, the highly varied composition of which can make it difficult to process.

Two thirds of the waste produced in Germany is already used to produce energy. The bulk of this energy, however, is produced from the incineration of waste and used wood [113, p. 34]. It is estimated that, in Germany, waste materials could theoretically be used to produce 8.11 billion m^3 biomethane [98]. This Diagram refers exclusively to waste materials which are not currently available for use in producing energy (Diagram 4.43).

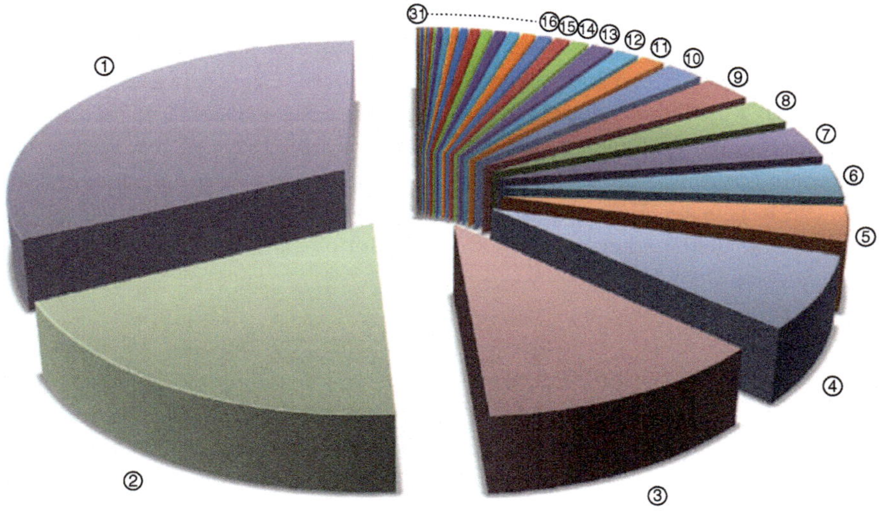

Diagram 4.43 Fermentation potential of waste; source: Biogasrat+, FvB, KTBL

After undergoing fermentation in a biogas installation, organic waste materials can be reused or processed into products such as high quality composts or fertilisers. Because raw materials can be reused in this way, feedstocks should largely be able to continue to be used for existing uses, competition for feedstocks should largely be avoided and it should be possible to maximise utilization of raw materials. Fermentation also avoids the release of greenhouse gases into the atmosphere associated with composting [96, p. 89].

This integrated waste utilization concept would need to be accompanied by a fundamental shift in the way the waste disposal industry operates—from a disposal-oriented process to supply-oriented planning for what would then be viewed as a secondary resource [96, p. 89].

What Is Organic Waste?

The current legislative environment (the Renewable Energy Act) and other factors mean that the use of organic waste materials for energy production needs to become a key pillar of the way the biogas and biomethane sector operates. Organic waste is necessarily produced during the manufacture of certain organic products and can be used for biogas production [102, p. 64]. Organic waste covers a very wide range of feedstocks, ranging from animal excrement, household waste, organic and green wastes and straw, to sewage sludge. The Renewable Energy Act promotes the use of biodegradable wastes, mixed municipal wastes and market wastes in particular for biogas production.

Animal Excrement

Animal excrements which can be used to produce energy in biogas installations include primarily wet manure from cattle, pigs and chicken and dung/litter. In Germany, these materials are particularly abundant in the North West and South, where livestock farming is widespread. This resource has the technical potential to produce 88,000 TJ per year, a relatively small value compared to its volume. This is due to the relatively low energy content of animal excrement.

Wet manure is largely fermented in small 75 kW biogas installations—either on large individual farms or as a joint venture by a group of smaller farms—and transformed into heat and electricity in local CHP plants.

In addition to energy production, fermenting animal excrement in biogas installations has other benefits. Fermented wet manure processed into digestate is a valuable mineral fertiliser for field application. The most important nutrients—70 % of nitrogen, 100 % of potassium and phosphorous—are retained in the digestate after fermentation, and the mesophilic process kills off most of the micro-organisms and bacteria in the manure. Consequently only 30 % of the nitrogen needs to be replaced through the use of mineral fertilisers, delivering significant savings on mineral resources. Increased use of digestate as a fertiliser also reduces greenhouse gas emissions by around 67 % compared to the use of farm manures and mineral fertilisers. It also reduces unpleasant odours arising from the decomposition of odour-producing materials during the fermentation process and avoids the methane emissions which would be produced if the wet manure were to be stored and not used to produce energy [95, pp. 17–18, 99, pp. 33–34; 42].

Fermentation of animal excrement thus makes an important contribution to protecting the climate and to fully utilising raw materials.

Organic and Green Wastes

The bio-waste ordinance (*Bioabfallverordnung*) defines organic waste as waste of animal, vegetable or fungal origin which can be broken down by micro-organisms, soil organisms or enzymes. Green waste includes any plant cuttings produced during garden, countryside or parkland management. It is consequently very heterogeneous in composition, making fermentation relatively difficult. Biogas production is easier if organic and green wastes are collected separately.

The potential offered by organic and green wastes varies significantly from region to region. More green waste is produced in rural areas, but home composting is also (still) more widespread in such areas. At the same time, urban areas produce more organic waste, but very little green waste. With a comprehensive collection system, 3.2 million Mg of organic and green waste could potentially be made available for fermentation per year [113, p. 35]. In the long term, it should even be possible to collect 7.5 million Mg per year [112, p. 26]. This, however, would require the introduction of universal household organic waste collection, something which is provided for—from 2015—by the new Waste Management Act (*Kreislaufwirtschaftsgesetz*). To fully exploit this potential, the introduction of universal household organic waste collection would need to be accompanied by a public information campaign to ensure that the public is aware of the need to separate different waste types (Diagram 4.44) [117].

Organic and green wastes can likewise be utilized both to produce energy and to produce other materials. After fermentation in a biogas installation, this waste can be composted and sold on faster while producing fewer noxious smells in one of around 1000 composting plants. The reduction in nutrient content is very slight, meaning that it can still

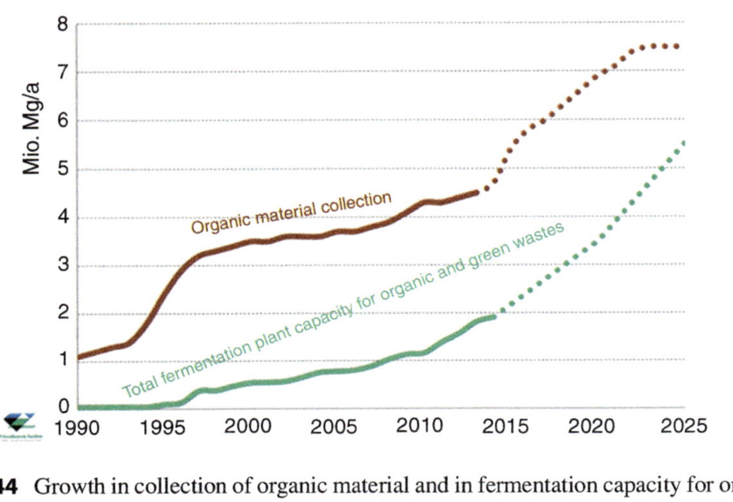

Diagram 4.44 Growth in collection of organic material and in fermentation capacity for organic and green materials in Germany since 1990 and projected future Diagrams; source: Witzenhausen-Institut

be used to produce high quality compost. Pre-fermentation is, however, only economically viable for volumes of waste upwards of 10,000 t per year [95].

Household Waste

The biogenic component of household waste is similarly heterogeneous in composition, ranging from organic and green wastes, paper and used wood, to textiles and rubber, collected via the domestic and commercial waste collection systems. The proportion of organic matter in household waste is expected to fall once separate collection of organic waste is made compulsory in 2015. Demographic changes are also likely to lead to a reduction in waste volumes. Household waste will, however, continue to contain waste materials suitable for fermentation in biogas installations. This would, however, require expensive upstream waste sorting, with the result that it is more efficient to burn it to produce energy in one of the 87 existing waste incineration plants [95, p. 27].

Straw

Straw—the part of the plant left after harvesting cereals and rape—is excellent for fermenting in biogas installations. Straw is produced in large quantities, but has a relatively low energy content. Despite its low energy content, in 2007 it had the technical potential to produce 110,00 TJ [101, p. 25].

The amount of straw produced depends on the area of land under cultivation and the type and mix of crops grown and is therefore somewhat variable. Biogas installations wishing to use straw as a feedstock for fermentation have to compete with traditional uses, including use as an animal litter and as a humus and nutrient source for the field. Although using straw to produce energy is better for the climate, it is an important, low-cost agricultural material for farms.

Sewage Sludge

Waste water treatment in the 9623 municipal waste water treatment facilities in Germany produces sewage gas [116, p. 3]. In around 1000 of these treatment facilities, this is used in CHP plants to generate electricity and heat, largely for internal use. Electricity generated from sewage gas and fed into the public grid is eligible for feed-in payments under the Renewable Energy Act. Sewage gas upgraded to biomethane is also eligible for inclusion in the biofuels quota. Using sewage gas to generate energy is only worth doing in very large treatment facilities. Upgrading to biomethane is not therefore undertaken to any significant extent [95].

(Gaining Access To) Waste Streams

In theory, a very large pool of biogenic waste materials is available to the biogas industry for fermentation. As described above, some waste is already used for other purposes, in part indeed for energy production, but most such waste would be able to be reused and competition for use of this waste should not, therefore, arise.

Leveraging this potential requires a fundamental transformation in waste disposal practices. The key priority under the Waste Management Act (*Kreislaufwirtschaftsgesetz*) is the avoidance of rubbish. A secondary objective is to reuse and recycle the waste that is produced. Only then should waste be used for energy production. It is therefore essential to develop logistical structures for collecting waste which cannot be recycled further, for delivering this waste for fermentation in biogas installations by the shortest route possible and finally, depending on the waste in question, to facilitate reuse of the digestate. Leveraging synergies and improving efficiency will require close collaboration between the agricultural, municipal waste management and energy sectors.

4.4.1.3 Renewable Raw Materials

Renewable raw materials and energy crops are specifically cultivated for conversion into energy in biogas installations. In Germany, around 2 million hectares of agricultural land is used to grow renewable raw materials. The Federal Ministry of Food, Agriculture and Consumer Protection believes that a further 2 million hectares could potentially be used for this purpose [94, p. 5]. Fermenting energy crops achieves the greatest possible biogas yields and produces a biogas with a very high methane content. For this reason energy crops have, to date, been the most widely used feedstocks in biogas installations. The 2014 Renewable Energy Act has, however, shifted the focus to waste materials.

Bioenergy is the only type of renewable energy able to deliver energy continuously and as demand requires. Electricity from biogas from renewable resources has a much smaller impact on the climate than electricity generated from brown or hard coal, irrespective of which of the various energy crops is used. Greenhouse gas emissions per kilowatt-hour for electricity generated from biogas produced using grass as a feedstock, for example, are 90 % lower than for energy produced from coal (Diagram 4.45) [94, p. 12].

What Are (High-Yield) Renewable Raw Materials?

Almost any plant with no or low wood content can be fermented in biogas installations. Plants used include in particular maize, grass silage, wholecrop cereal silage and sugar

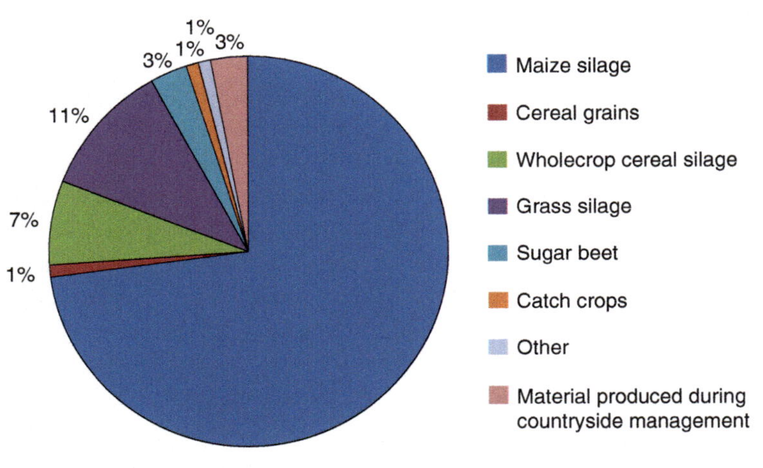

Diagram 4.45 Renewable raw materials as feedstocks in biogas installations by mass in 2012; source: FNR, DBFZ

beet. Due to its high per hectare energy yield and excellent fermentation qualities, the most widely used energy crop is maize. The 2012 Renewable Energy Act imposed a 60 % cap on the amount of maize (by mass) to be used for biogas production. For several years, the biogas sector has been working on developing alternative, equally high-yielding energy crops. This effort has already yielded significant successes, such as *Silphium perfoliatum*, sugar and fodder beets and wild plant mixes. Sustainable cultivation will require that the mix of renewable raw materials used varies by region and according to local conditions.

Annual and Perennial Crops

Energy crops are grown either as annuals or as perennials. Annuals are characterized by a crop rotation system, meaning that different crops, for example maize or various cereal types, are planted in a given field each year. Double cropping and polyculture are widely used, environmentally-friendly annual cropping systems. Double cropping permits two harvests in 1 year, for example by harvesting a winter crop in the spring, then sowing an energy crop such as maize or millet directly after the spring harvest for harvesting in October. The resulting diversified, year-round soil cover can prevent erosion and leaching of nutrients. Polyculture systems involve growing different plant species at the same time, allowing combinations that are optimized both for the soil and for biogas production.

Perennial energy crops are sown once only and can be used over a long period. They are not grown as part of a crop rotation system. The interval between harvests depends on the individual plant species [95, 107].

In Germany, the plants used to produce energy in biogas installations are predominantly annuals.

Sustainability Issues

As discussed above, fermenting renewable raw materials in biogas installations produces climate-neutral energy for the electricity, heat and motor fuels markets. This use releases only the CO_2 previously absorbed by the plants during the growing process. This represents a closed carbon dioxide cycle. In addition, because the soil is covered all year round, reducing erosion, it is able to store more CO_2.

Nonetheless, there has been much public debate over the use of renewable raw materials for energy production. The agricultural sector is already obliged by European and German regulations to grow energy crops using sustainable techniques. EU cross-compliance regulations set out clear requirements for humus balance and content, crop rotation systems, erosion protection and prevention of water pollution, aimed at discouraging monocultures. The 2013 reforms to the Common Agricultural Policy introduced the concept of 'greening', under which 30 % of direct payments to each farm would be linked to environmental improvements. This requires 5 % of field acreage to be treated as an ecological focus area. Farms farming more than 10 ha are required to grow two and farms farming more than 30 ha three different crops and permanent grassland must be maintained at the regional level. Germany's 'good professional practice' system (*Gute fachliche Praxis*) creates a national framework aimed at ensuring that companies in the agricultural sector (and other related sectors) adhere to the principles of animal and environmental protection, thus ensuring long-term soil fertility and land utility. Farmers also have an intrinsic economic interest in managing their fields sensibly and sustainably, in order to ensure that they can continue to be worked in the long term [95, 118].

4.4.1.4 Overview of Fermentation (Diagram 4.46)

Biogas is produced using a biological process. The core process involved in converting organic waste and renewable raw materials into energy in a biogas installation is fermentation. The fermentation process involves creating an anaerobic environment in which bacterial cultures from the raw materials fed into the system produce a mixture of gases, primarily methane and carbon dioxide [106, p. 11].

Biogas plants differ in how they are fed, the water content of the feedstock used, how the process is controlled and their operating temperature.

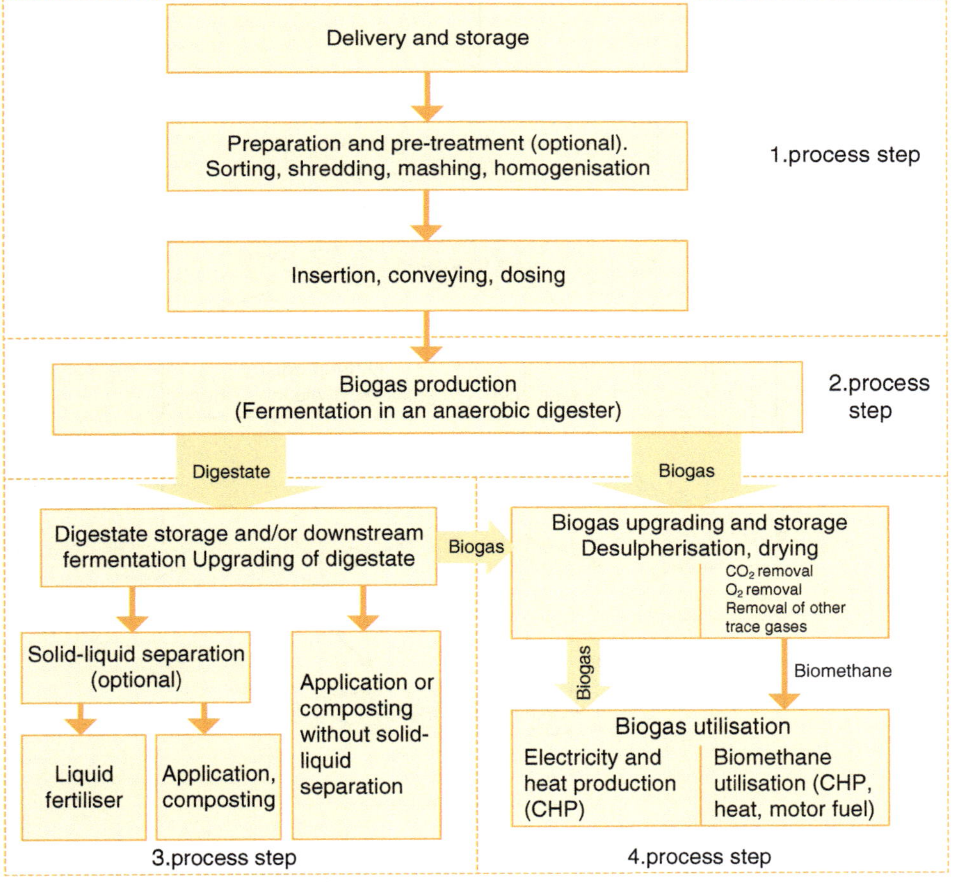

Diagram 4.46 General process flow for biogas production, source: FNR

Continuous and Discontinuous Fermentation Processes

Continuous processes involve regular addition of a feedstock. The process is heavily automated, ensuring that the quality of the gas produced remains constant.

Discontinuous processes are used for dry fermentation of stackable feedstocks in box fermenters. Percolate (the liquid extracted during digestate preparation) is added to the feedstock in closed boxes, resulting in biogas production. By operating multiple fermenters in parallel and feeding them at staggered intervals, this system can also deliver continuous gas production. The digestate produced in this process is also easier to prepare for composting, as it does not need to be dried (Diagram 4.47) [100, 111].

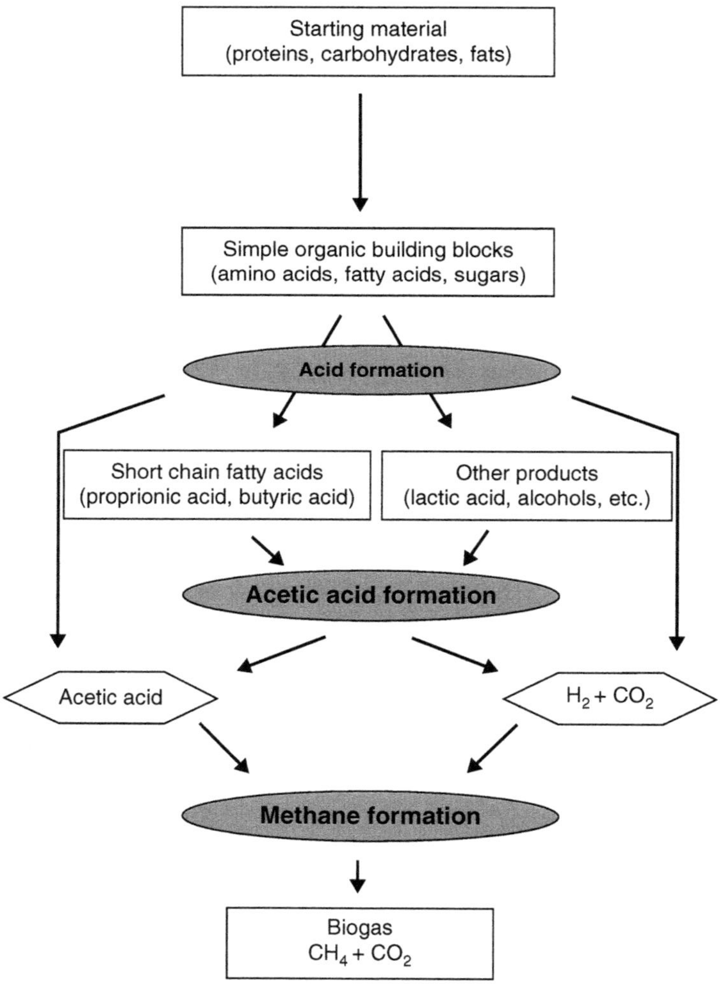

Diagram 4.47 Schematic representation of anaerobic digestion; source: FNR

Wet and Dry Fermentation

In practice, fermentation is considered to be wet if the dry matter content is around 12 %. Liquid (usually percolate) is added to the dry biomass so that it is capable of being pumped and stirred. Wet fermentation is always carried out as part of a continuous process. Dry fermentation may be carried out as either a continuous or discontinuous process. Dry fermentation is fermentation of substances with a dry matter content of 20–40 %. Dry matter contents greater than about 30 % are fermented using a discontinuous percolation process.

It should, however, be noted, that digester micro-organisms require a liquid medium to survive. Fermentation processes are classified as wet or dry on the basis of the dry matter content of the total feedstock used for fermentation. Adequate moisture content is always necessary for bacterial performance [106, p. 13].

Mesophilic and Thermophilic Process Temperatures

The process temperature is a major factor in determining gas yield from the fermentation process. Thermophilic operation, at a temperature between 55 and 66 °C, delivers a very high gas yield and digestion rate. To date, operation at thermophilic temperatures has only been used for continuous fermentation processes. Mesophilic fermentation processes, at temperatures of 33–37 °C, are more stable and are suitable for all fermentation types. A key advantage of high temperature thermophilic operation is that pathogens are killed during the fermentation process. Where fermentation is carried out under mesophilic conditions, to allow reuse of the digestate this must be carried out in a separate downstream process [100, p. 88, 111].

Upgrading to Biomethane

In order to use the biogas produced during fermentation as a substitute for or additive to natural gas, it must first be upgraded. Upgrading involves increasing the proportion of methane and removing non-flammable components. This usually involves desulfurizing, drying and compressing the gas and removing the CO_2 using any of a range of different processes, including chemical scrubbing, pressure swing adsorption (PSA) or water scrubbing. The gas is then odorized and the calorific value and pressure adjusted. Upgrading biogas to biomethane is only economically viable for large biogas installations or by pooling gas from a number of smaller installations (Diagram 4.48) [103, 115].

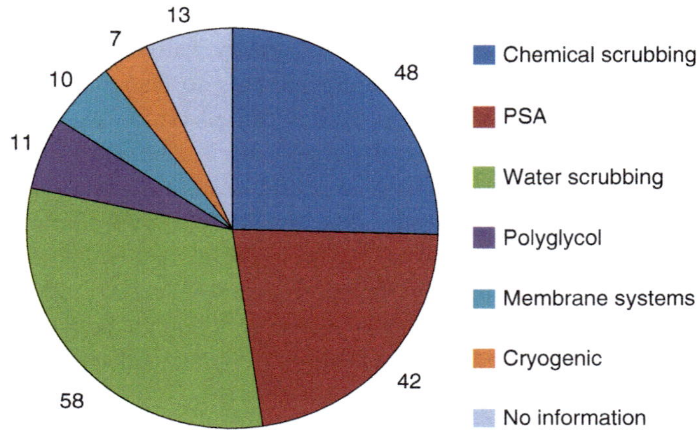

Diagram 4.48 Upgrading system use in Europe; source dena

Trade in Biomethane

The Gas Grid Access Ordinance (*Gasnetzzugangsverordnung*) guarantees priority connection of biomethane installations and continuous availability of the grid connection. Transport and sale of biomethane requires a 'biogas balancing group contract' (*Biogas-Bilanzkreisvertrag*) with the 'balancing group grid operator' (*Bilanzkreisnetzbetreiber*), which keeps a balance sheet for gas quantities. A feed-in contract regulates assignment to a balancing group and the quality of the biomethane fed into the grid, whilst an off-take agreement regulates off-take of the gas at the consumer. A key feature of the trade in biomethane is the requirement for a certificate of origin. This certificate records the installation in which the biomethane was produced and the feedstock used. This certificate is the basis for making the payments set out in the Renewable Energy Act, for inclusion under the Renewable Energies Heat Act and for inclusion in the biofuels quota. Since 2011, the German Energy Agency manages certification of gas origin and characteristics via the German Biogasregister:

> The system allows biomethane producers, traders and consumers to document gas volumes fed into the grid. The list of criteria imposes no restrictions on how the gas should be disposed off by producers or traders. Once information on production conditions has been verified by an independent auditor, the gas volumes can be assigned to a consumer and certification can be seamlessly documented from feed-in to off-take. This ensures that gas cannot be sold more than once [103, p. 35].

The Federal Office for Agriculture and Food is responsible for implementing sustainability criteria from Directive 2009/28/EC on the promotion of the use of energy from renewable sources and for the facility for double counting liquid and gaseous biofuels. To this end, it collects the relevant data and, depending on the application, forwards it to the grid operator (electricity) or central customs offices (fuel).

4.4.1.5 Biomethane as a Motor Fuel

Like natural gas, biomethane can be used as a motor fuel for cars, lorries and buses. In Germany, the most popular alternative fuel vehicles (in terms of new registrations) are vehicles powered by CNG (compressed natural gas). A total of 98,172 vehicles fuelled by natural gas or biomethane have been registered (2014 Diagram) [105]. The filling station network covers the whole country, with 919 filling stations, 181 of which offer biomethane derived from waste materials. In order to ensure reliability of supply and convenience for motorists, further expansion of this network is, however, required (Diagram 4.49) [109, 110].

The use of biomethane as a motor fuel is a relatively climate and environment-friendly alternative to conventional fuels. A 20 % biomethane/80 % natural gas mix emits 39 % less CO_2 than gasoline. Use of pure biomethane gives rise to a 97 % reduction in CO_2 emissions compared to using gasoline. Due to their low energy density, gaseous fuels do, however, need to be compressed to a high pressure or require more storage space. Cars are therefore equipped with pressurized tanks, which compress the gas to a pressure of 200 bar [104, 108].

Two different motor types are currently mass-produced for use in CNG vehicles. Bivalent vehicles have a gasoline tank in addition to the natural gas tank, allowing them

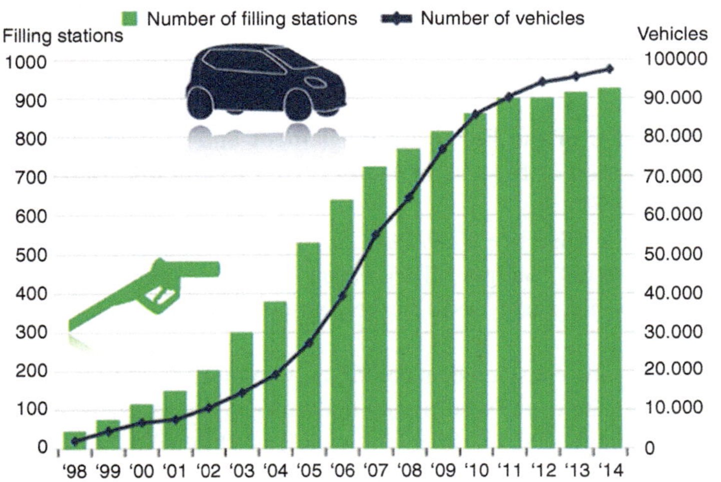

Diagram 4.49 Filling station and vehicle numbers 1998–2013; source: erdgas mobil

to run on gasoline should there be difficulties in obtaining CNG. Monovalent vehicles run on a single fuel only. Being optimized for a single fuel means they have more efficient engines, making them the better option from an environmental and climate-protection point of view. CNG vehicles are also financially attractive, as CNG is currently cheaper [108].

Natural Gas Vehicles and Infrastructure

At present, CNG vehicles make up around 1 % of total traffic. Germany has the second largest fleet of CNG cars after Italy and followed by Sweden and France. France, Italy and Sweden lead the way in terms of CNG-powered buses, followed by Germany and the Netherlands. This places Germany among the pioneers in both categories.

Germany's car industry has also once again shown its capacity for innovation, with a disproportionate number of German carmakers (Audi, Fiat, Mercedes, Opel, Seat, Skoda and Volkswagen) offering CNG vehicles as part of their standard range. The number of models available compared to conventional vehicles nonetheless leaves plenty of room for improvement. The steady growth in the range of vehicles available suggests a trend towards increased use of natural gas and biomethane-based CNG. This trend is also apparent in the commercial vehicle sector. Fiat, Iveco, Mercedes, Opel, Volkswagen, Kögel, MAN, Renault, Scania, Solaris and Volvo all offer CNG vehicles.

The number of CNG vehicles in circulation is tightly bound to the availability of CNG filling stations. Compared to the rest of Europe (with the exception of Italy), Germany already has a well-developed infrastructure.

4.4.1.6 Regulatory Framework

The Situation in Europe

Directive 2009/28/EC of the European Parliament and of the Council of 23 April 2009 on the Promotion of the Use of Energy from Renewable Sources (Renewable Energy Directive, RED)

The entry into force of Directive 2009/28/EC on the promotion of the use of energy from renewable sources (RED) obliges all European Union member states to increase the proportion of renewable energy used in the transport sector to at least 10 % by 2020 [Article 3(3)].

Biofuels will only be considered as contributing to meeting this target where defined sustainability criteria are met. These criteria must also be met to receive financial support for biofuel use [Article 17(1)]. For biomethane used as a biofuel to be included in biofuel quotas, it must also meet the requirements of the Renewable Energy Directive. Under Article 17(2–6), the following criteria must be met:

- Greenhouse gas emission savings from the use of biofuels compared to conventional fuels must be at least 35 %. From January 1, 2017, greenhouse gas emission savings from the use of biofuels must be at least 50 %. From January 1, 2018, greenhouse gas emission savings for biofuels produced in installations which started production on or after January 1, 2017 must be at least 60 % [Article 17(2)].
- Biofuels must not be produced from raw materials grown on land with high biodiversity value. This means land that in or after January 2008 had high biodiversity, irrespective of whether the land remains highly biodiverse, e.g. primary forest, areas designated for nature protection purposes and highly biodiverse grassland [Article 17(3)], land which stores large amounts of carbon dioxide, e.g. wetlands and "continuously forested areas" [Article 17(4)] and land which was peatland in January 2008 [Article 17(5)].
- Agricultural raw materials used for the production of biofuels must be produced in accordance with the requirements and standards given in part A and in point 9 of Annex II to Council Regulation (EC) No. 73/2009 of 19 January 2009 establishing common rules for direct support schemes for farmers under the common agricultural policy and establishing certain support schemes for farmers. They must also be produced in accordance with the minimum requirements for good agricultural and environmental condition defined pursuant to Article 6(1) of that Regulation [Article 17(6)]. Where cross-compliance requirements are met, biomass from land which was already under cultivation on January 1, 2008 is always considered to have been sustainably produce. This also applies to guarantees of origin for imported biomass for biofuel production or imported biofuels from third countries.

In addition, the European Commission undertakes to investigate the effects of biofuel production on land use, raw material prices and food security and to report on this to the European Parliament and Council every 2 years (Article 17(7), Article 23).

To ensure that sustainability criteria are met all the way along the biofuels production and supply chain, the Renewable Energy Directive requires economic operators to use a mass balance system operated by the member states [Article 18(1)], which:

- Enables consignments of raw materials and biofuels with differing sustainability characteristics to be mixed
- Ensures that it remains possible to identify the sustainability characteristics and sizes of the various consignments within a mixture
- Provides for the sum of all consignments withdrawn from the mixture to be described as having the same sustainability characteristics, in the same quantities, as the sum of all consignments added to the mixture

The sustainability of biofuels must be verified either by member states or using a voluntary certification system approved by the European Commission. Since July 19, 2011, the European Commission has approved 15 voluntary systems in use in the 27 member states. Approval is valid for 5 years.

As of March 2014, the following systems are currently approved by the European Commission:

1. **ISCC** (International Sustainability and Carbon Certification)
2. **Bonsucro EU**
3. **RTRS EU RED** (Round Table on Responsible Soy EU RED)
4. **RSB EU RED** (Roundtable of Sustainable Biofuels EU RED)
5. **2BSvs** (Biomass Biofuels voluntary scheme)
6. **RBSA** (Abengoa RED Bioenergy Sustainability Assurance)
7. **Greenergy** (Greenergy Brazilian Bioethanol verification programme)
8. **Ensus** (Ensus voluntary scheme under RED for Ensus bioethanol production)
9. **Red Tractor** (Red Tractor Farm Assurance Combinable Crops & Sugar Beet Scheme)
10. **SQC** (Scottish Quality Farm Assured Combinable Crops (SQC) scheme)
11. **Red Cert**
12. **NTA 8080**
13. **RSPO RED (Roundtable on Sustainable Palm Oil RED)**
14. **Biograce GHG calculation tool**
15. **HVO Renewable Diesel Scheme for Verification of Compliance with the RED sustainability criteria for biofuels**

As part of the sustainability certification process, all economic operators at all points in the production and supply chain are required to carry out independent audits of their suppliers and interfaces and to document these audits by means of sustainability certificates. Sustainability certificates are the basis for including these biofuels in biofuel quotas and for tax benefits.

As well as meeting requirements that raw materials used for biofuel production have been grown sustainably and to be recognized as contributing to the 10 % target biofuels, they must, as described above, also achieve a set greenhouse gas emission saving compared to a reference fossil fuel. Article 19 of the Renewable Energy Directive defines the procedure used to calculate this greenhouse gas emission saving. One option for demonstrating the required greenhouse gas emission saving is to use the default values for different raw materials listed in part A of Annex V to the Directive to calculate greenhouse gas emissions. The use of default values is dependent on member states reporting to the European Commission by March 31, 2010 that the cultivation of a raw material/feedstock in the relevant NUTS2 region is not associated with greenhouse gas emissions higher than the "disaggregated default values for cultivation" given in part D of Annex V to the Renewable Energy Directive.

Part A of Annex V to the Renewable Energy Directive gives default values for biomethane for

- Biogas from municipal organic waste as compressed natural gas: typical value 80 %, default value 73 %
- Biogas from wet manure as compressed natural gas: typical value 84 %, default value 81 %
- Biogas from dry manure as compressed natural gas: typical value 86 %, default value 82 %

Article 19 also allows greenhouse gas emissions to be calculated from actual values using the methodology set out in part C of Annex V to the Directive.

Actual greenhouse gas emissions should be calculated for all other feedstocks used in biomethane production. This is not straightforward, since biogas installations use multiple feedstocks (multi-input) to produce biomethane. The Renewable Energy Directive makes provision for mixing a range of different intermediate products, but nonetheless requires an individual greenhouse gas calculation to be made and evidence of greenhouse gas emission savings to be provided for each individual feedstock used. The use of a mean value covering the product as a whole is not permitted. Different greenhouse gas emission values can only be summed across feedstock mixes if a maximum greenhouse gas value has been specified for this step of the production process and is not exceeded. The use of a maximum value is intended to ensure that greenhouse gas values for feedstocks which do not meet greenhouse gas emission saving requirements are combined with greenhouse gas Diagrams for feedstocks which do meet these requirements, such that the requirement to be 35 % better than the reference fossil system is met.

Directive 2009/30/EC of 23 April 2009 Amending Directive 98/70/EC (Fuel Quality Directive, FQD)

The Directive on the quality of gasoline and diesel fuels (98/70/EC) requires suppliers of fuels (both fossil and biofuels) and electricity for powering road vehicles and mobile machinery and equipment ('Suppliers') to reduce life-cycle greenhouse gas emissions per unit of energy from fuel or electricity by 10 % compared to a baseline value by December 31, 2020 (Article 7a(2), Directive 98/70/EC).

Life cycle greenhouse gas emissions means all greenhouse gas emissions that can be assigned to the fuel during production, transport, distribution, processing and combustion (Article 2(6), Directive 98/70/EC). One component of life cycle greenhouse gas emissions is upstream emissions, meaning all greenhouse gas emissions occurring prior to transportation of the fuel to the refinery, for example emissions arising during transport of natural gas [Article 1(1)].

Six percent of the 10 % reduction target must be achieved through direct reduction in life cycle greenhouse gas emissions (Article 7a(2a), Directive 98/70/EC). This reduction can, for example, be achieved through the use of biofuels and electricity from renewable resources or by reducing greenhouse gas emissions arising during transport of crude oil or natural gas.

The remaining 4 % of the 10 % reduction target must be achieved by other measures described in the base directive (Article 7a(2b, c), Directive 98/70/EC).

Groups of suppliers can meet the reduction obligations jointly (Article 7a(4), Directive 98/70/EC). The Commission Directive sets out baseline values for reduction obligations and sets out rules for calculating life cycle greenhouse gas emissions for fossil fuels and electricity. The methodology for calculating life cycle greenhouse gas emissions for biofuels has previously been defined in the Fuel Quality Directive (Annex IV, Directive 98/70/EC). The baseline value for the greenhouse gas intensity of fuels is based on average fuel consumption in the EU in 2010. It is defined as 88.3 CO_2 equivalent per megajoule (gCO_2eq/MJ) (Annex II). By 2020, suppliers must reduce this Diagram by 6 %.

Suppliers must calculate life cycle greenhouse gas emission intensity using a standard methodology (Article 2(1), Annex I). The following factors are used in calculating CO_2 equivalence:

- Carbon dioxide (CO_2): factor 1
- Methane (CH_4): factor 23
- Nitrous oxide (laughing gas, N_2O): factor 296

Supplier calculations of life cycle greenhouse gas emission intensities must be based on baseline values rather than actual values. These baseline values are specified in the directive by raw material and transport type (Article 2(1), Annex I):

- Both for total life cycle greenhouse gas emission intensity
- And specifically for the intensity of upstream emissions

The actual life cycle greenhouse gas emission intensity of a fuel depends on many individual circumstances, such as the energy required to obtain the raw material and emissions during transport of the raw material. Emissions from the manufacture of machinery and equipment used during transport, production and refining are not taken into account (Annex Id).

For fuels with baseline values higher than the values for conventional oil production, the supplier may use actual values, where these are lower than the baseline values and can be verified using a methodology compatible with the ISO 14064 standard on quantification, reporting and verification of greenhouse gas emissions [Article 2(2)]. This applies, for example, to fuels extracted from oil shale. If provable reductions in upstream emissions—during gas or oil production—have been achieved, the supplier can deduct this from the baseline values [Annex I(3)].

Suppliers must report annually to each member state on the greenhouse gas intensity of the fuels and electricity supplied to that member state (Article 7a, directive 98/70/EG). The report must include the following details (Annex III):

- A unique identifier for the supplier, e.g. a VAT number
- The amount of fuel supplied
- The raw material source, e.g. conventional oil or oil shale, for each fuel supplied
- The amount of electricity supplied for road vehicles
- The greenhouse gas intensity of the fuel or electricity
- The reduction in upstream emissions [cf. Annex I(3)]
- The place of purchase of the fossil fuel or biofuel
- The place of origin of the fossil fuel, biofuel or electricity

Member states can permit suppliers to make use of data provided for other EU legislation-related purposes (Article 4, Annex III). Member states must provide an annual report on national fuel quality data to the Commission (Article 8(3), Directive 98/70/EC). Data must be transmitted to the European Environment Agency electronically (Article 5, Annex IV) and include the following details:

- Data from the company report in aggregated form
- The methodology used to calculate any greenhouse gas emissions which are lower than the values specified in Annex 1

Member states may make use of data that they have previously reported for other reasons (Article 5).

The Outlook for EU Regulatory Initiatives

Proposal COM(2012) 595 from 17 October 2012 for a Directive of the European Parliament and of the Council Amending Directive 98/70/EC Relating to the Quality of Gasoline and Diesel Fuels and Amending Directive 2009/28/EC on the Promotion of the Use of Energy from Renewable Sources

On October 17, 2012, the European Commission published its proposal for amending the Renewable Energy and Fuel Quality Directives. The objective of the proposal is to prevent indirect land use changes (ILUC) as a result of the use of biofuels. The draft proposal proposes the following:

- Biofuels will be taken into account for greenhouse gas emission reductions if they are produced in installations
  - which were in operation on July 1, 2014 and, compared to conventional fuels, until December 31, 2017, produce a 35 % greenhouse gas emission saving from January 1, 2018, produce a 50 % greenhouse gas emission saving
  - which came into operation after July 1, 2014 and produce a 60 % greenhouse gas emission saving compared to conventional fuels.
- Limiting the contribution made by renewable energy from conventional biofuels produced from agricultural biomass to the EU's 10 % target to a maximum of 5 % by energy content (new Article 3(4d), Directive 2009/28/EC).

- Introducing the ability to treat biofuels from certain waste materials as having four times their energy content for the purpose of meeting national and European biofuel targets.
- Introducing ILUC factors into reporting on EU member state biofuels policy greenhouse gas emissions.

Biofuels that do not produce a significant greenhouse gas emission saving when emissions as a result of indirect land use changes are taken into account and are produced from crops used for foods or feedstuffs will no longer be subsidized after 2020. For the 10 % target, the definitive measure is the energy content of the fuel used.

The European Commission proposal was debated in the European Parliament, where it was approved with some amendments at a first reading on September 11, 2013. On June 13, 2014, the Energy Council, an organ of the European Council involved in the legislative process, agreed the following key points:

- Biofuels will be taken into account for greenhouse gas emission reductions if they are produced in installations
  - which were in operation on July 1, 2014 and, compared to conventional fuels, until December 31, 2017, produce a 35 % greenhouse gas emission saving from January 1, 2018, produce a 50 % greenhouse gas emission saving
  - which came into operation after July 1, 2014 and produce a 60 % greenhouse gas emission saving compared to conventional fuels.
- Use of conventional biofuels in the transport sector: the maximum contribution to the 10 % target from conventional biofuels was set at 7 % of final consumption (new Article 3(4d), Directive 2009/28/EC).
- Use of advanced biofuels in the transport sector: member states will set their own targets for the share of final energy consumption from advanced biofuels for 2020. The reference value for this target is 0.5 % (new Article 3(4d), Directive 2009/28/EC, in conjunction with part A of Annex IX new).
- Introduction of the principle that renewable liquid or gaseous fuels of non-biological origin will be treated as containing four times their actual energy content and biofuels produced from feedstocks listed in parts A and B of Annex IX will be treated as containing twice their actual energy content.
- Member states should collaborate to prevent the provision of false information on biofuels by developing joint mechanisms for checking the traceability of raw materials along the length of the production and supply chain.
- The introduction of ILUC factors into reporting on EU member state biofuels policy greenhouse gas emissions annually on March 15.

Because the legislative process is based on Article 294 of the TFEU, the European Council and European Parliament must agree a joint position within 6 months of submission. Should they fail to do so, they must appoint a Conciliation Committee.

Some points of the agreement of the European Council are open to criticism. Allowing renewable liquid or gaseous fuels of non-biological origin to be treated as containing four times their actual energy content will allow the 10 % transport sector target to be achieved faster, but will reduce the actual proportion of biofuels placed on the market. The introduction of ILUC factors is open to criticism on two fronts. Firstly, the factors involved in and methodology to be used for calculating indirect land use changes are scientifically controversial. Secondly, it would be more sensible to work towards effective control mechanisms on indirect land use changes in relevant producing countries than to introduce a factor which evaluates biofuels in an undifferentiated manner while taking into account their specific origin. In addition, in Germany, from 2015 the Biofuels Quota Act will mean moving from a volume-based quota system to a greenhouse gas-based system which already takes into account the most important features of ILUC.

Proposal for a Directive of the European Parliament and of the Council on the Deployment of Alternative Fuels Infrastructure 2013/0012 (COD)
On January 24, 2013, the European Commission published its proposal for a Directive on the deployment of alternative fuels infrastructure. Under this proposal, by 2020 the number of CNG-powered vehicles should rise from the current one million to ten million. By the end of 2020, the following infrastructure should have been set up in each member state:

- A network of publicly accessible CNG refuelling points, with maximum distances between them of 150 km [Article 6(6)]
- A network of publicly accessible LNG refuelling points for heavy duty motor vehicles along the Trans-European Transport (TEN-T) Core Network at maximum distances of 400 km [Article 6 (3)]
- Publicly accessible LNG refuelling points for inland waterway transport in all maritime ports and by the end of 2025 in all inland ports in the TEN-T Core Network [Article 6 (1–2)]

On April 15, 2014, the European Parliament accepted a compromise proposal on the European Commission's draft Directive on the deployment of alternative fuels infrastructure. The compromise proposal still has to be formally approved by the European Council before the Directive can enter into force, probably in autumn 2014.

Key regulatory points for extending CNG mobility are:

- Member states will ensure through their national policy frameworks, that an appropriate number of publicly accessible CNG refuelling points have been set up by December 31, 2020 to allow the circulation of CNG motor vehicles in urban/suburban agglomerations, other densely populated areas and where appropriate within networks determined by the member state. [Article 6(6)]

- Member states will ensure through their national policy frameworks, that an appropriate number of publicly accessible CNG refuelling points have been set up by December 31, 2025 at least along the existing TEN-T Core Network to allow the circulation of CNG motor vehicles throughout the Union. [Article 6(6a)]
- Member states will ensure that CNG refuelling points for motor vehicles deployed or renewed [36 months from the date of entry into force of this Directive] comply with the technical specifications set out in Annex III.3.3.1. [Article 6(7)]
- Where appropriate, when displaying fuel prices at a fuel station, in particular for natural gas and hydrogen, comparison between the relevant unit prices shall be displayed for information purposes. Display of this information shall not mislead or confuse the user. [Article 7(3a)]

In principle, the Directive on the Europe-wide deployment of alternative fuels infrastructure must be viewed as positive, since market penetration of CNG vehicles is being significantly hampered by coordination problems, and the lack of a Europe-wide minimum infrastructure has meant that demand for CNG vehicles has failed to meet expectations (the 'chicken and egg' problem). The compromise hammered out between the European Parliament and European Council is a watering down of the European Commission proposal, which proposed minimum distances between refuelling points of 150 km in order to ensure that using such vehicles was broadly convenient and to enable the use of CNG vehicles for cross-border transport. The extent to which CNG vehicles will be adopted in practice is therefore heavily dependent on the design of measures set out in national policy frameworks.

The Situation in Germany

Biofuels Quota Act

The Biofuels Quota Act (*Biokraftstoffquotengesetz*) creates an obligation in the Federal Imission Control Act (*Bundesimmissionsschutzgesetz*) for a minimum amount of biofuels (biofuel quota) to be added to all gasoline and diesel fuels placed on the market from January 1, 2007, based on a company's total annual gasoline and diesel fuel sales volume, including the biofuels component.

The quota obligation can be met

- By the company itself or
- By a third party undertaking to meet this obligation (quota transfer)

either by adding the minimum amount of biofuel to fossil gasoline or diesel fuel or by placing a volume of pure biofuel corresponding to the minimum share on the market.

From 2015, the system changes from an energy-based biofuels quota to a climate-protection-based quota, aimed at reducing greenhouse gas emissions from biofuels. Under the Biofuels Quota Act and the Act Amending the Promotion of Biofuels (*Gesetzentwurf zur Änderung der Förderung von Biokraftstoffen*), quotas are as follows:

Diesel Fuel Quota 4.4 % from 2007

Gasoline Quota

- 1.2 % in 2007
- 2.0 % in 2008
- 2.8 % in 2009
- 2.8 % from 2010 to 2014

The amount of biofuel to be placed on the market is based on the energy content of the comparison fuel and not on volume. If the quota is exceeded in one calendar year, the excess may be carried forward to the next year.

Total Biofuels Quota as a Percentage of Fuel Use

- 5.25 % in 2009
- 6.25 % from 2010 to 2014

From 2015, quotas are based on the greenhouse gas emission saving from biofuels.

- 3.0 % greenhouse gas emission saving in 2015
- 4.5 % greenhouse gas emission saving in 2017
- 7.0 % greenhouse gas emission saving in 2020

Under Section 37a(4.1) of the Federal Imission Control Act, the biofuels quota can be met by the addition of biogas upgraded to natural gas quality (biomethane) to natural gas fuels. Biomethane must meet the requirements of the Biofuels Sustainability Regulations (*Biokraftstoff-Nachhaltigkeitsverordnung*) and the gas quality requirements for the use of natural gas as a fuel set out in the 10th Ordinance on the Implementation of the Federal Imission Control Act.

In addition, under the 36th Ordinance on the Implementation of the Federal Imission Control Act, certain biofuels count double towards meeting the biofuels quota set out in Section 37a of the Act. Under Section 7(1) of the 36th Ordinance on the Implementation of the Federal Imission Control Act, the following feedstocks count double towards the biofuels quota:

- Waste materials
- Non-food cellulosic materials
- Ligno-cellulosic materials

Outlook

On July 16, 2014, the German Cabinet approved a Twelfth Act Amending the Federal Imission Control Act (*Zwölftes Gesetz zur Änderung des Bundesimmissionsschutz-gesetzes*), which regulates the transformation of the biofuels quota system into a greenhouse gas-based quota system.

The act provides for the quota to be raised from 3 to 3.5 % in 2015 and 2016. From 2017, the quota will be reduced from 4.5 % to 4 % and from 2020 from 7 % to 6 %. Other key points in the act are:

- The introduction of an electronic register for companies subject to the quota.
- From 2016, biofuels placed on the market in the previous year by a company which was not subject to the quota will be able to be applied to the biofuels quota.
- Extension of the power to issue ordinances under Section 37d(1) to enable the regulator to issue ordinances to comprehensively regulate or substantiate the use of biomethane in meeting quota obligations and the system for certifying greenhouse gas emissions where biomethane is fed into the gas grid. In this context, there is a need to examine whether biomethane fed into the gas grid from a location outside the area of jurisdiction of the Federal Imission Control Act should be permitted to be included in the quota.

The 30 % greenhouse gas emission saving target has already been significantly exceeded (the saving is already more than 40 %), meaning that without a fundamental revision to these targets there will be no growth in biofuels. Biofuels quota and greenhouse gas emission saving targets should therefore be raised significantly. The EU Fuel Quality Directive, to which, in specifying quotas, the act refers, only sets out minimum quotas. EU member states are free to set their own more demanding quotas.

In addition, the current rules on transferral of biofuels quotas for pure biofuels placed on the market is not sufficiently flexible. Companies placing pure biofuels on the market are not 'companies subject to the quota' as set out in the Federal Imission Control Act. This means that they can sell the quota they have generated just once, and this must be done by March 31 of the following year. They are excluded from carrying the quota forward and are subject to price pressure from companies which are subject to the quota, since if the latter fail to purchase it, their quota expires and they lose a significant revenue component. Furthermore if by placing fossil fuels on the market, a company placing pure biofuels onto the market becomes subject to the quota, it can only transfer that portion of the quota pertaining to its obligations arising from the quantities of fossil fuels placed on the market.

Energy Tax Act (*Energiesteuergesetz*)

To encourage the use of biomethane as a fuel for powering cars and commercial vehicles, biomethane for fuel is exempt from fuel taxes until 2015.

Under Section 50 of the Energy Tax Act, tax exemptions for biofuels only apply if it can be demonstrated that the biomass used has been produced in accordance with specific environmental, social and sustainability requirements, as well as requirements pertaining to protection of the natural environment, and if the biofuel achieves a set greenhouse gas emission saving. Evidence that the biofuel meets these conditions must be submitted to the relevant central customs office. To prevent biofuels receiving duplicate subsidies, under Section 50(1.4) of the Energy Tax Act, only biofuels which are not included in the biofuels quota are exempt from tax.

To promote the development of biomethane as a fuel, tax exemptions for biomethane need to be continued beyond 2015, if necessary accompanied by the introduction of regular monitoring to assess market integration or a system to reduce exemptions once specific market shares are attained.

Biofuels Sustainability Regulations (*Biokraftstoff-Nachhaltigkeitsverordnung*)
The Renewable Energy Directive sets out sustainability requirements for the use of biomass to produce energy, which, in Germany, have primarily been transposed into national legislation through the Biofuels Sustainability Regulations. The Biofuels Sustainability Regulations set out requirements both for biomass and for documenting and certifying sustainability characteristics along the length of the production and supply chain. Certifying the sustainability of biomass is essential both for inclusion in the biofuels quota and for exemption from energy taxes.

These sustainability requirements also apply to gaseous biomass, i.e. biomethane, and encompass:

- Protection of the natural environment (Sections 4–6 of the Biofuels Sustainability Regulations)
- Sustainable agricultural cultivation (Section 7 of the Biofuels Sustainability Regulations)
- Achievement of a set greenhouse gas emission saving compared to reference liquid fossil fuels (Section 8 of the Biofuels Sustainability Regulations)

4.4.2 Synthetic Methane from Biomass

Franziska Müller-Langer, Marco Klemm, and Michael Schlüter

4.4.2.1 Introduction

Climate protection in the transport sector goes beyond the fields of application that have already been established. Biofuels are one way of reducing greenhouse gas emissions in the transport sector while simultaneously acting as substitutes for fossil fuels. Because the consumption of natural gas is on the rise worldwide [119–121], the production of natural

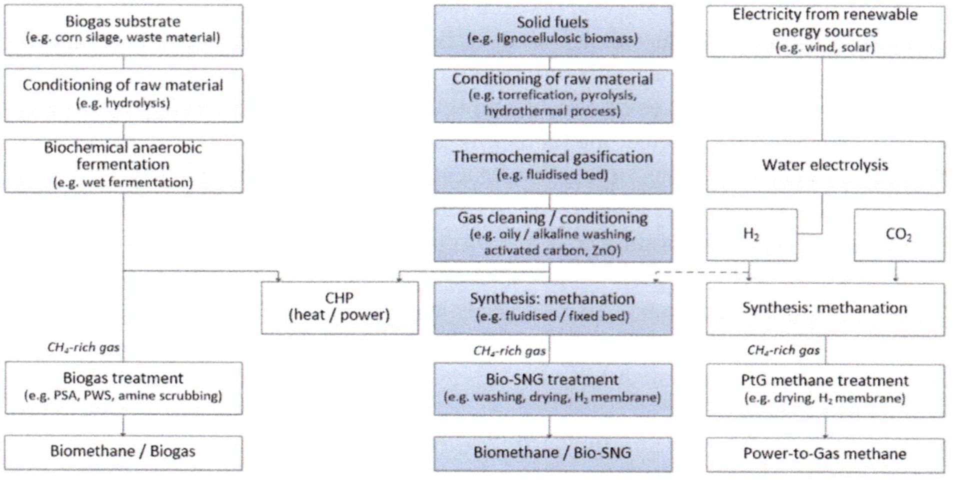

Diagram 4.50 An overview of the conversion chains of natural gas substitutes and how Bio-SNG fits in

gas substitutes made from renewable energies is becoming increasingly important. This is true not only in light of the overall objective of making a contribution towards environmental protection by reducing anthropogenic greenhouse gas emissions. It should also be noted that conditions are highly promising for producing and using natural gas substitutes thanks to many years of experience with combustion behavior and as a result of an extensive natural gas grid in many places. Moreover, by feeding natural gas substitutes into the natural gas grid, the location of consumption can be disconnected from the location of gas production.

The following conversion routes, which can partly be interlinked through synergetic effects, lend themselves as substitutes for renewable natural gas. These include (1) - so-called Power-to-Gas from renewable electricity, (2) conditioning of biogas produced through biochemical anaerobic fermentation and (3) Bio-SNG (synthetic natural gas) which is synthetically converted and conditioned using thermochemical gasification of biogenic solid fuels based on biomass conversion to biomethane (Diagram 4.50).

The principles of producing Bio-SNG shall be presented below followed by a classification based on technological, economic and ecological criteria. The chapter will end with a conclusion.

4.4.2.2 The Principles

The following observations are based on an overall process with a system structure as shown in Diagram 4.50, which simultaneously represents the boundaries of fuel production.

Raw Material Conditioning

Lignocellulosic biomass (e.g. wood as well as stalk material like straw, in specific ratios depending on the process) is primarily used in the production of Bio-SNG. Provision of the raw material entails delivering the biomass to the conversion facility, storage (i.e. storing a specific amount), conditioning (i.e. adjusting the quality of the raw material through mechanical or thermal pretreatment), and using suitable conveyor and sluice systems to feed the material into the gasification reactor. The plant technology required to do this is, on the whole, commercially available [122].

In contrast, most of the processes used for thermochemical raw material conditioning (for example torrefaction, pyrolysis, hydrothermal carbonization [123–127]) are currently still in the R&D phase. They serve to produce a gasified material which, compared to solid biomass that is only prepared mechanically, has a higher bulk and energy density, a lower water content and hence a higher energy content. This can also be achieved outside the fuel production plant. Significant for this are:

Gasification

Raw materials are converted into a high-calorific raw gas through endothermic gasification at high temperatures and through the stoichiometric use of a gasification agent containing oxygen (e.g. air, oxygen, steam). The composition of the raw gas depends on the gasification process (including the gasification agent, reaction conditions). In addition to the main components of CO, CO_2, H_2, CH_4 and H_2O, it contains various harmful components (e.g. particles, tars, nitrogen, sulfur and halogen compounds), which can require that the gas be laboriously purified and conditioned for catalytic fuel synthesis.

Gasification processes are significantly influenced by (1) the type of reactor (particularly moving bed and multi-stage reactors for fuel synthesis), (2) the way the process heat is supplied (allothermal vs. autothermal) and (3) pressure levels (atmospheric or pressurized), and have correspondingly different advantages and disadvantages [128–132].

The Bio-SNG plants are usually designed to use dual circulating fluidized bed gasifiers. These are usually allothermally run at up to 900 ° C and steam is used as the gasification agent [122, 132–134].

Gas Cleaning and Conditioning

With downstream processes in mind (for example the operating life of the catalysts), harmful components in the raw gas (e.g. particles (dust, ash, bed material), tars, nitrogen compounds (NH_3, HCN), alkalis (Na, K compounds), sulfur compounds (H_2S, COS, CS_2) and halogens (HCl, HCF) must be removed. The gas's extremely high purity requirements pose great technical challenges for gas purification technologies. For example, tar content may not exceed 1 $mg/m^3{}_{STP}$ in fuel synthesis, however it may reach around 50 $mg/m^3{}_{STP}$

when used in gas motors [129]. The predominantly multi-stage gas purification processes can be classified according to (1) temperature levels, (2) wet and dry processes such as filters and cyclones and (3) absorptive and adsorptive processes such as gas scrubbers, separators or sieves.

Some processes also serve to condition the gas (in other words to adjust the clean gas to the requirements of the synthetic natural gas process). This includes (1) adjusting the stoichiometric ratio of the gas components H_2 and CO, (2) removing CO_2 from the gas stream and (3) reforming light hydrocarbons (for instance CH_4, C_nH_m). The technologies suited for this are commercially available as part of conventional gas treatment processes and need only to be adapted to the properties of the raw or clean gas [122, 132, 134].

Methanation

Methanation is when the H_2 und CO contained in the syngas are catalytically converted into CH_4 and H_2O. The result is a methane-rich gas. Catalysts used for this (preferably nickel-based [134, 135]) typically operate at temperatures between 200 °C and 450 °C and pressures between 1 and 10 bar, which usually requires the syngas to be compressed beforehand [133, 135]. They are still in the testing phase for use in the methanation of biogenic gases. Because of their susceptibility to being deactivated by sulfur compounds (for example H_2S, COS), halogens and soot deposits, they place high demands on the quality of the gas (inter alia [135, 136]) and hence on upstream gas purification. As a rule, the necessary H_2/CO ratio of three in the syngas can only be achieved for the reaction control of methanation by way of gasification that uses water vapor as the gasification agent. Thus, missing hydrogen has to be supplied at the expense of carbon monoxide by way of a shift-reaction before or during the process [137].

Because methanation is a very exothermic process, the reaction heat that is generated has to be discharged. This can be done through a serial connection of multiple, commercially available, adiabatic fixed-bed or tube bundle reactors with intermediate cooling. Alternatively, technically more straightforward fluidized-bed reactors that have integrated bed cooling can be used. They enable heat to be easily extracted at high temperatures when the CO content in the syngas is higher than 10 vol.%. No prior stoichiometric adjustment of H_2/CO is needed and hence operations are almost quasi-isothermal [137, 138]. After fuel synthesis, the CH_4 content in the producer gas is approx. 40 vol.% and the CO_2 content is approx. 50 vol.% [139].

Gas Processing

In order to feed Bio-SNG into natural gas grids and use it in vehicles, a final gas processing step that uses CO_2 capture must take place, gas drying and—depending on the requirements of the natural gas grid in the area where it is being fed in—the addition, if necessary, of liquid gases (e.g. propane) to maintain combustion properties (see the

DVGW data sheets) [140, 141], DIN 51624 [142]). Furthermore, the gas must be compressed to the feed-in pressure if this hasn't already occurred in previous process steps. There is a series of processes that are commercially available to prepare the gas (for example pressure swing adsorption (PSA), amine scrubbing, pressure water scrubbing (PWS), Selexol/Purisol/Rectoisol processes, membrane separation, cryogenic processes) which are also partly used in biomethane plants [122, 132, 134, 143].

Glycol-based absorption processes (which use, for example, triethylene glycol (TEG) [143]) find application in the final gas drying process. An undesirable methane slip can occur, depending on the process used; this can be converted as a lean gas with corresponding exhaust gas after-treatment systems (for example with commercially available flox burners or through autothermal oxidation [144, 145].

4.4.2.3 Technological Classification

Technology development in the 1970s and 1980s focused on using syngas to produce SNG, however, there was a loss in commercial interest as a result of the price development of natural gas [146]. Both fluidized bed and fixed bed reactors with various currents were used for the synthesis. A large-scale SNG plant has been in operation in North Dakota (USA) since 1984 which uses coal gasification and methanation in a fixed bed reactor to produce SNG on the magnitude of around 1.5 GW [147, 148].

Today, however, fluidized bed reactors are often preferred for the synthesis of methane. Depending on the overall concept (by which electricity and, particularly, heat are generated as by-products depending on how the process is controlled), electrical net efficiency rates of more than 70 % can be achieved [122, 149]. The first demonstration plant to produce Bio-SNG of around 1 MW on the basis of biomass gasification and subsequent synthesis began operations in 2008 in Güssing (Austria). The commercial application of Bio-SNG will occur for the first time as part of the GoBiGas project, which plans to build an 80 MW Bio-SNG plant in Gothenburg by 2016. A demonstration plant is currently being commissioned at this site.

Table 4.13 contains an overview of the Bio-SNG plants installed at this time.

4.4.2.4 Ecological and Economic Classification

Successful market implementation of Bio-SNG—also in terms of competing with other alternative natural gas substitutes—requires added value in terms of lowering greenhouse gas (GHG) emissions along the supply chain and the costs related to this.

Greenhouse Gas Emissions Environmental effects like the effect of greenhouse gases are usually evaluated using so-called life-cycle analyses and ecological balance sheets. Greenhouse gas emissions for Bio-SNG range from 7 to 198 gCO_2eq/kWh depending on the overall concept [162] (our own studies reveal 40 to 126 gCO_2eq/kWh [163, 164]). These values are comparable to those for biomethane made from biogas. The biomass production and the conversion to SNG are decisive factors in overall emissions.

Table 4.13 Overview of existing and planned Bio-SNG-plants (DBFZ based on [150–160])

Location (city/ country)	Project/ Participant(s)	Plant capacity in MW[a]	Technology readiness level[b]	Operation, status
Saint-Fons/ Lyon/France	GAYA R&D Pilot Plant/GDF SUEZ	0.4	6 (demonstration plant)	Planned start 2015
Güssing/Austria	Biomassekraft-werk Güssing	1.0	6 (demonstration plant)	In operation since 2008
Gothenburg/ Sweden	GoBiGas I/Göteborg Energi	20	8 (commercial plant)	In operation since 2014
Gothenburg/ Sweden	GoBiGas II/Göteborg Energi	80–100	9 (commercial plant)	Planned start 2016
Malmö or Landskrona/ Sweden	Bio2G/E.ON	200	9 (commercial plant)	Planned start 2015

[a]Reference value SNG-output
[b]Technology readiness level (TRL) of the European Commission (1—basic principles observed, 2—technology concept formulated, 3—experimental proof of concept, 4—technology validation in lab, 5—technology validation in relevant environment, 6—demonstration in relevant environment, 7—demonstration in operational environment, 8—system completed and qualified, 9—successful mission operations) [161]

Greenhouse gases can be reduced by up to 87 % using the fossil fuel reference as per 2009/28/EC as a measure [165].

Costs An analysis of international studies of the last seven years on the cost to produce Bio-SNG reveals figures ranging from 4.7 to 12.8 EURct/kWh [162]. More recent studies have identified production costs of 8.9 to 10.2 EURct/kWh [122, 164], which are comparable to biomethane made from biogas [162, 163]. Experiences show that investment costs play a crucial role alongside annual full load hours for plant operation and biomass raw material costs. Depending on the overall concept, these are approx. 1410 to 2390 EUR per kW installed capacity [122, 164].

4.4.2.5 Summary and Outlook

The production of Bio-SNG using thermo-chemical biomass gasification enables a highly efficient supply of alternative natural gas substitutes. The first commercially run plants are starting up operations. Both the expected potential for reducing greenhouse gases and the production costs are comparable to those of biomethane made from biogas. In principle they can result in interesting synergies in the flexible integration of, for example, hydrogen from PtG and SNG (Diagram 4.50).

4.4.3 Synthetic Methane from Renewable Electric Energy

Reinhard Otten, Tobias Block, and Elias Hammer

CNG Vehicles: Are They Only a Shift from One Fossil Fuel Source (Crude Oil) to Another (Natural Gas)?

The perception of CNG mobility has changed significantly among experts and some policy makers over the last 5 years. Until around 2010, the CNG vehicle was simply considered a variant of the gasoline engine-driven motor vehicle which, instead of with gasoline, is fueled with another fossil energy source, namely natural gas, and therefore has some advantages when it comes to pollutant and CO_2 emissions, aside from drawbacks in weight, range and packaging. Biomethane as a renewable alternative to fossil natural gas, of course, has been known for quite some time and is established to a certain degree.

Still, this biogenous path was not a unique selling proposition on the surface, since the production and admixture of ethanol or biodiesel likewise provided non-fossil methods for traditional gasoline or diesel vehicles. The fact that biomethane, because of its particularly good efficiency per acreage (when obtained from energy crops) and the ease of production from organic waste, stands out among the biogenic fuels with respect to its ecological benefit has been described in the preceding Sects. 4.4.1 and 4.4.2, but is still little noticed in the public debate. One possible reason is that the biofuel associations, who advocate the continued funding of biomass-based energy sources, intentionally do not want to highlight any one of these sources so as not to create discrimination against others. And despite biomethane's excellent eco-balance, the amount of organic sources (energy crops, waste) for its production is naturally limited and the subject of fierce competition.

From the political perspective, the CNG vehicle was therefore primarily synonymous with the replacement of one fossil fuel by another. After all, this provided value in itself, since the predominant dependence of road mobility on crude oil has long created unease. At the same time, there are presently only a few countries where natural gas plays a truly big role as a fuel.

Yet, experts have demonstrated an entirely new appreciation for the significance of CNG mobility for several years now. Reasons for this trend include the above-mentioned potential of biomethane and the presumption that fossil natural gas reserves will significantly outlast oil reserves, but also the drastically growing share of fluctuating renewable energy sources, such as wind power and photovoltaic, and the development of Power-to-Gas technology, which can become a key element for the integration and continued expansion of these renewable energies. This will enable the CNG vehicle to become a partner and ideally complement to electromobility.

At first, this may sound strange. To understand what the production of green electricity—and with this, the basis for the truly environmental friendly operation of electric cars—could have to do with natural gas-operated vehicles requires a longer digression into the specialty of electric power supply, which is dominated more and more by the challenges of what is known as the energy turnaround.

Digression into the Field of Electric Power Supply and the Challenges of the Energy Revolution: Why Methane Plays a Key Role Again

In principle it holds true—at least for shorter distances—that the electric car is considered the torchbearer when it comes to mobility. And there are certainly good reasons: even considering charging losses and cooling/heating requirements in everyday use, it is at least twice as efficient as a comparable vehicle equipped with internal combustion engine. And—even more importantly—it can be operated already today with green electricity from renewable energies. At the same time, the mix of electricity in Europe is still heavily dominated by fossil or nuclear energy. Additionally, the expansion of renewable power sources, which we urgently need to establish a renewable basis for the electric car, is already slowing down in Germany due to the difficulties to integrate the increasing amount of fluctuating renewable energy into the grid. New power lines—to the extent they can be built today—can help, as can further fine-tuned load management ("smart grid"), which could even include electric vehicles.

These measures, however, have their limitations, and comprehensive studies show that they are only able to cushion the rising discrepancies between power supply and demand to a small degree. Quickly activatable back-up power plants will be unavoidable to fill the supply gaps left from time to time by wind and solar energy. There will always be general weather conditions during which, as the expansion of renewable energies advances, major surpluses will be recorded for weeks, which electricity traders will then not even be able to export abroad. After all, it can be assumed that Germany's neighboring countries will sooner or later expedite the expansion of wind and solar energy and have similar surpluses or have to compensate for shortfalls with back-up power plants. This scenario is something we must prepare for—if we are seriously aspiring to consistently expand renewable energies and plan to actually use the potential of e-mobility.

Methane: The Ideal Energy Source for Back-Up Power Plants in a Power System Dominated by Renewable Energies

So if there is one crucial capability in a power supply system with a high renewable energy component, it is that of swiftly, appropriately and safely compensating for the shortfalls that, by nature, result from the use of wind and solar energy with what are known as "back-up power plants." Nuclear power plants take at least 1 week to start up and shut down, and coal power plants generally take about 1–2 days. So they will only be able to contribute to stabilizing the frequency and voltage in the grid, as they have done in the past, by sensibly regulating their rotating masses, but essentially are not able to tailor their output much to demand. Both types of power plants, which due to the associated safety, environmental protection and logistics requirements in practice are always very large, are therefore often referred to as "base load power plants." In general, this is intended to express a particular quality ("reliable basic supply of electricity"), but the term also not so subtly indicates the lack of ability to respond to strong fluctuations in demand and compensate for the intermittent supply from wind and solar power. Gas-fired power plants, in contrast, are far better suited to meet these demands. They too, in principle, are "base load-capable," but even large plants can be brought from a standstill to maximum power in less than an hour, if necessary, and shut down just as quickly.

This can be accomplished even more quickly with decentralized, power-controlled co-generation plants, where electricity can be drawn within a matter of seconds. In addition, they contribute to the stability of the grid when intelligently controlled and linked ("swarm power concept") and are able to satisfy the demand for power even in regional networks, when it is not completely covered by renewable sources. As a result, they can considerably reduce the degree to which the power grid has to be expanded—the electric power is produced where it is needed.

Gas-fired power plants moreover have additional traits that make them suitable elements for the energy revolution:

- Gas-fired power plants require very little electricity of their own (about 1 % of the generated power, compared to 6–10 % for coal power plants and 6–16 % for nuclear power plants). Their so-called black-start capability is therefore excellent, i.e., the ability to start up during a general power failure, which is very important for the reliability with which electricity is supplied.
- Because of the low carbon content of methane and due to their better efficiency, gas-fired power plants produce the same amount of electricity as coal power plants at two to three times lower carbon dioxide emissions (43.5 % of the emissions of anthracite-fired and 33.2 % those of lignite-fired power supply) [166]. These values do not include the waste heat recovery that is possible, which in general is significantly easier to implement in the gas-fired power plants, which are generally smaller and decentrally located.
- There are options even today to supply gas-fired power plants with renewable energy. In Germany, approximately 4.4 % of electricity was produced from biogas in 2013, and approximately 10.6 % from natural gas [166, 167].

Interim conclusion: Methane as an energy source—which is also a source of energy for CNG vehicles—can and should play an important role as the energy turnaround progresses. The fact that, at present (2014), the opposite is true, namely that more cheap coal (to substitute nuclear energy) and less of the not-so-inexpensive natural gas (displaced by solar power) is being burned to generate power than in the prior years, is due to an outdated market design that does not adequately reward flexibility and low CO_2 emissions.

The message has already been heard by some policy makers: Gas-fired power plants are the ideal back-up partner for wind and PV power plants. If these, in the medium to long term, could then also be fed solely by renewable methane, supplying power from 100 % renewable energies would in fact be possible. Various studies have shown how this can work by storing excess electricity in the existing infrastructure, for example in simulations conducted by the Fraunhofer Institute IWES, which serve as the basis for publications by the German Federal Environmental Agency "Energy Goal 2050: 100 % renewable electricity supply" [168].

Storage of Green Electricity and Sustainable Energy Sources for Mobility: Two Key Challenges of the Energy Turnaround

This addresses the goal of not only managing phases that are marked by a shortage of electricity, but also rendering usable as much of the excess supply (e.g., when a lot of wind and sun are available, and consumption is low)—to support the aforementioned back-up power plants, for example, which is where the topic of Power-to-Gas (PtG) comes into play. The basis of the technology is the use of electricity to produce hydrogen from water by way of electrolysis. The hydrogen produced can be fed directly into the natural gas network, if mixed thoroughly; however, this is not possible and feasible everywhere and at all times because it must be ensured that the hydrogen concentration does not exceed 5 % at any of the withdrawal points in the natural gas network [169]. Due to the risk of equipment damage, many of these points are subject to even narrower limits, such as on CNG fueling stations (max. 2 % according to DIN 51624 [170]) or porous rock storage reservoirs (max. 1 %). So the option to have the hydrogen react with CO_2—in the presence of a catalyst and under certain pressure and temperature conditions—to form synthetic methane (Sabatier process, exothermic) is an interesting one. This would allow a substitute gas to be provided, which can be fed into the natural gas grid at any time and without restrictions. As a result, the two major energy infrastructures—power grid and natural gas network—can be bidirectionally connected to each other. One direction is common practice: electricity can be produced from gas (natural gas or biomethane). With PtG technology, the other direction now also becomes possible: electricity turns into methane. The natural gas grid could then become a buffer with huge capacity for the power grid. The PtG principle was developed collectively by the ZSW (Center for Solar Energy and Hydrogen Research) and the Fraunhofer Institute IWES (Institute for Wind Energy and Energy System Technology) in Kassel, Germany, and advanced to maturity in ZSW's laboratories in Stuttgart, Germany. The specific technology and development steps up to the first industrial application by automobile manufacturer Audi are described in Sect. 4.4.3.1.

But how and when can this technology be used in a meaningful way? Even today, wind power plants frequently must be taken off the grid because the power they produce is not offset by adequate demand in the particular network area. This is the case even though in 2013 in Germany on average only about 15 % of electric energy came from wind and solar. However on some days, the share accounted for significantly more than 50 %, resulting in looming instability and corresponding shutdowns in some areas of the network. At shares of 35 % (2020), 50 % (2030) and 80 % (2050) in the mix of renewable energies—as aspired by the German Federal Government—considerable "excess" amounts are therefore to be expected on a large number of days throughout the year. In certain regions—in electricity networks with high levels of PV feed-in from house rooftops or in regions with a large number of wind turbines, for example—renewable energies reach these levels in the mix already today. The Power-to-Gas plant of Audi AG, for example, is located in a power grid in which, over the course of the year, more than 80 % of electricity generated comes from renewable sources.

Currently, there are studies that state that rather than build and operate expensive storage facilities, it would be more economical in the next decades to simply not use these amounts of energy at all, i.e. to disconnect generation systems such as wind turbines and photovoltaic modules from the grid based on the situation. This ruthless calculation is generally based on the assumption that there will be no notable price increases for fossil energy and that people will accept a large number of additional wind turbines that are not allowed to feed into the grid for expanded periods and will not turn even in the best winds. These model calculations often only consider the electricity market and disregard the fact that very expensive technologies are already being introduced in the market in the mobility sector, and that large areas of land are already being cultivated with energy crops, so as to reduce the dependency on crude oil as an energy source and achieve comparatively small savings in CO_2 emissions.

In addition to the necessity to provide ecologically sound back-up power for the less reliable wind and solar power sources, it will certainly be useful to store the excess supply of electric energy. The resulting "surpluses" can then be used at a later time.

Comparison of Possible Energy Storage Systems

There are currently already a few storage options in the electric grid. These are either batteries (electrochemical storage), or primarily pumped-storage hydroelectric systems, i.e. artificial lakes into which water can be electrically pumped when there is excess energy or where water can be released downward through turbines when there is demand for energy (storage as position energy). These storage systems in existence today in Germany provide a maximum capacity of 0.04 TWh (=40 GWh). For illustrative purposes: Power consumption in Germany fluctuates between 30 GW (at night, on the weekend) and 80 GW (during the week, mid-day peak). From a strictly mathematical point of view, Germany could be supplied with electricity from these existing storage systems for 1 h at a low load of 40 GW. If the natural gas network, with its approximately 22 billion standard cubic meters of capacity (or 220 TWh of thermal energy), could be tapped as a reservoir, more than 100 TWh of electricity could be generated at the efficiency of more than 50 % of today's gas-fired power plants. With the assumed 40 GW of load, this would be sufficient for 2500 h, or 3 months, for Germany. The technically available storage capacity is significantly lower due to a minimum pressure in the natural gas network and the heat supply requirements; however, it becomes apparent that, with PtG, a storage capacity can be developed that is greater than what we have available today by three orders of magnitude.

Households, communication, water supply, business and traffic, etc. extensively depend on a reliable supply of power and gas. In addition, the energy market of the future will be dominated by variable sources such as wind and solar, and potentially uncertain fossil resource sources. The value of large accessible energy storage systems, which can be filled decentrally and flexibly with domestic renewable energy and used, is therefore evident.

Potential of the PtG Technology: How Many CNG Vehicles Could Be Supplied from PtG Plants?
Opinions vary widely on this question. Some experts say that the discrepancies between power supply and demand can largely be compensated for by other, more economical measures. Solutions that are cited include fine-tuned load management ("smart grid", among other things by including the storage capacities of electric cars), power-to-heat solutions, a comprehensive expansion of the power grid, and barrier-free globalization of the electricity market.

Others are of the opinion that the potential of these possibilities is being overestimated. Some of the arguments they list are as follows:

- So-called demand-side management, which refers to the balancing of electricity supply and demand, is extremely difficult and costly to organize and is further limited because by far not all consumers (in industry, business, IT, traffic, for example) have the necessary time flexibility.
- While power-to-heat, which involves the conversion of surplus electricity into heat by immersion heaters, is an economical option with high efficiency, it has limited potential. The existing water reservoirs are generally supplied directly by solar thermal systems, and demand for heat in households—the vast majority of which arises during the winter—will generally decline as a result of improvements in insulation promoted by policy makers. In addition, from an energy point of view, power-to-heat is a dead end and not a flexible solution such as PtG, where the stored energy is flexibly available both in terms of time and location for use as fuel, for re-conversion during times of low wind and sun, and for the demand-based conversion into heat.
- In light of the fact that energy sources (for example, offshore wind farms in the North Sea) and energy drains (for example, economically powerful regions in North Rhine-Westphalia and southern Germany) are physically located far apart, the expansion of the power grid is a fundamental component for the success of the energy turnaround, but is meeting with a lot of resistance and, also for economic reasons, is not likely to eliminate the existing and emerging bottlenecks at all voltage levels of the network.
- Optimizations at the international level appear to be even more difficult. Instead, present political developments (2014 EU elections) suggest that the EU member states are seeking solutions for greater national self-sufficiency rather than comprehensive international projects involving high coordination efforts and investment expenditure, which moreover are tied to extremely long-term obligations. But PtG solutions and their multifaceted energy vectors lend themselves precisely when decentralized solutions and regional (energy) independence rank near the top of the political agenda.

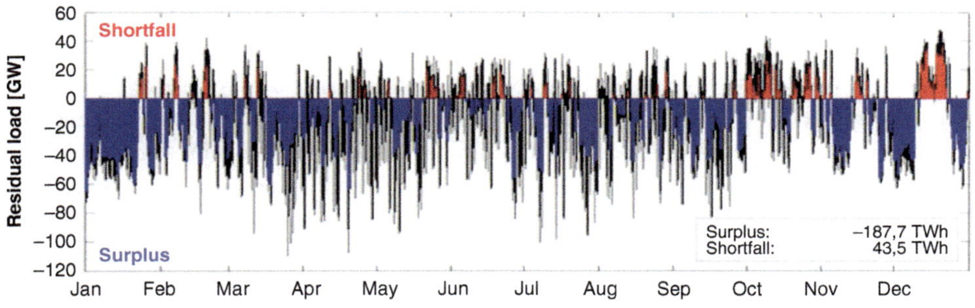

Diagram 4.51 Simulation of power deficits (residual load) and surpluses (negative residual load) in Germany at a share of renewable energies of 78 % and ideal power grid development—2050 forecast based on 2007 meteorological year [172]

Simulations conducted by the Fraunhofer Institute IWES (refer to Diagram 4.51) provide indications of the amounts of electricity to be expected in the future, for which no direct use exists in the power grid and which would consequently be lost by a temporary shutdown of the producer. In a scenario in which the installed renewable energy generation systems could directly cover a 78 % share in the mix of Germany's entire power supply, surpluses of 187 TWh would result in 1 year according to these simulations. This scenario already assumed an ideal—but very theoretical—expansion of the grid, in which bottlenecks in the transport of electricity are a thing of the past. So we talk about electricity for which no demand exists at any place in Germany. Assuming that slightly more than half, or approximately 100 TWh, of this energy could be used by PtG plants, approximately 60 TWh could thus be stored in the natural gas network in the form of synthetic methane and made available to the fuel market, for example, based on an efficiency of 60 % (only energy flow from power to gas, without waste heat recovery).

A CNG car with a fuel consumption of 4 kg/100 km (which in the case of H gas corresponds to approximately 6 l of premium on 100 km (39.2 mpg)) and 15,000 km (9320 miles) travelled per year requires 600 kg (1320 lbs) of methane per year, which is 7800 kWh. From a strictly mathematical point of view, the above calculated 60 TWh (60,000,000,000 kWh) from "surplus energy" produced in Germany (which would then no longer be excess energy) could be used to supply 8 million CNG cars, which is approximately 100 times more CNG cars than today.

This equates to approximately 20 % of cars on the roads in Germany now. So we are talking about a significant contribution to supplying the mobility sector with energy, and more specifically without having to add any wind turbines or PV power plants. Mind you, this is a theoretical value, since we do not know yet whether and when the re-conversion of stored energy could become more attractive economically than its use for mobility.

Some readers may have already wondered when the author will finally make up his mind whether synthetic methane generated from renewable energies ("renewable energy gas") is to be used for re-conversion or for CNG vehicles. After all, it can't be burned twice. This is correct, but at the same time highlights the major advantage of PtG technology: we do not have to make a decision today. As soon as synthetic methane is in the natural gas network, it is available for every conceivable use, and the willingness to pay for this renewable energy source will ultimately enable the decision to be made.

From an ecological point of view, the question could read: Is it more sensible to replace fossil natural gas in the energy sector or fossil crude oil in the mobility sector? The climate balance supports the second approach.

What Makes Renewable Energy Gas from PtG Plants So Attractive for the Mobility Sector? And Why Can CNG Mobility Become the "Locomotive" for the PtG Technology?

The first industrial PtG plant in the world was commissioned and is operated by the automotive manufacturer Audi (Diagram 4.52). Several arguments in favor of the use of synthetic methane as a fuel have already been hinted at in the preceding chapters. Here again are the reasons why practically all experts consider mobility to be the driver for PtG technology:

- Fossil energy forms are subject to high taxation in the mobility sector, at least in Europe. Renewable energy sources therefore have a better chance to gain a foothold than, for example, in the heating sector.
- The players in the mobility sector are faced with high CO_2 abatement costs. The leverage that exists in Europe in terms of legislation is the biofuel quota on the part of the mineral oil industry and the CO_2 fleet emission regulations of the EU, as well as national tax legislation based on CO_2 emissions for the automakers. Exactly the same amount of CO_2 is bonded during the production of synthetic methane as is emitted in its combustion, since CO_2 is the only carbon source in this process. Moreover, feeding it in regionally avoids the energy-consuming transport of fossil natural gas from abroad. In a quite conceivable scenario where these avoided CO_2 emissions are credited against the fleet figures of an automaker that proves that vehicles of its brand are powered with this fuel, the motivation to invest in the development and construction of PtG plants would certainly be quite high.
- With the natural gas grid, a transportation infrastructure is already in place for synthetic methane or hydrogen from PtG plants. No additional logistics chain from the production site to the fueling station needs to be developed.
- Every PtG plant produces renewable hydrogen, regardless of whether it is fed directly into the natural gas grid or first methanated by combining it with CO_2. This creates a crucial basis for turning fuel cell vehicles into a truly sustainable alternative to battery electric vehicles (BEVs) or those driven by internal combustion engines.

Diagram 4.52 Aerial view of the PtG plant (far right of the image) operated by Audi in Werlte (Emsland). The *left* and *center area* of the photograph show the biomethane plant of energy supply company EWE. The CO_2 that is captured by the plant is used in the PtG plant for the methanation of hydrogen. The waste heat of the PtG plant is utilized in various heat sinks of the biomethane plant (sterilization, CO_2 removal) (Source: EWE NETZ)

Outlook

Allow me to make one more important remark while taking a look into the future. The figure of 20 % of vehicles on the roads in Germany mentioned in the section above, which based on calculations, could be supplied entirely by PtG plants, is just the beginning. There are two important reasons for this.

The number was derived based on the efficiency of today's CNG vehicles. If hybridization or other efficiency measures were to come into play, the number of vehicles supplied by Power-to-Gas could be increased further. The share of synthetic methane in the fuel market could even double when imagining that, one day, a strong market penetration of battery cars and plug-in hybrids will perhaps allow 50 % of the distances traveled by car to be covered electrically. Additional big leverage—which albeit today is still very theoretical—could come from a growing number of fuel cell vehicles that are supplied directly from the hydrogen production of PtG plants. As a result of their high efficiency and the losses avoided in methanation, they can translate the available energy into approximately twice the mileage of CNG cars. Section 4.6 will explain why it is difficult to focus solely on this approach now.

The preceding calculations always only assumed the use of "excess" energy. This is due to the circumstance that, from today's perspective, the priority in the expansion of wind and solar energy clearly lies in the replacement of coal and nuclear power. However, in a broader perspective—which is valid even today for certain geographical locations—it is certainly legitimate to think about deliberately building wind turbines and PV power plants so as to produce fuel. The comparison to the production of biofuels as it is common today will help readers warm up to this idea. By combining PV fields and PtG technology (overall efficiency of pathway from solar energy to methane: approximately 10 %), about 20 times more fuel can be produced at the same level of land usage as with the cultivation of energy crops (overall efficiency max. 0.5 %). This factor increases considerably when basing the calculation on wind energy, which in the context of the fuel-tank-or-dinner-plate discussion uses practically no land at all, since wind turbines do not significantly impede the growing of crops. So from this perspective alone, it appears to make a lot sense to explore Power-to-Gas as a mobility topic, also because it remains to be seen what share battery electric and hydrogen-powered vehicles can in fact achieve in road traffic (see also Sect. 4.6). "Power-based fuels" therefore play a considerable role in scenarios of the German Federal Environmental Agency for the year 2050 when it comes to the aspired essentially complete reduction of greenhouse gas emissions [171].

Result

Power-to-Gas technology will be a necessary component of the energy landscape if the goal of converting power production in Germany as fully as possible to renewable energies is pursued, as it was consistently formulated by the last German administrations. It is still disputed at what share of green electricity the Power-to-Gas applications will become system-relevant for the power sector.

It is undisputed, however, that the crucial impetus for the development of PtG technology will initially come from mobility, once certain legal obstacles have been removed, such as the burden of end user duties. Huge potential for the mobility sector is apparent from a mere analysis of future "power surpluses." Still, in the broader perspective and with regard to the high dependency of the sector on crude oil, a discussion of the deliberate production of power-based fuels and the resulting avoidance of additional land usage by biofuels is also in order.

The development of PtG plants is an important door opener for the introduction of fuel cell vehicles and other hydrogen applications. CNG vehicles and the expansion of the network of CNG fueling stations are ways to lay the foundation for ensuring that PtG plants have good economic prospects and specific application even today. This, in turn, opens up entirely new avenues for storing electricity, swiftly continuing with the expansion of renewable energies, and, consequently, achieving ecological plausibility of electric vehicle drive systems. CNG mobility will therefore evolve into a partner for electric mobility in two respects: it will help supply ever larger amounts of green electricity and can complement electric mobility, almost with climate neutrality, in an area where electric cars today and in the long run are not likely to encounter broad customer acceptance: long distances.

4.4.3.1 Technical Realization of Power-to-Gas Technology (P2G®): Production of Substitute Natural Gas by Catalytic Methanation of H_2/CO_2

Michael Specht, Jochen Brellochs, Volkmar Frick, Bernd Stürmer, and Ulrich Zuberbühler

Motivation

Goal: "Power-to-Gas"

The original definition of Power-to-Gas was coined for the conversion of (renewable) electricity into a (renewable) gas, such as methane and/or hydrogen, in order to store this gas in the existing natural gas infrastructure and in order to be able to provide it to different consumer sectors at a later time [173–175]. The most important applications of P2G® in a renewable energy - based energy system are:

1. Long-term storage of renewable energy (RE) - for example for fluctuating excesses from photovoltaic/wind power installations
2. Stabilization of the power grid by the grid-controlled consumption of electricity and by feeding in electricity by reconverting the produced energy carrier
3. Partial transfer of the energy transport from the power grid to the gas grid
4. Production of sustainable fuels for mobility for which no substitutes are available elsewhere (with the "CH_4 mobility" entry market in the short term, the "H_2 mobility" market in the medium term and the provision of liquid hydrocarbons, e.g. for air traffic in the long term).

The opportunity for bidirectional coupling of the power grid and the gas grid lies in the convergence of the systems to a sustainable energy supply with electricity, heat and fuel, utilizing the existing grids for energy distribution and storage.

Goal: "Methanation"

The essential goals for the design of the reactor concept for methane synthesis are the temperature control in the reactor and the maximization of conversion for the production of a methane-enriched replacement gas (substitute natural gas (SNG) replaces natural gas with the same gas properties as the gas in the gas grid) produced from the reactants CO_2 and electrolytically produced H_2, which can be fed in the natural gas grid. Up until now, CO-based synthetic gases have been employed for methanation, however no CO-free gaseous reactants. Interest here has focused on the short-term industrial realization in a 6000 kW_{el} plant by the year 2013. This entails the construction of the "e-gas plant" for the automobile manufacturer Audi AG by the plant construction firm ETOGAS GmbH.

The objective of the research and development activities and the implementation following these activities is sustainable mobility based on the use of methane-powered vehicles. The work leading to the design of the "e-gas plant" is described in the following.

Power-to-Gas Technology for the Production of Substitute Natural Gas

Power-to-Gas is an approach to solving the problem of the (seasonal) storage of renewable energy. The fluctuating production of electricity from renewable energy, in particular from wind power and photovoltaic installations, serves the electrolytic production of hydrogen in the Power-to-Gas process, which is converted together with CO_2 to methane, the main constituent of natural gas, in a synthesis reactor and stored as substitute natural gas (SNG) in the natural gas grid. In the existing natural gas infrastructure, the chemical energy carrier methane produced from renewable resources is efficiently stored, distributed and provided for use as required. A particular advantage compared with other storage options is the use of the natural gas grid with its high storage and transport capacity. The Power-to-Gas process offers the option of merging the power grid and the gas grid to an integrated total system for the provision and storage of energy where required.

For long-term storage and the seasonal compensation of renewable energy, only chemical secondary energy carriers such as hydrogen and carbon-based fuels (e.g. substitute natural gas) that can be produced from different renewable energy sources come into question today. These represent the only foreseeable options at present for the seasonal storage of renewable energy and the reconversion of these energy carriers with a capacity in the order of TWh (for the example of Germany).

In modern gas and steam driven power plants or in decentralized cogenerating stations, SNG can be reconverted during times of high electricity demand and be used in industry or as a fuel - e.g. as "e-gas" in mobility. The principal process components of a Power-to-Gas plant are the electrolysis of water and the methanation process.

Electrolysis is available on the market as an alkaline system in the required MW_{el} performance class. However, this is not true for polymer electrolyte membrane (PEM) electrolysis and solid oxide electrolyte (SOE) electrolysis, which are still undergoing development. Due to the non-existent market, the available electrolysers are however not serial products, and are designed and produced as special constructions for the MW_{el} performance class, with correspondingly high specific costs. Besides the goal of improved efficiency, electrolysis research focuses particularly on the development of electrolytic block concepts and cost reduction by modularization of the hydrogen production plants.

The conversion of carbon oxides (CO, CO_2) by hydrogen to methane by Sabatier and Senderens has been known since the beginning of the twentieth century [176]. On a technologically relevant scale, the process of catalytic methane synthesis has been employed up to now mainly in coal gasification plants. Here, the CO-enriched synthesis gas resulting from the conversion of coal is converted to a methane-enriched product gas in a downstream methanation unit. The synthesis of methane is described by the following principal reactions:

Methanation reactions

$$3H_2 + CO \rightleftarrows CH_4 + H_2O(g) \quad \Delta H_R = -206\,kJ/mol \tag{4.2}$$

$$4H_2 + CO_2 \rightleftarrows CH_4 + 2H_2O(g) \quad \Delta H_R = -165\,kJ/mol \tag{4.3}$$

Water-gas shift reaction

$$H_2O(g) + CO \rightleftarrows H_2 + CO_2 \quad \Delta H_R = -41\,kJ/mol \tag{4.4}$$

In the methanation reaction, carbon monoxide and carbon dioxide are converted by hydrogen to methane and water vapour [Eqs. (4.2) and (4.3)]. The methanation reactions are coupled by the water-gas shift reaction [Eq. (4.4)]. According to prevailing academic opinion, the methanation of CO_2 takes place in two steps, the retro-shift reaction [the reverse reaction of Eq. (4.4)], followed by CO hydration [Eq. (4.2)]. The reaction mechanism for the direct methanation of carbon dioxide according to Eq. (4.3) is also under discussion [177].

The hydration of carbon monoxide and carbon dioxide is volume-reducing and strongly exothermic, so that the principle of Le Chatelier favors the methanation reactions at high pressures and low temperatures.

Specific Features for the Methanation of a CO_2-Based Synthesis Gas and Related Reactor Design Types

Methanation in combination with the P2G® process is characterized by the following specific features:

1. Methanation with CO-based synthesis gases has been technically realized and is available on the market. However, this is not true for H_2/CO_2-based gas educts.
2. The fresh CO_2-based synthesis gas fed to the SNG production process is stoichiometrically adjusted or exhibits a hyperstoichiometric H_2 fraction, with the goal of nearly complete CO_2 conversion.
3. CO_2 methanation places far greater demands on the catalyst than CO methanation.

Technologically, methane synthesis from largely stoichiometrically adjusted carbon oxide-hydrogen mixtures represents a major challenge, because on the one hand the greatest possible reaction conversion rate must occur in order to satisfy the country-specific feed-in specifications for synthetic natural gas. On the other hand, the exothermic reaction heat results in a pronounced rise in the temperature of the reactor, with negative consequences for the catalyst.

On the one hand, the educt gases for the methanation, which are characterized by nearly stoichiometric carbon oxide-hydrogen ratios, enable the direct production of SNG without downstream gas conditioning (CO conversion, CO_2 separation), but at the same time require a significant reduction of the CO_2 and H_2 concentrations in the product gas in order to comply with the specifications for substitute natural gas.

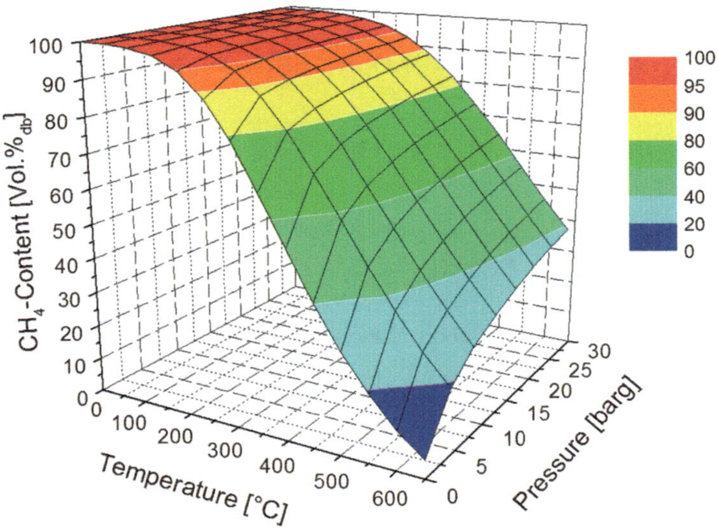

Diagram 4.53 Yield of methane (db: dry basis) in equilibrium as a function of temperature and pressure ([barg] = [bar$_{abs}$] − [bar$_{atm}$]); educt gas: 80 vol.% H_2; 20 vol.% CO_2 (db: dry basis)

The main challenge is the increased evolution of heat resulting from the high conversion rate of the reaction compared with non-stoichiometrically adjusted educt gases.

Calculations of the chemical equilibrium can be taken as the basis for the choice of the reaction conditions, as these allow predictions about the product gas composition for maximum achievable conversion with regard to the chosen process parameters. Here, the goal is a process control that delivers a product gas capable of feed-in following the removal of condensate. For the methanation employed in the P2G® process, obtaining a high CH_4 concentration in the resulting product gas with the least possible effort is of greatest importance. A basic prerequisite for a high rate of conversion is the optimal choice of the reaction conditions. Diagram 4.53 shows the methane concentration of the product gas that can be achieved in chemical equilibrium as a function of the parameters process temperature and process pressure. Theoretically, at low temperatures the production of a replacement gas with $CH_4 > 95$ vol.% is possible. Obtaining equilibrium requires an arbitrarily long residence time in the reactor system. With the use of a catalyst, the residence time can be reduced to a technically feasible level. The temperature at which the catalyst optimally performs is catalyst-specific and must be determined empirically.

In Diagram 4.54, the synthesis pressures required for the production of H-gas (95 vol.%$_{db}$ CH_4) and L-gas (90 vol.%$_{db}$ CH_4) according to the feed-in regulations for Germany are plotted [178, 179]. By way of example, for different stoichiometric numbers SN [see Eq. (4.5)] of an H_2/CO_2 mixture, equilibrium calculations for temperatures of

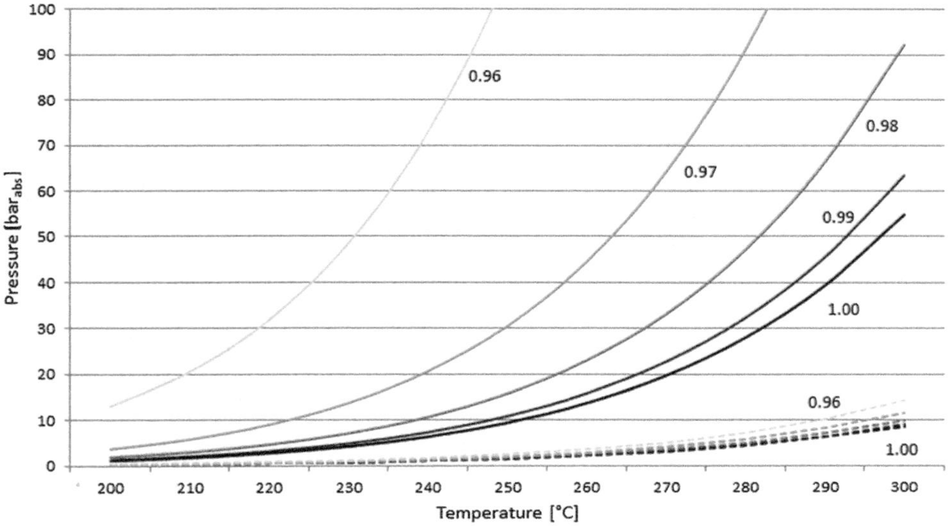

Diagram 4.54 Required process pressure for conversion under equilibrium conditions as a function of reactor outlet temperature for different stoichiometric numbers (SN) of a CO_2/H_2-mixture; H-gas (*solid lines*), L-gas (*dotted lines*)

200–300 °C were performed, specifying the product methane content [180]. Here, y_{H_2}, y_{CO} and y_{CO_2} are the volume fractions of the respective educt gas components.

$$SN = \frac{y_{H_2}}{3y_{CO} + 4y_{CO_2}} \quad [-] \tag{4.5}$$

It becomes clear that the minimum required synthesis pressure increases overproportionally with temperature. The objective of the process design is accordingly the lowest possible reactor temperature in reactor regions where the reaction has proceeded to a large extent (in the direction of the reactor outlet). This temperature is limited in the downwards direction by the start-up temperature and the conversion behavior of the catalyst employed. The choice of catalyst has a significant influence on the process pressure and thus on the energy requirements for the compression of the educt gas. In addition to the temperature, the educt stoichiometry has a substantial effect on the pressure. In the region of interest, low synthesis pressures are possible for stoichiometric numbers (SN) near 1.00. If the resulting product gas is to be fed into an L-gas grid, significantly lower synthesis pressures are required than for the production of H-gas. For example, if a catalyst has a minimum working temperature of 240 °C, a minimum pressure of 6.3 bar$_{abs}$ is required to maintain the limit value for CH_4 in H-gas. For L-gas the respective value can already be maintained at a pressure of 1 bar$_{abs}$. Here it must be noted that these considerations are completely valid only in chemical equilibrium. They represent the minimum requirements in regard to the process pressure. In practice, the pressure lies above the values determined here [180].

Different reactor systems can be employed for the methanation of carbon oxides. These must satisfy the requirements of sufficient heat control in the main reaction zone and exhibit a good reaction conversion rate. Within the scope of the P2G® process, the Centre for Solar Energy and Hydrogen Research (ZSW) investigates wall-cooled, catalyst-filled fixed bed reactors in configuration as tube bundle reactors and plate reactors (see below). These systems are designed so that sufficient heat dissipation is possible through the reactor wall. An important factor influencing this is the ratio of the cooling surface to the catalyst volume, as this directly influences the maximum temperature in the catalyst filling. Accordingly, in the case of a tube reactor, the reaction heat set free determines the maximum tube diameter. Possibilities for cooling medium for wall-cooled reactors are pressurized water, thermal oil or molten salts.

An important advantage of wall-cooled reactors is the precisely adjustable temperature profile for the catalyst filling, which is characterized in exothermic reactions by very high temperatures near the gas inlet zone and low temperatures at the filling end ($\approx$ cooling medium temperature). This impresses a temperature gradient on the reactor. Due to the high temperatures, very high conversion rates are obtained at the inlet zone. The low temperatures enable a favourable adjustment of the chemical equilibrium at the same time. This constellation enables nearly complete conversion in a single reactor stage [181].

The basic prerequisite for such operation is the efficient control of the reaction heat, as otherwise a reduced catalyst life must be expected. A suitable measure for influencing the temperature profile is a step-wise educt gas feed, which allows the distribution of the reaction heat over several filling regions. This reactor design was investigated by Wollmann et al. on the basis of the conversion of a diluted H_2/CO stream [182]. The authors were able to determine that, for the same educt gas feed-in rate, an educt gas stream fed in downstream results in a significant lowering of the upstream positioned temperature maximum.

In industrial applications, tube reactors are designed mostly as tube bundle reactors. With this design, the catalyst is inside the tube and the cooling medium (e.g. molten salt) is in the jacket area. A scale-up is realized by a greater number of tubes. The results of experimental investigations can therefore be transferred from a single tube to multiple tubes, provided that there is uniform gas distribution over the tubes in the stream, so that uniform pressure loss is ensured over the different catalyst beds.

Plate reactors represent an alternative reactor design. With this type of reactor, cooling plates subjected to a stream of pressurized water/steam are located at regular intervals in the catalyst filling. In the hot spot zone of the reactor, partial evaporation of the pressurized water occurs in the cooling plates, resulting in high heat transfer.

The essential features of the reactors utilized by ZSW are briefly described below. The tube reactors employed in the 250 kW_{el} and the 6000 kW_{el} P2G® plant (see section "Power-to-Gas Plant Design in the 25 kW_{el}, 250 kW_{el} and 6000 kW_{el} Performance

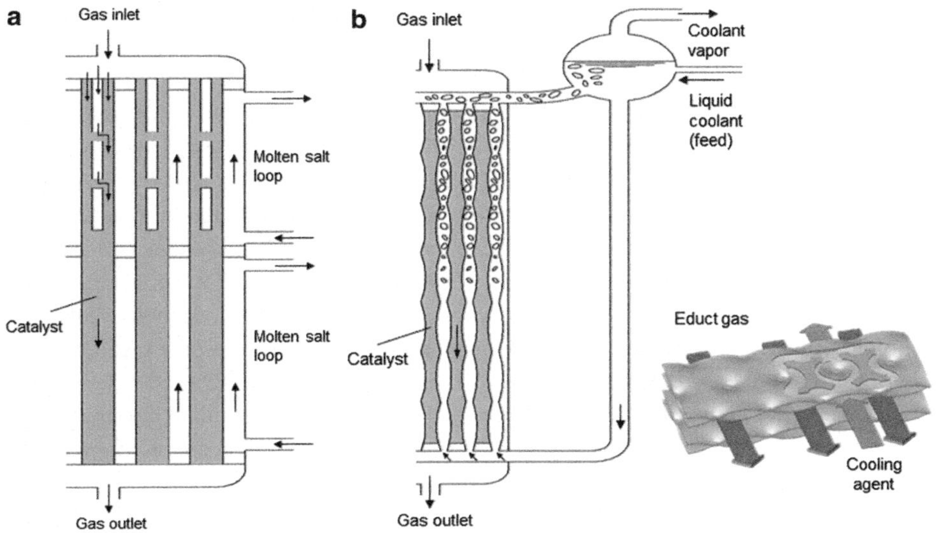

Diagram 4.55 (**a**) Schematic representation of the molten salt-cooled tube bundle reactor design with two separate cooling circuits and step-wise educt gas feed by a displacement pipe. (**b**) Schematic representation of the water/steam-cooled plate reactor design, consisting of thermal plates combined to form a heat exchanger packet (*graphic on right*: [183])

Classes") are characterized by a molten cooling medium (molten salt reactors). At least two separate cooling circuits impose a pre-set temperature profile on the reactor. Another special feature is the step-wise educt feed by a displacement pipe at the reactor inlet in order to limit the hot spot temperature. The design is illustrated schematically in Diagram 4.55a.

The plate reactor is based on an inexpensive design principle, in which pincushion-shaped thermal plates are formed in the production process by the expansion of spot-welded plates under high pressure. The thermal plates are combined to a heat exchanger packet containing the pressurized water/steam cooling medium. Between the thermal plates is the catalyst filling. The heat of reaction is dissipated by the partial evaporation of the pressurized water in the thermal plates in the hot spot region. The cooling circuit pressure is adjusted so that the required, temperature-dependent educt gas conversion is obtained at the reactor outlet. The principle is illustrated in Diagram 4.55b.

If the product gas composition at the reactor outlet does not meet the desired quality specification, it is possible to bring this into line with an additional separation process. Membrane gas separation processes are suited for this purpose. Thus, for example, if hyperstoichiometric process control with respect to H_2 is chosen, the excess hydrogen can be efficiently separated from the methane and recycled to the educt gas [184]. As the CO_2 conversion improves with the higher hydrogen content, this simultaneously increases the amount of methane produced.

Power-to-Gas Plant Design in the 25 kW_{el}, 250 kW_{el} and 6000 kW_{el} Performance Classes

At ZSW, two Power-to-Gas plants in the performance classes 25 kW_{el} and 250 kW_{el} were constructed. ZSW is also participating in a further 6000 kW_{el} plant within the scope of basic engineering, commissioning and plant monitoring. These performance figures refer to the electrical DC power consumption of the electrolysers in the design case.

The three plants are characterized by the following features: The 25 kW_{el} system is a research facility with two serially configured fixed bed methanation reactors with intercooler, as well as a product gas recycle loop. The 250 kW_{el} system is a research facility on a technically relevant scale, with variable configuration. Two different reactor systems (tube bundle and plate reactors) can be operated alone or in combination. A membrane gas conditioner enables methane enrichment in the product gas and the recycling of a hydrogen-enriched gas (permeate) to the educt gas. The 6000 kW_{el} plant is the world's first commercial Power-to-Gas plant which feeds methane into the natural gas grid. The design is based on the "once-through" concept in the form of a tube bundle reactor and feed-in to the L-gas grid. The designs of all three plants are illustrated in Diagrams 4.56–4.58.

The 25 kW_{el} Power-to-Gas Plant

ETOGAS GmbH, formerly SolarFuel GmbH, commissioned ZSW with the construction of the 25 kW_{el} plant. The container-integrated design enables operation directly at a biogas plant, utilizing real gases as educt. The design allowed the conversion of CO_2 from the "off-gas" of biogas plants with gas processing to generate bio-methane (e.g. via

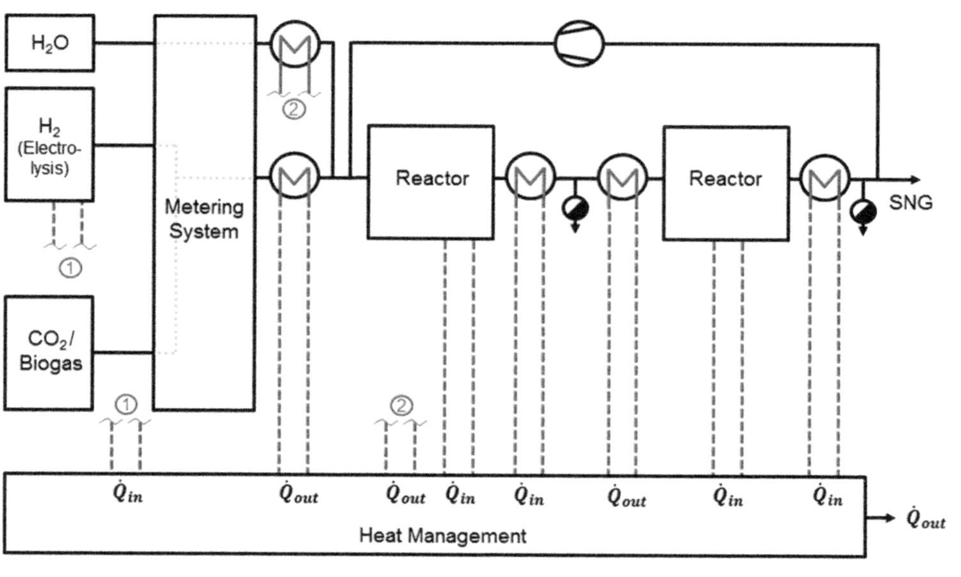

Diagram 4.56 Schematic illustration of the container-integrated 25 kW_{el} P2G® research plant facility

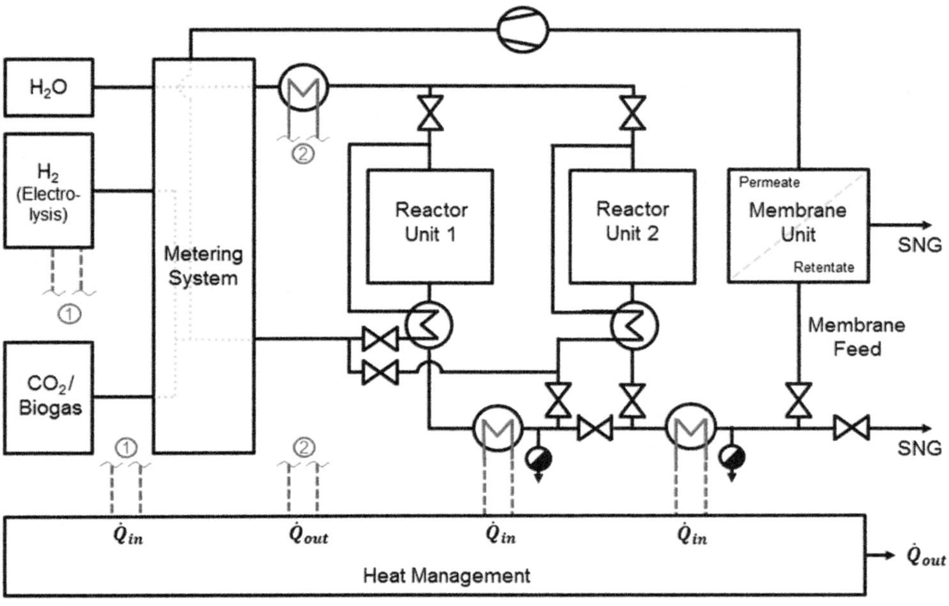

Diagram 4.57 Schematic representation of the 250 kW_{el} Power-to-Gas plant; Reactor 1: pressurized water-cooled plate reactor; Reactor 2: multi-zone, tube bundle reactor

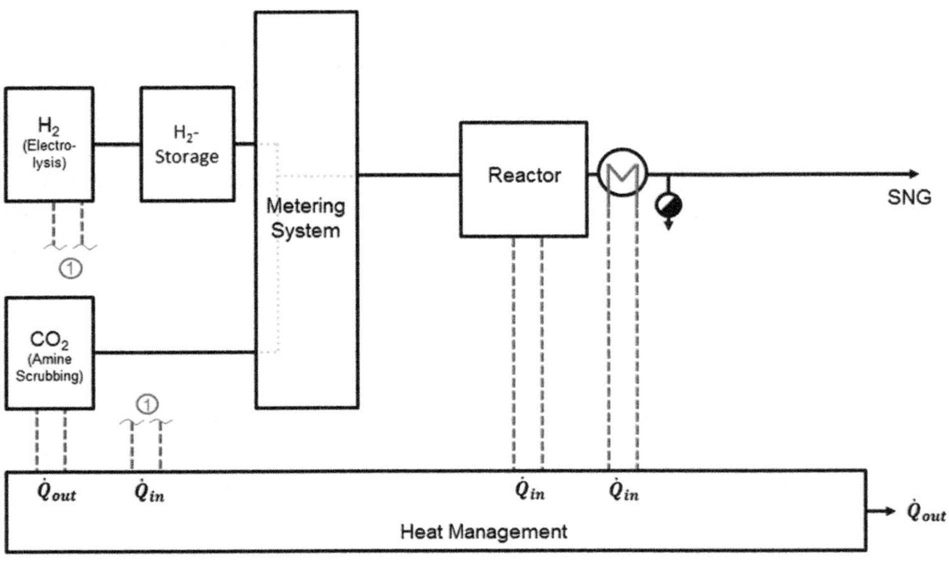

Diagram 4.58 Schematic illustration of the 6000 kW_{el} commercial Power-to-Gas plant of Audi AG in Werlte, Lower Saxony (Germany)

pressure swing adsorption or amine scrubbing) or the direct conversion of biogas by the addition of hydrogen. The plant was completed in 2009. With this plant, the operational capability of the technology was first demonstrated using real gases from biogas plants.

The production of hydrogen takes place in a high-pressure alkaline electrolysis with 25 kW_{el} input power. The educt gas streams are fed via a dosing device to a two-stage reactor system separated by a partial H_2O condensation stage. The synthesis unit design with intermediate condensation for the reduction of the moisture content increases the methane concentration of the product gas. A partial gas stream of the product gas is led via a "recycle loop" back to the first reactor stage to limit the temperature of the hot spot region. Deactivation of the catalyst due to carbon deposits is prevented by feeding water vapour to the educt gas.

The methane synthesis reactors are designed as tube reactors with a double jacket to temper the reaction chamber. A thermal oil is employed as the tempering medium. In addition to the high-pressure electrolysis and methanation stages, the container-integrated plant also incorporates the control and instrumentation electronics, including a fuelling module comprising a gas drying system, compressor, gas storage cylinders and fuelling device for natural gas - operated vehicles.

The 250 kW_{el} Power-to-Gas Plant

The construction of the 250 kW_{el} plant commissioned in 2012 took place within the scope of a publicly funded project [185]. The goal was the optimization of the technology for intermittent and dynamic operation and to point out cost reduction potential. The design is illustrated schematically in Diagram 4.57.

The plant includes a 250 kW_{el} high-pressure alkaline electrolyser and two different reactor systems that can be operated alone or in combination. Besides the proven tube bundle reactor known from the chemical industry, a pressurized water-cooled plate reactor is also employed. To temper the hot spot region, different approaches were realized. With the plate reactor, cooling of the principal reaction zone takes place by evaporation of water in the thermal plates, whereas in the multi-zone tube bundle reactor the temperature of the hot spot region is limited by the step-wise feed-in of the educt gas and molten salt cooling. With the multi-zone design of the molten salt cooling system, a precise temperature profile can be imposed in the methanation reactor.

To enhance the methane concentration, the plant can be operated as a two-stage synthesis process with a partial condensation stage or as single-step methanation with downstream gas upgrade. The processing of the methane-enriched product gas to substitute natural gas for feeding into the natural gas grid takes place via membrane technology by recycling the hydrogen-enriched gas (permeate) to the educt gas. This configuration enables hyperstoichiometric H_2 operation in the methanation reactor with nearly complete CO_2 conversion [184].

The 6000 kW_{el} Power-to-Gas Plant

The client for the 6000 kW_{el} plant in Werlte, Lower Saxony (Germany) is the automobile manufacturer Audi AG. The design and construction of the plant was carried out by the plant construction firm ETOGAS GmbH. Commissioning occurred in the fourth quarter of 2013. The plant is designed for the production of electricity-based substitute natural gas for feeding into the natural gas grid for use as sustainable fuel for mobility at natural gas filling stations.

The first commercial Power-to-Gas plant in the world with feed-in of substitute natural gas to the natural gas grid is characterized by a simple basic process plant conception in which no steam is dosed to the educt gas and the dried product gas is fed without further processing steps as replacement gas to the local L-gas grid. Methanation takes place in a molten-salt cooled tube bundle reactor with imposed temperature profile and single passage through the reactor ("once-through"). The temperature control in the hot spot region is implemented in the individual tubes by a step-wise educt gas feed and by the heat transfer medium.

The CO_2 required for the process is separated by amine scrubbing from the biogas of a waste material fermentation plant. Under full load, the P2G® plant produces a replacement gas volume flow of around 350 m^3_{STP}/h (without the bio-methane fraction, which is also fed to the gas grid). The feed-in of the replacement gas takes place to the local gas distribution grid (1.8 bar_{abs}). When the capacity of the distribution grid is exhausted (as in the summer, with low gas consumption) feed-in takes place to the transport grid (35–45 bar_{abs}).

Another feature of the 6000 kW_{el} plant is the intermediate storage of hydrogen. The hydrogen produced in the alkaline electrolysis with three electrolytic stacks is temporarily stored in a pressurized hydrogen tank at approximately 10 bar_{abs}. This enables methanation to be temporally decoupled from the intermittent operation of the electrolyser and results in a reduction of the number of start-up and shut-down ramps compared with electrolysis operation.

Results

Besides the construction and operation of Power-to-Gas plants in the performance classes 25 kW_{el} and 250 kW_{el}, experimental investigations on the hydrogenation of gases containing carbon oxides are being carried out at ZSW. Specifically, these include the screening of catalysts available on the market, investigations of their deactivation and the processing of product gases from methanation to a replacement gas by means of membrane gas separation technology (downstream membrane gas upgrade). This section discusses examples of experimental results.

Methanation Catalysts: Screening and Cycle Fatigue Resistance

Catalyst Screening

Before use in the reactor, catalysts available on the market are investigated with respect to their suitability for intermittent methanation operation. The investigations include the characterization of catalysts, for example with regard to start-up temperature and conversion behavior, as well as catalyst activation and the identification of deactivation mechanisms (thermal stability, poisoning and carbon depositions). Various reactors and analytical methods are employed for these investigations.

The investigations of the start-up temperature and the conversion behavior are carried out in a wall-cooled, oil-tempered tube reactor with approximately 80 ml filling volume. Diagram 4.59 shows examples from the results of commercially available, Ni-based catalysts for the start-up temperature ($T_{start\text{-}up}$) and the temperature measured in the hot spot region of the reactor ($T_{hot\ spot}$). The start-up temperature for describing the start-up behavior is determined by increasing the oil temperature in 10 °C steps. The temperature at which the filling bed temperature measured in the hot spot region is higher than the oil temperature (due to the exothermic reaction enthalpy) is referred to as the start-up temperature.

On the basis of the data determined in Diagram 4.59 ($T_{start\text{-}up}$, $T_{hot\ spot}$, y_{CH_4}), it can be stated that the catalysts 1, 3 and 5 are very well suited for CO_2 methanation. In addition to low start-up and working temperatures, these catalysts exhibit high temperature gradients in the hot spot region. Because of the exothermic character of the methanation reactions, this rise in temperature is an indicator of high catalytic activity. These results are confirmed in terms of the methane concentration obtained in the product gas. The test parameters do not reflect the respective optimal operating point of the catalysts, however they allow comparison under identical processing conditions.

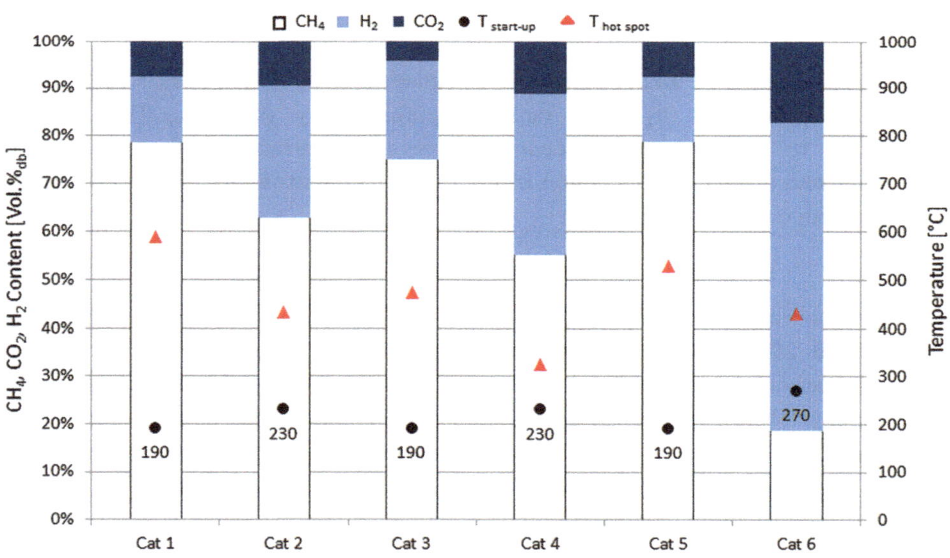

Diagram 4.59 Start-up temperature ($T_{start\text{-}up}$), temperature in the hot spot region ($T_{hot\ spot}$) and conversion behavior (resulting gas composition) of Ni-based catalysts available on the market (cat x) for CO_2 methanation; Educt gas: 80 vol.$\%_{H_2}$/20 vol.$\%_{CO_2}$; Basis of data$T_{start\text{-}up}$ and $T_{hot\ spot}$: SV = 2000 $l_{educt}/(l_{cat}$ h) and p = 7 bar_{abs}; Basis of data - gas composition: SV = 3000 $l_{educt}/(l_{cat}$ h); p = 1.5 bar_{abs}; $T_{cooling\ medium}$ = 300 °C

Catalyst Cycle Fatigue Resistance in Intermittent Operation

The catalysts employed for methanation in Power-to-Gas plants are subject to thermal stress as a result of intermittent and dynamic operation and the related frequent start-up and shut-down processes. The deactivation of the catalysts due to sinter effects leading to the reduction of the catalytically active surface was investigated in an oil-tempered micro-reactor [$V = 80$ ml; $SV = 4000\ l_{educt}/(l_{cat}\ h)$]. The cycle fatigue resistance of the catalyst was determined via automatic cyclic methanation operation with the starting conditions corresponding to the standby mode (temperature and pressure maintained in an H_2 atmosphere) and methanation conditions with H_2/CO_2 feed. Diagram 4.60 gives an example of the behavior of hot spot temperature and product gas quality over a few cycles. In Diagram 4.60, a cycle encompasses the start-up of methanation synthesis from the standby mode, as well as the shut-down of synthesis with return to the standby mode, maintaining the appropriate temperature with the tempering medium. Before beginning CO_2 feed-in, the temperature in the reaction chamber is identical to that of the tempering medium. The temperature of the tempering medium is adjusted so that it exceeds the start-up temperature of the catalyst. In the standby mode, H_2 streams into the reactor. As soon as CO_2 enters the reactor, the exothermic methanation reaction begins. This results in a pronounced increase in the temperature of the hot spot region, with the conversion of the educt gases. Upon reaching stationary methanation operation, the CO_2 dosing is discontinued and the reactor is flushed with hydrogen only. This causes the temperatures to fall to the level of the tempering medium (around 260 °C). After a brief holding time,

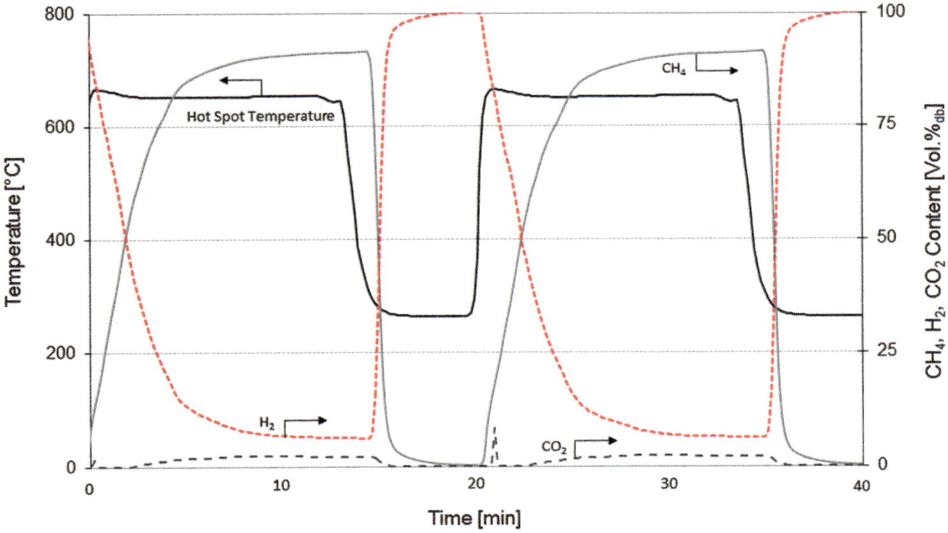

Diagram 4.60 Example of an experimental run for the investigation of the cycle fatigue resistance of catalysts during intermittent methanation operation (educt gas composition: 80 vol.$\%_{H_2}$/ 20 vol.$\%_{CO_2}$; $T_{cooling\ medium} = 260$ °C; $SV = 4000\ l_{educt}/(l_{cat}\ h)$; $p = 7\ bar_{abs}$)

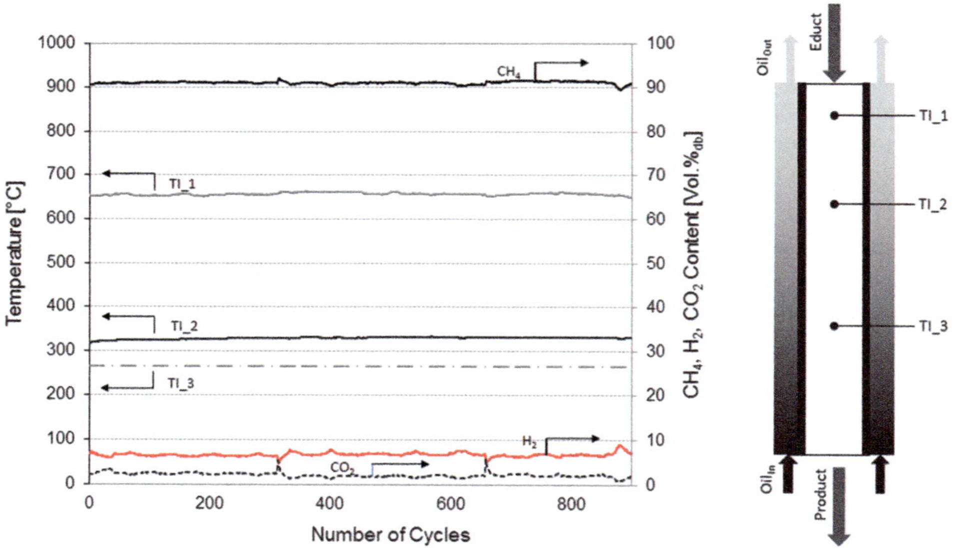

Diagram 4.61 Temperature profile and product gas quality as a function of the number of cycles (educt gas composition: 80 vol.%$_{H_2}$/20 vol.%$_{CO_2}$; $T_{cooling\ medium}$ (Oil) = 260 °C; SV = 4000 $l_{educt}/(l_{cat}\ h)$; p = 7 bar_{abs})

the process described above is automatically repeated until reaching the pre-defined total number of cycles. Diagram 4.61 shows the result of investigations over 900 cycles.

As Diagram 4.61 shows, no deactivation of the catalyst due to intense thermal stress was observed after 900 cycles. The temperatures in the reaction chamber and the quality of the product gas remained virtually constant. The thermal deactivation of the catalyst in the hot spot region due to sinter effects can in fact not be generally ruled out. Nevertheless, the results indicate that a high catalyst life can be expected. It must also be mentioned that the number of methanation cycles can be significantly reduced with temporary H_2 storage, further enhancing catalyst life (see Diagram 4.58). Frequent start-up and shut-down phases, such as those that occur in P2G® plants due to the intermittent supply of electricity, are typical, but do not seriously affect the catalyst's lifetime. Depending on the requirements of the power grid, the capacity of the temporary hydrogen storage and the resulting number of start-up and shut-down processes for methanation, a catalyst life of at least 1–2 years can be expected.

Results from the 25 kW_{el} P2G® Plant

Two different educt gases were investigated with the 25 kW_{el} P2G® plant:

- CO_2 separated from biogas and
- Purified biogas without the previous separation of CO_2.

The supply of the carbon dioxide required for methanation can take place from a number of sources. Examples are the extraction of CO_2 from the conversion of fossil energy sources [carbon capture and utilization (CCU) in place of carbon capture and storage (CCS)], in the production of lime/cement, in processes of the chemical industry and in the production of bio-ethanol. Particularly biogas plants come into question as sources of CO_2 for industrial Power-to-Gas plants. "Off-gas", consisting mostly of CO_2, arises during the processing of biogas to "bio-methane" (CO_2 separation). As this biogenic CO_2 is not burdened with climate-relevant emissions (CO_2-neutral cycle), it is well suited as an educt for methanation. Alternatively, CO_2 derived from biogas without previous separation can be used by combining the biogas with the H_2 in a methanation reactor unit.

The market entry of P2G® plants will take place where CO_2 is available "at no cost". On the one hand, this is CO_2 as the major constituent of biogas (as a gas mixture with a CO_2 fraction of 30–50 vol.%) or in the form of "off-gas", which consists primarily of CO_2 and is produced in the processing of biogas to "bio-methane" which can be fed to the natural gas grid.

In the anaerobic fermentation of biomass, crude biogas with the major constituents CH_4 (50–70 vol.%) and CO_2 (30–50 vol.%) is produced. Furthermore it also contains water vapour, the minor constituents H_2S, NH_3 and - depending on the type of pre-desulfurization - N_2 and O_2. Processing of the crude biogas to SNG takes place by the separation of water, the minor constituents and the major constituent CO_2, until the specified quality (replacement gas quality) is obtained. CO_2 is extracted in existing plants, for example, by pressure swing adsorption or by various scrubber systems. This CO_2 is then available as an educt for the production of SNG utilizing the P2G® process.

Experimental investigations on the conversion of gas mixtures containing H_2/CO_2 and $H_2/CO_2/CH_4$ have been performed with the 25 kW_{el} P2G® plant. Both variants were investigated with "real gases" from different biogas plants. The gases used as educts were biogas and "CO_2 off-gas" from a processing plant for the production of bio-methane from biogas. The educt gases were subjected to a fine purification process. Diagram 4.62a, b show the results of the SNG production from the educt gases "CO_2 off-gas/H_2" and "biogas/H_2". In both cases, the reactor was operated with comparable parameters (T: 250-550 °C; $p_{abs} = 7$ bar; Ni-based catalyst, space velocity SV: 2000-5000 $l_{educt}/(l_{cat}\ h)$).

Diagram 4.62a, b show that, with the use of both real gases - CO_2 off-gas from a pressure swing adsorption (PSA) process and biogas following purification, but without CO_2 separation - constant gas quality can be achieved over a longer period of time. The synthesis operation remained stable over the entire duration of the investigations. As a result of the gas composition and the combustion characteristics, the gas mixture produced was a replacement gas (L-gas) conforming to the German regulations DVGW G 260 and DVGW G 262 [178, 179]. Following drying, further gas conditioning is therefore no longer necessary for feeding to the L-gas grid.

The modelling/simulation of the two variants "Methanation of CO_2 off-gas" and "Methanation of purified biogas" in relation to the process efficiency is discussed in section "P2G® Process Efficiency".

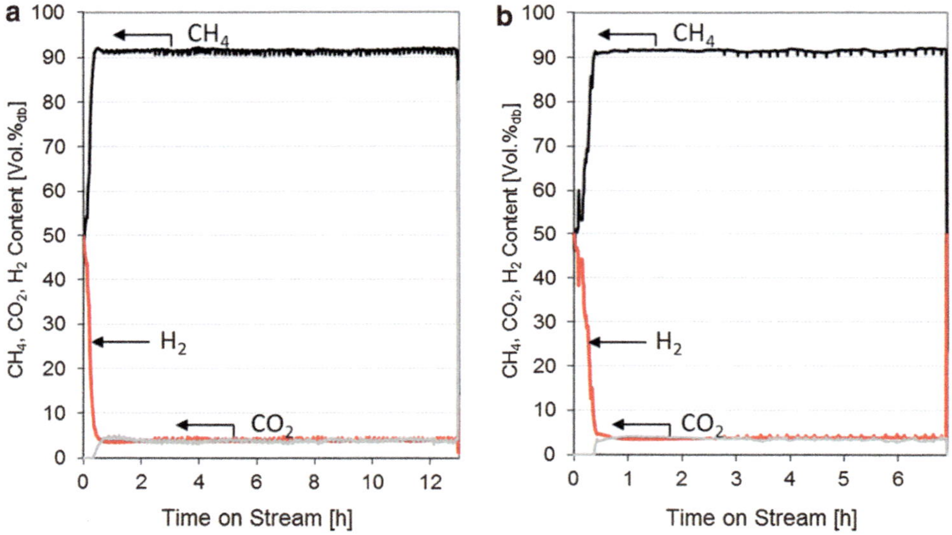

Diagram 4.62 (**a**) Product gas composition following the conversion of CO_2/H_2 in the 25 kW_{el} P2G® plant. As educt, the "off-gas" from a pressure swing adsorption (PSA) plant processed in a biogas plant to bio-methane was used. (**b**) Product gas composition following the conversion of purified biogas with H_2 in the 25 kW_{el} P2G® plant without previous separation of CO_2

Results from the 250 kW_{el} P2G® Plant

The results below refer only to a molten salt-cooled tube bundle reactor, as this was chosen for the 6 MW_{el} P2G® plant. The tube bundle reactors installed in the 250 kW_{el} and the 6 MW_{el} P2G® plant were designed based on experimental testing with a single tube. The primary goal was the transfer of the results for a single tube to a tube bundle reactor of the 250 kW_{el} P2G® plant. The results indicate that a scale-up from a single-tube reactor to a tube bundle reactor poses no problems. The modular construction of this reactor type allows adaptation to the required gas capacity through the number of reactor tubes.

Diagram 4.63 shows the example of the operational behavior for the tube bundle reactor of the 250 kW_{el} P2G® plant relative to repeated start-up and shut-down processes and fluctuating loads operated with a stoichiometrically adjusted H_2/CO_2 mixture - starting in the standby mode under an H_2 atmosphere. Upon reaching stationary operation conditions with constant product gas quality, the investigation of the fluctuating load between 100 % and 70 % on the basis of stoichiometrically adjusted educt gas stream ramps began. The standby mode under H_2 flushing was then started again and the entire sequence repeated.

Start-up from the standby mode requires more time than shut-down to the standby mode. This is due to the thermal inertia of the reactor system and the increased flushing time by the molar-number-reducing methanation. However, start-up and shut-down time are in the order of only a few minutes. The original gas composition is reobtained

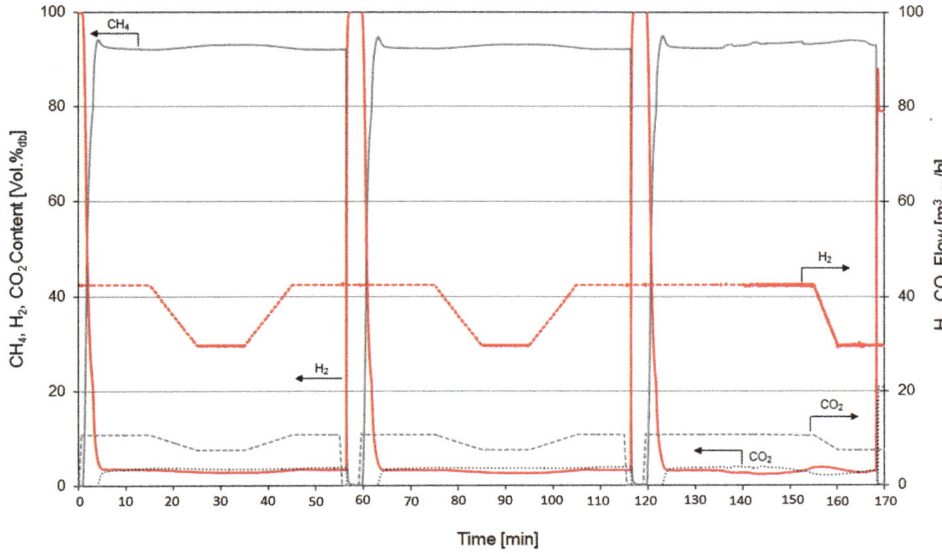

Diagram 4.63 Gas composition during load changes in the tube bundle reactor of the 250 kW_{el} P2G® plant as a function of the educt gas volume flow; T = 200 - 600 °C; p = 7 bar_{abs}; SV = 1365 $l_{educt}/(l_{cat}\ h)$

following repeated start-up and shut-down processes. The results also show that load changes between 100 % and 70 % have no significant influence on the product gas quality with the tube bundle reactor.

With the use of catalysts of the newest generation and optimal process control of the once through passage, the quality of the product gas meets the specifications of the German guidelines [178, 179] for feed-in as replacement gas to the German L-class natural gas grid without requiring additional downstream gas processing in addition to drying the gas.

Besides the definition of operational concepts and cost reduction options for P2G® plants, the production of a replacement gas for feeding to the H-gas grid with CH_4 fractions of typically > 95 vol.$\%_{db}$ is a major challenge. For moderate methanation pressures of < 10 bar_{abs} as in Diagram 4.54 with a "once-through" process, this is feasible only within certain limits. A promising option for downstream gas processing is the membrane gas separation process, in which the retentate is extracted as a replacement gas and the permeate is completely recycled to the educt gas stream for methanation (recycle loop).

One result of the process configuration in Diagram 4.57 with the "once-through" reactor concept in combination with membrane gas processing is shown in Diagram 4.64 for the tube bundle reactor. With the recycling of the permeate gas, the gas qualities of the retentate fulfil the requirements for feed-in as replacement gas to a gas group H grid.

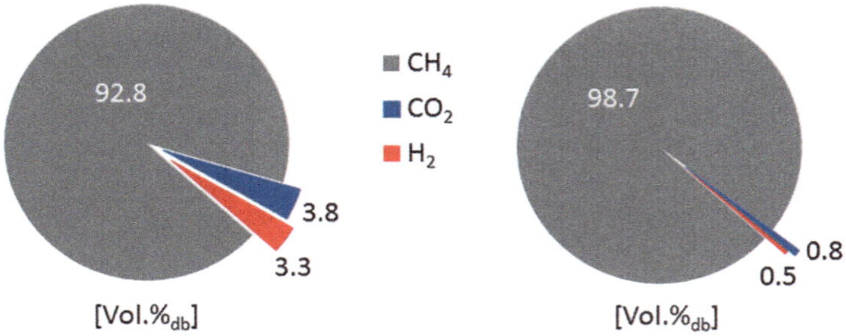

Diagram 4.64 Gas composition with the tube bundle reactor without membrane gas upgrade (*left*) (SV = 1365 $l_{educt}/(l_{cat}$ h); $p_{methanation}$ = 7 bar_{abs}) and with membrane gas upgrade (*right*) (SV = 1365 $l_{fresh\ gas}/(l_{cat}$ h); $p_{methanation}$ = 5 bar_{abs}; $p_{membrane}$ = 5 bar_{abs})

A comparative examination of the process chains with and without gas upgrade utilizing membrane technology indicates that, with downstream membrane gas processing, not only a high-calorific gas can be produced, but also the process pressure for methane synthesis can be lowered. The retentate gas produced fulfils the highest requirements for feed-in as a replacement gas for injection to the H-gas grid with limited process-related technical effort.

Results from Audi's 6000 kW_{el} E-gas Plant

Audi's 6000 kW_{el} e-gas plant in Werlte, Germany first supplied substitute natural gas (replacement gas) to the existing local L-class natural gas grid at the end of 2013. This plant was constructed with a molten salt-cooled tube bundle reactor designed for methanation with a product gas stream volume flow of 350 m^3_{STP}/h (see Diagram 4.55a).

Due to the start-up time required to reach stationary state conditions, the dynamic/intermittent operation of the methanation reactors in this industrial scale is viewed critically. The operational behavior in the start-up phase is therefore the object of particular interest in research. Diagram 4.65 shows an example of the tube bundle reactor used. Start-up from the standby mode under H_2 atmosphere takes place with a pre-selected temperature of the molten salt cooling medium with CO_2/H_2 feed-in from t = 0 s. Following the removal of condensate, the product gas composition is determined at the reactor outlet. After approximately 300 s the methane concentration in the product gas is > 90 vol.$\%_{db}$. Besides the CH_4 content, a further criterion for feed-in to the gas grid is a H_2 concentration of < 5 vol.$\%_{db}$. After barely a further 100 s, this requirement is satisfied so that following the removal of condensate and the dosing of air for adjusting the combustion properties, the product gas can be fed to the local L-gas grid after 400 s. In the stationary state, the e-gas plant delivers a maximum substitute natural gas yield of 3.3 MW_{SNG}.

A special feature of the e-gas plant is the temporary storage of 1200 m^3_{STP} of hydrogen. Switching the electrolysis off and simultaneously reducing the product gas yield of the methanation therefore allows further processing of the methanation for more than 1 h at

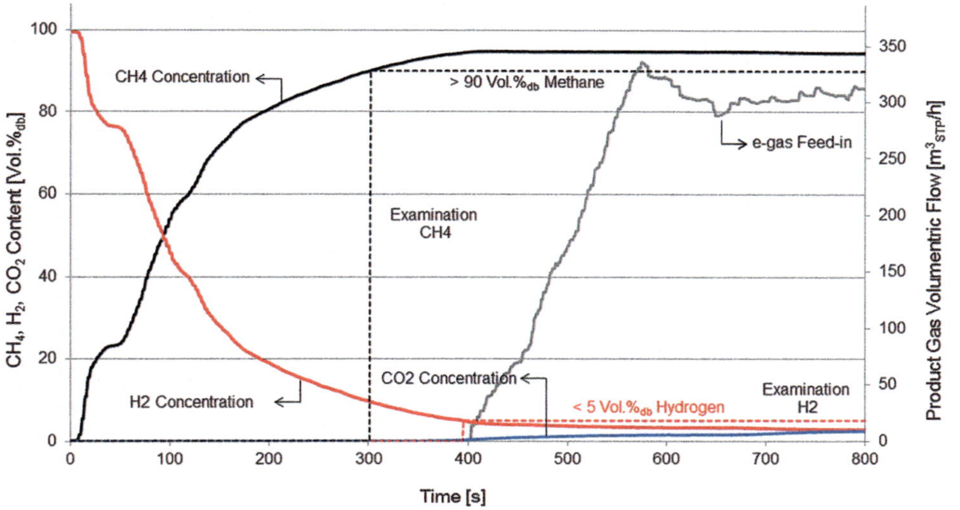

Diagram 4.65 Gas quality during start-up of the methanation reactor and feed-in of the product gas to the natural gas grid (T = 200 - 600 °C; p = 6.5 bar_{abs}; SV = 1500 $l_{educt}/(l_{cat}\ h)$) (Audi [186])

maximum filling level of the storage tank. If the electrolysis is started with the H_2 tank not completely filled, the tank is first filled before starting methanation. The temporary storage of H_2 enables a drastic reduction in the number of start-up and shut-down processes during methanation.

A further feature is the extraction of heat from the methanation reactor at a temperature level of < 200 °C to supply heat to the CO_2 separation for the amine scrubbing process. During operation of the e-gas plant, heat is provided to the CO_2 stripper (thermal CO_2 desorption). With the e-gas plant in the standby mode, the heat required for the CO_2 stripper is provided by the combustion of methane (e.g. the combustion of a partial stream of the biogas produced). This means a significantly higher bio-methane yield in methanation operation, as the combustion of methane can be dispensed with. The heat extracted for amine scrubbing is in the order of 400 kW_{th}. This extracted heat replaces 400 kW_{CH4} and must therefore be added directly to the plant efficiency. Furthermore, the e-gas plant supplies approximately 350 kW_{th} of low-temperature heat to the biogas plant (hygienization, fermentation heating). In relation to the overall consumption of electricity, including the peripheral current consuming devices, this results in an utilization level of around 60 % for the first Power-to-Gas plant on an industrial scale (Diagram 4.66).

P2G® Process Efficiency

For the determination of the process efficiency, the P2G® process was modelled at ZSW with the commercially available IPSEpro software. The heart of the IPSEpro software is an equation solver (Newton-Raphson algorithm), suited for solving implicit systems of equations. Taking the conservation of mass and physical relationships in chemical

Diagram 4.66 Methanation with Audi AG's 6000 kW_{el} Power-to-Gas plant in Werlte, Germany (molten salt-cooled tube bundle reactor; see Diagram 4.55a for the schematic representation of the design) (Audi [186])

reactions into consideration, element and material balances can be implemented in different models and complex processes can be interconnected.

Diagram 4.67 depicts the energy balance of the P2G® process for two selected cases (once-through process; without follow-on membrane gas separation):

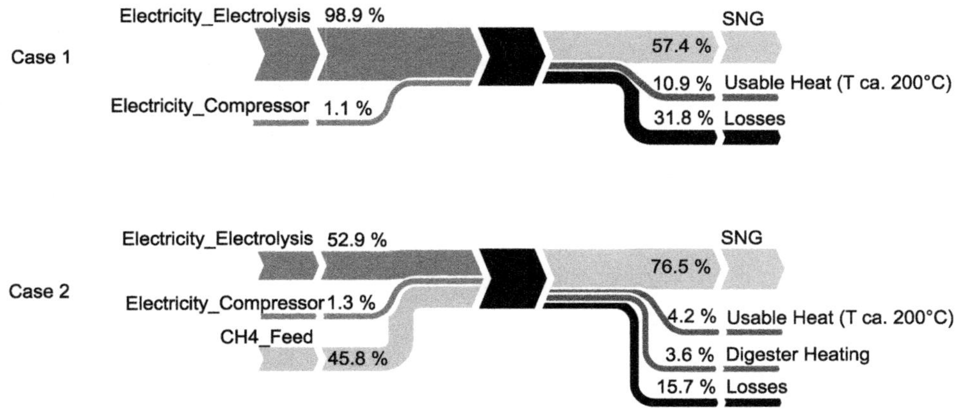

Diagram 4.67 Energy balances in the P2G® process for a stoichiometric CO_2/H_2 stream (case 1) and a stoichiometric biogas/H_2 stream (Case 2; biogas: 60 vol.%$_{db}$ CH_4, 40 vol.%$_{db}$ CO_2)

- *Case 1:*
  Methanation of a stoichiometric CO_2/H_2 gas stream
- *Case 2:*
  Methanation of a stoichiometric biogas/H_2 gas stream
  (Biogas: 60 vol.%$_{db}$ CH_4, 40 vol.%$_{db}$ CO_2)

The results of the process simulation are based on the data in Table 4.14. For calculation of the reaction conversion rate of the methanation, thermodynamic equilibrium was assumed at 8 bar$_{abs}$ and 260 °C.

It can be seen from Table 4.14 that for Case 1 (stoichiometric CO_2/H_2 stream) the composition of the product gas resulting from methanation fulfils the specifications for feed-in to the L-gas grid. For Case 2, the methane fraction of 95.8 vol.%$_{db}$ is significantly greater, so that an H-gas equivalent replacement gas is produced [178, 179].

For calculation of the chemical and thermal efficiency, in addition to the electricity demand of the electrolysis ($P_{el,electrolysis}$) and the chemical energy of the biogas ($P_{chem,educt\ gas}$), the most important electrical consumers, such as the CO_2/biogas compression and the product gas compression following methanation ($P_{el,compression}$) to achieve grid feed-in pressure, were also considered:

$$\eta_{chem} = \frac{P_{chem,product\ gas}}{P_{el,electrolysis} + P_{el,compression} + P_{chem,educt\ gas}} \quad [-] \tag{4.6}$$

$$\eta_{th} = \frac{\dot{Q}_{th,usable}}{P_{el,electrolysis} + P_{el,compression} + P_{chem,educt\ gas}} \quad [-] \tag{4.7}$$

Table 4.14 Basic data for the determination of the energy balances in the P2G® process (db: dry basis)

		CH_4 (vol.%$_{db}$)	H_2 (vol.%$_{db}$)	CO_2 (vol.%$_{db}$)	H_2O (vol.%)
Case 1	Educt gas	0.0	80.0	20.0	0.0
	Product gas	93.9	4.9	1.2	65.2
Case 2	Educt gas	23.1	61.5	15.4	10.0
	Product gas	95.8	3.4	0.8	50.9
Electrolysis					
Energy requirement		4.3	$kWh_{el}/m^3_{H2,STP}$		
Pressure		8.5	bar_{abs}		
Methanation					
Educt gas pressure		8.5	bar_{abs}		
Product gas pressure		8	bar_{abs}		
Temperature		260	°C		
CO_2/biogas pressure		1	bar_{abs}		
Usable heat temperature		Approx. 200	°C		
Gas grid feed-in pressure		16	bar_{abs}		

For Case 1, the conversion of a stoichiometric CO_2/H_2-stream to a product gas suitable for feed-in to the L-gas grid, a chemical efficiency of 57.4 % is obtained (Diagram 4.67). A quarter of the waste heat is generated at a temperature level of around 200 °C and can be utilized, for example, for the provision of a CO_2 stream from biogas by amine scrubbing. The resulting thermal efficiency is 10.9 %. For the case of direct biogas methanation (without CO_2 separation, in Case 2), a chemical efficiency of around 76.5 % and a thermal efficiency of 4.2 % at a temperature level of around 200 °C for the usable waste heat results. An advantage of the coupling of P2G®/biogas plants is that a part of the heat flow with $T < 200$ °C can be utilized to heat the fermenter.

The losses incurred during the conversion of electricity to chemically bound energy in the form of substitute natural gas occur both in the electrolysis ($\eta = 70$ %, corresponding to 4.3 $kWh_{el}/m^3_{H_2,STP}$) and in the methanation unit ($\eta = 83.2$ %). The optimization potential relative to the overall efficiency of the plant lies mainly in the electrolysis, as the efficiency of the methanation is determined by the reaction enthalpy of the CO_2 hydrogenation. Improving the efficiency of the electrolysis from 70 to 80 % (corresponding to 3.75 $kWh_{el}/m^3_{H_2,STP}$) leads to an improvement in the overall efficiency to > 65 % (Case 1). Taking the extracted heat at a temperature level of around 200 °C (e.g. for CO_2 separation by amine scrubbing) into consideration, an overall efficiency of > 75 % is possible without taking the low-temperature heat (e.g. for heating the fermenter) into account.

Costs of Electricity-Based Substitute Natural Gas

In relation to the economic constraints, the realization of Power-to-Gas technology represents a major challenge when transferring to a commercially viable scale. This is a consequence of the fact that the production of 1 kWh_{SNG} with an efficiency of around 60 % requires around 1.7 kWh_{el} of primary electrical energy. The inescapable conclusion is that the product "electricity-based substitute natural gas" must always be more expensive per kWh than the input electrical energy. Further important, sensitive influencing factors contributing to the costs of SNG are the investment costs and the annual number of hours of full utilization.

Cost Calculation

For calculation of the production costs for SNG, the parameters in Table 4.15 were chosen in order to describe the costs as a function of the mean electricity costs and the number of hours of full utilization during the year. Diagram 4.68a shows this calculation for

Table 4.15 Performance data and economic factors for the determination of SNG production costs

Performance data		
Chemical efficiency	57.40	%
Thermal efficiency	10.90	%
Auxiliary power (operation mode)[a]	1.10	%
Auxiliary power (standby mode)[a]	2.00	%
Number of hours full utilization[b]	Variable	h/a
Operating costs		
Procurement of electricity[b]	Variable	€ct/kWh_{el}
Procurement of CO_2[c]	0.00	€ct/kWh_{el}
Maintenance costs	3.00	$\%_{Invest}$/a
Revenues		
Substitute natural gas[d]	Resulting yield	€ct/kWh_{SNG}
Usable heat[e]	0.00	€ct/kWh_{th}
Financing		
Investment costs[b,f]	1000/2000	€/kW_{el}
Calculated interest rate[g]	5.00	$\%_{Invest}$/a
Amortization period	20	a
Residual value after amortization	0.00	€
Insurance costs	1.00	$\%_{Invest}$/a

[a]Percent of the installed electrical power
[b]Varies for the determination of the SNG production costs of as a function of sensitive influencing factors
[c]E.g. CO_2 from biogas plants with processing to bio-methane (no CO_2 costs)
[d]Result of the cost calculation
[e]Connection to a heat grid is not considered
[f]Plant costs only without site-specific costs for parcels of land and development
[g]Corresponding to return on investment when capital resources are invested

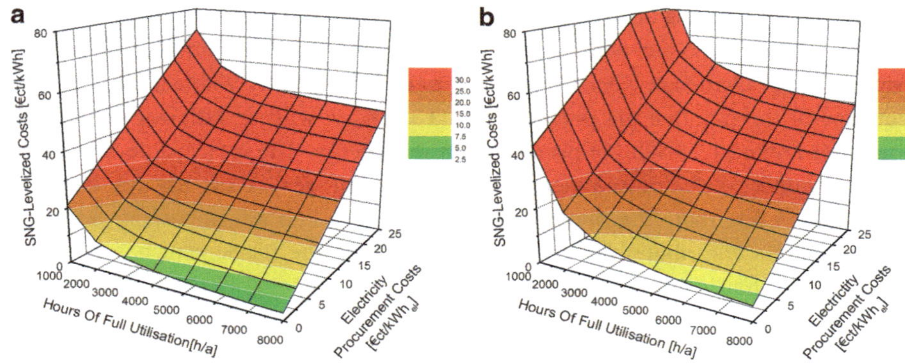

Diagram 4.68 (**a**) SNG production costs as a function of the annual number of hours of full utilization and the mean annual cost for procuring electricity for investment costs of 1000 €/kW$_{el}$. (**b**) SNG production costs as a function of the annual number of hours of full utilization and the mean annual cost for procuring electricity for investment costs of 2000 €/kW$_{el}$

investment costs of 1000 €/kW$_{el}$ and Diagram 4.68b for 2000 €/kW$_{el}$. The current investment costs of approximately 3500 €/kW$_{el}$ are higher. Nevertheless, the further development of this technology leads us to expect target costs of 1000 €/kW$_{el}$ within the next 10–15 years. These investment costs are based on plant sizes > 1 MW$_{el}$.

The performance data used as the basis for the cost calculation are oriented on the process efficiency of the P2G® plant of section "P2G® Process Efficiency", Diagram 4.67 (Case 1). The energy flow diagram for the methanation of a stoichiometric CO_2/H_2 gas stream indicates that, on the basis of the assumption of 100 kWh$_{el}$ (electricity), 57.4 kWh$_{SNG}$ (substitute natural gas) and 10.9 kWh$_{th}$ (usable heat) can be generated. Besides an auxiliary electricity requirement of 1.1 % during normal operation required for the peripheral auxiliary equipment (pumps, compressors, etc.), in standby mode around 2 % of the connected electrical load is required for tempering the electrolysis and the methanation reactor. Furthermore, the cost calculation for the costs of procuring electricity includes the operating costs for maintenance, as well as personnel costs and the costs of operational resources. On the other hand, the costs for the purchase of CO_2 are not considered here, because high CO_2 concentrations from the biogas process, which would otherwise be released to the atmosphere, can be utilized.

Diagram 4.68a, b show the influence of the annual number of hours of full utilization and the mean electricity procurement costs on the SNG production costs for a specific investment of 1000 €/kW$_{el}$ and 2000 €/kW$_{el}$ in a P2G® plant. To simplify reading the SNG production costs, the results are depicted with grid lines in intervals of 1000 h/a full utilization and 2.5 €ct/kWh$_{el}$ mean costs for the procurement of electricity. In addition, the SNG production costs are broken down using a color spectrum according to the legend in the diagram. These representations show that the SNG production costs increase with increasing costs for the procurement of electricity and decrease with an increase in the number of hours of full utilization.

For the case of specific investment target costs for the investment of 1000 €/kW_{el} in Diagram 4.68a, for electricity procurement costs 2.5 €ct/kWh_{el} and an annual number of hours of full utilization > 6000 h, operated together with bio-methane produced in fermentation, the SNG production costs are around 7.5 €ct/kWh_{SNG}. According to FNR [187], the production costs of bio-methane from biogas are in the order of 6–9 €ct/kWh_{SNG}. For a lower annual number of hours of full utilization or electricity procurement costs of > 2.5 €ct/kWh_{el}, the resulting costs of SNG production exceed the costs of obtaining bio-methane from biogas.

Conclusion

The P2G® process was first proposed in 2009 for the storage of renewable electricity in the form of substitute natural gas. The secondary energy carriers methane and/or hydrogen can be utilized as required for the reconversion to electric power or as an energy carrier for the mobility and heat consumer sectors.

Substitute natural gas could be fed to the natural gas grid for the first time as early as 2013 from the 6 MW_{el} plant (e-gas plant of the auto manufacturer Audi AG) on a technical scale. This shows that - in spite of the still existing optimization potential - the technology can be realized in the MW performance class today.

The following core statements can be derived from the empirical results of the P2G® plants in the 25 kW_{el}, 250 kW_{el} and 6000 kW_{el} performance classes:

- The production of substitute natural gas from CO-based synthesis gases is the state-of-the-art of this technology. However, due to the vastly lower reactivity of CO_2 compared with CO, this is not the case for H_2/CO_2 educt gases. In spite of this, synthesis from H_2/CO_2 using suitable catalysts with high selectivities and conversion rates is successful.
- The renewable fuel methane can be produced decentrally with P2G® plants of the MW_{el} performance class with considerably less technical effort compared with Fischer-Tropsch/methanol synthesis.
- The cycle fatigue resistance of the methanation catalyst was proven for up to 900 cycles (standby/full load) without loss of activity.
- In principle, substitute natural gas can be synthesized by a "once-through process", followed by drying of the gas, without subsequent gas processing (option for L-gas grids).
- In combination with membrane gas processing, the methane content can be increased up to 99 vol.$\%_{db}$ (option for H-gas grids).
- The start-up time of the methanation reactors from the standby mode to full-load operation is of the order of a few minutes. The same applies for the shut-down ramps.
- As a result of the short start-up and shut-down ramps, a P2G® plant can provide both positive and negative balancing energy for stabilization of the power grid.

- For the conversion of a stoichiometrically adjusted CO_2/H_2 stream, the chemical efficiency of the P2G® process relative to the required electrical energy input is currently < 60 %. However, the optimization potential for electrolytic hydrogen production allows us to expect an improvement of the chemical efficiency to > 65 %.
- When taking the heat extraction at a temperature level of around 200 °C (e.g. for CO_2 separation by amine scrubbing) into consideration, an overall efficiency of > 75 % is possible without taking the low-temperature heat (e.g. for heating the fermentation) into account.

For the present fuel market based almost entirely on crude oil, diversification of the resource base, including an increase in the proportion of renewable energy, is urgently required. The P2G® concept opens the way to a highly efficient new approach for the production of electricity-based fuels.

The calculated production costs for electricity-based methane must be seen in relation to the consumer prices that the customer in Germany pays today (2014) at the filling station. For fossil-based fuels this is approximately 18 €ct/kWh for gasoline, for hydrogen around 29 €ct/kWh and for compressed natural gas (CNG) 8 €ct/kWh. Even with the significant advantage of lower consumption obtained with an H_2 fuel cell vehicle compared with a combustion engine CNG vehicle, the costs of fuel cell powered vehicle are higher than for a CNG vehicle. Here, the state has possibilities for fiscal structuring which take account of the environmental effects of the different types of fuel.

Other factors of economic importance are the provision of system services in the electricity market (e.g. for balancing electricity) and the (seasonal) storage of electricity. Today, it is difficult to quantify factors such as reliability of supply and economic system optimization by the bidirectional coupling of the electricity and gas sectors. However, with increasing contributions of renewable energy, these will play an important role in future.

At the present time, end user surcharges in Germany (in particular the EEG surcharge) are levied on the procurement costs for renewable electricity for the production of electricity-based methane. When these surcharges are added to the mean costs for the procurement of electricity, considerably higher costs result for the production of SNG, and economic operational models cannot be demonstrated under the current conditions. On the other hand, Power-to-Gas technology is an energy conversion process and not the "consumption" of energy. Consequently, the end user surcharges have not been considered in Diagram 4.68a, b.

In spite of the higher costs of electricity-based SNG per kWh compared with the required electricity input, its production is one of the few options for the provision of regenerative fuels with per cent contributions in double figures for the mobility sector. Furthermore, it must be assumed that with the compulsory increased utilization of regenerative electricity production, the availability of inexpensive, weather-related fluctuating electricity will significantly increase at many hours over the year, so that in future SNG production can be used efficiently.

Acknowledgements
The authors wish to express their gratitude for financial support during the development of the P2G® technology.

The assembly and operation of the 25-kW_{el}-P2G® plant was financed by SolarFuel GmbH, renamed ETOGAS GmbH in 2013.

The assembly and operation of the 250 kW_{el} P2G® plant was supported by the Federal Ministry for Economic Affairs and Energy on the basis of a resolution of the German Parliament (funded project no.: 0325275).

The construction of the 6000 kW_{el} e-gas plant is an investment of the automobile manufacturer Audi AG.

The monitoring of the 6000 kW_{el} e-gas plant is supported by the Federal Ministry for Economic Affairs and Energy on the basis of a resolution of the German Parliament (project no.: 0325428).

4.4.3.2 Biological Methanation: Methane Production by Microbial Conversion of H_2 and CO_2

Joachim Krassowski and Oliver Jochum

Introduction

The chemical methanation described in the previous chapter occurs by heterogeneous catalysis in a gas-phase reaction.

As an alternative option to this type of methanation, the production of methane can be achieved through biological methanation, which utilizes natural metabolic processes of microorganisms. In contrast to the chemical methanation, the conversion of hydrogen and carbon dioxide into methane occurs naturally when specialized microorganisms are present. This biological pathway has been described scientifically since the mid-twentieth century and it is similar to the chemical methanation which is defined by the Sabatier reaction:

$$CO_2 + 4H_2 \rightarrow CH_4 + 2H_2O$$

The following sections provide an overview of the biological basics, the current state of technology, various process concepts for industrial implementation and the economic efficiency of biological methanation.

Biological Basics

Biological methanation is a biologically catalyzed reaction in which the conversion of carbon dioxide and hydrogen to methane occurs by metabolic processes of specific microorganisms. These microorganisms, so-called methanogenic archaea, belong to one of the oldest known living creatures. They combine hydrogen oxidation with the reduction of carbon dioxide in their energy metabolism to build up the cell internal energy carrier

ATP. In addition, the carbon dioxide serves as the sole carbon dioxide source of microorganisms to build up the cellular matter.

Pure methane, as a by-product of the metabolic processes, is released by the microorganisms. Due to these properties, the methanogenic archaea are also called chemoautotroph.

Methanogenic archaea are summarized into six orders. They are distributed according to their natural adaptations to various habitats with different temperature conditions. The temperature optima of various archaea reach from mesophilic (33–45 °C) through thermophilic (65–70 °C) to hyperthermophilic (65–85 °C) temperature values. The optimum pH value in each case is between 6.8 and 7.2 in neutral range [188].

Since these microorganisms absorb and utilize the reactant gases hydrogen and carbon dioxide only in solute form, the reaction always occurs in an aqueous sphere. In addition, a strictly anaerobic (oxygen-free) atmosphere is needed because oxygen causes a toxic effect on microorganisms. The natural occurrence of the species is therefore limited to oxygen-free aqueous habitats in which biomass degradation occurs, such as stomachs of ruminants, water sediments, swamps, rice fields and moors.

In their natural environment, the required carbon dioxide and hydrogen are typically formed by previous microbial degradation processes of biomass. The required transformation steps are carried out by different microorganisms, which provide the basic materials for the (energy) metabolism of methanogenic archaea.

The specific characteristics of methanogenic archaea have been technically used for decades for the production of biogas and sewer gas. The biomass degradation, which takes place in the digesters and septic tanks, is achieved in four consecutive steps, equivalent to the natural anaerobic biomass degradation of a variety of microorganisms. The final step of this process, an essential procedure in producing methane, is accomplished by methanogenic archaea through methogenesis. The archaea ensure a low hydrogen partial pressure by consuming the hydrogen which is released by the biological degradation of organic matter.

This is crucial for the activity of the acetogenic (hydrogen and carbon dioxide-producing) microorganisms in the anaerobic food chain. Both types of microorganisms have very close symbiotic relationships to each other and the methane content in biogas or sewer gas is limited by the hydrogen supply of the acetogenic microorganisms. The methane content depends on the type of biomass and varies between a maximum of 50–75 %.

In the context of laboratory experimentation, biological methanation involves the metabolic processes of methanogenic microorganisms caused by the addition of the feed gases hydrogen and carbon dioxide instead of biomass supply. Thus, since there is no limitation of hydrogen supply, in theory an almost complete conversion of the feed gases to methane can be achieved. Methane level about 95 % can be realized technically, so a biologically natural gas-equivalent can be produced by biological methanation. The procedural implementation of biologically catalyzed methanation is possible without high technical effort, due to the moderate process conditions. Furthermore, methanogenic archaea are very resistant to fluctuations of temperature and pressure, owing to their naturally acquired adaptations of extreme environments.

Similarly, impurities of the starting materials, for example traces of hydrogen sulfide (H_2S) in the reactant gases, are harmless according to current knowledge and could be nutritionally beneficial. Also on multi-day hunger phases without supplying substrates, the archaea react largely insensitively. The need for additional natural nutrients (for example, nitrogen, sulfur, phosphorus, potassium, calcium, magnesium, nickel, and others) varies among different methanogenic archaea.

Stage of Development

The biochemical production of methane from hydrogen and carbon dioxide was recently analyzed exclusively in laboratory scale. Table 4.16 summarizes scientific literature relating to biological methanation from 1978 to 2013 with reference to current process parameters and achieved productivity rates.

The scientific-based studies concerning the biological methanation offer a high diversity of experimental conditions and conclusions. They differ in procedure methods (batch or continuous, inoculum from monoculture or slurries, fermenter with or without immobilization of bacteria), operating parameters (temperature, pressure) and in relation to the composition of educt gases (ratio of volumes hydrogen/carbon dioxide and possible addition of hydrogen sulfide).

According to Table 4.16, researches between 1978 and 2010 were conducted in small-scale reactors of volumes ranging from 0.06 to 3 l. However from 2010 onwards, reactors of larger volumes between 7.8 and 80 l became the focus investigation. The reactors were implemented with a fermentative solution (inoculum), which was derived from slurries of different sources. Both digested sludge that contained a mixed culture of different archaea and monocultures were employed. Completely stirred-tank reactors and solid-bed reactors in which the concentration of microorganisms was increased by specific installations were engaged.

The experiments were carried out exclusively on mesophilic (37 °C) and thermophilic (65 °C) temperatures. The majority of experiments in these publications occurred in thermophilic conditions. Under these conditions a higher metabolic rate is theoretically expected.

The composition of educt gases corresponds in most cases to the stoichiometry of the underlying reaction equation.

The experiments were predominantly carried out at atmospheric pressure. Only a few attempts were executed at higher pressure. In addition, there are existing references which state that a heightened pressure causes an increasing cell concentration and thereby an increased production of methane.

The methane yields are stated in different forms. For this reason, the comparison of the previous results is difficult. Several studies refer the yields to the used working fermenter volume (e.g. [192, 193]) or to the used material for immobilization (e.g. [194, 195]). The yields are set in part as amount of substance and in part as volume. By volumes, there are

Table 4.16 Publications regarding biological methanation between 1978 and 2013

Source	Inoculum	Mode of operation	Immobilization	Temp. (°C)	Volume ratio H_2/CO_2	Reactor (l)	Medium	Volume flow (l/h)	Pressure (bar)	Productivity
Burkhard [189]	Sewage sludge	Batch	Packing Bioflow 40	37	40 %/60 %	27	n.a.	n.a.	1	1.3 l CH_4/h
Lee [190]	Sewage sludge	Continuously	PU sponge	35	40 %/60 %	7,8	n.a.	n.a.	1	1.05 l CH_4/h
Trechow [191]	n.a.	n.a.	n.a.	65	n.a.	10	n.a.	n.a.	1	150 l CH_4/h
Schmuck [192]	Activated sludge	Batch	None	37	40 %/60 %	3	n.a.	n.a.	n. a.	15 % CH_4, 80 % CO_2, 5 % H_2
		Continuously				80				10–20 % CH_4
Ako [193]	Sewage sludge	Continuously	None	37	80 %/20 %	1	0.1/d	n.a.	1	1.2 l CH_4/d
Yang [194]	Pond sludge	Continuously	Luffa	37	80 %/20 %	2.8	HRT 3 days	6	1	6.18 l CH_4/l/h
							HRT 6 days			11.96 l CH_4/l/h
Nishimura [195]	Pure culture	Batch	None	65	80 %/20 % + 500 ppm H_2S	2	1 l	30–120	1	max. CH_4: 12.5 l/l/h
		Continuously						120		~8 l CH_4/l/h
Nishimura [196]	Pure culture	Continuously	None	65	80 %/20 % + 500 ppm H_2S	2	1 l	300	1	~0.5 mol CH_4/l/h
									2	~1 mol CH_4/l/h
									3	1.28 mol CH_4/l/h
Yano [197]	Pure culture	Continuously	Hollow fibre	65	80 %/20 %	0.104	27 ml/h	1134 ml/h	1	70.6 % CH_4
								1588 ml/h		35.5 % CH_4
								2722 ml/h		17.8 % CH_4
Lundback [198]	Pure culture	Batch	None	37	75 %/25 %	0.15	0.075 l	n.a.	1	0.27 mol CH_4/mol H_2
Peillex [199]	Pure culture	Continuously	None	65	80 %/20 %	1.5	n.a.	60 l/l/h	1	12 l CH_4/l/h, 96 % CH_4
Peillex [200]	Pure culture	Batch	None	65	80 %/20 %	1.5	0.375 l	1–150 l/l/h	1	19.6 l CH_4/l/h, 50 % CH_4
		Continuously								2.3 – 3.2 l CH_4/l/h
Jee [201]	Pure culture	Continuously	Granular diatomaceous earth clay	65	80 %/20 %	0.136	18 ml/h	3.2	1	nach 25 h: 58 % CH_4 im Gas (540 ml/h)
Jee [202]	Pure culture	Continuously	Membrane reaktor	65	80 %/20 %	0.06	60 ml/h	0.76	1	0.75 ml CH_4/cm^2 contact area/h
			Ceramic reaktor			0.0755	16 ml/h	2.6		6 l CH_4/l ceramic/h
Ariga [203]	Sewage sludge	Batch	None	37	80 %/20 %	2	1 l	n.a.	1.01	~4 mol/m^3 (after 66 h)
									2.03	
									3.55	
Wise [204]	Sewage sludge	Batch	None	37	80 %/20 % + 0.3 – 1 % H_2S	5	2 l	42	1	25 mmol CH_4/g cell h 1.3–4 l/l h
				60					31	50 mmol CH_4/g cell h

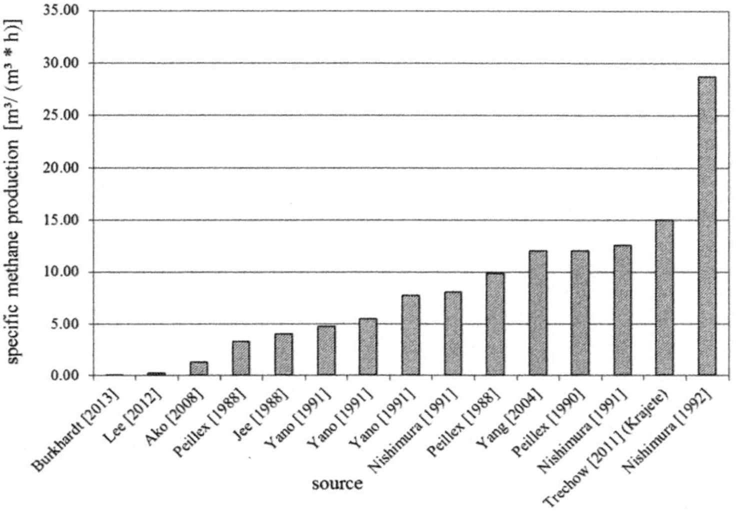

Diagram 4.69 Evaluation of different literature sources for space-time yield of the biological methanation

often missing detailed information about pressure and temperature conditions. In order to obtain a fair comparison of previous studies, the available literature was evaluated for methane productivity and scaled on a single size [CH_4 production in (m^3 CH_4/m^3 reactor volume/h)] as far as practicable (Diagram 4.69). As a result, a relationship between each achieved turnover rates and the measured methane concentration in the product gas can be assumed respectively, omitting the experimental conditions.

The values of determined space-time yields for methane from literature show a significant range from 0.05 to 28.7 m^3/m^3 reactor volume/h. A specific association between high productivity and the experimental conditions was not be deduced by these results.

It is striking to note that there is a high discrepancy in recent investigations. Burkhardt et al. and Lee et al. mark the lower end of the productivity with 0.05 m^3 CH_4/m^3 reactor volume/h or 0.13 m^3/m^3 reactor volume/h, in contrast Nishimura et al. marks the highest productivity with 28.7 m^3 CH_4/m^3 reactor volume/h [189–191].

Regardless of the detailed experimental conditions however, a relationship between each turnover rate achieved and the measured methane concentration in the product gas can be assumed. Even though there is insufficient information presenting exact compositions of the product gases is available, it should be noted that the range of reached methane concentration by continuous rudiments amounts to 18 % and 96 %. With the given data to the turnover rates, it can be derived that high turnover rates by biological methanation lead to a decrease in the methane concentration in the product gas.

Biological methanation shows up as a technology with very high potential on the basis of previous developments. However, there is need for research of ultimately recoverable performance. Furthermore additional processing steps to separate residual components out of the product gas are required.

Biological Methanation as a Central Component of Power-to-Gas Technology

The Power-to-Gas concept was developed [205] out of consideration of the increasing share of electricity production from volatile energy sources (wind, solar) and the associated need for integration of appropriate energy storage technologies in 2009. The concept describes the conversion of surplus electricity produced from renewable energies by electrolysis into hydrogen and subsequently methane through chemically catalyzed reactions. The converted methane is then fed into the existing natural gas infrastructure. Biological methanation is suitable to replace the chemical catalyzed reaction and serve as a central component of the process (Diagram 4.70). Due to the immense potential of the Power-to-Gas concept by storing surplus produced energy from volatile sources and the associated importance for the energy system in future, the current research in the field of biological methanation has a significant focus on integration into the Power-to-Gas concept.

Since 2012, different approaches have been developed to realize the process of biological methanation as a part of the Power-to-Gas concept on a larger technical scale. These approaches were based on research and development projects and were initiated by universities, research institutions and companies.

The plant size of methanation is primarily determined by the hydrogen flow rate of electrolysis. Due to the modular design of the electrolysis units, performances from a few kilowatts to several megawatts can be accomplished. The bioreactor is designed on the basis of the dimensioning of electrolysis unit, and is constructed as schematically shown in Diagram 4.71. The dimensioning of the bioreactor for biological methanation is very flexible and adaptable due to the simple construction. Thus, upscaling the reactor volumes to several 1000 l is possible for an industrial application.

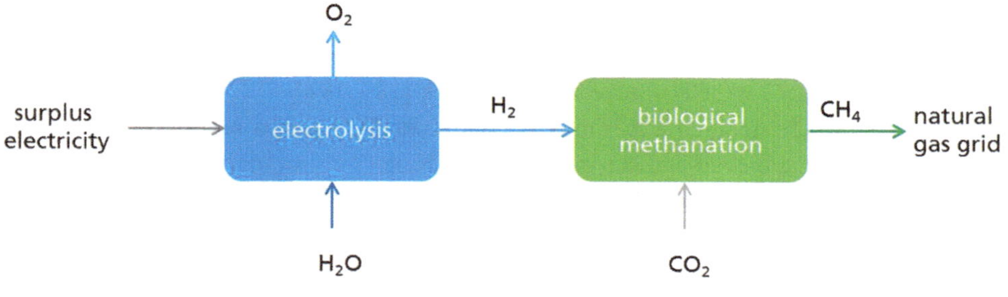

Diagram 4.70 Power-to-Gas concept

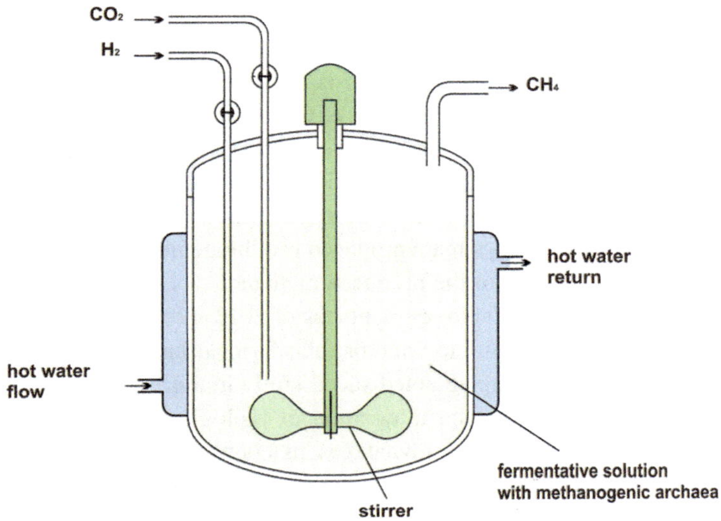

Diagram 4.71 Schematic view of a bioreactor with stirrer for biological methanation according to [206]

There are different ways to provide the required CO_2 volumetric flows which significantly depend on the installation site of the plant. The opportunities are shown below in terms of their advantages or disadvantages.

CO_2 Resourcing by Ambient Air

To provide the required volume of CO_2 stream, the CO_2 can be obtained directly from the atmosphere. Since ambient air only contains about 0.04 vol.% of CO_2, energy-consuming techniques are required to generate sufficiently large volume streams of CO_2 with high purity. Despite the advantage of a location-independent method of capturing CO_2, this technique, due to the high energy consumption, is not effective and thus not suitable for large-scale implementation of the biological methanation.

CO_2 Resourcing by CO_2 Emitting Industry Processes

In the metal industry, fossil energy carriers such as hard coal or lignite are used for (electric) power generation, CO_2 occurs as a major component of exhaust gas streams in large quantities and is emitted unused into the atmosphere. The exhaust gases pass various purification steps, so the volume streams of CO_2 are available almost in a pure form. The utilization of this CO_2 as a source for the methanation process has the advantage of being available in vast-quantities and of high purity. However, the Power-to-Gas concept will only be achievable if the storage of surplus electricity (from renewable energy) and carbon dioxide (from a power plant) is in close proximity to the natural gas grid. The large industrial and coal-fired power plant sites are generally not located in the vicinity of high potential wind and solar energy regions in Germany. Unfortunately, this method presents a locational disadvantage for the objectives of the Power-to-Gas concept.

CO_2 Resourcing by Biogas or Sewage Plants

The integration of the process with biogas or sewage plants provides special advantages. The carbon dioxide required for methanation is already highly enriched by the biogas volume stream: Biogas consists of up to 50 % of CO_2. There is also the possibility to use the entire biogas stream without prior concentration of CO_2 for the biological methanation. The hydrogen sulfide content in the resulting biogas volume stream functions as a source of nutrients for the microorganisms. The purification of the biogas is not necessary, because the CO_2 stream produced is of biogenic origin; this has a positive influence on the CO_2 emissions of the produced methane.

There are approximately 7850 biogas plants (as of 2013) [207] mainly in rural regions in Germany and the spatial vicinity to wind or solar farms is ensured in many cases. The Power-to-Gas concept can be implemented successfully in this combination.

There are various options for the integration of biological methanation in existing biogas plants that appear to create advantages in connection with the Power-to-Gas concept. The advantages of various dedicated techniques are explained in the following chapters.

Biological Methanation in Combination with Biogas Plants

The main advantage of integrated biological methanation is the use of the available CO_2 volume stream. In addition, the required microorganisms for the methanation process are already present in each biogas plant and can be taken directly for enrichment. If the biogas plant contains a unit for biogas upgrading, a connection to the natural gas grid already exists. Hence, additional investment costs can be avoided and the existing infrastructure can be used.

The integration of biological methanation with biogas plants is possible in three ways: in digester (section "Biological Methanation Directly in a Digester"), in a separate reactor at biogas plants with and without biogas upgrading (sections "Biological Methanation in a Separate Reactor" and "Biological Methanation in a Separate Reactor at Biogas Plants with Biogas Upgrading") or in combination with an in-situ electrolysis (section "In-Situ-Electrolysis and Biological Methanation").

Biological Methanation Directly in a Digester

Digesters and post-digesters are the central components of biogas production. The digester is charged directly with organic substrates, such as organic waste or renewable resources, and is the primary location of biogas production due to the basic degradation of organic material. The autotrophic and acetotrophic methane production steps occur here in a parallel and closely coupled way. The post-digester is usually charged with already degraded substrates from the fermenter, which are sometimes fed with a small proportion of fresh substrates.

The biogas of this process stage usually contains approximately 50 % CH_4 and CO_2. An obvious approach for the large-scale implementation of a biological methanation is the addition of H_2 directly into the fermenter or post-digester with the aim of increasing the concentration of methane in the biogas. Therefore, the biological methanation runs

Diagram 4.72 Methanation in biogas digester respectively post-digester

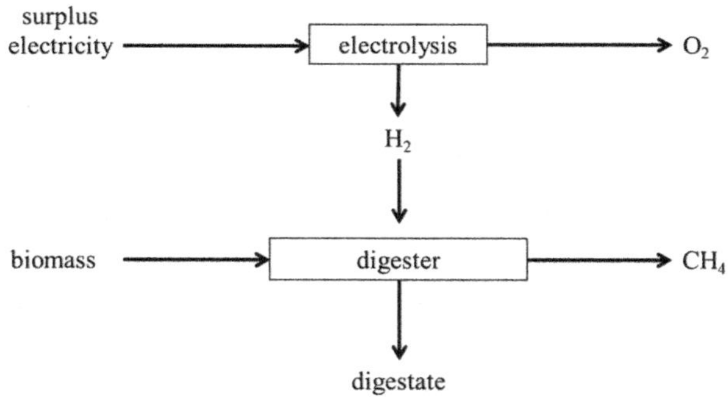

parallel to the microbial biomass degradation processes in the digester. Diagram 4.72 shows the process of this kind of methanation. The advantage of this option is that the methane producing microorganisms are already present in digesters of biogas plants and large quantities of carbon dioxide arise as a byproduct of fermentation. Furthermore, it can be assumed that the necessary nutrients for the autotrophic archaea system are also present in sufficient quantities.

Ideally, biological methanation should operate at maximum feasible hydrogen addition in order to produce maximum achievable product gas quality.

However, a relatively high hydrogen partial pressure in the digestors inhibits the degradation of biomass. Thus the addition of hydrogen is limited. CH_4 concentrations in the range of 98 % are required to feed the generated methane into the gas system and to use it as a fuel. Whether this value can be achieved within this concept without further purification technology has not yet been proven technically.

Biological Methanation in a Separate Reactor

The microbial conversion processes of hydrogen and carbon dioxide to methane can be realized relatively straightforward in a bioreactor, as shown in Diagram 4.71.

The cultivation of methanogenic microorganisms and the transformation processes take place in the same reactor, which are designed to be very variable in their dimensions. The introduction of feed gases and their solution into the digestor sludge is done via a special gassing appliance in combination with an agitator. This process is essential to yield the desired amount of methane.

For this system approach, as shown in Diagram 4.73, CO_2 and H_2 are fed to an independent biological methanation reactor.

The biology in this reactor can adapt specifically to the autotrophic environmental conditions, since no competing reaction of organic substrates occurs.

In addition, the influence of acetotrophic methane production in this system is excluded by a high partial pressure of hydrogen.

In principle, installing a CO_2 source is not necessary with this type of integration. The concentration of CO_2 can be compensated through a direct addition of raw biogas—with a composition of 50 % CH_4 and 50 % CO_2.

Diagram 4.73 Methanation in a separate reactor

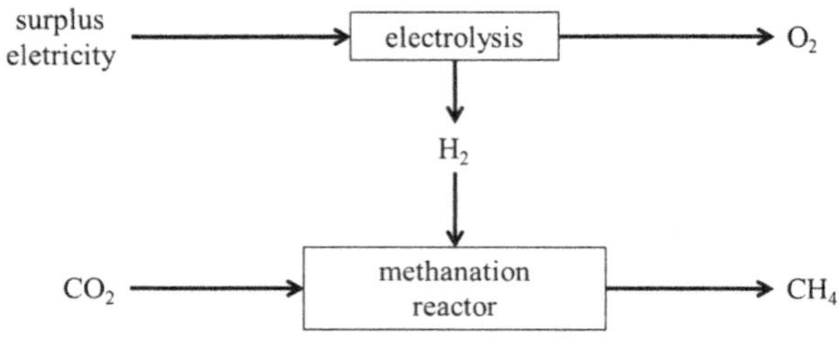

Due to the specialized reactor concept, significantly higher conversion rates and a higher CH_4 concentration are expected in contrast to the direct addition of hydrogen in the biogas fermenter. Implementations of this method are the subject of current research. Investigations by Krajete [191] that achieved the highest efficiency of all previously published investigations with such a system, indicate that there is a high potential for this concept.

Since this is a closed reactor system, fed with reactants as pure substance streams, further studies on the nutrient requirements of the microorganisms in the system are required. The maximum achievable concentration of methane coupled with simultaneous high turnover rates in this system has not been completely clarified.

Biological Methanation in a Separate Reactor at Biogas Plants with Biogas Upgrading

This particular system proposes an incorporation of a separate methanation reactor to biogas plants with downstream biogas upgrading. In Germany, there are approximately 151 biogas upgrading plants with an entry capacity of 93,650 Nm^3/h [208]. These are plants in which the CO_2 from the biogas plants is separated and biomethane with a methane concentration of up to 98 % is generated. The concentrated CO_2 stream can be used for biological methanation. In this concept, CO_2 and H_2 can be converted in an ideal ratio in the methanation reactor.

Moreover, different optimization paths can be implemented in this concept. In one version, the biological methanation simultaneously fulfils the function of a methane concentration unit for the raw biogas. A separate enrichment step can be omitted and the system can be dimensioned smaller overall. Alternatively, an existing biogas upgrading step can be used for the separation and recycling of unreacted H_2 and CO_2 from the biological methanation.

In-Situ-Electrolysis and Biological Methanation

A combination of the in-situ electrolysis and biological methanation in a system is also an emerging subject of current research. Most systems discussed previously use hydrogen in a gas phase, which is energetically less efficient to transform it into aqueous media, where the process of biological methanation occurs. As represented in Diagram 4.74, the in-situ electrolysis and methanation has the aim of transporting hydrogen production directly to the site of methane production.

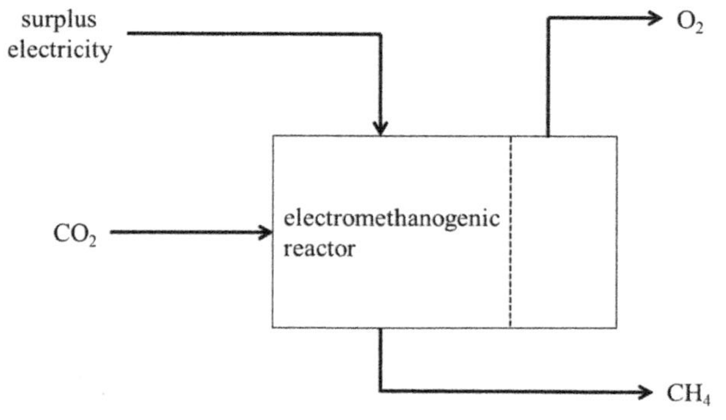

Diagram 4.74 In-situ-methanation

This system allows hydrogen ions to be directly utilized by autotrophic archaea in the electrically conducting media, in which the hydrogen is directly available for the microorganisms by its emergence [209, 210]. At the same time, CO_2 can be added in gaseous form into the reactor due to its relatively high solubility in water. It is expected that biological methanation reactors will work very efficiently using this method in the future.

However, there are still other factors subject to further investigations. For example, the ways of immobilization of microorganisms in conductive media or the nature of electron and mass transfer between microorganisms and CO_2.

Production Costs for Biological Methanation

The following economic approach of biological methanation on a large scale is based on theoretical assumptions, due to insufficient practical and application data. A reliable cost calculation is difficult; inaccuracies cannot be excluded. The following achievements aim at a broad overview of the essential costs and to permit an initial classification. The cost calculation was conducted as an example for a biological methanation in a separate reactor in combination with an existing biogas plant with a biogas upgrading unit. This system was chosen for the following reasons:

1. Economically feasible
   - Lower installation costs because connections to the gas grid of the biogas upgrading unit are readily available.
   - CO_2 source can be obtained directly from the upgrading unit of the existing biogas plant.
   - Locational advantage: renewable energy and biomass plants are located in close vicinity
2. Energetically efficient—reduce the energy waste in converting hydrogen from gaseous phase and back into aqueous media. Diagram 4.75 shows the process flow of the proposed scenario.

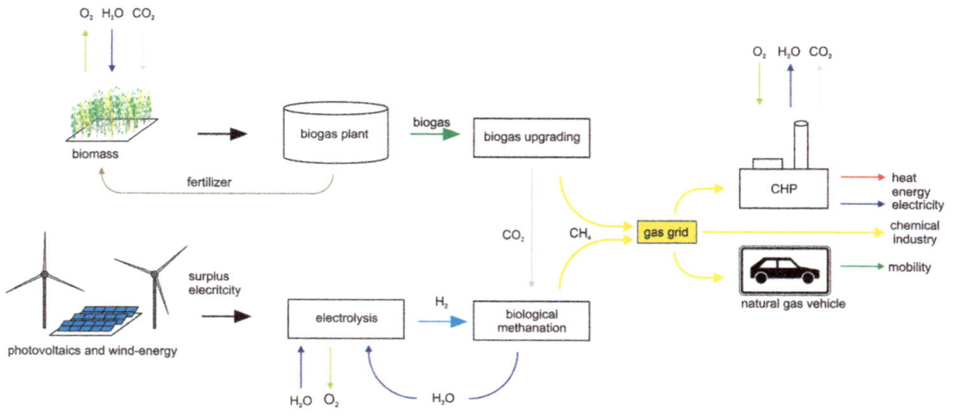

Diagram 4.75 Biological methanation in the Power-to-Gas concept

Hydrogen Production by Electrical Power

The use of electrolysis for hydrogen production is an established industrial process. The corresponding systems engineering has been achieved for small production volumes (<100 Nm^3/h) as well as for very large production volumes (>20,000 Nm^3/h).

Biological Methanation

Despite the promising results of specific product performance obtained in laboratory studies, there is still insufficient data of large-scale implementations of biological methanation. In the experiments with separate methanation, the average specific product performance that was achieved during experiments with a separate methanation reactor is about 5 m^3 methane/m^3 Reactor volume/h (Diagram 4.69). Using this value for the dimensioning of a large-scaled system, a production capacity of 500 Nm^3/h of biomethane and a fermentation volume by approximately 100 m^3 could be achievable. This calculation, based on investigations on a very small scale, may differ and cannot be considered reliable. Further, the large range of reactor systems used in different research projects shows that more studies are required.

Hydrogen and CO_2 Separation

The results of recent studies indicate that a high level of product performance cannot be simultaneously achieved with high product purity. Therefore, unreacted hydrogen and CO_2 from the product gas should be separated and repatriated to the process. The hydrogen and CO_2 separation provides a step based on already established technologies.

Assumptions for the Calculation of Production Costs

The production costs for biological methanation can be estimated for a simplified system under the following conditions:

- Electrolysis Power: 13 MW_{el}
- No consideration of gas storage for the reactant gases H_2 and CO_2
- No consideration of the dynamic behavior, such as potential performance degradation, caused by start/stop operation
- To calculate the costs of hydrogen generation, an efficiency of 70 % (energy-related) and a specific investment of 1000 €/kW_{el} are assumed. Except electricity, no operating costs for electrolysis are estimated.
- The investment costs are included with an annuity of 10 % in the annual costs.
- The educt gases are fed into the reactor with a stoichiometric ratio of 4:1 (hydrogen/ methane).
- The selectivity of the process to methane amounts to 95 %, the residual ratio of the reactants hydrogen and carbon dioxide is needed for biomass generation.
- 3 Nm^3 methane/m^3 reactor volume/h are assumed for the methane productivity.

  This value is below the average methane productivity in Diagram 4.69, which shows experimental results performed under laboratory conditions. Thus, realistic expected methane productivity should be reproduced in a bioreactor on industrial scale.
- The methane conversion in the bioreactor is assumed to be 84 % [202].
- Two reactor stages must be connected in series to obtain the desired residual hydrogen concentration below 10 vol.%. The total turnover is about 97.4 %. This results to the following gas composition at the reactor outlet:
  Methane: 87.9 vol.%
  Hydrogen: 9.7 vol.%
  Carbon dioxide: 2.4 vol.%
- Investment biological methanation: 950,000 €
- Operating costs biological methanation: 0.4 ct/kWh product gas

Determination of Production Costs

The following technical and economic data for the biological methanation can be deduced by previous assumptions:

- H_2 flow rate: 2565 Nm^3/h
- CO_2 flow rate: 641 Nm^3/h
- Product gas volume flow: 676 m^3/h (88 % CH_4, 10 % H_2, 2 % CO_2)
- Product gas power: 6.8 MW (energy-related)
- Size of reactor for biological methanation: 2 × 198 m^3
- Cost of capital: 95,000 €/a

Based on these numbers the energy-related production costs for the product gas at different electricity prices and full-load hours are calculated. Diagram 4.76 shows the results in graphical form. This clearly shows the influence of the full-load hours on the total costs. This is mainly caused by the high investment for hydrogen production in the electrolysis stage. Electricity costs also significantly affect the total production cost of the product gas.

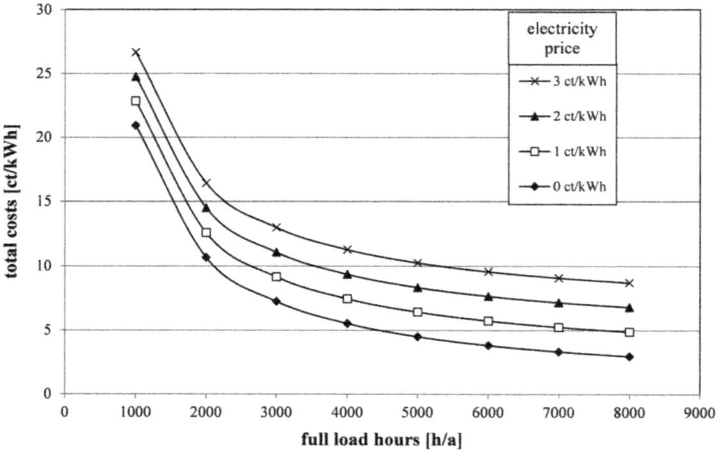

Diagram 4.76 Production costs for the product gas from biological methanation (energy-related) for a system with electrolysis power of 13 Mwel

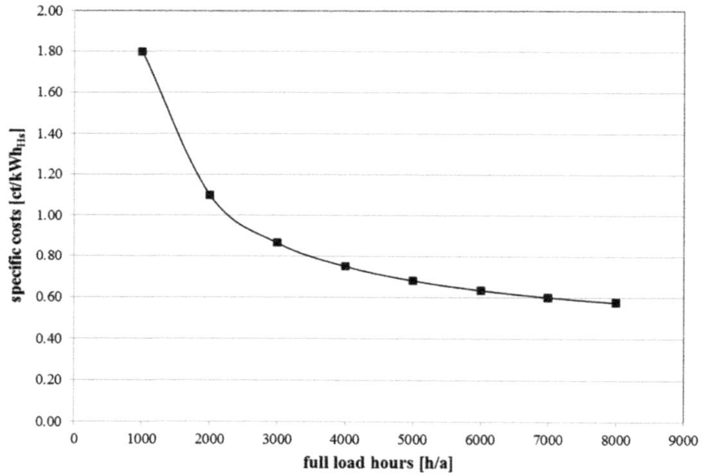

Diagram 4.77 Costs for biological methanation without hydrogen generation

If the overall production costs are divided into costs for hydrogen production and costs for biological methanation, it is clear that hydrogen generation accounts for 81–93 % of the total costs. The straight production costs for biological methanation as a function of the full-load hours are shown graphically in Diagram 4.77. The costs for electricity have no influence in this case, since they only affect the costs for hydrogen production. The total costs are mainly caused by the hydrogen production.

Summarizing the Consideration of Biological Methanation

The Power-to-Gas concept with the methanation and the resulting option to feed large quantities of methane into the natural gas grid provides an enormous potential for storing surplus energy, surpassing the possibilities of feeding hydrogen. The uses for methane are diverse and the existing natural gas network ensures a wider network of distribution. In addition to the heat generation and the re-conversion into electricity, methane is also commonly used as a raw material in the chemical industry. Moreover, the use of natural gas vehicles attracts more attention through scientific and economic studies and this will also play an important role in the future.

In the effort to increase environmental awareness in the mobility sector and the use of more fuel-efficient and environmentally technologies, biological methanation provides a regenerative fuel by the use of carbon dioxide from biogenic and hydrogen from renewable energy sources, with generated methane. This fuel is suitable for application in passenger and freight traffic due to its universal applicability.

First demonstration plants for the Power-to-Gas concept with biological methanation are in operation or in the planning stage [211]. To develop the existing potential, further research is needed. For example, the dynamic behavior of processes must be examined as well as the determination of robust data for integration into the overall energy system for continuous operation on an industrial scale. The foundation has been laid for the expected huge benefits that biological methanation provides in the Power-to-Gas concept for the future energy system.

4.5 Mixtures of Methane and Hydrogen

Helmut Eichlseder

Gaseous fuels, mostly being a mixture of gases containing methane and hydrogen, played a decisive role in the dawn of the internal combustion engine (then mainly in stationary application). Well known inventors of that time were Barsanti and Matteucci (1854) and especially Lenoir, who succeeded in the practical application of gas engines in 1860. Nikolaus August Otto's ambition was to improve the Lenoir engine. In 1876 he was able to provide evidence with coal gas as fuel (a gaseous mixture of 50 vol.% hydrogen and 25 vol.% methane) [212] and was granted his famous patent. The engine was subsequently modified for the operation with liquid fuel as well. For the mainly stationary applications of those times, gas remained the main fuel for the following years. Beside coal gas, supplied by the urban gas distribution stations and also called town gas, furnace gases (a side product of iron manufacturing in the blast furnace) and coal burning gases from charcoal piles were used.

The mobile application of gas engines is considerably older than that of Otto and Diesel engines. Therefore, so-called "storage gases" (carried in storage tanks on the vehicle with greater or lesser pressure) like natural gas, propane-butane mixtures and coal gas as well

Diagram 4.78 Viennese bus with gas tank trailer (operation with town gas, 1943) [213]

as "generator gas", produced by the gasification of solid fuels (charcoal, peat, brown coal or anthracite coal and especially wood), were used.

In special cases, gas was carried along unpressurized in voluminous rubber bladders in a roof extension or in a trailer (Diagram 4.78). A specialty for the use in airship engines was "Blau gas" (named after its inventor), a mixture of methane (37 vol.%), hydrogen (6 vol.%), ethylene, propylene, butylene and ethane, which corresponded to the density of air and thus did not result in the need to dump the (expensive) lifting gas, as the consumption of fuel during flight did not affect the lift forces.

For stationary engines, in addition to natural gas, various methane containing special gases—also using the gasification of solid fuels [214]—are investigated and applied nowadays, whereby the admixture of hydrogen [215] for the improvement of combustion relevant properties is being investigated and used in special cases.

Popular literature often describes the miraculous functional improvement of gasoline and diesel engines by adding small portions of on-board generated hydrogen. This

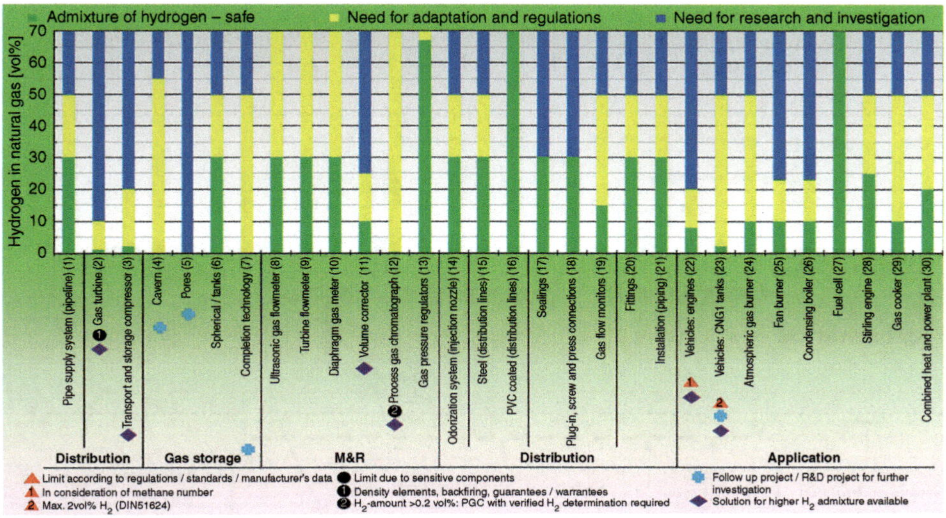

Diagram 4.79 Hydrogen tolerance of core elements of the gas supply system—hydrogen in natural gas [vol%] [216]

unrealistic potential does not correspond at all to the physical possibilities, but meets the wishful thinking of user and distributors of such on-board gas generators.

Today, other forms of mixed gas are more interesting for mobile application of a wider range, especially the admixture of hydrogen to natural gas or to biogas. In addition to the already described advantages of effective natural gas and methane production in Sect. 4.2, the direct admixture of hydrogen to natural gas creates further important benefits.

On the one hand, hydrogen from regenerative production can be stored directly, without any further processing, in the existing natural gas supply system. Together with storage and distribution possibilities, this poses a key technology for the future energy supply. Associated boundary conditions and opportunities for Germany can be described as follows: with a pipe system of 500,000 km it represents a huge storage system in itself and a nationwide network transporting twice the amount of energy of the electric power supply system. In addition, 20 % of the annual gas consumption can be stored in underground gas tanks [216] so that—with only a small concentration of hydrogen—huge energy quantities can be stored and transported. The question relating to the maximum permissible hydrogen concentration in the existing infrastructure (supply network, components, safety …) is being examined in several ongoing projects. The state of investigations can be seen in Diagram 4.79.

The depicted results of several current studies show that—except for some elements that are primarily integrated in the gas transportation system—the gas infrastructure shows a hydrogen tolerance of approximately 10 vol.%. Projects have been initiated in order to improve the scientific knowledge of some of the identified critical elements. In Germany, a mid-term tolerance of 10 vol.% can be assumed [216].

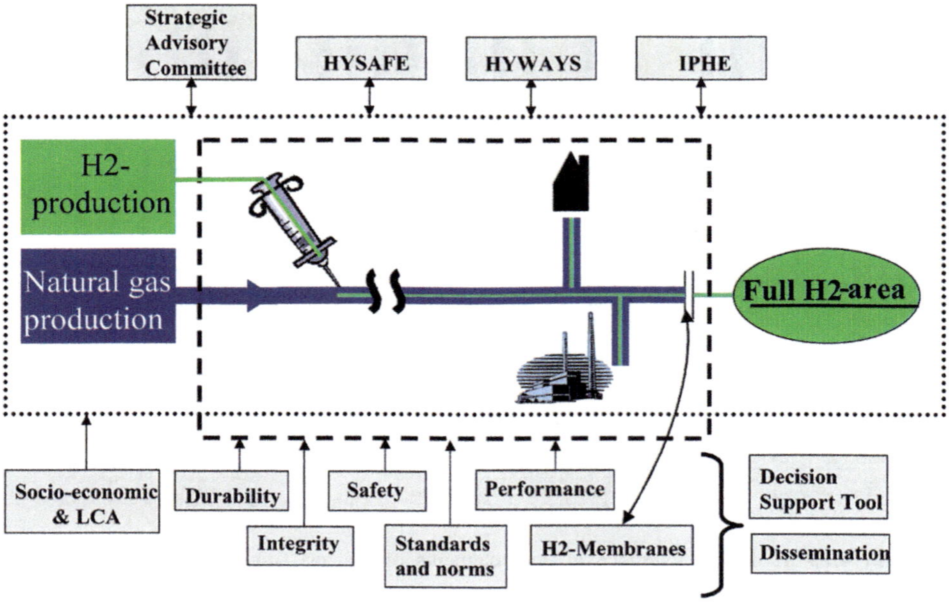

Diagram 4.80 Approaches for the use of the existing natural gas supply system [217]

Diagram 4.80 [217] shows an interesting view on approaches and possibilities for the use of the existing natural gas system for hydrogen.

Such use of regenerative hydrogen and the application in vehicle IC engines provides an appropriate bridging scenario, as widespread operations of vehicles with hydrogen-fuel cell drives or of battery-electric vehicles with the required infrastructure are not foreseeable in the next few years. This is especially true for emerging countries like China and India, which will likely show the highest growth in traffic-dependent energy consumption.

With the application of existing drive-train technologies with internal combustion engines, existing facilities can be used for vehicle production, maintenance and operation; and existing regenerative fuel can be gently introduced enabling a smooth transition. Thus CO_2 advantages of renewable fuels can be used according to the available amounts.

On the other hand, the admixture of hydrogen substantially improves some ICE relevant characteristics of natural gas and mixed gases containing methane. Concerning biogases and low energy gases, a substantial improvement of combustion characteristics is obtained enabling the efficient and functional proper operation of internal combustion engines with such fuels.

Due to the above mentioned reasons, several international projects are being carried out at the moment investigating and demonstrating the use of hydrogen-methane mixtures in different ratios in internal combustion engines for stationary and automotive application [218–225]. Table 4.17 gives an overview of the characteristics of natural gas—hydrogen mixtures for engine applications.

Table 4.17 Properties of natural gas-hydrogen-mixtures [223, 226, 231–233]

		Hydrogen	25 vol-% H_2 in CH_4	10 vol-% H_2 in CH_4	Methane	Natural gas[a] e-on [230, 231]
Wobbeindex[b] Wo (upper)	MJ/m^3	45.10	46.85	48.68	49.89	41.7
Wu (lower)	MJ/m^3	38.11	41.87	43.67	44.90	–
Density, liquid[c]	kg/m^3	71.0	–	–	422.8	–
At (boiling point)	K	20.2	–	–	111.4	78 ÷ 118
Density, gaseous[b]	kg/m^3	0.0838	0.5221	0.6097	0.6682	0.84 (0.7 ÷ 1.0)
Stoichiometric air demand	–	34.3	17.89	17.43	17.2	16
Calorific value (upper)	MJ/kg	142.0	59.09	56.81	55.63	51.9
Calorific value (lower)	MJ/kg	120.0	52.81	50.96	50.0	47.1
Energy density liquid[b]/ gaseous (300 bar)[b]	MJ/dm^3	8.52	–	–	21.1	–
	MJ/dm^3	2.50	8.10	9.72	10.90	–
Mixture calorific value[b]						
Mixture aspirated	MJ/m^3	2.97	3.15	3.16	3.17	3.15
Air aspirated	MJ/m^3	4.21	3.56	3.52	3.50	3.81
Ignition limits (theoretical)	Vol-%	4 ÷ 76 [223]	–	–	4.4 ÷ 15 [223]	4 ÷ 17
	λ	0.13 ÷ 10 [223]	–	–	0.6 ÷ 2 [223]	0.5 ÷ 2.3
Engine, upper[e]	λ	5.90	1.97	1.80	1.73	–
Auto-ignition temperature	K	858 [223]	–	–	868 [224]	848 ÷ 913
Minimum ignition energy	mJ	0.017 [223]	–	–	0.29 [223]	0.25
Diffusion coefficient	cm^2/s	55.1	–	–	18.3	–
Flame propagation speed of 25 %-Front v 0.25[b]	cm/s	5.0	–	–	1.7	–

(continued)

Table 4.17 (continued)

		Hydrogen	25 vol-% H_2 in CH_4	10 vol-% H_2 in CH_4	Methane	Natural gas[a] e-on [230, 231]
Laminar flame speed[b, f]	cm/s	233.2	45.6	40.8	38.3	–
Adiabatic combustion temperature[b, f]	K	2383.6	2233.6	2225.3	2220.8	2233
Quenching gap[c]	mm	0.64	≈1.82	≈2.01	2.13	–
Methane number		0	75 [222]	90 [222]	100	
Proportions by mass						–
c		0	0.726	0.739	0.749	–
h		1	0.274	0.261	0.251	–
Spec. CO_2-emission	g/kWh	0	180	191	198	200

[a]At: 273.15 K, 1013.25 mbar
[b]At standard conditions 298.15 K, 1013.25 mbar
[c]At 1013.25 bar
[d]At 298.15 K
[e]Single cylinder engine, n = 2000/min, WOT, SOI 320°CA, IP optimized
[f]CHEMKIN Simulation

The term Hythane® has been established for hydrogen-methane mixtures with a hydrogen share of max. 21 vol.% (a hydrogen share of 21 vol.% corresponds to an energy-related share of 7 % and a mass-related one of 3 %; according to other sources, the term Hythane® also includes mixtures with 8–32 vol.% hydrogen). A hydrogen share of up to 50 vol.% is called HCNG. In the following, H2NG is generally used for all mixtures including the volumetric content of hydrogen in percent: H2NG10, for example, specifies the mixture of 10 vol.% hydrogen and 90 vol.% methane.

Basically, both fuels are homogeneously mixable in each mixture ratio and can be stored together in a pressure tank. Materials apt for hydrogen can also be used for methane and their mixtures. Diagram 4.81 displays a comparison of energy densities of hydrogen and methane in gaseous and liquid condition. For further details please refer to [223].

Segregation of mixtures due to gravitational effects can be excluded due to the high diffusion tendency of hydrogen—the diffusion coefficient of hydrogen in air is three times higher than that of methane. Only at low temperatures is it possible that the thermal molecular motion is not sufficient to counteract the segregation due to the density

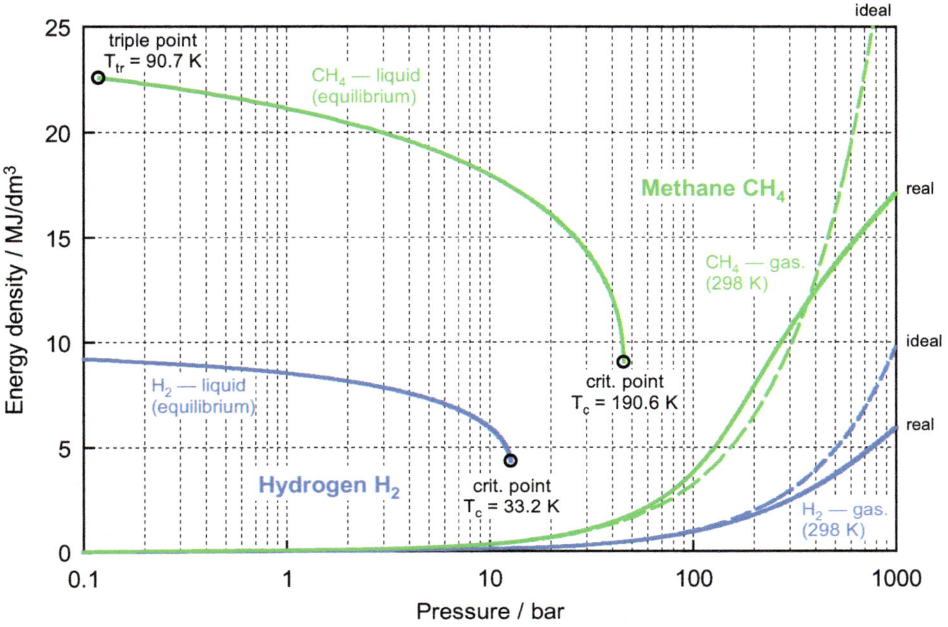

Diagram 4.81 Energy density of hydrogen and methane in compressed and liquid condition [223]

difference. A temperature decrease down to the boiling temperature of a gas component leads to its separation into a liquid phase and results in segregation. These temperature differences are, however, not reached in pressure tank systems [223].

For fuel injection in the engine, for gas burners in general and for safety reasons, the important question arises of how fast the combustion gas flows through the port cross-section and how much energy is transported therewith (Table 4.18).

A characteristic value for the energy flow rate at a critical state of flow, equal gas pressure and equal port cross-section is the (upper) Wobbe index Wo. Table 4.19 contains the Wobbe indices of methane, (Austrian) natural gas in different H2NG-mixtures and hydrogen.

In order to evaluate the suitability and properties of hydrogen-natural gas mixtures when applied in internal combustion engines, combustion specific parameters like ignition limits, ignition energy, auto-ignition temperature, flame speed etc. have to be observed. Even though the influence of different hydrogen ratios on efficiency and emission behavior is well documented in numerous publications, the availability of characteristic basic parameters is limited.

The properties of hydrogen-methane mixtures have been fundamentally investigated [226, 227, 228] at the Institute of Internal Combustion Engines and Thermodynamics at Graz University of Technology, performing both simulation and experimental

Table 4.18 Partial pressures, masses and energy contents in a tank at 350 bar and 25 °C [223]

	CH_4	H_2NG10	H_2NG15	H_2NG30	H_2NG50	H_2NG80	H_2
Vol% H_2	0	10	15	30	50	80	100
p_{H_2} (bar)	0	35	52.5	105	175	280	350
p_{CH_4} (bar)	350	315	297.5	245	175	70	0
$m_{ideal\ H_2}$ (kg)	0.00	0.28	0.43	0.85	1.42	2.28	2.85
$m_{ideal\ CH_4}$ (kg)	22.65	20.38	19.25	15.85	11.32	4.53	0.00
$H_{u\ ideal}$ (MJ/kg)	50.0	51.0	51.5	53.6	57.8	73.4	120.0
$H_{u\ ideal}$ (MJ/0.1 m^3)	1132	1053	1013	895	737	500	341
$m_{real\ H_2}$ (kg)	0.00	0.22	0.33	0.66	1.11	1.75	2.19
$m_{real\ CH_4}$ (kg)	20.94	18.85	17.80	14.66	10.47	4.19	0.00
$H_{u\ real}$ (MJ/kg)	50.0	50.8	51.3	53.0	56.6	70.7	120
$H_{u\ real}$ (MJ/0.1 m^3)	1047	969	929	812	655	420	263

H_u = Net Calorific Value. Net CV

Table 4.19 Volumetric calorific value and Wobbe index [223]

	CH_4	Natural gas in Austria	H_2NG15	H_2NG30	H_2NG50	H_2NG80	H_2
$H_{o\ vol}$ (MJ/Nm^3)	39.91	39.86	35.38	31.76	26.33	18.18	12.75
W_o (MJ/Nm^3)	54.00	53.01	52.01	50.02	47.48	44.87	48.66

$T = 0$ °C, $p = 1.013$ bar, Density air: $\rho_{Air} = 1.2929$ kg/m^3

H_o = Gross Calorific Value. Gross CV

investigations on a single-cylinder research engine as well as on a transparent engine. In the following, the results obtained are summarized and the impact of hydrogen admixture on mixture formation and internal combustion are described separately.

Mixture Formation Process

Experimental investigations on external mixture formation carried out on a transparent engine with optical access to the combustion chamber and to the intake manifold show negligible influence of the hydrogen concentration (Diagram 4.82). Diagram 4.83 shows the lambda distribution near the injection nozzle over crank angle during the first phase of mixture formation, in each case at the same start of injection [226, 227].

- In the first phase of mixture formation the differences are quite small. In the further course of the mixture formation process, the difference even between pure methane and hydrogen remains remarkably low, which could not be expected given the considerable difference in density (and diffusion coefficient—rather irrelevant in this case).

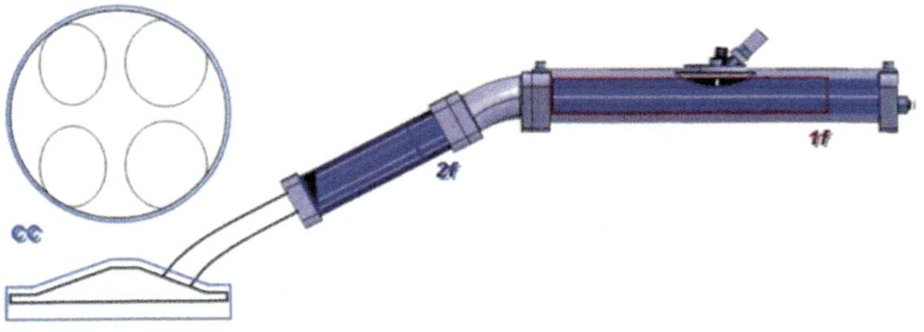

Diagram 4.82 Measuring points [226, 227]

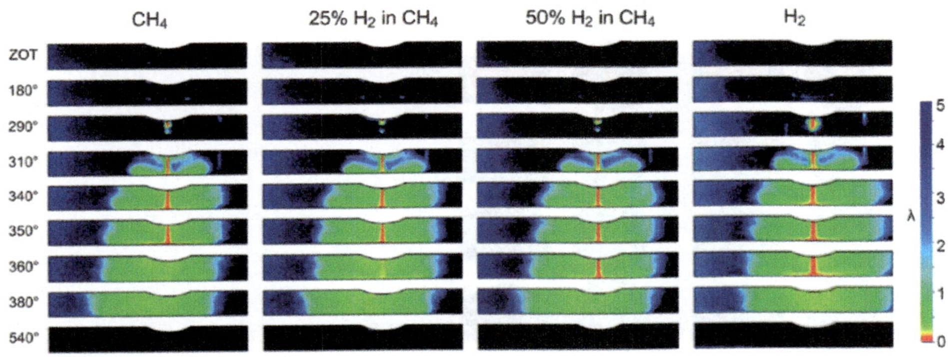

Diagram 4.83 Mixture formation in the intake manifold at position 1f, n = 1000/min, $\lambda = 1.3$, start of injection = 280 °CA ("early"), fuel pressure = 6 bar, mean effective pressures 5.4/5.4/5.3/4.2 bar (f.l.t.r.)

Diagram 4.84 depicts the charge composition in the combustion chamber during the intake and compression process for mixed gases with a hydrogen share of 10 % and 50 %. Initially, the nearly identical charge inhomogeneity independent of gas formation is evident; this is surprising with respect to the gaseous fuel and external mixture formation with a long mixing distance.

- As the hydrogen share of mixed gases in the natural gas supply system is clearly below 50 % and as the Wobbe index and thus the injection duration show similar values, it can be stated that within the observed range for mixed gas the impact on mixture formation is quite low.

Combustion

The specific values of pure methane and hydrogen differ significantly in one fact that causes an influence on the ignition and combustion process to be expected: The auto-ignition temperature with 595 °C and 585 °C respectively are quite high for both gases, but the minimum ignition energy of hydrogen with 0.017 mJ is more than a magnitude smaller than of methane (0.29 mJ). The high auto-ignition temperature strongly indicates

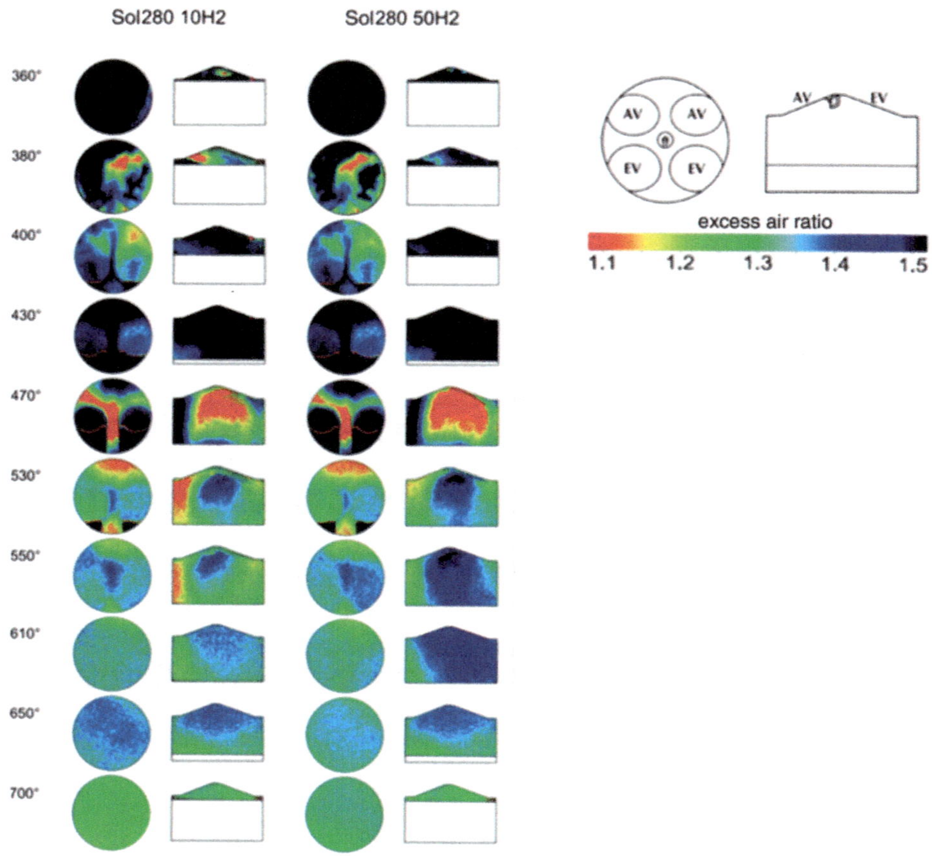

Diagram 4.84 Mixture formation in the combustion chamber [226]

that even for fuels containing hydrogen, only processes with spark ignition or pilot injection are suitable.

- Reaction-kinetic considerations of determining combustion relevant parameters with the software package CHEMKIN by Sandia National Laboratories showed a strong decrease in the ignition delay time during the auto-ignition process; an admixture of 4 vol.% hydrogen reduced the ignition delay time by nearly two-thirds.
- The laminar flame speed does not show a linear behavior with increasing hydrogen share. With a small hydrogen share, the increase of laminar flame speed is practically not visible, only at a share of more than 50 % it strongly increases, see Diagram 4.85.

The analyses of cylinder pressure characteristics and rate of heat release carried out on a single-cylinder 500 cm^3 research engine confirm, as expected from reaction-kinetic

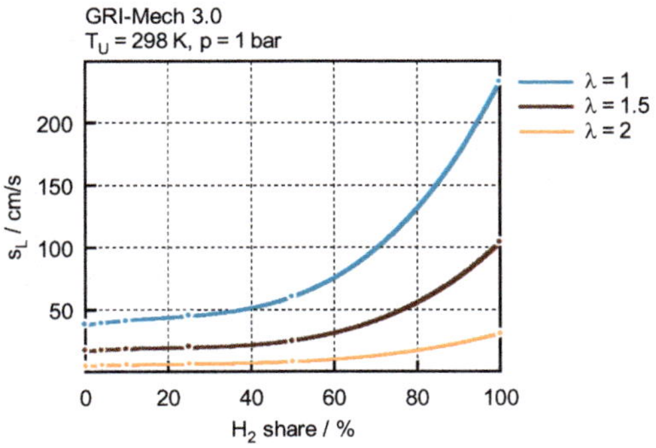

Diagram 4.85 Laminar flame speed as function of hydrogen share [226]

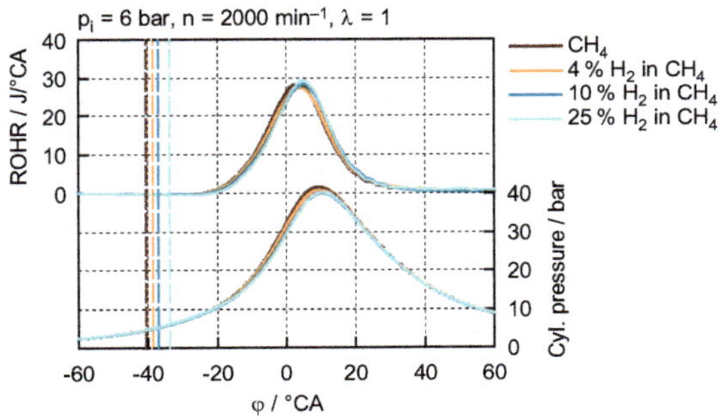

	IP °CA	ID [1] °CA	DOC [2] °CA	MFB50% °CA
25 % H_2	−33.7	12.9	26.3	5.7
10 % H_2	−37.1	14.1	25.3	5.3
4 % H_2	−38.9	15.1	28.9	4.5
Methane	−40.6	16.6	26.7	4.0

[1] Delta ignition signal/zero crossing heat release rate

[2] Delta MFB90%/MFB10%

Diagram 4.86 Pressure and rate of heat release: fuel variation with related values for ignition timing, ignition delay, combustion duration and 50 % mass fraction burned point [226]

simulations, the impacts of changed ignition delay and flame speed, see Diagram 4.86. Based on different ignition timing, relevant for the nearly constant position of the 50 % mass fraction burned point, the influence of the accelerated ignition process is evident. Combustion duration, however, varies moderately as expected from the laminar flame speed differences in the investigated mixture range.

Diagram 4.87 shows the (surprisingly good) correlation between the simulated laminar flame speed and the turbulent flame speed as well as the combustion duration measured in the transparent engine. An essential property for practical engine operation, which is clearly distinct in hydrogen, is the tendency to backfire and abnormal combustion. This reflects the behavior of flame speed, as these anomalies are inconspicuous up to a hydrogen share of 50 %.

Investigations also carried out on a single-cylinder 500 cm^3 research engine concerning the possible increase of lean operation limit showed that this can only be achieved with a hydrogen share of more than 20 %.

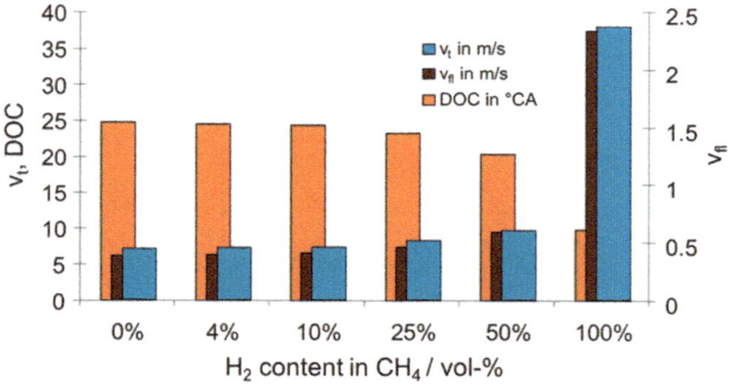

Diagram 4.87 Laminar flame speed vfl (simulation), turbulent flame speed vt (measurement on the transparent engine at n = 1000/min) and combustion duration (measurement on the thermodynamic research engine at n = 2000/min), $\lambda \approx 1$, WOT [228, 229]

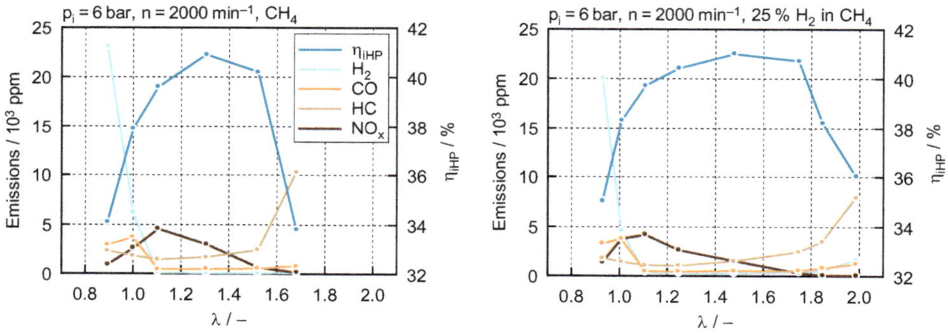

Diagram 4.88 Operation with CH_4 (*left*) and 25 % H_2 in CH_4: emissions and high pressure efficiency, $p_i = 6$ bar, n = 2000/min [228]

A very positive effect of increasing hydrogen share is that the amount of unburned fuel in the exhaust gas is considerably smaller at $\lambda > 1$ and that the loss of efficiency is also smaller with increasing air-fuel ratio, see [228] and Diagram 4.88 respectively.

The influence of the fuel composition on the single losses can be assessed by means of a thermodynamic analysis of losses and is presented in Diagram 4.89 [226]. It can be seen that at the boundary conditions of $\lambda = 1.3$, 2000/min and $p_i = 6$ bar the indicated thermal efficiency slightly decreases with increasing hydrogen share, resulting from higher wall heat losses and from a slightly lower efficiency of the ideal engine, which cannot be completely compensated by faster combustion. Only with the use of wider ignition limits (compare right column, $\lambda = 1.5$ instead of 1.3) a noteworthy improvement of efficiency can be achieved.

For further details and comparing optical investigations please refer to the doctoral theses [226, 227] as well as publications [228, 229].

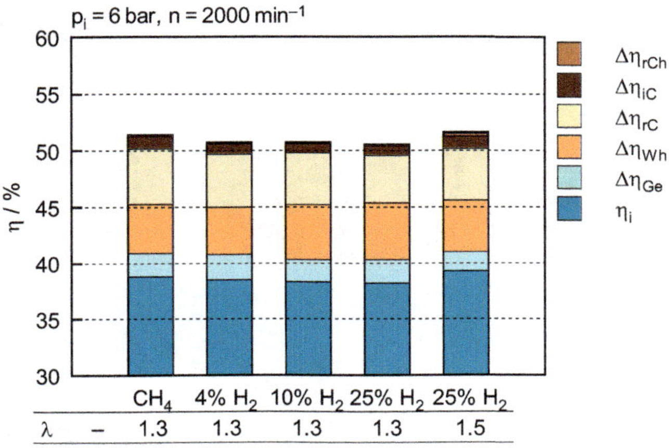

Diagram 4.89 Thermodynamic analysis of losses [226]

It is not surprising that investigations on a natural gas passenger car with 25 % hydrogen admixture, as described in [225], reach comparable conclusions. It deserves mentioning here that studies of a commercial vehicle diesel engine [219] with internal gas mixture formation and diesel pilot injection arrive at the same results; mixed gases with a small hydrogen amount of 10 % show minor impacts on combustion but with a share of 20 % upwards clear changes like reduced combustion durations, significantly reduced HC and CO (plus PM) emissions as well as slightly higher NOx-emissions and considerably increased combustion stability can be observed.

For confirmation of identified functional properties of external mixture formation of a passenger car spark ignition engine, as well as for the verification of the practicability and acquisition of operational experience, the Institute for Internal Combustion Engines and Thermodynamics at Graz University of Technology together with the HyCentA Research GmbH built up a prototype vehicle which was approved for road traffic [223, 224, 226, 230].

The baseline vehicle Mercedes Benz E200 NGT is serially designed for a bivalent operation with natural gas or gasoline. The adaptation enabled an operation with natural gas, hydrogen and a mixture of these two with variable composition without affecting the possible gasoline operation. This concept demonstrates the possibility of a gradual introduction and use of hydrogen, beginning with conventional fuels followed by natural gas up to mixed gas operation in compliance with available hydrogen supply (Diagram 4.90).

Therefore, gas specific components of the fuel supply were replaced by hydrogen suitable ones, focusing on hydrogen resistant materials as well as the high pressure level and increased safety requirements. Additionally, the adaptation of the engine was essential, especially with regard to gas injectors, intake manifolds and engine management system and their application. The most important changes on the vehicle affect the exchange of the original pressure gas cylinder by pressure tanks for 350 bar hydrogen as well as a novel electronic gas safety system (Diagram 4.91).

Diagram 4.90 Prototype vehicle for hydrogen/natural gas/mixed gas and gasoline operation [223]

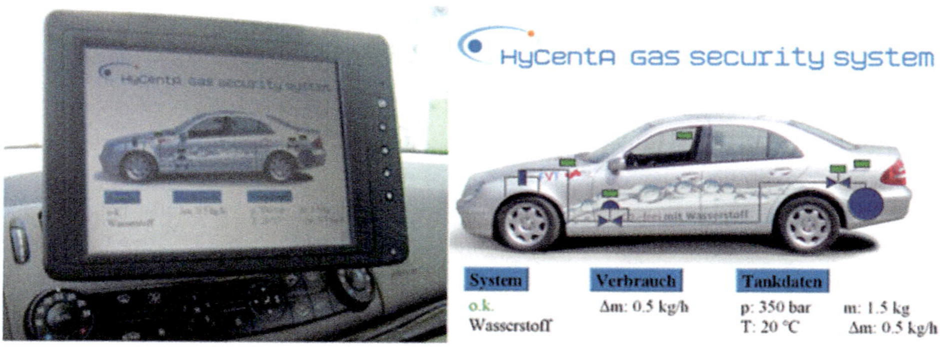

Diagram 4.91 Gas security system [223, 230]

Beside extensive data acquisition, logging function and visualization, this safety system handles the switch-over between gas and gasoline operation in driving mode, offers multiple leak detection as well as extensive safety measures leading to an unrestricted type approval for general road traffic. The determined characteristics of gas operation, based on research engines, could be confirmed with the full engine. The prototype vehicle succeeded in demonstrating the practical evidence of the fuel-suitability of mixed gases for automotive applications and in pointing out the bridging technology for the gradual use of renewable hydrogen.

4.6 Overall System Comparison: Methane-Based Mobility Compared to Other Drive Concepts and Energy Sources

Reinhard Otten and Hendrik Gosda

Methane as fuel—as the chapters above have shown, the topic has many different facets, and from the perspective of a world where more than 99 % of vehicles are equipped with internal combustion engines and in over 90 % of cases use crude oil-based energy sources, using methane as an energy source for vehicles is quite obviously a step in the right direction. Still, as the opportunity for a breakthrough also appears likely in Europe, many market players and decision-makers already consider CNG and LNG vehicles an outdated alternative. The first battery- or hydrogen-operated electric vehicles are now available in the market, bearing witness to a carbon-free future with zero emissions and tank-to-wheel efficiencies that, at first glance, make any car running on an internal combustion engine a relic.

This chapter therefore juxtaposes methane-based mobility with these "new," even less established drive concepts. First, however, as an interim conclusion of the preceding pages of this book and as a starting basis for the comparison, a SWOT analysis (**S**trengths; **W**eaknesses; **O**pportunities; **T**hreats) of methane as an energy source for mobility shall be provided.

4.6.1 SWOT Analysis for Methane as a Fuel

Strengths of Methane as a Fuel

- Almost no harmful by-products, therefore extremely suitable for clean, efficient combustion
- Hydrocarbon with the highest content of hydrogen, therefore comparatively low CO_2 emissions (an advantage of approximately 23 % compared to gasoline at the same efficiency)
- Highly knock-proof, resulting in efficiency advantages over gasoline engines
- Large fossil fuel reserves
- Very well developed worldwide infrastructure (transport and storage)
- Universal energy source with established sales markets in the sectors involving heat, electricity and (regionally varying) mobility
- Even today, high percentage of renewable sources (Germany in 2013, for example: 20 % biomethane in natural gas for road traffic) [234]
- Low costs due to low commodity prices and also lower taxation in many countries.

Weaknesses of Methane as a Fuel

- More complex to store in the vehicle (pressurized tanks) than liquid fuels
- Energy density nonetheless lower by approximately a factor of 5 (at 200 bar), accordingly high space requirement of the tanks or lower range
- Subjective reservations toward natural gas ("explosive, dangerous, difficult to refuel")
- Low density of fueling station network; in many countries, no comprehensive infrastructure at all
- Relatively high investment and operating costs for fueling stations
- "Chicken and egg problem": fueling station density will grow depending on the number of vehicles, however, vehicles are only attractive when the fueling station density is good
- Tank-to-wheel efficiency of natural gas engines is slightly worse than that of diesel engines and significantly worse than electric motors operated by batteries or fuel cells
- In case of Power-to-Gas production: Expense and loss of efficiency due to methanation process at PtG plants

Opportunities of Methane as a Fuel

- Development of regional natural gas markets into a global market has started (LNG transport by ship, worldwide construction of LNG terminals), resulting in lower prices and reduced dependence on individual supply sources
- Given preference over diesel when entering low-emission zones due to lower NOx emissions
- In the course of the boom of LNG logistics: new applications of natural gas for mobility uses, for example, long-distance trucks and shipping
- Great potential of renewable sources: biomass (energy crops, organic waste, manure) and synthetic methane or hydrogen admixture from Power-to-Gas plants (use of cheap green electricity from fluctuating sources)
- Synergies with hydrogen: miscibility ("hymethane"), extraction of hydrogen from natural gas, hydrogen as an intermediate in the methanation in PtG plants.

Risks of Methane as a Fuel

- No methane-powered vehicles may enter zero-emission zones and no benefits from zero-emission legislation
- Subjective or objective supply bottlenecks in certain regions due to the use of the natural gas supply for political leverage
- High greenhouse gas effect from leaking methane (factor of approximately 25 compared to CO_2): methane-based mobility is discredited due to improper handling of leaks in pipelines or of evaporative emissions at LNG terminals and fueling stations

– Environmental risks during the mining of unconventional fossil natural gas resources (fracking, exploitation of methane hydrates in the ocean): discreditation of methane as an energy source in general.

4.6.2 Criteria for the Evaluation of Different Drive Systems

Looking at studies on road mobility in the future, the decisive success criteria with respect to the type of engine always appear to be the same:

– Low CO_2 emission of the vehicle
– Low local emission of pollutants
– High efficiency
– Low energy costs.

Better studies at least manage to include vehicle costs, and the really good ones even consider system costs of the associated infrastructure.

Ultimately, these add up to relatively similar scenarios: battery drives and fuel cells will win the race sooner or later, and the market penetration rate only depends on how quickly these drive systems can be offered at attractive prices based on anticipated economies of scale, and how extensively public subsidies (interestingly enough, strictly private investments are rarely mentioned) will provide charging stations or hydrogen fueling stations.

In terms of the fuel cell, forecasts are distinguished by remarkable consistency: for approximately 40 years now, predictions have been made with clockwork regularity that the breakthrough of the hydrogen car is about 10 years away.

The Human Factor: Attractiveness Is a Crucial Criterion for the Customer
If the above-mentioned criteria were in fact sufficient, the CNG car, for example, would have accounted for a very large market share for some years now: its efficiency is better than that of gasoline engines, emissions of pollutants and CO_2 are less than both gasoline and diesel engines, the purchase price is comparable to a diesel vehicle, and the entire compelling package brings joy at every fill-up because savings can easily range from 30 to 50 %.

And yet the market share of natural gas-powered vehicles in Europe is less than 1 % or, in one or two countries that offer particularly favorable general conditions, in the low single-digit percentage range.

The considerable financial advantages quite obviously were only sufficient for a small number of customers to also tolerate average ranges of 300–400 km (186–248 miles) per tank filling and a relatively fragmented network of fueling stations (in Germany, approximately 940 out of a total of approximately 14,000 stations), and, first and foremost, to take the step toward the unknown. When new car customers are asked why they did not select a natural gas vehicle, they cite the following reasons:

- Loss of convenience due to lack of range, combined with low fueling station density ("If I purchase an expensive new car/company car, I don't want to lose time on refueling and have to plan every trip based on refueling availability")
- Lack of awareness ("I didn't even think of it; natural gas was not on my radar, and I also don't know anyone who drives such a car")
- Skepticism in terms of maturity and reliability of the technology and, as a result, in terms of value retention ("Natural gas vehicles are still relatively new in the market; surely they still have some initial problems and lose a lot of value")
- Lack of knowledge and uncertainty on the part of the dealer and lack of road testing opportunities ("When I spoke to the dealer, he didn't really know much about it and advised me to purchase a diesel—I was able to immediately take it for a test drive and pick up a demo car inexpensively")
- Uncertainty regarding the refueling process ("Refueling is so different; I don't know if I can manage that")
- Apprehension about gaseous fuel ("Natural gas—even worse, at 200 bar—is too dangerous for me; I don't feel like driving around with such a bomb in the car")
- Fear of complete loss of mobility ("With a natural gas car, I would be afraid to be stranded; after all, you can't refuel from a canister")
- Skepticism in regard to the expertise of repair shops ("If I ever have a problem with the car, I'm sure the shop doesn't know what to do with such an unusual car").

So when asking the question about the future powertrain mix, it could be a mistake to neglect the customers and their reasons for purchasing. Confidence in the technology and a certain degree of market penetration ("critical mass" in the sense of awareness, familiarity) appear to play a major role, and the discerning new car customer in particular, who does not shy away from tacking a few thousand euros on to the price for suitable extra comfort options and for whom, very often, the employer provides the car as part of the compensation package, will not tolerate any disadvantages in practical use.

So it already becomes evident at this point that the fuel cell-powered car, which suffers from almost every one of the actual and perceived weaknesses of the CNG vehicle in significantly exacerbated form, will have a very difficult time when it comes to customer acceptance. Yet, acceptance is extremely important if real progress is to be made with regard to sustainability, for only an environmentally sound vehicle that is in fact sold, and replaces one that is less environmentally friendly, is indeed eco-friendly.

Battery electric vehicles also need to overcome some of the above-mentioned hurdles on the part of the customer; the added difficulties here are charging options for the battery, which do not exist yet for every user, and—depending on the use of the vehicle—the charging times.

As an interim conclusion, it should be noted here that—in addition to efficiency, greenhouse gas emission in the product life cycle, local emissions and overall costs—two essential criteria must also be considered: customer acceptance and practicability of the energy supply.

To illustrate why road mobility with CNG vehicles is of interest in this context and can play an important supporting role in paving the way, especially for electric and hydrogen mobility, the following sub-chapters will provide a qualitative comparison of these drive concepts, in conjunction with the respectively possible provisioning of renewable energy. The benchmark will be CNG mobility based on Power-to-Gas technology, which was described in Sect. 4.4.3. It will be explained why it would be wrong, in the view of the authors, to hamper this integral approach in the hopes of making the big leap in efficiency, right now and for everyone, with the battery-operated car and fuel cell. We will shed light on misconceptions and disprove theses that engineer a competition between CNG and electric or hydrogen mobility, instead of pointing out the synergies. Finally, the intent is also to clarify that, during times of fluctuating energy sources, efficiency is not everything, but merely *one* element of the eco-effectiveness we need to achieve the climate protection targets across sectors.

4.6.3 Road Mobility with Renewable Energy: The CNG Vehicle in Combination with Renewable Methane from Power-to-Gas Plants as the Benchmark

Ever since vehicle manufacturer Audi presented the project on the construction of the first industrial Power-to-Gas plant worldwide and on the use of the produced synthetic methane in CNG vehicles of its brand in May 2011, the idea of storing "excess" power in the form of hydrogen or methane in the natural gas network has evolved from an academic niche to a core topic of the energy industry. Hardly any convention in the industry goes by where the topic of Power-to-Gas is not on the agenda, and approximately 20 smaller and larger PtG projects have been developed since in Germany alone. Audi's 6-MW plant equipped with a sophisticated heat management system (for a description, refer to Sect. 4.4.3.1) has been feeding synthetic methane ("Audi e-gas") into the natural gas network since 2013, and even utility companies such as E.ON or Thüga Group have meanwhile commissioned—albeit smaller and simpler—plants, which produce pure hydrogen and feed it into the natural gas network.

The industry agrees that using stored energy for mobility applications arguably promises the greatest chances of achieving economical operation with the new technology within the next few years. According to today's forecasts, use of the stored gas for re-conversion to electricity appears to become economical only when the share of renewable energies in the power mix has far surpassed the 50 % threshold [235]. Still, this does not even seem to be the problem, since we have such a long way to go in terms of mobility before we reach these orders of magnitude, considering that, at most, approximately 6 % of energy forms used in road traffic comes from renewable sources. What's more, the ecological effect of these biofuels (essentially bioethanol and biodiesel) is highly controversial, and even ethical concerns are raised due to the competition with food production. Only biomethane produced from waste, which is used in CNG vehicles, is spared criticism—also due to its excellent greenhouse gas balance.

In any case, the attempt to find truly convincing alternatives to crude oil—which for its part increasingly comes from sources and production methods which are extremely objectionable because they pose a danger to the environment (key words: oil sands, deep sea, arctic, fracking, etc.)—is certainly commendable.

In fact, the Power-to-Gas concept, which has since been picked up by a number of countries, is very compelling in many respects. Here is the list of advantages discussed in Sect. 4.4.3 again in the form of key words:

- Storage and use of daily and seasonal power surpluses is possible in the form of hydrogen and methane, providing access to lower energy prices at the energy exchange
- Contribution to the stabilization of the power grid
- Bidirectional coupling of power and gas network, thereby tapping a reservoir of vast capacity (Diagram 4.92)
- Use of existing, well-developed infrastructure
- Diverse options for use of the stored energy, including provisioning of almost climate-neutral long-distance mobility
- During methanation of the energy, CO_2 waste gas flows are bonded
- A large number of possible synergies with bioenergy and sewage treatment plants
- Unlike biofuels of the first generation, no competition with food production.

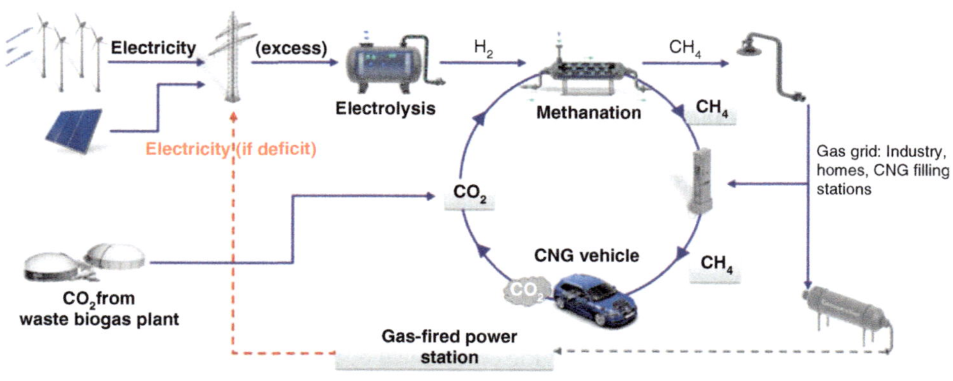

Diagram 4.92 Schematic of Power-to-Gas: Surplus electricity can be converted into hydrogen and/or synthetic methane and then stored in the natural gas grid. Audi ensures that the energy vector is set toward mobility in the energy balance by the so-called e-gas fuel card. A closed circuit also results for CO_2 in the energy balance: Carbon dioxide emissions of the CNG vehicle do not exceed the amount that was previously bonded during methanation (Source: AUDI AG)

Nonetheless the combination "Power-to-Gas + CNG vehicle" is subject to considerable criticism, especially as regards the loss of energy in the overall chain from the energy source to the wheel ("well-to-wheel"). A theoretical calculation for passenger cars, which is based on the New European Driving Cycle (NEDC) and does not consider the potential use of waste heat in the individual process steps and in the vehicle, in fact shows the following well-to-wheel efficiencies. It goes without saying that these also impact the overall economics [236]:

- Battery electric vehicle: 52 %
- Fuel cell-vehicle: 20 %
- CNG vehicle: 12 %.

The decisive factor for the differences is the particular efficiency of the respective vehicles. The battery-powered vehicle, which draws electricity directly from the grid, additionally benefits from the fact that no conversion losses from electricity to hydrogen or methane occur in the upstream chain.

The full extent of the efficiency advantages of the fuel cell over the internal combustion engine assumed here has not yet been proven in practice. Initial experiences with hydrogen buses in Switzerland, for example, show that an efficiency gain of only approximately 20 % was achievable over traditional diesel buses, with the energy for the requisite nightly heating of the drive unit not having been included in the calculation here. As a result, the energy-related advantages of the fuel cell system compared to diesel vehicles equipped with hybrid drives are not yet tangible [237].

Of course, it must also be noted that considerable improvements in efficiency are still to be expected as development progresses. However, this also holds true for natural gas engines: truck manufacturer Scania has developed engines that, at the best efficiency point of approximately 40 %, rank just a few percentage points behind even the best Euro-6 diesel engines, and exhibit similarly good values in the partial load range as well [238].

Still, compared to the described alternatives, efficiency remains the Achilles heel of CNG vehicles operated with synthetic methane. The following sub-chapters explain, however, that this certainly is not a deal breaker.

4.6.3.1 The Comparison to the Battery Electric Vehicle

Given its lack of local emissions and excellent drive efficiency, the electric vehicle ranks near the top on the mobility wish list of many politicians. In comparison to our benchmark, which is the CNG vehicle operated with methane from PtG plants, the following theses or assertions are frequently voiced, questioning the plausibility of the CNG/PtG option.

Claim No. 1
"Tesla has shown that the electric car is also suitable for long distances; other vehicle concepts will not be needed any more in the future."

Indeed, since U.S. firm Tesla has been able to achieve actual ranges of more than 300 km (186 miles) with its series-produced, solely battery-operated sports sedan and, as a result of extremely powerful rapid chargers, also provides long-distance capability without unreasonable constraints, this thesis is becoming increasingly more popular. It is still incorrect, though; at least for the foreseeable future. The reason is that Tesla relies on an unsustainable concept in two respects.

The first problem concerns the car itself. It is equipped with a lithium-ion battery pack weighing approximately 700 kg (1540 lbs), which can store a total of up to 85 kWh of energy. This not only costs a lot of money, but also puts the concept in jeopardy because of environmental concerns.

Member of the BMW Board of Management, Development, Dr. Herbert Diess made the following comments in this regard during the introduction of BMW electric vehicles as part of the Vienna Engine Symposium in May 2014 [239]: "Today's customary energy storage technologies and range targets of 300–400 km (186–248 miles) make vehicle concepts heavy and expensive. Moreover, vehicles equipped with storage systems of more than 50 kWh lose their ecological legitimacy. Since lithium-ion batteries require a tremendous amount of energy in production, these vehicles are no longer able to compensate for the additional CO_2 emissions generated in the production phase over their life cycle."

The other problem relates to the rapid chargers. While the concept of exclusive charging stations with 90–135 kW of power pursued by Tesla is conceivable for a small number of luxury-brand cars, it is not readily scalable. It will be difficult for customers to accept a 20-min charging time for a distance of 200 km (125 miles) in the long run. And if two or three other customers are already in line at the station, charging can turn into a very tedious matter. During peak hours, lines are already very long at today's fueling stations along highways. The flow of energy during a fueling process for a diesel vehicle corresponds to approximately 20,000 kW, that is, about 200 times that of a Tesla quick charger. So you can imagine how many charging stations would have to be provided along the highways for electric long-distance mobility to work flawlessly with a massive rollout. It would also be questionable to what degree modifications to the electrical transmission networks would be necessary to reliably deliver the requisite electric power to the highways.

In terms of the vehicle, it remains to be seen how well traction batteries, or electric vehicle batteries, will tolerate frequent rapid charging in addition to the normal aging process of the cells.

Ultimately, anything but the conceptual compatibility of the electric car with the needs of the electricity grid would be counterproductive. But battery-operated cars used for long-distance travel would be charged, to a large degree, at times when power demand is already high. As a result, they are hardly able to make an overall positive contribution to "demand-side management."

What remains undisputed is that the electric car offers many advantages when used for short distances, when long charging times are possible at night at home, or during the day at work, and it should and will acquire significant market share. Even now, there are millions of second or third cars that rarely travel more than 100 km (62 miles) a day. A very large number of these vehicles could be replaced with an electric car. Still, the customer base will remain limited for a long time to come. Not every city or suburban dweller has a guaranteed parking space where the car can be charged, and again, the question of sufficient power of the local grids becomes a problem once a certain number of vehicles are on the road. Last but not least, electric cars will also remain significantly more expensive in the medium term than comparable conventional passenger cars. The approximately 10,000-euro difference found in the market today, which still does not cover manufacturers' costs at present, is not likely to disappear completely in the next years.

So we have to remain realistic. Yes, vehicles operated with green energy and moderate battery sizes up to a maximum of 30 kWh are an excellent and environmentally friendly alternative when used for short distances, if the service life of the battery and vehicle is sufficient, and when they replace a conventionally driven vehicle. It will still take a long time, however, until they will completely displace vehicles with internal combustion engines from the market. This assessment is supported by virtually all available studies on the future of the passenger car market. On average, these studies arrive at the conclusion that, by 2030, approximately 22 % of all newly registered passenger cars worldwide will derive their power entirely (BEV) or partially (PHEV) from electricity, and that 85 % will be equipped with an internal combustion engine, although to a large degree with varying degrees of electrification ranging from full hybrids to range-extender vehicles [240].

The forecast market shares in the order of 20 % by the year 2030 will come to a significant degree from legislation, which will mandate a certain quota of zero-emission vehicles, such as in individual states of the USA, for example, that will follow the path of Californian legislation. Plug-in hybrids will account for a considerable share in the mix of electric vehicles. When short distances have to be covered, they offer all the advantages of an all-electric car, but minimize crucial purchasing barriers—aside from the price and the presumed high loss of value—on account of their long distance-capable drive systems with internal combustion engines (regardless of whether they are diesel, gasoline or natural gas engines).

Claim No. 2

"It is much better to use surplus electricity directly in electric cars, instead of going through the trouble of converting it into methane, only to burn it in cars with inefficient internal combustion engines."

It would be foolish to contradict this statement. It goes without saying that the direct path from the power grid to the car battery should be given preference. The question is just how much surplus electricity can electric cars in fact store. In principle, this amount is limited by

- The actual number of electric cars in existence
- The actual battery sizes of these cars (if they are plug-in hybrids, the rechargeable batteries are rather small)
- The technical feasibility of charging these batteries via smart-grid control units in a way that is in fact compatible with the needs of the grid, i.e. controlled externally depending on the demand in the power network at certain times.
- The willingness of users of these vehicles to utilize these smart-grid options and charge the battery at appropriate off-peak times, i.e. not as soon as they arrive at home and want to plug in their vehicle, for example, but rather starting at 1 a.m. in the morning; this could mean, for example, that the car might not be available for a spontaneous trip in the evening, and mobility would be considerably limited.

Yet even in an "ideal scenario," where the entire passenger car population (approximately 42 million) in Germany has been replaced with battery cars, and all of these cars are organized via a smart grid and connected to the grid, the power availability would be exhausted in a very short time (order of magnitude: a few hours [241]). The anticipated seasonal excesses, which would occur sometimes over a few weeks when the share of wind and solar energy is high, could only be stored on a very small scale in the traction batteries. The idea is nonetheless correct—electric cars should most definitely be charged when there is an excess of electricity. The real capacity that can indeed be tapped, however, does not compare to the potential that the Power-to-Gas technology has to offer.

Result

Two key conclusions can be drawn from the critical analysis of the two claims:

1. It would be wrong to limit ourselves solely to electric mobility and its potential (which is quite significant in the medium and long term) when it comes to motorized private transportation. In terms of long distance-capable passenger cars, the internal combustion engine will remain widely used—giving more leverage to fuels that are very low in carbon, such as methane, or allow even climate-neutral operation of the vehicles because CO_2 is bonded during their production phase (biomethane, synthetic methane from PtG plants). CNG vehicles can thus superbly complement the gradual

progression of electrification of the vehicle market, and they can do so at comparatively low—or even "negative"—CO_2 abatement costs.

2. If we want the electric cars and plug-in hybrids that are added to the vehicle population to be able to apply their positive climatic effect, considerable progress is required in two areas. For one, action must be taken which is aimed at more than compensating for the "ecological rucksack" of the vehicles during their useful life. This requires a long life cycle and a high recycling rate of all components, notably the traction batteries. Secondly, the share of renewable energies in the energy mix must be increased further—or else the emission of greenhouse gases is only shifted from the tail pipe to the power plants. Power-to-Gas concepts can make a significant contribution in this regard (see Sect. 4.4.3), and they, in turn, can already gain a foothold today in the energy landscape if CNG vehicles are available in the market as users of energy.

4.6.3.2 The Comparison to the Fuel Cell Electric Vehicle (FCEV)

In many scenarios of the future, mobility with fuel cell vehicles plays a major role. This is certainly understandable, since the fuel cell vehicle—apart from water vapor—is entirely emission-free in its operation, just like the battery electric vehicle. Additionally, there are two key advantages:

- The higher achievable range
- The considerably lower time needed to replenish the energy storage system.

Some experts even consider it to be the better alternative to the battery vehicle. Both advantages, however, come with the trade-off of very high complexity. To significantly surpass the energy density of the lithium-ion battery, the hydrogen must be stored at very high pressures in the vehicle. For commercial vehicles, 350 bar (5076 psi) has become established as a widely common standard, while a pressure of 700 bar (10,150 psi) has been adopted for passenger cars allowing for smaller tank diameters. This results in very high complexity (and very high costs) for the energy storage system in the vehicle—in addition to the likewise high costs for the fuel cell stack, battery and power electronics—and for the fueling stations as well, which ultimately must provide pressures of 850–900 bar (12,300–13,000 psi) to achieve sufficient overflow velocities. This is a major challenge with respect to the quality of materials and sealing between the individual components, most notably in the case of hydrogen, where the molecules are the smallest in the realm of nature. The complexity for cooling the components and for measures to prevent icing (during expansion of the gas during the refueling process) is significant. The cost for even low-capacity hydrogen fueling stations today is well above one million euros, which exceeds the current price of natural gas fueling stations by a factor of 4 or 5 [242].

In addition to the hurdles described in Sect. 4.6.2 in terms of acceptance by the customer, hydrogen mobility—looking at the overall costs for vehicles and the infrastructure—will have a difficult time becoming a valid alternative. At the same time, the gap in the market for fuel cell vehicles is becoming ever smaller: the range and rapid charging capability of battery vehicles will continue to improve, while the conventional vehicle equipped with internal combustion engine will continue to advance. Individual hydrogen mobility faces strong competition from plug-in hybrids in particular, which combine the local-emission-free design of the electric car with the unrestricted long distance-capability of the conventional vehicle. In terms of price, plug-in hybrids will also be less expensive in the long run than fuel cell vehicles can ever be. Above all, however, they can draw on established infrastructure (socket at home and gasoline, diesel or CNG fueling stations).

Still, hydrogen has one more huge systemic advantage: it can be used to store fluctuating energy forms such as solar and wind, which brings us back to the field of synergistically linking energy management and mobility, which is a distinguishing factor of our benchmark "CNG vehicle + synthetic methane from Power-to-Gas plants." The following claims, which are intended to demonstrate the superiority of hydrogen as an energy source and of the fuel cell vehicle as a means of transportation, are frequently made:

Claim No. 1
"The methanation of hydrogen in Power-to-Gas plants represents expensive loss of energy. The natural gas network can absorb large amounts of hydrogen, without exceeding the limits for hydrogen concentration."

In principle, this statement is not incorrect. The methanation stage can indeed be dispensed with, and pure hydrogen can be fed into the natural gas grid, where no problems arise from the hydrogen concentration in the network because the quantities fed are negligibly small compared to the flow rate. However, the number of locations in the natural gas grid where this is possible in practice with industrial-scale PtG plants is not arbitrarily high. After all, there are quite a few end users and service providers for whom the grid operators in Germany must guarantee lower hydrogen concentrations than the 5 %-level that initially represents the maximum upper limit according to the regulations of the DVGW (German Technical and Scientific Association for Gas and Water) in the interest of the technical compatibility of the network components. In addition to the almost 1000 CNG fueling stations alone in Germany, this also includes industrial customers, such as manufacturers of safety glass, which use natural gas as feedstock or a process medium. Among the service providers, operators of huge porous rock storage reservoirs could incur significant damage at hydrogen concentrations above 1 %.

Following an intensive dialog between stakeholders, the German Federal Network Agency therefore developed a position paper, in which the following provision is proposed [243]:

> After determining hydrogen-sensitive applications in the network, the network operator must establish, based on binding information provided by the hydrogen feed-in party as part of the application for connection to achieve the planned feed-in as well as based on the necessary data from its own grid, whether and to what extent hydrogen can reach sensitive applications or grid coupling points to downstream or (in the case of return-feed) upstream networks. If this is the case, the network operator, if necessary in coordination with an affected customer, must review whether the application or the customer in question (storage reservoir or power plant operator, for example) will be able to tolerate an elevated hydrogen content, or an elevated amount of hydrogen, in isolated cases (example: very brief instances where the threshold is exceeded); however, an affected customer is not obligated to accept a briefly elevated hydrogen content if damage cannot be ruled out with even very brief excess levels. If this is not the case, the network operator is required to calculate the scope in which hydrogen can be fed in at the requested connection point.

This clearly shows that the possibilities of feeding pure hydrogen into the grid are generally limited, and the times of operation of such plants may potentially be severely restricted by the network operator as a result of the direction and speed of flow in the natural gas network.

Surprisingly, it is suggested time and again that feeding pure hydrogen into the natural gas grid could then also provide for corresponding hydrogen applications to come into effect. This is not the case, of course, because it is mixed with natural gas. As far as mobility is concerned, feeding hydrogen into the natural gas network initially only benefits CNG vehicles, and perhaps LNG vehicles. It is therefore useless to become mired in ideological trench warfare when it comes to the question of the meaningfulness of methanation; this is about practical considerations in the individual case, which require a pragmatic approach.

Still, it is often argued that a Power-to-Gas plant with a methanation stage could not possibly pay off simply because the synthetic methane, which comes out at the end of the efficiency chain, would entail excessively high costs per kilometer in a vehicle with an internal combustion engine. A study conducted by the German Aerospace Center (DLR) [244] for example, which was compiled for the German Federal Ministry of Transport, calculates fuel costs (without taxes) of 12 €/100 km (62 miles) for a CNG car operated with PtG methane. At this price, this energy pathway would not be competitive in the long run. This calculation, however, was based on average electricity costs of 8.5 ct/kWh and 4000 h of operation of the PtG plant per year. This price corresponds approximately to the overall costs of 1 kWh of wind power, so at first glance the approach seems comprehensible if it were not for the fact that the fundamental idea of Power-to-Gas is disregarded, which is to convert electricity into gas and store it precisely when the price at the energy exchange is low. In 2012, the average power spot price at the energy exchange for the least expensive 4320 h (which is approximately half of the year) was 3.03 ct/kWh [244]. On weekends with a lot of wind and sun, prices at the exchange even go negative; on August 17, 2014, for example, the values were minus 2 ct/kWh for hours, with swings to minus 5.9 ct/kWh. Negative prices are symptoms of a lack of flexibility in production and electricity demand, and hopefully a temporary event. It is clearly and easily comprehensible that a further expansion of renewable energies, whose marginal costs are close to 0 ct/kW since they require no fossil resources, will always lead to recurring phases of very low electricity prices—even if flexibility options, such as Power-to-Gas, were to be expanded quickly and help reduce the gaps between supply and demand. As a result, the operating costs for vehicles that are supplied with energy from PtG plants can be much lower than was calculated in the above-mentioned study.

In addition, it generally holds true that, when energy is available in abundance, the importance shifts away from efficiency and toward the ability to render the energy practically usable, even though this statement requires a mind shift by the efficiency-driven engineer. Nature is an interesting teacher in this regard: plants store sunlight by way of photosynthesis, and with the help of CO_2, in the form of biomass. This is done with an efficiency in the order of 1 %. Nevertheless, photosynthesis is undoubtedly a model for success.

Claim No. 2

"Fuel cell vehicles are technologically so mature that they can replace vehicles with internal combustion engines."

Anyone who has had the chance to test drive one of the modern limited-lot hydrogen cars is inclined to agree with this statement. The drive system, including the cooling system and high-pressure storage technology, operates so quietly and inconspicuously that it almost makes you believe you are driving a battery-operated electric car. Still: for a fuel cell car to make sense for the customer and the environment, a lot more development work is needed. Four important items shall be highlighted here:

- Costs
  Even at high quantities and the corresponding economies of scale, manufacturing costs would still be much too high. Despite tremendous successes in development, the approximately 30 g of platinum required per passenger car remains relatively high [246]. A theoretical widespread use of fuel cells in vehicles and co-generation plants would therefore further drive up the price of platinum. In a scenario for 2050, which postulates a 50 % share of the fuel cell drive system in new passenger car registrations, production of platinum that is approximately 12 times higher than that of 2012 would be needed for these vehicles alone, unless the platinum content can be significantly reduced. Current studies arrive at the prognosis that a broad-based use of fuel cell vehicles is only possible if a suitable substitute can be found for platinum as a resource [246]. Regardless, a fuel cell vehicle will always be relatively expensive; in addition to

the fuel cell stack, high costs are also associated with the battery, power electronics, and most notably of the leak-proof 700 bar (10,150 psi) high-pressure storage tanks.

- Service life

  The long-term durability of the battery stacks—preferably over the service life of the vehicle—remains a major challenge and is subject to conflicting objectives in the criteria related to cost, efficiency, temperature resistance and performance. The problem for the customer is similar to that of the traction battery of the electric vehicle: when an expensive component has to be replaced after 8 or 10 years, it generally constitutes a total economic loss because the repair costs—assuming, at a minimum, cost-covering prices for replacement parts and labor—exceed the residual value of the vehicle.
- Life cycle assessment

  As a result of the many materials—platinum, CFRP high-pressure tanks, copper windings, and large numbers of electronic components with heavy use of rare metals—which are either highly energy-intensive or obtained through high consumption of resources, the ecological rucksack of the extremely complex fuel cell vehicle carries a heavy weight at today's stage of development. To achieve, in sum, an ecological benefit compared to a conventional car, emissions would have to be very low during the service life. Hydrogen from natural gas and its liquefaction for transport purposes are therefore not an option if true progress is to be achieved. Only if the hydrogen is produced from renewable energies and transported in the gaseous state, or produced directly at the fueling station, can the eco-balance turn positive. Another prerequisite is that the life cycle of the vehicle does not end after only 8 years as a result of the above-mentioned total economic loss. The useful life must be long enough to more than compensate for the ecological rucksack from production.
- Safety

  In contrast to methane, it is almost impossible to odorize hydrogen, making leaks in the tanks, piping and connections very difficult to detect. A problem for accident scenarios involving fire is the circumstance that the extremely hot flame that develops as a result of the predictable hydrogen blowout from an overheated tank is virtually invisible in daylight and gives off almost no heat (methane: yellowish flame, considerable heat emission). Before marketing H_2 vehicles to the masses, appropriate technical and organizational measures have to be developed to ensure that, in an emergency, no increased safety risk exists for passengers and first responders. The fact that a tank that has 700 bar (10,150 psi) of internal pressure theoretically poses a greater risk than a tank with 200 bar (2900 psi) is presumably a problem that has little practical relevance due to available safety precautions. However, it remains to be seen whether potential customers share this assessment.

Claim No. 3

"An infrastructure of H_2 fueling stations, and later pipelines, will follow automatically, based on demand, once a large enough number fuel cell vehicles are on the roads."

The chicken and egg problem is well-known from the CNG vehicles market: Sales of cars are slow because there are not enough fueling stations, and since only a small number of cars on the roads are CNG vehicles, it is not worthwhile investing in one of the relatively expensive fueling stations. Consequently, attempts to achieve a threshold value at which the tug-of-war between supply and demand is a non-issue are underway in those countries where the government provides incentives to help establish H_2 mobility by offering subsidies for the development of infrastructure and vehicle fleets. The fact that it will not be easy to persuade private customers to purchase a hydrogen car has already been addressed. But even if a large fleet would have become established as a result of statutory quotas or extensive subsidies, the question that arises is who is supposed to operate and maintain the hydrogen fueling stations.

A look at the situation found with CNG vehicles proves worthwhile. Germany has almost 1000 CNG fueling stations and some 100,000 CNG vehicles. Fueling station operators say that there should be at least 200 regular users for every CNG fueling station to be able to cover operating and maintenance costs and for the investment of approximately 250,000 € to pay off. Given the model offensive that automakers are embarking on, this target could be achievable for the majority of locations in the coming years, but some operators are now relinquishing their locations. The situation is therefore at a tipping point, but the threshold value needed to overcome the chicken and egg problem could be surpassed with a little perseverance on the part of politics and operators, as has been successfully demonstrated in Italy, for example. This trend has been, and still is, driven by the low price of natural gas, which in addition to lower market prices than crude oil products also benefits from tax relief.

Hydrogen fueling stations command an investment that is 5 times higher, consume significantly more electricity due to the low energy density of hydrogen by volume, the much higher pressure level (700 bar instead of 200 bar in the vehicle), and the resultant very high cooling needs, and necessitate significantly more intensive servicing and maintenance for the leak-proof high-pressure components and the complex technology in general. So far, there are no conclusive concepts of what a business case for hydrogen fueling stations could look like after the funding projects expire. At the same time, however, the price for hydrogen must be very inexpensive to ensure a continued flow of customers who purchase an expensive fuel cell vehicle.

In light of these conflicts, discussions have now turned to a new technology that operates based on cooled, but still gaseous hydrogen in order to get away from the extreme pressures and pave the way for new possibilities for the economical operation of H_2 fueling stations [247]. Still, it is questionable whether this pathway, now the fourth following cryogenic liquid storage and 350/700-bar pressure storage, will gain a foothold in light of the focus on the 700-bar option for passenger cars.

In sum, the creation and self-sufficient management of an H_2 fueling station infrastructure for a maximum pressure of 700 bar in the vehicle seems to be a huge challenge. From today's perspective, this is only conceivable if comprehensive hydrogen mobility is mandated by law (registration quotas, driving restrictions for other concepts).

There is one more point that should be made: H_2 fueling stations could produce hydrogen on-site by way of an electrolyzer and can therefore be a direct, renewable alternative to our benchmark—synthetic methane for CNG vehicles. Apart from the fact that these H_2 fueling stations also have the problem of high pressure levels, they offer a few basic advantages. They can produce hydrogen when electricity is cheap. They do not lose any energy during the conversion into methane and can supply fuel cell vehicles, which are favorable in terms of efficiency, directly and without the need for complex transportation by special-purpose tanker trucks. In fact, strictly from a mobility perspective, this is the ideal Power-to-Gas configuration. Still, there is a big problem, namely the difficult challenge of reconciling supply and demand. For one, this approach is not immune either to the aforementioned chicken and egg problem (the fueling station is already there, along with the expensive electrolyzer, but buyers are few and far between) and secondly, it is not unlikely that demand for hydrogen is slow during times where power supply is high or, conversely, that electricity is very expensive over days and weeks, while a large number of vehicles need refueling. Admittedly, on-site pressure storage tanks can mitigate the problem, but they cannot fundamentally solve it.

It becomes evident at this point that the classic Power-to-Gas idea has been substantially lost in this concept. After all, the key was, in principle, to be able to store and transport the large excess amounts of electricity produced in the future, and to use the extensive capacity of the natural gas network for this purpose. Simulations show that there will be phases of surpluses, at times lasting for weeks, which will alternate with periods where back-up power plants have to produce electricity and will ideally avail themselves of the storage reservoir. Small hydrogen tanks, which are full after just a few hours and able to give off their energy only in one direction (local mobility), are able to make only a small contribution to this problem. And very large hydrogen tanks would surely far exceed the investment expenditure of an H_2 fueling station—unless other consumers are available locally, such as a co-generation plant, which would allow for the costs of such a storage tank to be shared.

Result

Hydrogen can be produced from various primary energy sources, including renewables. Hydrogen burns without greenhouse gas emission. And hydrogen permits fast fueling of vehicles and is clearly superior to the battery vehicle in this regard. So it would certainly be a mistake to completely dismiss a carrier of energy that offers such crucial advantages.

Admittedly, the previous points do not appear to be very optimistic for the fuel cell vehicle, at least not in a scenario that is open to different types of technology.

But a look at the drawbacks leading to such an unfavorable prognosis reveals one application option that is already being widely pursued and where hydrogen mobility can prove itself and mature in practical experience: City bus fleets. For the following reasons, these fleets are much more likely to achieve progress and avoid unnecessary disappointments:

- No chicken and egg problem: due to their high demand for energy per kilometer and the long distances they cover on a daily basis, a manageable number of buses can ensure relatively good utilization of an expensive hydrogen fueling station.
- Zero emission at the perfect location: Since the buses generally operate in urban areas, they can use their emission advantage (pollutants and noise) to tip the balance at precisely the right location.
- Almost no competition from battery vehicles: as a result of the typical long distances covered every day in line operation, battery buses are only conceivable for very specific applications; high daily mileages would make the required batteries too heavy and the charging times too long.
- Large units, low costs: The costs for the attained ecological benefit are considerably lower because investments are made in large units, which are used more intensively. This is far more efficient than the investment in passenger cars with low energy conversion, sitting idle for 23 h a day.
- Local support: The operator of a bus fleet can make technicians available at any time, who can immediately take care of defects.
- No private investment risk: The operators of city buses are generally owned by public authorities and can make a decision based on democratic processes, without individuals having to assume risk.

4.6.4 Summary, Outlook and Possible Synergies

If one were to provide a very simple summarizing response to the question as to why, in the comparison of the renewable energy-based mobility concepts discussed here, the authors think so highly of the CNG vehicle concept that is supplied with synthetic methane from PtG plants, the answer could read as follows:

Because it already works today and sensibly ties the mobility and energy sectors together.

Individual approaches that are divorced from reality may arrive at different results. Anyone who looks solely at mobility may perhaps wish for a vehicle pool that consists only of battery and fuel cell cars, which are supplied in each case with renewable energy. People are entitled to have this wish and work on it, and one day it might come true. But

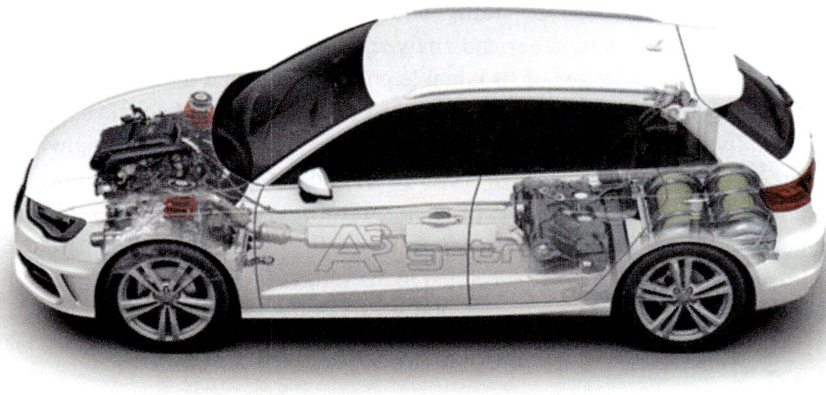

Diagram 4.93 Cutaway view of the Audi A3 Sportback g-tron, a CNG vehicle: The bivalent engine design (fuel tank in front of the rear axle, lightweight CNG pressure tanks in CFRP/GFRP hybrid design behind that) improves the acceptance of the natural gas-operated passenger car because mobility remains ensured when no CNG fueling station is available. As confidence in the CNG infrastructure grows, a switch can be made to monovalent natural gas concepts—an important step for customers to accept the special characteristics of fuel cell cars also (Source: AUDI AG)

anyone seeking to directly enforce this desire, and rejects other, complementary solutions, overlooks three mistakes in this scenario:

1. It is entirely unrealistic in the foreseeable future and does not help further the objective of making actual progress in protecting the climate, here and now.
2. It does not offer any solution for the seasonable storage of large amounts of renewable energy, neglecting one of the big challenges of the energy revolution.
3. It ignores the realization that major changes must grow organically within the circumstances of today's (infra)structures.

People who want to bring about positive change should always focus on these two questions: what is desirable, and what is feasible?

And what can "what is feasible"—which, in itself, already constitutes a step in the right direction—look like in order to ultimately achieve "what is desirable"?

In the specific example of mobility, the answer consists of two theses:

Thesis No. 1
Efficient solutions can only be found when cross-sector thinking displaces individual approaches.

The discussion about electric mobility already showed that the energy industry, automotive industry, and city and traffic planning sectors must cooperate much more closely than in the past. Anyone desiring a change in energy policy not only when it comes to power generation, but also in traffic and heat applications, is well advised to contemplate all the subjects together and analyze the entire system. The close interrelationships between the challenges of sustainable power supply and mobility were outlined in detail in Sect. 4.4.3 and the preceding sections of this chapter. It became apparent that methane—ideally supplied by Power-to-Gas plants—and motor vehicles designed for this energy source can play a key role. However, this should not give rise to the impression as if this technology alone could solve the problems of road mobility. Electric vehicles, with their high efficiency and the potential—yet to be confirmed in practical experience—of using hydrogen as fuel, will surely be needed in the future. It is therefore important not to obstruct the way there, but to pave it. This brings us to the second thesis:

Thesis No. 2
The CNG vehicle and its possible energy pathways open up not only synergies between different economic domains, but also clear the way for other sustainable forms of mobility.

Section 4.4.3 already described that electric and natural gas mobility form a beautiful symbiosis, both in terms of sustainable energy provisioning (as a result of Power-to-Gas as a key technology) and with respect to customer requirements (short distances/long distances). However, the following determination is also correct: Power-to-Gas technology and the natural gas vehicle pave the way for hydrogen mobility. The reasons for this are as follows:

1. Every Power-to-Gas plant—regardless of whether with or without methanation—produces hydrogen, which can be tapped directly for hydrogen vehicles, while the remainder, for which no consumer is available at that time, can be fed into the natural gas network, either in pure form or synthesized into methane. This solves the above-described dilemma of hydrogen fueling stations with on-site production, which in practice will have difficulty balancing supply and demand. In the long run, PtG plants, which will emerge decentralized in many areas over the course of the energy revolution, could be connected via hydrogen pipes, forming the basis for an H_2 infrastructure. Experts agree that the economic success of PtG plants, and with this their continued development, will depend on the successful marketing of the fed-in energy in mobility. This is where the biggest returns per unit of energy could be achieved, since energy prices are high and the incentives for lowering CO_2 emissions are the greatest. Still, significant demand for synthetic methane, which will enable long-distance mobility that is almost climate-neutral, cannot be achieved without a sufficient number of CNG vehicles in the market. So the simple formula is this: without CNG vehicles no PtG plants, and without PtG plants no effective hydrogen infrastructure. What will help the success of Power-to-Gas technology is its capability to grow organically. Since every single plant can operate independently and produce hydrogen and/or methane at virtually any quantity, plants are able to respond flexibly to demand for electricity using hydrogen and methane as energy sources. In addition, it is not necessary to

construct a huge pipeline network right from the start, as would be required for broad-based hydrogen mobility. Developing such an extensive hydrogen pipeline network as a prerequisite, in contrast, would represent a huge investment risk for governments who are already burdened with high debt.
2. If it should not be possible to translate methane into success in mobility, despite the tremendous advantages this energy source has to offer, it most certainly will not be possible to do so with hydrogen, a far more problematic energy source. Vehicles and fueling stations are expensive, the energy density is even lower, and there is no possibility to carry gasoline as a back-up fuel and, with this, to mitigate the problem of the fragmented fueling station network. This is why the CNG vehicle (initially in bivalent form, later monovalent) is so important as a forerunner for the fuel cell vehicle. People must first gain confidence in fuels that are not named gasoline or diesel and also cannot be added to the tank using a canister. Anyone who fails to get dealers, sales agents, mechanics and customers accustomed to gaseous fuels and their traits will have a difficult time generating significant sales volumes for fuel cell cars.

In summary, it can be stated that the CNG vehicle is not only an option for lowering greenhouse gas emissions that is practicable and economical here and now, but also a key element in the energy turnaround and a perfect partner for the electric vehicle. Moreover, in the medium to long term it could prove to be the decisive door opener for the broad-based use of fuel cell technology in passenger cars as well.

References

References for Section 4.1

1. Blesl M, Beestermöller R, Kuder R, Fahl U (2013) Energie- und gesamtwirtschaftliche Auswirkungen veränderter Rahmenbedingungen auf die Nutzung von Erdgas in Deutschland. Studie für das Zentrum für Energieforschung Stuttgart (ZfES), Stuttgart
2. Schlesinger M, Lindenberger D, Lutz C (2014) Die Entwicklung der Energiemärkte – Energiereferenzprognose. Studie im Auftrag des Bundesministeriums für Wirtschaft und Technologie, Basel
3. Andruleit H, Bahr A, Babies HG, Franke D, Meßner J, Pierau R, Schauer M, Schmidt S, Weihmann S (2013) Energiestudie 2013. Reserven, Ressourcen und Verfügbarkeit von Energierohstoffen. Bundesanstalt für Geowissenschaften und Rohstoffe (BGR), Hannover
4. Remme U, Blesl M, Fahl U (2007) Global resources and energy trade: an overview for coal, natural gas, oil and uranium. Forschungsbericht/Institut für Energiewirtschaft und Rationelle Energieanwendung. Universität Stuttgart, Stuttgart
5. Berner HJ, Bargende M (2000) Innovative Fahrzeugantriebe—Erdgas als alternativer Kraftstoff—ein Überblick. VDI-Gesellschaft Fahrzeug- und Verkehrstechnik, Dresden
6. Bargende M (2002) Monovalente Erdgasmotoren: Der Schlüssel zur CO_2-Emissionsminderung. Management Circle-Fachkonferenz "Erdgasfahrzeuge", Stuttgart

7. Friedrich C, Quarg J, Berner H-J, Bargende M (2001) Leistungs- und Emissionspotenzial eines aufgeladenen 1-Liter-Erdgas-Motors. 4. In: Stuttgarter Symposium: Kraftfahrwesen und Verbrennungsmotoren, Stuttgart
8. Cerbe G (2008) Grundlagen der Gastechnik: Gasbeschaffung, Gasverteilung, Gasverwendung, München
9. Turns S (2000) An introduction to combustion: concepts and applications. McGraw-Hill, Boston, MA
10. Berner H-J (2007) Ausgewählte Aspekte der Entwicklung aufgeladener PKW Erdgasmotoren—Strategie und Werkzeuge. Vertiefungsseminar, 5. Management Circle Fachkonferenz, Frankfurt
11. Bargende M, Berner H-J, Bohatsch S, Ferrari A, Hofmann B (2008) Ein strahlgeführtes CNG-Brennverfahren für minimale CO_2-Emissionen. Der Antrieb von morgen: MTZ Konferenz, München
12. Bargende M (2014) Grundlagen der Verbrennungsmotoren, Vorlesungsumdruck, IVK Universität Stuttgart, 3. Auflage Stuttgart
13. Menon V, Komarneni S (1998) Porous adsorbents for vehicular natural gas storage: a review. J Porous Mat 5:43–58
14. Bohatsch S (2011) Ein Injektorkonzept zur Darstellung eines ottomotorischen Brennverfahrens mit Erdgas-Direkteinblasung. Dissertation Universität Stuttgart, Stuttgart
15. VDI (2006) VDI-Wärmeatlas. Berlin
16. Bullis K (2010) &la;heise.de,&ra; Technology Review. [Online]. http://www.heise.de/tr/artikel/Erdgas-Transport-leicht-gemacht-1075371.html
17. Bohrmann G, Kuhs WF (2006) Gashydrate der Meeresböden: Methanspeicher im Ozean, Geofokus Artikel. GMit 23:8–18
18. Pischinger R (2009) Thermodynamik der Verbrennungskraftmaschine. Wien
19. Latscha H, Kazmaier U, Klein HA (2008) Organische Chemie: Chemie-Basiswissen II. Berlin
20. Joos F (2006) Technische Verbrennung: Verbrennungstechnik, Verbrennungsmodellierung, Emissionen. Berlin
21. Schmid D (2014) Vorlesungsmanuskript Motorische Verbrennung. IVK Universität Stuttgart, Stuttgart
22. Zeuch T (2003) Reaktionskinetik von Verbrennungsprozessen in der Gasphase: Spektroskopische Untersuchungen der Geschwindigkeit, Reaktionsprodukte und Mechanismen von Elementarreaktionen und die Modellierung der Oxidation von Kohlenwasserstoffen. Dissertation Universität Göttingen, Göttingen
23. Kallinen K, Haerkoenen M, Pitkamnen M (2004) Advanced catalysts for CNG engines. SAE-Paper Nr. 2004-28-0028
24. Deutsches Zentrum für Luft- und Raumfahrt e.V. (2007) Abschlussbericht—Erdgasmotor mit innerer Gemischbildung (EDI), Stuttgart
25. Gülder ÖL Correlations of laminar combustion data for alternative S.I. Engine. SAE-Paper Nr. 841000
26. Ulinski M, Moore P, Elia M, Metghalchi M (1998) Laminar burning velocity of methane-air-diluent mixtures. In: Proceedings of the ASME internal combustion engine division, Port Lauderdale, Florida
27. Metghalchi M, Keck J (1977) Laminar burning velocity of iso-octane-air, methane-air and methanol-air at high temperature and pressure. In: Proceedings of Eastern Section of the Combustion Institute, Hartford, CT
28. Warnatz J, Maas U, Dibble R (2010) Combustion. Berlin
29. Arbeitsgemeinschaft für sparsamen und umweltfreundlichen Energieverbrauch e.V. (1992) BHKW und Methanzahl. Hamburg

30. DIN 5162: Kraftstoffe für Kraftfahrzeuge—Erdgas—Anforderungen und Prüfverfahren
31. DVGW (1993) Gasbeschaffenheit, Technische Regeln Gase, Arbeitsblatt G 260, DVGW
32. Arbeitsgemeinschaft für sparsamen und umweltfreundlichen Energieverbrauch e.V. (2000) BHKW und Methanzahl. Kaiserslautern
33. BDEW Bundesverband der Energie- und Wasserwirtschaft e.V. (2009) Gastechnik Zahlen, Daten, Fakten, Berlin

References for Section 4.2

34. BP Statistical Review of World Energy: http://www.bp.com/content/dam/bp/pdf/Energy-economics/statistical-review-2014/BP-statistical-review-of-world-energy-2014-full-report.pdf. Accessed Jan 2015
35. Andruleit H et al. Energy Study 2013 – Reserves, resources and availability of energy resources (17), Bundesanstalt für Geowissenschaften und Rohstoffe (BGR): Hannover. Accessed Jan 2014
36. BDEW Bundesverband der Energie- und Wasserwirtschaft e. V.: www.bdew.de/internet.nsf/id/DE_Bilder-und-Grafiken. Accessed Aug 2014
37. Deutscher Verein des Gas- und Wasserfaches e. V.: www.dvgw.de. Accessed Aug 2014
38. Technische Regel – Arbeitsblatt DVGW G 260 (A) "Gasbeschaffenheit", Hrsg. DVGW, Bonn, Februar 2013, ISSN 0176-3490
39. Richtlinien für die Beschaffenheit des Gases, GWF 82, 11.11.1939, p 745
40. EASEE-gas (European Association for the Streamlining of Energy Exchange-Gas). www.easee-gas.eu. Accessed Aug 2014
41. Graf F, Bajohr S (2013) Biogas – Erzeugung, Aufbereitung und Einspeisung. DIV Deutscher Industrie Verlag, München. ISBN 978-3-8356-3363-6
42. U.S. Energy Information Administration (EIA). (2011) World Shale Gas Resources: an initial Assessment of 14 Regions outside the United States. Washington, DC. www.adv-res.com/pdf/ARI%20EIA%20Intl%20Gas%20Shale%20APR%202011.pdf. Accessed Aug 2014
43. U.S. Energy Information Administration (EIA) (2013) Technically recoverable shale oil and shale gas resources: an assessment of 137 shale formations in 41 countries outside the United States. Washington, DC. www.eia.gov/analysis/studies/worldshalegas/pdf/fullreport.pdf. Accessed Aug 2014
44. Asociacion Espanola De Companias De Investigacio, Exploracion Y Produccion De Hidrocyburos Y Almacenamiento Subterraneo (ACIEP) (2013) Evaluación preliminar de los recursos prospectivos de hidrocarburos convencionales y no convencionales en España. http://aciep.com/publicaciones-oficiales. Accessed Aug 2014
45. Andruleit H et al (2013) Energiestudie 2012 – Reserven, Ressourcen und Verfügbarkeit von Energierohstoffen (16). Bundesanstalt für Geowissenschaften und Rohstoffe (BGR), Hannover
46. Polish Geological Institute National Research Institute (PGI) (2011) Assessment of shale gas and shale oil resources of the lower Paleozoic baltic-podlasie-lublin basin in Poland. www.pgi.gov.pl/en/dokumenty-in/doc_view/769-raport-en.html. Accessed Aug 2014
47. Johnson C, Boersma T (2013) Energy (in) security in Poland – the case of shale gas. Energy Policy 53:389–399
48. Horsfield B, Schultz HM (2010) Shale gas research: the way forward in Europe. Oilfield Technology (4):14–18
49. http://s.hswstatic.com/gif/oil-drilling-derrick.gif. Accessed Dec 2014

50. Cerbe G et al (2008) Grundlagen der Gastechnik. Hanser, München. ISBN 978-3-446-41352-8
51. Homepage of Wirtschaftsverband Erdöl- und Erdgasgewinnung e. V. (WEG). www.erdoel-erdgas.de. Accessed Aug 2014
52. Fox J (2010) Gasland, Film, USA. www.youtube.com/watch?v=mmf_JYrEhyw. Accessed Aug 2014
53. Frimmel FH, Ewers U, Schmitt-Jansen M, Gordalla B, Altenburger R (2012) Toxikologische Bewertung von Fracking-Fluiden. Wasser und Abfall 14(6):22–29
54. Frimmel FH, Gordalla B, Ewers U (2013) Unkonventionelle Gasgewinnung in Deutschland – Was bedeutet das für das Wasser? GWF – Gas/Erdgas 154(1–2):80–86
55. Umweltbundesamt (2012) Umweltauswirkungen von Fracking bei der Aufsuchung und Gewinnung von Erdgas aus unkonventionellen Lagerstätten – Risikobewertung, Handlungsempfehlungen und Evaluierung bestehender rechtlicher Regelungen und Verwaltungsstrukturen, Gutachten, Dezember 2012. www.umweltbundesamt.de/publikationen. Accessed Aug 2014
56. Umweltbundesamt (2014) Umweltauswirkungen von Fracking bei der Aufsuchung und Gewinnung von Erdgas insbesondere aus Schiefergaslagerstätten - Teil 2: Grundwassermonitoringkonzept, Frackingchemikalienkataster, Entsorgung von Flowback, Forschungsstand zur Emissions- und Klimabilanz, induzierte Seismizität, Naturhaushalt, Landschaftsbild und biologische Vielfalt, Gutachten, Juli 2014. www.umweltbundesamt.de/publikationen. Accessed Aug 2014
57. Andruleit H et al (2009) Energiestudie 2009 – Reserven, Ressourcen und Verfügbarkeit von Energierohstoffen. Hrsg. Bundesanstalt für Geowissenschaften und Rohstoffe (BGR), Hannover, Dezember 2009
58. Kidnay AJ, Parrish WR (2006) Fundamentals of natural gas processing. CRC Press, Boca Raton, FL, ISBN 978-0-8493-3406-1
59. Exxon Mobil Production Deutschland GmbH (2009) Die Aufbereitung von Erdgas, Broschüre, Hannover. www.exxonmobil.com/Germany-German/PA/Files/about_what_upstream_technology_Erdgasaufbereitung.pdf. Accessed Aug 2014
60. Haynes WM et al (2014) Handbook of chemistry and physics (95th edn). CRC Press, Boca Raton, FL. Internet Version: www.hbcpnetbase.com. Accessed Aug 2014

References for Section 4.3

61. Air Products LNG Technology brochure. http://www.airproducts.de/~/media/Files/PDF/industries/lng/en-lng-technology-brochure.pdf
62. Linde LNG Technology. http://www.linde-engineering.com/internet.global.lindeengineering.global/en/images/LNG_1_1_e_13_150dpi19_4577.pdf
63. Kidnay AJ, Parrish WR (2011) Fundamentals of natural gas processing. CRC Press Taylor & Francis Group, 2. Auflage
64. Mokhatab S, Mak JY, Valappil JV, Wood DA (2013) Handbook of liquefied natural gas. Gulf Professional Publishing, 1. Auflage
65. Schier S (2008) Das Snohvit-Projekt – Erdgasverflüssigung nördlich des Polarkreises; GWF 149, Nr. 2
66. Fasold, HG (2011) Long-term gas supply for Europe – Pipeline projects and LNG-chains; GWF International (1):S. 22–28

67. Okamura T (2008) Future forecast for life cycle greenhouse gas emissions of LNG and city gas 13A. IGRC, Paris
68. IEA World Energy Investment Outlook (2014) International energy agency, www.iea.org
69. ARI Inc. and ICF Int. (2008) Greenhouse gas life-cycle emissions study: fuel life-cycle of US natural gas supplies and international LNG. Report prepared for Sempra LNG, 10, November 2008. http://www.cpuc.ca.gov/NR/rdonlyres/430BCFA0-37C6-4E62-AD2C-7C3A34013181/0/ARI_LCA_NOV_10_08.pdf (08/2014)
70. PACE (2009) Life cycle assessment of GHG emissions from LNG and coal fired generation scenarios: assumptions and results; Feb 2009. http://www.lngfacts.org/resources/LCA_Assumptions_LNG_and_Coal_Feb09.pdf (08/2014)
71. Barnett (2010) Life Cycle Assessment (LCA) of Liquefied Natural Gas (LNG) and its environmental impact as a low carbon energy source; Dissertation Okt
72. IGU Word LNG Report – 2014 Edition, www.igu.org (03/2015)
73. Gas Infrastructure Europe, www.gie.eu (03/2015)
74. Eurogas Statistical Report, 2014, www.eurogas.org (03/2015)
75. www.fnb-gas.de/de/netzentwicklungsplan/netzentwicklungsplan.html (03/2015)
76. www.entsog.eu (03/2015)
77. Grundlagen der Gastechnik, Hrsg. G Cerbe, 7. Auflage, Carl Hanser Verlag, München, Wien (2008)
78. Gas2energy.net – Systemplanerische Grundlagen der Gasversorgung, Hrsg. J Mischner, H-G Fasold, K Kadner, Oldenbourg Industrieverlag GmbH, München (2011)
79. Gasdruckregelung und Gasdruckregelanlagen, Hrsg. B Naendorf, 4. Auflage, Deutscher Industrieverlag (2014)
80. Gasmessung und Gasabrechnung, Hrsg. U Wernekinck, 4. Auflage, Oldenbourg Industrieverlag GmbH, München (2012)
81. Odorierung, Hrsg. K Pritsching, 3. Auflage, Vulkan-Verlag (2010)
82. Mastering future challenges with gas innovations (2014) www.dvgw-innovation.de (03/2015)
83. Untertage-Gasspeicherung in Deutschland. Erdöl, Erdgas, Kohle 130(11):S. 402–412 (2014)
84. Einfluss von Biogas und Wasserstoff auf die Mikrobiologie in Untertagespeichern, DGMK-Forschungsbericht 756 (2013)
85. Westphal K (2014) Die internationalen Gasmärkte: Von großen Veränderungen und Herausforderungen für Europa. Energiewirtschaftliche Tagesfragen 64(1/2):47–50
86. Golden Rules for a Golden Age of Gas World Energy Outlook Special Report on Unconventional Gas, www.iea.org (02/2013)
87. IEA World Energy Outlook (2014) International energy agency, www.iea.org
88. Global Vision for Gas – The Pathway towards a Sustainable Energy Future, Studie International Gas Union, Juni 2012
89. Energiefahrplan 2050, Mitteilung der Kommission an das Europäische Parlament, den Rat, den Europäischen Wirtschafts- und Sozialausschuss und den Ausschuss der Regionen, Brüssel Dezember 2011, ec.europa.eu (02/2013)
90. EU-Kommission: EU energy, transport and GHG emissions trends to 2050 Reference scenario 2013 (16.12.2013), ec.europa.eu (02/2014)
91. BP Energy Outlook 2035, www.bp.com (03/2015)
92. BP Statistical Review of world energy June 2014

References for Section 4.4.1

93. AC Biogas: agri.capital erwirbt Biomethananlage Pliening von der r.e Bioenergie. Münster 2012
94. AEE, Agentur für Erneuerbare Energien: Anbau von Energiepflanzen. Umweltauswirkungen, Nutzungskonkurrenzen und Potenziale. Berlin 2013, S. 5 und S. 12
95. AEE, Agentur für Erneuerbare Energien: Reststoffe für Bioenergie nutzen. Potenziale, Mobilisierung
96. Baur F (2010) Effiziente Nützung von Biomasse – Reststoffe, Nutzungskonkurrenzen und Kaskadennutzung. In: FVEE – Themen 2010 (2010), S. 88–92
97. BDEW, Bundesverband der Energie- und Wasserwirtschaft e.V.: http://www.bdew.de/internet.nsf/res/1500F0BC37AA46C1C1257BCC004CA1E7/$file/BDEW-13-00025_Erdgas_Kapitel_6.pdf (2014).
98. Biogasrat+ e.V., Fachverband Biogas, KTBL 2013
99. Biogasrat+ e.V.: Biogas und Landwirtschaft. Berlin 2011, S. 33 f., S. 42
100. Dany C (2013) Das Pferd von hinten aufzäumen. Biogas J 1:86–92
101. DBFZ, Deutsches BiomasseForschungsZentrum: Biomassepotenziale und Nutzungskonkurrenzen. Berlin/Leipzig 2013, S. 25
102. DBFZ, Deutsches BiomasseForschungsZentrum: Endbericht – Globale und regionale räumliche Verteilung von Biomassepotenzialen. Status Quo und Möglichkeiten der Präzisierung. Leipzig 2010, S. 64
103. dena, Deutsche Energie-Agentur: biogaspartner – gemeinsam einspeisen. Biogaseinspeisung in Deutschland und Europa – Markt, Technik und Akteure. Berlin 2013, S. 35
104. dena, Deutsche Energie-Agentur: Erdgas und Biomethan im künftigen Kraftstoffmix. Handlungsbedarf und Lösungen für eine beschleunigte Etablierung im Verkehr, Berlin 2011
105. erdgas mobil: Erdgasfahrzeuge im Aufwind: Neuzulassungen steigen im ersten Quartal 2014 um 20 Prozent. 2014
106. FNR, Fachagentur Nachwachsende Rohstoffe: Biogas. 9. Auflage, Gülzow 2013
107. FNR, Fachagentur Nachwachsende Rohstoffe: Leitfaden Biogas. Von der Gewinnung zur Nutzung. 6. Auflage, Gülzow 2013
108. FNR, Fachagentur Nachwachsende Rohstoffe; BMELV: Biokraftstoffe. 3. Auflage, Gülzow 2007
109. gibgas GmbH (2014): http://www.gibgas.de/Fahrzeuge/Historie/Statistik (25.06.2014)
110. gibgas GmbH (2014): http://www.gibgas.de/Tankstellen/Service/Statistik (25.06.2014)
111. Kern M, Sprick W (2011) Bioabfallvergärung in Deutschland – Potenzial, Technik und Perspektiven. In: 5. Biomasse-Forum 2011, S. 61–74
112. Raussen T, Kern M (2014) Stand und Perspektiven der Bioabfällvergarung in Deutschland. Biogas Atlas 2014(15):19–28
113. Schüch A, Daniel-Gromke J, Liebetrau J, Nelles M (2014) Stand und Perspektiven der Abfall- und Reststoffvergärung in Deutschland. Biogas J 2:34–38
114. Verordnung über die Verwertung von Bioabfällen auf landwirtschaftlich, forstwirtschaftlich und gärtnerisch genutzten Böden (Bioabfallverordnung – BioAbfV) 2013
115. FNR, Fachagentur Nachwachsende Rohstoffe: Bioenergie die vielfältige erneuerbare Energie. 5. Auflage, Gülzow 2013
116. DWA, Deutsche Vereinigung für Wasserwirtschaft, Abwasser und Abfall e.V.: 25. Leistungsvergleich kommunaler Kläranlagen. Hennef 2012, S. 3
117. BMU, Bundesministerium für Umwelt, Naturschutz und Reaktorsicherheit; UBA, Umweltbundesamt: Ökologisch sinnvolle Verwertung von Bioabfällen. Anregungen für kommunale Entscheidungsträger. Berlin, Dessau-Roßlau 2012
118. NABU Schleswig-Holstein: Die sogenannte „Gute fachliche Praxis". URL: http://schleswig-holstein. nabu.de/themen/landwirtschaft/hintergrund/14504.html (2012)

References for Section 4.4.2

119. IEA (2013) World energy outlook. International Energy Agency, Paris
120. BMWI (2014) Monitoring-Bericht nach § 51 EnWG zur Versorgungssicherheit bei Erdgas. Bundesministerium für Wirtschaft und Energie, Berlin
121. European Commission (2003) World energy, technology and climate policy outlook. European Commission, Brussels
122. Müller-Langer F (2012) Analyse und Bewertung ausgewählter zukünftiger Biokraftstoffoptionen auf der Basis fester Biomasse. DBFZ Report Nr. 9, ISSN: 2190-7943, Leipzig
123. Bergman PCA, Boersma AR, Kiel JHA, Prins MJ, Ptasinski KJ, Janssen FJJG (2005) Torrefaction for entrained-flow gasification of biomass. Energy Research Centre of the Netherlands (ECN), ECN-C-05-067, Petten
124. Kietzmann F, Klemm M, Blümel R, Clemens A (2013) Demonstrationsanlage für HTC-Kohle in Halle – abfallwirtschaftliche Ziele und technische Umsetzung. 25. Kasseler Abfall- und Bioenergieforum, Kassel, 16–18 Apr 2013
125. Klemm M, Schmersahl R, Kirsten C (2012) Biofuels: upgraded "new" solids. In: Meyers RA (ed) Encyclopedia of sustainability science and technology. Springer, New York
126. Funke A, Ziegler F (2010) Hydrothermal carbonization of biomass: a summary and discussion of chemical mechanisms for process engineering. Biofuels Bioprod Biorefin 4
127. Mitchell P, Kiel J, Livingston B, Dupont-Roc G (2007) Torrefied biomass – a foresighting study into the business case for pellets from torrefied biomass as a new solid fuel. University of Aberdeen, ECN, Doosan Babcock, ITI Energy, Conference All Energy
128. Kaltschmitt M, Hartmann H, Hofbauer H (Hrsg) (2009) Energie aus Biomasse: Grundlagen, Techniken und Verfahren. 2. Auflage. Springer, Berlin
129. Knoef H (2005) Handbook biomass gasification. BTG biomass technology group, Enschede
130. Vogel A (2006) Technologische und verfahrenstechnische Untersuchungen. Schriftenreihe "Nachwachsende Rohstoffe", Band 29, Analyse und Evaluierung der thermo-chemischen Vergasung von Biomasse. Landwirtschaftsverlag Münster
131. Karl J (2004) Wirbelschichtvergasung – Stand der Technik. Schriftenreihe "Nachwachsende Rohstoffe" Band 24, Münster
132. Klemm M (2012) Biomass gasification for rural electrification, small scale. In: Meyers RA (ed) Encyclopedia of sustainability science and technology. Springer, New York
133. Hofbauer H, Rauch R, Fürsinn S, Aichernig CH (2006) Energiezentrale Güssing – Energiezentrale zur Umwandlung von biogenen Roh- und Reststoffen einer Region in Wärme, Strom, Bio-SNG und flüssige Kraftstoffe. Schriftenreihe 79, Technische Universität Wien, Repotec, Berichte aus Energie- und Umweltforschung, Wien
134. Kiendl I, Klemm M, Clemens A, Herrmann A (2014) Dilute gas methanation of synthesis gas from biomass gasification. Fuel 123:211–217
135. Haldor Topsøe A/S (2009) From solid fuels to substitute natural gas (SNG) using TREMP™. Brochure Haldor Topsøe
136. Mozaffarian M, Zwart RWR (2003) Feasibility of biomass/waste-related SNG production technologies. Energy Research Centre of the Netherlands (ECN), ECN-C-03-066, Petten
137. Urban W, Althaus W (2005) Analyse und Bewertung der Nutzungsmöglichkeiten von Biomasse – Band 3: Biomassevergasung, Technologien und Kosten der Gasaufbereitung und Potenziale der Biogaseinspeisung in Deutschland. Untersuchung im Auftrag von BGW und DVGW, Fraunhofer-Institut für Umwelt-, Sicherheits- und Energietechnik (Fraunhofer UMSICHT), unveröffentlichter Endbericht, Oberhausen
138. Biollaz S, Seemann M, Schildhauer TJ, Stucki S (2005) Methan-aus-Holz: Phase 1, Erzeugung von synthetischem Erdgas aus Holz zur Nutzung als Biotreibstoff (CNG/CMG).

Schlussbericht, Paul Scherrer Institut (PSI) im Auftrag des Bundesamtes für Energie Bern, Gazobois SA, Villigen
139. Wokaun A (2003) Ecogas – Teilprojekt: Methan aus Holz. Schlussbericht, Paul Scherrer Institut (PSI), Villigen
140. Deutsche Vereinigung Des Gas- Und Wasserfaches E.V. (2000) Arbeitsblatt DVGW-G 260: Gasbeschaffenheit. Deutsche Vereinigung des Gas- und Wasserfaches e.V., Technisch-wissenschaftlicher Verein, Bonn
141. Deutsche Vereinigung Des Gas- Und Wasserfaches E.V. (2004) Arbeitsblatt DVGW-G 262: Nutzung von Gasen aus regenerativen Quellen in der öffentlichen Gasversorgung. Deutsche Vereinigung des Gas- und Wasserfaches e.V., Technisch-wissenschaftlicher Verein, Bonn
142. Din Deutsches Institut Für Normung E. V. (2008) Kraftstoffe für Kraftfahrzeuge – Erdgas – Anforderungen und Prüfverfahren (DIN 51624:2008-02). Deutsches Institut für Normung e.V., Berlin
143. Urban W, Girod K, Lohmann H (2008) Technologien und Kosten der Biogasaufbereitung und Einspeisung in das Erdgasnetz – Ergebnisse der Markterhebung 2007-2008. Fraunhofer-Institut für Umwelt-, Sicherheits- und Energietechnik (Fraunhofer UMSICHT), Oberhausen
144. E-FLOX GMBH (2014) Schwachgasbrenner. Renningen. http://www.e-flox.de/Produkte/Schwachgasbrenner/tabid/60/Default.aspx (letzter Zugriff: Oktober 2014)
145. Haase Energietechnik AG: Autotherme Oxidation für Abluft und Schwachgase: VocsiBox®. Broschüre, Neumünster, FE-366/6.2002 RD
146. Ullmanns Encyclopedia of Industrial Chemistry (2002) Methanation and methane synthesis. Gas production, vol 15, 6. überarb. Ausgabe, Wiley-VCH
147. Perry M, Eliason D (2004) CO2 recovery and sequestration at Dakota Gasification Company – a progress report. Gasification Technologies Council, Arlington
148. U. S. Doe (2006) Practical experience gained during the first twenty years of operation of the great plains gasification plant and implications for future projects. U. S. Department of Energy Office of Fossil Energy
149. Rönsch R (2011) Optimierung und Bewertung von Anlagen zur Erzeugung von Methan, Strom und Wärme aus biogenen Festbrennstoffen. DBFZ Report Nr. 5, ISSN: 2190-7943, Leipzig
150. Risø DTU; DGC Dansk Gastenisk Center A/S (2010) Bio-SNG potential assessment: Denmark 2020. Risø-R-1754 (EN) Report, Roskilde
151. Chandal M, Williams E (2009) Synthetic Natural Gas (SNG): technology, environmental implications, and economics. Climate Change Policy Partnership
152. Sudiro M, Bertucco A (2010) Synthetic Natural Gas (SNG) from coal and biomass: a survey of existing process technologies, open issues and perspectives. Natural Gas, Primoz Potocnik (Hrsg), InTech
153. Kopyscinski J, Schildhauer TJ, Biollaz SMA (2010) Production of synthetic natural gas (SNG) from coal and dry biomass – a technology review from 1950 to 2009. Fuel 89:1763–1783
154. European Biofuels: http://www.biofuelstp.eu/bio-sng.html (last downloaded October 2014)
155. Specht M, Zuberbühler U (2013) Power-to-Gas (P2G®): Innovative Systemlösung zur Integration erneuerbarer Energien. dena-Energieeffizienzkongress, Berlin, 25.11.2013
156. Van Der Drift B, Biollaz S, Waldheim L, Rauch R, Manson-Whitton CH (2012) Status and future of bio-SNG in Europe. IEA-bioenergy conference, Vienna
157. Göteborg Energi. http://www.goteborgenergi.se/English/Projects (last downloaded 31.08.2014)
158. Gaya Project. http://www.projetgaya.com/en/calendar/ (last downloaded 25.02.2015)
159. GÖTEBORG ENERGI AB, Göteborg Energi's biogas plant GoBiGas is now fully operational, Press release 18.12.2014.
160. E.ON Sverige AB. https://www.eon.se/om-eon/Om-energi/Produktion-av-el-gas-varme-och-kyla/Bio2G/Lokalisering-tillstand-och-samrad/ (last downloaded 27.02.2015)

161. European Commission (2011) High-level expert group on key enabling technologies. European Commission, Brussels
162. Müller-Langer F, Majer S, O'Keeffe S (2014) Benchmarking biofuels – a comparison of technical, economic and environmental indicators. Energy Sustain Soc 4:20
163. Müller-Langer F, Rönsch S, Weithäuser M, Oehmichen K, Seiffert M, Majer S, Scholwin F (2009) Erdgassubstitute aus Biomasse für die mobile Anwendung im zukünftigen Energiesystem. Endbericht zum Forschungsvorhaben FZK 22031005, Fachagentur Nachwachsende Rohstoffe e. V.
164. Rönsch S, Zeymer M, Majer S (2010) Treibhausgasvermeidungskosten von synthetischem Methan und Methanol aus Biomasse und Braunkohle. Chem Ing Tech 86(10):1678–1689
165. European Union (2009) Directive 2009/28/EC of the European Parliament and of the Council of 23 April 2009 on the promotion of the use of energy from renewable sources and amending and subsequently repealing Directives 2001/77/EC and 2003/30/EC. Official Journal of the European Union used by inland waterway vessels and repealing Directive 93/12/EEC. Official Journal of the European Union

References for Section 4.4.3

166. Umweltbundesamt: Energieziel 2050: 100% Strom aus erneuerbaren Quellen. Umweltbundesamt, Reihe Broschüren/Faltblätter 2010 (www.umweltbundesamt.de/sites/default/files/medien/publikation/long/3997.pdf)
167. Icha, P.: Entwicklung der spezifischen Kohlendioxid-Emissionen des deutschen Strommix in den Jahren 1990 bis 2013. Umweltbundesamt, Reihe Climate Change 2014 (www.umweltbundesamt.de/publikationen)
168. Sunbeam GmbH: Website "Multitalent Biogas", Artikel "Biogas: das Multitalent für die Energiewende" 2014 (http://www.multitalent-biogas.de/branche/biogas-das-multitalent-fuer-die-energiewende)
169. DVGW, Deutscher Verein des Gas- und Wasserfaches e.V.: DVGW Arbeitsblätter G260 (Gasbeschaffenheit) und G262 (Nutzung von Gasen aus regenerativen Quellen in der öffentlichen Gasversorgung), 2008
170. Beuth: DIN 51624, Kraftstoffe für Kraftfahrzeuge – Erdgas – Anforderungen und Prüfverfahren, 2008. Norm abrufbar unter http://www.beuth.de/de/norm/din-51624/104466710
171. Purr K et al (2014) Treibhausgasneutrales Deutschland im Jahr 2050. Umweltbundesamt, Reihe Climate Change (http://www.umweltbundesamt.de/publikationen/treibhausgasneutrales-deutschland-im-jahr-2050-0)
172. Sterner M et al.: Die Speicheroption Power-to-Gas: Ausgleichs- und Integrationsmaßnahmen für EE. Fraunhofer IWES 2011

References for Section 4.4.3.1

173. Specht M, Baumgart F, Feigl B, Frick V, Stürmer B, Zuberbühler U, Sterner M, Waldstein G (2010) Storing bioenergy and renewable electricity in the natural gas grid, AEE Topics 2009. Renewable Energy Research Association (FVEE), Berlin, p 69
174. Specht M, Brellochs J, Frick V, Stürmer B, Zuberbühler U, Sterner M, Waldstein G (2010) Speicherung von Bioenergie und erneuerbarem Strom im Erdgasnetz (Storage of bio-energy and renewable electricity to the natural gas grid). Erdöl Erdgas Kohle 126(10):342

175. Specht M, Zuberbühler U, Baumgart F, Feigl B, Frick V, Stürmer B, Sterner M, Waldstein G (2011) Storing renewable energy in the natural gas grid as methane via Power-to-Gas (P2G): a renewable fuel for mobility. In: Proceedings of the 6th conference "gas powered vehicles – the real and economical CO_2-alternative", p 98, Stuttgart, Germany, 26th/27th October
176. Sabatier P, Senderens J-B (1902) Nouvelles synthèses du méthane (New method for the synthesis of methane). Comptes Rendus hebdomadaires des séances de l'academie des sciences, Paris, p 514
177. Zagli E, Falconer JL (1981) Carbon dioxide adsorption and methanation on ruthenium. J Catal 69:1
178. Deutscher Verein des Gas- und Wasserfaches e.V. (DVGW) (2011) Arbeitsblatt DVGW G 262 – Nutzung von Gasen aus regenerativen Quellen in der öffentlichen Gasversorgung (German Technical and Scientific Association for Gas and Water (DVGW): work sheet DVGW G 262 – utilization of gases from regenerative sources in the public gas supply), Bonn: Wirtschafts- und Verlagsgesellschaft Gas und Wasser mbH
179. Deutscher Verein des Gas- und Wasserfaches e.V. (DVGW) (2012) Arbeitsblatt DVGW G 260 – Gasbeschaffenheit (German Technical and Scientific Association for Gas and Water (DVGW): work sheet DVGW G 260 – gas quality). Wirtschafts- und Verlagsgesellschaft Gas und Wasser mbH, Bonn
180. Frick V (2013) Erzeugung von Erdgassubstitut unter Einsatz kohlenoxid-haltiger Eduktgase – Experimentelle Untersuchung und simulationsgestützte Einbindung in Gesamtprozessketten (Production of substitute natural gas from carbon oxide containing educt gases – Experimental investigation and simulation-supported inclusion in overall process chains), Berlin: Logos Verlag GmbH, Dissertation University of Stuttgart
181. Seglin L, Geosits R, Franko B, Gruber G (1974) Survey of methanation chemistry and processes. In: L. Seglin (ed) Methanation of Synthesis Gas, Advances in Chemistry Series No. 146. American Chemical Society, Washington, DC, p 1
182. Wollmann A, Benker B, Keich O, Bank R (2009) Potenzialermittlung eines modifizierten Festbett-Rohrreaktors (Determination of the potential for a modified fixed bed tube reactor). Chemie Ingenieur Technik 81(7):941
183. DEG (2014): DEG-Engineering GmbH, Gelsenkirchen, Germany
184. ZSW (2013) Patent application PCT/EP2013/071095
185. BMU (2011) Public Fund of the Federal Ministry for the Environment, Nature Conservation and Nuclear Safety in Germany (BMU), funded project no.: 0325275 A-C
186. Audi (2014): Audi AG, Ingolstadt, Germany
187. FNR (2014) Agency for renewable resources (FNR). http://biogas.fnr.de/daten-und-fakten/ (date of retrieval: 15.09.2014)

References for Section 4.4.3.2

188. Leuchtenberger A (1998) Grundwissen zur mikrobiellen Biotechnologie: Grundlagen, Methoden. Verfahren und Anwendungen Chemie in der Praxis, Teubner, Stuttgart, Leipzig
189. Burkhardt M, Busch G (2013) Methanation of hydrogen and carbon dioxide. Appl Energy 111:74–79
190. Lee JC, Kim JH et al (2012) Biological conversion of CO_2 to CH_4 using hydrogenotrophic methanogen in a fixed bed reactor. J Chem Technol Biotechnol 87(6):844–847
191. Trechow P (2011) Windstrom treibt Erdgasfahrzeug an; VDI- Nachrichten

192. Schmuck S, Keser B, Mietzel T (2010) Biological Sabatier process and its potential for the electricity industry
193. Ako OY, Kitamura Y et al (2008) Steady state characteristics of acclimated hydrogenotrophic methanogens on inorganic substrate in continuous chemostat reactors. Bioresour Technol 99 (14):6305–6310
194. Yang Y, Zhang ZY et al (2004) Continuous methane fermentation and the production of vitamin B12 in a fixed-bed reactor packed with loofah. Bioresour Technol 92(3):285–290
195. Nishimura N, Kitaura S et al (1991) Growth of thermophilic methanogen KN-15 on H_2-CO_2 under batch and continuous conditions. J Ferment Bioeng 72(4):280–284
196. Nishimura N, Kitaura S et al (1992) Cultivation of thermophilic methanogen KN-15 on H_2-CO_2 under pressurized conditions. J Ferment Bioeng 73(6):477–480
197. Yano T, Aoki K, Nagai S (1991) Kinetics of CH_4 production from H_2 and CO_2 in a hollow fiber reactor by plug flow reaction model. J Ferment Bioeng 71(3):203–205
198. Lundbäck KM, Klasson KT et al (1990) Kinetics of growth and hydrogen uptake by Methanobacterium formicicum. Biotechnol Lett 12(11):857–860
199. Peillex J-P, Fardeau M-L, Belaich J-P (1990) Growth of Methanobacterium thermoautotrophicum on H_2 and CO_2: high CH_4 productivities in continuous culture. Biomass 21 (4):315–321
200. Peillex J-P, Fardeau M-L et al (1988) Growth of Methanococcus thermolithotrophicus in batch and continuous culture on H_2 and CO_2: influence of agitation. Appl Microbiol Biotechnol 29:560–564
201. Jee HS, Nishio N, Nagai S (1988) Continuous CH_4 Production from H_2 and CO_2 by Methanobacterium thermoautotrophicum in a fixed-bed reactor. J Ferment Technol 66 (2):235–238
202. Jee HS, Yano T et al (1987) Biomethanation of H_2 and CO_2 by Methanobacterium thermoautotrophicum in membrane and ceramic bioreactors. J Ferment Technol 65(4):413–418
203. Ariga O, Okumura T et al (1984) Biomethanation under elevated pressure. J Chem Eng Jpn 17 (6):577–582
204. Wise DL, Cooney CL, Augenstein DC (1978) Biomethanation: Anaerobic fermentation of CO_2, H_2 and CO to methane. Biotechnol Bioeng XX:1153–1172
205. Sterner M (2009) Bioenergy and renewable power methane in integrated 100% renewable energy systems: Limiting global warming by transforming energy systems Erneuerbare Energien und Energieeffizienz – Renewable Energies and Energy Efficiency, Vol. 14, Kassel University Press, Kassel
206. CIEC Promoting Science (2013) Chemical reactors, http://www.essentialchemicalindustry.org/processes/chemical-reactors.html
207. Fachverband Biogas (2013) Branchenzahlenprognose für die Jahre 2014 und 2015, http://www.biogas.org/edcom/webfvb.nsf/id/DE_Branchenzahlen/$file/14-11-25_Biogas%20Branchenzahlen_Prognose_2014-2015.pdf
208. Edel M, Blume A et al (2014) Branchenbarometer Biomethan: Daten, Fakten und Trends zur Biogaseinspeisung, Berlin
209. Cheng S, Logan BE Electromethanogenic reactor and processes for methane production WO 2009/155587 A3, 2009
210. Logan BE Grot S et al Bio-electrochemically assisted microbial reactor that generates hydrogen gas and methods of generating hydrogen gas US 7491453 B2, 2005
211. Deutsche Energie- Agentur GmbH (dena): Power-to-Gas: Eine innovative Systemlösung auf dem Weg zur Marktreife, Berlin, 2013

References for Section 4.5

212. Schnürle A (1939) Die Gasmaschine. List-Reihe Heft 5, Wien 1939
213. NN Kraftstoff CNG (2011) – Fahren mit Erdgas: 20 Jahre Erdgas-Fahrzeuge in Österreich. In: Zeitschrift der österreichischen Vereinigung fur das Gas- und Wasserfach. Forum Special 5
214. Herdin G, Zankl M (2011) Die Nutzung von Synthesegas im Gasmotor. Beitrag zur 13. Tagung "Der Arbeitsprozess des Verbrennungsmotors", Graz
215. Utz J, Smaling R (2005) Verbesserte Verbrennung durch Wasserstoffanreicherung – Ein vielversprechendes Konzept für den homogenen Magerbetrieb. MTZ 10:785–791
216. Müller-Syring G (2012) Entwicklung von modularen Konzepten zur Erzeugung, Speicherung und Einspeisung von Wasserstoff und Methan ins Erdgasnetz. DVGW Forschung 2012
217. Florisson O (2006) The Value of the existing natural gas system for hydrogen, the sustainable future energy carrier (progress obtained in the NATURALHY project). 23rd World Gas Conference, Amsterdam
218. Ma F, Yu W, Liu H, Wang J, Zhao S (2007) Experimental study on thermal efficiency and emission characteristics of a lean burn hydrogen enriched natural gas engine. Int J Hydrogen Energy 32:5067–5075
219. McTaggart-Cowan GP, Jones HL, Rogak SN, Bushe WK, Hill PG, Munshi SR (2006) Direct-injected hydrogen-methane mixtures in a heavy duty compression ignition engine. SAE Paper No. 2006-01-0653, 2006
220. Munshi S Medium/Heavy duty hydrogen enriched natural gas spark ignition IC-Engine operation. 1st International Symposium on Hydrogen Internal Combustion Engines. VKM-THD Mitteilungen, Heft 88, Institut fur Verbrennungskraftmaschinen & Thermodynamik, Technische Universitat Graz 2006
221. Munshi S, Nedelcu C, Harris, J et al (2004) Hydrogen blended natural gas operation of a heavy duty turbocharged lean burn spark ignition engine. SAE Paper No. 2004-01-2956
222. Wang J, Huang Z, Fang Y, Liu B, Zeng Z, Miao H, Jang D (2007) Combustion behaviours of a direct-injection engine operating on various fractions of natural gas-hydrogen blends. Int J Hydrogen Energy 32:3555–3564
223. Eichlseder H, Klell M (2012) Wasserstoff in der Fahrzeugtechnik – Erzeugung, Speicherung, Anwendung. 3. Auflage, Wiesbaden
224. Eichlseder H, Klell M, Schaffer K, Leitner D, Sartory M (2009) Potential synergies in a vehicle for variable mixtures of CNG and hydrogen. SAE 2009-01-1420 in: Hydrogen IC Engines, SP-2251, ISBN 9780768021479, SAE International 2009
225. Bach C, Butler T, Huber M (2013) Fuel consumption and emissions investigation on a HCNG operated passenger car, EMPA Dubendorf
226. Skalla C (2011) Gemischbildung und Verbrennung von Wasserstoff/Methan-Gemischen in Pkw-Anwendungen. Dissertation, TU Graz
227. Haslacher R (2011) Entwicklung optischer Messverfahren zur Analyse von Brennverfahren fur PKW Gasmotoren. Dissertation, TU Graz
228. Skalla C, Haslacher R, Eichlseder H (2009) Mischgase mit Wasserstoff – verbrennungs-relevante Eigenschaften und Eignung für die motorische Anwendung. 9. Internationaler Kongress Motorische Verbrennung, München
229. Skalla C, Haslacher R, Leitner D, Eichlseder H (2009) Fahrzeugkonzepte für Wasserstoff – Mischgase als Bruckentechnologie. ATZ/MTZ Konferenz Gasfahrzeuge, Stuttgart 2009
230. Klell M, Eichlseder H, Sartory M (2010) Variable mixtures of hydrogen and methane in the internal combustion engine of a prototype vehicle – regulations, safety and potential. Int J Veh Des 54(2010):2
231. DVGW, Deutscher Verein des Gas und Wasserfaches: Richtlinie G260

232. E.ON Hanse AG, THB – Technische Hinweise und Bestimmungen (Gas) fur das Netzgebiet der E.ON Hanse AG, Stand Mai 2007
233. Fukuda M, Korematsu K, Sakamoto M (1981) On quenching distance of mixtures of methane and hydrogen with air. In: Bulletin of JSME Vol 24. Nor. 193, Paper No. 193-13

References for Section 4.6

234. dena, Deutsche Energie-Agentur (2014) Energiesteuerermäßigung für Erdgas und regeneratives Methan im Verkehr. Position paper. July 29, 2014
235. Hermann H et al (2014) Prüfung der klimapolitischen Konsistenz und der Kosten von Methanisierungsstrategien. Study, Öko-Institut e.V. www.oeko.de/oekodoc/2005/2014-021-de.pdf
236. Gosda H (2014) Ökonomische und ökologische Analyse von Technologiekonzepten zur Reduzierung der Treibhausgasemissionen im Mobilitätssektor. Final thesis at the TU Braunschweig, Braunschweig
237. Seraidou N (2014) Brennstoffzellenpostauto. Ein emissionsfreier Antrieb für unsere Umwelt. VÖV 2014 Symposium
238. Stojanovic Z (2014) Scania's high performance methane gas engines for CNG/LNG. Presentation at the NGVA Europe V show & workshops, Brussels
239. Diess H (2014) BMW i3 und BMW i8. Zeitgemäße Antriebsstränge für eine neue Freude am Fahren. 2014 Vienna Engine Symposium
240. Audi (2013) Future Powertrain Mix 2030. Interne Analyse von 20 Studien zum zukünftigen Pkw-Antriebsmix, Ingolstadt
241. IWES-Fraunhofer (2010) Modellierung einer 100% erneuerbaren Stromerzeugung in 2050. Technical report, Fraunhofer Institute
242. Lenz HP (2011) Die Rolle der Infrastruktur für das Automobil der Zukunft. In: ATZ extra 16, pp 112–117
243. Federal Network Agency (2014) Positionspapier zur Anwendung der Vorschriften der Einspeisung von Biogas auf die Einspeisung von Wasserstoff und synthetischem Methan in Gasversorgungsnetze. http://www.bundesnetzagentur.de/cln_1431/DE/Sachgebiete/ElektrizitaetundGas/Unternehmen_Institutionen/NetzzugangundMesswesen/Gas/Einspeisung_v_H2_u_synth_CH4/Einspeisung_v_H2_u_synth_CH4_node.html
244. DLR, ifeu, LBST, DBFZ: Power-to-Gas (PtG) im Verkehr – Aktueller Stand und Entwicklungsperspektiven. Short study for the BMVI 2014. http://www.bmvi.de/SharedDocs/DE/Anlage/UI-MKS/mks-kurzstudie-ptg.pdf?__blob=publicationFile
245. Block T (2013) Entwicklung eines Konzepts zur Modellierung einer PtG-Anlage im intermittierenden, wirtschaftlich optimalen Betrieb. Master thesis at the University of Flensburg, Flensburg
246. Bernhart et al (2013) Fuel cells – a realistic alternative for zero emission? Roland Berger Stuttgart, Munich, Frankfurt. http://www.rolandberger.de/media/pdf/Roland_Berger_Fuel_cells_20140113.pdf
247. Brunner T (2013) Hydrogen fuel cell technology. Fuel cells and hydrogen joint undertaking. BMW Group. http://www.fch-ju.eu/sites/default/files/3-BRUNNER%20%28ID%20193389%29_0.pdf

Vehicle Development for Natural Gas and Renewable Methane

5

Manfred Adolf, Michael Bargende, Michael Becker, Thorsten B. Bender, Matthias Budde, Albert Ebner, Florian Feix, Günter Figer, Peter Heine, Andreas Jauss, Timm Kehler, Mahir Tim Keskin, Eduard Köhler, Andreas Kufferath, Winfried Langer, David Lejsek, Claudia Petersen, Ulrich Philipp, Ayhan Sarikaya, Rolf Sauerstein, Michael Schaarschmidt, Alexander Schenk, Peter Volz, Sascha Weiske, Florian Winke, Holger Winkelmann, Helge Wollenhaupt, and Klaus Wunderlich

M. Adolf • A. Schenk
BorgWarner Emissions Systems, Moerikestraße 155, 71636 Ludwigsburg, Germany

M. Bargende (✉)
Forschungsinstitut für Kraftfahrwesen und Fahrzeugmotoren Stuttgart (FKFS), Pfaffenwaldring 12, 70569 Stuttgart, Germany

Institut für Verbrennungsmotoren und Kraftfahrwesen (IVK), Universität Stuttgart, Pfaffenwaldring 12, 70569 Stuttgart, Germany
e-mail: Michael.bargende@fkfs.de

M. Becker • M. Budde (✉)
BorgWarner Corporate Advanced R&D, Moerikestraße 155, 71636 Ludwigsburg, Germany
e-mail: mbudde@borgwarner.com

T.B. Bender • A. Jauss (✉) • A. Sarikaya • H. Winkelmann
Adam Opel AG, Friedrich-Lutzmann Ring T1-03, 65423 Rüsselsheim, Germany
e-mail: Andreas.jauss@de.opel.com

A. Ebner • P. Heine • P. Volz • H. Wollenhaupt (✉) • K. Wunderlich (✉)
Daimler AG, Mercedesstraße 137, 70327 Stuttgart, Germany
e-mail: Helge.wollenhaupt@daimler.com; Klaus.k.wunderlich@daimler.com

F. Feix, M.A. • T. Kehler (✉) • C. Petersen • M. Schaarschmidt, M.B.A.
erdgas mobil GmbH, Neustädtische Kirchstraße 8, 10117 Berlin, Germany
e-mail: Timm.kehler@erdgas-mobil.de

R. van Basshuysen (ed.), *Natural Gas and Renewable Methane for Powertrains*, Powertrain, DOI 10.1007/978-3-319-23225-6_5

5.1 Natural Gas Storage and Handling Systems

Andreas Jauss, Holger Winkelmann, Thorsten B. Bender, and Ayhan Sarikaya

Natural gas vehicles are a mature technology already available today for solving the environmental problems caused by the automobile. The advantage is based on the features of the main fuel content methane. While keeping a minimum fuel quality standard, it doesn't matter to the vehicle technology whether the Methane is several million years old or just a few days. There is complete compatibility between fossil and renewable sources for this most simple of all hydrocarbons. Its highest possible ratio between hydrogen and carbon and the resulting features considering *production/availability (fossil & renewable)*, the *justifiable effort in the fuel system* and the *engine and environmental features* make methane an ideal vehicle fuel.

The notation "Methane Vehicle" would be the precise term. In the following however, the historically established term "Natural Gas Vehicle" will be used—the use of methane out of renewable sources is explicitly included. The following chapters are valid for *bivalent* passenger vehicles up to 2.8 t and light commercial vehicles up to 3.5 t gross vehicle weight with natural gas tanks in an under-floor installation (natural gas and gasoline operation). *Monovalent* natural gas vehicles (solely natural gas operation) have not established themselves so far. From the customer's point of view, natural gas fuel systems should have the following features:

- Safety
- Reliable and failure-free operation

G. Figer (✉)
AVL List GmbH, Hans-List-Platz 1, 8020 Graz, Austria
e-mail: Guenter.figer@avl.com

M.T. Keskin (✉) • U. Philipp (✉) • F. Winke (✉)
Institut für Verbrennungsmotoren und Kraftfahrwesen (IVK), Universität Stuttgart, Pfaffenwaldring 12, 70569 Stuttgart, Germany
e-mail: Mahir-tim.keskin@ivk.uni-stuttgart.de; Tim.keskin@web.de; ulrich.philipp@ivk.uni-stuttgart.de; Florian.winke@fkfs.de

E. Köhler (✉)
formerly KSPG AG, Neckarsulm (Germany), Konradusstraße 10, 74821 Mosbach, Germany
e-mail: Eduard.koehler@web.de

A. Kufferath • W. Langer • D. Lejsek (✉)
Robert Bosch GmbH, Postfach 300240, 70442 Stuttgart, Germany
e-mail: David.lejsek@de.bosch.com

R. Sauerstein • S. Weiske
BorgWarner Turbo Systems, Kaiserstr. 1, 67292 Kirchheimbolanden, Germany

- Satisfying real life driving range in comparison to other types of fuel and natural gas vehicle competitors
- No constraints in terms of interior and luggage compartment space, as well as payload.

The fulfillment of the first two criteria is expected from an OEM. Beside the excellence of the natural gas powertrain, the two latter features are crucial in terms of market success. Hence the integration of the natural gas system into the vehicle architecture is an essential challenge, storage space is required for approximately four times the fuel volume for the equivalent energy content compared to premium gasoline (natural gas storage at 3000 psi/ 200 bar). In order to fulfill the driving range requirements, accommodation of the natural gas system should be done in the advanced vehicle development phase or it should be at least *package protected*.

For the vehicle classes mentioned above, high-pressure storage of the methane in its gaseous state has been established (*compressed natural gas* = *CNG*). It is provided by the natural gas grid and compression up to the vehicle filling pressure is done at the filling station. With pressure storage, the gaseous state is present under all vehicle operation conditions, merely some minor ingredients of the natural gas can condense into the liquid state under certain conditions, hence limitation is needed [1]. This especially affects higher hydrocarbons (propane, butane, compressor-oil from filling stations). The choice of material needs to be suitable to the natural gas quality present in the particular market. CNG systems—for example suitable for the European market—might be problematic in other global regions due to non-limited fuel ingredients, e.g. in terms of inner corrosion of the tanks caused by sulfurous compounds.

Beside the storage the handling of the fuel, the feed on demand with a certain quality and the desired injection pressure for the engine, is the task of the natural gas system. Supercharging is recommended for retrieving the optimal efficiency of the engine, when using methane as a fuel, due to its very high knocking resistance and its elevated air demand/stoichiometric fuel ratio. This goes along with a strong variability of the inlet manifold pressure. Hence, for sequential natural gas injection into the inlet manifold (*MPFI*) an adaptation of the injection rail pressure is recommended (*MAP adaptation*). The actuator is the pressure regulator of the natural gas system, which continuously adapts the rail pressure to the present inlet manifold pressure. Prior to natural gas injection into the engine, the necessary quality—as far as it can be influenced by vehicle systems—needs to be provided. Typical related topics are filtration of particles or the separation of liquid up to high-viscosity ingredients from the fuel.

5.1.1 System Overview and Legal Requirements

The following lists important legal requirements for natural gas systems in passenger cars—without the claim for global completeness. They are relevant for type approval of natural gas components, the overall natural gas fuel system and vehicles and are an important design guideline. They are documented and available, hence will not be dealt with in detail into here.

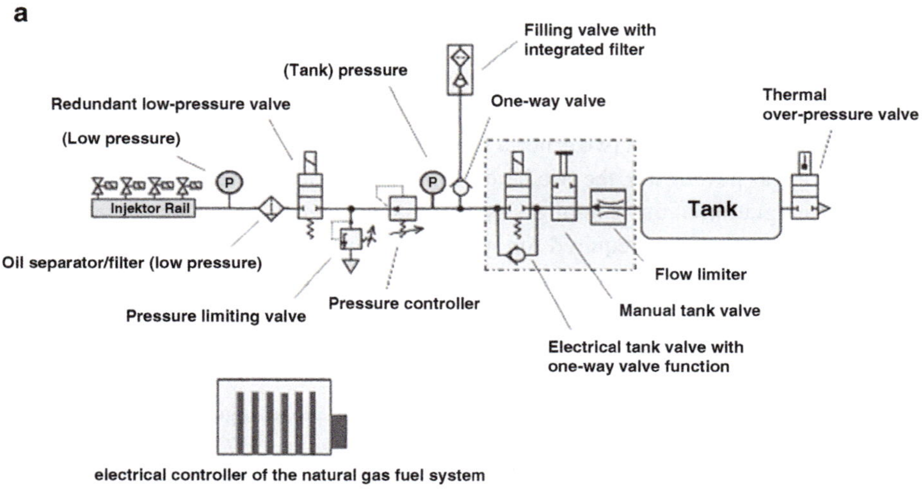

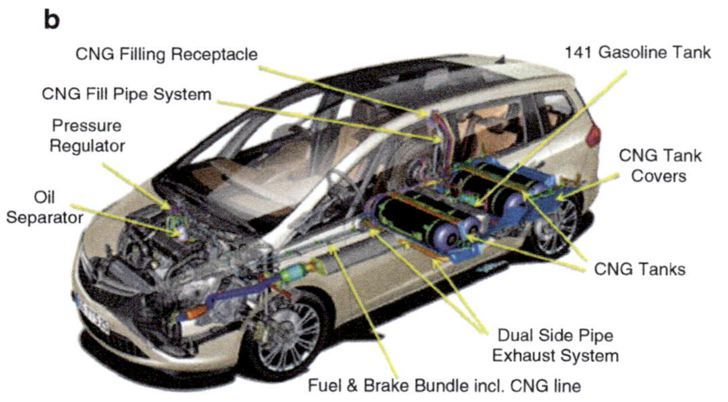

Diagram 5.1 Nature gas fuel system (**a**) block diagram (**b**) in vehicle

Field experience, manufacturer specific in-house testing and validation are mandatory for safe design, see Sect. 5.1.3. The type approval requirements alone are not sufficient. Where indicated, the type approval requirements need to be clearly surpassed in order to provide safe operation over the lifetime. Diagram 5.1 shows the structure of an underfloor natural gas system.

Depending on the legal requirements or the OEM-concept, the basic installation (Diagram 5.1a) can be extended by other components. These can be additional overpressure fuses (*P-PRD*), thermally triggered fuses (*T-PRD*), high pressure filters, redundant high pressure shut off valves as well as manual shut off valves etc.

Essentially two global regions with different natural gas system regulations can be distinguished: *North America* and *Europe*. The basic difference is the permitted nominal storage pressure. In North America (USA and Canada) the nominal value is 3600 psi at 70 °F (248 bar at 21 °C), whereas in Europe according to *ECE R110* it is 200 bar at 15 °C. Many other states have adapted *ECE R110* such as for example Switzerland, Norway, Tunisia, Turkey, The Russian Federation, Belorussia, The Ukraine, Uzbekistan, Japan, Australia, New Zealand, The Republic of Korea, Malaysia, Thailand and The Republic of South Africa.

Europe

In Europe, the *ECE R110* is relevant for type approval of natural gas system components and systems of serial production vehicles. Fulfillment of the *ECE R115* is required for system approval of single registrations of conversions or after-market installations. However the natural gas system components here need to be authorized as well to be in conformity with *ECE R110.*

North America

In North America—in contrary to Europe—the self-obligation of the manufacturers governs the development and testing of components and systems according to the acknowledged rules of technology if they shall be entered into the market. In the USA the *Federal Motor Vehicle Safety Standards (FMVSS)* are the legally binding requirements for natural gas vehicles:

- *FMVSS 301* Fuel System Integrity
- *FMVSS 303* Fuel System Integrity of Compressed Natural Gas Vehicles
- *FMVSS 304* Compressed Natural Gas Fuel Container Integrity

In addition, the following industrial standards also have to be considered as acknowledged rules of technology and need to be incorporated in the design and validation of components and systems:

- *ANSI/CSA NGV2* American National Standard for Compressed Natural Gas Vehicle Fuel Containers
- *ANSI PRD1* American National Standard for Pressure Relief Devices for Compressed Natural Gas Vehicle (NGV) Fuel Containers
- *ANSI NGV3.1/CSA12.3* American National Standard/Canadian Standards Association Standard for Fuel System Components for Compressed Natural Gas Powered Vehicles
- *ANSI NGV1* American National Standard for Compressed Natural Gas Vehicle (NGV) Fueling Connection Devices

In addition, the *Standard Nr. 52 of the National Fire Protection Association (NFPA)* was integrated into the set of requirements. Similar to the *ECE R110*, there are defined requirements for components and systems for serial and aftermarket installation. This standard however goes beyond *ECE R110* and includes requirements for filling devices for *Compressed Natural Gas (CNG)*, for *Liquefied Natural Gas (LNG)* and for *hydrogen (gaseous & liquefied)*. For Canada—similar as for the USA—following standards are legally binding:

- *CMVSS 301.2* Canada motor vehicle safety standards—CNG Fuel System Integrity
- *CSA B109* Natural gas for vehicles installation code
- *CSA B51*, Boiler, pressure vessel and pressure piping code—part 2 requirements for high-pressure cylinders for the on-board storage of natural gas as a fuel for automotive vehicles
- *CSA B108* Compressed natural gas for fueling stations installation code

Rest of the World

The following international valid ISO Standards apply to the rest of the world:

- *ISO 11439* Gas cylinders—high pressure cylinders for the on-board storage of natural gas as a fuel for automotive vehicles—second edition
- *ISO 14469* Part 1–3 road vehicles compressed natural gas (CNG) refueling connector
- *ISO 15500* Part 1–19 road vehicles—compressed natural gas (CNG) fuel system components

In many areas of the world, the European or the North American Standards are worked into the regional regulations or they are referenced. Despite extensive compliance, it is recommended to obtain detailed information about the valid regulations in the intended markets. Even (smaller) deviations—e.g. for the permissible gas permeation, the required lifetime or the durability testing—need to be considered.

5.1.2 System Components

Natural gas components and systems are safety relevant; hence related DFMEA and SFMEA are to be performed in the concept phase of development. The following gives a description of the most important components in contact with natural gas.

5.1.2.1 Filling Receptacle

In general, the filling of natural gas takes place through a filling receptacle standardized according to NGV 1 (see Sect. 5.1.1)

The couplings of the filling station are quick-connected to the receptacle by form fit. A discharge of natural gas into the environment practically does not take place—to achieve this the pressure in the filling line is fed back into the filling station right after the filling process. Coarse particles (ca. >50 μm) are pre-filtered within the filling receptacle. The filter body can consist of sintered metal, metal mesh filter or sheet metal with micro orifices formed by lasing. Downstream, there is a check valve that avoids the return flow of natural gas into the environment. Different shut off body shapes can be used (ball type, piston type, etc.) in application with elastomer sealing. The shut off force can be applied to these bodies by spring or magnet. The first shows increasing shut off force with increasing opening and forms an oscillatory system in conjunction with the shut off body mass. As during the filling process a wide range of flow and pressure is passed through, natural frequencies can occur, damaging inner components and causing noise and vibration issues. Especially the elevated noise excites an impression of minor value to the customer. Within the filling receptacle and the subsequent filling line high fluid mechanical forces can occur. Therefore it is recommended to perform coupled CFD- and FEA-simulation to identify (oscillation) load in the filling path and check for fatigue strength. Later design validation has to take place using real hardware.

The shut off force of the magnetic solution diminishes with the square of the opening distance; therefore stimulation of oscillation is mitigated to a large extent and flow resistance is reduced due to the wide opening.

A robust design and monitoring of the filling receptacle production is required as failure can lead to multiple errors in the natural gas system, e.g. by flaking particles. Excessive flow dynamic load at the elastomer with the consequence of abrasion, component dislocation or other damaging needs to be avoided by design measures (Diagram 5.2).

5.1.2.2 High Pressure Piping and Connectors

For connection to high-pressure natural gas piping, metallic compression fittings have been established although other concepts are under investigation for cost reasons. The high-pressure pipes have a self-centered and gas tight connection by friction and plastic form fit. The form fit avoids ballistic accidents with pipes being under high inner pressure, as they cannot be pressed out of the opening of the cap nut.

Diagram 5.3 shows the initial screwing by a 450-degree turn with plastic form fit. For service, the connection must be able to be re-connected at least 20 times (follow the

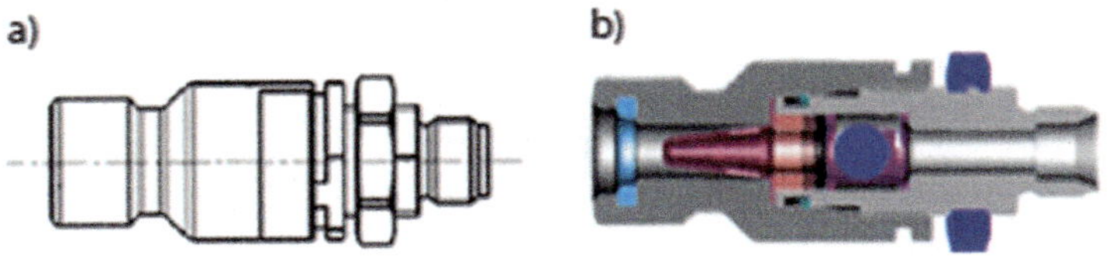

Diagram 5.2 (**a**) Passenger vehicle filling receptacle according to *NGV1* (**b**) Section of the *NGV1* receptacle [2]

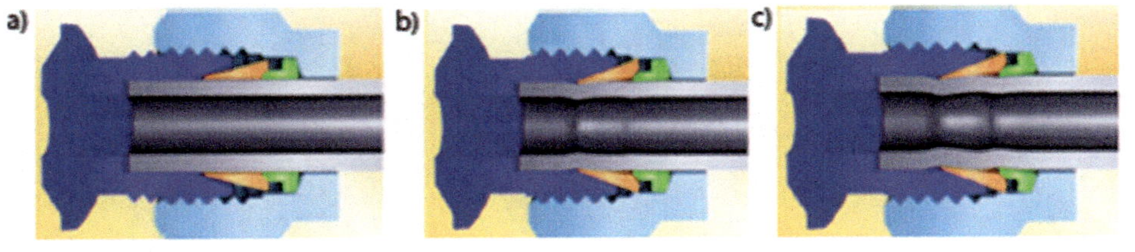

Diagram 5.3 Initial screwing of a clamp ring pipe connector [3]. (**a**) Hand-tight, (**b**) 360°-turn, (**c**) 450°-turn with final plastification/sealing

different re-connection procedure). Cold seizing—typical for stainless steel—is avoided by silver coating of the thread flank.

The shape of the pipes shall be designed to guarantee a repeatable tension-free installation. Multiple neighboring tanks e.g. can be connected by hook-shaped intermediate piping which has just one installation direction without the need to dismount the tanks themselves. Due to the high corrosion resistance, normally the material AISI 316Ti (1.4571) is used for the piping and AISI 316L (1.4404) for the connectors, see Sect. 5.1.3.4.

During the filling process a high natural gas mass flow is created. In order to reach an optimized filling mass, the mass flow must be distributed evenly into the different tanks. This is even more valid if the filling process is interrupted significantly earlier prior to the regular end—at that point the pressure level in the different tanks can vary a lot and during start of the vehicle undesired noise can occur, as the pressure will equalize between the tanks. CFD simulation is helpful in order to investigate the influence of parameters for the flow distribution such as number of pipe bending, pipe section or inlet geometry from the tank valve into the tank on a qualitative basis. A later verification must be done by real filling testing.

5.1.2.3 Tank Valve (Diagram 5.4)

The tank valves serve to fill the tanks and to feed the natural gas to the engine as well as to shut off the tanks when the natural gas system is not in use. Furthermore, the tanks are

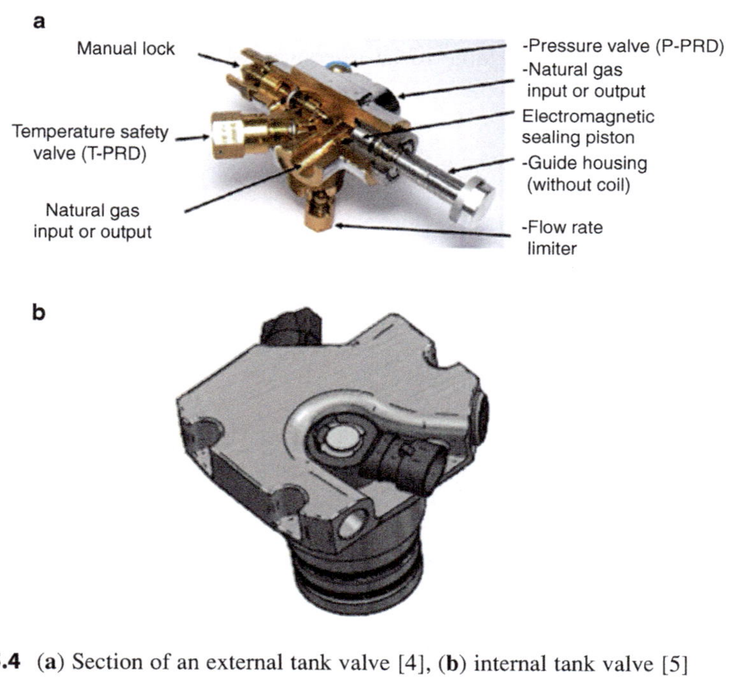

Diagram 5.4 (**a**) Section of an external tank valve [4], (**b**) internal tank valve [5]

protected by the valve assembly from impermissible temperatures (T-PRD) and—if necessary—pressure (optional P-PRD).

The assembly of the tank valves with the tanks has high safety relevance. Beside the stored chemical energy, the tanks carry a large volume of pressurized gas with a serious amount of potential energy. Beside other measures, a sudden release of this gas due to a tank burst must be prevented by the safety concept of the tank valves.

During filling, the gas flow passes a check valve in the tank valve assembly that stops the tank being emptied by a back flow. This check valve can be positioned in a separate filling channel. In this case, the withdrawal of gas from the tank is achieved by a separate feed channel and a shut off valve, which normally is actuated by a solenoid.

Often the check valve and the shut off valve are realized in a combined design within just one channel for tank filling and engine feed. The solenoid valve needs to be able to open against a pressure difference at the level of the highest operation pressure of the tank. Due to the required natural gas mass-flow for the engine operation a suitable inner diameter of the valve opening section is mandatory. However, the pressure force at the valve opening section cannot be overcome by usual solenoids. Therefore, a very small pilot opening equalizes the pressure between the tank and the pipe system towards the engine. After that, the solenoid force is able to open the full section.

The dynamic filling flow can cause high loads within the tank valve. A durable sealing function needs to be guaranteed for which reason it shall be investigated by the above named CFD/FEA-simulation plus real filling cycling in order to check for life-time tightness. If high loads are found on the sealing material that cannot be handled by the usual elastomer (HNBR, Viton), then PFA (Perfluorakoxy—Fluoride Plastic) can be a solution.

For service or in case of emergency, a manual shut off valve is present, which is positioned within the tank valve in a place where the tank can be closed independently from the open or close position of the solenoid valve. Its layout is made for the maximum number of opening and closing cycles to be expected in the field (e.g. for service reason).

Towards the inner of the tank, the tank valve has a flow limiter, which in case of a significant leakage in the piping system, e.g. a pipe rupture, limits excessive mass flow down to a minimum. In order to achieve this, a spring loaded shut off body is used which is pulled into closing position by the high flow forces present in such a case. After shut off, the static pressure difference between the inner tank and the environment keeps the shut off body in the closed position.

As a safety device, a thermal fuse (T-PRD, see Sect. 5.1.2.5) that triggers at a temperature of ca. 110 °C and releases the tank contents into the environment without closing again is mandatory by law.

An optional over-pressure fuse (P-PRD *Pressure-Triggered Pressure Relief Device*) irreversibly triggers within a window of over-pressure relative to the regional nominal pressure—therefore in the USA at higher pressure level compared to Europe. It works as a "fast fuse" in conjunction with tanks with good heat conductivity in case of fire scenarios—see Sect. 5.1.3.3. So called burst discs are in use for this consisting of certified Nickel sheet metal which after the initial pressure load is formed to the shape of a rotation-symmetric calotte shell fixed at its circumference. The triggering/rupture of this membrane releases the total tank contents into the environment.

External and internal designs for tank valves can be distinguished. Internal valves offer advantages in case of crash and fire, see Sects. 5.1.3.2 and 5.1.3.3.

Even for the internal tank valves it is recommended that—after shut off by the manual valve—the solenoid valve and the piping connectors are accessible for service, or if necessary, exchange. During in-field service, the tank valve must not be separated from the tank. Should this be necessary, it is absolutely mandatory that the tank is without inner pressure (trained personnel, follow documented working instructions—danger of fatal accident!).

5.1.2.4 Tank

Currently cylindrical tanks with rounded endings can be type-approved for pressurized natural gas in vehicle applications. At least one end has an opening to accept the tank valve. The tanks represent up to 60 % of the overall cost of the natural gas system (in case of composite). Four different designs need to be distinguished from each other. In the area of passenger cars, Type 1 and Type 4 tanks have been established, which will be described in the following (Diagram 5.5):

Natural Gas Tank Type 1

Type 1 natural gas tanks are made out of metal and are homogeneously isotropic—usually made from heat-treated steel. Pipe material for example can be rotated around its longitudinal axis and guided rolls close the pipe ends by warm forming with or without leaving an opening for the tank valve (*Spinning Process*). Due to the material compression, the wall thickness at the end increases, although it is not necessary, as the favorable stress distribution in this area does not require it. Therefore, tanks are also drawn from plate material or formed out of blocks (*Billets*) by Reverse Impact Extrusion followed by a hot roll process of the cylindrical section. Even after these processes, the valve side needs to be closed by *Spinning*, however the bottom side is thin walled, light and without the issue to close the area gas tight by *Spinning* if no opening is required.

Due to the simple geometry the *Membrane Theory of the Shells* [12, 13] reduces in the first approach towards the so-called *Vessel Formula* [14]. With this, the stress in the

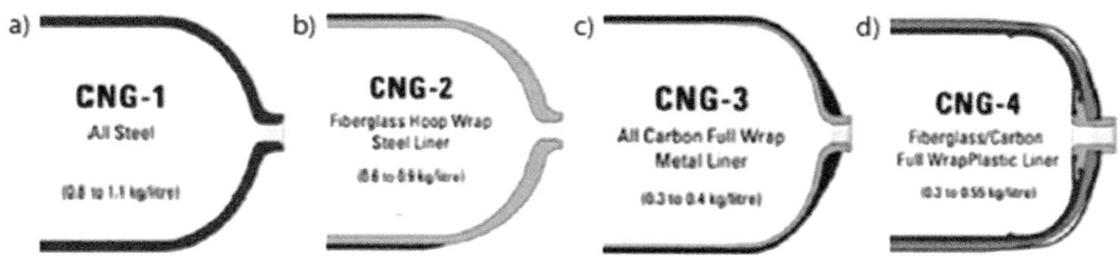

Diagram 5.5 (**a**) Type 1 metal, (**b**) type 2 metal/part-composite, (**c**) type 3 metal/composite, (**d**) type 4 plastic/composite [6]

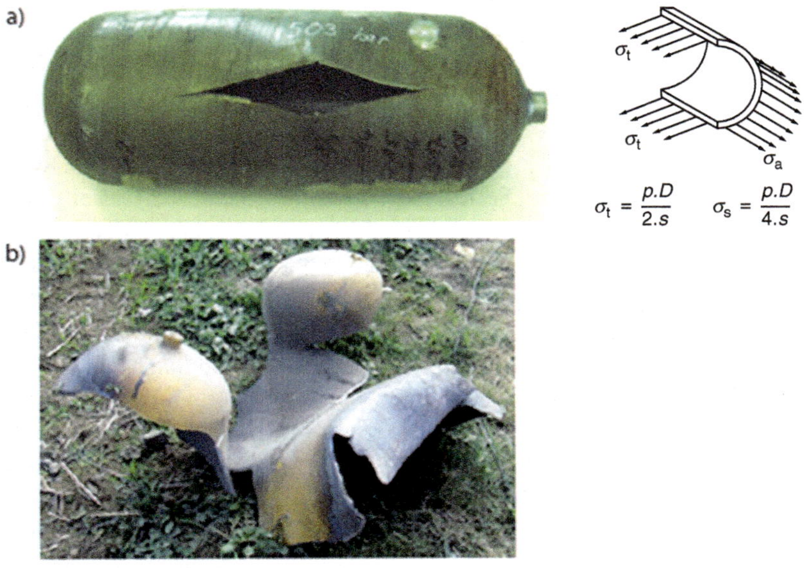

Diagram 5.6 Bursting: (**a**) incompress. medium of low energy (water), (**b**) compress. medium of high energy (natural gas—*localized fire*)

cylindrical area in the circumference direction shows double the value compared to the longitudinal direction, hence bursting by inner pressure is initiated by a longitudinal rupture, see Diagram 5.6. The stresses in the almost spherical calotte are approximately the same in all directions and of the level as the longitudinal stress in the cylindrical area.

The energy absorption in case of a crash with strain of rupture above 10 % for the heat-treated steel is an advantage. One time abrasion on the steel tank in an accident scenario is not critical. The thermodynamic filling efficiency is favorable—the high heat conductivity guides the compression heat quickly towards the environment. Therefore the mass of natural gas in the tank at the end of the filling process is comparably high. The heat transfer towards the T-PRD in case of a fire scenario occurs essentially via the high conductive tank shell and the quickly heated filling mass of gas, which then starts to circulate in the tank—see Sect. 5.1.3.3. Permeation does not occur via the tank shell. This can be essential if there is a low threshold for hydrocarbon-permeation (such as in the USA) and in bivalent vehicles the existing gasoline system does not leave a big buffer for the additional natural gas system. The high weight (ca. 0.9–1.0 kg/l with steel) and the potential corrosion sensitivity are a drawback.

Natural Gas Tank Type 4

Fiber reinforced Type 4 plastic tanks eliminate the disadvantages of Type 1 tanks: High weight and corrosion risk. This is "bought" by higher manufacturing costs driven by the reinforcement fiber, especially the carbon fiber. Despite the cost situation, the advantages compared to Type 1 prevail—however the specific features have to be taken into account. In passenger vehicles—if the application allows (permeation)—Type 4 composite tanks have been established due to their low cost plastic liner compared to the metal liner of Type 3 tanks.

Pressurized gas tanks with their distinctive and symmetrical inner load and low number of outer load applications utilize the features of reinforcement fibers ideally. Low tank weight of about 0.35 kg/l can be reached. With this technology, tank sizes can be realized that result in competitive driving ranges compared to other propulsion systems while steel tanks of this size would create fatigue issues in the vehicle body mounting area. Furthermore, the low tank weight means less weight in the tank brackets, which incidentally are cheaper. Overall the result is lower fuel consumption, better CO_2 emission and improved driving dynamics. As an example, with a 7-seater mini-van, Diagram 5.1, it was possible to reduce weight by 85 kg compared to steel tank application of the same volume, resulting in a lower inertia mass class for tailpipe emission testing and a CO_2 reduction of ca. 3–4 g/km in NEDC.

The liner—the gas-tight inner shell of the composite tank—is made of plastic such as HDPE or Polyamide for Type 4 tanks. The aim is a minimized methane-permeation. Such liners are normally manufactured by blow forming. Depending on the diameter of the initial pasty hose out of the extruded liner plastic, material overlapping in smaller diameter end areas of the blow formed liner shell can occur. In these so-called *Pinch Lines* the material overlapping gets welded during blow forming. The gas tightness of the pinch lines is very reliant on manufacturing parameters as for example polyamide-granulate humidity during extrusion or material temperature during blow forming. They are potential areas of problems. It is therefore recommended to manufacture liners without pinch lines with a continuous smooth surface—means without geometric disruption. The blow forming tooling and process need to be adapted to this (*Pinch Line Free Liner*).

Usually both ends have an opening with a metallic thread insert—the so-called boss—to accept the tank valve, or at the opposite side a closing plug, (this second opening is necessary as a fixation for the later fiber wrapping process). This plug can be equipped with a T-PRD, see Diagram 5.9. It has to be considered that the joining area between liner plastic and metal boss is not exposed to the inner pressure of the tank. If it is in the area of the inner pressure, this seam can be separated as the pressure front creeps into it, which can lead to external leakage. Therefore, it is recommended that e.g. the tank valve sealing are in touch with the liner surface in order to seal against the environment already there.

To carry the inner pressure load, the liner is wrapped in different directions with multiple layers of reinforcement fiber that are normally embedded in thermosetting plastic. Numerically controlled winding machines process the wrapping of the liner. It is turned around its longitudinal axis and resin-impregnated fiber is laid down in defined positions as they are guided by a numerically controlled support. Typical thermosetting systems for embedding fiber are Epoxy- or Polyvinyl Ester-resin. To carry the high load in circumference direction according to the *Vessel Formula* fiber wrapping with small pitch—so called *Hoop Wraps*—are done in the cylindrical area of the tank. The load in longitudinal direction is supported by so called *Helix Wraps* with very large pitch that are also wrapped around and therefore cover the liner ends up to the end bosses. Special attention needs to be applied to the transition between the cylindrical and the ball shaped end areas—the so-called shoulders—in terms of layout for handling accidents or crash load. In this area, the hoop wraps end and the helix wraps alone need to cover the large circumference here where they are packed toward the boss ends. Hence the shoulder area has an abrupt reduction in wrapping structure leading to a change in stiffness with locally increased stress. This becomes complicated because the high circumference load from the cylindrical area still disturbs the nearby shoulder as it is not completely faded away. For design validation drop tests towards the shoulder area with subsequent hydrostatic burst tests are performed. If necessary additional impact protection has to be applied.

Composite out of carbon fiber (*CFC, C-fiber*) and/or glass fiber (*GFC, G-fiber*) have established themselves for load support. For Type 4 tanks, *High Strength* C-fibers are preferred with a high static tensile strength (ca. 4900 MPa), outstanding cycling resistance and low density of about 1.8 g/cm^3 [15]. This is in contrast to type 3 tanks where *High Modulus* C-fiber is regularly used (to cover metal liner fatigue). Natural gas tanks out of pure CFC from underbody installation in passenger car use over 12 years under German environment conditions have been examined and there was practically no degradation in burst pressure. One disadvantage is the low tolerance in terms of over loading, e.g. during a vehicle crash with contact to other hard components. This is caused by the brittle nature of C-fiber with a comparably low strain to rupture of ca. 2.1 %, see Sect. 5.1.3.2.

G-fiber has a significantly higher toughness with a strain to rupture of ca. 5 %. A disadvantage is its higher specific weight of ca. 2.5 g/cm^3 and the degradation of the initial tensile strength (ca. 2000 MPa) over its lifetime. Therefore G-fiber can be used in addition to C-fiber in order to elevate the fracture toughness under sudden impact load, see Sect. 5.1.3.2. C- and G-fiber can be wrapped alternating within one layer (*Hybrid Wrapping*) or an inner C-fiber wrapping is protected by an outer G-fiber wrapping.

In contrast to metallic barriers of type 1 to type 3 tanks, the liners of type 4 tanks are subject to permeation, which has to be minimized. The permeation dictates the operation conditions of type 4 tanks. Rapid pressure decrease needs to be avoided so that there is no expansion of the methane that previously diffused into the plastic structure of the tank. Otherwise the so-called *Liner Buckling* into the inner of the tank volume can occur. This can happen under service conditions when discharging of the tanks is done too fast as the service personal is used to it from previous type 1 tanks. Even sealing elastomer underlays potential damage by sudden decompression of the dissolved methane leading to local internal ruptures of the material. Therefore pressure cycling gradients needs to be kept low in order to protect the plastic and elastomer in use in a way that a damage-free back diffusion of the methane out of the material is possible. To minimize permeation *Multi-Layer* designs with almost permeation-free barriers, e.g. EVOH (*Ethylene-Vinyl-Alcohol-Copolymer*) as used in gasoline tanks, are under investigation. The different materials are coextruded as concentric layers and blow formed into the liner shape. The diffusion mentioned above is a lot less, however it needs to be taken into account that the back diffusion through the tight barrier is very slow. If methane which once passed this barrier does not permeate away to the environment fast enough there might be even more effects such as *Blistering* between the co-extruded layers or the *Liner Buckling* as mentioned above.

Due to the full plastic design, the heat conductivity of type 4 tanks compared to type 1 steel tanks is at a lower level of about two times the power of ten (factor perpendicular to the fiber: CFC ca. 10, GFC ca. 200, HDPE ca. 100). The thermodynamic filling efficiency is therefore not as good. At the same pressure level at the end of the filling process, a higher gas temperature—causing a lower filling mass—will be reached. The surface of the tank does not experience such high temperature swings. Condensation of air humidity—such as on type 1 tanks—only occurs under extreme cycling. The consequences of the different heat conductivity and the different degradation process in fire scenario compared to steel are discussed in Sect. 5.1.3.3.

Composite tanks need to be protected from permanent UV radiation, which is the case with underbody vehicle installation. For pickup trucks with composite tanks mounted on the bed, the customer must be informed that the tank cover should not be permanently removed, which is also valid as it is the protection from the payload in crash events. Similar to this, the overall logistic chain from the tank manufacturer to the OEM production line needs to be monitored in terms of intermediate storage of the tanks.

Natural Gas Tank Type 3

Type 3 tanks have a gastight metal liner barrier—most of the time made from aluminum—which is also fully wrapped by fiber composite. The statements on composite given for type 4 are also valid for type 3, except the increased use of stiffer *High Modulus* C-fibers in order to mitigate liner fatigue problems during hydraulic pressure cycling in type approval testing. The real number of cycles during a vehicle lifetime is much lower. As a counter measure, the so-called *Autofrettage*- process is performed during which the tank is hydraulically over-pressurized just below the damage threshold. The plastic setting of the (Aluminum-) liner causes a residual compressive stress in its material as it is compressed by the composite mantel after *Áutofrettage*. The inner gas pressure in operation needs to overcome the compressive stress first so that at the end there is a lower level of tensile stress present in the liner material, which has a positive effect on fatigue.

Metal liners are more expensive than plastic liners, which is the reason that type 3 has been established in heavy commercial vehicles such as busses etc. rather than passenger cars. However type 3 liners are a tight permeation barrier, which enable its application in bivalent vehicles according to the strict US-*Shed-Test* thresholds. As with the other tank types with a metallic hull in contact with the natural gas, the application of type 3 has to be investigated depending on the regional natural gas quality in terms of inner corrosion.

Diagram 5.7 Corrosion of a Typ-2-Steel/Glass Fiber-Tank [8]

Natural Gas Tank Type 2

For this type, the composite wrapping does not cover the overall tank shell but only the cylindrical area. There is the risk of corrosion in a capillary gap between the inner metal hull—mainly made from steel—and the outer composite, see Diagram 5.7. The longitudinal load according to the *Vessel Formula* still needs to be carried by the metal body, as the *Hoop Wraps* in the cylindrical area are effective by form fit only for the circumference load. In OEM passenger car underbody installation this tank type has been barely used.

Free Conformable Structural Tanks

Free-formable tanks for pressurized gas according to the *Membrane Theory of the Shells* or welded assemblies of the load-carrying tank hull are currently not foreseen for type approval. Due to an improved utilization of the given vehicle package space up to 35 % more gas mass can be stored [7]. Corresponding design which "works in computer simulation" need to be assessed including the manufacturing process and all kind of loads in vehicle application in terms of its safety features. This needs to be done very conservatively, as any risk cannot be accepted. Hence these designs have not yet found their way into the type approval documents or into production vehicles. Furthermore, the cost situation is disadvantageous, especially in conjunction with composite solutions (Diagram 5.8).

5.1.2.5 Thermally Triggered Pressure Relief Device (T-PRD)

Thermal fuses (T-PRD *Thermal-Triggered Pressure Relief Device*) have to release the total methane content of the tank into the environment when a temperature threshold (normally 110 °C) is exceeded in order to avoid a tank burst event.

Eutectic meltable solder as well as glass bulb filled with liquid are established T-PRD. Meltable fuses should not be positioned within the channel to be opened for pressure relief as re-solidification may occur due to the cooling effect of the discharging gas flow with a subsequent plugging of the channel by remaining solder. The correct triggering with permanent opening of the discharge channel as well as investigation of the creep behavior at elevated temperature under tank pressure cycling need to be proven by testing.

Without doubt, digital triggering is reached with the liquid-filled glass bulb, see Diagram 5.9. As a consequence of exceeding the temperature threshold, the inner liquid expands and cracks the glass bulb, which then releases a shut off body in the discharge

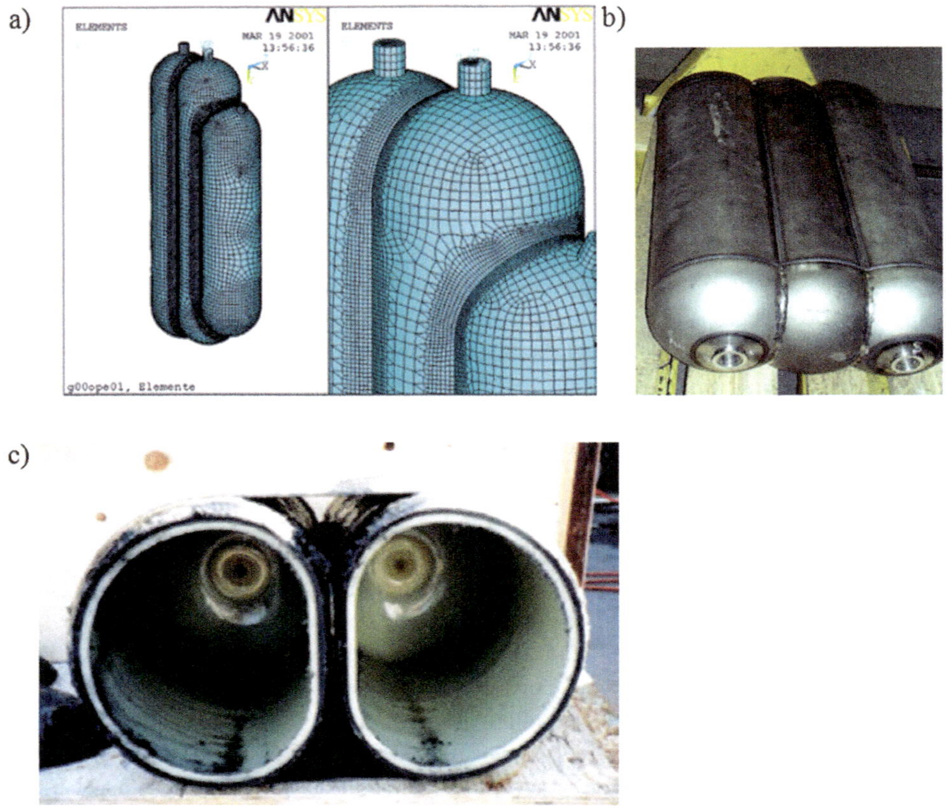

Diagram 5.8 (**a**) FEA conformable tank [7] (**b**) prototype [7] (**c**) conformable composite tank [8]

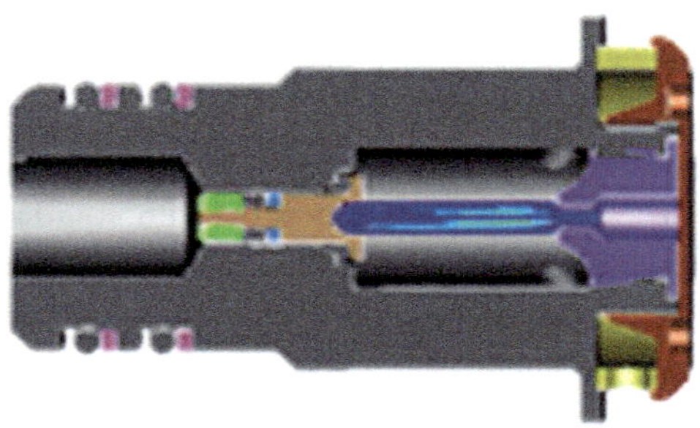

Diagram 5.9 Thermal fuse in thermo-plug [2]

channel. As a quality check, it is recommended to perform an axial load test with each glass bulb at the end of their manufacturing process with a load level beyond those which can be reached in the later vehicle application.

5.1.2.6 Pressure Regulator

The term pressure regulator is only correct for mechanical regulators, as for example diaphragm or piston regulator types. Electrical triggered regulators are simply regulator control elements, since the control algorithm is embedded in the controller. The pressure tapping point is located in the injector rail. Nevertheless, the term pressure regulator is established for all type of devices and will be used in the following. Pressure regulators adjusting the natural gas injection pressure for the engine need to cover the whole range of tank storage pressure up to the maximum required mass flow. To achieve the maximum range, the natural gas mass should be used down to the lowest possible tank storage pressure. Generally a pressure transducer is installed in the high-pressure fuel interface to support the fuel level indication and storage system diagnosis. A particle filter and a high-pressure shut-off valve containing a pilot valve function (similar to the tank valve) are incorporated in the high-pressure fuel interface as well.

Double stage regulators are most common in compressed natural gas vehicles. In mechanical regulators, both stages are based on mechanical pressure reduction. In so-called electrical regulators the second stage consists of a valve with a pulse width modulation triggered solenoid, regulated by a control module. The mid stage pressure downstream of the first regulator stage reduces the supplied tank pressure down to around 20–30 bar. The second stage executes the fine adjustment of the desired injector rail pressure (Diagram 5.10). With (turbo) charged engines mechanical or electrical regulators have a variable outlet pressure between 200 and 1000 kPa (2–10 bar). A spring-loaded safety valve prevents outlet overpressure exceeding thresholds of about 15 bar. The valve releases the overpressure temporarily, safely to the atmosphere. The internal leak rate of the regulator in the closed state should not exceed the sum of all injector internal leak rates, to avoid pressure build-up between the regulator outlet and the injector rail during vehicle parking. Elevated injector rail pressures could prevent injector opening.

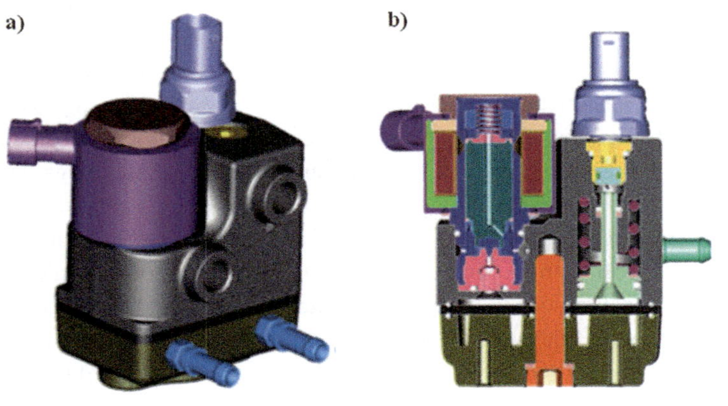

Diagram 5.10 (**a**) Electronic pressure regulator [4], (**b**) section of double stage electrical regulator [4]

Mechanical regulators represent the closed loop control in a single unit. The difference between the atmospheric pressure and the injector operation pressure is subjected to spring-loaded diaphragms/pistons (in naturally aspirated engines). Coupled to the diaphragm/piston is a seated valve, which regulates the outlet pressure by opening or closing. With (turbo) charged engines—when the injection pressure shall exceed the strong fluctuating manifold pressure by a certain difference value—a reference pipe transfers the manifold pressure instead of the atmospheric pressure to the regulator diaphragm/piston (MAP-Adaptation). Diaphragm regulators experience hardly any friction forces in contrast to piston regulators and reveal in general a better control performance. They are more bulky, which has a negative effect on narrow engine compartment packaging.

Electronic regulators demonstrate the best pressure regulation performance when coupled with a controller that is capable of supporting individual PID programming. The pressure on the downstream side can be adjusted independently to the engine load. This enables changes in the fuel quality (heating value) to be compensated for, which can be identified by the lambda control, by regulator outlet pressure adaptation. This kind of regulator can be used for many engines with different nominal power without changing its size, if the maximum mass flow still fits. The adaptation is limited to calibration parameters. The device itself is very compact, since the mid stage pressure regulation is done by a mechanical piston regulator stage, without incorporating any drawbacks and also the PWM-controlled valve is a compact solution. The latter substitutes in addition the high-pressure side shut off valve—preventing gas flow in electrically unpowered conditions—which is a common component of mechanical regulator assemblies. Because of the integration in the controls architecture, the electronic regulators are mainly found in OEM vehicles, but offered in increasing numbers.

5.1.2.7 Filter- and Oil Separator

Regional occurring malfunctions of natural gas injectors due to liquid oil residues, out of the fuel storage system, as well as the desire to supply particle free gas to the engine have led to this component. Oil sludge can build up in the injectors. The consequences range from deviation from the desired exhaust quality—the lambda controller demands a homogeneous injector function in narrow limits at multiple cylinder engines—to driving performance faults and non-drivability. Oil has been observed in compressed natural gas storage systems with high mileage and its source are the refueling station gas compressors. Analysis of the gas contamination in the supplied CNG during the refueling procedure throughout German refueling stations using filter and adsorption separation has revealed oil content up to 150 ppm in single cases. Insufficient compressor maintenance and prolonged operation time—leading to high compressor temperature according to high demand—will result in elevated amounts of lubrication oil entering the compression chamber. This oil content mostly evaporates into the compressed gas during the process. The oil vapor and some liquid oil aerosols will be supplied to the vehicle storage system during refueling. The evaporated oil content will condense only when subjected to a significant decompression of the gas. On the one hand, this happens when the pressure

Diagram 5.11 Oil separator/filter section [11]

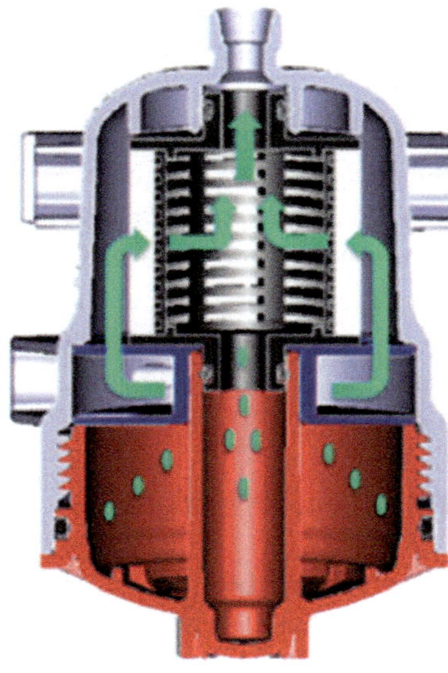

drops in the storage system—the reason why oil residue is found inside the containers—on the other hand with high system pressures only downstream of the pressure regulation resulting in the above mentioned consequences. Oil separation with the least effort is therefore possible between the pressure regulator and the engine, for example in two separation stages [11]. In the first stage the oil will be separated gravimetrically then the natural gas flow will be strongly deflected and directed into the second stage. There filter media separates remaining oil aerosols as well as particles. Downstream of this component, the gas is oil- and particle free (Diagram 5.11).

5.1.2.8 Low Pressure Hose and Piping

Flexible low pressure piping supplies the natural gas demand into the injector rail of the engine. It isolates the engine movement from the vehicle body mounted components such as the oil separator/filter unit or the pressure regulator. The stiffness of the piping and of the fixation points—beside the fixation on the vehicle site location of the interfacing components—need to be designed such that the transfer of vibration to the vehicle and the fuel storage system is avoided.

For example, these vibration frequencies could be working noises of the CNG-injectors leading to elevated noise levels noticeable by the customer. The flexible low-pressure piping has to withstand the thermal and mechanical impacts resulting from the closeness to the engine. Its construction in general is a multilayer design, consisting of an inner, middle and an outer layer. The inner layer is responsible for keeping the permeation low; therefore HNBR is a common material.

The intermediate layer, a mesh out of Polyamide/Nylon, Polyester or Cotton needs to guarantee working pressures up to 30 bar. To protect against mechanical and temperature impact and to withstand chemical impact from media like oil, grease, battery acid or others and from the environment (UV, Ozone) the outer layer out of NBR (for example) completes the construction.

To assure sufficient space for other components, low-pressure piping can be manufactured as preformed parts. The leak tight interface may be achieved by using barb connectors and hose clips or by using pipe connections with conical metal seals and nuts. When barb connectors and hose clips are used, the minimal slip force requirement at the maximal temperature and maximal working pressure has to be obeyed. If ferule connections are used the connection between the metal interface and the hose material is done by compression and this interface has to be validated with cyclic pressure pulsation at the different required temperatures as the operation environment requests.

5.1.3 Vehicle Integration

5.1.3.1 Packaging

The accommodation of an underfloor natural gas system ideally takes place during the concept phase of the vehicle development. In regular underbody structures, the transverse arrangement of the natural gas tanks in front and behind the rear axle has established itself. For cost reasons, the aim is to use a minimum number of tanks, valves and piping, *whereby no restriction in usability for the customer must occur*. A one-tank system is often insufficient as the necessary driving range requirements might not be met or the vehicle usability is restricted (e.g. in the luggage compartment). Large tank diameters are ideal as the storage volume increases by the square of it. Experience and initial simulation are used in order not to let the tank package envelop reach into crash deformation zones. The tank system including its suspension needs to be investigated for natural frequency in order to not suffer resonance damage during later durability validation. The same is valid in terms of *fatigue*—especially the mounting points at the vehicle body need to be monitored.

In the underbody area, strap mounting is preferred to neck mounting. This is especially valid in terms of composite tanks (type 4) as in crash events, loads may be concentrated in the sensitive area of the end bosses with a neck mounting. Strap mounting allows the tanks to slip a bit under high crash load, so that they get out of the way. The length of each pair of the circumference half straps, which are needed to carry the tank, should be identical as this avoids the tank twisting during assembly and later issues with the piping connection.

Short side pipes exhaust systems have proven themselves—they end in front of the rear axle and enable up to 15 l increased overall tank volume. Ideally they are packaged in the non-visible area. However in some global regions, they cannot be type approved.

The protection covers for the tanks should be fixed by quick connectors for easy access during inspections. Appropriate testing needs to make sure that there is no undesired accumulation of snow, ice, dirt/dust or stones possible within the covers. Suitable measures must guarantee that no stones can be clamped between tank surfaces or tanks and other components, otherwise vehicle vibration can lead to damage of the corrosion protection (type 1) or of the composite structure. Such gaps can be closed e.g. by closed-cell foam material.

In terms of interaction of the natural gas system with the rest of the vehicle it is recommended to perform System-FMEA.

5.1.3.2 Crash Safety

During and subsequent to vehicle collision there must be no danger originated from the natural gas system. The assessment - done by standardized testing - shall reveal that there will be no risk due to the potential energy stored within the natural gas high pressure tanks or by any leakages. Some of the legally required tests are drawn for assessment of the natural gas system, see Diagram 5.12a–c for the EU27+ states. Other global regions potentially require other standards, as for example, the relevant US rear impact is also valid for natural gas vehicles (70 % overlapping, 1361 kg deformable barrier, 88 km/h).

The vehicle manufacturers can define further crash testing for assessment of the natural gas system, see Diagram 5.12d. Here the position of the impact pole is guided towards the most sensitive spot of the natural gas system, which usually is at the side of the tank valve (identification from previous variation in crash simulation). Furthermore—see Diagram 5.12c—testing can be modified and performed including a trailer hitch in order to check for interaction with the rear natural gas tanks.

In the field of natural gas systems, it is recommended to not just fulfill the crash-testing requirements borderline, but to make provision for sufficient safety margin for real life accidents, see Diagram 5.13.

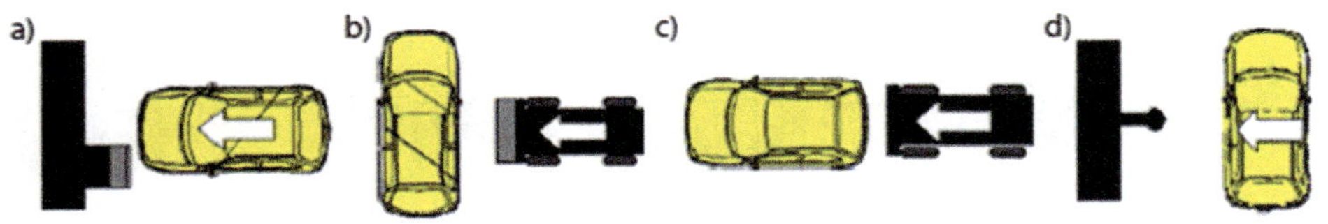

Diagram 5.12 Crash-testing with natural gas vehicles. (**a**) 65 km/h—40 % overlapping, deformable barrier, (**b**) 50 km/h—950 kg, deformable barrier, (**c**) 54 km/h—1800 kg, rigid barrier, (**d**) 29 km/h—10 inch post, vehicle on *Flying Floor*, (**a–c**): Legal requirement in EU27+: A, (**d**) recommendation: modified pole testing

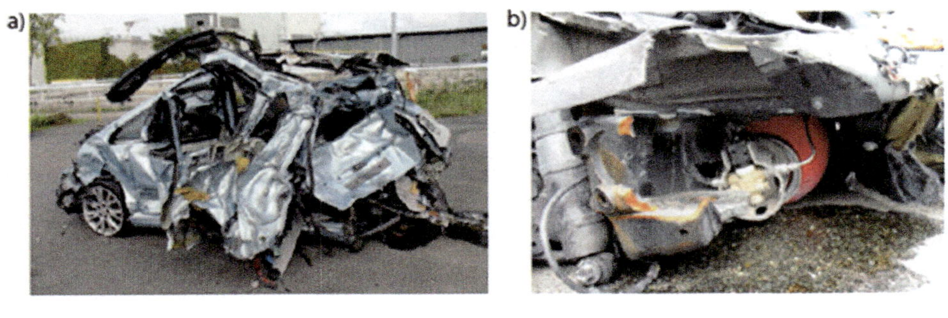

Diagram 5.13 (**a**) 40-to-truck impact onto unoccupied natural gas vehicle (**b**) no harm caused by the natural gas system

High Pressure Fuel Lines and Connectors

Natural gas high-pressure fuel lines made from stainless steel have a wide ductile range beyond yield strength, which is why they display favorable behavior in crash events. Optimization can be done in the area of the connectors especially at the interfaces to the components such as valves or pressure regulator, which can break away and become leaky at low bending load [9]. Diagram 5.14b shows the relocation of the sealing geometry and screwing into the component housing in order to avoid foreign contact bending.

Tank Valves

Section 5.1.2.3 shows external and internal tank valves. The external tank valves are derived from design solutions used in the technical gases industry and the safety relevant components have an exposed position outside of the robust tank shell. Depending on the OEM safety concept, specific structures for the external tank valves can be applied to protect against crash impact, see Diagram 5.15a, as all safety devices of the tank valve still need to be functional in a post-crash scenario [16].

For example, the solenoid valves must not be deformed so that they can close when no longer supplied with electrical power—see Sect. 5.1.3.5. Temperature or over-pressure fuses must not be damaged so that they do not trigger prematurely and still safeguard against a potential subsequent fire scenario. Internal tank valves integrate these safety

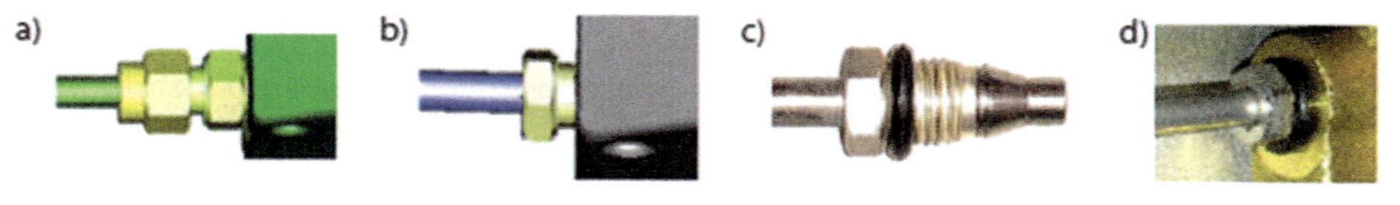

Diagram 5.14 (**a**) Commercial available natural gas fuel line connector, (**b–d**) crash-optimized automotive variant

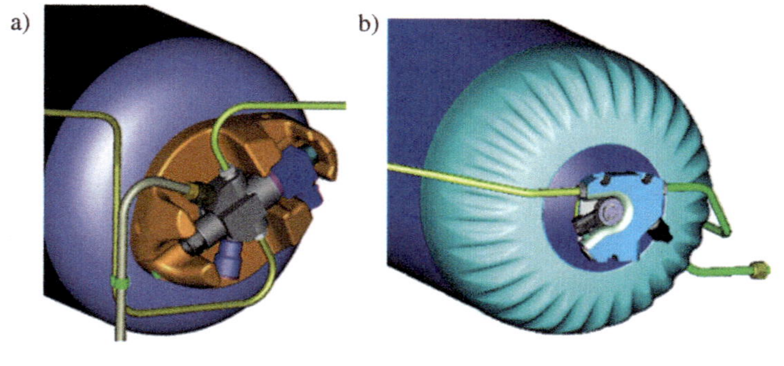

Diagram 5.15 (**a**) External tank valve including crash protection [4, 16], (**b**) crash-optimized internal tank valve [5, 10]

relevant components widely into the contour of the tank shell within the large interface opening and therefore do not require additional crash protection, see Diagram 5.15b. Therefore, there is also improved heat conduction from the tank into the valve with advantages for the duration till the T-PRD triggers in the case of fire for Type 1 tanks. For application with composite tanks, sufficiently large access for hot gas from the environment towards the T-PRD has to be foreseen.

Tank

Type 1 steel tanks are crash resistant including abrasion load, see Diagram 5.13 for example. Crash simulation modeling is very well established. Composite tanks are coming into use in increasing numbers for which crash safety has to be ensured. Simulation modeling of composite tanks integrated into the vehicle crash modeling is under development based on real single layer material testing as a basis for a multi-layer finite-element-analysis modeling. After verification based on real tank hardware, these virtual models can be integrated into the overall vehicle crash models, see Diagram 5.16. The aim is to obtain statements of layer failure under crash load for a safe tank design. Ideally the number of tank tests could be limited to verification only.

The *Fall-Tower-Test*, Diagram 5.17, is used to examine the impact toughness of pressurized gas tanks. The derivation of the impact energy can be gained from critical load cases out of crash simulation such as, for example, concentrated contact forces on the tank by a trailer hitch. Typical features for the composite fiber appear as already named in Sect. 5.1.2.4: Pure CFC-Tank design is not recommended in underfloor installation of passenger vehicles as they tend to brittle fracture when exposed to concentrated crash

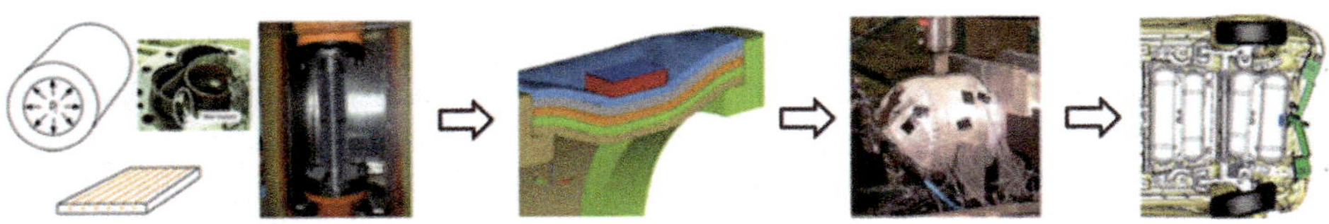

Diagram 5.16 Development of a crash simulation model for composite pressurized gas tanks [18]

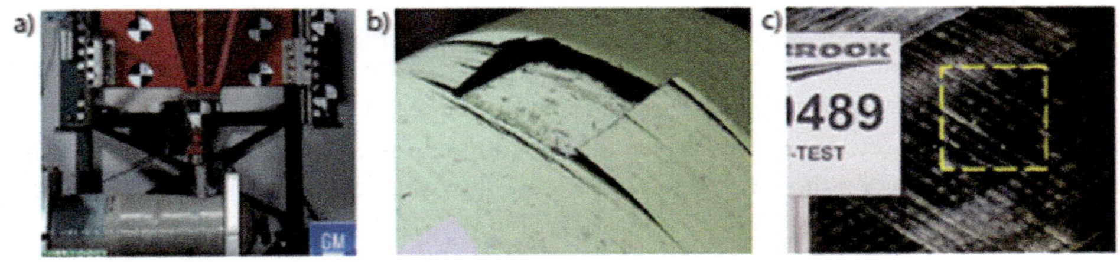

Diagram 5.17 (**a**) Fall tower impact, (**b**) pure cfc-wrapping, (**c**) cfc/gfc-hybrid wrapping

load, see Diagram 5.17b. A remedy could be a combination of CFC/GFC-structure, so-called Hybrid Wrapping, see Diagram 5.17c or a layered structure with GFC to the outside.

The axial load on composite tanks can be assessed using component testing which is relevant in side crash scenarios such as Diagram 5.12b or especially Diagram 5.12d with its significantly deeper impact into the vehicle structure.

Composite tanks are very sensitive in terms of abrasion such as loaded slipping on road surface. Diagram 5.13b shows the case as a consequence from a wheel breaking off under very high crash load with subsequent hit to the ground and slipping in the area of the tank dome. Steel tanks do not require special protection for this, however composite tanks should have at least metallic slide protection in order prevent exposure to abrasion. This for example can be integrated into the mounting system of the tank [17]. Furthermore the *Leak-Before-Burst* criteria during abrasion load should be tested with composite tanks. During this, leakage is permitted but it must not lead to a burst event. Appropriate safety precautions need to be arranged when performing such testing.

5.1.3.3 Fire Protection

Vehicle fire can lead to bursting of CNG tanks if there is no sufficient fire protection. Such cases have a high percentage on failure of CNG tanks [19]. Therefore vehicle manufacturers should focus on fire impact on CNG tanks with the same seriousness as for crash behavior, for example. The fuel system performance under fire scenario is discussed in the following.

Tank Valve and Thermo Plug

The position of the *T-PRD* in the tank valve or within the *Thermo-Plug*, see Diagram 5.9, should be adjusted to the heat conduction characteristic of the CNG tank in a fire scenario in order to guarantee a minimum triggering time. Steel and composite tanks differ significantly in this respect, see further down below. The material and cross section of the tank valve housing must enable quick heat conduction towards the thermal fuse. In this respect for example, aluminum has advantages over brass and significant advantages over stainless steel. Initial thermal simulation must be verified using real testing, see below.

For both kinds of safety devices (T-PRD and P-PRD) it is recommended to guide the excess flow towards the road surface after triggering so that it is dissipated. If the excess flow is guided in a different direction and is ignited in this post-accident scenario, this can lead to a several meters long jet flame which endangers first responders or objects close by. The excess flow guiding pipe/device which directs the gas flow towards the road surface must not be deformed by gas flow rebound or plugged by foreign material (such as dirt, ice formation, etc.). On the one hand, the complete pressure discharge of the tank shall occur speedily, on the other hand, the excess flow should not be too strong—see the diagram in Diagram 5.19b.

Localized Fire

During type approval, typical CNG tank sizes for passenger vehicles need to pass fire testing while being exposed to fire over their total length. In doing this, the T-PRD within the tank valve is rapidly heated and triggered in order to avoid the potentially fatal burst event. In reality, there have been cases of localized fire on the tank surface during which the T-PRD did not trigger prior to bursting [20, 21]. During FMEA different scenarios can be found leading to localized fire load on CNG tank surfaces, such as fire from gasoline accumulation in lane grooves after accident, fire of an overturned vehicle with the tank valve lying in the water of a roadside ditch, wheelhouse burning, etc. A complicating factor is that vehicles might be significantly deformed after accidents when fire can occur, see Diagram 5.13 for example. The hot gas flow from fire might not be guided along the contours as under the defined conditions in a vehicle burn test. This means that the temperature field and load of a fire in a post-accident scenario cannot be predicted. Due to this the following methodology takes the tank surface as the system boundary into account for a hard burn testing criteria as CNG tanks in the vast majority survive the initial crash event of accidents. If this criteria is passed, one can assume that a more severe fire scenario will not occur in a vehicle after an accident. *The methodology provides a good overview of the events happening with the different tank design and safety concepts under fire load.* The assessment is separated into the particular procedures *Test-to-Failure* and *Test-of-Safety-Concept.*

Test-to-Failure

A certain number of the same CNG tanks (type, size, batch of manufacturing) with different filling pressures are exposed to localized fire load up to the point of destruction. The result is a pressure borderline over time under fire load that the tank design can bear without any safety device. That means the tanks are neither equipped with P-PRD nor T-PRD during this kind of test procedure. The positioning of the fire depends on the later planned position of the T-PRD's in serial production. The burst event—or any other kind of destructive failure—is deliberately caused. The emerging pressure wave or tank shrapnel are life threatening, hence the testing needs to be done in a locked area in compliance with a related safety distance and usage of protective shelter. Gasoline is the fuel with the most realistic temperature gradient over time and is also the most practicable one during test execution. When a tank burst occurs, there is only the gasoline pan and the tank bracket that are destroyed. The localized gasoline fire—in a window with a length of 25 cm in longitudinal direction of the tank—is placed at the maximum distance to the position of the T-PRDs intended for the later production. The fire is ignited from the remote and protected area. Beside the fire window, the tank is shielded from the flames by sheet metal protection, see Diagram 5.18b.

With just one T-PRD in the tank valve, the fire position would therefore be at the opposite end of the tank, whereby type 1 tank fire load would only be applied up to the end of the cylindrical contour, as the rather ball shaped end is subjected to lower mechanical stress and does not fail prior to the cylinder area. With type 3 and type 4, the tank ends will

Diagram 5.18 (**a**) Localize fire load, (**b**) test assembly of a composite-tank, (**c**) assembly with *wind shielding* on military training area

be under fire load as the fiber wrapping is of less thickness there and the degradation mechanism of composite is critical earlier (see below). If T-PRDs are foreseen/necessary at both ends of the tank—i.e. in the tank valve and the opposite *Thermo-Plug*—the fire window will be positioned right in the middle. Diagram 5.18 shows the principle test setting. The tank receives U-shaped wind shielding using heavy concrete blocks. Wind is the essential disturbance in order to reach a steady temperature load on the tank—therefore the efficiency of the wind shielding needs to be checked (wind speed measurement just prior testing) and improved if necessary. Prior to a test, the weather forecast needs to be taken into account; no appropriate wind shielding is realistically possible for wind speeds exceeding a certain level. The U-shaped test configuration leads to a so-called chimney effect of the flames on the tank so that the rear 180° of the tank circumference are subjected to the fire load. The height of the tank above the gasoline level needs to be adjusted in a way that the maximum exothermal load is effective at half of the height of the tank (850–920 °C). Temperature measurement in this flame area close to the tank hull gives information if the single test can be accounted. Diagram 5.19a shows the pressure progression of several single burn tests and the resulting borderline curve for a type 1 steel tank example. The burst point at the zero point of time, hence without any fire load, is the result of a hydrostatic burst test with water filling. Type 1 steel tanks typically show a strong decrease in this so-called *Sufferable Pressure Curve* in the left area of the diagram that is caused by the degradation of the heat treatment due to the temperature load. Reaching the borderline curve means bursting—therefore T-PRD and P-PRD need to trigger and pressure-relief the tank at an appropriate distance prior to this curve when the same testing is done with tanks fully equipped with these safety devices.

Failure Mode Type 1

Type 1 tanks remain gas-tight and the failure mode is always burst in correctly performed fire tests. A rapid pressure increase in the gas contents is typical due to the outstanding heat conduction of the tank shell. At the same time however, the heat treatment of the steel quickly degrades without loss of wall thickness, which is in contrast to composite tanks. Other than in hydraulic burst testing with water filling, bursting with natural gas filling releases a tremendous amount of energy, see Diagram 5.6. Measured initial speed of the ball shaped burst pressure wave is at about 1440 km/h.

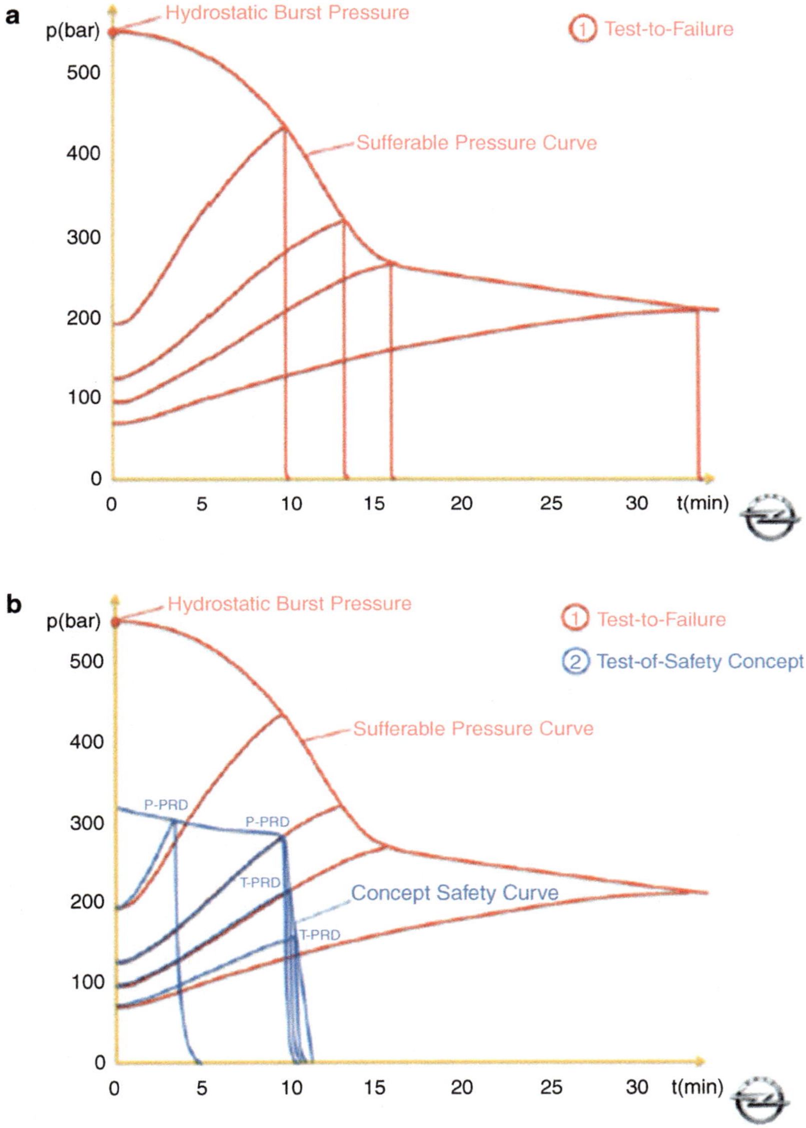

Diagram 5.19 (**a**) Natural gas tank failure tests and sufferable pressure curve, (**b**) Test of safety concept on a natural gas tank

The subsequent event is usually the combustion of the released natural gas as a fireball. The initial crack of the tanks occurs in longitudinal direction such as in hydraulic burst testing, however the crack spreads out further driven by the pressure front of the expanding gas and can even lead to dismantling of the tank shell into several pieces of shrapnel. After the burst event, testing staff need to stay in the protected shelter for an appropriate time as tank shrapnel can still fall down from a great height impacting into ground. Simple walls are not sufficient for protection as the fragments can fall back to earth in an almost vertical direction.

Failure Mode Type 3

Due to its metallic liner, type 3 tanks also stay gas-tight until they burst. The difference compared to type 1 is the degradation of the composite tank shell. Reinforcement fiber is excavated bit by bit by pyrolysis of the matrix resin and therefore no longer carries load. Practically, this mean a loss of wall thickness until the remaining fiber suffers an overload fracture. The bursting is less caused by increase of the inner pressure as the composite acts like a thermal isolator and the pressure level is significantly lower compared to type 1 burn testing—it is mainly the named effect of the excavated fiber as the resin burns away.

Failure Mode Type 4

The effect off degradation of the composite portion is analog to type 3 tanks. However the behavior of the plastic liner is significantly different, leading to diverse failure modes. On the one hand, type 4 can fail by bursting; on the other hand, the liner can melt so that the tank is no longer gas-tight and just burns as the gas filling is squeezed through the fiber wrapping. Furthermore, both failure modes can occur for one tank, which means that it can start to burn due to its own gas contents and still can burst with some time delay. This means that a type 4 tank burning with a typical natural gas flame on the surface and without the T-PRD trigger is not an all-clear message, as it still can burst!

Test of Safety Concept

Within this particular procedure, the same tanks (type, size, batch of manufacturing) are burned starting with the same initial pressure level as for *Test-to-Failure*, however equipped with the full set of safety devices such as T-PRD and P-PRD. Diagram 5.19b shows the earliest triggering of the T-PRD occurs after more than 9 min. This shows that high initial filling pressure of steel tanks can only be safeguarded by P-PRD in this kind of *Localized Fire* as for the T-PRD triggering the heat conduction needs to conquer the distance from the fire area first, which under these conditions is not in time. At the zero point of time, the *Concept Safety Curve* just shows the triggering pressure of a cold P-PRD (burst disc). The decrease of the P-PRD triggering pressure over time traces back to the loss of stability of the Nickel burst disc when being exposed to heat. While type 1 steel tanks are protected by P-PRD, as well as T-PRD, the release pressure of these safety devices comes closest to the *Sufferable Pressure Curve* when initial filling pressure is in the medium range. The necessary safety distance from the borderline curve needs to be considered in the direction of the existing pressure gradient in this area. A safety distance criteria which takes this into account can be the minimum length of the hypotenuse of the averaged grade triangle created between the triggering point of the safety device and the tanks burst point on the *Sufferable Pressure Curve*—in each case gained while testing with the same initial filling pressure. This takes the remaining time duration as well as the pressure difference still to be overcome till bursting into account for the safety criteria. Different burn testing—even with the same starting parameters—show variation as can be seen in the two tests with the highest initial filling pressure in Diagram 5.19b. This needs to be considered for the formulation of an appropriate safety distance to the *Sufferable Pressure Curve*.

Vehicle Burn Testing (Diagram 5.20)

The vehicle burn test is used to verify the serial natural gas fuel system. Typically a gasoline pan is placed below the vehicle and completely covers the area of the natural gas storage system. If the test procedure described under *Localized Fire* has been passed, no deviation is expected, as within the vehicle test the distribution of hot gas towards the T-PRD is promoted. Other burn testing at a vehicle level can be done e.g. in order to verify the described *Localized Fire* methodology (for example representing arson of the vehicle interior or somewhere external etc.). The natural gas system must represent the production status. The same safety requirements as named under *Localized Fire* need to be met. The vehicle fire shall proceed as long as it needs till there is no remaining pressure in the natural gas tanks as it was relieved by triggering safety devices. The excess flow flames are typically checked and documented using camera monitors, however this should not be used for definite determination of a pressure-free condition in the tanks. Due to this, the vehicle needs to be equipped with steel ropes/chains so that it can be overturned to the side from a protected area at the end of the test. The natural gas storage system is now visible and each tank must be shot from a safe and protected distance by suitable hard-core ammunition (e.g. 7.62x51 NATO) to guarantee the pressure free condition. Only after that can staff be allowed to approach the burned vehicle. This must be done with breathing protection in order to avoid harm from poisonous smoke gas. The on-site inspection and later assessment of the disassembled natural gas tanks shall include the check of the safety

a) b)

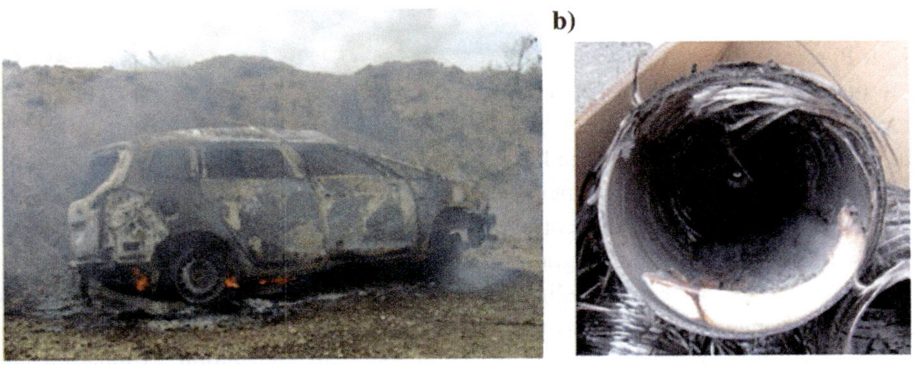

Diagram 5.20 (**a**) Vehicle burn testing, (**b**) section of a type 4 natural gas tank from vehicle burning

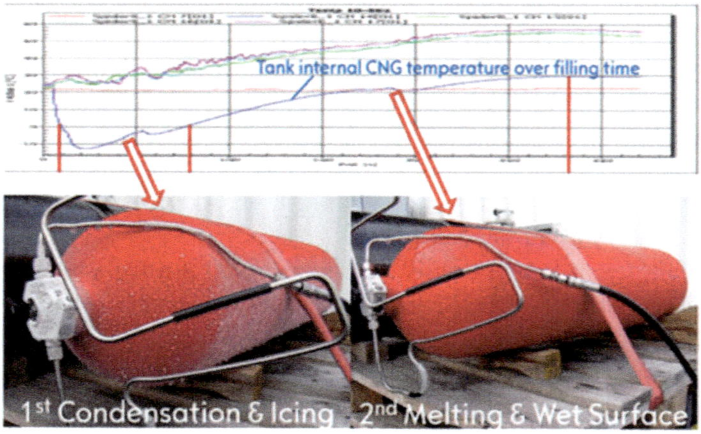

Diagram 5.21 Icing and condensation of water during the fast filling of a type 1 cylinder

channels of the T-PRD and P-PRD in order to determine which fuse has been triggered and if all triggering did take place as intended.

5.1.3.4 Corrosion Protection

The corrosion protection for natural gas fuel systems has particular importance as there are rarely technical standards available for underfloor installed systems operated in northern latitudes, see Sect. 5.1.1. In contrast to normal underbody components, natural gas components are subjected to thermodynamic cycles that lead to increased corrosive load. As an example, Diagram 5.21 shows the water condensation (sweating) on a type-1 tank surface during a fast filling cycle. Hence the evaluation criteria of normal underbody components are not sufficient in corresponding corrosion tests—the aggravation of test methods is necessary. Progressive corrosion on natural gas systems may become safety relevant, which underlines the importance of their consideration.

High Pressure Piping and Connectors

The material AISI 316Ti (1.4571) is in use for piping systems and AISI 316L (1.4404) for fitting systems, see [22]. Alloys with nickel and mainly molybdenum increase the resistance against pitting and stress corrosion cracking. By connection to other components consisting of different material such as aluminum valve housing, the potential differences can lead to galvanic corrosion in a humid environment which is present in the vehicle underbody area, see Diagram 5.22a.

Contact corrosion can be eliminated, for example by separating the electron-conducting connection of the paired metals by coating (preferably of the less noble metal). For aluminum, this can be a non-conductive anodization. Another measure is the avoidance of electrolytes between the metals, for example by sealing, see Diagram 5.22b.

a

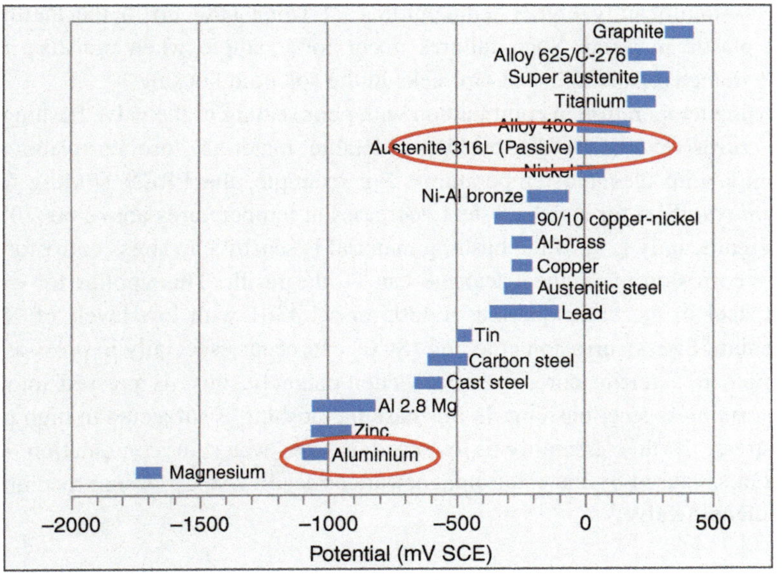

b

Diagram 5.22 (**a**) Potential differences of different metals [24] (**b**) isolator/sealing for a connection of aluminum/stainless steel

Tank Valve

The design of the solenoid coil and the therein-positioned guiding bushes for the inner shut off piston must receive special attention. In case of leakages in the bushing, there is the risk of successive depletion of the entire fuel mass, even if the inner piston is in the closed position. In an underfloor installation, a high combined load is caused by corrosion (for example salt water in the winter time) and self-heating by the solenoid. Both loads occur cyclically and superimposed. Proper magnetic coils generate at full-onboard voltage temperatures in the range of 85–90 °C (185–194 °F). Faulty solenoids generate—depending on how close they come to the short circuit by decreasing ohmic resistance—significantly higher temperatures; at worst case up to the melting of the solenoid plastic material. Such failures occur for example when moisture penetrates through a defective solenoid seal or cracks in the solenoid housing.

Increasing temperature in combination with poor sealing of the valve bushing will lead to high corrosive load at the bushing. Metallic materials lose corrosion resistance significantly with elevated temperatures. For example, the PREN (*Pitting Resistance Equivalent Number*) for stainless steel decreases at temperatures above 60–70 °C (140–158 °F) significantly [23]. If the bushing material is sensitive to stress corrosion cracking or pitting corrosion, permanent leakage can be the result. This applies for example to stainless steel in the basic qualities 1.4305 and 1.4301 with low levels of Nickel and Molybdenum. Stress corrosion cracking (SCC) can occur especially in press-assemblies, for example if a ferrite core, used for better magnetic flux, is pressed into an outer austenitic stainless steel bushing. In this case the bushing is subjected to high permanent tensile stress. If this assembly is exposed to salt water in combination with high temperatures, caused by the electromagnet, it can lead to SCC. Design recommendation for the solenoid valve:

- Effective sealing of the coil, the electrical connection and the piston-bushing also in combination with temperature cycling
- Design of the coil housing robust against temperature cycles
- Reduction of thermal load by peak-and-hold control of the solenoid valve. Full voltage only to open the valve; reduced voltage to keep it in open position
- Bushing made from high corrosion resistant material, stress free-annealed (e.g. 1.4529)
- Bushing should not be loaded by tensile stress caused by press fit. Bushing should be made out of one piece (one piece design)—designed for the magnetically less conductive, high corrosion resistant bushing material

Recommendation for laboratory corrosion testing of tank valves with combined cyclic corrosion load:

- Valve screwed into a thick-walled pressure test body of small inner volume at full nominal pressure by nitrogen filling, see Diagram 5.23
- Salt spray cycles with intermediate drying in order to elevate the salt concentration within the remaining humidity
- Electrical cycling (opening/closing) of the solenoid, cycling time must represent the full temperature swing

The above-mentioned design properties need to be ensured by appropriate quality assurance in the serial production.

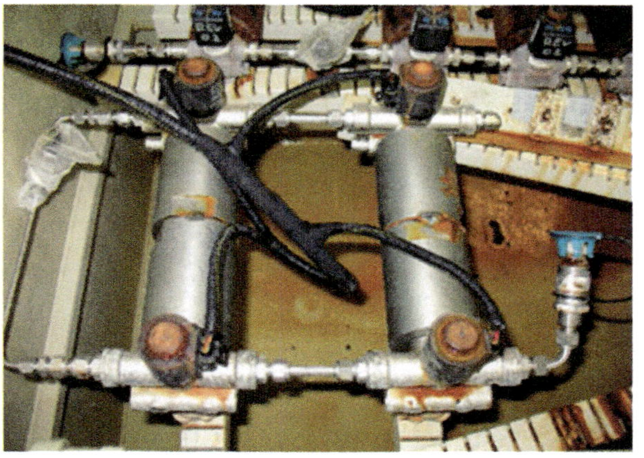

Diagram 5.23 Corrosion test for tank valves: cyclic electrical/ thermal loads overlapped with cyclic corrosion loads

Type 1 Tanks

These are sensitive to corrosion, especially steel tanks with low nominal diameter and small wall thickness, since the typical layered corrosion quickly leads to a high percentage in reduction of the wall thickness and thus the margin of safety. Since these low diameter tanks also underlie the full manufacturing process they are economically inefficient relative to the storage volumes gained. The thermodynamics of the refilling process and the good heat conduction of the type 1 tank shell in general leads to comparably high temperature swing on the tank surface with humidity condensation effects. This contributes significantly to corrosion loads, see Diagram 5.21. Correlation to the field could be generated in laboratory cyclic corrosion testing without refilling only by a significantly excessive number of cycles compared to other underbody components. Accordingly, the OK criteria must be adapted. With this, the salt spray load cycles intermit with drying cycles. Non-cyclic corrosion tests under constant salt spray application have been found to be not very effective. Furthermore real vehicle corrosion tests with all the thermodynamic effects of the natural gas system are recommended. Internal corrosion of type 1 tanks has to be assessed—taking the regional qualities of natural gas into account—based on regular internal inspection [27]. In Europe internal corrosion is, compared to external corrosion, no major issue due to the protection by traces of filling station compressor oil that enter the tanks with the natural gas and due to the absence of some critical corrosive ingredients in the fuel. With other global gas qualities the use of type 1 tanks needs to be investigated for the region case by case (for example hydrogen sulfide with water, sulfurous acid and sulfuric acid leading to corrosion of steel).

Numerous pressurized gas tank manufacturers globally are proficient in the production of basic type 1 steel tanks. However this is not the case for the outer corrosion protection that is mandatory for vehicle applications in the area of elevated corrosive environments. The background is that these tanks have a history in the application of industrial gas; typically they are not exposed to a corrosive environment as it takes place in some vehicle applications. In particular, the corrosion load by salt water is completely missing. The

a)

b)

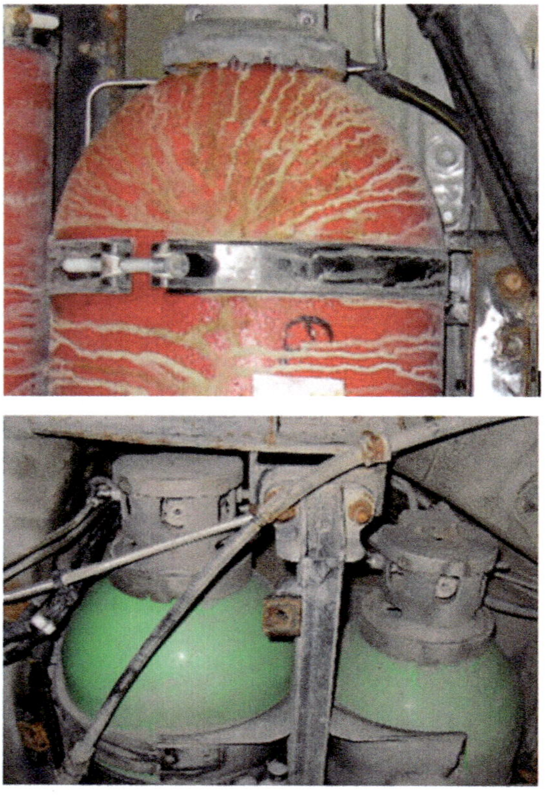

Diagram 5.24 (**a**) End of cyclic vehicle corrosion test—not passed, (**b**) test passed

type approval guidelines in present form provide no sufficient remedy—the tests to assess the tank corrosion protection are not representative by far. Therefore currently only a few manufacturers are able to provide type 1 natural gas tanks that meet the corrosion requirements of the vehicle OEMs. This is only possible with appropriate quality control of the coating process flow and its specifications. For example, the meticulous observation of the cleaning process result of the crude steel surface can be named as well as its immediate subsequent procession into the coating process without any dwell. So far only a few type 1 tank manufacturers invested in corresponding coating facilities with hermetic sealing to the environment (Diagram 5.24).

Type 3 and 4 Tanks

The metal liners of type 3 tanks may be subjected to internal corrosion caused by extremely poor gas qualities. This is not the case in correlation to minimum gas standards [1]. Type 3 and 4 tanks require the galvanic isolation of the carbon fibers and the aluminum of the liner or the boss, as there is a potential difference between these two materials. The cases with stress corrosion cracking of glass fiber tanks in the 1990s have been solved meanwhile by the use of boron-free fiber material. Type 3 and especially

Diagram 5.25 End of cyclic vehicle corrosion test with a type 4 tank—free from corrosion

type 4 tanks in under-floor vehicle applications do not tend to corrode in general, see Diagram 5.25.

5.1.3.5 Controlling, Regulation and Diagnostic of the Natural Gas System

Diagram 5.26 shows the typical components of a natural gas system consisting of tank valves, high-pressure regulator, pressure sensor and the pressure- and temperature sensor, at the natural gas injector rail of the engine with its gas injectors. Other temperature sensors and a manual fuel mode switch (natural gas/gasoline) may be optionally present. If no manual fuel mode switch is present, the control module performs the fuel mode switching automatically. The control and regulation functionality is embedded in a control unit; most of the time it is integrated into the engine control module (ECM). A separate CNG control unit, which is in close communication with the ECM is also possible.

Diagnostics on the Natural Gas System Malfunctions in the electrical/electronic system can be detected during vehicle operation via the *On-Board Diagnostics* and are documented in the malfunction memory. Where appropriate, a warning with a recommended course of action can be displayed to the driver. The sensor signals can be used for plausibility checks performed by the ECM software. For example, this allows the detection of non-measurable, potentially safety-relevant malfunctions (e.g. by *Blind-Tank-Diagnosis*). Plausibility checks can also be used for driver information, such as the quality of natural gas, which can be obtained via the lambda control circuit of the engine. *Off-Board Diagnostics* can be performed via the vehicle's diagnostic connector with conventional test equipment. They include both the readout of the ECM malfunction code memory as well as active device control testing for functional checks.

Leak Detection High Pressure Side Natural gas systems have to be sealed. Leaks can be detected via plausibility checks and, if appropriate, can be displayed to the driver with a remark to park in an open-air area only. The diagnosis can be initiated by a short-term opening of the tank valves before the engine is started. The high-pressure line is then applied with the tank pressure and sealed between the tank valves and the inlet valve of the pressure regulator. The (temperature compensated) pressure gradient is observed by the high-pressure sensor and checked for plausibility with estimated to thresholds.

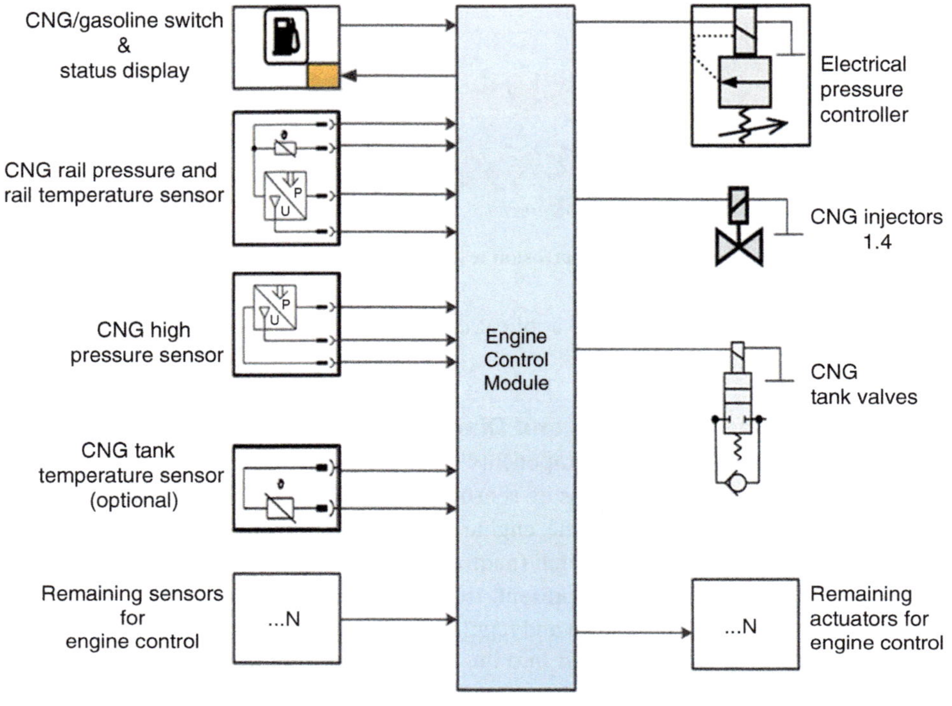

Diagram 5.26 Electrical block diagram of a natural gas fuel system

This test is aborted if the engine is started during this pre-start diagnosis and it is replaced by the *Continuous Leak Diagnosis* algorithm, which is performed in parallel to driving. Thereby the temperature compensated pressure gradient is set into relation to the maximum possible pressure gradient based on the fuel consumption of the engine. If this is not plausible, a forced safety fuel mode switch to the gasoline mode is triggered because of a potential leakage—all valves in the gas system will be closed. Furthermore, there is a corresponding action warning displayed to the driver. The leak diagnosis only lasts for one ignition key cycle and is therefore restarted with the next cycle.

Leak Detection Low Pressure Side This diagnosis is basically done in the same way as for the high-pressure side, but in the area between the pressure regulator and the natural gas injectors. In case a malfunction is detected, the default action is a forced fuel mode switching into the gasoline mode.

Blind-Tank Diagnosis [25] This diagnosis identifies faulty (e.g. valve stuck closed) tank valve solenoids. If a faulty solenoid is not identified or is ignored (reduced range not recognized) the corresponding tank is still filled but not emptied. Together with several filling cycles and ambient temperature increase from winter to summertime, critical P-PRD threshold pressure in the tank can be reached. If there are P-PRD's as part of the tank safety system, they may be triggered depending on their calibration, which then is an undesired event. The diagnosis software uses the estimated CNG consumption of the engine and calculates the expected high-pressure drop gradient of the intact system. This calculated value is set in relation to the measured pressure drop (both temperature compensated). If the pressure decreases faster than expected—depending on the value of the gradient—one or more blind tanks with a defective solenoid valve can be identified. Similar to this, blind-tank diagnosis of individually controllable tank valves can be done by the device control check, while the engine is running and measure the pressure drop and check it for plausibility. A blind tank is then quickly identified. The principle difference between the leakage diagnostics is that the blind tank diagnosis is evaluated during the emptying of the complete tank system, i.e. not only during just one key cycle.

Control of the Solenoid Valves Tank and high-pressure valves should be driven directly from the output driver of the control unit in order to simplify the monitoring. The valves must be closed in power off condition. Therefore, the positive supply voltage of the valves should be switched directly by the control device (HSD *High Side Driver*). A circuit where the ground is controlled (LSD *Low Side Driver*) should be avoided, as a grounding failure in an accident scenario would prevent the mandatory *Close-without-Voltage* function of the valves. A control by *PWM-* method (*Pulse Width Modulation*) offers the advantage of a *Peak-and-Hold* operation to reduce power consumption (up to 50 %), temperature load, thermo-capillary effects and potential corrosive load. So overall the solenoid durability is fostered. When a forced fuel mode switching to gasoline operation is triggered, the controller switches the solenoid valves off forcing them into the closed position. This is also the case for regular gasoline operation or key cycle off. In addition to the above mentioned diagnosis or security properties, further checks can be implemented, which can cause a forced fuel mode switching such as excessive pressure in the injector rail or an unexpected higher engine torque compared to the driver's request.

System Characteristics in Accident Situation

To prevent danger due to the natural gas system in the case of an accident, the system should be closed upon detection of such scenarios. This is the case when airbags are deployed (front/side impact), belt tensioners have been activated or a vehicle roll over is detected. Previously, a safety delay waited until the vehicle speed fell below a defined threshold. A reactivation of the system should only be possible by means of a new key cycle—the initial leak diagnosis will start accordingly.

HMI-Strategy

The *Human Machine Interface (HMI)* describes all the information path of the natural gas system to the driver. Accordingly the following features are included:

CNG Level Indicator This transfers the temperature compensated pressure signal into a fuel gauge indication. The information corresponds to the still usable residual fuel mass in the gas tank. The indication can be provided by either single fuel gauges—the indication switches between gasoline and natural gas filling level then—or two separate fuel gauges. Analog as well as digital displaying methods are in use.

Board Computer The board computer calculates and displays fuel consumption and remaining driving range. In driving conditions, the consumption can be displayed e.g. in kg/100 km or in kg/h when the vehicle is stopped. With information derived from the engine's lambda signal, the specific energy content of the natural gas filling (gas quality) and subsequent the displaying of the energy consumption over distance or time is possible. The latter is interesting in case the pricing of natural gas at the filling stations is changed from mass to energy units in the future.

Driving Mode Indication Some manufacturers allow the customer to manually switch between natural gas and gasoline mode, while others provide automatic switching starting from the natural gas mode. The respective operating mode can be optionally displayed on the fuel mode switch and on the driver information display. If manual switching is denied—e.g. due the fault in CNG fuel system—this can be reported back by e.g. rapid blinking of the LED on the fuel mode switch.

Text Messages on the Instrument These can include messages about the condition of the natural gas system. In conditions without consequences to the continuation of the driving operation, they usually appear briefly (*Pop-Up*). In case of safety or potential walk home issues, the message remains until the driver manually confirms it. Examples are low CNG fuel level or a leakage in the CNG system.

CNG Filling Stations in Navigation System For planning of the travel route or in the case of a low natural gas filling level the natural gas filling stations should be stored in the navigation system. Through the navigation these filling stations can be reached with a minimum time effort.

5.2 Passenger Car Gasoline Engines for Natural Gas

5.2.1 Bivalent and Monovalent Systems

Michael Bargende

Wikipedia [26] defines *bivalent engines* as vehicle propulsion units that are able to use two different types of fuel. This would need to be supplemented with the statements that both fuels must also be transported within the vehicle and that a switchover from one fuel to another is possible without a hardware change.

Bivalent engines are not to be confused with hybrid drives, where at least two independent or coupled drive aggregates are present.

Neither are dual-fuel systems described here. Dual-fuel systems also have two different fuel types, typically a flammable and a difficult-to-ignite fuel, which are carried in separate tanks. However, both fuels are burnt together, whereby the substitution rate can vary. This can go so far that the flammable fuel is only used as a "chemical spark plug". In this case, the system is described as a pilot injection principle.

The bivalent systems focussed on here describe systems that carry both fuels separately, whereby either one fuel or the other can be transported to the combustion chamber and burnt.

Typically, fully-fledged mixture formation systems are available for both fuels, which permits unrestricted operation with either of the two fuels.

The motivation to transport and burn two different fuels from two separate tank systems mainly results from two reasons: (1) There is no widespread coverage of supply of one of the fuels, so that if refuelling is not possible, the vehicle can continue using the other fuel as a replacement, or (2) one of the fuels is not to be or cannot be used for cold start and warm-up, since either the exhaust aftertreatment system cannot be brought to the conversion temperature quickly enough or a complete re-certification is to be avoided, in order to keep the costs for a retro-fit low. It is also conceivable that a reliable engine start cannot be guaranteed with one of the fuels at very low temperatures or due to the implemented mixture formation system and the engine must therefore be warmed up with the other fuel.

The latter motivation means that a complete exhaust gas certification procedure does not necessarily need to be conducted for the bivalent engine to be approved for road use. This assumes that once the catalytic converter has reached its peak conversion performance, it is permissible to switch over to the second fuel. Note that it must be guaranteed that the emission of regulated harmful pollutants by an engine at operating temperature is below the appropriate limits for either fuel.

Examples of such a design are bivalent engines that can be run with gasoline or autogas (LPG). LPG operation is typically only activated once the coolant temperature has reached 60 °C. This means that every cold start and warm-up is performed using gasoline. The vehicle cannot be started with LPG, although exclusive operation with gasoline does not pose any difficulties. This threshold of 60 °C coolant temperature has two effects. One is that, for lambda = 1 controlled engines with three-way catalytic converters, the exhaust aftertreatment system has reached its full conversion temperature and the emitted pollutants are below the legislative limits. Secondly, since the engine is relatively warm, good and consistent LPG mixture preparation occurs in the inlet manifold. This used to be of major importance when so-called gas-mixers were used and no sequential inlet manifold injectors were implemented.

Both reasons make a retrofit of gasoline engine driven vehicles with so-called autogas systems relatively low-cost, since a full (in Europe: European) emissions test is not required for every engine/vehicle combination.

A further example is the bivalent operation with gasoline and ethanol (E100) or methanol (M100) in so-called flex-fuel vehicles (FFV *Flexible Fuel Vehicle*). Due to the high evaporation enthalpy of alcohol, a confident and low-emission cold start using pure alcohol below an external temperature of 15 °C is not easily achievable, such that here, too, the possibility is used of either starting with gasoline and only switching to ethanol once a certain engine temperature has been attained, or, if no second tank is available, to mix gasoline or up to 15 % special volatile hydrocarbons into the alcohol [27]. In the latter case, these vehicles are no longer powered by bivalent engines, but are FFVs, since only one type of fuel is carried by the vehicle, albeit one that can be varied from one filling procedure to the next.

Bivalent natural gas engines were also started with gasoline until just recently and only switched to natural gas once a certain engine temperature had been reached. The reason for this was mainly due to the approximately 150 K higher light-off temperature of the three-way catalytic converter. Starting with gasoline meant that the emissions limits could be more easily fulfilled. This was predominantly applicable to Europe and all other regions of the world that had adopted the European emissions legislation. In Europe, the so-called THC (total hydrocarbons) are limited, whereas in the USA—and in all regions in the world that have adopted the American emissions legislation—the so-called NMOG (non-methane organic gases) or NMHC (non-methane hydrocarbons) are limited. The reason for the US legislation is that it limits precursor substances (NMOG + NO_x) for the generation of ground-level ozone. In the USA, methane emissions are thus classified as greenhouse gases and not as pollutants and converted to CO_2 emissions.

In the meantime, bivalent natural gas engines have entered the market that either only start with natural gas and warm-up or for example only start with gasoline every seventh cold start and after every time gasoline is filled up and at very low temperatures.

The customer driving a bivalent natural gas vehicle expects that he or she is able to drive unrestrictedly with both fuels. This means that both the engine response, i.e. the torque response during an acceleration phase, as well as the full load characteristics and hence the power available, together with the drivability with respect to engine vibration and noise when driving with either fuel are practically identical. This is also true for the mileage of the vehicle, in other words, the lifetime of the engine.

Injecting natural gas either into the inlet manifold or into the cylinder during the intake stroke causes a significantly higher volume displacement and thus lower filling due to the much lower density compared to gasoline vapour (approximately a factor of 5). At partial load with quantitative control, this causes a beneficial dethrottling and thus lower charge exchange, in contrast to full load, where the lower charging causes an approximately 8 % performance loss compared to gasoline operation, which must be accepted. This typical nominal performance loss of approximately 8 % is not solely caused by the lower charging, but also due to the lack in natural gas of the mixture cooling effect present in gasoline caused by the released enthalpy of evaporation, which effectively increases charge.

In principle, forced induction can compensate for this loss in performance through increased boost pressure. However, this is only possible if the turbocharger layout is not transferred unchanged from the gasoline engine, but is modified in such a way that the higher boost pressure from the TC is delivered without exceeding the component temperature tolerance limits on the turbo side, both concerning the outlet valve and the turbine itself. Favorable conditions prevail if a so-called Active Waste Gate can control the desired higher boost pressure when running with natural gas by late opening of the waste gate; whereas when running in gasoline mode, a lower maximum boost pressure is required, mainly due to knock reasons, but possibly also due to limits caused by reaching the maximum permissible transmission input torque.

In contrast to LPG operation, where the possibility of retrofitting existing vehicles is relatively easy, a retrofit of a CNG system for a bivalent CNG-gasoline operation is very seldom. This has several reasons, not only because of the lower attractiveness due to the current ratio of only 1000 natural gas fuel stations in Germany compared to more than 6000 LPG fuel stations, but also due to the considerably higher effort required for the retro-fit.

One of the reasons for this is the significantly more complex tank technology. LPG tanks are typically filled under 5–10 bar pressure [28]. Since the fuel can be stored in a liquid state, the tank can be constructed relatively small. Examples are known where the tank fits in the spare tyre recess [28].

Handling an LPG pressure tank and the legislative regulations are relatively simple due to the relatively low pressures.

CNG natural gas vehicles require pressure tanks with filling pressures of 200–250 bar in order to guarantee an adequate range when in natural gas mode. These tanks, as well as the pipework leading to the engine and the pressure reduction stages, are subject to substantially stricter safety regulations. In addition they are more voluminous, such that placement in the boot of the vehicle means severe limitations in the remaining space available.

Apart from the tank and pipework restrictions, converting a vehicle from gasoline operation to natural gas requires component changes in order to be able to guarantee the identical lifetime as for gasoline operation.

This affects in particular the inlet valves and seats that are protected in gasoline operation by the wear-prevention additives present in gasoline. These additives can also be employed with LPG, but not with CNG. The solution lies in other, hardier materials that do not require the protection afforded by the additives. Under certain circumstances, reinforced pistons may be necessary, since due to the higher anti-knock properties of natural gas, the mixture can be ignited earlier at full load and thus the peak pressures are higher. In addition, natural gas produces significantly fewer deposits on the piston than when in gasoline mode. These deposits form a type of thermal insulation, so that when in natural gas operation, the piston surface temperature in conjunction with the higher wall heat flows due to the earlier combustion can increase so significantly that both mechanically and thermally more stable pistons are required.

In addition, it can be expected that further tightening of the emissions limits will make a new formulation of the catalytic converter coating necessary to guarantee a timely converter light-off with regard to methane emissions.

Concerning the higher anti-knock attributes of natural gas, it should be noted that the resulting possible increases in compression ratio are not or are only partially implementable because of the continuing requirement to guarantee unrestricted operation in gasoline mode.

This is particularly true for engines with turbochargers where, compared to the otherwise structurally identical naturally aspirated engine variants, the compression ratio has been reduced.

However, this effect seems to be losing in importance. The current bivalent natural gas engines from Daimler, the M270NGT and the M274NGT, have a compression ratio of 11:1, despite the turbo charger [29].

Higher compression ratios are problematic in any case, since even when operating with natural gas, the quality of natural gases with lower knock resistance must be taken into account (for example L-gases with propane/butane blends) and a load intervention to control knock should be avoided. Depending on the stroke/bore ratio and valve angle prevailing in the base engine, the form of the combustion chamber may be so unfavorable that compression ratio higher than 11:1 leads to prolonged combustion and an increase in cycle variability. These effects counter the increase in efficiency from an increase in compression ratio such that a compression ratio of 11:1 is effectively the best compromise between the two opposing effects.

This trend also originates from the fact that the base engine is increasingly a highly optimized, directly injected, turbocharged gasoline engine with high power density. By directly injecting the gasoline and the resulting improved use of fuel evaporation cooling, combined with favorable combustion chambers thanks to the corresponding stroke/bore ratio, valve angle and minimized fire land volume, the compression ratio could be increased due to the lower knock sensitivity.

In addition to this, there is no longer competition regarding the positioning of the gasoline injectors and natural gas injection valves in the inlet manifold in bivalent systems. Gasoline operation can be realized with direct injection and natural gas operation can use appropriately positioned injectors in the inlet manifold. There may be limitations due to heat entry into the gasoline fuel injectors when running with natural gas. This can lead to a refractioning of the fuel and thus to a coking up of the injector in non-flushed operation. Concepts involving centralized injector locations are fundamentally more thermally critical for these reasons.

Good dynamic results can be achieved even with inlet manifold injection if the natural gas injectors are optimally positioned. The relevant effect of optimal mixture cooling via fuel evaporation that occurs with gasoline direct injection, above all at full load, does not exist with natural gas, so that the better mixture homogenization with inlet manifold injection in comparison with direct injection is the main focus, together with the avoidance of complexity and the high costs associated with natural gas direct injection valves.

Monovalent systems are so highly optimized for operation with natural gas that, if at all intended, running the engine with gasoline is only possible with limitations concerning rated power and, under certain circumstances, smoothness and responsiveness of the engine.

Opel coined the phrase "monovalent-plus concept" [30]. In this case, the engine has been rigorously optimized for natural gas operation. If necessary, the engine can also be run with gasoline that can be stored in an approximately 15-l tank. Running the engine with gasoline is limited both by the possible range due to the small tank and also due to the rated power.

This concept also provides the manufacturer with the advantage of a simplified certification for gasoline operation.

To date, bivalent natural gas vehicles have had the reputation of being sub-optimal when in natural gas mode; whereby only monovalent engines could exploit the full potential of operating with natural gas.

The four crucial aspects, only improvable with monovalent engines, were:

Mixture Formation

A consequence of the competition between inlet manifold gasoline injection and inlet manifold gas injection is a sub-optimal solution concerning the positioning of the gas injection valves, since normally the original position of the gasoline injectors is kept in order to avoid the necessity of a complicated calibration of the ignition advance angle and the cold start and warm-up strategies for gasoline injection. In particular the cold start and warm-up strategies are decisive here, since gasoline is typically used to start and the catalytic converter light-off is achieved with gasoline and only then is the system switched over to gas operation. This results in a positioning of the gas injector valves relatively far from the inlet valve for space reasons. This makes it very difficult, above all in dynamic operation, to guarantee an exact stoichiometric mixture. In monovalent operation, the gasoline injector valve can be replaced with a gas injector valve at the identical location, whereby the exact stoichiometric gas-air mixture is positively influenced.

The Compression Ratio

Natural gas is, in principle, very inflammable and hence knock proof. However, if butane and/or propane are blended in, as is the case for certain L-gas qualities, then knock can indeed occur and anti-knock control must be employed. This means that, in particular for turbocharged engines whose compression ratio in gasoline mode must be reduced as compared to naturally aspirated engines, monovalent natural gas engines can be designed to have substantially higher compression ratios. The limit for the increase in compression ratio is reached when the free flame surface for a roof-shaped 4-valve combustion chamber in proximity to the TDC position of the piston is so narrow that no high turbulent flame propagation speed is possible and thus the combustion duration and the cycle variability increase to such a level that the gain in efficiency from the increased compression ratio is over-compensated.

The stroke/bore ratio for a monovalent natural gas engine can be designed more in the direction of long stroke, as long as the base engine does not have to be a gasoline engine (perhaps rescinding several hundred revolutions of nominal engine speed), whereby the limit for increasing the compression ratio is reached later. Realistically, compression ratios above 13:1 are not practical, since the peak pressure (see below) increases too much. A maximum compression ratio of 12:1 seems to be a good compromise for monovalent natural gas engines.

Peak Pressure

Typically, gasoline engines have been designed according to the mechanical stability of the base engine for a maximum peak pressure of below 100 bar, partially also up to 120 bar, since higher pressures, even with turbocharging, did not occur. Turbocharged, monovalent natural gas engines with high compression ratios can easily exceed this limit due to their high knock rating. Average cylinder peak pressures of between 150 and 160 bar are no problem, whereby individual working cycles can even reach 190–200 bar peak pressure due to cycle variability. To date, these peak pressures could only be run with base engines that were specifically designed for gas operation, otherwise changes to the ignition timing delay and/or boot pressure reduction and/or compression ratio reduction had to ensure that the mechanical limits of the specified engine would not be exceeded in monovalent natural gas operation.

Ignition Voltage Requirements

Natural gas requires a higher ignition voltage than conventionally used with gasoline, due to the reluctance of the methane molecule to ignite. This leads to a higher thermal loading of the ignition coil and the electrode burn-up of the spark plug is more pronounced. The latter can be compensated for by using improved materials (such as platinum) or special spark plugs (if not available as standard in the form of lifetime spark plugs). The higher thermal loading of the ignition coil can become critical if, for example, pencil-type ignition coils are used in the base engine, which have very little reserves due to the tight installation space. In monovalent engines, the complete ignition system can be optimized for use in natural gas operation, without having to consider a fully-fledged gasoline mode.

Today, since mainly direct injection gasoline engines now provide the basis for natural gas engines, the competitive situation for both mixture formation systems has largely eased. A typical configuration consists of direct gasoline injection, either with lateral or central injection position and a sequential inlet manifold gas injection. However, a purely monovalent engine design would lead to a direct injection of natural gas. Despite a range of investigations into the advantages of direct gas injection, particularly in stratified mode [31], but also in homogeneous mode [32], there is to date no series production. The main reason surely lies in the difficulty of mass-producing a natural gas injector that reaches the lifetime of a gasoline injector and in particular remains gastight towards the combustion chamber when the engine is switched off over the lifetime of the engine. Hence bivalent operation no longer has any restrictions compared to a monovalent design with regard to mixture formation, disregarding the costs for dual fuel systems.

The trend towards increasingly higher engine performance per litre due to turbocharging and downsizing, in order to further reduce the specific fuel consumption, leads to a litre performance of over 100 kW/l for mass-produced engines. These relatively low volume engines with 800–1200 cm^3 total capacity display even as three-cylinder engines relatively small combustion chambers with short flame paths. In combination with long stroke designs and the possibility of mixture cooling with direct injection due to gasoline evaporation, this leads to the possibility, even in turbocharged gasoline mode, of achieving compression ratios of 11:1 or even higher.

This eliminates the necessity of a monovalent design to be able to increase the compression ratio, as the limit has been practically reached in bivalent operation. The same is true for the peak pressure. Due to the design in gasoline operation for average pressures > 25 bar in order to reach the nominal performance > 100 kW/l at moderate engine speeds around $n = 5500$ rpm, peak pressures of well over 100 bar at full load in the area of maximum torque must be tolerable by the base engine.

The requirements concerning ignition voltage in gasoline mode are becoming increasingly stricter due to turbocharging and high compression ratio. The requirement to retard ignition when knocking combustion occurs makes it particularly necessary to have reserves, since the ignition timing at high cylinder pressure must be relocated towards ignition TDC and hence even higher pressures. This results in the ignition systems being designed in such a way that even turbocharged natural gas operation with high compression ratio, but typically higher advanced timing requirements can be reliably ignited.

In summary, the conclusions that can be drawn from this discussion are that practically all technical reasons that so far have supported a monovalent instead of a bivalent design do not, or soon will no longer, apply. This is mainly due to the fact that in terms of design, gasoline engines are approaching a level that used to represent a measure for an increase in efficiency for natural gas engines.

Three arguments remain for a monovalent design of natural gas vehicles:

- The introduction of natural gas direct injection. Further advantages of efficiency could be employed, as described in [31] and [32].
- The costs for the vehicle and development. A monovalent design can completely waive a second fuel system, including mixture formation components, control unit calibration and emissions certification.
- The range. Depending on the packaging situation within the vehicle, a larger natural gas tank (perhaps even divided into several smaller tanks) can be mounted by eliminating the gasoline tank, which would enable a significant, customer-relevant increase in vehicle range.

5.2.2 Thermodynamics

Mahir Tim Keskin and Michael Bargende

Thermodynamic analysis is an ideal tool to assess the advantages and disadvantages as well as the potential and limitations of running passenger car Otto engines with natural gas. The results show the inherent differences to gasoline operation due to the different material properties, the consequences for engine design and the peculiarities to be noted when simulating and analyzing gas engines.

5.2.2.1 Thermodynamic Basics for Combustion Engines

As is the case for all combustion engines, the calculation of natural gas engines is mainly based on three fundamental equations: the first law of thermodynamics (conservation of energy), ideal gas law and mass balance. The equations can be formulated in a manner appropriate to combustion engines as follows [33]:

$$\frac{\mathrm{d}U}{\mathrm{d}\varphi} = \frac{\mathrm{d}Q_\mathrm{b}}{\mathrm{d}\varphi} + \frac{\mathrm{d}Q_\mathrm{w}}{\mathrm{d}\varphi} - p\frac{\mathrm{d}V}{\mathrm{d}\varphi} + \sum\left(\frac{\mathrm{d}m}{\mathrm{d}\varphi}h\right) \tag{5.1}$$

$$p \cdot V = m \cdot R \cdot T \tag{5.2}$$

$$\frac{\mathrm{d}m}{\mathrm{d}\varphi} = \sum\left(\frac{\mathrm{d}m}{\mathrm{d}\varphi}\right)_{\mathrm{E/A}} \tag{5.3}$$

A detailed analysis of the individual terms of the equations enables initial statements to be made on the areas where natural gas and gasoline-driven engines differ: First, the differing calorific material properties have an impact on inner energy and enthalpy in the first law and on the individual gas constants in the ideal gas law. In addition there are the indirect dependencies for example the effect of different reaction kinetics on the combustion process and hence on pressure and temperature. Differences can also be expected regarding wall heat compared to gasoline operation due to different heat transfer properties, density and viscosity, but, as far as assessable, these have a subordinate role in the total system, such that proven models of diesel and gasoline engines can be used further.

5.2.2.2 Primary Influences of Natural Gas Operation

In the following, a primary influence denotes those influences that originate only from the change in fuel, hence they occur even with an unchanged engine design and ultimately result from the differing fuel characteristics. First findings can be reported by observing a "perfect engine", while in a real process assessment further correlations will become clear.

Ideal Comparison Process

Assuming a constant volume cycle, the thermal efficiency for an ideal motor is well-known to be dependent only on the compression ratio ε and the isentropic exponent κ. Whereas the compression ratio is only indirectly influenced by the fuel used and can thus be left constant within the context of the primary influences defined above, there is a direct dependency between the isentropic exponent and fuel (this is frequently ignored and the material properties for air at average temperatures are used instead; this is however not a mandatory condition for consideration as an "ideal engine"). It can be described at first as the quotient of isobaric and isochoric heat capacity, whereby this ratio can also be described using the degrees of freedom of the observed gas molecules [34]:

$$\kappa = \frac{c_p}{c_v} = \frac{f+2}{f} = 1 + \frac{2}{f} \tag{5.4}$$

The number of degrees of freedom depends primarily on the size of the molecule and the temperature, whereby increasing molecule size and temperature are also logically associated with an increase in the degrees of freedom. It can be directly inferred from this that the methane molecule, which is much smaller compared to iso-octane as a typical gasoline component, has fewer degrees of freedom and thus possesses a higher isentropic exponent—an attribute that has a fundamentally positive effect on process efficiency, see Diagram 5.27. As an aside, it is also worth noting the degrees of efficiency that

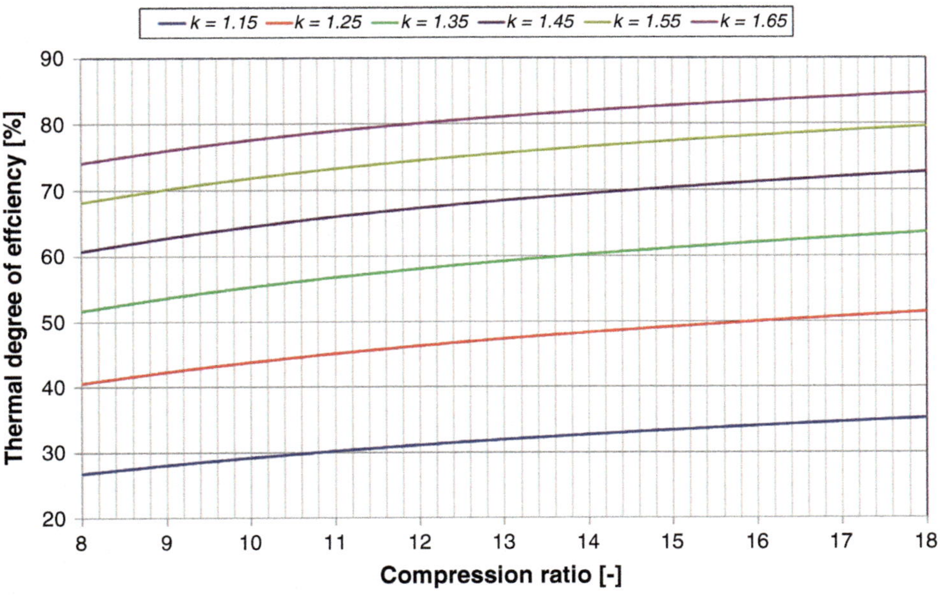

Diagram 5.27 Thermal efficiency of the constant volume cycle depending on isentropic exponent and compression ratio

are theoretically possible with working media with a high proportion of noble gas ($f = 3$).

For engine operation, it is not the material properties of the fuel itself which are relevant—there is a very significant different between $\kappa_{Methane} \approx 1.20$ as opposed to $\kappa_{Octane} \approx 1.03$ (each at 300 °C and 2 bar) [35]—but the values for the utilized air-fuel mixture. If one assumes a mixture of ideal gases, the isentropic exponent of the mixture can be calculated using the heat capacity of the mixture, that itself results from the linear combination of the individual heat capacities multiplied with the according share of the mixture:

$$\overline{\kappa} = \frac{\sum x_i \cdot c_{p,i}}{\sum x_i \cdot c_{v,i}} \tag{5.5}$$

If one uses the molar material properties and mixture shares, the first part of Eq. (5.4) and the dependency

$$c_v = \frac{\mathfrak{R}}{\kappa - 1} \tag{5.6}$$

then the isentropic exponent of the mixture can hence be stated as

$$\overline{\kappa} = \frac{\sum x_i \cdot \frac{\kappa_i}{\kappa_i - 1}}{\sum x_i \cdot \frac{1}{\kappa_i - 1}} \tag{5.7}$$

whereby x_i represents the molar fraction. Assuming a stoichiometric mixture composition and an oxygen content in air of 21 % results in the following isentropic exponents for each material—again at 300 °C and 2 bar—[35] of:

$$\kappa_{Methane-air} \approx 1.36$$
$$\kappa_{Octane-air} \approx 1.33$$

These values, also representative for gasoline and natural gas (even when considering different compositions of natural gas, for example H-gas or L-gas, only very small deviations from the value of pure methane occur), are in a similar area. However, these minor differences in calorific characteristics result in an increase of approximately 6 % in an ideal process assuming a compression ratio of eps = 10. This, however, is immediately to be put into perspective as it is obviously a very simplified view, particularly of the calorifics (ignoring, among others, the change of gas composition by the combustion process, the temperature dependency, the dissociation).

The isentropic exponents determined here are of practical importance for the thermodynamic zero-line determination. If one assumes that for a cylinder temperature close to the wall temperature, the polytropic compression can be considered as adiabatic and reversible, then for external mixture preparation depending on the fuel used, different recommendations result for the polytropic exponents:

$$n_{\mathrm{Natural\ gas}} \approx 1.36$$
$$n_{\mathrm{gasoline}} \approx 1.33$$

These theoretical values fit well with the values found in literature in which lower values from 0.01 to 0.02 units are mentioned using more exact observations, for example $n_{\mathrm{Natural\ gas}} = 1.34$ [36]. The fuel used for internal mixture formation is irrelevant at normal injection times for the polytropic exponents, such that these correspond to the values for air and can be selected in the range 1.37 to 1.4 [33, 37].

Independently from the process management in addition to the previous observations, a statement can also be made concerning the expected CO_2 emissions. Whereas in Otto engines using gasoline a reduction in CO_2 emissions can mainly be achieved only via an improvement in the efficiency, a change to natural gas results in a reduction in emitted CO_2 with unchanged efficiency due to the significantly better C/H atom ratio with comparably low effort. This results directly from the reaction equations with corresponding standard enthalpy formation and is shown using the example of alkanes in Diagram 5.28. This gives an advantage of approximately 20 % for methane over

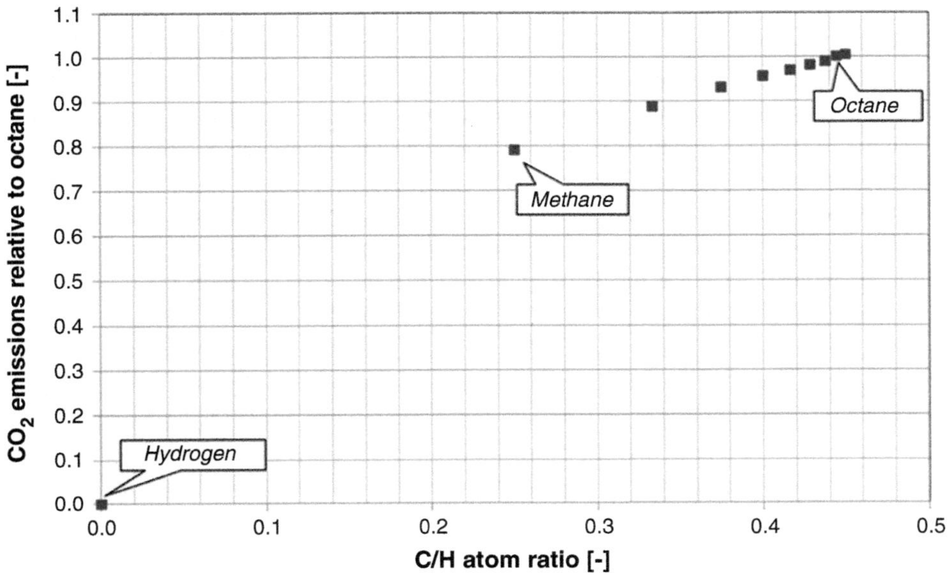

Diagram 5.28 CO_2 emissions for identical energy release and complete combustion relative to octane as a function of C/H atom ratio

n-octane related to the same energy release (this means for identical load, if one assumes identical degrees of efficiency).

Overall, a significant reduction in CO_2 emissions and a slight improvement in the degree of efficiency can be expected due to the fundamental characteristics of methane for an otherwise unchanged engine.

Perfect Engine

A temperature dependency of the material properties is considered for the perfect engine in contrast to the ideal thermal cycle. In most cases, a linear dependency of the heat capacity and an independency of the individual gas constants from the temperature are assumed, which is a reasonable approximation that neglects dissociation at high temperatures, see Diagram 5.29.

If one linearly approximates the data of a calorific approach for combustion gases, for example from [38], then the values shown in Table 5.1 can be determined for gasoline and methane at $\lambda = 1$ (these values do not necessarily follow from the calorific approach used, but depend especially on the temperature window over which the linear approximation is effected). For the sake of completion, the lower heating value and the stoichiometric air requirement are also listed.

The following attempts to consider a comparative calculation between methane and gasoline according to the Seiliger process. It is based on the mutual boundary conditions shown in Table 5.2 and a Seiliger process with isochoric-isobaric heat release [39]. Table 5.3 shows the state values and Diagram 5.30 shows the *p*-*V* diagrams.

Most noticeable are the significantly lower air and fuel masses for methane at the starting point, whereby an observation of the perfect engine clearly reveals the filling losses for external mixture formation. This has a negative effect on the achievable rated power but is positive for the partial load efficiency via the dethrottling. It should be noted that due to the higher heating value of methane the lower mass is largely compensated for, so that the 14 % lower fuel mass is contrasted with a significantly lower reduction in added heat of approximately 4 %.

As can be expected, the higher isentropic exponent of the methane-air mixture leads to a slight increase in the compression final temperature and pressure. This is contrasted with a less pronounced increase in temperature and pressure during combustion due to the lower heat contribution. In total, the calculation with methane results in an advantage in degree of efficiency of approximately 0.5 % compared to that of gasoline; thus the advantage of degree of efficiency is considerably less than in the ideal process. If the air-to-fuel ratio for gasoline is increased to such a degree that either the heat supply or the indicated work is identical in both cases, this advantage is further reduced to 0.3 % and 0.4 %. The peak temperatures for methane are slightly higher than for gasoline. Overall, the differences in a perfect engine are comparatively slight.

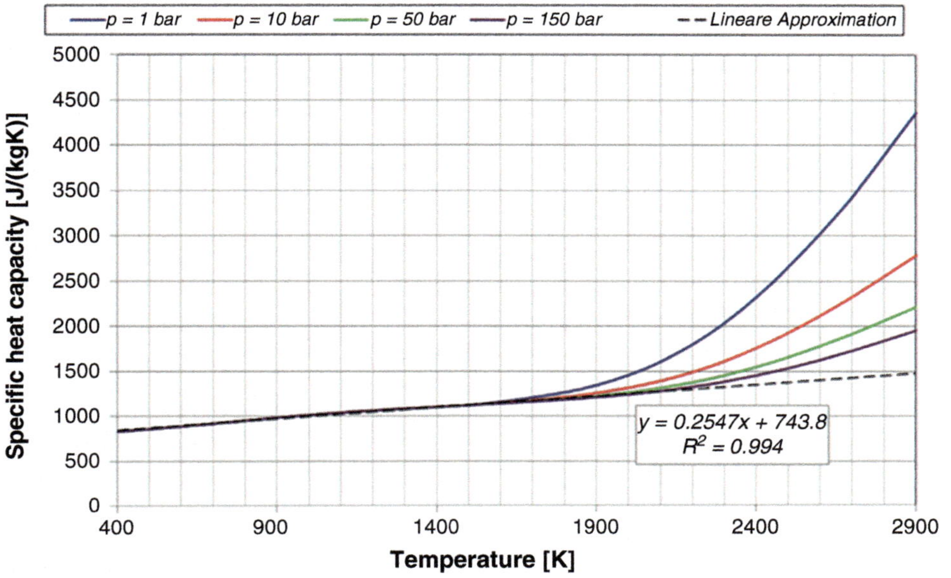

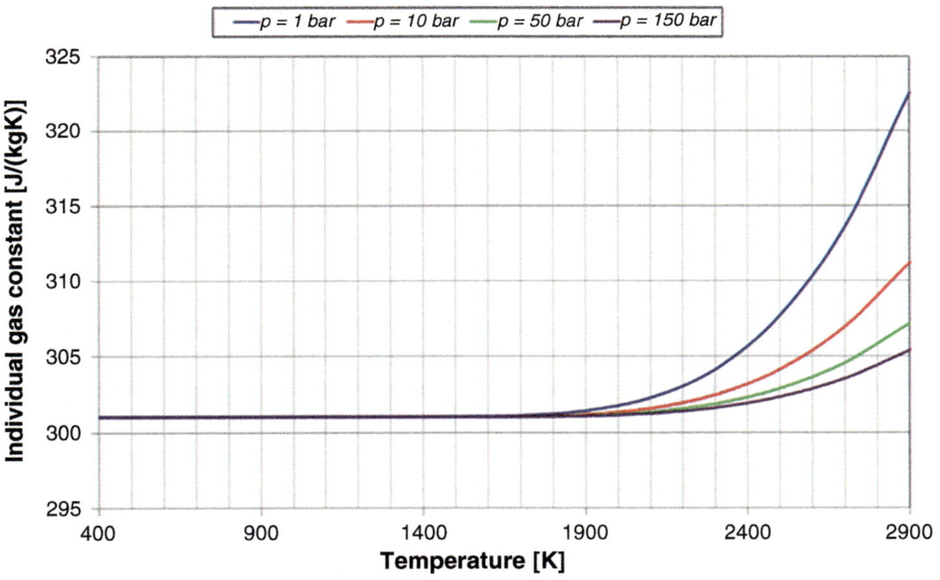

Diagram 5.29 Specific heat capacity at constant volume and individual gas constants for stoichiometric methane-air mixtures

Table 5.1 Data used for the calculation as a perfect engine

	Gasoline ($C_8H_{15.63}$)	Methane (CH_4)
c_v(T) (J)	720.4 + 0.2427 T/(K)	7443.8 + 0.2547 T/(K)
R (J/kg K)	288.7	301.2
H_u (MJ/kg)	42	50
L_{st} (–)	14.7	17.2

Table 5.2 Boundary conditions for the calculation as a perfect engine

Start pressure p_1 (bar)	1
Start temperature T_1 (K)	300
Compression ratio ε (–)	10
Stroke volume V_h (dm^3)	0.5
Lamda λ (–)	1
Proportion of iso-volumetric added heat x_v (–)	0.1

Table 5.3 State points for the calculation as a perfect engine

	Pressure [bar]		Temperature (K)		Mass (mg)	
State points	Gasoline	Methane	Gasoline	Methane	Gasoline	Methane
1	1.0	1.0	300	300	Air: 600.6	Air: 581.0
2	22.2	22.4	667	671		
3	86.2	85.7	2586	2572	Fuel: 40.9	Fuel: 33.8
3*	86.2	85.7	2905	2887		
4	5.9	5.9	1772	1755		

Real Process Calculation

Simplified assumptions such as the neglecting of wall heat losses, the specification of isochoric-isobaric heat release and the highly simplified calorifics are not required when performing a real working-process calculation, so that realistic temperatures and pressures result from the calculation and enable more reliable statements to be made.

In this manner regarding the calorifics (using [38]), the gaseous composition of the combustion gas can be derived taking the dissociation into account.

The latter has a slightly stronger influence on methane, as shown by the specific inner energy trace in Diagram 5.31, whereby the differences overall are low. By way of example, Diagram 5.32 shows the results of a chemical equilibrium calculation for methane and octane smoke. The higher H_2O fraction and the lower CO_2 fraction for methane as opposed to octane are clearly recognizable. This is correspondingly true for the notable amount of dissociation which occurs from around 2000 K. With methane this results in lower CO equilibrium concentrations for example. Regarding NO, the differences are less conspicuous at first glance due to the logarithmic scaling of the

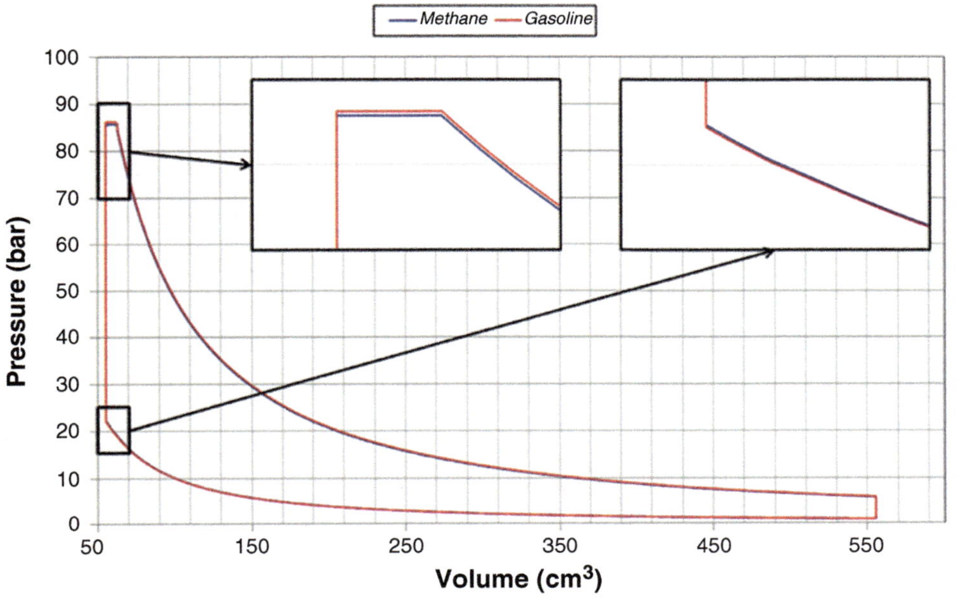

Diagram 5.30 p-V diagrams for the calculation as a perfect engine with gasoline and methane

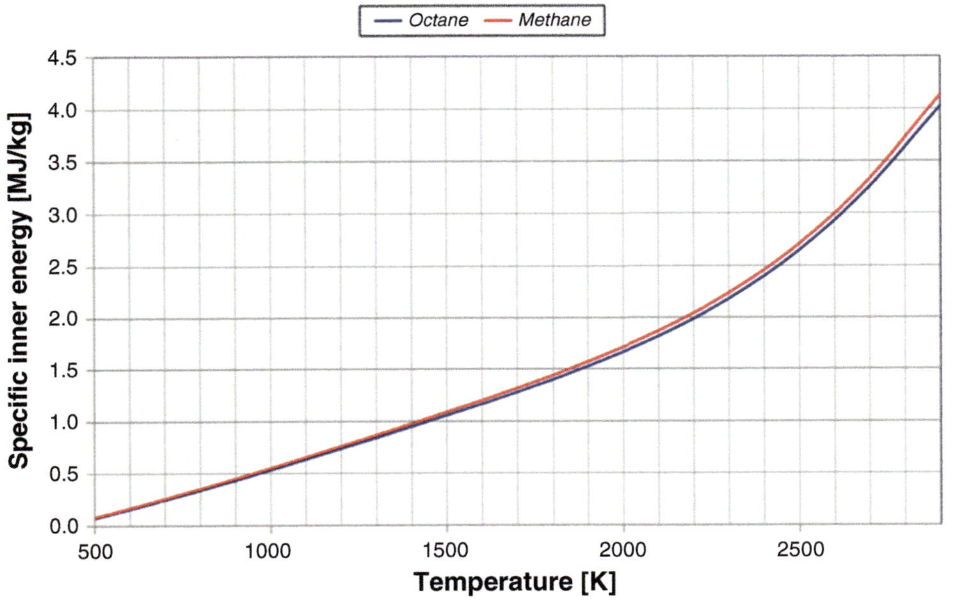

Diagram 5.31 Specific inner energy according to [38] for octane and methane (at $p = 1$ bar, $\lambda = 1$)

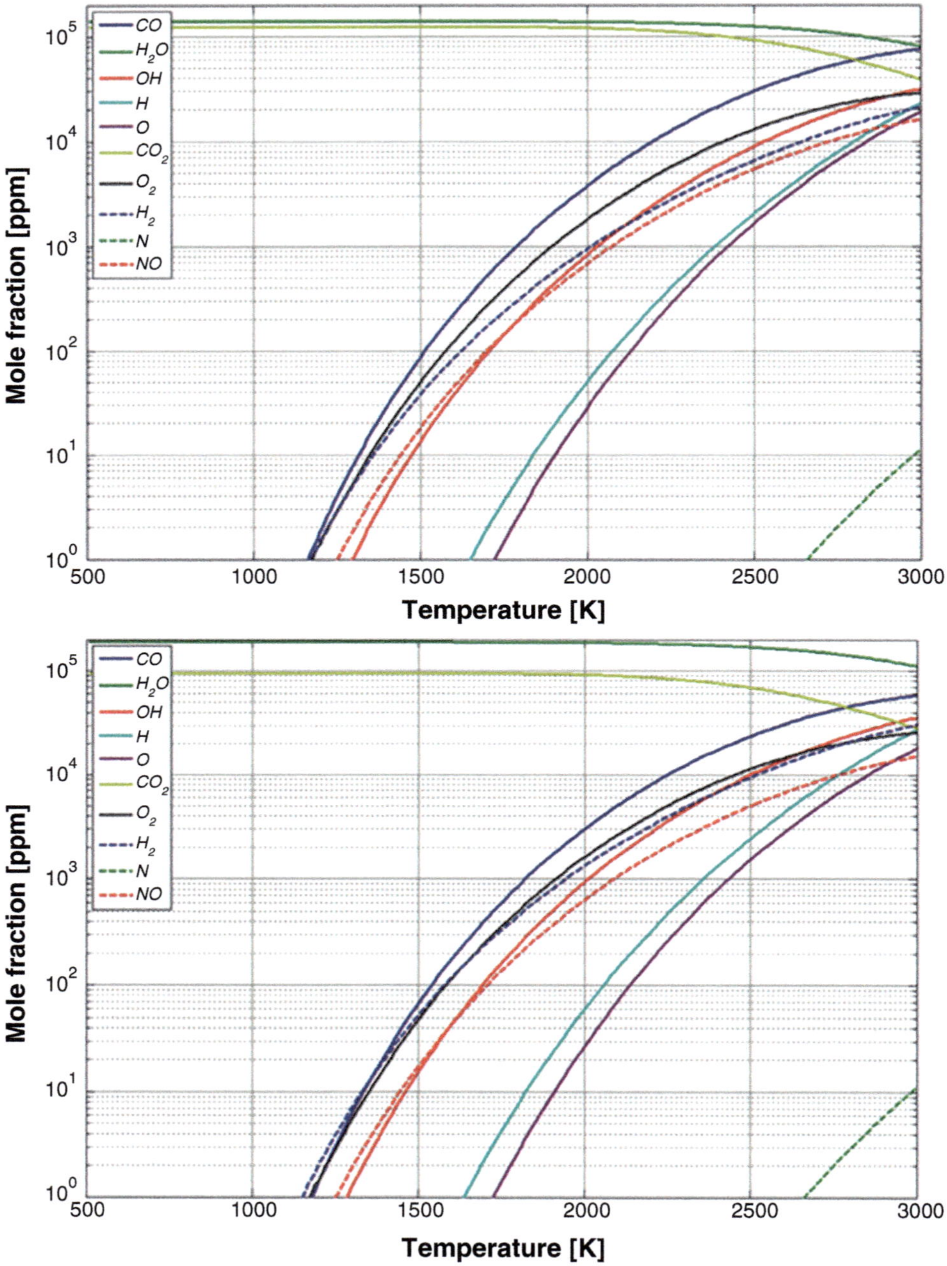

Diagram 5.32 Results of an equilibrium calculation according to [38] for octane (*above*) and methane (*below*); N_2 is outside of the represented area

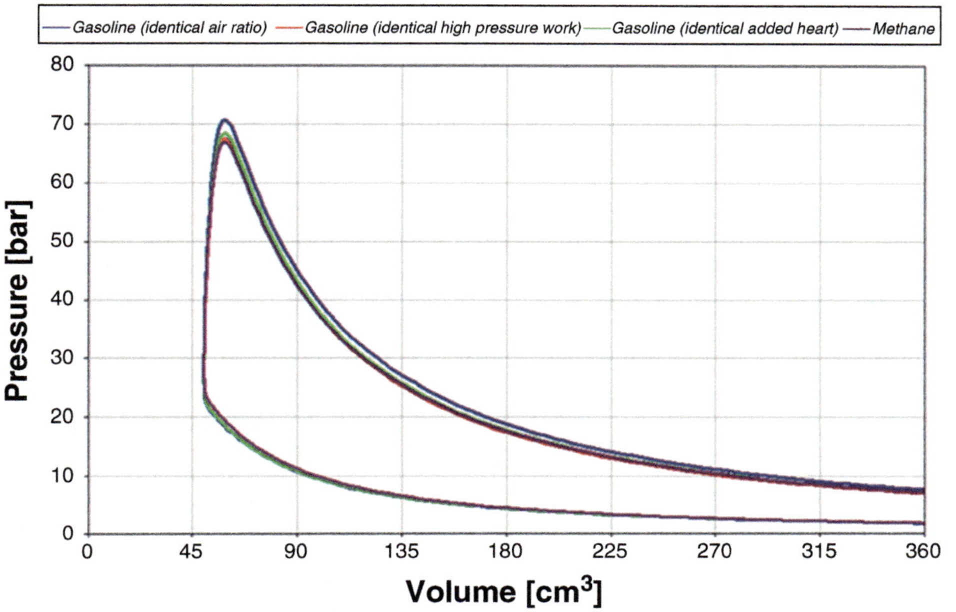

Diagram 5.33 *p-V* diagram for methane compared to gasoline with the same air ratio, identical heat addition and identical indicated high-pressure work

diagram; however the NO equilibrium concentrations are also notably lower for methane and at 2500 K lie more than 7 % under the octane values.

If the high-pressure part is simulated under the same boundary conditions as the perfect engine and assuming the same Wiebe heat release for both fuels, the results are qualitatively similar to those in the previous sub-chapter, compare with Diagram 5.33. Thus the worse filling at the same pressure and temperature is confirmed, as wells as the higher compression final temperature and the lower peak temperature. The latter observation flips into the opposite just as for the perfect engine after adaptation of the air-to-fuel ratio for gasoline for the same heat addition or same indicated work. This is primarily due to the higher compression final temperature, as shown by a comparison of identical compression final temperatures in Diagram 5.34. It is notable that the high-pressure efficiency for the same indicated work, in contrast to the calculation with the perfect engine, is now slightly worse for methane than for gasoline.

Diagram 5.35 shows an additional aspect in which the nitrogen oxide emissions were calculated according to Zeldovich [40]. In all observed comparative cases, methane lay more than 20 % below the values for gasoline. This is an even larger advantage than the observation of the NO equilibrium concentrations. This difference results from a combination of different factors.

On the one hand the effect of the equilibrium concentrations of other species on the formation of nitrogen oxides is accounted for in the Zeldovich mechanism—so for

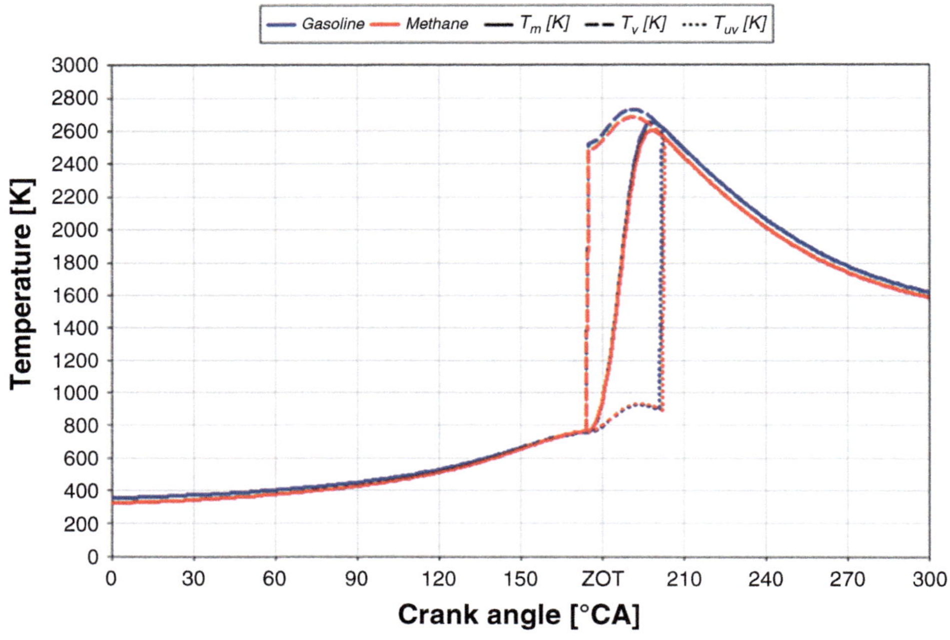

Diagram 5.34 Temperature trace for methane compared to gasoline with identical heat transfer and final compression temperature

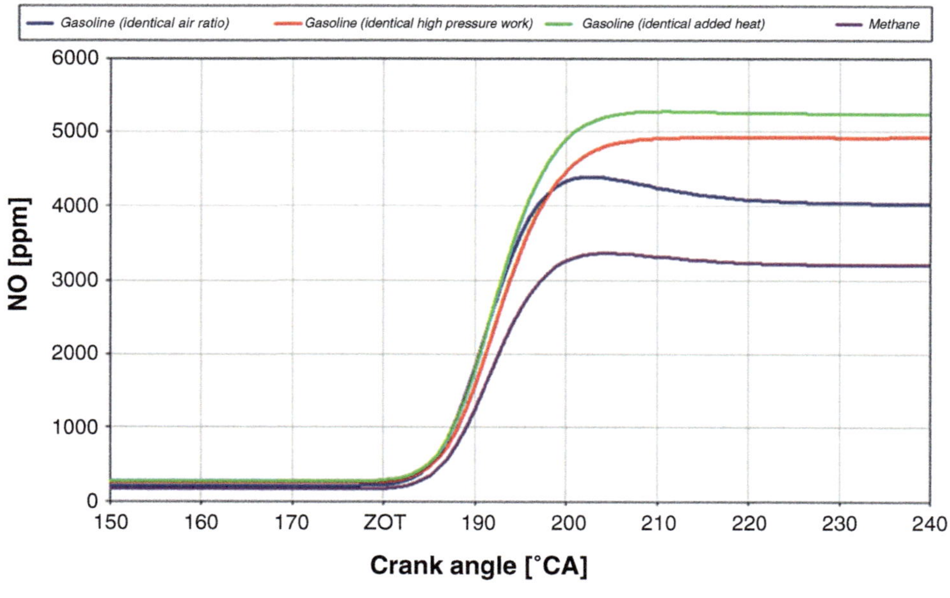

Diagram 5.35 Nitrogen oxide emissions according to Zeldovich for methane compared to gasoline with identical air-to-fuel ratio, identical heat supply and identical indicated high-pressure work (residual gas content 5 %)

example in the case of methane, both the low concentration of oxygen radicals and the significantly higher concentration of hydrogen radicals in the extended Zeldovich mechanism hamper the formation of NO molecules.

On the other hand the changed pressure and temperature trace also effects the nitrogen oxide emissions. This effect makes the higher contribution: as shown in post-processing comparison calculations determining the equilibrium concentrations, the difference between methane and gasoline at identical pressures and temperatures over time would be just 10 %.

In summary, it can be stated that with an unchanged engine design, there are only comparatively small difference between operating with methane or with gasoline. This makes the following even more interesting: Observing the secondary influences by modifying the engine design.

5.2.2.3 Secondary Influences of Natural Gas Operation

Note: Unless otherwise stated, the results in this chapter are based on the results of a real process calculation analogue to section "Real Process Calculation".

Compression Ratio

As mentioned previously, the compression ratio in conjunction with the isentropic exponent is the only factor influencing the thermal degree of efficiency in an ideal constant volume cycle. Diagram 5.36 shows its increase as a function of compression ratio. It is

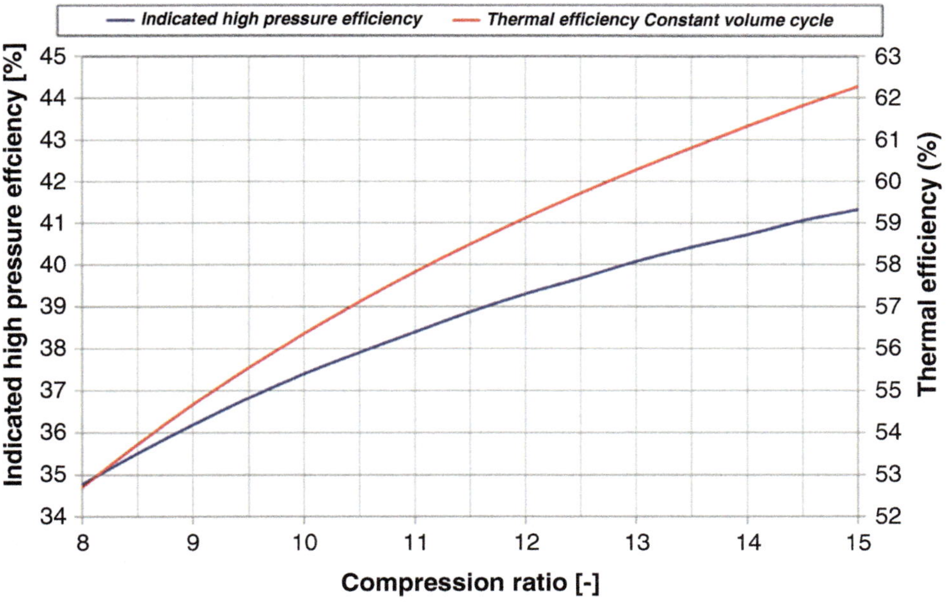

Diagram 5.36 Thermal efficiency and indicated high-pressure efficiency as a function of compression ratio

evident that a clear gradient is available for the typical ranges of compression ratio ($\varepsilon = 9...12$) for gasoline-driven engines also under realistic boundary conditions, even though it is lower than in an ideal process due to wall heat losses, amongst other causes. The influence of the changed compression ratio on the combustion process is neglected due to the use of the Wiebe model for heat release; in reality, an increase in combustion duration is possible in case of an unfavourable form of the combustion chamber, so that the efficiency gradient flattens out even more.

Whereas the compression ratio in gasoline-driven Otto engines is limited due to the tendency to knock—even with direct injection—, operating the engine with natural gas offers the potential to significantly increase the compression ratio and hence the degree of efficiency due to the considerably higher knock resistance. For example, if one were to assume an increase in compression ratio from $\varepsilon = 9$ to $\varepsilon = 11$ via simple modifications to the piston, which can be realized with a reasonable amount of effort and can be considered a realistic increase according to [41], then this would result in an improvement in the indicated efficiency of approximately 6 %. If one were to assume a turbocharged engine, the expected possible increase in the compression ratio compared to the gasoline-driven variant is even clearer; an increase of 3 units results in an improvement in efficiency of almost 9 % [42]. This value is significantly higher than the differences caused by primary influences, such that an increase in compression ratio can be understood as an important basis to run an engine with natural gas at high efficiency.

MFB50

The tendency for Otto cycle gasoline-driven engines to knock not only limits the maximum possible compression ratio, but also leads to the necessity of retarding ignition by the knock controller in certain operating conditions, particularly at low speeds.

Depending on the resulting position of mass fraction burned (50 %), this can lead to a deterioration in efficiency that, as shown in Diagram 5.37, is only relatively weakly dependent on the compression ratio, but increases progressively with retarded timing. This effect is amplified again by the increasing combustion duration due to the late combustion timing—this is considered in all following calculations by the use of a quasi-dimensional combustion model. Thus for MFB50 positions 10° crank angle after the optimum position, the deterioration is only approximately 5 % whereas for 30° CA considerably more than 25 %.

The effect of MFB50 on the efficiency is thus in the same order of magnitude as that of the compression ratio; for example the efficiency is the same for $\varepsilon = 9$ and optimum MFB50 as it is at $\varepsilon = 11$ and an MFB50 point 18° CA after TDC.

By keeping an MFB50 point that is optimally efficient in natural gas operation, i.e. with unchanged compression ratio, considerable reductions in fuel consumption can be expected in many operating points that lie in the same order of magnitude as those that can be achieved through an increase in compression ratio.

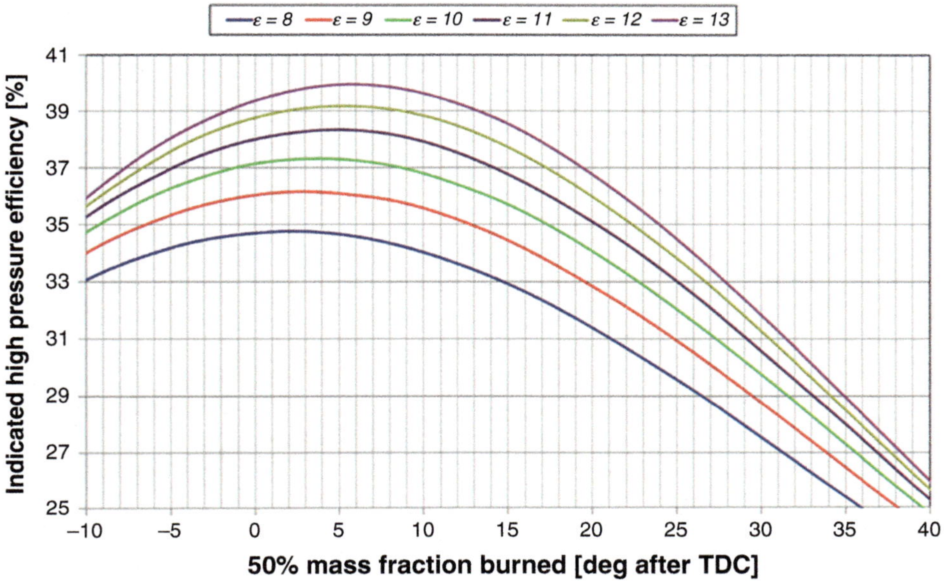

Diagram 5.37 Indicated high-pressure efficiency as a function of compression ratio and MFB50 %

Peak Pressure

Both an increase in compression ratio and the omission of retarded combustion timing do not only improve efficiency, they also cause an increase in peak pressure and the pressure gradient. This not only brings with it disadvantages regarding acoustics, but also increases strength requirements for engine components, which can have negative effects on engine friction due to the consequently increased piston mass.

Since an increase in peak pressure is only possible to a limited degree and the knock border, as described, is much higher for methane, delaying combustion to limit peak cylinder pressure can be relevant in natural gas engines as it is in principle for diesel engines, particularly if the engine design originated from a gasoline engine and is turbocharged. Diagrams 5.38 and 5.39 show that the peak cylinder pressure is largely linearly dependent on both the compression ratio and on the MFB50 point, whereby the start pressure, in a simplified view, acts as a multiplier.

The flattening of the traces to the horizontal plane for late positions of MFB50 mean that the peak pressure is no longer reached during combustion but at the end of the compression phase. As a rule of thumb, it can be derived from the diagram that assuming an efficiency-optimal MFB50 point, the combustion must be retarded by approximately 5° CA if the compression ratio is increased by one unit and the previous peak pressure is not to be exceeded.

Diagram 5.40 shows that normally such a process still has efficiency-bringing advantages even at full-load: A higher compression ratio for the same peak pressure

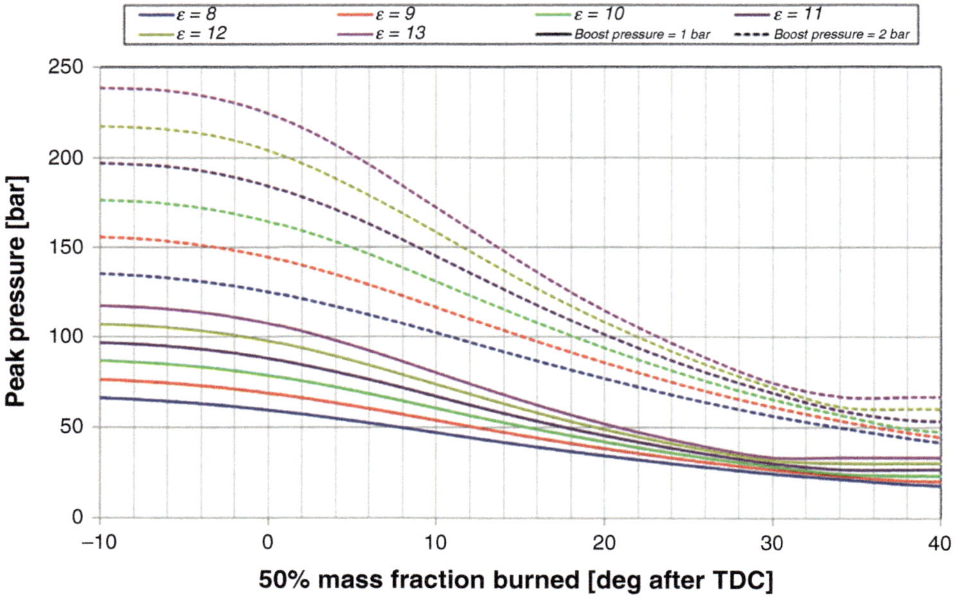

Diagram 5.38 Peak pressure as a function of MFB50 for various boost pressures

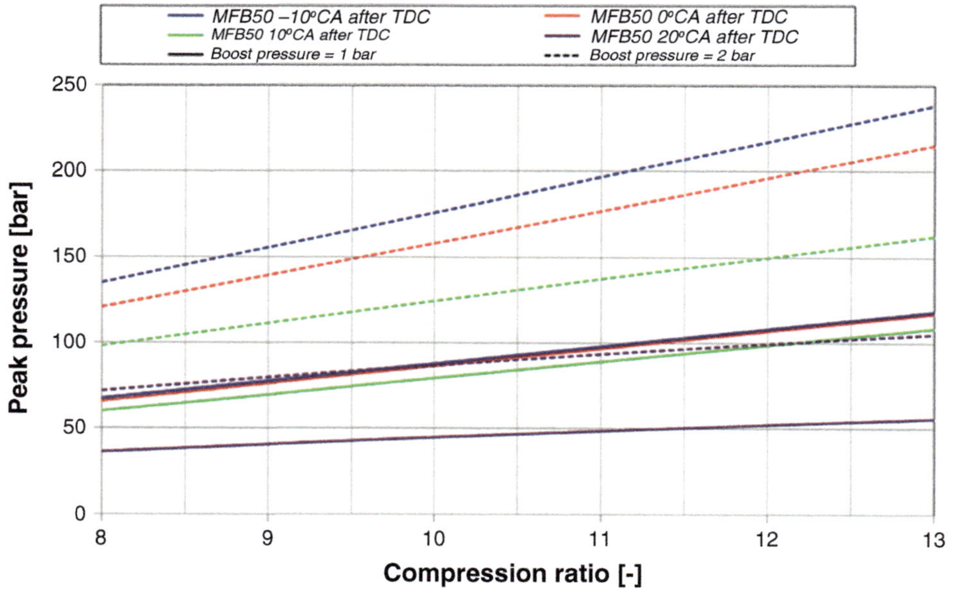

Diagram 5.39 Peak pressure as a function of compressions ratio for various boost pressures

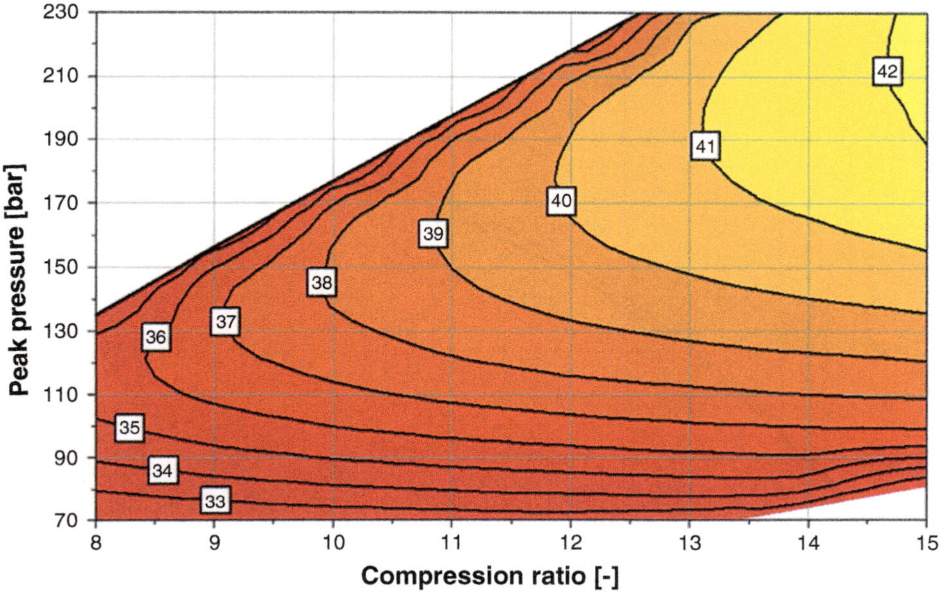

Diagram 5.40 Indicated high-pressure degree of efficiency [%] as a function of maximum permissible peak pressure and compression ratio

continues to yield benefits. However, depending on the permitted peak pressure, these are considerably lower than without the peak pressure limitation. So for example, with a permitted peak pressure of 110 bar and a boost pressure of 2 bar, increasing the compression ratio from $\varepsilon = 9$ to $\varepsilon = 11$ allows only to achieve an advantage of roughly 2 % due to the necessary combustion late shift (representing only a third of the original potential, compare section "Compression Ratio"). At very low permitted peak pressures, an increase in the compression ratio can even have a negative effect, so for example for a permissible peak pressure of 90 bar, the optimum compression ratio would be $\varepsilon = 14$. For thermodynamic reasons it is thus worthwhile designing the engine for natural gas operation for higher pressures to increase the indicated efficiency.

Taking the boost pressure into account as a further influencing factor on peak pressure, the overall picture is extended by an additional component. It can clearly be seen in Diagram 5.41 that an increase in the compression ratio or boost pressure only makes sense if the correspondingly high peak pressure is permissible. This confirms for example that the benefits of a higher compression ratio diminish with decreasing permissible peak pressure and in extreme cases—here with permissible peak pressures under 100 bar—can flip into disadvantages.

In the example shown, a moderate increase in boost pressure from 2.5 bar to 2.6 bar results in deteriorating degrees of efficiency of up to 1.5 %, provided that peak pressures over 185 bar (at $\varepsilon = 14$) and 200 bar (at $\varepsilon = 15$) cannot be permitted. Since this is also

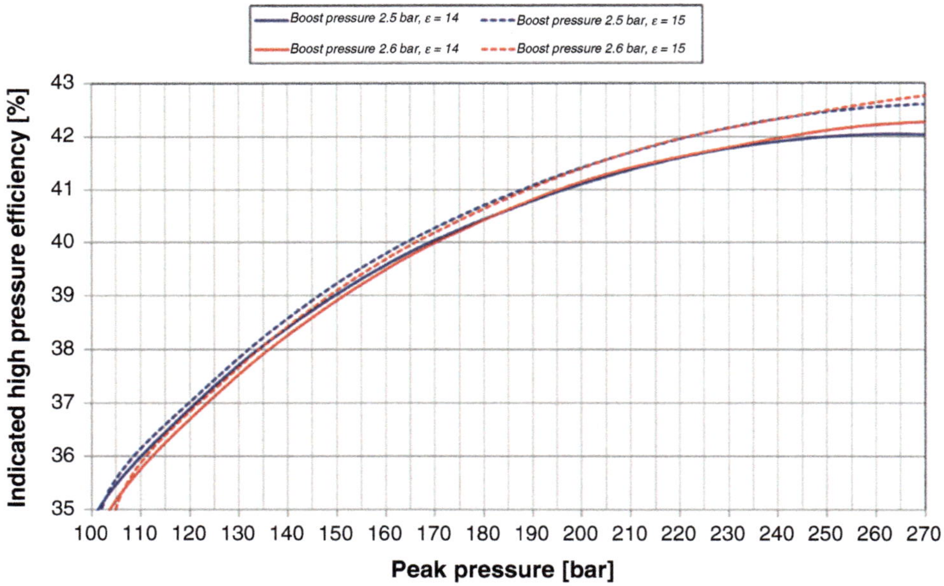

Diagram 5.41 Indicated high-pressure degree of efficiency as a function of permissible peak pressure for different combinations of boost pressure and combustion ratio

associated with an increase in charge of approximately 4 %, the losses in efficiency can be taken into account in order to achieve the highest possible power. However, other situations are conceivable under other boundary conditions where the worse efficiency can no longer be compensated for via the gain in charge.

These dependencies are principally very similar to those of a diesel engine, so that it is also conceivable that different power levels for natural gas engines can be realized simply on the software side via different calibration.

Exhaust Gas Temperature

Maintaining the optimum MFB50 point and the increase in compression ratio also have an effect on the exhaust gas temperature, refer to Diagrams 5.42 and 5.43. As expected, the measures designed to increase efficiency lead to a reduction of the cylinder temperature when the exhaust valve is opened, despite the increase in peak temperature. Maintaining the most efficient MFB50 point has a clear effect on the temperatures, but the increase in compression ratio also leads to a noticeable reduction of the exhaust temperature.

This results in serious consequences, particularly for turbo-charged engines. From the point of view of component protection, a reduction in exhaust gas temperature at full load is beneficial and especially desirable due to the lack of possibility to use mixture enrichment to do so when running in natural gas mode. However, this simultaneously reduces the available exhaust gas enthalpy for the turbine, which itself has feedback effects on the boost pressure and must be considered when designing the turbo-charger.

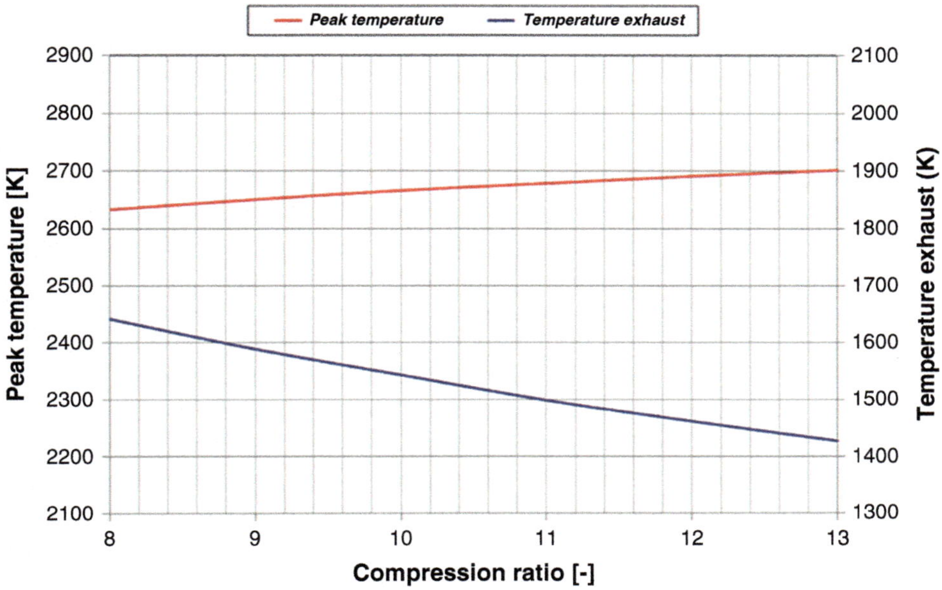

Diagram 5.42 Temperature as a function of compression ratio (for the MFB point with optimum efficiency)

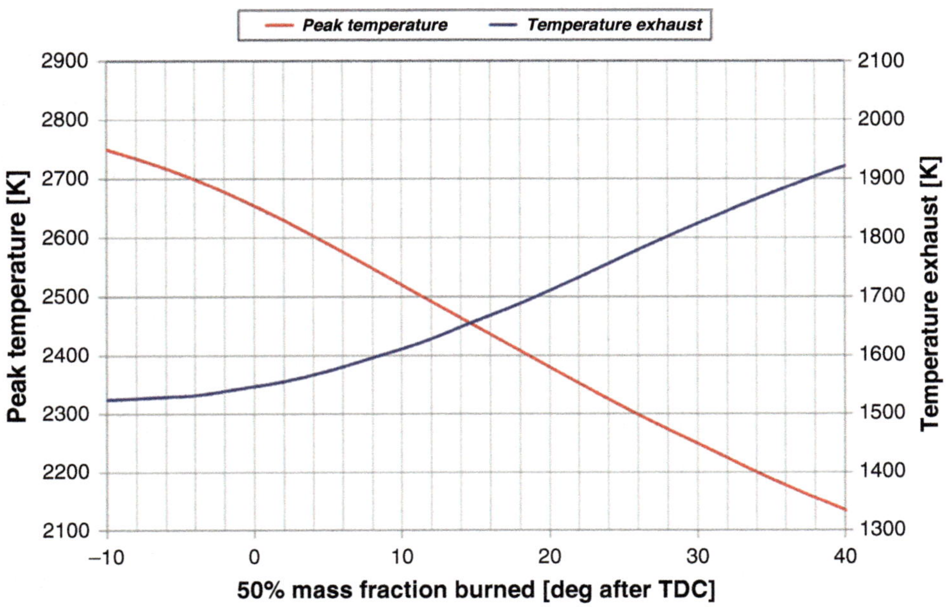

Diagram 5.43 Temperature as a function of MFB50 (for $\varepsilon = 10$)

Furthermore, the sinking exhaust gas temperatures also have significant effects on the exhaust gas aftertreatment systems. More detail on this is provided in the chapters devoted to turbo-charging and exhaust emissions.

Mixture Formation

Just as in gasoline operation, both external mixture formation with fuel being introduced into the inlet manifold and inner mixture formation with fuel being introduced directly into the cylinder are possible in natural gas operation. The second possibility—the direct injection of natural gas—places substantially higher demands on the injector [42, 43], but also a number of advantages. For example, the disadvantages caused by the gaseous aggregate state of the fuel regarding cylinder filling that can be up to 8 % at full load [42] can be avoided if the injection is performed after the inlet valve has closed.

Simultaneously, the direct injection can also lead to an increase in turbulence and thus be used to optimize the combustion. This can also be used to create the possibility of implementing a stratified combustion process, where complete dethrottling can be achieved in large partial load areas. This permits the realization of far-reaching fuel consumption saving potential compared to stoichiometric homogenous operation, see Diagram 5.44.

Compared to gasoline operation, the advantages of the direct introduction of fuel can be evaluated in a similar fashion. It is to be noted however, that there are a few differences. For example, since there is no evaporation of fuel in natural gas operation, there is no possibility of cooling the inside of the cylinder, which can be considered as being non-critical due to the much higher knock resistance. It is also worth noting that the direct injection of natural gas, in contrast to gasoline injection, leads to an increase in pressure as a result of the increase in charge mass. The effects on cylinder temperature can be different according to the injection timing: while normally an increase in temperature follows the increase in mass and pressure, when injecting very late (where the injection needs to occur with correspondingly high pressures) a reduction in temperature is conceivable since the fuel temperature can be then significantly lower than the cylinder temperature, refer to Diagram 5.45.

5.2.2.4 Simulation of Gas Engines

Quasi-dimensional combustion models make it possible—assuming sufficient knowledge of physical-chemical influence parameters—to also represent a change in the fuel type, i.e. only certain model parameters need to be modified without having to change the model itself. For the simulation of natural gas engines, this means that proven models used for gasoline engines can, in principle, also be used for natural gas operation.

The Entrainment model has, amongst others, proven itself capable of simulating laminar turbulent flame propagation with positive ignition and will be introduced briefly [44]:

The basic assumption, as shown in Diagram 5.46, is of a hemi-spherical flame front moving with a propagation speed perpendicular to the flame front away from the spark

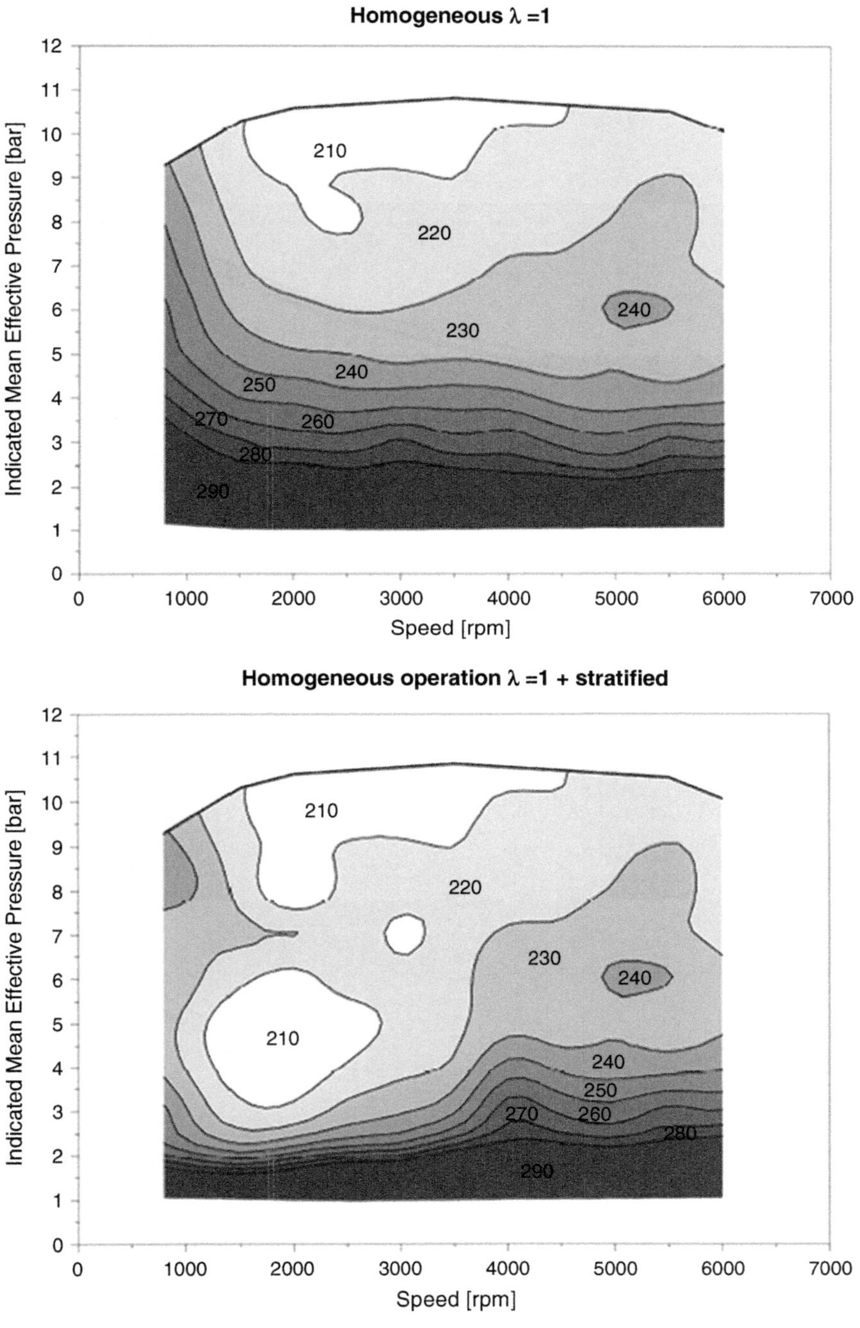

Diagram 5.44 Indicated specific equivalent consumption in an engine map in purely stoichiometric homogenous operation and with additional implementation of stratification (from [42])

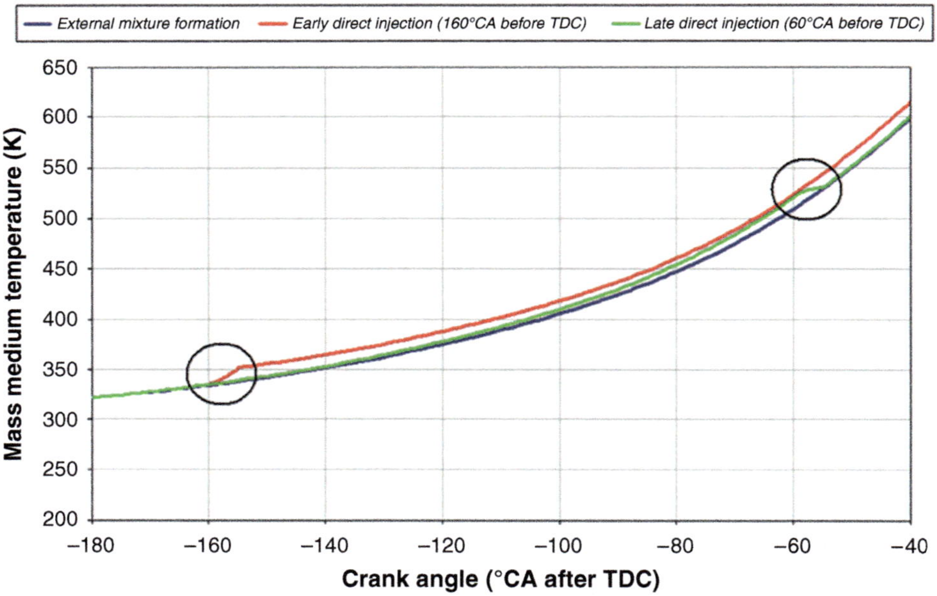

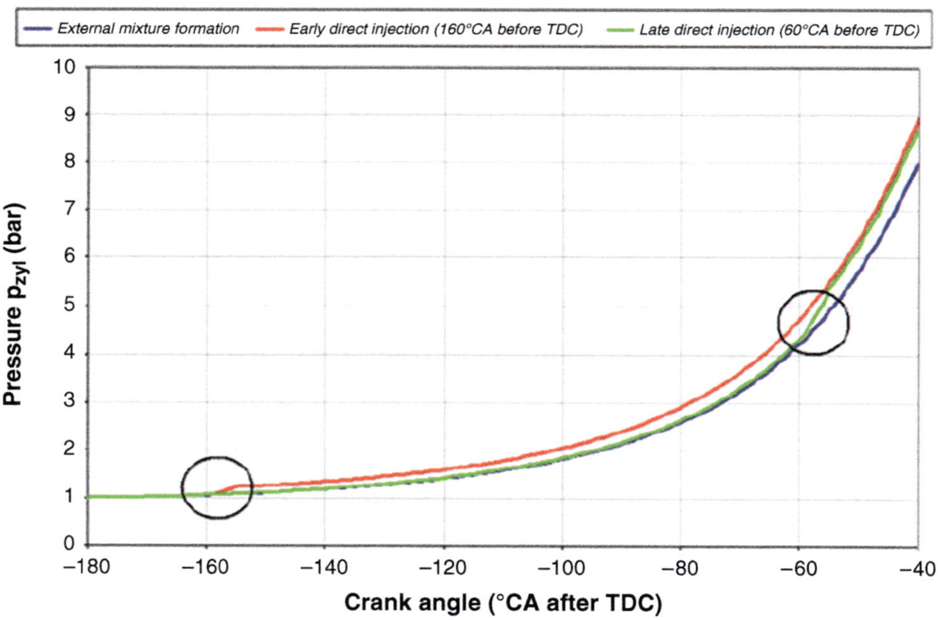

Diagram 5.45 Pressure and temperature during compression for earlier and later direct injection and also for external mixture formation

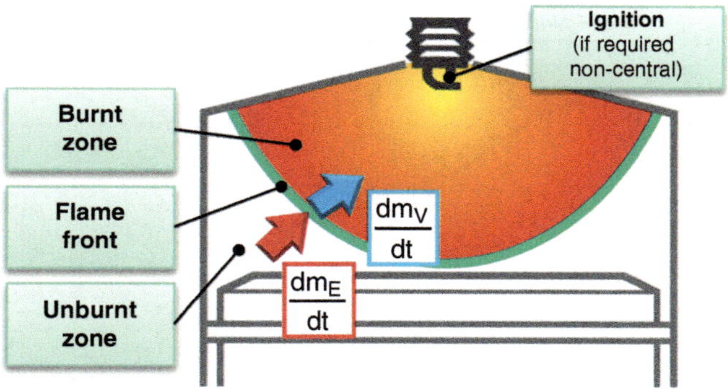

Diagram 5.46 Schematic drawing of the Entrainment model, see [45]

plug. The position of the spark plug is normally slightly off-centre to compensate for the not quite perfect hemi-spherical propagation that occurs in reality and thus to avoid an unrealistically synchronous impingement on the cylinder walls. The combustion chamber is thus divided into three areas: Unburnt zone, burnt zone and the flame front that divides the two areas. The flame front does not appear thermodynamically and is allocated to the unburnt zone so that the model corresponds to a two-zone calculation.

The speed at which the flame front globally penetrates the unburnt zone is modeled as the sum of laminar combustion speed and isotropic turbulence speed.

$$u_{\mathrm{E}} + u_{\mathrm{Turb}} + s_{\mathrm{L}} \tag{5.8}$$

Resulting in the mass introduced into the flame zone by:

$$\frac{\mathrm{d}m_{\mathrm{E}}}{\mathrm{d}t} = \rho_{\mathrm{uv}} \cdot A_{\mathrm{F}} \cdot u_{\mathrm{E}} \tag{5.9}$$

The sought after combustion process results from the introduced mass and a characteristic combustion time τ_{L}:

$$\frac{\mathrm{d}m_{\mathrm{v}}}{\mathrm{d}t} = -\frac{\mathrm{d}m_{\mathrm{uv}}}{\mathrm{d}t} = \frac{\mathrm{d}m_{\mathrm{B}}}{\mathrm{d}\varphi} \cdot \frac{\mathrm{d}\varphi}{\mathrm{d}t} = \frac{m_{\mathrm{F}}}{\tau_{\mathrm{L}}} \tag{5.10}$$

The characteristic combustion time τ_{L} is formulated for the complete laminar combustion of a turbulent vortex of Taylor length l_{T} over the laminar flame speed s_{L}:

$$\tau_{\mathrm{L}} = \frac{l_{\mathrm{T}}}{s_{\mathrm{L}}} \tag{5.11}$$

The Taylor length l_{T} results from the integrated length and turbulence speed of the current calculation step, see [44].

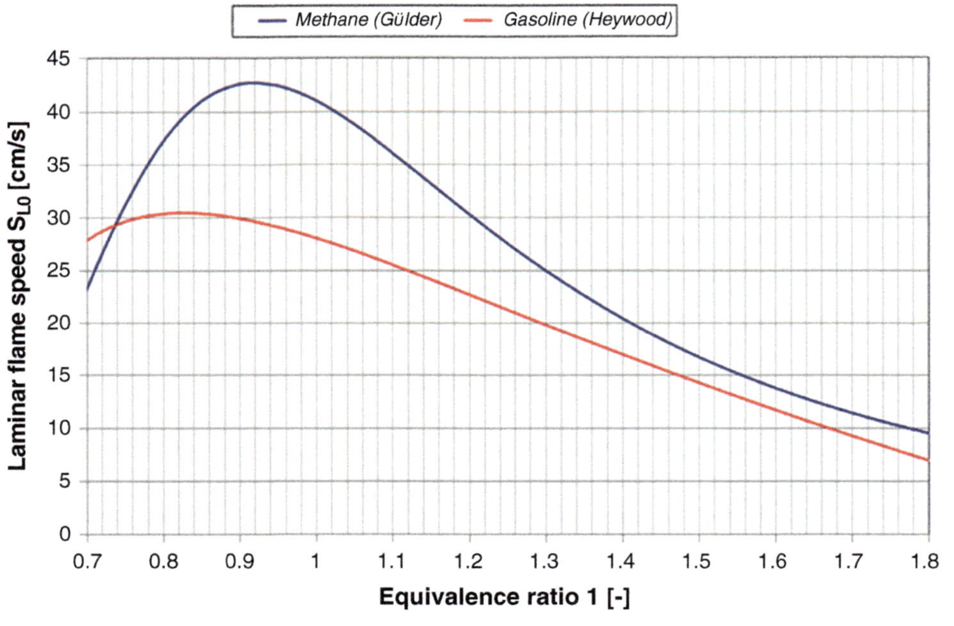

Diagram 5.47 Laminar flame speed for methane and gasoline according to Gülder and Heywood under reference conditions (298 K, 1 bar)

A homogenous, isotropic k-ε model according to [33, 46] can be used to describe the turbulence. Normally, the only parameter to be adjusted in the scaling factor C_k for the global turbulence level, while all other parameters can normally be kept engine independent.

In order for the Entrainment model to be useable for natural gas engines, the fuel characteristics that are different to gasoline must be considered. This is true for the laminar flame speed, values for various fuels can be taken from approaches from literature [47, 48]. As shown in Diagram 5.47, the laminar flame speed for methane is higher than for gasoline, so that even taking the differences and uncertainties between different authors into account, it can be safely assumed that methane does not burn significantly slower than gasoline. When interpreting the diagram, it is worth noting that on the one hand the diagram only shows the relationships under reference conditions—different authors and fuels can easily arrive at different gradients over temperature and pressure—and on the other hand, even for the same fuels, the values in literature can generally display differences in the order of 5—10 cm/s.

Furthermore, a changed "flame stretch" can be assumed for different fuels [49]. This can be incorporated into an Entrainment model via a fuel-dependent adjustment of the Taylor length. Diagram 5.48 shows an example of the simulation for an operating point for a change in fuel type. It is to be noted that the combustion speed can vary according to the composition of the natural gas [50].

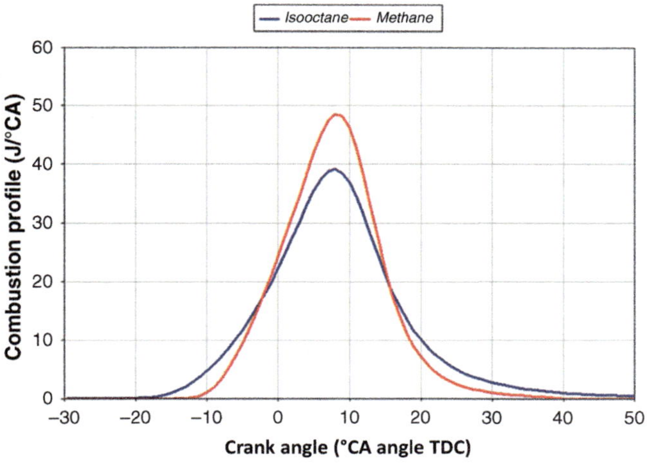

Diagram 5.48 Simulation of an operating point with methane and gasoline under the same boundary conditions, ignition timing adjusted for optimal efficiency

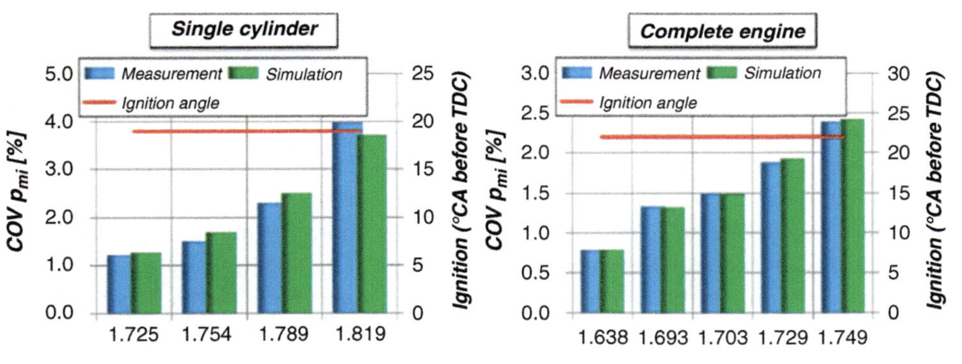

Diagram 5.49 Simulation and measurement of cycle-to-cycle variations on a single cylinder engine and a perfect engine operating with natural gas, from [45]

A model for the simulation of cycle-to-cycle variations was developed on top of the Entrainment model [45]. This model is also directly applicable for natural gas fuels. Example results for the prediction of the lean burn border for a large gas engine are shown in Diagram 5.49.

5.2.2.5 Holistic View

As shown by the previous observations, natural gas as a fuel for Otto engines behaves in many respects similarly to gasoline, so that proven analysis and simulation tools can be used with slight modifications. Pressures and temperatures in unmodified engine designs lie in similar ranges, whereby similar efficiencies can be expected. The more favourable chemical characteristics cause—particularly regarding CO_2—improved exhaust raw emissions and, via the increased knock resistance, offer the potential of modifying the engine's design to achieve a significant increase in efficiency. It can be beneficial to design the engine for high peak pressures that lie well above those for normal gasoline-driven Otto engines in order to take full advantage of this potential; and direct injection in conjunction with stratified combustion can be used to their advantage. All in all, natural gas offers a number of advantages over gasoline from a thermodynamic point of view that speak for the continued use in future vehicle concepts.

5.2.3 Mixture Formation and Engine Control

David Lejsek, Winfried Langer, and Andreas Kufferath

5.2.3.1 Mixture Formation

Introduction

Due to their gaseous nature, the processes during injection and mixture homogenization of natural gas differ fundamentally from the effects during the injection of liquid fuels. The physical properties of natural gas have to be considered when designing the injector, the calibration of the injection timing and the injector positioning.

The focus during engine calibration is on the requirements of the respective combustion system. The non-existence of wall film formation for the gaseous fuel is particularly advantageous. This allows for a by far more flexible positioning of the injector and is advantageous for the dynamic operation of the engine. Even at low ambient temperature, the mixture homogenization is much better than for liquid fuels, but the icing of individual components such as the pressure regulator must be considered. This effect can be controlled using appropriate calibration measures [51].

During injection into the intake manifold, a reduction in cylinder charge may be experienced due to the displacement effect of the gas. This leads to a power reduction of approximately 10 % under otherwise constant engine boundary conditions. This power reduction can be partially compensated for if a charging system is present. During engine calibration, the compressibility of the natural gas must be considered. A supercritical flow condition during injection must be maintained for exact fuel metering to the combustion chamber. Due to the constant injection pressure, this constraint is dependent on the conditions in the intake manifold and the combustion chamber. Especially for charged engines, additional measures during design and control of the injector have to be implemented.

Due to the current trend of SI-engine downsizing, the use of natural gas is a very attractive measure to reduce CO_2-emissions. Beside conventional intake manifold natural gas injection, additional injection technologies are currently being developed and combustion systems with direct natural gas injection are being investigated. Here engine measurements show potential for a further increase in efficiency of the internal combustion engine and hence an additional reduction of CO_2-emissions.

Components for Natural Gas Injection

The physical properties of the injected natural gas lead to special requirements upon the corresponding component designs. All moving parts of the injector need to be designed with non-lubricated guidance. The small density and the increased pressure losses in the flow path require larger flow cross sections and specially designed throttle areas are required to achieve the required flow rate.

Diagram 5.50 shows a current natural gas injector for intake manifold natural gas injection. Apart from the principle function, the only parts it shares with the injector version for gasoline intake manifold injection is the shape and the electrical injector control, meaning that the natural gas injector can be very easily integrated into existing engine systems. All functional parts are specially designed for the use of natural gas. The application spectrum ranges from the classical mounting position in the intake manifold with a designated injector for each cylinder (Multi Point Injection) to the use of a single injector before the throttle and the use of multiple injectors mounted to the intake manifold of large natural gas/diesel-engines.

The gas passes through the natural gas injector displayed in Diagram 5.50 in the longitudinal direction. In non-actuated cases, a return-spring keeps the valve seat shut at the bottom of the valve. This valve seat was designed as an elastomer-steal flat seal to

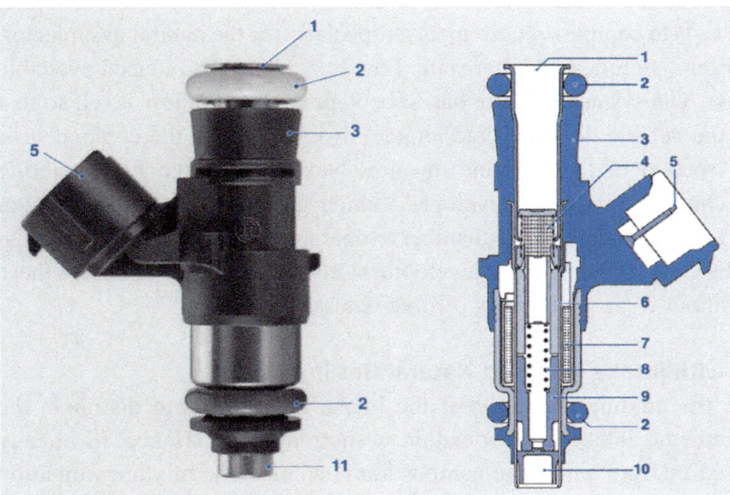

1 Pneumatic connector
2 Gasket
3 Injector body
4 Filter
5 Electrical connector
6 Sleeve
7 Solenoid
8 Spring
9 Anchor with elastomeric seal
10 Valve seal
11 Nozzle

Diagram 5.50 Natural gas injector (Bosch)

reduce injector leakage. The elasticity of the valve seat prevents a second, undesired opening of the injector during the closing process due to the damping nature of the elastomer.

These features improve the injection quantity tolerance. The resistance of the solenoid is 8.5 Ohm at an operation voltage of 14 V and hence permits injector operation with the standard power stage. To achieve the highest possible injector flow rate, the design flow path through the injector reduces the pressure losses inside the injector. The valve is operated supercritically. At the throttle area, the flow reaches sonic speed so the injector basically behaves as an ideal nozzle. Hence the exiting natural gas mass is proportional to the injection time. This is especially important for an exact fuel metering and facilitates easy engine control unit calibration. The smallest flow cross section and hence the governing throttle position is positioned downstream of the valve seat and hence reduces the influence of the manifold pressure on the mass flow rate.

The demand of a supercritical flow translates into the requirement that the rail pressure at the smallest injector cross section has to be twice that of the manifold pressure. For charged engine types, the manifold pressure is up to 2.5 bar. This leads to a required system pressure of 7 bar taking all pressure losses at throttling positions into account. This required pressure level and the gas tank volume define the driving range of a natural gas vehicle.

Corresponding to the gasoline direct injection, the natural gas can also be injected directly into the combustion chamber of the internal combustion engine. This procedure leads to additional advantages, especially the avoidance of cylinder charge disadvantages and hence leads to increased torque generation. In this way, the specific performance values of modern engines can be met with gasoline direct injection as well as with natural gas injection. Compared to the intake manifold natural gas injection, such a direct natural gas injection system leads to complex requirements, especially for the natural gas injector. The injector has to offer a significant higher static flow rate due to the reduced available injection time window. The system pressure has to be kept at a certain low level, so that the driving range of the vehicle does not deteriorate. Also, the size of the component is limited to the cross section of the gasoline injector. Next to the high fuel metering accuracy, a sealing concept has to be developed, which guarantees a lifetime secure operation of the injector under combustion chamber temperatures. A complete new design of the injector with the corresponding high development effort is required to ensure these injector properties without wear while using oil-free natural gas.

Mixture Formation with Intake Manifold Natural Gas Injection

The primary goal of the mixture formation at the intake manifold is to distribute the natural gas evenly into the intake airflow within a short distance. Hence an already sufficient homogenized mixture enters the combustion chamber. The mixture formation is supported by an additional mixing process during the flow across the intake valves. The mounting position of the natural gas injector at the intake manifold and direction of the

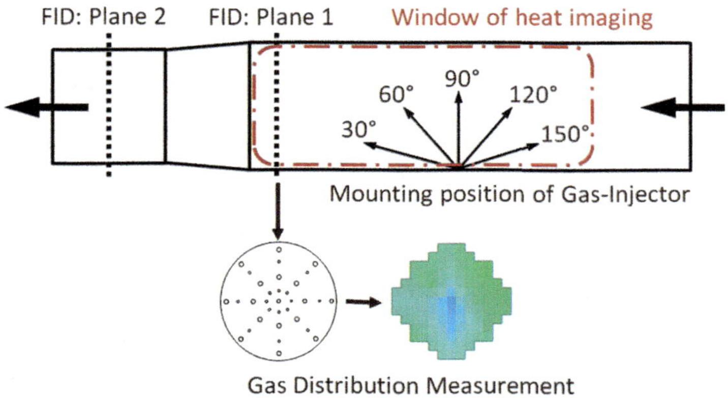

Diagram 5.51 Schematic description of a flow test bench with mounting position for the injector, measuring planes for the FID and a viewing window for the infrared camera [52]

gas jet are of particular importance for the success of the homogenization. This was shown by experiments on the component test bench that evaluated the fundamentals of mixture formation with gaseous fuels.

Diagram 5.51 shows the setup of a test procedure to quantify the homogenization during natural gas injection in dependency of the injector mounting position.

The injector can be mounted at different mounting angles respective to the main flow direction in a test bench. The intake manifold, which is made out of acrylic glass, is streamed with an adjustable air mass flow rate. The air is conditioned to a certain pressure and temperature in order to emulate the operating conditions of a naturally aspirated engine. The cross section of the intake manifold is examined in two measuring planes, a near and a far position with respect to the injector mounting position, for the HC-concentration using a flame ionization detector. The results are time resolved homogenization levels in the two measurement planes. To assess the mixing quality more closely, the test bench is equipped with an infrared camera. For a better contrast, a heated plate is installed behind the intake manifold in the viewing direction of the camera. Additionally, the acrylic glass pipe is heated up before the measurement by hot air flow to reduce disturbing influences. The injected cool gas can accurately be visualized and displayed in pseudo-colors using this testing procedure.

As an example, Diagram 5.52 shows the results of the FID-measurement for different mounting angles of the gas injector at a part load operation point. Displayed are the λ-values in the two measuring planes in a time oriented order as well as the maximum homogenization of the corresponding measurement. This λ-value is gained by a statistical evaluation of the single HC-concentrations at the measurement points in the measuring planes 1 and 2. The difference between the individual values and the mean value is evaluated. A homogenization index of 100 % means perfect mixing. Significant differences in the distribution of the HC-concentration can be seen in the first

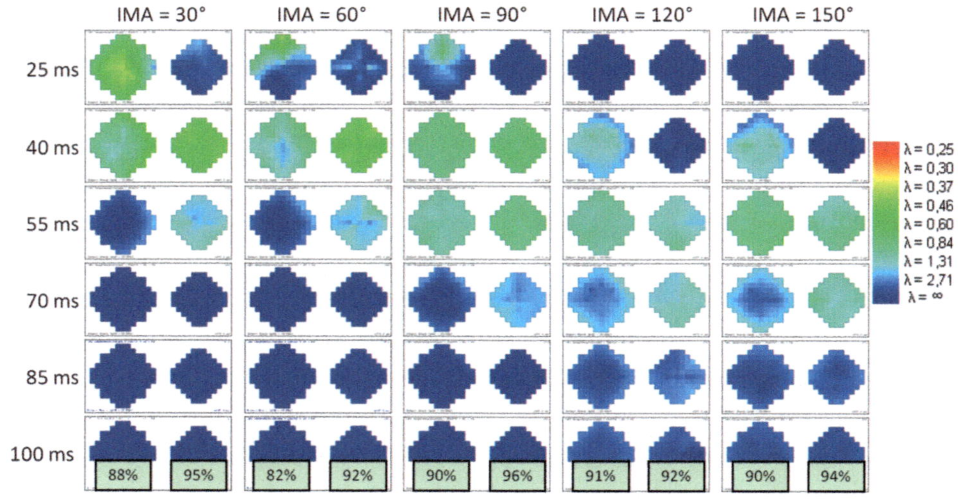

Diagram 5.52 λ-distribution figures as well as the max. homogenization index for different injector mounting angles (IMA) at the flow test bench, $m_L = 13.6$ kg/h, $p_S = 0.685$ bar: left plane 1, right plane 2 [52]

measurement planes. The measurement in the cross section of the second measurement planes shows notably smaller differences favored by the longer time for mixture homogenization at this measurement location. From the test it can be stated that an injection angle smaller than 90° leads to good mixture homogenization. Mounting angles that lead to an injection opposed to the main flow direction lead to very good homogenization values at later points in time. Also, such a mounting position leads to higher cylinder charge reductions and hence to a power reduction which is shown by engine test measurements.

The evaluation of the infrared camera images is done in multiple steps. First the original image is transformed into a monochrome image. By subtraction of the background, an enhanced image is created which shows just the injection process. Finally the final image is created by a pixel multiplication. Diagram 5.53 shows the enhanced images of a perpendicular injection, an injection direction flow direction and an injection against the main flow direction are display for one part load operation point.

The impact of the gas injector mounting position can be clearly identified. The injection in opposite direction of the main flow direction homogenizes the mixture very well. Also, the injection perpendicular to the main flow leads to a fast and efficient mixing of the cylinder charge. The injection in the main direction of flow also homogenizes sufficiently, but slightly retarded. The method of infrared imaging confirms in all examined operating points the results of the experiments to determine the degree of homogenization.

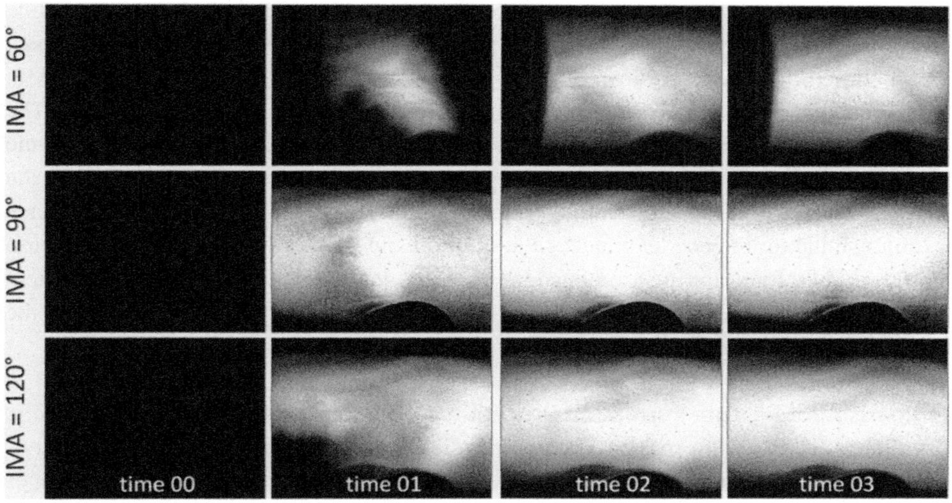

Diagram 5.53 Infrared images of intake manifold natural injection for different injector mounting angles (IMA) for different points of time, $m_L = 13.6$ kg/h, $p_S = 0.685$ bar [52]

Mixture Formation for Natural Gas Direct Injection

There are also experiments to examine mixture formation for the direct injection of natural gas into the combustion chamber. These tests can be split up in two parts, pure 3D-CFD-calculations and engine testing with components, which are only useable under lab conditions and do not represent the development status of series production parts. The boundary conditions during direct natural gas injection are different to the conditions of natural gas intake manifold injection. The injection time is shorter since the injected fuel injected cannot be advanced for a complete working cycle. After intake valve closure, pressure and temperature increase due to compression. The effective flow cross section of the injector has to be significantly increased compared to the intake manifold injection version. Additionally the injection pressure needs to be raised to ensure a supercritical flow condition.

To evaluate the mixing processes and to do first optimization steps, mainly 3D-CFD-simulations are used at this early research step, since video endoscopy of the combustion chamber is only meaningful by using complex tracer methods for the natural gas. For validation purposes of the complex phenomena, spray chamber imaging of the gas injection is required. Therefore the natural gas or an inert substitute gas is injected into a defined environment and the flow pattern is visualized using optical measurement procedures such as the Schlieren-technique. Subsequently a calibration and validation of the 3D-CFD-simulation is done. The calibrated model is then transferred to simulations with for engine operation typical boundary conditions. Examples for such an approach can be found in [53] and [54]. The results of such 3D-calculations show that good homogenization of the mixture in the combustion chamber can be achieved by a narrow, oriented directly down to the piston surface spray layout. This is in contrast to gasoline direct injection, where the direct wetting of the piston surface must be avoided. A compact gas cloud homogenizes significantly better than wide spread jets. First engine tests with natural gas direct injection basically confirm the results of the 3D-CFD-calculations. The homogenization of the cylinder charge is not the single optimization parameter. Due to the strong displacement effect of the natural gas and the increased injection pressure, the cylinder charge motion is substantially influenced by the spray impulse. This can enhance or diminish the turbulence level needed at ignition and thus elongate or shorten the burn duration. Therefore, during combustion concept engineering for natural gas direct injection, the interaction of the injection with the in-cylinder charge motion and the spray layout must be considered.

Comparison of Mixture Formation Procedures Under Engine Operating Conditions
The choice of mixture formation has an immediate impact on the combustion concept of the engine and hence on its characteristics. In [55, 56], the advantages and disadvantages of different combustion system concepts with internal and external mixture formation are methodically evaluated on an unchanged test engine. The effects of different combustion system concepts on the drivability of a sub-compact class vehicle are discussed. The results cover gasoline as well as natural gas as fuel. In the following, the engine test results of the natural gas injection in comparison to the state of the art gasoline direct injection at a turbocharged SI-engine are illustrated more closely.

Combustion Concepts with External Mixture Formation
Today, the injection of natural gas into the intake of each cylinder is the classic version of the use of this fuel in SI-engines. It is in many cases offered by OEMs off-the-shelf. The injection into the intake upfront of the throttle is critical mainly due to emission reasons. Reasons are especially inhomogeneities during mixture formation under dynamic operating conditions. Central gas injection is also outdated because of slow response behavior. Gas injectors that are specially optimized for the operation with natural gas are used for fumigation, Diagram 5.50 The combination with gasoline direct injection as a bi-fuel concept is state-of-the-art for current powertrains with natural gas intake manifold injection. With bi-fuel systems, the flexibility of use and the range of such cars increases substantially. Moreover, mass series production engines today can be adapted to natural gas operation with little effort. Disadvantageous is that these engines must allow the operation with gasoline and thus have to feature a compression ratio common for gasoline engines due to knocking. So often the full efficiency potential of natural gas operation is not completely used. An alternative are SI-engines that are optimized to natural gas use featuring increased compressed ratio and an improved mechanical stability but have a reduced power output during gasoline operation.

The impact of natural gas injection with external and internal mixture preparation on the full load behavior of a SI-engine is shown in Diagram 5.54. The full load torque curve when using gasoline direct injection is also displayed. The different mixture formation

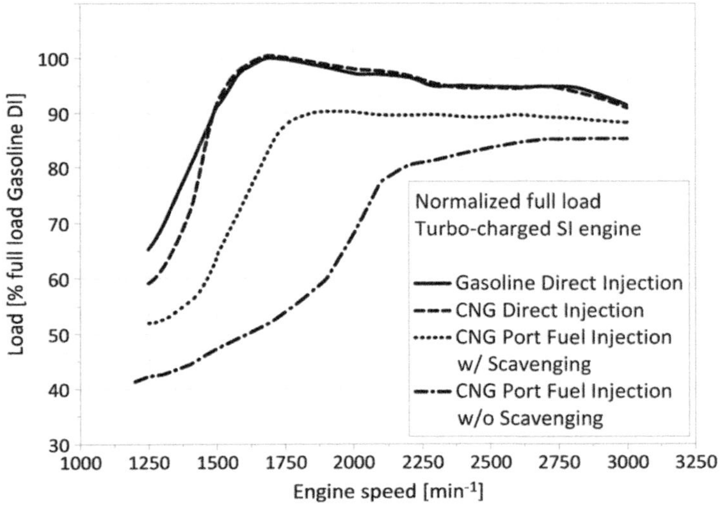

Diagram 5.54 Normalized full load behavior with external and internal natural gas injection as well as gasoline direct injection for a turbocharged SI-engine at low engine speeds [56]

procedures were examined on the same test engine without making changes. The maximal boost pressure is identical for all mixture formation procedures.

In Diagram 5.54, the decreased cylinder charge of the pure intake manifold natural gas injection (without scavenging) is clearly visible. The decrease in cylinder charge is caused by the displacement effect of the natural gas injected into the intake manifold. The volume captured by the natural gas reduces the cylinder charge by approximately 10 % in comparison to gasoline direct injection. Due to the reduced mass flow rate through the engine, the turbine power output is also decreased. Hence the achievable boost pressure is limited, especially at low engine speeds. Hence the maximum torque output of the natural gas engine with intake manifold injection is approximately 50 % lower than when using gasoline direct injection. This characteristic in the low end torque area is responsible for the significantly reduced power generation of the natural gas engine and for a restricted dynamic and drive away behavior compared to gasoline engines with the same displacement. The rated power of the natural engine is reached at the same engine speed but is reduced by 10 % due to the cylinder charge reduction.

To increase the reduced mass flow rate and hence increase the achievable torque level at low engine speeds the method of purging air though the engine (scavenging) can be applied for natural gas intake manifold injection just as is known for gasoline direct injection. Here the intake and exhaust cam timings are chosen in a way to generate a large valve overlap. Hence a purging of fresh air charge from the intake manifold to the exhaust manifold is possible during gas exchange if a certain pressure difference is given. This procedure increases the mass flow rate through the engine and shifts the operation point of the turbine towards higher efficiencies. This results in a higher boost pressure of the compressor and hence a higher achievable engine load. Since with natural gas intake manifold injection, not pure air but a gas-air mix is scavenged, the scavenging degree during full load operation is limited by the maximum permissible HC-raw emissions. If the full conversion rate of the catalytic converter is utilized, 90 % of the gasoline direct injection maximum torque level can be reach by natural gas intake manifold injection and scavenging, Diagram 5.54.

The potential of scavenging can be further increased by injection timings while the intake valve is opened [open valve injection (OVI)], which prevents a scavenging of fuel-air mixture. But the time available for injection is significantly reduced by this approach. The natural gas mass flow rate has to be increased to inject the required fuel mass for full load during this reduced time anyway. This can be achieved for example by the use of two injectors for each cylinder. Also, an advancement of the existing gas injector towards higher static mass flow rates can be a solution.

Another possibility can be used with current bi-fuel-engines. Since these engines are mostly equipped with gasoline direct injection, the low end torque area can be operated in gasoline mode. Such a mixed operation strategy requires additional calibration effort, and depending on the driving behavior, leads to an increased gasoline consumption of the natural gas vehicle.

Overall, from a current point of view, the natural gas intake manifold injection technology offers substantial potential with respect to increasing the specific loads of the engine. With an attractive cost to usability ratio, the customer acceptance of natural gas vehicles can be increased and hence the market share can be enlarged.

Combustion Concepts with Internal Mixture Formation

The direct injection of natural gas into the combustion chamber of a SI-engine almost leads to the torque characteristics of the gasoline direct injection, as shown in Diagram 5.54 only at very low engine speeds the torque level is decreased due to displacement effects and the missing cooling of the cylinder charge. Hence by the use of internal mixture formation, the drivability of a vehicle with turbocharged SI-engine and gasoline direct injection can be established while still offering the CO_2-emission advantages of natural gas. Additional advantages of the natural gas direct injection are especially during catalytic converter heating. As tests with prototype components showed [56], increased combustion stability as well as higher temperatures inside the catalytic converter are possible in comparison to natural gas intake manifold injection, while decreasing the overall HC-raw emissions. This is especially important during the early phase after cold engine start, since the catalytic converter light of temperature, which is 150 °C higher than for gasoline, can be reached even faster by direct gas injection, see Sect. 5.2.7.

The advantages of direct natural gas injection are in contrast to the high development effort of a series production CNG-direct injection valve. The requirements upon such a component are a lot more complex than for an intake manifold injection valve. For operation in the scavenging regime, very high natural gas mass flow rates have to be realized in the available space of the gasoline direct injection valve. Additionally, the demand for high leak tightness and a very good fuel metering quality at temperatures of the combustion chamber wall over the whole life span have to be mentioned.

The advantages of the natural gas direct injection can be met in the medium term by the use of revised intake manifold injection technologies. Especially the low-end torque behavior of current SI-engines with gasoline direct injection can be reached cost-efficiently by the use of open valve injection or a gasoline-natural gas mixed operation mode. The procedures are constructive for the transition from bi-to monofuel systems to increase the customer acceptance of natural gas vehicles. If the natural gas direct injection valve can be mass-produced, natural gas direct injection is the best choice for monofuel engine operation with an optimal design of the engines towards maximal compression ratio and mechanical stability.

5.2.3.2 Bi-Fuel Engine Management System

Introduction

In a CNG bi-fuel system, conventional engine management systems are extended to support the additional CNG fuel system. The functionality required to control the CNG components is realized by either an additional CNG engine control unit or as an integrated extension of a "gasoline"-ECU to a bi-fuel ECU (Diagram 5.55). Robustness is increased and space is saved by avoiding the addition of a separate NG-ECU with its own set of electrical connectors.

A separate CNG-ECU is most often used in small-volume projects or with CNG-retrofit systems. One must be aware that, regardless of how the engine management system is realized, other engine management functions are also active during CNG operation. These functions must be evaluated for operation with CNG fuel and may have to be adapted.

Functions of the Engine Control Unit

The engine control unit regulates the air and fuel delivery to the engine according to the operating conditions and the torque request from the driver and the accessories (e.g. A/C compressor). Adjustments for CNG fuel delivery control might include only adjusting for the density and the stoichiometric ratio of the CNG fuel. With this adjustment, the engine would at least run and deliver an amount of torque that roughly matches the amount

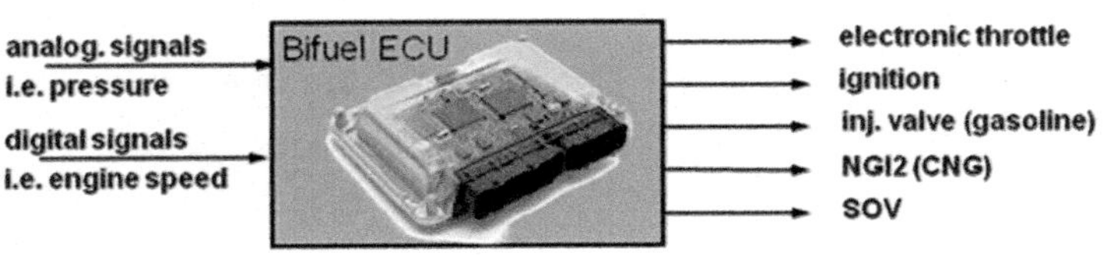

Diagram 5.55 Bi-fuel engine control unit (Bosch)

requested. However, the requirements of a modern spark-ignited engine are myriad and extend from ensuring precise torque delivery, to control of the air to fuel ratio of the intake charge to ensure catalytic conversion of the emissions, to diagnosis, all the way through calculation of driving range based on the fuel level.

The functions and diagnoses of the original engine control unit are specially targeted toward gasoline. For a careful consideration of the influence of the fuel type on these functions and diagnoses, detailed knowledge and analysis of the engine management functions is necessary.

One may assume that the functional architecture of modern engine control units is modular and the calculations are physically based. Therefore extensions of existing functions and evaluation of their influences can be executed in a structured manner. The high-level structure of the Bosch bi-fuel engine control unit is shown in Diagram 5.56.

First off, one considers fuel independent modules are general support functions such as system control, communication functions, i.e. CAN, and the diagnostic system. These functions control only the sequence of events, the communication to external units and provide the structure of error messaging. The special content of the messages and diagnosis results are delivered by the corresponding subsystems. Also fuel independent data are the determination of the operating data, like camshaft angle position, throttle angle, air mass temperature, pressure and knock information.

Functions are only independent of the fuel in use if their variables, and the preceding variables on which they depend, have no impact of the fuel or the impact is removed.

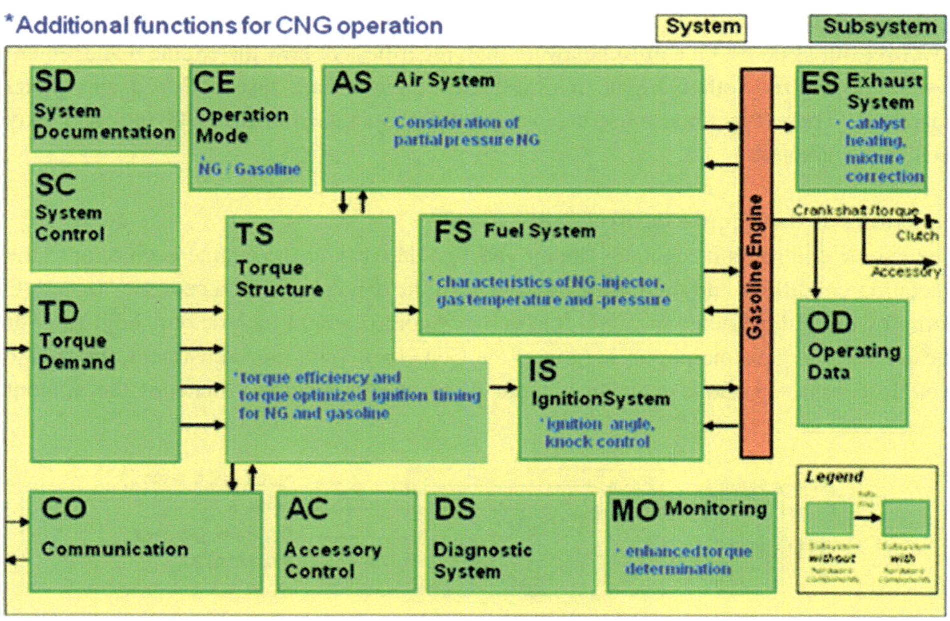

Diagram 5.56 Example of the functional structure of a Bi-fuel engine control unit (Bosch)

As an example, the physical quantity of torque is fuel independent. This is achieved by using fuel-specific thermodynamic efficiencies in the calculation model. Obviously the operating conditions of engine speed and temperature are fuel independent. It must still be considered that functions based on a sensor have a diagnosis. For instance the engine temperature is checked during warm up by a model value based on power, ignition efficiency and optionally additional values. However the impact of the fuel on this diagnosis as a result of its combustion efficiency is small enough in comparison to the other effects that it can be neglected.

With the objective of maintaining constant torque during the fuel change between gasoline and CNG operation, the target torque value shall be independent of the fuel, even if the maximum torque of each fuel is different or smaller in the CNG operation mode. The resulting dead space in the accelerator pedal beyond the maximum torque is more acceptable, compared to changing the target torque value during fuel change, which would need to be smoothed by filters.

In the "torque structure" module, the target values for air mass, fuel mass and ignition angle are derived from the target torque and the fuel specific combustion efficiency. Fuel specific combustion efficiency is affected by various conditions like fuel quality, homogenization of mixture, inert gas, mixture, tumble and swirl, lambda and ignition angle. To obtain maximum efficiency during engine operation and also precise torque control, all parameters and maps of the torque calculation should be calibrated separately for each fuel.

The module "air mass system" regulates the desired air mass using the electronic throttle body. The injection of the gaseous methane fuel into the intake manifold occupies space, which results in a partial pressure and increases the manifold pressure in the case of constant air mass. In the case of subsonic air mass flow, meaning the pressure ratio over throttle is more than a factor of two, the throttle has to be opened in order to achieve the same mass flow. In case of a stoichiometric mixture the partial pressure of the fuel methane (G20 or H-Gas) is around 10 % of the partial pressure of the air mass. For L-Gas (G25 or methane with 14 % nitrogen) it will be around 20 %. In case of a camshaft control for exhaust gas rate EGR, the parameters should also be fuel specific to consider the compatibility of exhaust gas on ignition and combustion mixture with methane.

The "ignition system" provides the ignition angle for the maximum efficiency at the current operating conditions. Therefore the corresponding parameters should be fuel specific. Also the knock adaptation should be fuel-specifically calculated to avoid knocking during fuel change from CNG to gasoline operation.

The module "fuel system" calculates the injection time according to the required fuel mass and the current CNG rail pressure and temperature. Due the separate fuel system with injectors of different mass flow characteristics, a separate calculation is required. Methane is always gaseous under the environmental conditions in the injector and in the manifold, and unlike gasoline, cannot build up a wall film in the manifold. Therefore no fuel enrichment factors for dynamic load conditions or during cold start und warm up should be transferred from the gasoline system. The ignitability of a mixture is limited to an air-to-methane mass ratio between 0.8 and 1.4 (relative to a stoichiometric mixture), which depends somewhat on the combustion chamber. Exceeding these limits could increase the HC-emissions as a result of effects ranging from poor combustion up to misfire, which would be also be perceptible as a reduced drivability.

Engines with gasoline injection use mixture adaptation to correct the tolerances of the air mass and fuel system. For a precise mixture control during the change of fuel, it is recommended to calculate the mixture correction factor specific to each fuel. This would offer a fuel specific mixture diagnosis and the mixture control during fuel change only has to correct for short-term effects like temperature.

In addition to the tolerances of the CNG fuel system, which are comparable to the tolerances of the gasoline system, the gas quality (composition) must be considered. This can be done by increasing the range of the mixture adaptation factor or by including a separate gas quality factor that is adapted during specific conditions after CNG refueling.

The emissions control system must consider the different exhaust gas temperatures and differing exothermal reactions in the catalyst due to fuel specific effects on conversion rate of the catalyst. To achieve at least the same emissions results as in the gasoline operation mode, an adjustment of the fuel specific exhaust gas parameters is required. Additionally the water content of the exhaust gases of methane combustion is higher. To protect the lambda sensor from being damaged by thermal shock resulting from contact with condensed water during warm up after engine start, the sensor heating must be adjusted for each fuel. Additionally, the characteristic curve of lambda sensor is sensitive to hydrogen in the exhaust gases due to its higher velocity. Due to the higher hydrogen content of exhaust gases during operation with methane in the rich-mixture range, the characteristic curve of the lambda sensor should be adjusted specifically for each fuel.

The control of CNG tank valves and the corresponding diagnosis is also a part of the CNG fuel system module. The CNG tank valves are safety relevant components, therefore the principal control is defined by Regulation ECE-R110 [57]. Additionally a stuck-open tank valve would not affect normal operation but could be dangerous in case of an accident. Therefore a diagnostic function is recommended that checks from time to time that the valve can be successfully closed. This could be done by closing the tank valves during idling of the engine and checking for the expected pressure drop.

Additionally, an external leakage diagnosis is useful. Depending on the position of pressure sensors and shut off valves in the CNG system, one can monitor for a pressure drop while the engine is off, which would indicate leakage. During engine operation only the comparison between consumed fuel and reduction of the tank fuel level could be done for diagnosis. However, due to the fact that the temperature of the CNG in the tank is not known, the accuracy of such a diagnosis is low. Use of a tank temperature sensor would improve such a diagnosis.

Due to the torque control function of the engine control unit, torque monitoring functionality is implemented as a safety measure. Due to the self-contained architecture of the torque monitoring functionality of the ECU, modification of fuel injection outside the engine control unit could render the monitoring function ineffective or could result in false-positives in error detection.

For secure and effective implementation of bi-fuel engine control, solely adaptation of the fuel injection system is not sufficient. The functionality of the engine control unit should be thoroughly analyzed and evaluated to avoid negative impacts on drivability, emissions and diagnosis.

5.2.4 Ignition

Matthias Budde, Michael Becker, Manfred Adolf, and Alexander Schenk

5.2.4.1 Introduction: Requirements for Ignition Systems in Natural Gas Operation

The fundamental requirements for ignition systems in natural gas engines do not differ largely from those in gasoline engines: For both engine types, the ignition system has to be designed for reliable inflammation in all operation conditions combined with long lifespan for all components. Due to the high stability of methane molecules, more energy is required for the ionization of the air-fuel-mixture in natural gas operation compared to gasoline operation. Together with an inadequate design of the ignition system, this fact can lead to a reduced lifespan.

Natural gas is well suited as fuel for combustion systems with high charge dilution by air (lean combustion) because of its increased ignition boundaries compared to gasoline ($0.7 <$ air-fuel-ratio AFR < 2.1 in homogenous mixtures without charge motion [58]). Diagram 5.57 shows the operation range of state-of-the-art natural gas engines (incl. stationary large bore engines). The limiting region of engine-damaging knock can be avoided by running leaner mixtures with increasing load. For too lean mixtures the operation range is limited by misfire. Lean combustion and its equivalent, the dilution by external cooled exhaust gas (exhaust gas recirculation EGR), represent another obstacle for the ignition system because of the increased amount of inert gas.

Further requirements for the spark plugs are created by higher inflammation and combustion temperatures in natural gas operation because of the lack of evaporation cooling and the more efficient combustion position closer to top dead center. Therefore, cold spark plugs should be used ideally to reduce both electrode temperatures and wear. Due to state-of-the-art bivalent operation with natural gas and gasoline, spark plug electrodes are more likely to be matched for gasoline operation with the critical conditions for cold start, sooting and quenching. This leads to stronger electrode wear in natural gas operation and shorter maintenance intervals. However, state-of-the-art spark plugs with double-precious-metal reinforcement (iridium-platinum) at the electrodes allow for runtimes over 60,000 km. In natural gas operation the demand for ignition voltage is typically increased by 2 kV which has no major effect on electrode wear, but the discharge

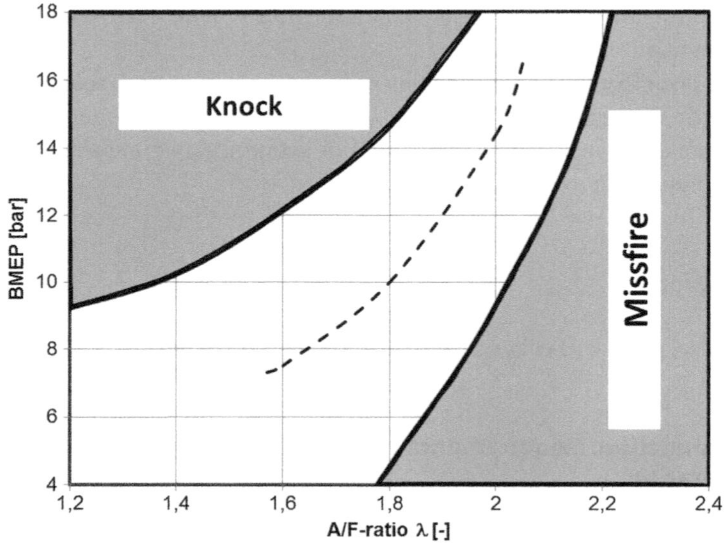

Diagram 5.57 Possible operation range of natural gas engines with lean mixtures

current is limited by the ignition coil. In stationary natural gas engines with monovalent lean operation, cold spark plugs with an increased amount of wear-resistant electrode material are used.

5.2.4.2 Classification of Ignition Systems for Natural Gas Engines

A broad variety of natural gas engines can be found in the market: from small single-cylinder engines in micro combined heat and power plants (e.g. HONDA CHP module with 1 kW electrical power) to the world's largest natural gas engine Wärtsilä 50SG with 18 MW electrical power generated in 18 cylinders. Below there is a classification of current natural gas engines and its typical ignition components:

Micro Combined Heat and Power plants

In the area of micro plants for the combined heat and power generation for one or two family houses, modified single-cylinder gasoline engines are used with natural gas that are originate from emergency power unit applications. The ignition systems are also taken over; they usually are inductive systems with standard spark plugs such as also used in passenger car applications.

Passenger Car Natural Gas Engines

In passenger car applications, gasoline engines are usually adapted for the requirements of natural gas operation (which is also true for mini combined heat and power plants, e.g. the module of the company "Lichtblick", which uses a Volkswagen passenger car natural gas engine). The vehicles are sub-divided into monovalent and bivalent vehicles: Monovalent engines are designed for pure natural gas operation with little use of gasoline fuel during cold start and as emergency reserves [59]. The gasoline fuel is stored in a separate tank with a size of maximum 15 l for monovalent vehicles. Bivalent engines are designed for both fuels, gasoline and natural gas, and in principal the type of fuel can be switched during operation. In conclusion, the ignition system can be optimized for natural gas operation in monovalent vehicles whereas in bivalent vehicles a compromise for both fuels is considered. The ignition coils are taken over from the gasoline base engines, the spark plugs (usually with a 14 mm thread) have a colder design for monovalent vehicles and the electrode distance is decreased. For both, monovalent and bivalent vehicles, the center and mass electrodes are equipped with precious metal reinforcement from platinum and/or iridium alloy.

Natural Gas Engines for Commercial Vehicles and Small Combined Heat and Power Plants

Existing commercial vehicle Diesel engines are used as the basis for light to heavy-duty vehicles and small combined heat and power plants for economic reasons [59]. For the implementation of the Otto-type combustion system, some adaptations for mixture formation and of course ignition are necessary in these engines. The Diesel injector, for example, is replaced by the spark plug in the cylinder head. To minimize the integration of the ignition system into the existing series Diesel ECU, a special 24 V ignition coil control module is used. Inductive systems for ignition coils such as in passenger car applications are typically used in the commercial vehicle sector whereas in combined heat and power plant engines capacitive ignition systems are commonly applied. The spark plugs are mainly so-called J-gap spark plugs with a 14 mm thread with platinum or iridium reinforced electrodes. For mobile applications, spark plugs are often derived from passenger car spark plugs, whereas in combined heat and power plants special industrial spark plugs with high precious metal contents and electrode gaps between 0.2 and 0.4 mm are generally used.

Medium-Size and Large-Bore Stationary Natural Gas Engines

Large-bore stationary natural gas engines as they are used in combined heat and power plants or as compressor power units in gas pipelines are usually based on Diesel engines from off-highway, railroad or marine applications and are specifically adapted for stationary operation with natural gas. Engines for compressor power units are designed for a high reliability to ensure stable pipeline operation at any time. Therefore, these engines run on moderate mean pressures and with easily-ignitable mixtures, which enables the use of relatively simple ignition systems (e.g. industrial spark plugs without precious metal reinforcement) and are replaced preemptively. On the other hand, engines for power generation are designed for the highest possible electrical efficiency and therefore run on very high mean pressures. For engine out NO_x reduction, lean combustion is used resulting in high requirements for the ignition system in terms of high ignition voltage demand and high ignition energy. Additionally, the spark plug maintenance intervals should be as long as possible and be in line with oil change intervals.

For this engine size class, three concepts of ignition system according to the engine bore are in place:

- Up to approx. 150 mm bore diameter, open combustion chambers with standard industrial spark plugs can be applied without slow combustion or misfire events.
- Up to approx. 200 mm bore diameter, some producers use pre-chamber spark plugs with a passive pre-chamber. The larger bore diameters lead to longer flame distances resulting in the need for higher ignition energies and larger ignition volumes. The concept of pre-chamber spark plugs separates the region around the spark plug electrodes from the main combustion chamber using a pre-chamber that is located around the spark plug. This mini combustion chamber is ignited conventionally by the spark between the electrodes. The combustion enters the main chamber via overflow holes as torch jets because of the emerging pressure drop between pre-chamber and main chamber and leads to large-volume ignition in the main chamber.
- Above 200 mm bore diameter, gas-fueled pre-chambers are used [60]. This concept uses a separated pre-chamber with its own gas supply and spark plug and is linked to the main chamber only via overflow channels. At ignition time, a relatively rich and easily ignitable mixture is provided leading to high combustion temperatures in the pre-chamber and makes the use of special industrial spark plugs with extreme cold electrodes necessary. With this concept, the torch jets entering the main chamber through the overflow holes also serve as large-volume ignition source.

However, this classification is not valid in every case. The engine MWM TCG 2032, for example, with a bore diameter of 260 mm would fit in the category gas-fueled pre-chamber, but uses a pre-chamber spark plug instead. Capacitive ignition systems connected via an ignition rail to the ignition control unit and via ignition cables to the spark plugs are commonly used in all three concepts. Spark plugs are always special industrial plugs with 18 mm thread (or 7/8″ in the US market) in this engine size class. Apart from such exotic species, special solutions can be found e.g. direct installed coil-on-plugs in the oil reservoir under the cylinder head cover in some Caterpillar engines. For the architecture of the complete ignition system, several configurations can be found in the market. Usually the ignition control unit is designed as a separate unit taking over other functions such as knock control.

Besides these three concepts, other methods for igniting the air-gas-mixture can be found in medium-size to large-bore stationary natural gas engines. Diesel pilot injection (e.g. MAN Diesel Natural Gas engine 32/40 DF) or the PGI method (Performance Gas Injection) are two examples that are not explained further here. In addition, several high frequency corona systems are currently in development that also deliver, as a matter of principle, a higher ignition volume.

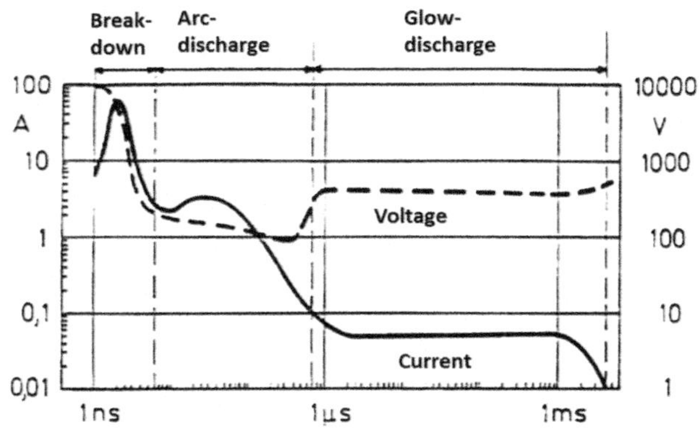

Diagram 5.58 The phases of spark discharge according to [61]: Time dependent characteristic of current and voltage for a transistorized ignition coil (TIC)

5.2.4.3 Function and Layout of Various Ignition Systems

This chapter explains in detail today's most commonly used systems with spark plugs and ignition coils and in addition a high-frequency corona ignition system.

Ignition

The spark formed by the ignition at the spark plug can be divided in three consecutive discharge types with significantly different energetic and plasma-physical characteristics (Diagram 5.58).

The voltage at the spark plug increases until the streamer discharge formed in the electrical field reaches the opposite electrode. Then breakdown occurs and the impedance of the electrode pathway decreases drastically. The current increases quickly due to the discharge of the spark plug's scattering capacities. Very high temperatures are created in the conductive channel by the dissociation and ionization of atoms and molecules. The spark then turns into the arc phase with very low voltage where the current is determined by the discharge of high-voltage capacities. A hot spot is created at the cathode by strong electron emissions (burn spot) and cathode material is evaporated leading to strong erosion of electrode material. The channel temperature decreases to approx. 6000 K. The plasma expands due to heat conduction and diffusion and the beginning exothermal reaction leads to an advancing flame front.

With currents below 100 mA the transition to glow discharge happens. In a transition phase, the change between arc and glow discharge can happen multiple times depending on mixture variations between the electrodes. During glow discharge the voltage increases again—the electron current is supported by striking ions—and the channel temperature is down to approx. 3000 K. The temperature is below the melting limit and the electrodes are now mainly atomized by striking charge carriers [62].

During this process, the coil as energy storage discharges completely into the discharge channel. The spark stalls when the voltage falls below the limit for sustaining the channel.

The rest of the energy dies in the secondary side of the ignition coil. Due to significant heat losses at the electrodes, the energy contained in the spark plasma is much lower than the electrical energy routed to the spark plug. For today's commonly used transistorized ignition coils (TIC), the glow discharge is mainly responsible for inflammation while the reliability of ignition increases with peak current and discharge duration [62].

Transistorized Ignition Coil

Coils used in ignition systems switched by a transistor without distributors are formed by a closed magnetic circuit made from laminated silicon steel plates with concentric stacked primary and secondary coils and filled with epoxy resin. By turning on the primary-side current, energy is stored inductively in the magnetic circuit's air gap. After breaking the primary side current by the transistor, a voltage is built-up in the secondary coil until breakdown at the spark plug. The breakdown voltage is a result of the cylinder pressure and the electrode distance at the spark plug. The maximum achievable voltage is limited by the shutdown voltage (terminal voltage in the transistor) and the transmission ratio from secondary to primary side in the coil (Diagrams 5.59 and 5.60).

After breakdown, the energy discharges via the secondary coil in the spark. During this glow discharge (discharge duration), the secondary side voltage remains limited to the value of the burning voltage and is nearly constant.

State-of-the-art ignition coils can deliver a voltage higher than 40 kV. However, this high voltage cannot be used in today's engine operation as the isolation durability of spark

Diagram 5.59 Layout of an ignition coil

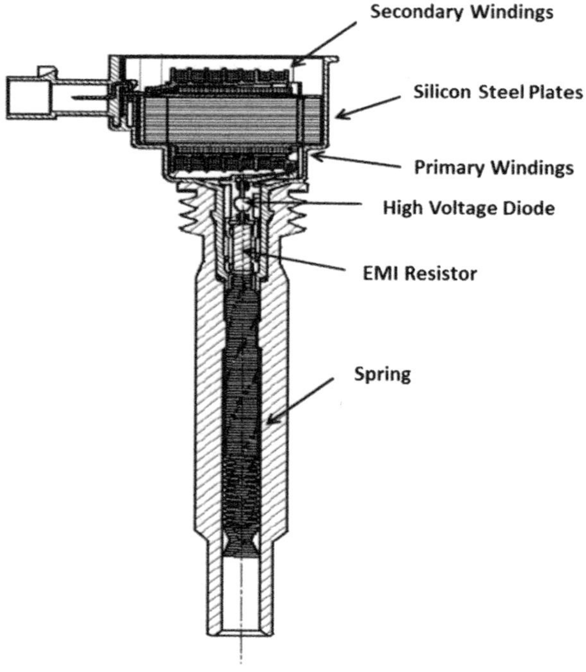

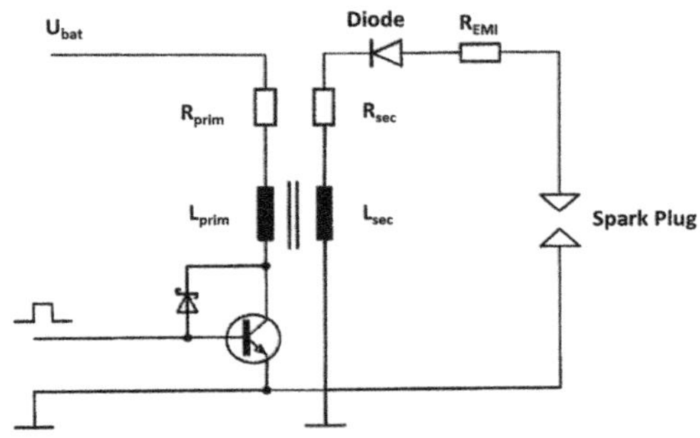

Diagram 5.60 Layout of a transistorized ignition system

plugs prevents the use of engines with higher ignition pressures and therefore higher breakdown voltage. The output or burning energy is determined by the measurement of discharge duration and the spark current at constant burning voltage. Together with the transmission ratio and the coil's shutdown current the maximum spark current (glow current) on the secondary side coil is defined. The spark duration can be varied in wide boundaries by the choice of the storage inductivity and the operation point of the magnetic circuit.

The coupling between primary and secondary side of the ignition coil is more than 90 %. Due to transfer losses and the resistances in the circuit (secondary side coil impedance, interference suppression resistance in coil and spark plug) only 50 % of the electrically stored energy in the primary circuit are used at the spark plug. The cylinder conditions (pressure, temperature, charge motion etc.) determine together with the electrode distance the burning voltage during the spark duration. High load or charging increase the burning voltage and decrease the spark duration (coil discharge duration).

The standard coil ignition has been widely established because of its good cost-value ratio [63].

Dual Coil Ignition System

In applications with external EGR, a continuous spark is a beneficial support for the inflammation even enabling re-ignition of blown-out sparks. In a dual coil ignition system, two TICs are operated alternately; they are decoupled by high voltage diodes. This way, an oscillating spark current can be set that does not fall below a predefined current level. Compared to a multi-spark ignition system where one coil is charged and discharged multiple times, no interrupts for re-charging appear in a dual coil ignition system. However, in order to shorten otherwise excessively long re-charging periods, an additional DC-DC-converter is necessary for coils with sufficient high ignition energy.

High-Voltage Capacitive Discharge Ignition System (CDI)

In a CDI, the energy is temporarily stored in a capacitor and the necessary high voltage is created by a fast low-loss transformer while switching. The inductive charging of the capacitors for voltages of 100 V and more is done by an engine flywheel.

These ignition systems feature an extremely fast voltage increase (some kV/μs) and are therefore very resistant against shorting due to spark plug fouling. Disadvantageous is the very short burning duration of approx. 100 μs leading to possible misfire events in inhomogeneous mixtures and the high spark current leading to increased spark plug wear.

In stationary large-bore engines, CDI are charged by the installed power grid. The achievable high ignition voltage and the fast voltage rise are advantages here whereas the burning duration is of lower importance.

High Frequency Corona Ignition System

Contrary to standard ignition systems using a thermal plasma in the spark for the air-fuel-mixture inflammation, the high frequency or corona ignition system principle is based on the corona gas discharge. A corona discharge is the ionization of an electrically non-conductive fluid in the vicinity an electrical conductor; Diagram 5.61 shows the corona discharge at an electrode with 5 ignition tips in air at environmental pressure. The up to 1000-fold larger ignition volume compared to a spot-shaped spark ignition is clearly visible. This behavior is called the volume ignition effect. Photo-ionization and field emissions lead to the corona discharge.

One basic condition for the creation of a corona discharge is the existence of strongly inhomogeneous electrical fields. While in a homogeneous electrical field an electrical breakdown happens instantaneously with reaching the ignition voltage, a stable discharge without electrical breakdown is established in a current-limited circuit as it is formed in field-weak regions of a strongly inhomogeneous field. These strongly inhomogeneous fields can be created by field enhancements at very asymmetric electrodes. This is usually achieved by the use of tip electrodes in corona ignition systems forming a point-plane arrangement in the combustion chamber with electrical conductive engine components (piston, cylinder head, valves) as electrical ground potential. Diagram 5.62 shows the corona discharge at a single electrode tip at environmental pressure and at 10 bar. Obviously, the corona discharge is pressure sensitive, meaning the volumetric extension is smaller at higher pressures with otherwise constant boundary conditions. However, this

Diagram 5.61 Corona discharge in air at environmental pressure

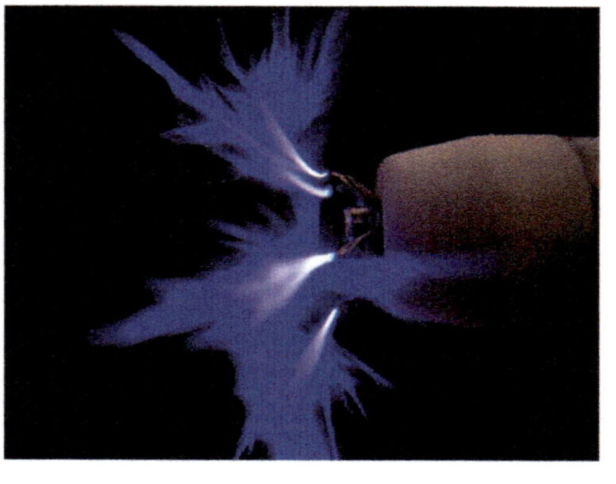

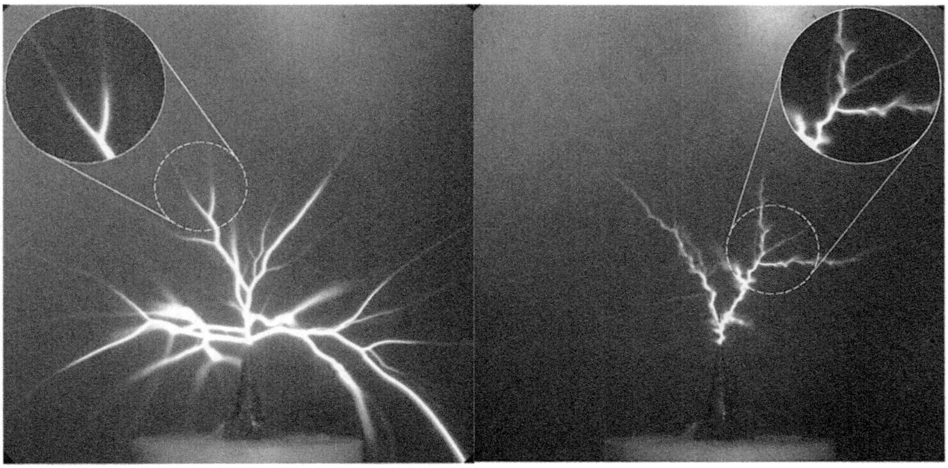

Diagram 5.62 Corona discharge in air at environmental pressure (*left*) and at 10 bar (*right*) [64]

effect is partly compensated for by the modified shape of the corona. Coming from a rather straight shape, the single streamers are more likely to deliquesce and turn into a helical structure with increasing pressure. Additionally, they modify their position in space more often leading to larger ignition volume when integrated over the ignition duration.

BorgWarner's corona ignition system "EcoFlash" uses a high frequency alternating voltage in the low megahertz region for the generation of high local field strength at the electrode tips. This alternating voltage is generated in a control unit containing the high frequency generator and some DC-DC transformer stages as well as the complete logic for open- and closed-loop control. The generated voltage of some 100 V is transferred to the igniter via a coaxial cable. The igniter is based on the principle of an LC-series-oscillating-circuit and increases the voltage by a factor of 100 at the electrode tips when activated at its resonance frequency. This results in local electrical field strength of up to 300 kV/mm decreasing drastically with increasing distance from the electrode. Diagram 5.63 shows the top view of the ignition system and the equivalent circuit diagram of the igniter. The igniter contains a coil as inductance and a ceramic-metal-feedthrough acting as the capacitance of the LC-oscillating-circuit. Therefore, the resonance frequency as well as its electrical efficiency is determined by the layout of the igniter.

Some engine testing results in natural gas operation using a spark plug with a transistorized ignition coil or a high frequency corona ignition system in a 4 cylinder 2.0 l passenger car engine with 115 kW rated power are shown below.

As in gasoline operation, the de-throttling of the engine in natural gas operation also leads to fuel consumption reduction in part load. This can be achieved by exhaust gas recirculation or by keeping the exhaust valves open during the intake stroke (valve overlap). However, the inflammation is complicated by the higher content of inert

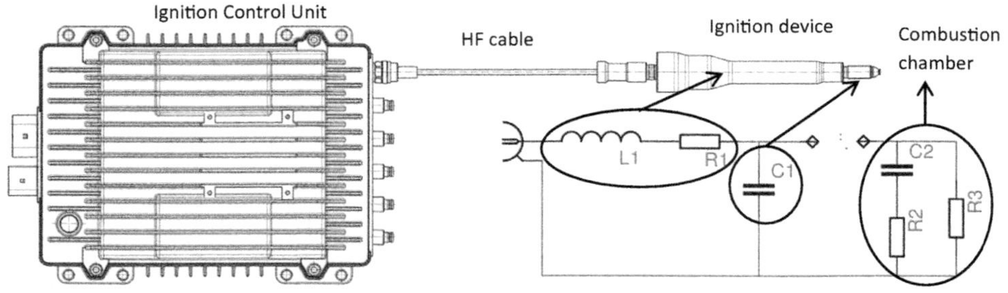

Diagram 5.63 Corona ignition system "EcoFlash" (*top view*) and equivalent circuit diagram of igniter and combustion chamber

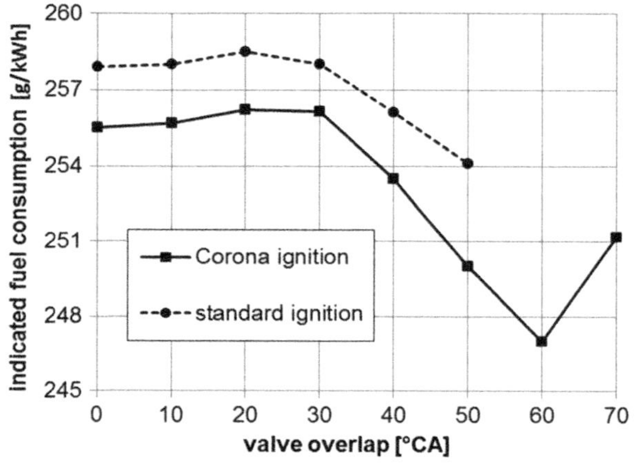

Diagram 5.64 Indicated fuel consumption over valve overlap for the corona ignition system and a standard ignition system (load point 2000 rpm, 2 bar BMEP) [65]

gases. Diagram 5.64 shows the potential for fuel consumption reduction with increasing valve overlap in natural gas operation for a load point of 2000 rpm and 2 bar BMEP. By using a spark plug with a transistorized ignition coil, a reduction of approx. 1.5 % in indicated fuel consumption can be achieved at the limit of the combustion stability compared to the base without valve overlap. By using a corona ignition system (here: EcoFlash) this advantage can be raised to approx. 4.3 % by further de-throttling and higher residual gas capability enabled by the improved inflammation stability by means of the volume ignition effect.

For keeping the emissions legislation limits, a fast catalyst heating is also favorable in natural gas operation. This is complicated by lower exhaust gas temperatures and higher catalyst operating temperatures compared to gasoline operation (400–500 °C compared to 250–350 °C for gasoline). High exhaust gas temperatures for fast catalyst

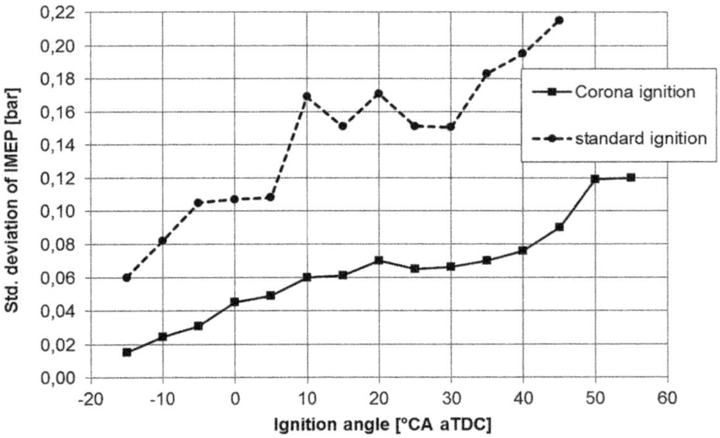

Diagram 5.65 Standard deviation of IMEP over ignition angle for the corona ignition system and a standard ignition system (catalyst heating load point: 1200 rpm, 1 bar IMEP) [65]

light-off need to be achieved by ignition delay in natural gas operation, too. Due to a large volume change per degree crank angle (higher piston velocity) for late combustion, lower combustion stabilities as a matter of principle have to be compensated by short burn delays. Diagram 5.65 shows the standard deviation of the indicated mean effective pressure over ignition angle for a typical catalyst heating load point (1200 rpm, 1 bar IMEP) in natural gas operation. Due to the volume ignition effect of the corona ignition system, significantly higher combustion stability can be achieved over a wide range of the ignition timing. Therefore, the usable ignition angle range is increased having a positive effect on the available exhaust gas enthalpy for catalyst heating and the fuel consumption.

5.2.4.4 Summary and Perspective

Natural gas engines set higher requirements for the ignition systems than comparable gasoline engines. Appropriately designed, state-of-the-art ignition systems can comply very well with those requirements. New ignition systems such as the corona ignition system are in development to utilize the full potential of the fuel natural gas, e.g. by increased charge dilution.

The currently foreseeable introduction of 48 V vehicle electrical systems will also have a positive influence on ignition systems for natural gas operation. The biggest value can be realized for multi-spark or dual coil ignition systems through omission of the DC-DC-converter. The higher vehicle system voltage leads to shorter recharge periods and therefore allows for higher ignition energies. For high-frequency corona ignition systems one DC/DC-converter stage can also be reduced according to the circuit design.

5.2.5 Charging in Natural Gas Operation

Matthias Budde, Michael Becker, Rolf Sauerstein, and Sascha Weiske

5.2.5.1 Introduction: Fuel Specification—Effects on Combustion Systems

Natural gas as a fuel is becoming more and more important for the compliance with climate targets for CO_2 emissions reduction. While in the past mainly bivalent naturally aspirated engines were introduced into the market, there is now a change towards monovalent engines optimized for CNG usage. Limited gasoline operation is also possible in these concepts to enable sufficient mobility in regions with low natural gas fueling capacities. In recent times concepts were introduced combining downsizing, turbocharging and CNG allowing for another significant step towards energy efficient propulsion. Besides these applications in the field of passenger cars, the use of natural gas in commercial vehicles and stationary engines has been established to some extent where mainly CI engines were adapted for natural gas usage (trucks, coaches) or landfill gas (stationary engines for electric power generation). With the adaptation of the engine design and the necessary materials in the area of intake (fuel injection in intake manifold/upstream turbo-compressor) and in the area of hot gases, fundamental experience has been gained that can be used synergistically in the area of passenger car engines. Below we will explain in detail the specialities of the combination of natural gas and turbocharging.

When using natural gas as fuel, there are some additional challenges for the engine's charging system. The mixture heating value is lower and the air requirement is higher. Both aspects can be compensated for through higher charging rates, which is enabled by the higher knock resistance of natural gas. Additionally, the combustion can be shifted to a more efficient position closer to top dead center leading to lower exhaust gas temperatures and thus to a cost-saving potential for the turbine material compared to gasoline operation. As a matter of principle, a gaseous fuel always has the disadvantage of a lack of combustion chamber cooling due to evaporation leading to higher component wear and lower volumetric efficiency, which is partly compensated for by the facts mentioned earlier. The disadvantage of a lower laminar burning velocity can be compensated for with a dedicated design of intake ports and combustion chamber (e.g. masking of intake valves) for an increase in charge motion. This again leads to a lower volumetric efficiency that must be compensated for by charging. Diagram 5.66 shows a comparison of fundamental parameters of natural gas (CNG, quality H) and RON 95 gasoline fuel.

Natural gas as a fuel for combustion engines underlies less stringent regulations as gasoline or diesel fuel. In the German fuel station grid, different gas qualities can be found with methane numbers from approx. 75 to approx. 90. Below we will use knock resistant quality H gas when talking about natural gas operation. Further impacts of the different parameters of CNG versus gasoline fuel can be found in the literature [67, 68].

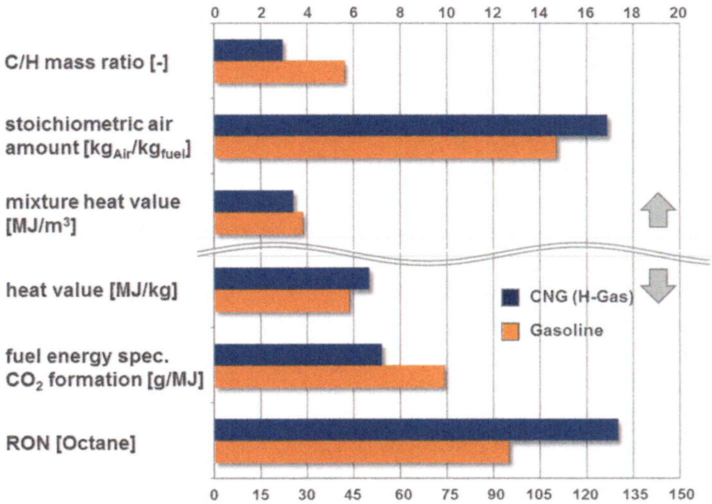

Diagram 5.66 Engine relevant fuel parameters of natural gas and gasoline fuel [66]

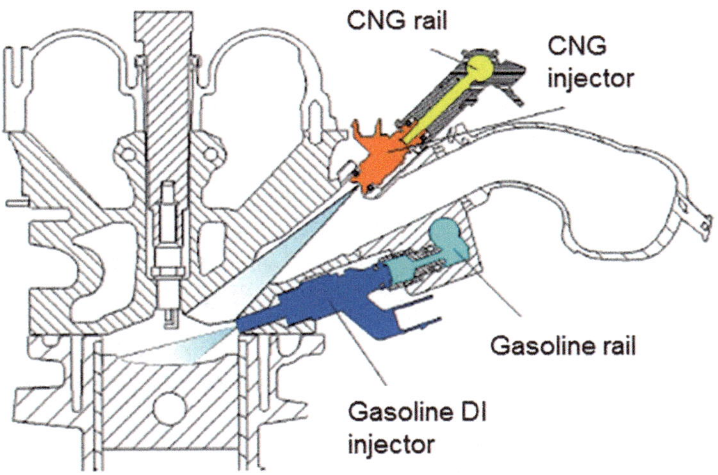

Diagram 5.67 Fuel injection system of a bivalent natural gas engine [69]

5.2.5.2 Turbocharger Matching for Natural Gas Operation

Due to the limited availability of natural gas and improved cold start behavior for gasoline operation, today's natural gas engines are usually designed as bivalent models for both natural gas and gasoline operation. State-of-the-art bivalent engine concepts therefore feature a dual injection system with direct high-pressure injection for gasoline and intake manifold injection for natural gas (e.g. natural gas engine based on Volkswagen's EA211 as in Diagram 5.67). The gaseous injection of natural gas into the intake manifold leads to

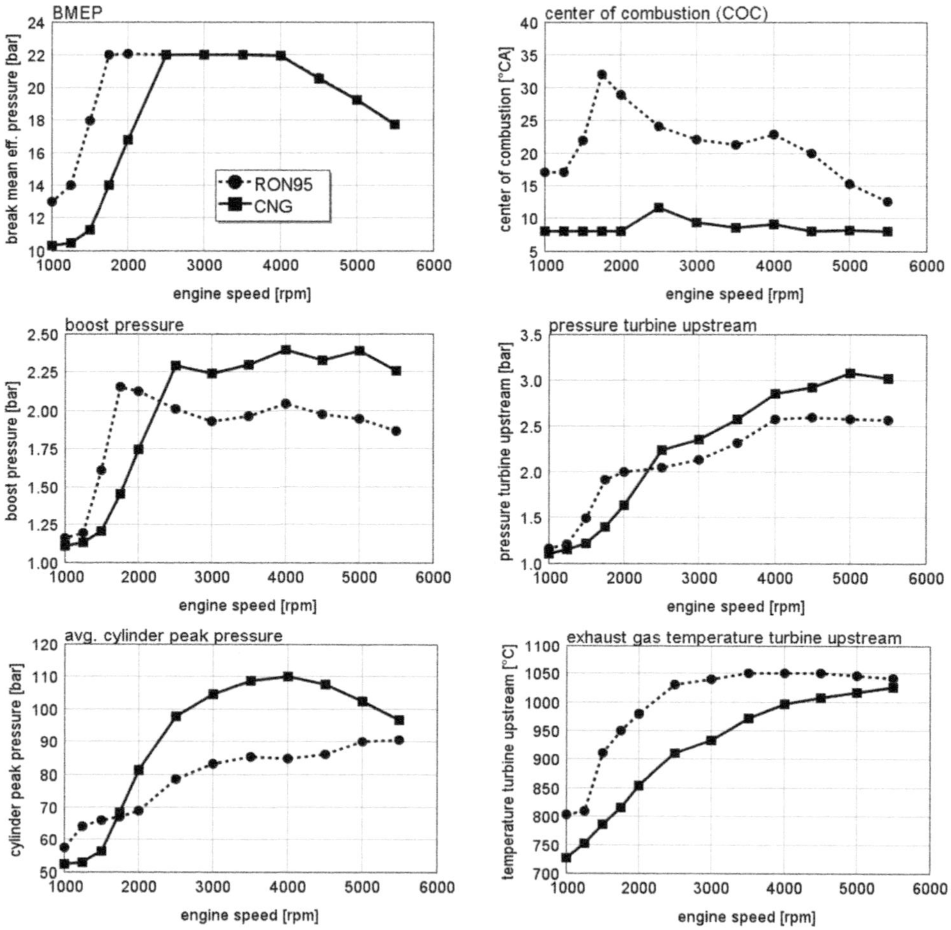

Diagram 5.68 Full load comparison between gasoline and natural gas operation for bivalent engine layout

a decrease in air efficiency depending on the fuel quality. This has to be compensated for by charging. Direct injection systems for natural gas are currently in development.

Possibilities to Charge a Bivalent Engine Concept

A simulation stud compared a bivalent engine concept with a monovalent natural gas engine in terms of effects on the charging system. An engine with 85 kW/l effective power and a maximal brake mean effective pressure of 22 bar was defined as the basis for the potential assessment of turbocharged natural gas engines.

In a full load comparison between gasoline and natural gas operation with identical turbo charger layout, the necessary compromises of the air path and mixture formation design can be seen (Diagram 5.68). For the same rated power level of 85 kW/l, loss in low-end torque of up to 35 % is found, which is unacceptable for drive-away and expected elasticity behavior. The decreased exhaust enthalpy level in natural gas operation in combination with the necessity to avoid scavenging leads to a decreased boost pressure level at low-end torque and therefore unachieved torque targets even with an optimum position of the combustion. Improvements could be made by the introduction of natural gas direct injection into the combustion chamber.

In gasoline operation, significant late positions of combustions are necessary to avoid knock and to keep a maximal scavenging rate of 8 % (resulting from the requirement to keep a minimum air-fuel ratio of 0.95). This increases the exhaust temperature level. The higher knock resistance of natural gas is beneficial here and leads to a decrease of the exhaust temperature level of 125 K at comparable engine power. Enrichment is necessary in gasoline operation above speeds of 3500 rpm because of reaching the turbine inlet limit temperature of 1050 °C. This is an advantage for natural gas operation where lower exhaust gas temperatures even at rated speed do not reach the turbine inlet limit temperature leading to significantly lower fuel consumption in engine map regions outside the current NEDC. The trend of the boost pressure and the resulting exhaust backpressure over speed demonstrate the higher requirements for the charging system by higher pressure ratios over compressor and higher expansion numbers over turbine for natural gas operation. Both numbers are still in the normal operation range of a turbo charger. However, an increase in BMEP can lead to a re-configuration of the aerodynamic components.

Further differences between gasoline and natural gas operation can be demonstrated with a load sweep at a constant engine speed of 4000 rpm (Diagram 5.69).

The higher knock resistance of natural gas enables an efficient position of the combustion close to top dead center even at high loads leading to an increase of the engine inner efficiency. Cylinder peak pressure and exhaust gas temperature rise with increasing load (Diagram 5.69). After reaching the knock limit, it is necessary to retard the ignition when the load is increased further. This is the case at 18 bar BMEP and above for

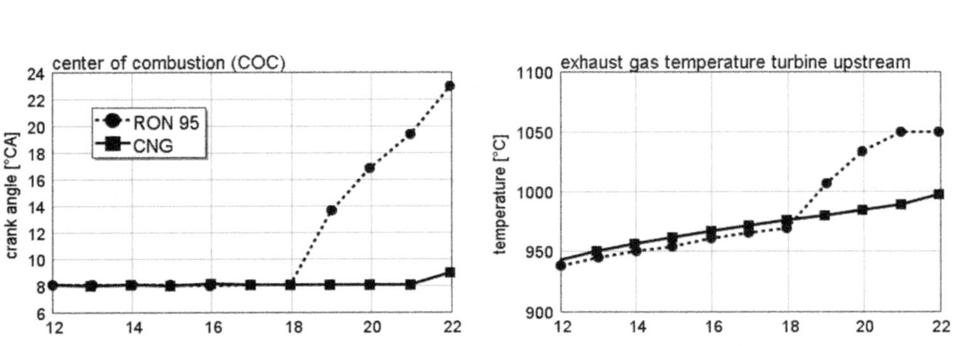

Diagram 5.69 Combustion parameter for gasoline operation vs. natural gas operation for bivalent engine layout

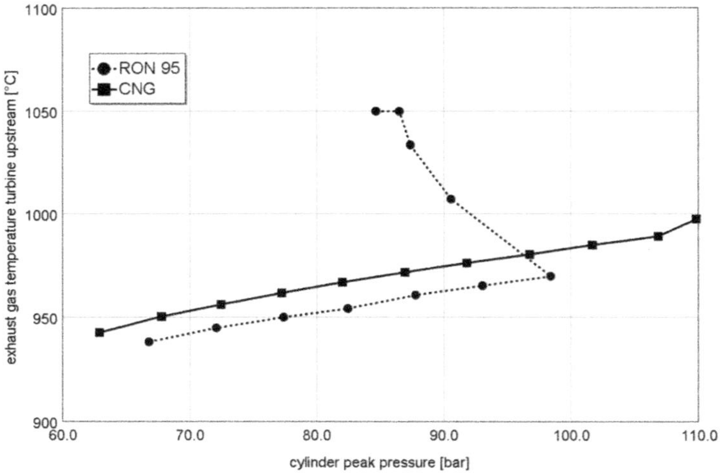

Diagram 5.70 Relationship of cylinder peak pressure and exhaust gas temperature upstream turbine for a load sweep at 4000 rpm

gasoline operation. On the other hand, the position of center of combustion can be kept at 8 °CA aTDC up to a load of 21 bar BMEP in natural gas operation, then the ignition delay also becomes necessary due to knock or reaching the mechanical limit of cylinder peak pressure. Furthermore the earlier ignition time leads to an increase in cylinder peak pressure and a reduction of the temperature level at the end of the expansion when the exhaust valve opens in comparison to gasoline operation.

The benefit of reduced exhaust gas temperatures with natural gas is partially compensated for by the lack of component cooling due to enrichment [70]. The comparison of cylinder peak pressure and exhaust gas temperature shows the ratio of increase of mechanical engine load and decrease of exhaust gas enthalpy relevant for turbo charging compared to gasoline operation (Diagram 5.70). The increasing energy density with load and speed in combination with the engine's peak pressure capability defines the achievable effective power of a bivalent engine in stoichiometric conditions. With the assumption of a stoichiometric combustion up to an exhaust gas temperature of 1050 °C, a reduction in exhaust gas temperature of up to 80 K is achieved along the load sweep for natural gas operation. At the same time, an efficient combustion close to TDC is possible for natural gas operation due to the higher knock resistance up to the engine's mechanical limit. It is desirable to increase the engine's mechanical limits for the benefit of efficient combustion parameters at higher engine loads.

The gaseous injection of natural gas in the intake manifold leads to a significant decrease of volumetric efficiency depending on the fuel quality that has to be compensated for with higher boost pressures by the charging system (Diagram 5.71). For single-staged turbo charged engines, this can lead to power reduction especially in high altitude

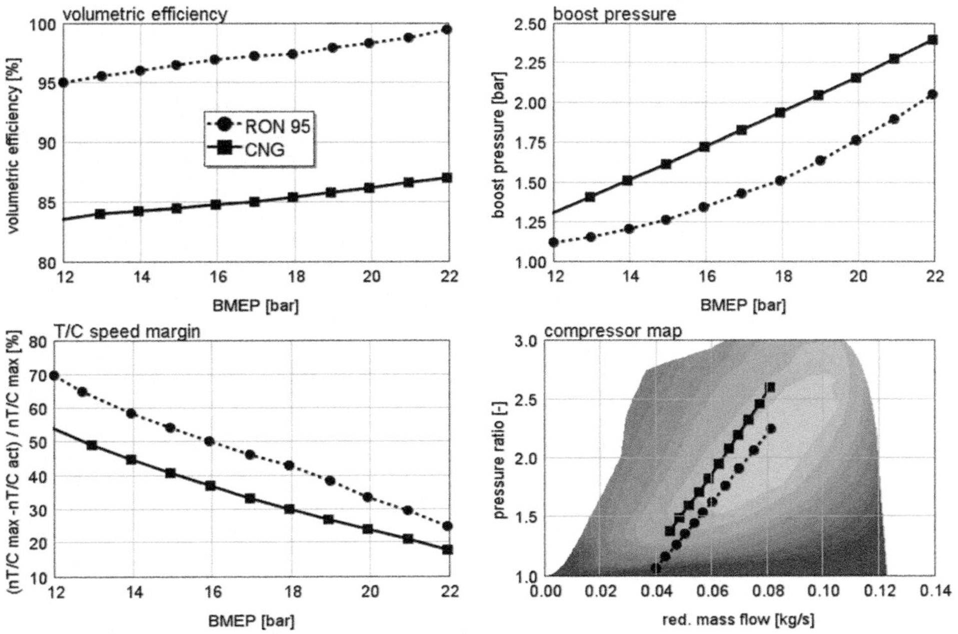

Diagram 5.71 Comparison of charging-relevant parameters for a load sweep at 4000 rpm

operation due to earlier reaching of the turbo charger speed limit and could be compensated for by direct fuel injection into the cylinder.

Different gas qualities result in different potential for fuel consumption reduction. In low to medium part load conditions there is almost no need for ignition delay and efficient combustion positions close to TDC are achievable. Additionally, the engine can already be de-throttled at low loads due to the gaseous injection of natural gas into the intake manifold. In the mentioned load sweep, a fuel consumption benefit of 7 % in low and medium part load and up to 16 % for full load conditions in bivalent natural gas operation compared to gasoline operation is achievable.

Different charging concepts can be applied to compensate for the torque loss at low speeds in natural gas operation. Besides a variable turbine geometry, two-stage charging systems consisting of a combination of two turbo chargers or a combination of one turbo charger and a supercharger are possible. In both two-stage combinations, the permanently engaged base turbo charger can be enlarged regarding its flow capacity leading to lower pumping losses at rated power and lower intake temperature levels for gasoline operation. The use of a regulated two-stage turbo charger can already increase the torque from idle speed upwards [71, 72]. On the other hand, there is an increase in thermal inertia for the charging unit. This can be improved by a compact design of regulated two-stage turbo chargers with decreased exhaust surface areas from turbine inlet to catalyst inlet

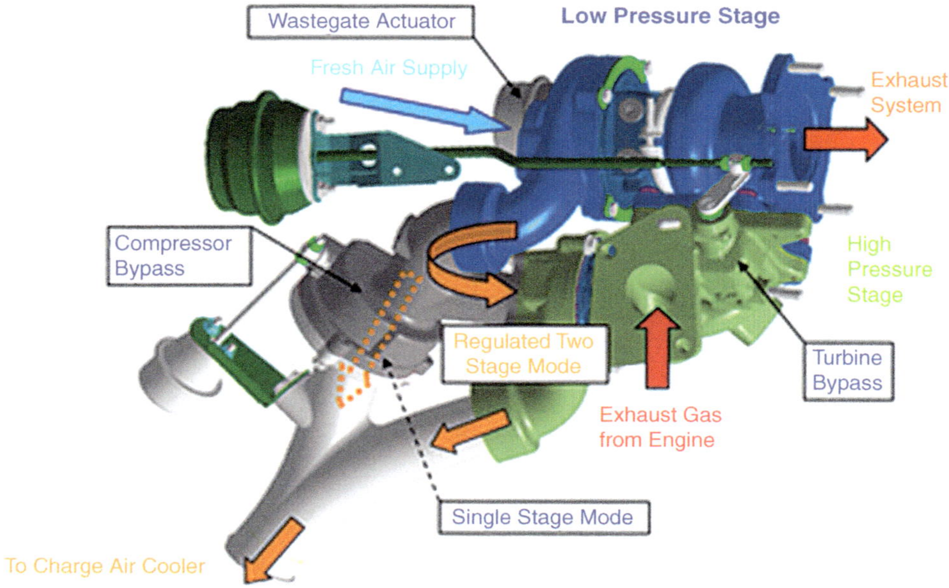

Diagram 5.72 Compact design of a regulated two-stage turbo charger [72]

(Diagram 5.72). Superchargers as second stage avoid the increase of exhaust system thermal inertia but suffer from additional necessary drive power and therefore an increase in engine load. Additionally, the efficiency of superchargers decreases disproportionally high with decreasing size and complicates their application in small displacement engines. Additionally, the placement in the front-end accessory drive is necessary, complicating the application in critical package situations. For efficiency reasons, switchable mechanical components have to be used with combined turbo-supercharger concepts to minimize parasitic losses in naturally aspirated operation.

A significant advantage of two-stage charging with turbo charger and supercharger and especially with two-stage turbo chargers is the independence from electrical power—relevant for frequent acceleration or dynamic drive behavior. Additionally, very high boost pressure levels can be achieved by two-stage turbocharging as the maximum pressure ratio for turbo chargers is three times higher than for superchargers.

Another variant for multi-stage charging is the use of an electrical accessory charger (eBOOSTER™) in combination with the base turbo charger. Contrary to screw-type or G-type superchargers, the turbo-type eBOOSTER™ is driven electrically und therefore has the potential to use electrical energy gained by recuperation or stored in the battery. Contrary to standard superchargers, the maximal achievable compression ratio is 1.7 limiting the total achievable degree of charging. But it could be sufficient for single natural gas engines and must be analyzed in detail for the specific engine.

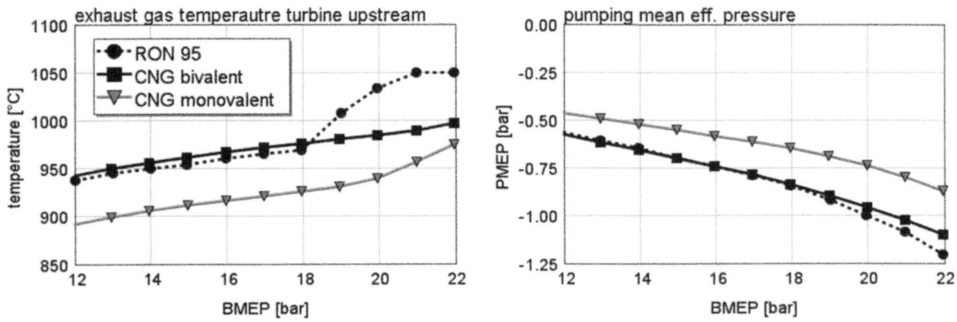

Diagram 5.73 Comparison of bivalent gasoline/natural gas operation with monovalent natural gas operation with 1.5 units increased compression ratio and VTG turbine in load sweep at 4000 rpm

Potential for Optimization in Monovalent Natural Gas Engines

The biggest reduction potential for CO_2 can be achieved in monovalent natural gas operation. Because of the natural gas' higher knock resistance, the compression ration can be increased by 1.5–2 units. Along the earlier discussed load sweep at 4000 rpm, this measure leads to a further increased cylinder peak pressure and according to the larger expansion volume with identical valve timing to an additional decrease of the exhaust gas temperature of 50 K compared to the bivalent natural gas operation (Diagram 5.73). The variability in the swell behavior of a VTG turbine (as used in this case) with an optional bypass offers the additional potential of decreasing the pumping losses by 23 % reducing the required boost pressure level at identical engine power.

At full load, the deficits in low-end torque in bivalent natural gas operation are reduced significantly by the increase in compression ratio and the use of a VTG turbine (Diagram 5.74). The torque deficit of 10 % compared to gasoline operation cannot be balanced in this example due to limitation by the compressor's surge line. Here, the use of a so-called high-flow-compressor with a wider air mass flow range could be beneficial.

The turbine's reduced absorption capability due to the closure of the VTG blades leads to an increase in pumping losses in speed ranges lower than 2000 rpm in monovalent natural gas operation. The appearing shift of the position of the center of combustion leads to an increase of the exhaust gas temperature with simultaneous increase of the cylinder peak pressure level. From 3000 rpm onwards, a reduction in pumping losses by 27 % is possible by the increase of usable turbine capacity.

Diagram 5.74 shows the necessity for significant ignition time delay for the monovalent layout with the increase of the compression ratio. Investigations are necessary to determine whether the possibility of transforming the higher compression ratio of the monovalent layout into a reduced effective compression ratio using Miller or Atkinson valve timing is beneficial.

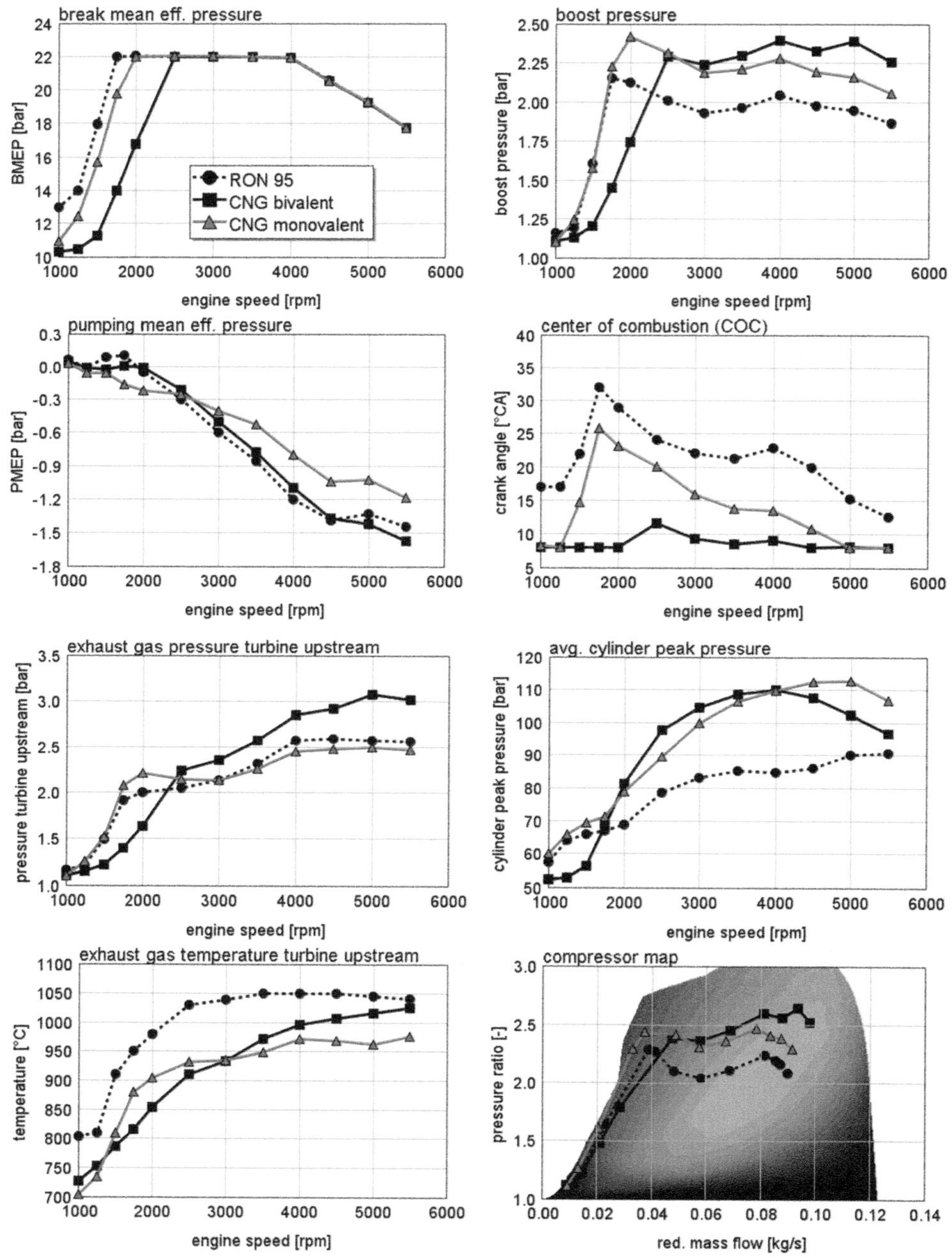

Diagram 5.74 Full load comparison between bivalent gasoline/natural gas operation and monovalent natural gas operation with 1.5 units increased compression ratio and VTG turbine

At rated power, the ideal timing for the center of combustion can be achieved. The exhaust gas temperature of the described monovalent natural gas layout does not exceed the 980 °C limitation and is therefore appropriate for the cost-efficient use of the VTG technology, even at high specific engine power. For the described concept of a monovalent natural gas engine with a 1.5 units increased compression ratio and a VTG turbine, 2–4 % fuel consumption savings depending on speed and load point can be realized in addition to the potential of a bivalent natural gas engine compared to the base gasoline operation.

5.2.5.3 Material

Turbochargers for natural gas operation can generally be made from the same materials as those for gasoline operation. However, it becomes apparent that austenitic cast iron is more preferable than ferritic material due to its higher temperature resistance. Special attention has to be paid to the turbine and particularly on the exhaust gas temperatures and composition. As exhaust gas temperatures can, according to the combustion system, be much higher than diesel-like exhaust gas temperatures, mainly temperature-resistant alloys with high nickel contents are used for the turbine wheel and the turbine housing. For gasoline applications these turbines are already in series production and are durable up to exhaust gas temperatures of 1050 °C. Alloys for higher temperature limits are currently in an experimental stage and could be interesting for natural gas operation for high specific loads because of the lacking possibility to enrich to reduce exhaust gas temperatures.

When using biogas, special attention needs to be paid to the exhaust gas composition as biogases can contain higher amounts of sulfur. Sulfur reacts during the combustion and in the hot exhaust gas to sulfur dioxide and further to sulfuric acid. This acid acts highly corrosively in the exhaust system and in the turbine, which is unfavorable for the durability. Special applications with increased contamination or variable gas composition should also be checked for components that might lead to a corrosive attack. This affects particularly large bore engines with gas injection upstream from the compressor where, due to leakage flows via the compressor-side shaft seal ring to the bearing housing, the thrust bearing can be in direct contact with the gas mixture. Finally, for highly dynamic operation and frequent waste gate actuation, attention has to be paid to the seating region to avoid material fatigue.

5.2.5.4 Summary and Perspective

A very efficient powertrain can be created by the combination of natural gas as fuel and turbocharging. The following aspects are relevant for the optimization of the complete system:

- The high knock resistance of natural gas allows for early and efficiency-optimized ignition angles leading to low exhaust gas temperatures (exhaust gas enthalpy) compared to gasoline.

- The peak pressure capability of the base engine in combination with the chosen compression ratio determines the load and speed threshold above which the ignition angle needs to be retarded from the most efficient timing.
- The gaseous injection of the fuel into the intake manifold leads to a decrease in air efficiency that needs compensation by charging.
- Because of the gaseous injection, no enrichment for the limitation of the exhaust gas temperatures is possible. Therefore, the achievable power is limited by the exhaust gas temperature the respective components can tolerate.

These characteristics require a dedicated matching of the thermodynamics and the mechanical layout of the turbo charger to achieve the power targets in bivalent operation. Usually, an adaptation of the compressor and turbine pairing is made, leading to compromises in both gasoline and natural gas operation. To increase the exhaust gas temperature resistance of the turbocharger's hot gas leading components (turbine housing, turbine wheel, heat shield, waste gate) up to 1050 °C helps to keep the power density in natural gas operation. Two-stage charging can fulfill the requirement of high low-end-torque, either as a combination of turbo charger with supercharger, or as a regulated two-stage turbocharging concept as it is established in diesel engines.

However, the biggest CO_2 reduction potential can be achieved with monovalent natural gas engines. A high compression ratio optimized for natural gas can be combined with a VTG turbine to gain further efficiency potential. Thus, exhaust gas temperatures can be kept below 980 °C allowing for a cost-attractive use of the VTG technology. On the compressor side, high-flow compressor wheels permit a wider air mass flow range and thus a higher low-end-torque while still achieving the targets for rated power.

For both bivalent and monovalent natural gas engines, new combustion systems offer promising solutions:

- Combustion systems with Miller-/Atkinson valve timing can compensate for the necessary delay of ignition timing for high compression ratios by adaptating the intake valve closing timing.
- The use of low-pressure or high-pressure EGR in medium to high part load can contribute to a further de-throttling. The experiences made with diesel and now also with gasoline engines promote the fast availability.

The use of the discussed solutions promises a further significant increase in efficiency and therefore CO_2 reduction.

5.2.6 CO_2 and Efficiencies

Helge Wollenhaupt, Klaus Wunderlich, and Albert Ebner

Abbreviations

BMEP	Brake Mean Effective Pressure
BSFC	Brake Specific Fuel Consumption
DI	Direct Injection
ECU	Engine Control Unit
HRF50	50 % Heat Release Fraction
NEDC	New European Driving Cycle
NG	Natural Gas
PFI	Port Fuel Injection
RON	Research Octane Number
VCR	Variable Compression Ratio
WLTP	Worldwide harmonized Light vehicles Test Procedures
WOT	Wide Open Throttle

Reduced operating costs and hence lower total costs of ownership are the most encouraging aspects in choosing a natural gas vehicle from a customer point of view. For manufacturers however, it is mainly the significant reduction in CO_2 emissions compared to conventional powertrains that makes natural gas as an engine fuel so attractive. CNG has the potential to sustainably reduce fleet fuel consumption and contribute to achieving ambitious exhaust emission targets in order to cope with increasing challenges from regulations around the globe. This potential, however, can only be exploited if the infrastructure can cope with the sale of a considerable number of natural gas vehicles [73].

The reduction in CO_2 emissions achieved by operating a spark-ignited engine with methane instead of gasoline is mainly based upon the special physical and chemical properties of methane:

- Favorable calorific properties (C/H ratio, calorific value)
- Higher knock resistance

The advantage of the calorific properties of methane can be utilized quite easily without the need to fundamentally change the mechanical components of an existing spark-ignited engine. When natural gas was introduced as fuel for vehicle applications, the fuel systems of existing gasoline engines were usually replaced or supplemented by a separate or additional CNG systems. Only very slight mechanical adaptations were made to these engines, that is only where required to cope with the special demands of the CNG combustion (modified valve seats for example). Fundamental engine parameters, however, such as displacement, compression ratio and with these the basic ECU calibration were rarely changed compared to the original gasoline engines that were used as a starting point.

Along the course of subsequent spark-ignition engine development towards smaller engine displacement *(downsizing)* combined with higher levels of boosting, direct injection *(DI)* systems and higher compression ratios for improved overall efficiency, the sensitivity of the fuel for irregular combustion phenomena has become increasingly important. With its considerably higher knock resistance compared to gasoline, methane has the potential to push overall efficiency of spark-ignited engines into unrivalled regions and therefore increase its effective CO_2 benefit compared to conventional fuels even further.

With regards to mixture formation, port fuel injection *(PFI)* is still state of the art for fuels in gaseous state of aggregation. Supplementary to a different or additional fuel system, further fundamental modifications to a modern turbocharged gasoline engine are necessary to make as much use of the higher knock resistance of methane as possible without causing considerable disadvantages in torque characteristics, driving experience or exhaust emissions compared to the existing gasoline DI combustion process. Engine displacement, turbocharger layout as well as the maximum combustion peak pressures the engine should be able to resist are only a few among the most relevant parameters that need to be taken into consideration. For future concepts, direct injection of gaseous fuel in the combustion chamber *(CNG-DI)* is expected to give new perspectives and pave the way for additional improvements. This technology, however, is still being developed.

Natural gas powertrains based upon diesel engines—running solely with CNG with spark ignition or in dual-fuel mode by substituting only a share of diesel—exploit the high knock resistance of methane fairly well. This is due to the relatively high compression ratios and high permitted combustion peak pressures originally incorporated into diesel engine design. Instead, other challenges, for example in the area of exhaust temperature management, exhaust gas treatment (see Sect. 5.2.7) and fairly unusual topics for diesel engines such as knock detection and antiknock control are to be discussed here.

It is fair to state that the use of compressed natural gas for passenger car applications comes with a set of disadvantages as well, which are mainly caused by the lower energy density of CNG storage compared to liquid fuels. The pressurized-gas vessels contribute to additional mass of the vehicle—increasing with higher CNG capacities for optimized mileages—, as well as restrictions in packaging, luggage and passenger compartments and possibly even disadvantages in the aerodynamic setup and the structural safety concept of the vehicle. Most of these aspects do have a direct or at least an indirect negative effect on fuel consumption. Because of this, the advantages in energy consumption and CO_2 emissions natural gas provides compared to conventional fuels is reduced accordingly. Therefore the target must be on the one hand to minimise these disadvantages with appropriate vehicle concepts and on the other hand to compensate by choosing appropriate engine parameters, that is by specifically optimizing the engine for CNG operation. In doing so, attractive and competitive vehicles can be provided on the market—even in comparison with other types of energy storage. Nevertheless, the still limited availability of CNG fuel stations, even in rather attractive markets for mobile use of natural gas, requires carrying at least a small amount of conventional liquid fuel (reserve) in the CNG vehicle.

5.2.6.1 Material Properties

As the first and simplest molecule within the group of alkanes, methane is the hydrocarbon with the highest number of C–H bonds in relation to the number of carbon atoms (Diagram 5.75), hence methane has the smallest C/H ratio. For that reason, methane has special properties concerning the specific energy content and the generation of CO_2 by reacting chemically with oxygen. In Table 5.4, a comparison in material properties is given between methane and isooctane (2,2,4-trimethylpentane), which is often used as reference substance for gasoline and applied for the definition of the octane number (RON).

Diagram 5.75 Methane molecule (CH_4) (source: wikimedia.org)

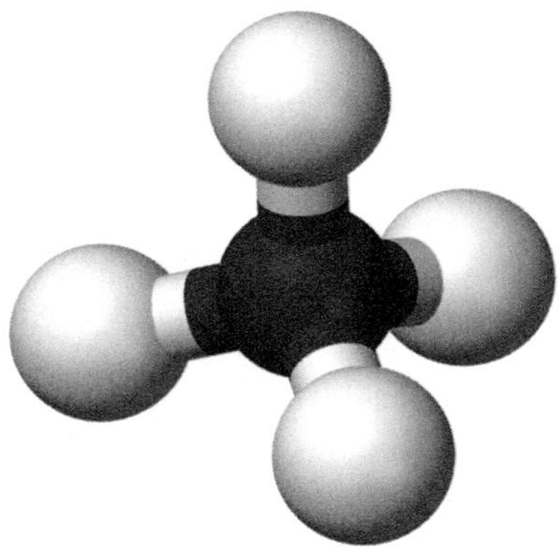

Table 5.4 Properties of methane (CH_4) and isooctane (2,2,4-trimethylpentane, C_8H_{18}) (source: wikipedia.org)

	Methane	Isooctane (2,2,4-trimethylpentane)
Total formula	CH_4	C_8H_{18}
Structural formula	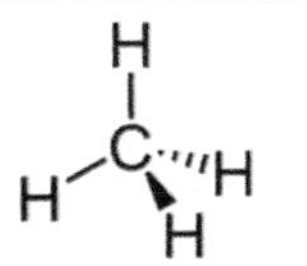	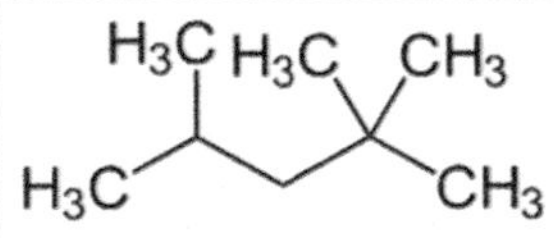
Molar mass	16.04 g/mol	114.23 g/mol
Calorific value	50.0 MJ/kg	44.9 MJ/kg
Octane number (RON)	130	100 (per definitionem)

A theoretic derivation of the chemical reactions of these substances with oxygen and the correspondent mass of released CO_2 is given as follows:

The ideal reaction of isooctane with oxygen occurs by this chemical equation:

$$2C_8H_{18} + 25O_2 \rightarrow 16CO_2 + 18H_2O$$

The released mass of CO_2 and the mass of the reactant C_8H_{18} are therefore in the following proportion:

$$n_{CO_2} = 8 \cdot n_{C_8H_{18}} \text{mit } n = \frac{m}{M} \quad \frac{m_{CO_2}}{m_{C_8H_{18}}} = \frac{8 \cdot M_{CO_2}}{M_{C_8H_{18}}} = \frac{8 \cdot 44.01\,\text{g/mol}}{114.23\,\text{g/mol}} = 3.08$$

In consideration of the calorific value of isooctane ($H_{uC_8H_{18}} = 44.9\,\text{MJ/kg}$), the generated mass of CO_2 relative to the released amount of energy is:

$$W_{C_8H_{18}} = H_{uC_8H_{18}} \cdot m_{C_8H_{18}} = H_{uC_8H_{18}} \cdot \frac{m_{CO_2}}{3.08} \quad \frac{m_{CO_2}}{W_{C_8H_{18}}} = \frac{3.08}{H_{uC_8H_{18}}} = 68.6\,\frac{g_{(CO_2)}}{MJ_{(CH_8H_{18})}}$$

In the ideal and complete combustion of isooctane, a total mass of 68.6 g CO_2 is generated per each 1 MJ of released energy.

For comparison, here is the chemical equation for the ideal reaction of methane with oxygen:

$$CH_4 + 2O_2 \rightarrow CO_2 + 2H_2O$$

The released mass of CO_2 and the mass of the reactant CH_4 are in the following proportion:

$$n_{CO_2} = n_{CH_4} \text{mit } n = \frac{m}{M} \quad \frac{m_{CO_2}}{m_{CH_4}} = \frac{M_{CO_2}}{M_{CH_4}} = \frac{44.01\,\text{g/mol}}{16.04\,\text{g/mol}} = 2.74$$

In consideration of the calorific value of methane ($H_{uCH_4} = 50.0\,\text{MJ/kg}$), the generated mass of CO_2 relative to the released amount of energy is:

$$W_{CH_4} = H_{uCH_4} \cdot m_{CH_4} = H_{uCH_4} \cdot \frac{m_{CO_2}}{2.74} \quad \frac{m_{CO_2}}{W_{CH_4}} = \frac{2.74}{H_{uCH_4}} = 54.8\,\frac{g_{(CO_2)}}{MJ_{(CH_4)}}$$

In the ideal and complete combustion of methane, a total mass of 54.8 g CO_2 is generated per each 1 MJ of released energy.

Comparing the specific mass of CO_2 generated with methane compared to isooctane, the following relative reduction in CO_2 emissions can be derived:

CO_2 emission of **methane** relative to **isooctane**:

$$1 - \frac{54.8\, g_{(CO_2)}/MJ_{(CH_4)}}{68.6\, g_{(CO_2)}/MJ_{(C_8H_{18})}} = -20.1\,\%$$

The gravimetric fuel consumption, on the contrary, is only a function of the ratio of the calorific values:

Gravimetric fuel consumption of **methane** relative to **isooctane**:

$$1 - \frac{44.9\, MJ/kg}{50.0\, MJ/kg} = -10.2\ \%$$

Hence the relative reduction in CO_2 emissions by combusting methane instead of isooctane is about twice the corresponding relative reduction in gravimetric fuel consumption. This is due to the lower C/H ratio of methane.

In order to measure and compare different fuel types concerning their effect on CO_2 emissions, knowing the corresponding calorific parameters is mandatory. Thus an analysis aiming at CO_2 reduction can only be as accurate as the knowledge of the effective calorific values. As natural gas is a natural product, however, its exact composition and calorific parameters might differ considerably depending on its origin as well as location and time of extraction from the supply system. Besides pure methane, natural gas usually contains other hydrocarbons such as ethane, propane, butane and pentane in different concentrations. Other ingredients are mostly inert gases like nitrogen and helium which do not affect CO_2 emissions directly. In addition, raw natural gas contains a considerable amount of CO_2 when extracted, which is normally removed during production. Nevertheless, residual fractions of carbon dioxide from natural sources as well as remaining CO_2 carrier gas from biogenically produced methane might affect the measurable CO_2 emissions negatively.

5.2.6.2 Thermodynamics and Combustion Process

For a thermodynamic fuel comparison of natural gas and gasoline at engine operation, not only the physical fuel properties are of special importance, but also the combustion process and particularly the mixture preparation method upon which the combustion process is based. Although spray-guided direct injection *(DI)* might be considered state-of-the-art for spark-ignited gasoline engines, CNG engines available on the market are still equipped with external mixture preparation concepts, i.e. with port fuel injection *(PFI)* or insufflation to be more precise. Based on technology that is currently available, a proper fuel comparison between natural gas and gasoline is therefore also a comparison of different mixture preparation methods.

In order to properly compare gasoline and natural gas engine operation from a thermodynamic point of view, the same characteristic parameters need to be taken into consideration as for comparison of direct *(DI)* and port fuel injection *(PFI)* [74–76]. The effective efficiency of the real engine process can be split into an internal efficiency and several fractions of losses:

$$\eta_e = \eta_{th\,v} - \Delta\eta_{BV} - \Delta\eta_U - \Delta\eta_K - \Delta\eta_{LW} - \Delta\eta_R$$

with η_e, effective efficiency of the real engine process; $\eta_{th\,v}$, internal efficiency of the ideal engine process (Otto/constant-volume cycle); $\Delta\eta_{BV}$, losses due to non-ideal combustion; $\Delta\eta_U$, leakage losses; $\Delta\eta_K$, heat losses; $\Delta\eta_{LW}$, charge-exchange losses; $\Delta\eta_R$, friction losses.

Different mixture preparation concepts have an effect on at least four of these parameters ($\eta_{th\,v}$, $\Delta\eta_K$, $\Delta\eta_{LW}$, $\Delta\eta_R$). Charge-exchange losses, for example, are normally reduced with direct injection relative to port injection due to an effect of improved dethrottling [75]. Insufflation of gaseous fuel in the manifold, however, has a dethrottling effect at part-load conditions as well. Both effects need to be determined carefully at each operating point in order to compare both operation modes properly. A similar situation arises from the analysis of friction losses. At gasoline DI operation, on the one hand, considerable drive losses are normally seen at the high-pressure fuel pump, which furthermore depend strongly on engine load or fuel flow rate respectively. Natural gas operation, on the other hand, causes increased mechanical friction losses at high engine loads due to higher combustion pressures (see also Sect. 5.2.6.3), causing increased piston and bearing stresses to which the engine might be submitted accordingly. Thus a thermodynamically correct comparison of both operation modes is valid only under the specific boundary conditions of each operating point.

For a more global comparison of natural gas and gasoline operation within the same engine type, for example to condense the differences in fuel consumption between both operation modes into a single number, it appears to be sufficient to refer to the calorific values only. It is worth emphasizing here that the calorific properties of the fuels are in effect additional to the particular engine efficiencies. Thus the differences in gravimetric fuel consumption and CO_2 emissions of methane relative to isooctane, as derived from the calorific properties in Sect. 5.2.6.1, apply for the same effective engine efficiency η_e, where thermodynamic and mechanical losses altogether are identical for both operating modes.

5.2.6.3 Knock Resistance

With progressively higher engine load conditions, another fuel property becomes increasingly important which is also strongly affected by the molecular structure: the considerably higher knock resistance of methane relative to gasoline (see also Table 5.4).

Compared to a spark-ignited engine designed for operation with gasoline only, the compression ratio can be increased for operation with natural gas on principle, improving internal efficiency of the combustion process considerably. Diagram 5.76 shows the

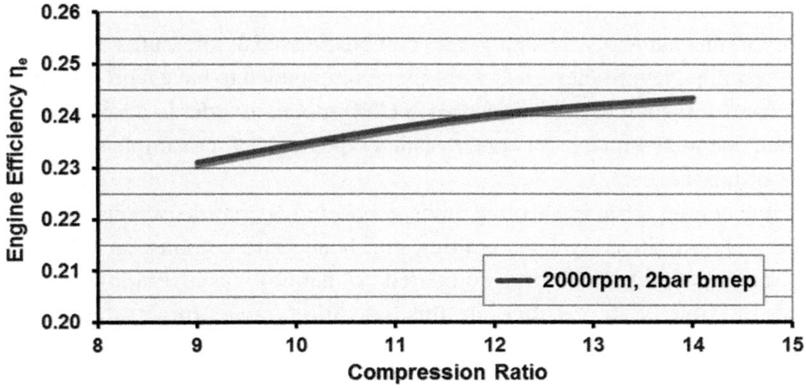

Diagram 5.76 Effective engine efficiency of a 2.0 l gasoline engine at 2000 rpm, 2 bar bmep

simulated effective engine efficiency of a 2.0 l gasoline engine as a function of compression ratio at an engine speed of 2000 rpm and at 2 bar bmep.

A noticeable increase in compression ratio of a spark-ignited engine, however, cannot be necessarily transferred into a practical application easily, as this would cause a multitude of interfering effects that need to be analyzed closely. Besides an increase in combustion pressures with higher mechanical and thermal stress for various components, an improvement in internal efficiency along with a higher compression ratio results in lower exhaust gas temperatures, thereby affecting exhaust gas aftertreatment (see Sect. 5.2.7) and also the operating conditions of a turbocharger. Another substantial question concerning the definition of the compression ratio is whether or not an engine concept is aiming at running purely with natural gas (monovalent) or able to be operated optionally with both types of fuel (bivalent; gasoline and natural gas mode). With regards to the significantly different antiknock properties of these fuels, the answer to that question might substantially influence fundamental engine design parameters. In addition, engine layout is also affected considerably by the vehicle concept for which the powertrain is designed. The vehicle concept with its tank capacities, in turn, is strongly influenced by the availability of CNG fuel stations. Therefore it is very important that further improvement in CNG infrastructure on the one hand, and continuing development in engine and vehicle technology on the other hand occur as closely associated as possible. It is worth mentioning that engine concepts with variable compression ratio (VCR), for example, have the potential to solve this trade-off resulting from two types of fuel with significantly different knock properties with which an engine should be able to be operated.

Natural gas engines with compression ratios valid for bivalent operation, or at least for optional operation with gasoline in a reserve mode, are expected to be able to run in CNG mode with highly efficient ignition timings even under boosted engine operating conditions. The high knock resistance of natural gas allows for early combustion timings without the risk of provoking irregular combustion phenomena, provided that the engine is capable of mechanically resisting the combustion peak pressures these operating conditions require. Nevertheless, it is strongly recommended to have a properly calibrated antiknock control system in operation even in CNG mode, in order to avoid damage to the engine when operated with a less knock resistant type of gas, for example with a relatively high share of butane.

If the same engine is run in gasoline mode at boosted operation, especially in the lower rpm range, only significantly later ignition and hence late combustion timings can be achieved due to knock limitation compared to natural gas operation. The thermal efficiencies feasible with gasoline in this operation range might thus be reduced significantly compared to CNG mode.

In order to better illustrate these differences, an 'ignition timing efficiency' might be defined, representing the reduction in thermal efficiency of an actual operating point or ignition calibration setting relative to an efficiency-optimized timing, which might not be feasible due to knock limitation. Engine torque modelling in a modern ECU, for example, is calculated as a function of ignition timing by essentially the same principle. The determination of an ignition timing efficiency for CNG and gasoline mode respectively is a fairly simple method to assess and compare the relative effect of the fuels' knock resistance on thermal engine efficiency.

Diagram 5.77 shows the average combustion peak pressure, combustion timing (HRF 50 % position) as well as the effective ignition timing efficiency under boosted WOT

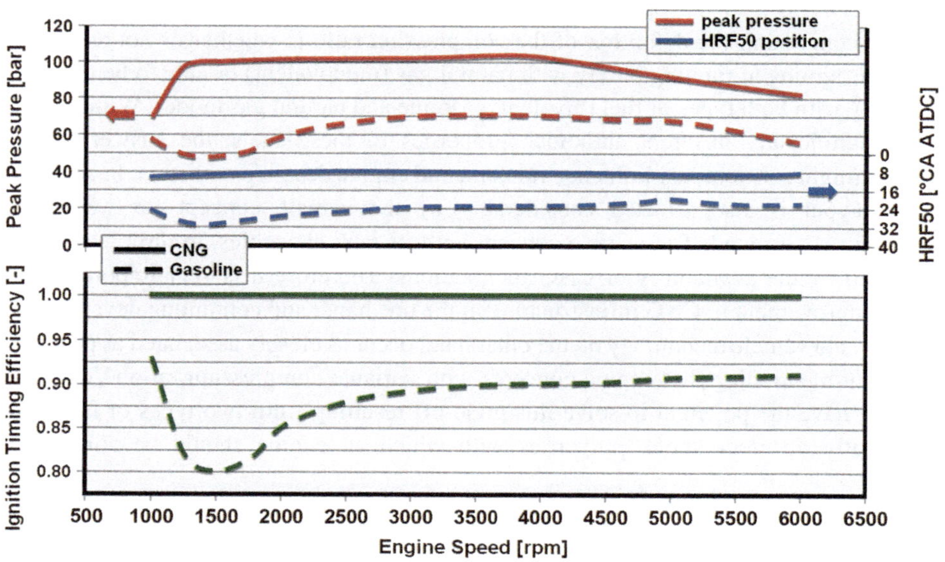

Diagram 5.77 Average combustion peak pressure, HRF50 position and ignition timing efficiency of the Mercedes-Benz 'M274 Natural Gas' engine under WOT conditions [73, 77]

conditions of the Mercedes-Benz 'M274 Natural Gas' engine in CNG and gasoline mode. In CNG mode, an efficiency-optimized combustion timing can be adjusted even at maximum torque operation—hence the ignition timing efficiency is always optimal.

The illustrated ignition timing efficiency in gasoline mode is considerably lower, representing the disadvantage in thermal efficiency due to knock limitation relative to the earlier ignition achievable with CNG. This disadvantage needs to be compensated for by additional charge (i.e. boost pressure) and hence additional fuel and energy in order to provide the same torque output. Comparing both operation modes on the same engine, the ignition timing efficiency is a measure which comprehensively illustrates the impact of the fuels' knock resistance on overall engine efficiency. It is worth emphasizing that these differences between gasoline and CNG are in effect additional to the different calorific properties and hence CO_2 emissions derived in Sect. 5.2.6.1.

5.2.6.4 Downsizing and Downspeeding

The advantages of combining turbocharging, improved internal cooling due to direct injection and smaller engine displacement in order to reduce mechanical friction have substantially affected recent gasoline engine development. Development continues further along the path of progressive engine *downsizing* towards higher boost rates and increasing specific power and torque output, complemented by *downspeeding*, i.e. load-point shifting towards lower engine speeds and higher loads.

Due to its high knock resistance, methane or natural gas appears to be the ideal fuel for that purpose and therefore has the potential to improve overall engine efficiency even further (see Sect. 5.2.6.3). Operating small-capacity engines with natural gas, however, is associated with several disadvantages too. First, the mechanically permissible maximum combustion peak pressure of a spark-ignited engine originally designed for gasoline operation is exceeded fairly easily when operated in CNG mode with fuel-efficient early ignition timing, especially if the compression ratio has been increased relative to the original gasoline engine. In order to restrict the maximum combustion pressures, engine efficiency at high torque output might have to be deteriorated by ignition calibration without actually obtaining the knock limit. In other words, the thermodynamic benefit of the high knock resistance of natural gas at high specific engine load is not fully exploitable because of mechanical reasons in this case. Furthermore, the gaseous state of aggregation of the fuel at mixture preparation eliminates the possibility to use mixture enrichment, that is, internal cooling by the vaporization enthalpy of additional fuel, in order to protect engine components from overheating at high specific power output. In case of external mixture formation (port fuel injection), disadvantages in transient engine response compared to liquid fuel have to be accepted, even in the case of turbocharged engines. This is due to displacement of intake air in the manifold by gaseous fuel during transient torque build-up. Moreover, a torque increase by means of *scavenging* at low engine speeds is also not possible in this case, as not only air would be purged into the exhaust system but also unburnt fuel. The latter disadvantages, however, will not apply for internal mixture preparation *(CNG-DI)*. In a nutshell, it can be stated that *downsizing* in combination with natural gas, especially with external mixture formation, is not as straightforwardly applicable as with direct injection of a liquid fuel.

In choosing engine displacement and compression ratio in favor of the specific demands of natural gas and in combination with adapted transmission gear ratios, improvements in overall powertrain efficiency can be achieved with CNG by means of *downspeeding* even better. For example, the CNG passenger car engines 'M270/274 Natural Gas' introduced in 2013 by Mercedes-Benz feature a combination of a relatively large engine capacity, an increased compression ratio and a smaller turbocharger compared to the corresponding gasoline engine platform. This layout enables the engine to be operated in CNG mode with stoichiometric mixture and efficiency-optimized ignition timing in the whole speed and load operating range, under compliance of mechanical pressure and temperature limits accordingly. A comparison of specific fuel consumption of this engine in gasoline and natural gas operation mode is shown in Diagram 5.78.

The minimum specific fuel consumption *(BSFC)* in gasoline mode occurs in a medium engine speed and load range, as efficiency and fuel consumption tend to deteriorate at high engine load due to knock limitation. Note that the fundamental characteristic of this BSFC map is representative for spark-ignited gasoline engine in general.

In contrast to that, engine efficiency in CNG mode with its efficiency-optimized ignition timings even under WOT conditions tends to improve steadily along with engine load, as combustion is not knock limited in this case. With regards to engine friction being primarily a function of engine speed, minimum specific fuel consumption with CNG occurs at very low engine speeds and at maximum torque.

Focusing on the minimum BSFC operating point with gasoline and comparing CO_2 emissions in CNG mode accordingly (based upon measured fuel consumption and calorific fuel properties as shown in Sect. 5.2.6.1), operation with natural gas instead of gasoline corresponds to a reduction in CO_2 emissions of approximately 24 %, i.e. at the same engine speed and load-point. By means of additional *downspeeding*, i.e. load-point-shifting towards lower engine speeds, overall engine efficiency and hence CO_2 emissions with natural gas can be improved even further. In the shown example, CO_2 emissions with CNG turn out to be lower by about 28 % compared to the minimum BSFC operating point with gasoline, at the same effective engine power output. This number is considerably higher than the expected reduction in CO_2 emission derived from the calorific properties of the fuels only (see Sect. 5.2.6.1). This is basically due to the higher knock resistance of natural gas being exploited here also (see Sect. 5.2.6.3).

As the early ignition timings used with natural gas even at high loads help to significantly reduce cyclic deviations in the combustion process (see also Diagram 5.77), engine operation feels particularly smooth in CNG mode even in the low-end torque area, without any negative effect on comfort or driving experience. In combination with comparably high transmission and final-drive gear ratios as well as appropriate shifting strategies, the fuel-saving potential of *downspeeding* with natural gas can be exploited under real-life driving conditions very straightforwardly.

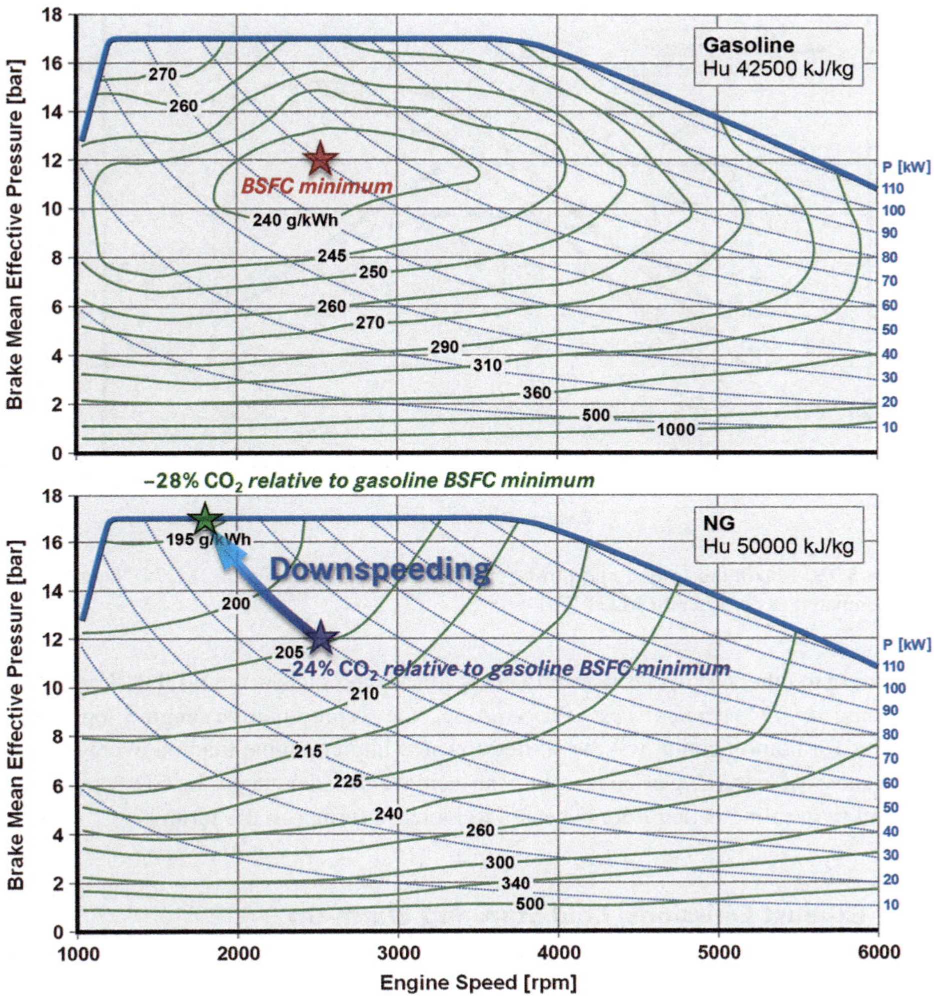

Diagram 5.78 Specific fuel consumption of the Mercedes-Benz 2.0 l 'M274 Natural Gas' engine [73]

For that reason, the natural gas variants of the Mercedes-Benz M270/274 engine family have been designed in particular to provide high low-end torque. The maximum torque output under steady-state conditions is available at a low engine speed of 1250 rpm. It is worth emphasizing that this is achieved in CNG mode with port fuel injection and hence without scavenging. The maximum torque output of this engine at this speed has been increased by 59 % compared to its predecessor M271, Diagram 5.79, whereas the homologated CO_2 emissions in the NEDC were reduced by 22 % with the same vehicle.

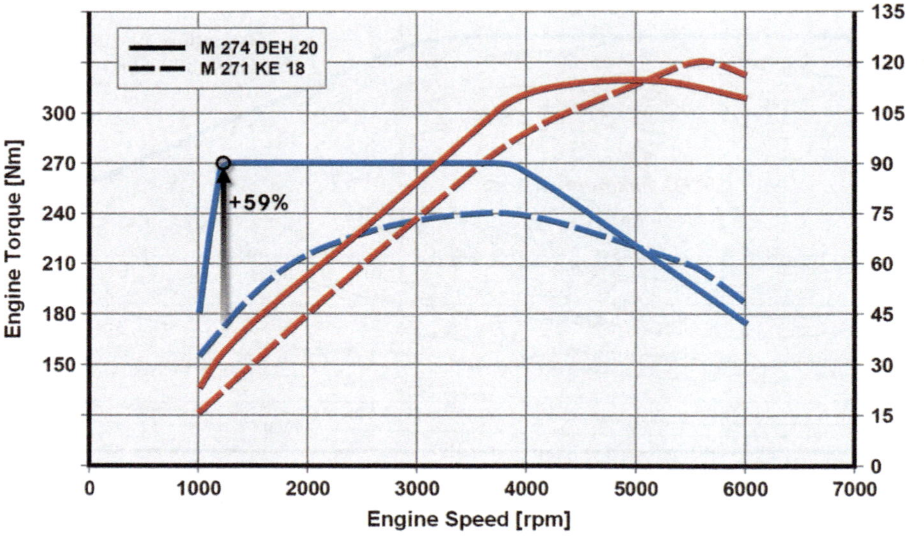

Diagram 5.79 Maximum torque and power curves of the Mercedes-Benz 'M274 Natural Gas' engine compared to predecessor M271 [73]

Referring to other test cycles currently being introduced such as the WLTP *(Worldwide harmonized Light vehicles Test Procedures)*, the representative engine operating conditions for homologation will be shifted towards higher engine load on average. The attractiveness of natural gas concerning the achievable advantages in CO_2 emissions compared to gasoline is therefore expected to increase further in the future.

5.2.7 Exhaust Emissions, Cold Start and Warm-Up

Klaus Wunderlich, Helge Wollenhaupt, Albert Ebner, Peter Heine, and Peter Volz

Abbreviations

ADAC	'Allgemeiner Deutscher Automobil-Club e.V.', largest German Automotive Club
DI	Direct Injection
Euro 4	European Exhaust Emission Statute, valid since 01/2005
Euro 5	European Exhaust Emission Statute, valid since 09/2009
Euro 6	European Exhaust Emission Statute, valid since 09/2014
GWP	Global Warming Potential
HRF50	Heat Release Fraction 50 %

NEDC	New European Driving Cycle
FTP-75	US EPA Federal Test Procedure
RON	Research Octane Number
TSP	Thermal Shock Protected
TWC	Three Way Catalyst

Natural-gas-powered engines are characterized by considerably lower CO_2 emissions due to the combustion of naturally, biogenically or synthetically produced methane—the predominant component of natural gas. Concerning the pollution of the atmosphere with waste emissions affecting global warming however, additional challenges are raised by the potential slip of incompletely burnt residuals of methane through the exhaust system.

Hydrocarbon emissions produced during the combustion of gasoline are composed of various hydrocarbon structures with differing chain lengths, substances with unsaturated bonds (olefines), as well as aromatics and aldehydes. The hydrocarbon emissions generated in the combustion of methane, however, consist mainly of pure CH_4, usually by a proportion of more than 90 %.

It is worth emphasizing in this respect that the climate-changing effect of CH_4 is about 25 times higher than the effect of CO_2. In U.S. greenhouse gas emission legislation for 'passenger cars' and 'light duty trucks' for example, CH_4 and N_2O emissions are being accounted for as non-limited emission components whose greenhouse impact is assessed by a GWP (*global warming potential*) penalty factor. Methane is given a GWP penalty of $25 \times CO_2$, if the emission limit of 0.03 g/mi is exceeded. For N_2O emissions over 0.01 g/mi a GWP penalty of 298 is given relative to CO_2, which, however, is irrelevant for homogeneous combustion with stoichiometric mixture.

For that reason, the exhaust emission aftertreatment of natural gas engines, especially with regard to methane slip, is of utmost importance.

5.2.7.1 Catalyst and Exhaust Gas Aftertreatment

Natural gas, biogas or synthetically produced methane gas normally consists of more than 90 % pure methane. The accompanying components are mostly inert gases such as nitrogen or carbon dioxide.

The methane molecule is symmetrical and is characterized by its saturated and inactive chemical properties. Thus the chemical bonds between carbon and the hydrogen atoms do not tend to break up easily, whereas especially the unsaturated double bonds in the hydrocarbon structures of gasoline emissions show a much more pronounced tendency to thermally break up and consequently build free radicals. In addition, it is worth mentioning that the chemical binding energy of a C–H bond (435 kJ/mol) is considerably higher than the binding energy of an atomic C–C bond (300 kJ/mol). Both facts contribute to the stable and inactive chemical properties characteristic for the methane molecule. For that reason, noticeable catalytic conversion rates of methane emissions on the catalyst surface normally do not occur at temperatures

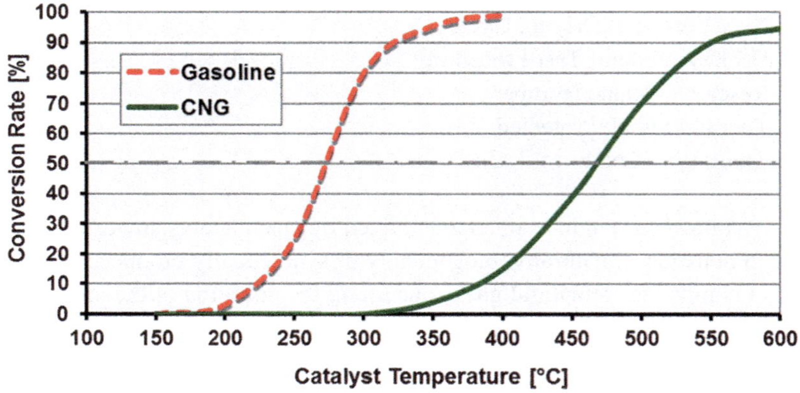

Diagram 5.80 Catalytic conversion of hydrocarbon emissions from gasoline and CNG combustion

lower than 400 °C. Gasoline hydrocarbon emissions, on the contrary, tend to be converted at about 250 °C, Diagram 5.80.

Furthermore, the higher amount of water being produced in methane combustion tends to shift the catalytic conversion even further towards higher temperatures. The maximum conversion rate of a three-way catalyst *(TWC)* system, especially under thermally-aged conditions, might therefore not occur below 550 °C.

In addition to the restricted catalytic conversion of untreated methane emissions, the smaller ratio of carbon atoms in the emissions of methane combustion leads to a reduced amount of carbon monoxide (CO) in the raw emissions compared to gasoline. This tends to dampen the occurrence of the NO_x redox reaction on the catalyst surface:

$$2CO + 2NO \rightarrow 2CO_2 + N_2$$

Therefore the optimal lambda working point in CNG operation mode with activated closed-loop lambda control appears to be around 0.996, which is adjusted and controlled precisely by means of the postcatalyst oxygen sensor. An example of postcatalyst emissions at constant-speed driving in a CNG-powered passenger car is shown in Diagram 5.81 as function of air/fuel ratio.

As natural gas vehicles are usually based upon or derived from existing gasoline vehicles, the catalyst volume is mostly taken over from the corresponding gasoline engine as well. Therefore the space velocities in the catalyst, i.e. the normalized flow rates of the exhaust emissions along the catalyst surface, are normally comparable. The longer the emission stays in contact with the catalyst surface, i.e. the lower the space velocity, the better the efficiency of the catalyst conversion process. As the conversion rates with natural gas engines are shifted considerably towards higher temperatures, this is usually

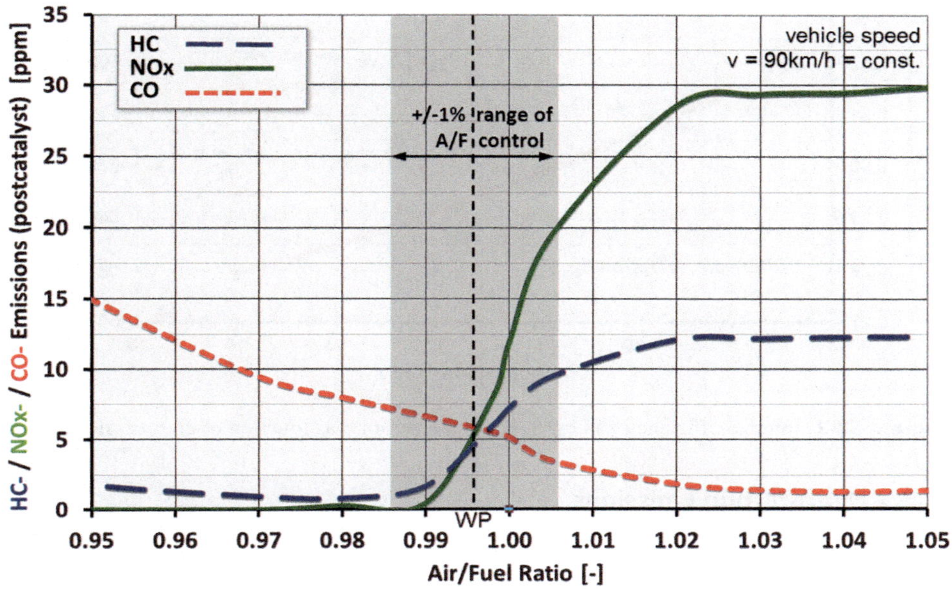

Diagram 5.81 Operation point of A/F control and measured emissions at CNG operation

compensated for by increased precious-metal loading of the catalyst with palladium or platinum respectively.

A frequently discussed general lean-mixture operation of natural-gas-powered engines in order to exploit an additional decrease in fuel consumption has turned out not to be beneficial. This is due to NO_x emissions and further reduction in exhaust gas temperatures. Thus further discussions in this chapter are related to stoichiometric mixture in combination with a TWC system exclusively.

5.2.7.2 Pollutants and Exhaust Temperature

The high knock resistance of methane with a RON of about 130, as described in Sect. 5.2.6, does enhance the possibilities to thermodynamically optimise the spark-ignited engine compared to gasoline, for example by means of *downsizing* with turbocharging and increased compression ratios.

Yet, an increase in internal combustion efficiency due to a higher compression ratio is in conflict *(trade-off)* with the requirements concerning exhaust emission aftertreatment. With regards to the HC and NO_x emissions and the exhaust gas temperature level required for a proper catalytic conversion of these pollutants, the challenges concerning emission aftertreatment clearly increase alongside a progressing compression ratio. An increase in compression ratio by 5 units, for example, corresponds to a decrease in exhaust gas temperature by about 70 K in the same engine operating point, as shown in Diagram 5.82.

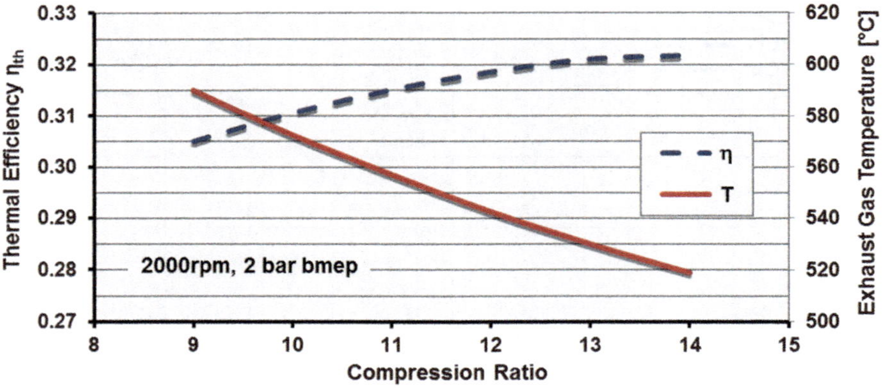

Diagram 5.82 Internal efficiency and exhaust gas temperature as function of compression ratio

5.2.7.3 Hydrocarbon Emissions

One of the main effects responsible for the generation of the hydrocarbon (HC) emissions is flame quenching on the cold walls of the combustion chamber and in the squish and piston crevice areas, therefore referred to as wall or crevice quenching. An increase in compression ratio is normally accompanied by a considerable rise in hydrocarbon emissions due to an unfavorable chamber geometry concerning the volume/surface ratio, the distance between flame initiation and piston crown, as well as bigger quenching areas.

In addition to these challenges arising from an emission point of view, the occurrence of irregular combustion phenomena, even with a highly knock resistant type of fuel, tends to increase significantly along with a rise in compression ratio. Correspondingly, higher required ignition voltages also cause an increase in electrode wear and a higher risk of shunt firing on the insulator of the spark plug. It can be stated that the geometrical compression ratio sensibly achievable with a spark-ignited combustion process appears to be limited to a maximum of about 13.5–14.

The efficiency-optimized early combustion timings adjustable with methane as fuel, even in combination with high compression ratios, naturally lead to an early end of combustion. Hence the burnt gas temperatures at the time of exhaust valve opening turn out to be relatively low, impeding a secondary reaction of incompletely burnt hydrocarbons during the subsequent exhaust phase, Diagram 5.83.

5.2.7.4 NO_x Emissions

Another harmful pollutant is nitrogen-oxide emission, normally summarized under the collective term NO_x, the generation of which is strongly affected by combustion temperature. NO_x formation, as described in the Zeldovich reaction process, is an endothermic oxidation of nitrogen, which can be described by this chemical equation:

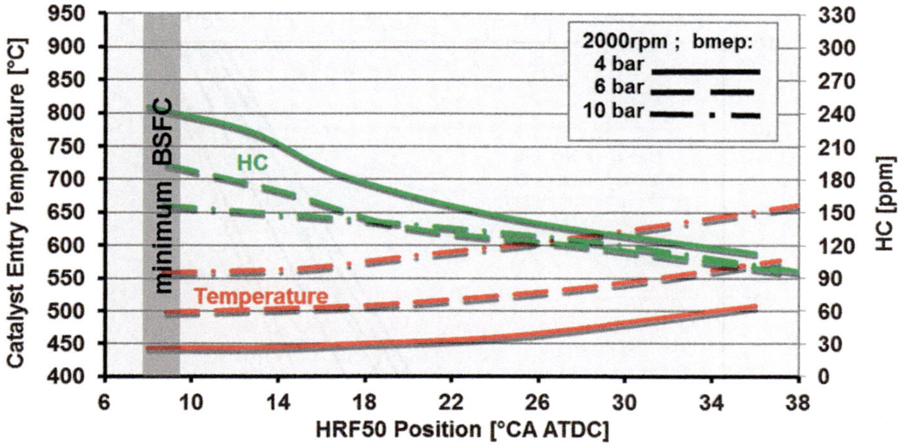

Diagram 5.83 Untreated HC emissions and exhaust temperature versus combustion timing

$$N_2 + O_2 \Leftrightarrow 2NO$$

This gross equation is the result of a more complex reaction mechanism based upon a series of individual elementary reactions. Besides nitrogen oxide, small quantities of nitrogen dioxide and nitrous oxide also happen to be generated in this process. As those quantities are rather negligible, however, the generation of nitrogen-oxide emissions might be described in a simplified way by the Zeldovich mechanism:

$$N_2 + O \Leftrightarrow NO + N$$
$$O_2 + N \Leftrightarrow NO + O$$

A thermal equilibrium in the concentration of nitrogen oxide, as shown in Diagram 5.84 in dependence of temperature and mixture ratio, is hardly reached in the combustion chamber, as the corresponding reaction kinetics happen rather slowly in relation to the quickly changing temperature, pressure and local mixture conditions. Note that the formation of nitrogen oxides can be reduced considerably by exhaust gas recirculation and lowered combustion peak temperatures accordingly.

The formation of nitrogen oxide emissions shows the tendency to increase with higher combustion temperatures generated by higher engine load, which is also reflected in exhaust temperatures. Analysing the behavior within the same engine operating point, the formation of NO_x is increased by earlier ignition timings that might be adjusted in order to optimise engine efficiency. This has an increasing effect on combustion temperature as well. The exhaust gas temperatures, however, are relatively low in this case, which is due to an extended expansion time until the beginning of the exhaust phase, as shown in Diagram 5.85.

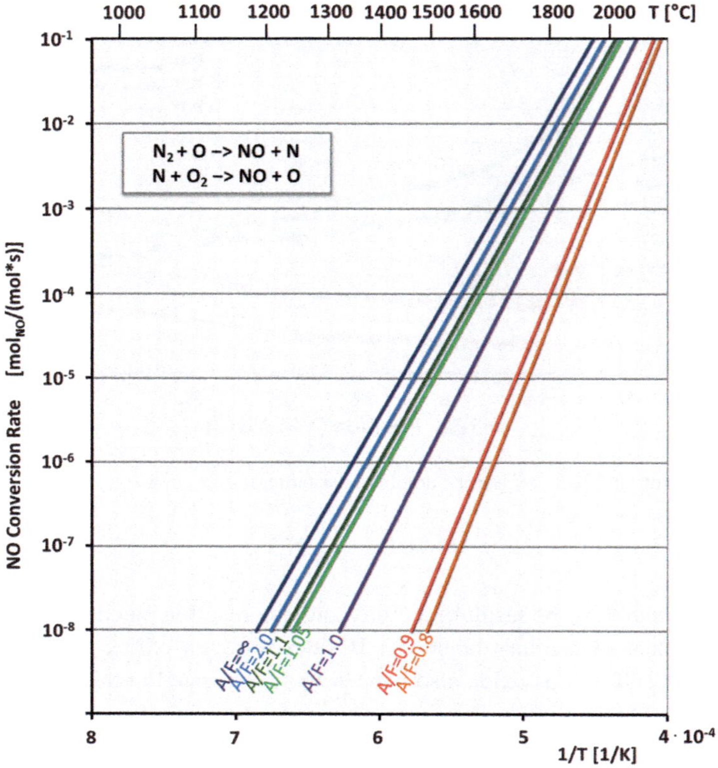

Diagram 5.84 NO formation rate during methane combustion by Zeldovich [78]

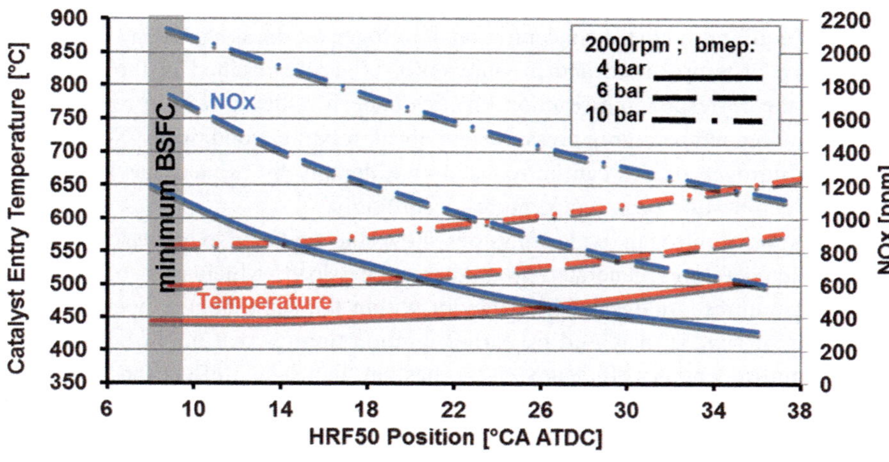

Diagram 5.85 Untreated NO_x emissions and exhaust temperature versus combustion timing

5.2.7.5 Exhaust Gas Temperature

Besides a given catalyst surface with precious-metal load as well as space velocity and lambda control respectively, the main parameter substantially defining the effectiveness of the exhaust-emission aftertreatment system is the exhaust gas temperature.

The aforementioned challenge arising from the chemically stable properties of the methane molecule and consequently the high temperatures required for a sufficient catalytic conversion is intensified further by an additional technological component: turbocharging.

Turbocharging has become an inherent part of recent gasoline engine development, which is characterized by a progressing degree of *downsizing*. Especially in case of CNG engines with external mixture formation (PFI), disadvantages in volumetric efficiency due to the partial pressure of gaseous fuel in the manifold and consequently reduced torque and power output of the engine can be quite easily compensated for by means of turbocharging. The secondary expansion of burnt gas in the exhaust turbine, however, tends to reduce the exhaust gas temperature by another 120–140 K, dependent on engine operating point.

It is worth mentioning, on the other hand, that the reduction in exhaust gas temperature has some advantages as well, as this might help operation of the engine without the need for mixture enrichment across the whole performance map. This is of particular importance especially in regards to the limited tolerance of methane combustion towards richer than stoichiometric mixtures and also concerning the lack of liquid fuel and its vaporization enthalpy typically used to reduce combustion temperatures.

5.2.7.6 Cold Start and Catalyst Heating

When the first Euro 4 natural gas vehicles were introduced on the market, the PFI nozzles used for CNG were derived from gasoline injectors with only marginal modifications. The increase in needle lift or nozzle cross-section, as required for sufficient gas flow rates, did have a negative effect on cold start performance of these injectors. The cold start therefore had to be done normally in gasoline mode with these CNG engines. This also allowed for an efficient catalyst heating by means of mixture enrichment in combination with a secondary air system. Once the required catalyst temperature of about 400 °C and the upper end of the dew point on the oxygen sensor—as required for closed-loop lambda control—were obtained, engine operation finally could be changed into CNG mode. At that time, gasoline mode was allowed until the fourth urban driving cycle within the relevant emission certification test (NEDC).

With introduction of the Euro 5 emission regulation, the allowed portion of gasoline operation within a NEDC test in CNG mode was restricted to a total of 60 s. By means of improved cold start performance of the CNG injectors, and with regards to quasi-monovalent CNG vehicle concepts with gasoline tank reservoirs smaller than 15 l, it became possible and necessary to perform the engine cold start and warm-up phases in CNG mode exclusively.

In order to comply with the more stringent Euro 6 emission regulation, it has furthermore become crucial to have the exhaust emission systems, consisting of catalyst as well as pre- and post-catalyst oxygen probes, ready for operation very quickly.

Downward pressure on engine manufacturing costs as well as the introduction and distribution of gasoline direct injection, allowing for efficient catalyst heating strategies without the need for secondary air, caused secondary air systems to disappear from the market. This evolution has been transferred also to the CNG variants derived from these gasoline engines accordingly.

Alternatively, lean-mixture operation during cold start and warm-up is commonly applied for natural gas engines. As the maximum rate of combustion is shifted towards leaner mixtures with methane compared to gasoline, CNG combustion can be run considerably more stable under lean mixture conditions (e.g. with an A/F ratio of about 1.07). This contributes significantly to reducing hydrocarbon emissions. As the fuel is injected in a gaseous state of aggregation, wetting of the cold combustion-chamber walls and therefore the generation of unburnt emission components is reduced substantially. Furthermore, the lower combustion temperatures cause NO_x raw emissions to drop as well.

In addition, engine warm-up is carried out with later ignition timings, reducing combustion temperatures further. The exhaust gas temperature, on the contrary, is considerably increased due to the late combustion timings—with the burn-out enduring into the exhaust phase. Hence the untreated exhaust emissions of the pollutants NO_x and HC are reduced and the catalyst is heated up more quickly (see Diagrams 5.83 and 5.85).

The higher proportion of water released in methane combustion relative to gasoline requires a robust oxygen sensor concept with sufficient protection against imminent damage from water droplets. Lambda probes with a coating on the sensitive heated zirconium oxide element—so-called *TSP* sensors—can be heated up quickly after engine cold start. Lambda probe operability, as required for closed-loop mixture control, might be achieved with this type of probe before the actual dew point stop on the sensor is obtained.

Pilot controlled mixture formation might therefore be abandoned prior to the second drive-off within the first urban cycle of the NEDC, in favor of closed-loop lambda controlled mixture preparation within the ideal lambda scope.

5.2.7.7 Particulate Emissions

With introduction of the Euro 6 emission regulation, an additional harmful substance is being limited, bringing the direct injection gasoline engine in line with diesel engines accordingly: the limitation of the particle number emissions. The associated permissible limit value for particle numbers will be tightened in two steps to 6×10^{11} particles per kilometre by 2017. Natural gas vehicles already comply with that limit today or actually tend to undershoot that number of particles by a considerable proportion, even though engines with external mixture formation are not required to comply with this regulation.

Normally the particle number emissions determined with natural gas engines are close to the limit of detection.

In a medium-term perspective, it is expected that CNG direct injection concepts will be introduced corresponding to direct injection of gasoline. Then natural gas vehicles will have to comply with the particle number regulation as well. Particulate exhaust emissions with CNG-DI concepts are expected to undershoot the permissible limit similarly to external mixture preparation.

The low level of particulate emissions is mostly based on the fact that no wall wetting in the combustion chamber occurs with CNG during engine start, warm-up or acceleration phases due to the gaseous state of aggregation. A subsequent vaporization of fuel from the chamber walls and consecutive formation of inhomogeneous mixture zones, which might become the source of particles, does not happen here as a matter of principle. Only a small contribution to particulate emissions is expected from the oil film on the chamber wall in this case.

Diagram 5.86 compares the particle number emissions of the Mercedes-Benz 'M270/274 Natural Gas' engines with piezoelectric gasoline direct injection and CNG port injection in both operation modes [77]. The highly dynamic piezoelectric injectors allow for multiple gasoline injections within a combustion cycle. In conjunction with an appropriate injection strategy, involving up to five injections per cycle, the number of particle emissions emitted in gasoline mode also remains considerably below the permissible limit. This is due to reduced spray-penetration depth and improved mixture formation resulting in a lower risk of wall wetting in the combustion chamber. The spray

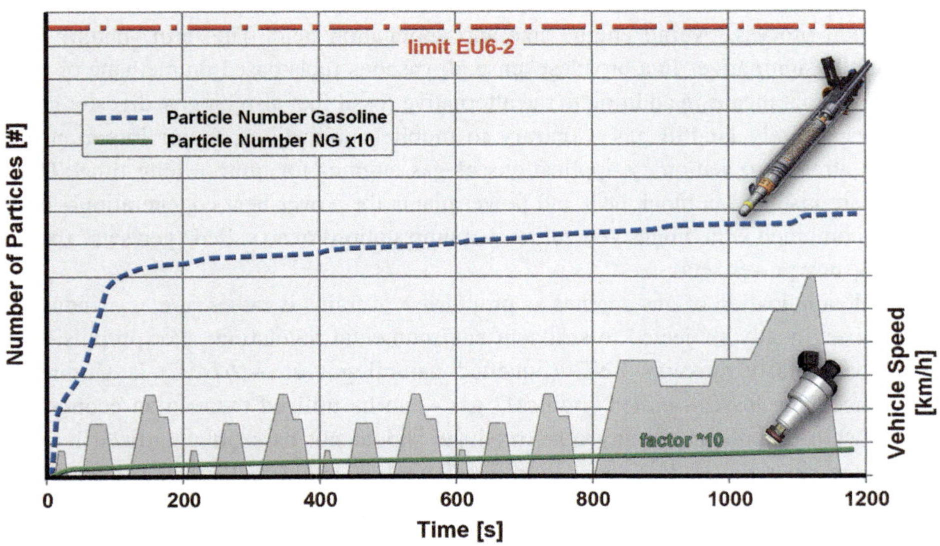

Diagram 5.86 Number of particulate emissions of the Mercedes-Benz 'M270/274 Natural Gas' engine in the NEDC [77]

kinetics, especially at the final injection just before spark timing, help increase the local turbulences and contribute to a good combustion and burn-out behaviour [79, 80].

In a nutshell, it can be stated that natural gas engines are characterized by considerable advantages in exhaust emissions, despite the difficult challenges arising from the special properties of methane concerning its effect on global warming *(GWP)*. In the *ADAC EcoTest*, for example, whose criteria also take CO_2 and pollutant emissions beyond standard certification cycles into consideration, natural gas vehicles are usually found among the top scorers in the ranking list. Especially by using methane generated from regenerative sources, the natural gas operated engine represents the combustion engine technology with the lowest level of exhaust emissions currently available. The introduction of CNG direct injection *(CNG-DI)*, which is still in development, will provide further advantages, for example concerning improved catalyst heating strategies with very late ignition timings or regarding the avoidance of methane slip under engine operation with valve-overlap conditions.

5.2.8 Thermal and Mechanical Load of Natural Gas Spark Ignition Engines[1]

Eduard Köhler

5.2.8.1 Stationary and Mobile Application Fields of Gas Engines

The specific layout of gas engines depends on the type and quality of the used gas, combustion process, overall engine size and application field. The term "natural gas" colloquially summarizes in a broadest sense all gaseous fuels based on methane or mine gas (CH_4). That means in addition to the alternative fossil fuel also biogas, digester gas or dump respectively landfill gas. Contrary to mobile applications, major importance is attached already to stationary applications of gas engines for quite a long time. Those engines are installed in block heat and power plants for power-heat cogeneration—often directly combined with biogas generation—, pump stations or so called "gen sets" (power or emergency power sets).

Mobile application of gas engines as propulsion of trains is rather rare, e.g. industrial locomotives locally subjected to stringent environmental regulations. For propulsion of ships—particularly tankers—LNG (liquefied natural gas at −167 °C) is gaining in importance. The losses—called "boil-off" gas—can be utilized there in an economical way for ship propulsion. In any case, methane should not pass off unutilized into the atmosphere since its GWP (Global Warming Potential) is unequally higher than that of carbon dioxide (CO_2).

[1] Eduard Köhler, Mosbach (KSPG AG, Neckarsulm, retired end of 2013).

For motor vehicles LNG seems to be an improper fuel due to its boil-off losses during longer downtimes. There is the consideration now nevertheless using LNG as alternative fuel for commercial truck transport on long distances because of its higher energy storage density compared to CNG.

For road traffic purposes CNG (compressed natural gas under a pressure of 200–250 bar) is preferred so far. The launch of CNG in the domain of passenger cars, trucks/commercial vehicles and busses only hesitantly went on over a long time. The automakers are now increasingly respecting however the potential of CNG within the scope of their strategies to reduce CO_2 fleet emission figures. The very favorable molecular weight ratio of carbon to hydrogen (C/H: 12:4·1 = 3:1) respectively CNG (in reality 3.7 to 4:1) results in a CO_2 emission reduction by about a quarter compared to gasoline, although there is an increased fuel consumption owed to the low fuel mass density.

Taking the liquid fuel LNG, a comparison gets rather easy if the efficiency and equivalent output of the engine are approximated. LNG's mass density in liquefied condition is distinctly lower compared to gasoline, but its net calorific value is higher. From this fact results an increased volumetric fuel consumption already mentioned. But the much more favorable CO_2 equivalent enables this alternative fuel gaining an advantage regarding the specific emission in question:

Fuel parameters	LNG	Gasoline RON 98	Volumetric fuel consumption ratio:
Net calorific value (MJ/kg)	50	41	$\frac{\dot{V}_{CNG}}{\dot{V}_B} = \frac{750 \times 41}{424 \times 50} = 1.45$
Mass density (liquid) ρ (kg/m³)	424	750	CO_2 benefit of CNG:
CO_2 equivalent (kg/l)	1.17	2.32	$\frac{CO_{2-CNG}}{CO_{2-B}} = 1.45\frac{1.17}{2.32} = 0.731$ respectively 27 %

Automakers offer now more and more motor vehicles dedicated to CNG, besides passenger cars for alternative operation also light commercial vehicles/trucks for distribution of goods and commuter busses/busses for public bus transport. Aspects of environmental protection, the possibility of particulate matter free driving in green environmental or exclusion zones of inner cities are prevailing over the reduced cruising range resulting from the lower energy storage density of natural gas. Increased engine efficiency, comparably low priced fuel and tax relief are incentives for German car drivers. The higher purchasing price is confronting a favorable pay-off period.

5.2.8.2 Passenger Car Engine Concepts for Natural Gas

According to state of the art, natural gas mass-production engines are predominantly applying the fuel mixture compressing principle. According to the classification, they are spark ignition gas engines (well defined from diesel gas engines and real gas diesel engines [81]). There are two fundamentally different conceptions of combustion, which can be distinguished by the combustion-air ratio λ:

- Engines operated with stoichiometric fuel rate ($\lambda = 1$) equipped with three-way catalyst.
- Lean-burn engines ($1.66 \leq \lambda \leq 1.74$) showing low raw exhaust gas emission within the combustion-air ratio in question.

With increasing excess air the laminar flame speed of natural gas is distinctly higher than that of gasoline. This qualifies natural gas for lean-burn combustion processes. The output deficit can be compensated by turbo- or supercharging. Passenger car engines are operated using a stoichiometric combustion-air ratio allowing the application of a regular three-way catalyst including lambda (oxygen) control. On the other hand, the operation of biogas engines in block heat and power plants with catalysts is nearly impossible without purification of the gaseous fuel. After all, there are traces of gas like hydrogen sulfide (H_2S) and siloxanes (organic silicon compounds building solid deposits (silicon dioxide: sand) during combustion) which are tremendously affecting the service live and thus making the catalyst operation uneconomic. For that reason these gas engines are operated by feeding a lean mixture. The excess of air decreases the combustion respectively exhaust gas temperature in a way that the emission limit values e.g. of the German regulations "TA Luft" (Bundes-Immissionsschutzgesetz: Technische Anleitung zur Reinhaltung der Luft—TA Luft) can be met.

The spark-ignited gas engine characterizes control of quantity of mixture, external fuel-mixture generation and premixed combustion. The gas injection pressure has to be higher than the intake respectively charge air pressure. As in case of a stoichiometrically operated engine a homogeneous mixture of gas is generated and ignited by spark plug ("direct ignition" to be distinguished from pre-chamber ignition at lean operation) comparable to a conventional spark-ignited engine. Also regarding spark-ignited gas engines there are intensive efforts to benefit from the advantages of modern fuel direct injection technology in soon future.

5.2.8.3 Influence of Fuel Quality

Natural gas is available in fluctuating quality as H ("high") and L ("low") gas. In Germany the "DVGW-Arbeitsblatt G260" (Deutscher Verein des Gas- und Wasserfaches e.V.) is mandatory. CNG provided by the not area-wide gasoline station network has to meet in addition the German industrial standard DIN 61425 (existing since 2009). The adjustment of differences in fuel quality and consequently varying operating conditions of the engine accomplishes the electronic control unit (ECU) of the engine by means of signals from the lambda oxygen sensor and knock sensor system, thus changing the injection time of the gas injection valves and ignition timing.

The composition of natural gas is different by region. The admixture of ethane, propane, butane and ethene by the gas suppliers for purpose of control of the calorific value of the gas-air mixture respectively Wobbe index (characteristic value of thermal load of gas burners) influences the anti-knock properties. These hydrocarbons on the one hand are increasing the calorific value of the gas, on the other hand they are decreasing the methane number—the indicator of knock resistance of the gas comparable to the octane number of gasoline—at the same time. For the combustion in the engine the methane number is more important than the real content of methane or the calorific value. This number directly influences the ignition timing as well as the boost pressure in case of super- or turbocharged engines and thus the engine output and efficiency.

Regenerative biogas produced in biogas plants—a mixture consisting for the most part of methane but also carbon dioxide, water vapor and nitrogen as well as small amounts of hydrogen sulfide, halogens and siloxanes—is increasingly gaining importance. The big gas engines e.g. of block heat and power plants are specifically designed for this purpose. Feeding in the gas supply system however necessitates the purification in sour gas treatment plants. For that reason the particular requirements of unpurified (low grade) biogas are of no relevance for passenger car engines. "E-Gas" produced by regenerative electricity, e.g. an approach pursued by the automaker AUDI, is in contrast pure synthetic methane. This is optimally suited for CNG for passenger car engines particularly designed for this gaseous fuel.

5.2.8.4 "Bi-Fuel" ("Bivalent") and "Mono-Fuel" ("Monovalent") Engine Operation

Passenger cars to be operated with CNG are currently "bi-fuel" cars to a large content with respect to the still scarce CNG infrastructure of gas stations. Instead of "bi-fuel" there is the term "bivalent" too. "Bi-fuel" has to be distinguished from "dual fuel" which refers to diesel gas engines ignited by a small amount of pilot diesel fuel. The quantity of diesel fuel could be increased there however to 100 % respectively mere diesel operation. In case of "bi-fuel" the switch over between CNG and gasoline is done by a switch actuator or automatically. But the basic layout for gasoline operating mode impedes the full utilization of the potential of CNG. The compression ratio has to be aligned with the lower knock resistance of gasoline. A variable compression ratio would be appropriate [82], but the launch of series production of this technology is not foreseeable. The lower calorific value of the gas-air mixture by 8 % (H gas) has to be respected too in addition. The unfavorably changing conditions of gas operation mode are generally increasing the thermal and mechanical load of the engine. The fact that those engines hardly show an output deficit in CNG operating mode today is complicating the situation.

In contrast, "monovalent" vehicles (also called "mono-fuel") are to a large extent to completely optimized for the natural gas operating mode, and they are capable to scoop out better or entirely the available potential. In case of a so-called "quasi monovalent" CNG operating mode passenger cars have an optional small gasoline tank with a volume of less than 15 l. Under the directive of current EU regulations they are fiscally classified as "monovalent" natural gas vehicles. Then regarding motor vehicle tax they are assessed according to the lower CO_2 emissions of the CNG operating mode. Spark ignition passenger car engines only dedicated to CNG are still under development.

5.2.8.5 Subsequent Natural Gas Retrofitting of Passenger Series Production Engines

Compared to LPG (liquefied petroleum gas, a mixture of propane and butane) the subsequent CNG retrofitting of series production passenger cars is of distinctly minor importance until now. In any case "retrofitting" is not the correct term, because it's actually a matter of enhancement for "bivalent" operating by conversion kits available for some spark ignition engines. The subsequent conversion of passenger car diesel engines into "bi-fuel" spark ignition engines is not relevant with respect to the comprehensive technical interventions concerning the engine. The usual modifications concern

- On the side of the vehicle an additional compressed-gas tank with shut-off and tank as well as back-pressure valves,
- A peripheral engine equipment like gas pressure regulator, in case of central fuel-mixture generation gas mixer (e.g. venturi injector), in case of single fuel-mixture generation (port gas injection) gas distributer, intake manifold with gas injectors and
- An adaption of electronic engine control unit (ECU). Besides this, pressure and temperature sensors are also required.

An optional engine intervention is usually restricted to the exchange of the valve seat rings. Such subsequently retrofitted engines are otherwise hardly prepared in particular for CNG operating mode. They are designed for gasoline operation and depend for that reason on the cooling effect and lubricating properties of the liquid fuel. Defects of in general not correspondingly adapted mechanical components therefore cannot be definitely excluded. In spite of careful workmanship by certified retrofitting car repair shops a subsequently retrofitted passenger car can turn out to be more susceptible to failure. In practice once in a while there occur problems with the gas pressure regulator as well as switching-over between gasoline and CNG. The vehicle dynamics are adversely affected by the loss of power and excess weight of the pressure tanks. Natural gas passenger cars therefore have the reputation to be particularly environmentally beneficial in fact and cost-efficient but a little "lazy".

5.2.8.6 Changed Operating Conditions of Natural Gas

Due to the very high knocking resistance natural gas is particularly applicable for the combustion in the spark ignition engine. The knock resistance corresponds to an octane number (ROZ: research octane number) of 120–135 (depending on the composition of the gas) compared to 95 respectively 98 (100) ROZ of unleaded premium gasoline. According to that fact there are good preconditions for

- Distinctly raising the compression ratio more or less and
- Operating the engine at full load with optimum efficiency to a large extent.

The latter requires the adaption of the ignition map to the extended possibilities of the alternative CNG operating mode. The gas quality possibly noticeably varying from gas station to gas station has to be definitely recognized by the engine sensor system to avoid engine damage. Exploitation of the potentials inevitably means that the engine is exposed to higher load. On top of that the gaseous state of the fuel, as it will be explained more detailed in the next chapter, has an unfavorable effect on

- Temperature and
- Wear of components.

Here—due to "dry" combustion—as already mentioned above, in particular

- Missing cooling and
- Lubricating properties

of the gaseous fuel become additionally noticeable. Overheating and shortcoming of lubrication can cause engine damage if there are no engine-engineering precautions. Tightened operating conditions also result from the fact that "bi-fuel" engines cannot abstain from downsizing trend too.

5.2.8.7 Increased Thermal and Mechanical Load Resulting from Tightened Operating Conditions

Thermal Load

In case of "monovalent" natural gas operating the compression ratio of naturally aspirated engines can be raised to about $\varepsilon = 13.5$, that of turbo- or supercharged engines to about $\varepsilon = 12$. For biogas operating—not relevant for passenger cars—even $\varepsilon = 15$ is possible. This ability is based on the high amount of inert carbon dioxide (CO_2) of biogas. It was mentioned already that there are rarely (status 06/2014) passenger cars only for "monovalent" CNG operating on the market at present. Therefore there are really found compression ratios in the range of $\varepsilon = 9.6$–11.0. It's known furthermore that in the advanced development divisions of the car makers also compression ratios of about $\varepsilon = 12.0$ for turbo- respectively supercharged engines are tested.

By reason of the high knock resistance a natural gas engine shows the potential also to be stoichiometrically operated under full load condition at or near the optimum crank angle for the center of combustion mass. Thus the optimum efficiency can be kept beyond the medium rpm range and where appropriate up to the rated speed with a favorable center of combustion mass at 8–10° crank angle after firing top dead center, as this is shown in Diagram 5.87 [83, 84].

With increasing efficiency the exhaust gas temperature is decreasing thereby due to the larger expansion of the combustion gases in the cylinder. For purpose of protection of the components there is in general a temperature limit of 850–950 °C. In case of turbocharged

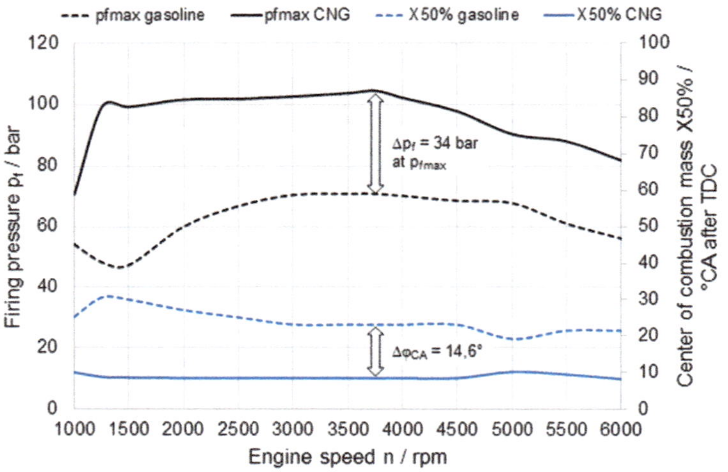

Diagram 5.87 Firing pressure p_f and center of combustion mass X50 % at wide-open throttle depending on engine speed n and both gasoline and CNG operating mode of an inline 4 spark ignition engine (compression ratio $\varepsilon = 11.0$, turbocharged engine) (data from [83])

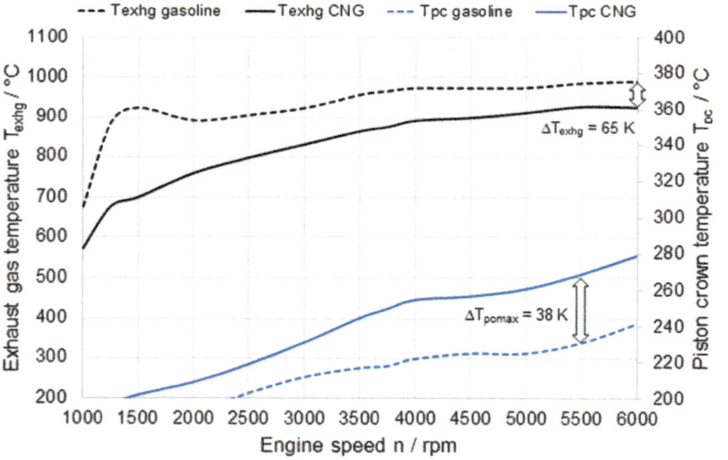

Diagram 5.88 Exhaust gas T_{exhg} and piston crown temperature T_{pc} at wide-open throttle depending on engine speed n and both gasoline and CNG operating mode of an inline 4 spark ignition engine (compression ratio $\varepsilon = 11.0$, turbocharged engine) (data from [83])

spark ignition engines the exhaust gas temperature can rise however over 1000 °C. An efficiency optimized CNG operating decreases the exhaust gas temperature compared to gasoline operating, as this is shown in Diagram 5.88 [83] for example. The potential of reduction of exhaust gas temperature can be specified with about 150 K [84]. In the case in question at full load and low rpm more than 200 K are achieved. At medium rpm the reduction of temperature drops down to about 100 K. At high rpm there remain 65 K anyway, whereupon at gasoline operating mode the temperature reducing wide-open-throttle enrichment comes into effect. On the other hand this is causing the considerable excess of fuel consumption.

The combustion temperature depends besides on the load also on the instant of time of the energy conversion. It is in principle increasing with the spark advance—acceleration of energy conversion—and leaning of the mixture till the knock limit. CNG distinguishes regarding its inflammation and burns, compared to gasoline at comparable engine operating condition, slower by trend [84]. Hence the maximum combustion temperature is first of all lower too. For gasoline 2000 (until 2500) °C are according to textbook. The reference value of CNG is about 1900 °C. On the other hand this value is more or less increasing due to tightened CNG engine operating parameters. For the upper limit there no reliable figures are available.

Thermal problems do not only result from a higher combustion temperature, rather from missing internal cooling. The thermal load in the combustion chamber is higher since an already gaseous fuel cannot achieve a noteworthy cooling effect as this is the case for gaseous fuel by abstraction of the enthalpy of evaporation. Thus the wide-open-throttle enrichment applied for gasoline operation at higher engine speed for purpose of protection of components is rather ineffective. This concerns in particular the intake ports in the cylinder head, the inlet valves, the valve seat rings, the pistons, the cylinder walls, but finally the total heat balance of the engine.

Diagram 5.88 shows amongst others, concerning the increase of component temperature of CNG operating mode at full load, an exemplary comparison of piston temperature [83]. The temperatures are relating to the "piston crown" overall. That the reason why the exact location can neither be localized nor the temperature level figures can be consequently rated. Significant however is the CNG related raise of temperature, at low rpm of about 20 K, at higher rpm up to 38 K. In contrast, the maximum piston temperatures measured by means of a different high performance engine are enabling a differentiation concerning piston combustion bowl, 1st piston ring groove and piston pin boss, Diagram 5.89. CNG operating mode is causing a raise of temperature also here, whereupon the local temperature limits are reached respectively exceeded. The available

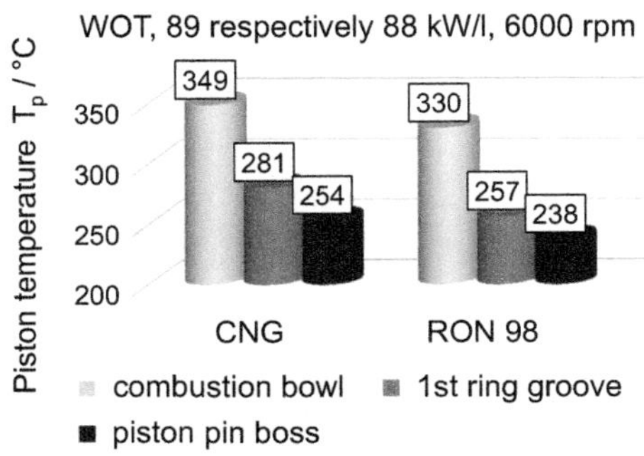

Diagram 5.89 Maximum piston temperatures T_p at both gasoline and CNG operating mode by way of example of a spark ignition engine with small swept volume and high power density (data: KS Kolbenschmidt GmbH, D-74172 Neckarsulm)

examples verify the CNG related higher thermal load on the components with engine specific differences as respected.

Mechanical Load

The mechanical load of the engine is determined by the level of the firing pressure (maximum cylinder pressure) and the variation in time of the combustion. The more fuel energy is converted near firing top dead center, the higher is the firing pressure, so much more rampant is the raise of pressure (pressure gradient $dp_{cyl}/d\alpha$) and harder the combustion noise too. Exemplary pressure histories for CNG engine operating mode can be found e.g. at [85]. For the crank drive—pistons, con rods, crankshaft and the associated plain bearings—the firing pressure is a primary layout characteristic.

In case of subsequent retrofitting the engine generally gets no raise of compression ratio. Without adaption of the ignition map the firing pressure would drop at CNG operating mode. Depending on spark advance and increase of compression ratio the firing pressure is more or less intensely rising, as this is illustrated by a schematic draft in Diagram 5.90. In case of modern "bi-fuel" engines the compression ratio is individually aligned with the maximum tolerable value for gasoline operation. Concerning the turbocharged engine (base engine with supercharger) described at [83] this is raised from $\varepsilon = 9.8$ auf 11. At a comparable operating point CNG increases the firing pressure

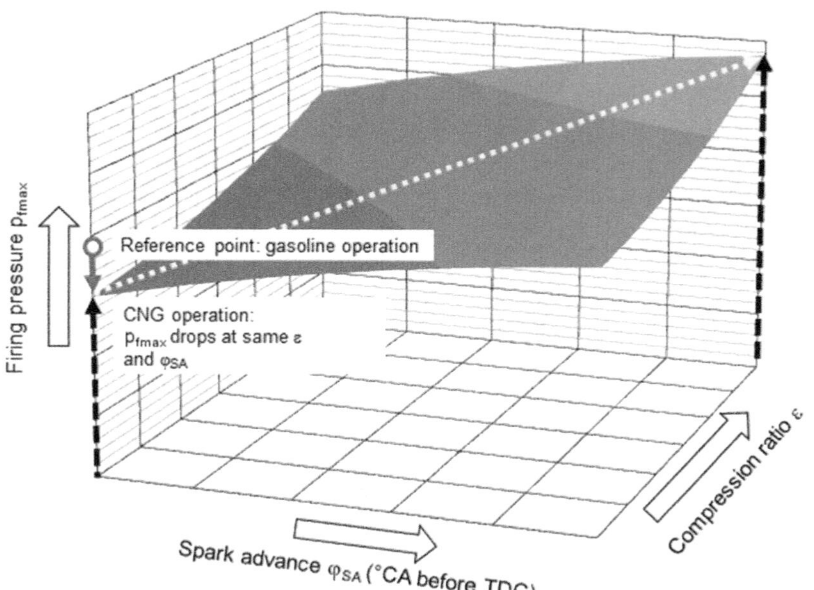

Diagram 5.90 Schematic draft of the effects of a CNG specific adaption of compression ratio ε and spark advance φ_{sa} (optimization of the center of combustion mass) on firing pressure p_{fmax} (reference point: firing pressure at gasoline operation of base engine)

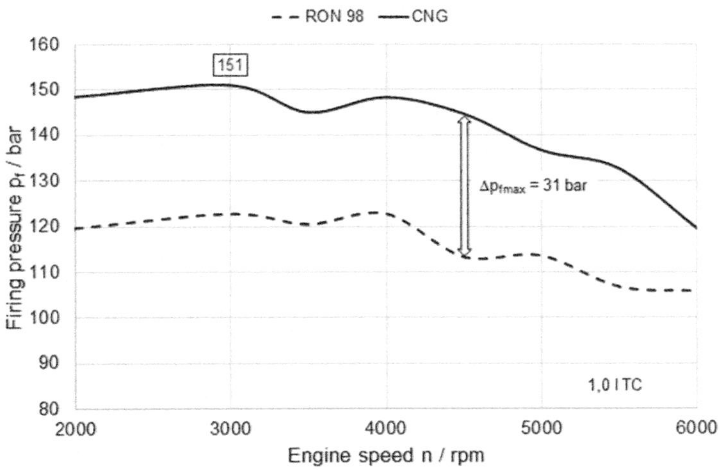

Diagram 5.91 Firing pressure p_f at gasoline and CNG operating mode depending on engine speed n by way of example of a spark ignition engine with small swept volume and high power density (data: KS Kolbenschmidt GmbH, D-74172 Neckarsulm)

due to the spark advance for an efficiency optimized center of combustion mass e.g. by about 30 % respectively 34 bar compared to gasoline, Diagram 5.87. Regarding the current "bi-fuel" engine generation there is evidence of a tendency limiting the firing pressures to 100–120 bar, because the pistons of many conventional spark ignition engines may currently reach its maximum loading at about 100 bar.

In case of high density of engine power CNG is reaching soon a firing pressure level which was still subject to passenger car DI diesel engines a few years ago, Diagram 5.91. The small turbocharged spark ignition engine with a moderate compression ratio of $\varepsilon = 10$ (not verifiable description) still under advanced development can equalize the specific output of 88 kW/l in CNG operating mode. The firing pressure is rising to 151 bar thereby. The firing pressure increase is 31 bar. Regarding another, in this case supercharged test engine [86] with a compression ratio of $\varepsilon = 12.5$ the firing pressure is limited to 140 bar. A firing pressure increase beyond this level seems not to be planned for the next engine generation.

Higher component temperatures and intensified mechanical load on components don't cause the evidently increased wear of some components on its own. Missing hydraulic properties of liquid fuels are also of major importance. The wetting with finely distributed fuel droplets is e.g. capable decisively damping the impact of the valve heads on the valve seat rings, the "Achilles heel" of gas engines. In a natural gas engine there occurs increased mixed and boundary (nearly dry) friction. Gasoline is characterized by low viscosity in fact, nevertheless this still contributes to lubrication. Lubricating and detergent/dispersant additives cannot be added to gaseous fuels. Detergents prevent respectively reduce deposits residues from combustion in the intake ports, in the valve and combustion chamber area and on the spark plugs. Gas engines in principle are tending to stronger deposits and thus increasing risk of coking. This can cause a raise of temperature, finally uncontrolled combustion in terms of knocking and glow ignition and thus engine damage. There is an option however to admix dosed liquid additives to the intake air.

The engine lubricating oil is also subject to unfavorable influence of gas operation mode. This is thermally higher loaded and requires a better thermal stability for that reason. In any case it is in direct contact with the blow-by gases containing acid and acidiferous substances. Lube oil is progressively ageing by oxidation, nitration und sulfation. The latter are chemical reactions of lube oil with nitrogen oxides respectively certainly negligible sulfur in case of CNG. Oxidation respectively nitration (as well as sulfation in case of sour gas) cause thereby also lube oil acidification with corresponding corrosive potential. Corrosion due to lube oil acidification can concern the whole oil-circuit system even including the disintegration of paper filters. Finally the increased condensate formation has to be respected. Natural gas contains more hydrogen than gasoline. Thus proportionally more water vapor is arising from combustion process and condensing. The unavoidable "blending" with lube oil is favoring the internal engine corrosion.

According to [87] nitration occurs in the lube oil of gas engines—indeed a broad interpretable term—and contributes to an increase of viscosity. In case of natural gas operating mode no lube oil dilution by fuel input is happening. In case of gasoline direct injection engines this has to be considered as an annoying evil in fact, but in this way it has a compensating effect. Burning lube oil also tends to the formation of ashes depending on the specific additivation and decisively contributes to a deposit and thus coking problem.

Accelerated ageing of lube oil is shortening the service interval, a competition criterion and for the acceptance of CNG not without meaning. An overall quantification of the practical effects of thermal destabilization and corrosive potential in particular by CNG is difficult however. There are no actual statements of the carmakers available. A guideline for the oil service interval could be 15,000 km (with reference to [88]) respectively 1 year. Hence this is reduced by a half.

5.2.8.8 Layout of Passenger Car Spark Ignition Engines for Natural Gas Operating Mode in Accordance with Corresponding High Load

Just the natural gas propulsion technology can be characterized by a moderate technical input compared to other alternatives. The stoichiometric CNG operating mode of passenger car spark ignition engines requires no fundamental conceptual modifications, however an engine-engineering upgrade including interventions into engine mechanics concerning the well known weak points [82, 89]. The higher thermal and mechanical load is tightening the requirements concerning adequate endurance strength, the more so, as there are rarely deductions with respect to the density of engine power in the meantime. Correspondingly high are the development efforts for a CNG engine version, although this will be derived from a conventional spark ignition engine for economic reasons.

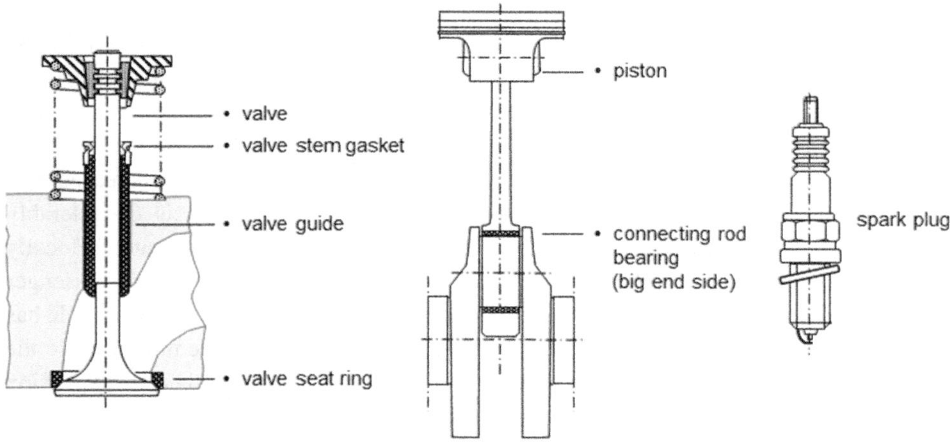

Diagram 5.92 Engine components primarily affected by the CNG related upgrade and to be adapted to the higher thermal and mechanical load respectively specific requirements of the gaseous fuel

Diagram 5.92 shows engine components anyway concerned. The carmakers meanwhile see a big challenge developing spark ignition engines with optimum efficiency by exploitation of the CNG potentials. The requirement of necessary as well as optional adaption measures will be highlighted in the following sections.

Engine Castings

(a) Cylinder head

The adaption of the compression ratio is shifted to the comparably low cost piston as far as possible. One reason also to review the design of the combustion chamber and the intake ports in the cylinder head is the CNG specific optimization of fuel-mixture formation and combustion. Even at high gas speed on the gas injectors there is only a moderate penetration of the intake flow due to the low mass density of the gas. The fuel-mixture generation therefore needs more charge motion compared to a liquid fuel. The CFD flow simulation is an important development tool in an early stage of concept to guarantee the fuel-mixture generation and equal charge of all cylinders. The four valves technology in combination with a "pent roof" combustion chamber geometry, typical for spark ignition engines, and a diesel typical shape of the intake ports supports a fast and as complete as possible combustion. That way e.g. [90] reports about a test engine with intake ports designed as swirl and charging ducts. The purpose is "the increase of turbulence in the combustion chamber in case of reduced stability of combustion due to high degree of EGR for reduction of nitrogen oxide at optimum engine efficiency. The swirl of the intake flow is intensified thereby at low rpm and switched-off charging duct in a way well known from diesel engines.

For reasons of thermal and with respect to high cylinder pressure of diesel engines in any way high mechanical load, CNG does not force to call off the substitution by nearly 100 % of grey cast iron (GJL) by aluminum. Depending on the type of the aluminum alloy however there occurs already a loss of hardness and strength in a temperature range of 160–200 °C increasing with runtime [91]. Diagram 5.93 shows by example of the Al alloy $AlSi_6Cu_4$ how the tensile strength initially increased by a T6 heat treatment (solution heat treatment, quenching, ageing) is considerably reduced after an ageing process over 500 h at 250–300 °C, temperatures locally attained or exceeded in the combustion chamber of today's spark ignition passenger car engines. At CNG operating mode a local temperature increase of about 30 K has to be anticipated without instantly concluding necessary corrective measures. In the shrink fit zone around the valve seat rings their comparably small thermal expansion and the favorable heat conduction of aluminum—contrary to grey cast iron—in principal have an relieving effect. There are verbal statements however concerning damage of the aluminum material around the valve ring seat because those became too hot.

In case of need the available repertoire of design measures to improve the cylinder head cooling—from optimization of the areas streamed by the coolant (size of heat exchange surface, wall thicknesses, flow cross sections/local flow speeds) over increase of the coolant flow rate to a different flow through concept (e.g. cross flow cooling [83])—is sufficient. Regarding water jacket sand cores of cylinder heads however the conflict of goals between cooling and manufacturing technology aspects (avoidance of scrap by broken sand cores) is getting worse.

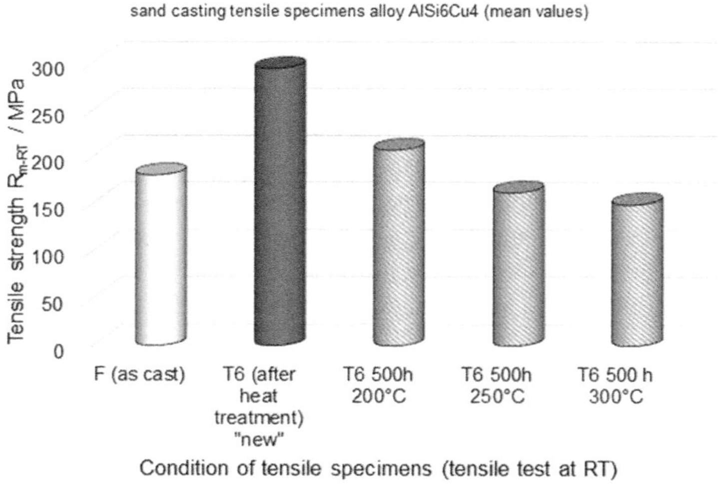

Diagram 5.93 Loss of strength of aluminum alloys due to long-term thermal load by way of example of the alloy $AlSi_6Cu_4$ (tensile strength $R_{m\text{-}RT}$ at room temperature of sand casting specimens in as cast condition, after heat treatment and after artificial ageing at 200, 250 and 300 °C over 500 h [91])

In case of high thermal load more expensive warm hardening AlSiMg primary alloys [e.g. $AlSi_7Mg$ respectively $AlSi_{10}Mg(Cu)$] [distinguished from cost-saving AlSiCu secondary alloys (e.g. $AlSi_6Cu_4$)] are applied. They are preferred due to their all together more beneficially assessed material profile. The choice of a cylinder head alloy including heat treatment and options of beneficial influence on microstructure (e.g. microstructure modification) however always mean a compromise regarding high static as well as dynamic high temperature strength also after long runtime, low thermal fatigue, high ductility, very good thermal conductivity, but also good castability and only slight hot tearing tendency [92]. Besides a design to withstand high load respectively corresponding stresses and a proper material, the ingate and feeder as well as densener/mold chilling technologies are essentially contributing to good performance characteristics. The chilling of the mold on the fired side of the cylinder head effects a very fine microstructure with a DAS (SDAS (secondary) dendrite arm spacing) of about 20 μm. This is reducing the increased risk of thermo-mechanical fatigue (TMF) with cracks occurring in the land between the valves respectively those and the spark plug thread respectively—as the case may be—the fuel injector boreholes.

(b) Engine block

Regarding engine blocks of passenger car spark ignition engines the rate of substitution of grey cast iron by aluminum is about 50 %. In addition, here are some aluminum engine blocks of passenger car diesel engines (e.g. BMW, Volvo). Since these diesel engines are withstanding a firing pressure load of at least 180 bar, the question of the appropriate material may rarely be at issue in context to CNG too. Also rough-textured distinguishing features like closed- or regarding manufacturing advantageous open-deck design may be of minor importance, the more as e.g. the Volvo passenger car diesel engines are representing the diesel untypical open-deck design. This issue is anyway a matter of controversial debate among experts.

If the engine block of the base engine should turn out as insufficient concerning strength and rigidity, there is—comparable to the cylinder head—a repertoire of design, material and casting process related measures available. A conservative dimensioning, increase of the land between the cylinder bores, may prohibit itself, because this would break open the "common part philosophy". A more structural rigid two-parts design—an engine block upper part and a bedplate bolted together with this—can be considered in contrast, if it is compatible with the existing production facilities. The cooling of the land between the cylinder bores can be improved too, but this could be a challenge considering the small land of "Siamese cylinder bores". Regarding cell sections, bracing ribs and camber the particular restrictions of the casting process have to be respected. For purpose of lightweight design considerable gains of rigidity have to be combined with only a few extra weight. Sometimes a spark ignition engine has a diesel engine companion. In this case there it's worth to check whether it would make more sense to build the CNG version on the basis of the diesel engine block.

For engine blocks of "bi-fuel" engines a material with higher strength—AlSiMg primary alloys instead of AlSiCu secondary alloys or grey cast iron of superior grade (GJV)—may not be required in general. Regarding some Al engine blocks the full material strength potential is barely tapped. Here a strength increasing T6 heat treatment can be applied. In contrast, cost-saving aluminum high pressure die casting (HPDC) is rather limited regarding heat treatment with respect to casting process implications (included gas pores under very high injection pressure damaging the casting under high solution heat treatment temperature).

The current limited experiences make it difficult finally assessing the suitability of different cylinder surface technologies for CNG. In case of grey cast iron no real problems have become known. This statement also covers increased wear, e.g. caused by corrosion. Regarding aluminum cylinder surfaces containing a high amount of silicon there is probably as good as no CNG experience available, since these are mostly found in engines installed in passenger premium or sports cars. There is no knowledge about the subsequent retrofitting of those engines. The galvanic nickel-silicon carbide dispersion coating, well-proven at racing and motorbike engines, could not succeed at passenger car series production engines. It has turned out that they cannot be applied worldwide by reason of highly sulfur containing gasoline in some regions of the world and connected corrosion phenomena. The novel thermal spray coatings based on differently alloyed carbon steel (process denominations APS: atmospheric plasma spray, wire arc spray (LDS in German), PTWA: plasma transferred wire arc) are just recently finding a more broadly based application.

(c) **Cylinder head gasket**

State of the art as generally known is the multilayer steel gasket (MLS). This technology is characterized by operational reliability, that means advantages concerning locally increased contact pressure per unit area (in this specific case this means load per unit length), reduction of bolt forces and thereby caused cylinder distortion, reduced so called "head lift" as well as even more equally distributed contact pressure. Higher firing pressure at CNG operating mode is equivalent to larger sealing joint oscillation ("head lift"), the gasket has to be fit in case of need. This is accomplished by a specific design of the "stopper" increasing the contact pressure around the cylinder bore and influencing the "head lift" as well as adding at least one further functional layer. The gasket manufacturers are following up mostly different design concepts. Those gaskets can actually have up to five layers (as a sum of crimped functional layer and not crimped distance layers). The particular composition of the gasket complies with the pressure level, the rigidity conditions and other specific circumstances of the particular engine. Due to the very high firing pressure level of diesel engines there are proper gasket designs principally available, whereupon the today's possibility of integration of "head lift" and temperature sensors into the gasket are of interest just for gas engines.

Powertrain

(a) Piston

The piston is a central component concerned by natural gas operation mode. The raise of compression ratio requires design measures, since the reduced compression volume falls partly or completely on the piston. Aside from the valve pockets in the piston crown typical for four valve engines, in case of port fuel injection the pistons show a flat piston crown or a shallow combustion bowl with relatively big diameter. The reduction of respectively omitting the combustion bowl comprises a potential for raising the compression ratio, concerning the exemplary pistons in Diagram 5.94 e.g. from $\varepsilon = 9.8$ of the base engine to 11.0. The shape of the combustion bowl of the "bi-fuel" engine in question was adapted by reducing its diameter and depth. If this is not sufficient, a slightly projecting piston crown within the spherical combustion cap in the cylinder head is the simplest but—due to the disadvantageous shape of the combustion chamber—not the best measure with respect to combustion.

Port fuel (gas) injection (PFI) of "bi-fuel" engines is increasingly combined with gasoline direct injection (GDI). Then for the shape of the piston crown—depending on operating with homogeneous fuel mixture or stratified fuel-air charge (FSI), in the latter case also on the requirements of the specific combustion process—an individual compromise has to be found. Regarding spark ignition gas engines derived from diesel engines—certainly rarely relevant for passenger car engines—for purpose of reduction of the compression ratio the piston combustion bowl has to be increased and shaped like a tub. This could have a negative impact on the piston compression height.

Diagram 5.94 Adaption of the combustion bowl in the piston crown for raise of combustion ratio from $\varepsilon = 9.8$ to 11.0 for "bi-fuel" operating mode by way of example of the pistons of the inline 4 spark ignition engine Mercedes Benz M270 (data: KS Kolbenschmidt GmbH, D-74172 Neckarsulm)

To support the fuel-mixture formation by charge motion, a sufficiently large squish area is provided generating an inward directed squish flow when the piston is approaching the top dead center. Furthermore a consequent reduction of the clearance (dead space), in particular a top land of small height, is recommended. But this corresponds to the state of the art of up-to-date spark ignition engines anyway. This is accelerating the combustion, contributing to a complete combustion of the gas-air mixture and reducing the fuel consumption as well pollutant emission.

The current piston design of spark ignition engines [93] is based on a more compact, more stability giving, stress minimizing layout and high-temperature resistant piston alloys (e.g. KS 309, MAHLE M142, M145). There is a very good compromise regarding low mass, rigidity, shape related fatigue strength and unremarkable piston noise. This design shows sufficient potential, at least for current "bi-fuel" engines and those still under development. The piston skirt is asymmetrically shaped and has a low friction coating. The piston skirt side walls have a convex shape, are inclined and supporting the piston crown by ribs on the outside. Thus a better bracing by a reduced piston pin length can be achieved. Low piston compression height, undercut ring land and reduced wall thicknesses are backing the lightweight design concept. CNG specific reinforcement can increase the total piston mass by 5–10 % [89]. Thereby it has to be respected that the increased mass in the piston crown area can only be insufficiently compensated by the adaption of the inner piston contour.

The fatigue strength of the piston is based on a sufficiently rigid and strong piston pin in a bearing piston pin boss. A "swimming" piston pin support (to be distinguished from a piston pin clamped in the con rods' small eye) is imperatively recommended. If the firing pressure is increased, roughly estimated characteristic numbers for the piston pin boss and pin are indicating the amount of rework, e.g. increasing the piston pin length [94] and/or the piston outer diameter as well as correcting the ratio of inner to outer diameter. An already very small piston compression height can be opposed to an increase of the outer diameter, because it could become necessary in particular to reinforce the 1st ring land loaded by the cylinder pressure. Finally, not only the fatigue strength of the piston pin is improved, but also the contact pressure in the piston pin boss is relieved. Also a "trapezoid con rod" serves this purpose, but the current piston design is already tailored for that. At last high temperature-resistant piston alloys are cable to raise the loading capacity of the piston pin boss by about 30 %. An optional DLC coating [DLC: diamond-like carbon (diamond like, friction and wear reducing carbon coating)] is available as further measure for prevention.

An evaluation of the fatigue strength by FE calculation is illustrating further need of design reworking by means of local safety factors, identified by superposition of thermal and mechanical stress, temperature distribution and temperature specific fatigue strength diagrams of the piston alloy. Local critical temperatures of the component e.g. on the piston crown, ring land area and piston pin boss have to be respected in particular. Here again it should be referred to the comparison of temperatures in the Diagrams 5.88 and 5.89. A CNG specific increase of temperature on the piston crown, as it results from Diagram 5.88 [83], lets also expect a corresponding high temperature in the 1st ring groove, Diagram 5.89. The engine in question showing a high power density is exceeding here the local temperature limit of 270 °C by 11 K.

With increasing temperature and shortage of lubrication, favored by "dry" combustion, the ring groove side faces are generally affected by the phenomenon of "micro welding". The punctual welding of a piston ring and ring groove side face increasingly damages the same. Leakage and consequently a strongly rising amount of blow-by gas are very unpleasant results. That's the reason why for CNG operating mode a reinforcing protection of the ring groove, e.g. hard anodizing of the 1st and optionally 2nd ring groove, are mandatory [94]. Unprotected ring grooves can be a risk already from a temperature of 240 °C.

At a temperature of 350 °C, definitely possible on the piston crown, also high-temperature resistant piston alloys are suffering the loss of their advantages they are showing in the temperature range of 250 to far beyond 300 °C. That way they have an about 25 % higher fatigue strength at 300 °C [93]. As Diagram 5.89 outlines, CNG is raising the temperature on the center of the piston crown of the engine in question to the limit. The increase of temperature is 19 K compared to gasoline, regarding another engine even 34 K. The many degrees of freedom (compression ratio, spark advance, efficiency of piston cooling, natural aspiration/turbo charging/ supercharging, power density, comparison based on "bi-fuel" or conventional engine) are not admitting overall statements about the absolute levels of the CNG related piston temperatures. It can be acted on the assumption anyway that the temperature in the 1st piston ring groove of "bi-fuel" engines is raised by CNG by 10–30 K, the piston crown temperature by 20–40 K.

Characteristic for high-temperature resistant piston alloys is the increase of the copper and nickel content as well as e.g. small additions of vanadium and zirconium. The adjustment of material strength and hardness as well as volume stabilization are effected by solution heat treatment, quenching and annealing. Above 250 °C the intermetallic Al_2Cu phases originating from precipitation hardening are losing their importance. More temperature-stable, hard phases building nanocrystalline precipitations are taking over a primary function in case of these specific alloys [93]. "Bi-fuel" engines are partially equipped with forged pistons today [94]. The higher material strength and ductility by manufacturing conditions compared to cast pistons is based on a homogeneous microstructure achieved by the forging process. The benefits are diminishing however in the high temperature range. That's the reason why piston cooling measures generally are of big importance. CNG can necessitate the adaption of the oil squirting nozzles installed on the crankcase to a higher flow-rate [94]. If the spray cooling is no longer sufficient, in an exceptional case an expensive cooling gallery piston including further design features of diesel

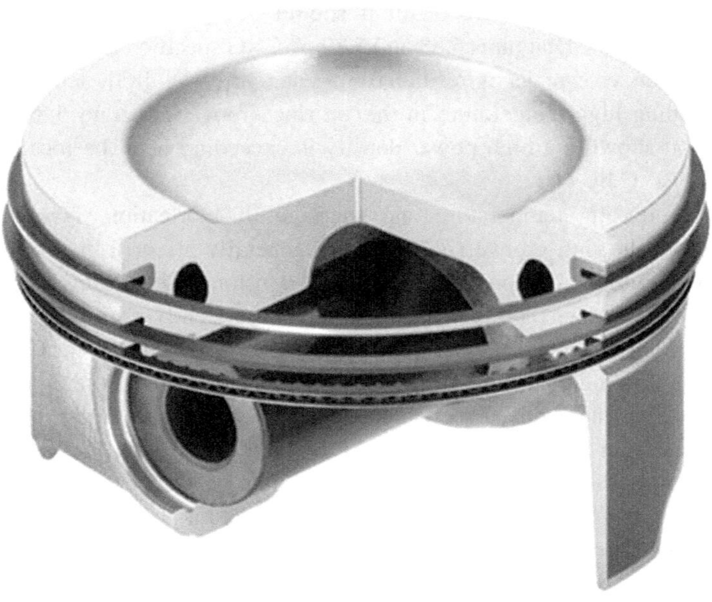

Diagram 5.95 Spark ignition engine pistons with ring carrier and cooling gallery (illustration: KS Kolbenschmidt GmbH, D-74172 Neckarsulm)

pistons have to be considered for spark ignition engines too. Diagram 5.95 exemplary shoes such a cooling gallery piston with ring carrier.

The piston ring pack of spark ignition engines typically characterizes a rectangular ring as top ring and a "three-part" oil control ring, both from steel. In between there is often a less expensive piston ring from grey cast iron as taper faced ring or stepped taper faced (scraper) ring, in the latter case with a bevel or recess on the inner top edge affecting a so called "positive ring twist" which applies more contact pressure to the lower edge of the running surface. A CNG related higher gas force on the piston ring back of the compression rings is increasing the piston ring friction an thus also the wear of the running surface. This can be counteracted by a superior base metal grade and/or a reinforcement of the running surface, in particular of the top ring, whereupon the available references of the automakers are restricted to overall statements [94]. Steel rings are gas nitrided anyway. Coatings of the running surface are deposited galvanically or by thermal spray coating processes. In particular for diesel engines there is made brisk use of a repertoire of wear resistant surface coatings based on chrome, molybdenum, metal-ceramics, DLC, diamond etc., so that—apart from cost aspects—there is no lack of bulk production options. In case of high-output gas engines there are references that the piston ring manufacturers are increasingly applying chrome nitride (CrN respectively CrCN with embedded carbon) in combination with the PVD process (Physical Vapor Deposition).

(b) Connecting rod and crankshaft

CNG is increasing the mechanical load on connecting rods and crankshaft. Since the firing pressure is on a level between conventional spark ignition passenger car and DI-diesel engines, the CNG operating mode is no new challenge however. But depending on the increase of firing pressure a reworking of the crank drive beyond the pistons can become necessary. The cross sections of cast con rods can be scaled up, or as the case may be, cast con rods can be substituted by forged ones. In contrast, an increase of the con rod width would involve extensive modifications. The trapezoid con rod already mentioned above does not only relieve the piston pin boss but also on the con rod side the particularly high loaded bearing bush in the con rods' small end.

Not only a higher firing pressure but also a stronger combustion due to CNG related spark advance are increasing the dynamic load of the crankshaft. A higher torsional vibration excitation results from the torque characteristics significantly influenced by the gas forces. Today the ordinary calculation of the crankshaft dynamics by multi-body system simulation gives information about the effects. In case of need there is the fall back to the experiences with diesel engines, that are torsional or combined torsional-bending vibration dampers. A strength deficit can possibly be compensated already by an alternative strength increasing finishing treatment of the fillets respectively transition radii between the journals and crank webs. Regardless of the cost a material with higher strength can be applied, as the case may be nodular graphite iron can be substituted by forged steel. In case of determined bearing width only the journal diameters can be increased. But due to the evident increase of friction losses this is very unpopular today. Also in case of a rigidity deficit a substitution of nodular graphite iron by steel with a by 25–30 % higher Young's modulus instead of more favorable dimensions may be considered when indicated. A forged crankshaft improves strength and rigidity at the same time.

(c) Plain bearings

The plain bearing manufacturers are showing the load limit of a certain plain bearing type by means of the acceptable average contact pressure. The current absolute load limits of plain bearings only appropriate for main bearings are 90 MPa, if appropriate for main as well con rod bearings 130 MPa and for con rod small end bearing bushes 180 MPa average contact pressure. If these key figures are exceeded scaling-up would become necessary with known consequences however. Regarding con rod bearing and small end bearing bush CNG can demand subsequent improvement. On the con rod big end part certainly more expensive "sputter" bearings [89] known from diesel engines can be applied, on the bearing cap part high-grade three-layer bearings. Furthermore the plain bearings of modern passenger car engines have to meet more severe stop-start requirements in the meantime. Here also in case of mixed friction particularly wear resistant sliding layers e.g. based on PAI (polyamide-imide) with dispersed solid lubricant particles are applied.

Also the "sweet gas" CNG implies an increase, but not overall comprehensible corrosion risk due to acidification with accelerated oil aging and increased water input of the lube oil. This is posing the question for the corrosion-resistance of current plain bearing materials. First of all the entry into force of the German "Altfahrzeug-Verordnung" (EU end-of-live vehicles directive) may be remembered, which is prescribing unleaded plain bearing materials for passenger car engine applications. Therewith leaded bronze based CuPbSn materials have lost their applicability. With lead a solid lubricant is omitted. The amendment has enforced the development of alternative plain bearing materials e.g. based on AlSnCu, AlSnSixx, AlZnBi, for highly loaded con rod plain bearings e.g. on CuZnxx, for con rod small end bushes also on CuSnZn (xx stands for certain further relevant alloying constituents). On the part of the plain bearing manufacturers the newly developed AlSnCu based bearing materials are declared to be insensible against oil corrosion. Regarding coppery alloys zinc has hardening and corrosion resistance improving properties. Thus a good corrosion resistance is also attested to brass plain bearing materials. "Sputter" sliding layers deposited on the bearing metal are consisting e.g. of $AlSn_{25}Cu_{2.5}$ and are considered as corrosion-resistant too. Con rod small end bearing bushes get a thin zinc coating for corrosion protection again. The equivalent coefficient of thermal conductivity of aluminum-based plain bearings is about 50 W/m/K, that of copper based plain bearings 75–80 W/m/K.

Valve Train

(a) Valve timing

Natural gas is also influencing the valve timing. It is well-established to modify the valve cam in such a way that the "touchdown" speed of the valves (the final about 0.2 mm valve lift in the pre-cam angular range [89]) is reduced. This is a supporting measure to lower the particularly critical valve seat wear. The low speed scavenging (increase of the "low-end" torque respectively kneeling of corresponding speed) cannot be applied in case of CNG port injection concerning the unavoidable loss of cylinder charge [82, 89]. Also the reduced residual gas compatibility of the natural gas combustion has to be respected.

Just for the natural gas operation mode a valve timing according to Miller or Atkinson is adequate. According to Atkinson the intake "stroke" is stretched by late closing of the intake valves, and thereby a part of the intaken fuel-air mixture is pushed back into the intake system. The shorter compression stroke enables a higher compression ratio. In contrast, according to Miller the inlet valves are closed early during the intake stroke. Due to the intermediate expansion there occurs an inner cylinder charge cooling effect ("miller effect"). The combustion temperature is decreased, the distance to the knock limit and the efficiency are increased.

The owing to the smaller cylinder charge quantity dropping off output can be compensated by turbo- or supercharging.

(b) Valve train components (camshaft, valve train mechanism)

Natural gas operation mode is not forcing measures concerning the valve train mechanism, if the base engine shows a trusted layout. Only abstaining from hydraulic valve-clearance compensation—hardly found at passenger car engines today—would to be reconsidered with respect to the valve setting problem by increased valve seat wear. But by reason of accelerated lube oil aging however wear protection measures on the contact points of the valve train mechanism, reinforcement of push rod and lever contact areas could turn out as reasonable. As for the camshafts, these are cast e.g. from different cast iron materials (procedurally hard chilled cast iron, nodular cast iron with following induction hardening of the cams), forged from steel (with following hardening of the cams) or constructed in terms of particular light weight versions. Thereby cams from hardened and annealed roller bearing steel ($100Cr_6$, hardness > 52 HRC) or sintered PM steel are applied [95]. In case of real need there are many options to optimize the valve train.

(c) Valves

In contrast to the cooling of the inlet valves by the intake flow the outlet valves are particularly thermally loaded by the hot exhaust gas. Missing internal cooling of natural gas intensifies the thermal impact. In case of subsequent retrofitting furthermore the exhaust gas temperature can unnecessarily rise by means of a retarded combustion caused by an unadjusted ignition map. Otherwise the exhaust gas temperature drops, Diagram 5.88. A reliable quantification of the valve temperature increase is difficult. The +10 K specified at [83] on the exhaust side could be assessed as the not representative result of two opposing effects, lower exhaust gas temperature but missing internal cooling.

With increasing temperature the risk of hot corrosion, high temperature oxidation and scaling is coming up. As is well known, valves are manufactured from corrosion-resistant, heat- and wear resisting steels [95–97]. High component temperature and shortcoming of lubrication let rise above all the fretting in the valve seat area. Diagram 5.96 shows the temperature profile in the valve head respectively valve seat area according to [98] in a schematic description. Regarding hardenable martensitic inlet valve steels (e.g. X85CrMoV18-2/material No. 1.4748) a temperature of 600–650 °C may not be exceeded, because otherwise the induction hardening in the valve seat area is losing its effect [96]. The reason is that engine operation is equivalent to an extended annealing process. More heat-resisting austenitic outlet valve steels (e.g. the commonly used steel X50CrMnNiNbN21-9/material No. 1.4882) can also be applied on the inlet side in case of need. Instead of monometallic valves a martensitic-austenitic material combination for valve stem and head in terms of friction welded bimetallic valves are the rule, since the top end face and the recess area of the valve base have to be hardened for reason of wear resistance [97].

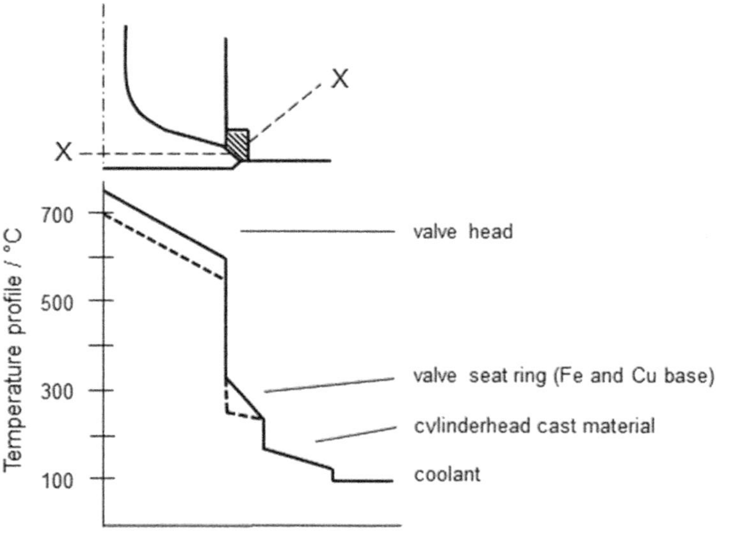

Diagram 5.96 Temperature profile on the valve head respectively seat area according to [98] as schematic draft (actually higher temperatures occurring on the outlet valve side)

On the exhaust side 800 up to more than 1000 °C are reached. On this temperature level, besides more heat-resisting steels, the valve seat area requires a plating (armored protection) from special nickel or cobalt alloys [96] (e.g. registered trademarks Stellite, Tribaloy, Eatonite, etc., plated by means of the submerged arc welding on process). Regarding spark ignition engines of high power density plating can be respected as a standard. In case of need the inlet valves can get an armored protection too for CNG operating mode. The valve head could alternatively be manufactured from extremely heat-resisting nickel-base alloy as e.g. Nimonic 80 A or Inconel 751. But regarding the cost aspect this remains widely reserved for big ship diesel engines with heavy crude oil operation. As is well-known, hollow and by two-thirds sodium filled valve stems improve the heat dissipation due to the liquefaction of the filling material and thus the so-called "shaker" effect. The achievable temperature reduction is 80 to 150 K [95].

Valve lightweight design is reducing the forces in the valve train and thus friction and wear. It has a favorable effect on valve train vibrations, after-opening and -closing of the valves and thus on the fretting in the valve seat area. The lightweight design options vary from hollow valve stems and heads (until now no market penetration for cost reasons [97]) to the reduction of the valve stem diameter, for passenger car engines to a minimum of 5 mm [89]. In case of shortcoming of lubrication also the seizure tendency of austenitic steels in the valve guides has to be particularly respected. Hard chrome plating of the valve stem or nitriding respectively carbonitriding (salt bath nitriding) of the whole valve is recommended [89]. The need of a polish finishing depends on the process applied in practice.

(d) **Valve seat rings**

Valve seat rings have turned out to be the "Achilles heel" of gas engines. They are dissipating 75 to 80 % of the heat transmitted through the valves [98]. Depending on their material temperatures below 300 up to more than 400 °C are reached on the outlet side. Missing internal cooling and shortcoming of lubrication of CNG are increasing the fretting (wear effected by friction) of the valve seat rings in a critical way. Due to "dry" combustion there is no lubrication at all by the gas on the inlet side, on the outlet side combustion residues can provide a limited solid lubrication. Here there should be referred to the importance of balanced, major deposits avoiding ashes and their appropriate composition regarding the solid lubrication by a specific additivation of the lube oil.

The mechanical load of the valve seat ring is caused by the valve spring force, the gas force acting on the valve head, the impact force during the valve closing and the friction forces also in context to the free—or for CNG operating mode not recommended—enforced valve rotation. Regarding natural gas not only the forces mentioned above, but also the gas force and thermally related larger distortion of the valve seat ring and the deflection of the valve head as well as relative movements are stimulating the wear. In the valve seat area there emerge adhesive, abrasive, corrosive and tribo-chemical phenomena as well as plastic deformation [97]. Thereby punctual micro-welding occurs in the valve sealing face. When releasing again there is a following removal of material. That's the reason why the valves are gradually engraving more and more, Diagram 5.97. Thus the setting of the valves is going on and on. The tappet clearance (finally in case of a hydraulic valve-clearance compensation too) is depleted bit by bit and the valves are getting leaky. The consequence is the phenomenon of "blowing", the rapidly increasing thermal damage of the valves.

Diagram 5.97 Valve with engraved valve seat area (illustration: MS Motorservice Deutschland GmbH, D-71679 Asperg)

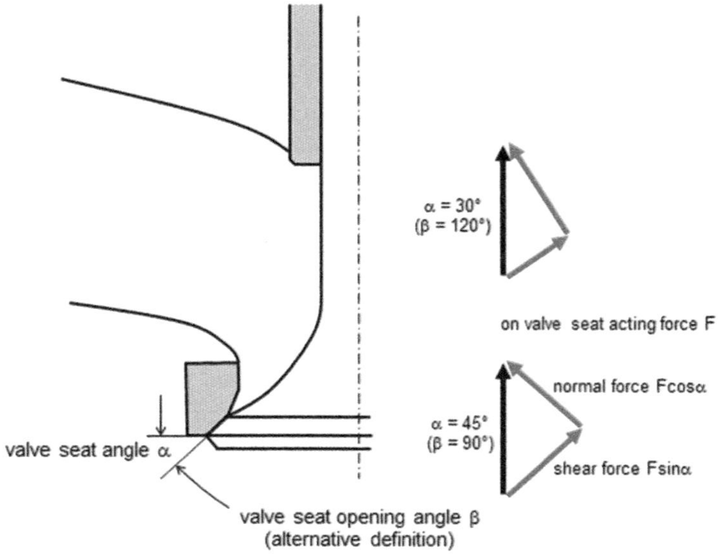

Diagram 5.98 Normal and shear force on the valve seat depending on the valve seat angel α

Besides material related measures "gimmicks" by design have turned out to be beneficial. These mainly concern a favorable modification of the valve seat geometry. A smaller seat angle α, that means a lager opening angle $\beta = 180° - 2\alpha$ (there is no consistent definition!), is in fact increasing the normal force $F_N = F \cos\alpha$ acting on the valve seat. In contrast, the shear force $F_S = F \sin\alpha$, is assessed to be more unfavorable, and thus the friction force as well as the relative movement are reduced, Diagram 5.98 [98]. Therefore it's proved and tested to reduce the valve seat angle α for CNG at least on the outlet side from 45 to 30° (20°) [89, 95–97]. It's useful too, to optimize the relevant parameters like seat height, seat width [95] respectively seat surface (increase of the impact respectively heat transfer surface) and valve head edge height (increase of rigidity) within the given scope [97]. Further recommendations concern the abstaining from the angle deviation (small difference between the angles on the valve seat and valve seat ring side) and the relocating of the valves relative to the combustion chamber. Also on the inlet side the more heat-resisting valve seat rings of the exhaust side can be applied. For a familiar mileage also in case of "dry" combustion in gas engines special materials have been developed. Higher macrohardness alone is not the crucial issue [97]. The micro-hardness of the hard phases is important at the same time. Valve seat rings for gas operation have to be all in all more heat-resisting, temperature-resistant and wear-resisting. Regarding the combination of the valve seat ring and the valve head material the latter should have the higher hardness. For purpose of not too much weakening the valve head about two-thirds of the wear should be shifted to the valve seat ring.

Valve seat rings for truck/commercial vehicle and bigger engines are cast from high-alloyed cast iron materials or chrome steels applying the centrifugal casting process. The centrifugal force contributes to a more compact microstructure. But valve seat rings are also individually cast in lost (sand) molds. The increased wear resistance is based on the accumulation of appropriate hard phases. These are existent in terms of a distinct, closed mesh of composite and special carbides in a—e.g. in case of hardened cast iron—mainly martensitic matrix [99]. Special carbides are building alloying elements like Cr, Mo, V, W, Nb and Ti. Composite carbides is building the Fe matrix material itself with the carbide building elements. Chrome in particular improves the corrosion-resistance. In case of cast iron the equally distributed solid lubricant is provided by precipitated graphite.

Regarding the bulk production of passenger car engines mainly PM materials based on sintered powder metal alloys are applied. Their advantage is the combination of a high potential of wear reduction with a competitive manufacturing process [100]. That way hard and soft, high- and low-melting materials in powder form can be compacted to a high performance composite material [99]. At high sintering temperature a steel matrix—the quality is depending on the provided basic powder—with dispersed hard particles is generated by diffusion and recrystallization. Carbide building elements like Cr, W, V, Mo are building again with also added carbon the required hard phases [97, 99]. Furthermore Ni and Si can be alloyed. Just with respect to "dry" combustion the composite still gets a copper phase (optional with tin) for purpose of improving the heat conduction, embedded solid lubricants (no lead due to the EU end-of-life vehicles directive) and e.g. an acrylic resin filler material [100], provided that the manufacturing process related pores are not closed by copper infiltration.

Major attention is paid to the hard particles of such special materials. High-melting tungsten respectively tungsten carbides are particularly contributing to the superior material properties. Diagram 5.99 exemplary shows the microstructure of such a PM material especially developed for "dry" combustion. There are different material approaches however, e.g. with alloys based on nickel and cobalt [97, 100]. In this

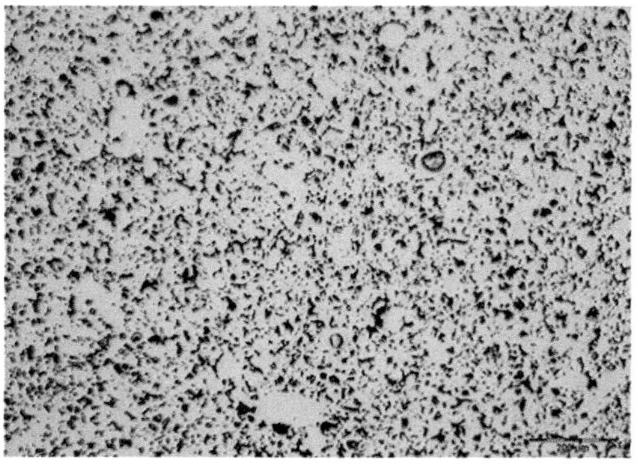

Diagram 5.99 Multi-phase microstructure of a valve seat ring from sintered PM material: Here hard particles from tungsten and other carbides in a predominantly martensitic matrix (illustration: material HT + of MS Motorservice Germany GmbH, D-71679 Asperg)

case the wear resistance is based on the hard, temperature-resistant intermetallic Co-Mo-Cr-Si-Phase. According to [100] molybdenum silicides are mainly contributing to this hard phase. The amount of hard particles is restricted in practice by the minimum strength of the sintered compound.

As solid lubricants manganese sulfide (MnS), molybdenum disulfide (MoS_2), calcium fluoride (CaF_2), graphite and chromium sulfide (CrS) can serve [97, 100]. There are efforts to generate solid lubricants instead of addition of powder in terms of a precipitating phase during the sintering process [100]. The machining tool wear, inversely proportional to the hardness of the material, which can amount to obviously more than 45 HRC, is another important criterion. Solid lubricants also turn out to be beneficial in this way. By closing the pores by means of infiltration a tool wear increasing interrupted cut can be avoided furthermore.

(e) **Valve guides**

The mechanical load of the valve guide is influenced by the impact force of valve train mechanism, the rigidity of the cylinder head and not least by its own length. Valve guides are centering the valve in the valve seat, supporting the lateral forces acting on the valve stem and dissipating the remaining 20 %, in case of sodium-filled hollow valve stems 30 %, of the heat transmitted through the valves [97]. The stem clearance is subject to a conflict of objectives, the more so as this has a direct influence on gas sealing, lube oil supply and thus also oil consumption, heat dissipation and wear. This can require a newly adjustment for gas engines within tight limits. Finally valve guide materials are in demand which have improved self-lubricating properties, even if the valve stem is never completely free of lube oil.

Valve guides consist of gray cast iron, non-ferrous metals or PM materials. The corresponding manufacturing processes are sand casting, continuous casting respectively extrusion molding and sintering, also in this case enabling a best possible adaption of the material to the specific requirements. Non-ferrous metals cannot cope with natural gas in spite of their good heat conduction and sliding properties. Cast iron materials, not applied for passenger car engines in fact, have a pearlitic matrix. Add-on of phosphorus is increasing the wear resistance in terms of a ternary Fe-P-C hard phase, the so-called steadite or also phosphide (eutectic) mesh [97]. Alloying elements like chrome are increasing the quality of the cast material. Solid lubricant of gray cast iron is the precipitated graphite again.

Regarding passenger car engines PM valve guide materials are substituting non-ferrous materials to a large extent at least on the exhaust side. Basis here is a mainly pearlitic matrix and a hard phase from steadite or cementite too [97, 100]. For natural gas the properties can be adapted to the requirements by powder from a higher-alloyed steel as well as appropriate add-ons and dosage [100]. Copper and tin respectively bronze, carbon respectively graphite and special solid lubricants like molybdenum disulfide (MoS_2) or manganese sulfide (MnS) are serving here for purposes comparable to those of valve seat rings.

Diagram 5.100 Valve stem gasket with double sealing lips (illustration: SKF GmbH, D-97421 Schweinfurt)

(f) **Valve stem gasket**

In order to face the shortcoming of lubrication of the valve stem valve stem gaskets with double sealing lips are applied [94]. The second sealing lip, also designated as backpressure sealing lip, had been integrated in addition for a better sealing against the exhaust gas flow and thus for reducing the blow-by [97]. By means of the second lip also lube oil in the area between both lips is retained and thereby the valve stem lubrication is necessarily improved, Diagram 5.100. In any case, also for natural gas operation the temperature of the valve stem guide on the location of the valve stem gasket should not substantially exceed 150 °C to prevent damage.

Gas Injectors

Also for the gas injectors in the inlet ports (PFI: Port Fuel Injection) the gaseous fuel is a challenge. Only the decompression of the gas from the system to manifold pressure effects a certain cooling-down. Moreover the missing lubricating and damping properties of the gas have to be respected regarding the design of the seat geometry and the closing mechanism. The CNG inlet port injection of current "bi-fuel" engines is increasingly combined with gasoline direct injection (GDI). If switched to CNG, the gasoline injectors are missing the cooling-down effect by gasoline flowing through. Due to the high thermal load then critical temperatures can be easily reached [82]. The increase of the heat dissipating areas can necessitate a modified sealing concept for the injector. But the injector can be set back for purpose of projecting as little as possible into the combustion chamber. An unwanted consequence can be coking at gasoline operation [82]. Regarding gas injectors there is still a demand for development. The components available on the market (status 2011) still had—at least recently—not enough flow rate for high cylinder specific output in the upper speed range [101].

Crankcase Ventilation

With the firing pressure the amount of blow-by gas and their temperature are also rising. The crankcase ventilation must still have the ability under these more difficult constraints too, to maintain the required small negative pressure in the crankcase. During the combustion of natural gas more water vapor has to be disposed. That's the reason why the crankcase ventilation is more susceptible to freezing at minus temperatures. This has to be duly respected regarding the ventilation concept and its accordant adaption.

Exhaust-Gas Turbocharger

External fuel-mixture formation—currently still characterizing the state of the art of the CNG technology—means a loss of cylinder charge as generally known due to displacement by the not negligible gas volume. Just in case of "bi-fuel" engines, the horsepower loss can hardly be compensated only by raise of the compression ratio. Therefore and in particular because of the high knocking resistance natural gas operation is predestined for exhaust-gas turbocharging. For adaption to the smaller mass flow rate the turbocharger is exchanged, e.g. for a device which is laid-out for an about by 20 % reduced swept volume of conventional spark ignition engines [83]. A smaller rotor diameter is moreover improving the engine response. The improvement of low-end torque (deficiency of torque at low rpm) requires multistage turbocharging [82].

The coolant cooled exhaust manifold integrated into the cylinder head is state of the art for spark ignition engines in the meantime and serves also for component heat protection at wide-open throttle [89]. For the application of a VTG turbocharger (turbocharger with variable turbine geometry) for spark ignition engines the exhaust gas temperature has to stay distinctly below 1000 °C. At reference [89] 950 °C, at [83] 926 °C are not exceeded, Diagram 5.88. But [83] specifies the lay-out temperature of the turbocharger with 1050 °C, since, when the engine is alternatively operated with gasoline, 1000 °C are nearly reached. Because CNG—exploiting the efficiency potential—is adequately reducing the exhaust gas temperature, in case of "monovalent" CNG operating the application of a VTG turbocharger could be possible according to [84]. How representative this exhaust gas temperature level shall be for the next engine generation has to be waited and seen.

The turbocharger bearing housings of current spark ignition engines are coolant cooled. At extreme high exhaust gas temperature a revised design using highest-heat resisting materials for the turbine housing and rotor can hardly be went round [102]. The further development of the turbine housing provides heat-resisting, thin-walled steel casting or sheet steel structures. Less component mass is lowering also the "thermal inertance" relevant regarding CNG (see catalyst). Downsizing enforced by CO_2 fleet emission targets is promoting the turbocharge application for spark ignition engines. There persists the question whether an expensive turbocharging technology will find acceptance or the exhaust gas temperature has to be finally limited.

Engine Cooling Systems

In view of a distinctly higher mechanical and thermal load by CNG operating mode very high specific outputs for series production engine are currently still avoided (status 2014). According to [82] a reasonable upper limit for CNG operating mode preliminarily seems to be at about 75 kW/l. Diagrams 5.88 and 5.91 are already referring to the next engine generation. The specific output of the concerning engine with small swept volume is nominally 88 kW/l. An equivalent high power density at thermally less advantageous operating conditions, raised demand for cooling and increasing application of turbo-/supercharging are implying a high input of heat into the engine cooling circuit and lube oil and so asking for a thereupon laid-out engine cooling system.

(a) **Engine cooling circuit**
The repertoire of measures (intensification of the local heat dissipation, increase of the coolant flow rate up to the increase of the vehicle radiator) need no particular comments here. The specific measures concern the integration of the gas pressure regulator into the cooling circuit to avoid freezing. Regarding big gas engines a specific coolant additive for raising the boiling points is admixed to the thermally highly loaded coolant.

(b) **Oil cooler**
An oil cooler with more cooling capacity can become necessary. Reference [94] e.g. reports about an upgrade of the oil cooler by two plates. This measure is explained by the increased heat input into the lube oil due to intensified piston cooling. For big engines besides the main oil circuit cooling a shunt (secondary) oil circuit cooling can be additionally considered.

(c) **Charge-air respectively fuel-mixture cooler**
Current "bi-fuel" engines have a conventional charge-air cooler as mandatory in combination with turbo-/supercharging. The raise of cooling capacity mentioned at [101] can be explained by the progression of the power density of the CNG demonstrator compared to the base engine. Fuel-mixture cooler built-in before the turbocharger are only applied for big lean-burn engines based on diesel engine.

Spark Plugs, Ignition

CNG is harder to inflame compared to gasoline. That's the reason why a by some kV higher ignition voltage is necessary. With respect to the ignition coils the increase of ignition voltage however should be kept within a limit. The CNG-air mixture nevertheless should be reliably ignited. Further requirements concern the prevention of hot corrosion, decrease of electrode wear and primarily the avoidance of thermal overloading of the spark plugs. Therefrom noble metals are applied, e.g. a center electrode with a thin tip from an iridium alloy and a small plate inserted into the ground electrode [103]. The former has a diameter of only 0.6 mm and thus needs less ignition voltage according to the manufacturer instruction, Diagram 5.101. Its shape should support an optimum propagation of the flame front.

The heat rating of the spark plug, its heat capacity respectively its ability for heat dissipation is adapted to the higher thermal load of the CNG operating mode by design measures at the spark plug insulator and heat transfer surface as well as an appropriate choice of material. Colloquially phrased, "colder" spark plugs are necessary. As is well known, these have a short insulator around the center electrode with a small insulator root area and conduct more heat by the spark plug thread. A core from copper in the center electrode improves the heat dissipation. For the heat rating there are no consistent designations of the spark plug manufacturers. Further preventions concern measures for reduction of deposits by the combustion and an optional coating for improved corrosion protection.

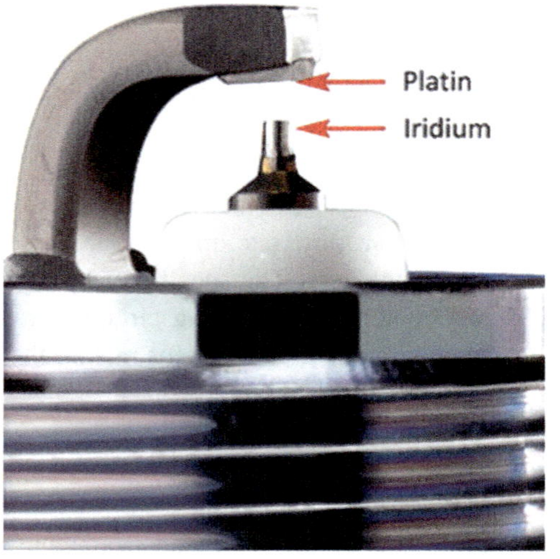

Diagram 5.101 Electrodes of a spark plug optimized for gas operation (illustration: NGK Spark Plug Europe GmbH, D-40880 Ratingen

Catalyst (Catalytic Converter)

In case of natural gas operation the component heat protection due to the cooling effect by wide-open-throttle enrichment is omitted. This also concerns the catalyst. An optimum efficiency combustion otherwise lets drop the exhaust gas temperature. An exhaust gas turbocharger has the effect of an additional temperature drop (temperature reduction below up to far more than 200 K). For the three-way catalyst therefore not primarily the exhaust gas temperature altitude currently is a problem. In fact, different natural gas qualities are causing variations of the ease of ignition with the consequence of unnoticed misfire. Also the required higher ignition voltage is contributing to this. The overfeeding with methane is overheating the catalyst. This phenomenon seems to afflict conventional ceramic catalysts more than for this application specifically laid-out metal catalyst substrates [104]. The minimum sulfur content of CNG is lowering the conversion of the catalyst only a little. Deposits of ashes related to not qualified lube oil additives are impacting the long-term function on the other hand.

Regarding "monovalent" natural gas operation the HC emissions consisting of unburned methane to a large extent are a particular challenge. The high activation energy of methane requires a "light-off" temperature of about 450 °C, an increase of 50–100 K compared to gasoline operation. After cold start regarding "catalyst heating" the thermal inertance of an exhaust gas turbocharger is not insubstantial. The optimization of the three-way catalyst for the methane conversion is based on an increased noble metal loading with modified composition. Faster heating-up also presupposes a modified design and a further reduced distance between precatalytic converter and engine. "Bi-fuel" engines don't know this problem by the way, since the cold start is automatically carried out in the gasoline operating mode.

Under long-term wide-open-throttle operation the three-way catalyst is also subjected to the risk of overheating. As is well known, above 900 °C accelerated ageing is beginning, 1200 °C are giving rise to destruction. "Bi-fuel" engines are reaching the higher exhaust gas temperature when operated with gasoline, Diagram 5.88. In case of "monovalent" CNG operating mode of engines with high power density the reduction of the exhaust gas temperature has to be seen in shortening the duration of combustion ("compact combustion" [101]). Insufficient adaption of the ignition map respectively a deficient engine base for that, as this may be on hand in case of subsequently retrofitted engines, result in too hot exhaust gas as already mentioned above.

Lubricating Oil

Lube oils are classified according to API, ACEA, ILSAC, GLOBAL and other standards. Thereto there are specific releases by the carmakers and engine manufacturers. The petroleum industry offers gas engine lube oils for "monovalent" natural gas operation (see e.g. [105]). For these there are releases by the big diesel engine manufacturers who have also gas engines in their product portfolio. The general standardization of the lube oils for gas operation is still in preparation. Regarding "bi-fuel" engine operation the lube oil has to meet the requirements of both fuels. In case of subsequent retrofitting by repair shops specialized on this job, the carmakers are rather reserved with recommendations. Users of such cars therefore are often unconfident with the choice of the properly lube oil. A short overview of the fundamental facts are given e.g. by [106].

Lube oils for gas operation need a high thermal stability and a low affinity to oxidation respectively nitration. Contrary to unpurified (low-grade) biogas etc. CNG is free from sulfur and other contaminations to a large extent. The danger of acidification (indicator TAN: Total Acid Number) with the risk of corrosion going along as well as the resulting necessity for neutralizing additives with alkalinity provision (indicator TBN: Total Base Number) are correspondingly relativized. Gasoline also contains purifying substances, so-called dispersants. For gaseous fuels the lube oil has to balance their deficiency by appropriate additivation. But ashes building additives with purifying, anti-oxidative, anti-corrosive, neutralizing and wear reducing effects are contributing to increased formation of deposits just in gas engines. The additives are based on organometallic active agents as well as phosphorus and sulfur. Calcium containing dispersants and wear reducing, zinc containing ZDDP [(or also ZnDTP) Zinc-di-alkyl-di-thiophosphate] are in particular in the focus. There is a need of lube oils containing only few sulfur and phosphorus. They should burn with lowest ashes residues as possible measured by the sulfate ashes originating during combustion (organic residues according to DIN 51575 in Germany). Such lube oils are labeled with the additional designation "Low-SAPS" (SAPS: sulfate ashes, phosphorus, sulfur) or only "Low-Ash". Since there cannot be abstained from the specific effect of additives, new additive complexes have been developed. Lube oils poor in ashes are not completely novel. At the same time they are serving as protection for the catalyst too.

There is the proof that certain deposits of ashes—quasi as solid lubricant—are definitely helpful for gas engines to reduce the valve seat wear. For that reason a few ashes but of certain consistency and slight depositing affinity should develop. According to product information of the petroleum industry fully synthetic fuel-efficient engine oils with "Low SAPS" properties respectively CNG specific additivation in the range of API SN/CF respectively ACEA C1-10 to C4-10 are coming off best for spark ignition passenger car engines. Their superior grade base oils show a lower coking affinity and are thermally higher resistant than mineral oil. By reasons of the complex issues only the carmakers can give recommendations based on engine release test runs. In case of high thermal load of the lube oil there is also the option to increase the lube oil volume for stretching the service interval. This is increasing the absorptive capacity of the lube oil and correspondingly delaying its ageing.

5.2.9 Acoustics of Natural Gas Powered Vehicles

Ulrich Philipp

5.2.9.1 Natural Gas Engines

Engines designed for monovalent or bivalent operation with natural gas are normally derivatives of conventional, liquefied fuel powered engines. In the bus and truck sector, a diesel engine is typically used as a basis, whereas in the passenger car sector, a gasoline engine is used as a base. Accordingly, acoustic comparisons for these two concepts must be made with the corresponding base. Since the base engine is normally unchanged, any differences in acoustic behavior are due to the changed combustion excitation. Adapting the engine with regard to turbocharging and fuel supply also causes additional acoustic effects.

These are:

- Noise emanating from the turbocharging and air loading system due to the increased charge pressure required to compensate for the reduced aspirated air mass caused by natural gas injection.
- Noise emanating from the injectors caused by pressure pulsations, the much larger stroke of the jet needle and the lack of hydraulic damping of gaseous fuels. Diagram 5.102 shows the frequency spectrum of the noise level measured at 1 m distance for different gas engines of different power compared to a diesel engine with a power rating in the lower range of the comparison field. Even so, the noise of the diesel engine is higher across the mid-range of frequencies by approximately 15 dB(A). However, it must be noted that this comparison does not show measurement results from vehicle engines, but from engines for block-type cogeneration plants. Running an engine designed for Otto fuels with 765 cm^3 with natural gas results on the one hand in an average power loss of 13.7 % despite adjustment of the ignition timing to an optimal efficiency, Diagram 5.103, and on the other hand in an improvement in specific fuel consumption of 17.7 %, see Diagram 5.104. The power difference is caused by the lower cylinder charge, since CNG has a lower density compared to fuel vapour.

Other causes are the lower flame propagation velocity and the higher ignition delay for CNG combustion. This leads to an acoustically beneficial lowering of the pressure rise,

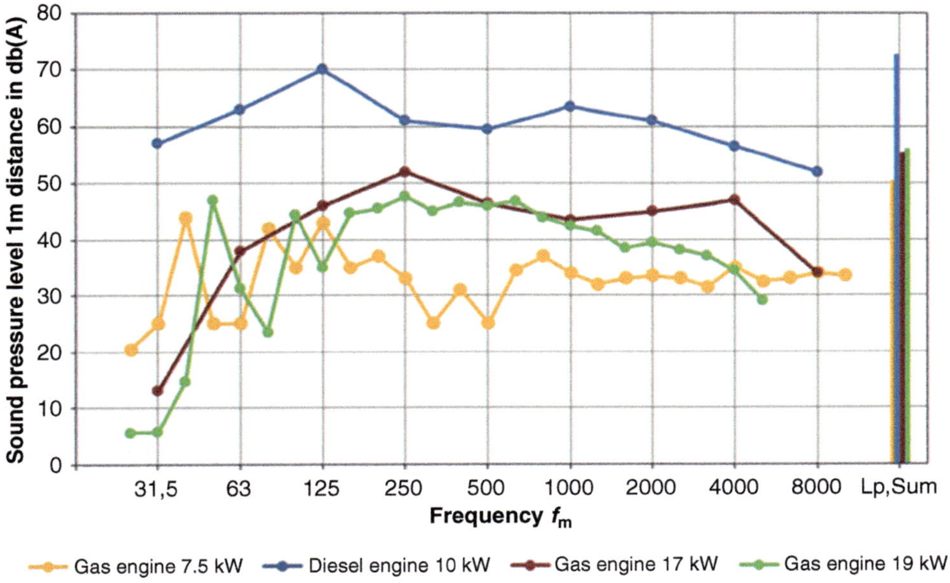

Diagram 5.102 Spectral noise levels [1 m distance in dB(A)] of various diesel and gas aggregates [107]

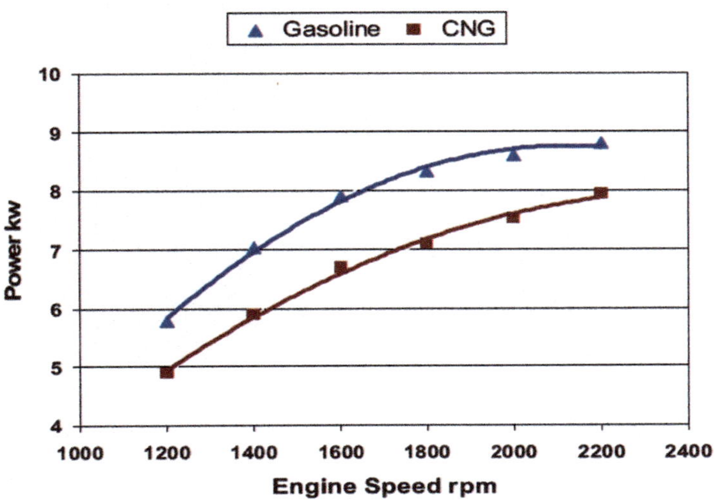

Diagram 5.103 Full-load curve as a function of engine speed for natural gas and gasoline [119]

but also to a reduction in mean pressure [119]. The average noise pressure level at 1 m distance at 2000 rpm (speed with maximum efficiency) is correspondingly reduced by 0.5 dB(A) and at 2200 rpm (rated speed) by 2 dB(A) (Diagram 5.105).

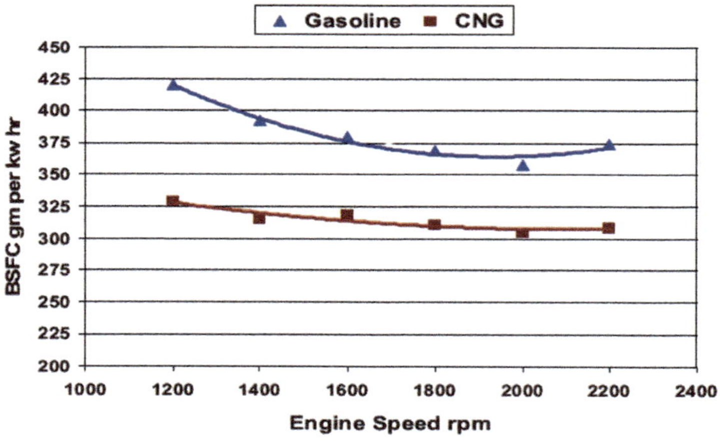

Diagram 5.104 Specific fuel consumption at full-load as a function of engine speed for natural gas and gasoline [120]

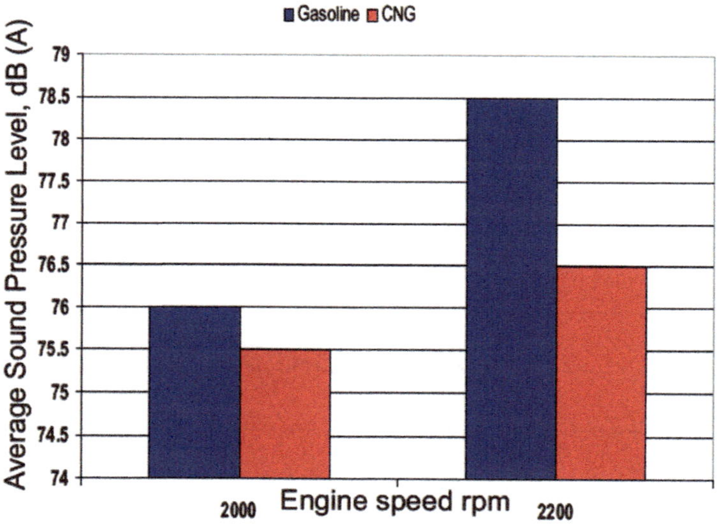

Diagram 5.105 Influence of fuel on the average noise pressure level at 1 m distance [108]

5.2.9.2 Combustion Noise

Mixture formation in natural gas engines is better when compared to engines that are operated with liquid fuel, since the gas is more completely mixed with air and distributed more evenly throughout the cylinder. As a result of this, CNG engines are fundamentally smoother and quieter [117, 118]. A further advantage resulting from the better mixture formation is lower specific fuel consumption, since the fuel cannot escape in to the crankcase via the lubricating film, as is the case for liquid fuels [110]. This occurs particularly with rich mixtures, which do not occur with CNG engines anyway, since the cooling effect caused by the evaporation of the liquid fuel is not present and the maximum of the laminar flame propagation velocity is close to $\lambda = 1$. These findings apply to retrofit and/or bivalent engines that do not have additional adaptations such as higher compression ratio to run with natural gas.

The actual combustion noise is determined by applying FFT (*Fast-Fourier-Transform*) to the combustion chamber pressure signal and transforming it into the frequency domain, where the resulting combustion chamber pressure spectrum is weighted using a frequency-dependent curve, the so-called structural attenuation level. This structural attenuation level is actually an engine-specific value. In order to be able to compare different engines and/or combustion processes, a standardized curve is used in indicating systems or in special devices that measure combustion noise (noise meters). A further weighting in the frequency domain is the A-weighting, which considers the frequency dependence of the human ear's sensitivity. This is considered when determining the structural attenuation index. The final calculation of the overall level over the frequency range results in a single characteristic value for combustion noise that can be directly related to the sound pressure level of the corresponding engine. As can be derived from the engineering units for combustion chamber or sound pressure (1 Pa = 10–5 bar), the noise attenuation of an engine contains several orders of magnitude, so that the structural attenuation index is stated in decibel [dB(A)] (Diagram 5.106).

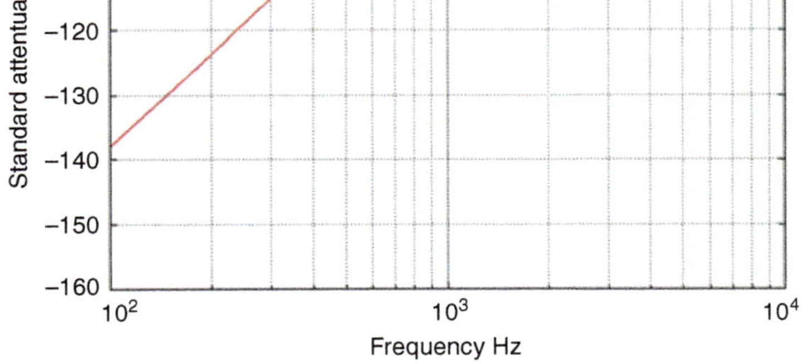

Diagram 5.106 Standard structure attenuation curve of a combustion engine [108]

The different frequency ranges of the combustion chamber pressure spectrum are determined by various characteristic values of combustion chamber pressure, see Diagram 5.107. The peak pressure is responsible for low frequencies; as the frequency increase, the first and second derivations of combustion chamber pressure begin to dominate, while the high frequency range is excited by cylinder pressure oscillations.

The combustion excitation is not only transmitted within the engine structure and finally radiated from the surface, but additional noise components emerge in conjunction with the mechanics of the engine, if caused by alternating forces clearance is traversed in component pairings such as piston-cylinder or bearings. This process creates impulse excitation that is slightly dampened by the lubricating film. Special methods using correlation and coherence techniques [108] can separate direct and indirect combustion noise and mechanical noise components, which are due to different mechanisms of excitation.

The comparison of combustion noise between natural gas and gasoline operation with optimal timing of 50 % MFB (compare Sect. 5.2.2) and identical indicated mean pressure

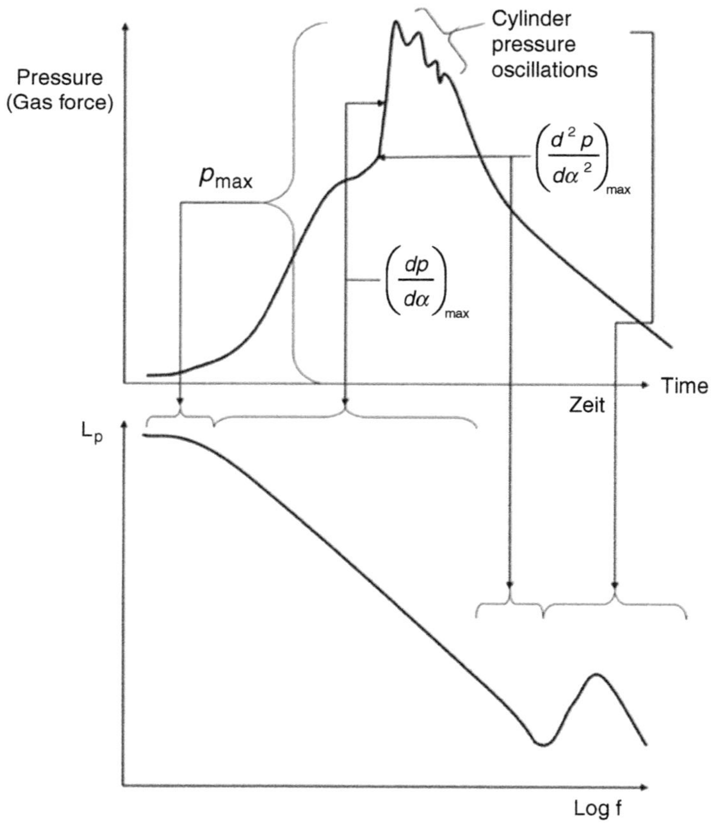

Diagram 5.107 Relationship between combustion chamber pressure characteristics and the combustion chamber pressure spectrum [108]

leads to a calculation for methane of 81.89 dB(A), which is a combustion noise approximately 3 dB(A) higher than isooctane [78.76 dB(A)]. Diagram 5.108 shows the corresponding combustion chamber pressure signals and Diagram 5.109 shows the combustion noise spectra.

Diagram 5.110 shows combustion noise level, combustion chamber maximum pressure and the first derivation of combustion chamber pressure calculated with methane over ignition timing. The typical earlier ignition timing for natural gas engines due to the high knock resistance (and also beneficial for reasons of efficiency) leads to a significant increase in combustion noise level and the maximum value for the first derivation of the combustion chamber pressure.

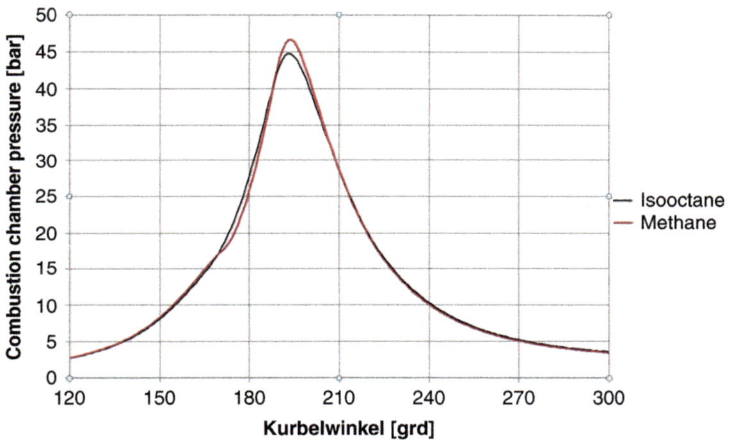

Diagram 5.108 Calculated combustion chamber pressure signal for isooctane and methane

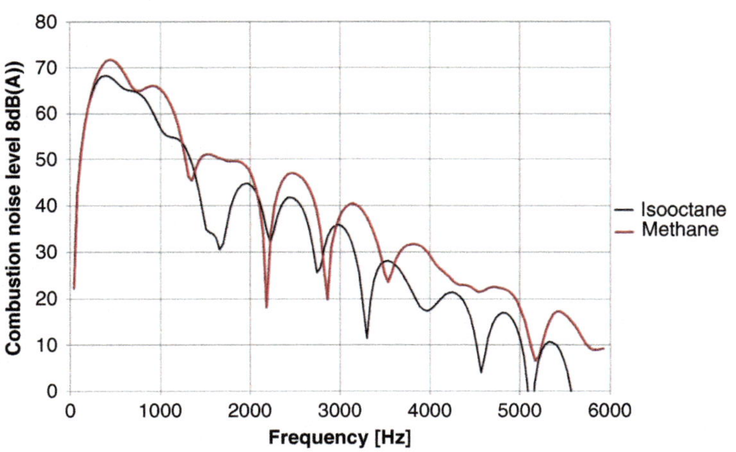

Diagram 5.109 Calculated combustion noise spectrum for isooctane and methane

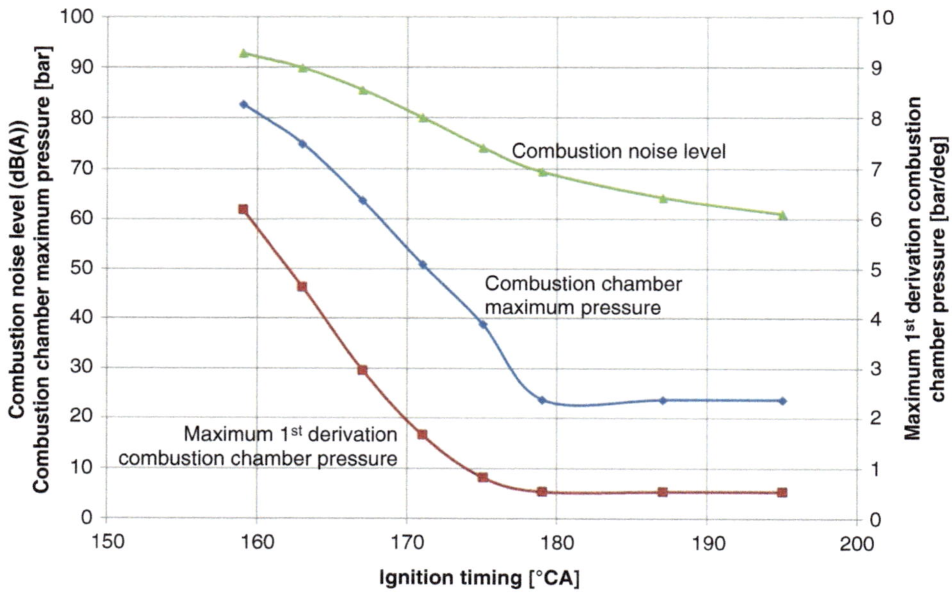

Diagram 5.110 Combustion noise level, maximum combustion chamber pressure and 1st derivation of combustion chamber pressure with methane while varying the ignition timing

5.2.9.3 Turbocharging and Charge Air System

The reduced cylinder charge due to the injection of natural gas into the intake manifold can be compensated and even over-compensated with turbocharging. If a naturally aspirated engine is modified to take an exhaust gas turbocharger, then the turbocharger has additional damping effects on the pulsations in the inlet and exhaust system.

The average damping value of a turbocharger for a diesel engine is shown in [111] to be 10 dB, [112] gives values for the ignition order of 20 dB. Experimental investigations on a four-cylinder engine show the damping of intake noise to be 12 dB and more at lower frequencies [113]. The damping of exhaust noise by the turbine is quantified as 15 dB, whereby particularly the low frequencies are heavily damped. This can be seen using the engine order as an example in Diagram 5.111. The higher damping caused by the turbocharging weakens the load and speed dependency of the orifice noise level, whereby the characteristic engine orders are less succinct compared to normally aspirated engines. According to [112], the damping effect due to turbocharging in stratified operation of gasoline engines with direct injection can be positive in as far as the fundamentally increased noise excitation at the intake orifice is partially damped and compensated for and, due to the damping effect of the turbine in the exhaust tract, it is perhaps possible to reduce the exhaust system volume.

On the other hand, the additionally required intake air can lead to acoustic effects as they are known from stratified charge engines with dethrottling at part load [112, 114], Diagram 5.112. The increase in the exhaust side gas exchange excitation in stratified mode as opposed to homogenous operation is estimated in [112] to be 4 dB. A further measure to

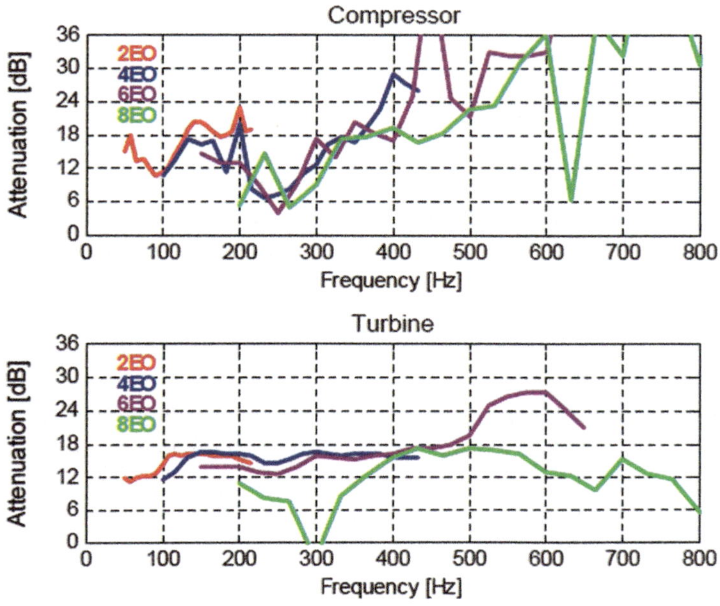

Diagram 5.111 Measured damping of the intake noise (*above*) and exhaust noise (*below*) [113]

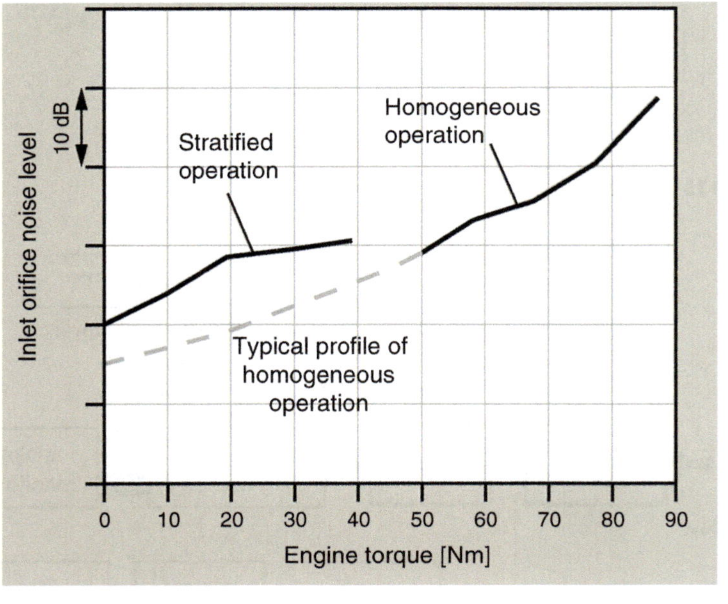

Diagram 5.112 Intake orifice noise level as a function of load at constant speed [114]

improve NVH behavior is, as is usual with engines using liquid fuels, to manufacture the inlet manifold out of synthetic material, which also leads to an additional weight saving [112].

5.2.9.4 Injector Noise

The pulse width control by the ECU produces rapid opening and closing of the injector. The needle stroke is 4–6 times higher than for liquid injectors and the hydraulic damping factor of liquid fuel is not present. These effects lead to noticeable injector operation noise [120]. In addition, whistling noises can be diagnosed from the vicinity of the injectors when slowly accelerating in low gears from 30 to 70 km/h. Various measurements have identified the cause as gas pulsations in the connecting channels between the injector and the intake manifold (Diagram 5.113).

Based upon calculations performed using CFD, this whistling effect can be remedied by reducing the jet inside diameter by 40 %. The noise measurements carried out according to the layout in Diagram 5.114 resulted in the improvements shown in Diagram 5.115 in the frequency range from 3 to 20 kHz.

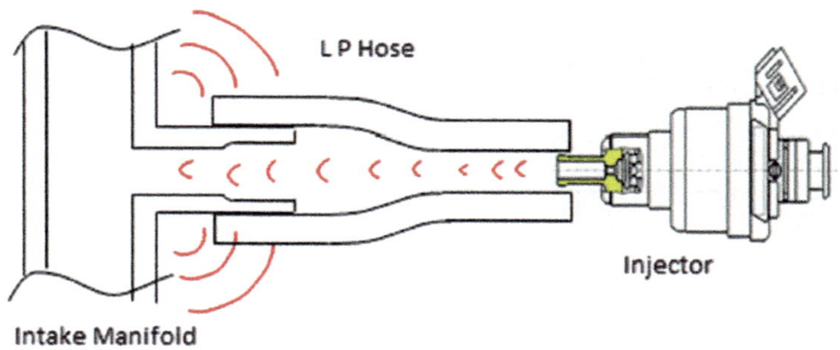

Diagram 5.113 Gas flow from the CNG injector [120]

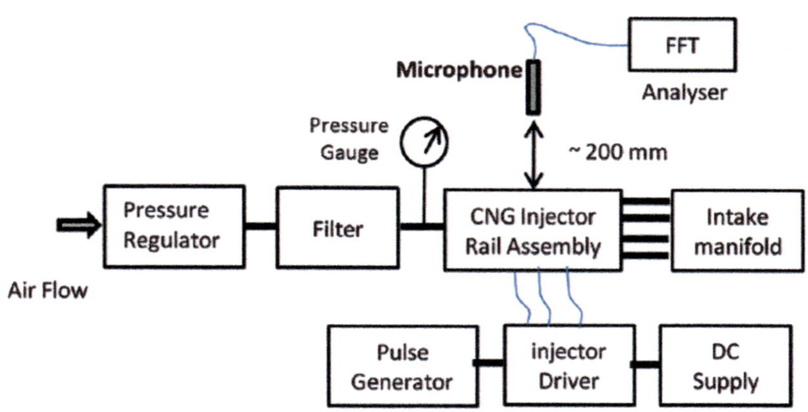

Diagram 5.114 Schematic view of the noise measurement setup [120]

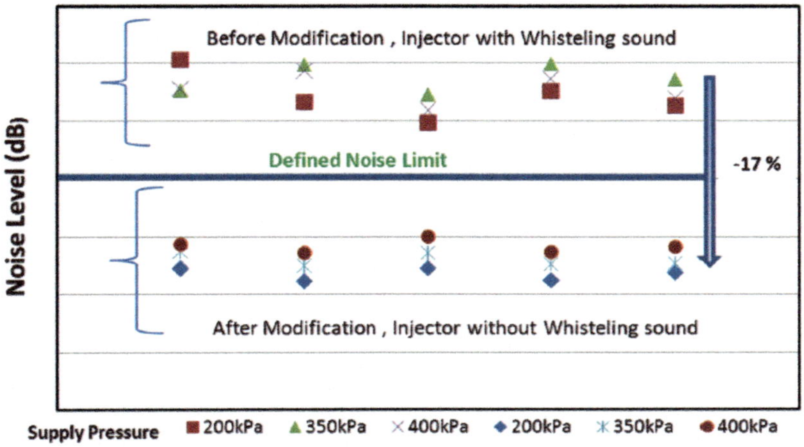

Diagram 5.115 Comparison of noise levels before and after injector optimization [120]

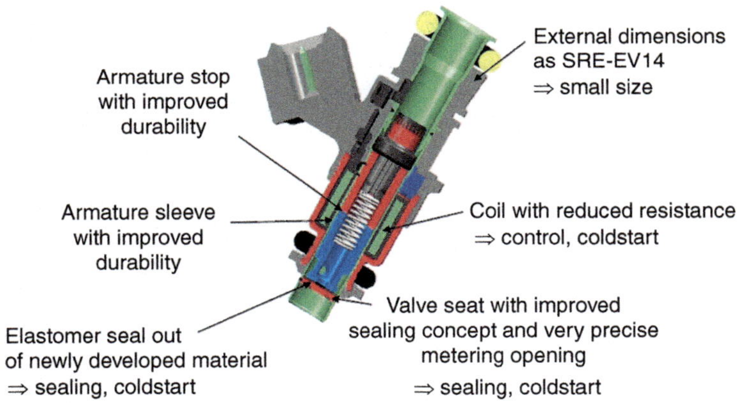

Diagram 5.116 Layout of a natural gas injector NGI2 from BOSCH [116]

In order to reduce the noise excitation impulses in the jet needle seat, it is located in an elastomeric disc as shown in Diagram 5.116 that simultaneously assumes the role of seal.

Noise also emanates from the flow of gas out of the injector nozzle into the intake manifold, since the gas expands, as opposed to the constant volume flow of liquid fuel. An invention is described in [119] that makes the gas flow through a porous shell with floor after exiting the injector nozzle (Diagrams 5.117). Diagrams 5.118 and 5.119 show the gas distribution 2.4 and 3.8 ms after injection with and without this shell. The formation of pulsed waves is reduced or eliminated by the distribution of the gas in the transverse

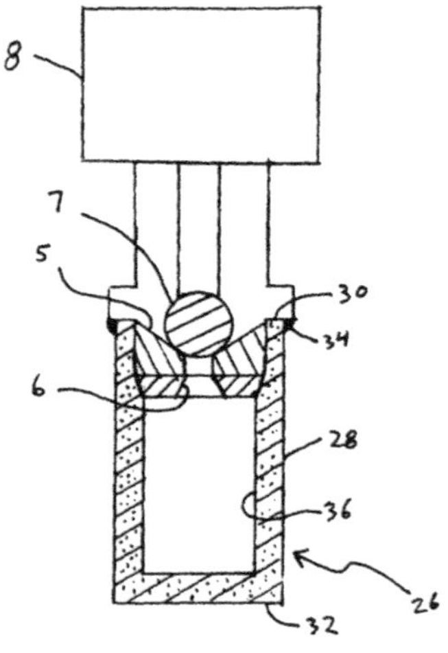

Diagram 5.117 Natural gas injector with porous shell at the injector orifice [119]

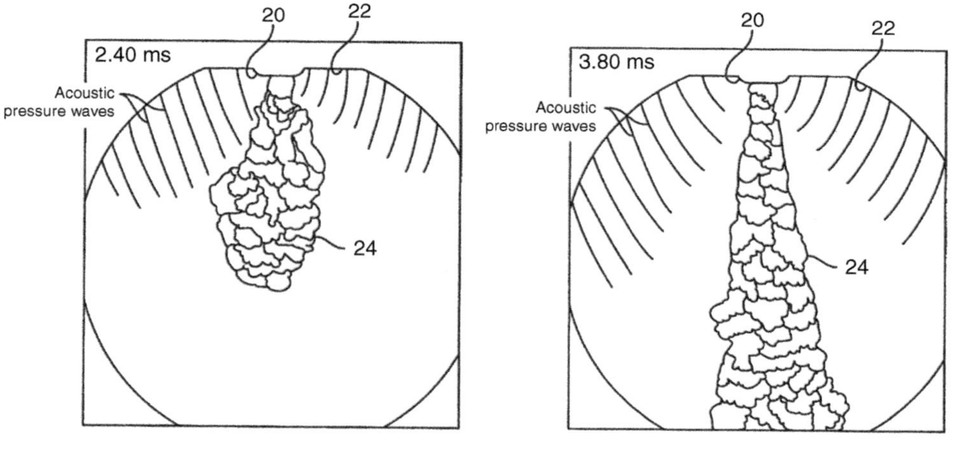

Diagram 5.118 Gas distribution in the inlet manifold [119]

direction. A reduction in noise of 4.7 dB(A) results from using a shell with a length of 25 mm manufactured out of stainless steel 316 l and with a pore width of 40 μm, while a shell manufactured from bronze and a length of 12 mm still results in a 2.8 dB(A) lower noise level.

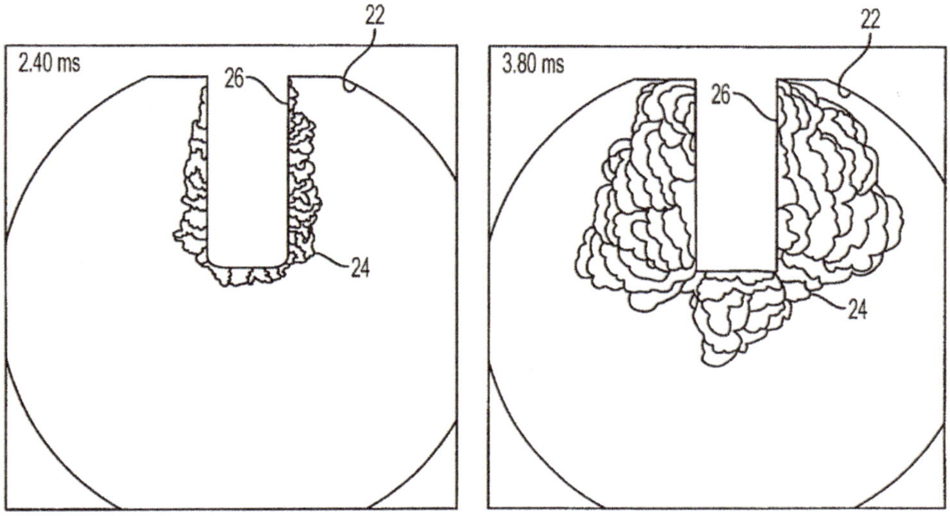

Diagram 5.119 Gas distribution in the inlet manifold with device to reduce noise [119]

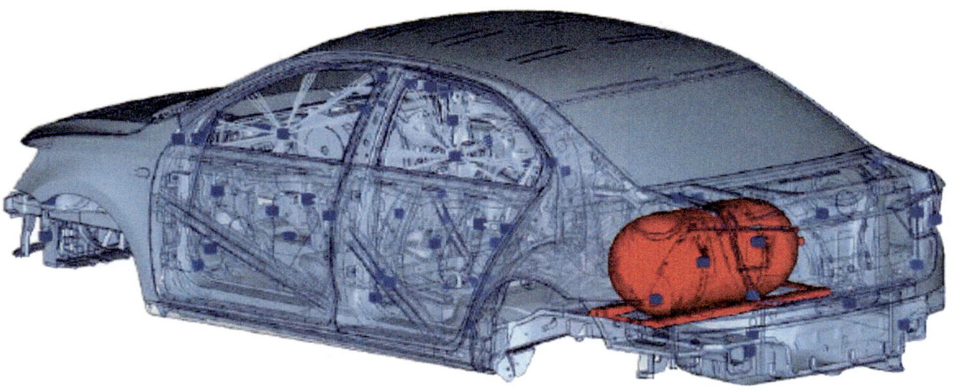

Diagram 5.120 FEM model of vehicle structure in order to calculate NVH behavior [121]

5.2.9.5 Overall Vehicle

The installation of the CNG fuel system in the vehicle has considerable influence on the vibration behavior of the vehicle structure and hence NVH and also on safety-relevant vehicle structures concerning crash behavior for rear impact due to the additional components, particularly the relatively heavy fuel tank see Diagram 5.120. A *Multidisciplinary Design Optimization* (MDO) process is described in [121] which optimizes vehicle structure with respect to NVH and crash safety with the target conflict

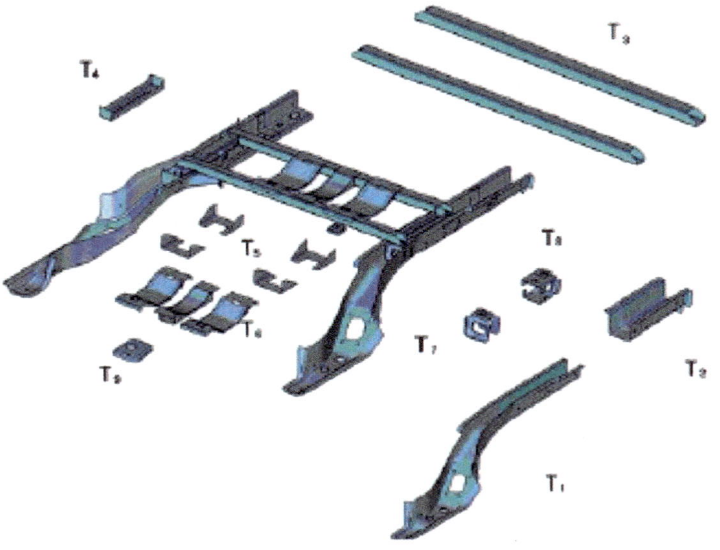

Diagram 5.121 Components for the optimization process [121]

of light construction (Diagram 5.121). The target with respect to NVH was to leave the natural frequency of the sub-system of 23 Hz unchanged, which considering the increased mass also requires an increase in component stiffness.

5.3 Engine Concepts for Commercial Vehicles

Günter Figer

5.3.1 Introduction

Natural gas engines represent an economically interesting powertrain solution as an alternative to diesel engines for use in medium and heavy-duty trucks. The combination of long-term low pricing of natural gas, together with national and local subsidies for powertrain solutions that reduce greenhouse gas emissions are often convincing arguments for fleet operators to force the conversion from diesel to natural gas powered propulsion. Furthermore, the continuously growing road transport sector is leading to increasing CO_2 emissions despite ever more efficient powertrain technologies. In order to reduce the increase of CO_2 emissions, the EU has initiated appropriate programmes whereby the emission of greenhouse gases are to be reduced by 6 % by the year 2020

[122]. Based on the approximately 25 % lower greenhouse gas potential of natural gas, and the largely greenhouse gas-neutral bio-methane (see Chap. 4), both of these gases are thus important elements in a strategy to achieve these emission targets. The lower gas price and the more favorable CO_2 footprint are the most important drivers for a significant increase in the registrations of natural gas powered commercial vehicles. An annual growth rate of between 16 and 18 % has been forecast for the future [123].

Regarding the legal framework for heavy-duty commercial vehicles powered by natural gas engines, the emission levels have been gradually reduced in the same manner as for diesel engines. This reduction applied to NO_x, particulate mass and later particulate number as the critical items for diesel engines, and in addition CH_4 emissions and non-methane hydrocarbons for natural gas engines. While in Europe methane emissions have been limited a few years ago, US legislation is currently limited to the regulation of greenhouse gases. This takes into account that from a certain threshold methane has a more than 20-fold higher greenhouse effect than CO_2, hence the CH_4 limit for the USA is shown as a dashed line in Diagram 5.122.

The greatest challenges for commercial vehicles with gas engines are the limited range compared to a diesel truck, the limited power density and a currently unsatisfactory brake thermal efficiency. In addition to this, service intervals must be extended and the currently high vehicle system costs in terms of acceptable pay-back period must be reduced. CNG (Compressed Natural Gas), but in particular LNG (Liquefied Natural Gas) are relevant for commercial applications. Both fuels consist mainly of methane; the technical difference between the two fuels is however the different methods of storage. As described in Sect. 5.1, CNG is stored in cylindrical high-pressure tanks at pressures between 200 and 250 bar. Liquid LNG, which is stored below its boiling point of 111 K (−162 °C), uses highly insulated cryogenic tanks, see Sect. 5.3.2.4. Both storage systems have in common that the energy density is less than diesel and hence the range of the vehicle is reduced by comparison (see Table 5.5). LNG is stored in liquid form at densities up to three times higher than storage in gaseous form and enables vehicle ranges of between 900 and 1000 km for heavy-duty vehicles with maximum tank volumes. CNG tank systems can currently achieve ranges of between 450 and 500 km.

Maximum energy efficiency is particularly important because of the limited vehicle range compared to diesel fuel. Whereas the simplicity of diesel engine conversion was an important factor for previous engine technologies, the focus of new engine development

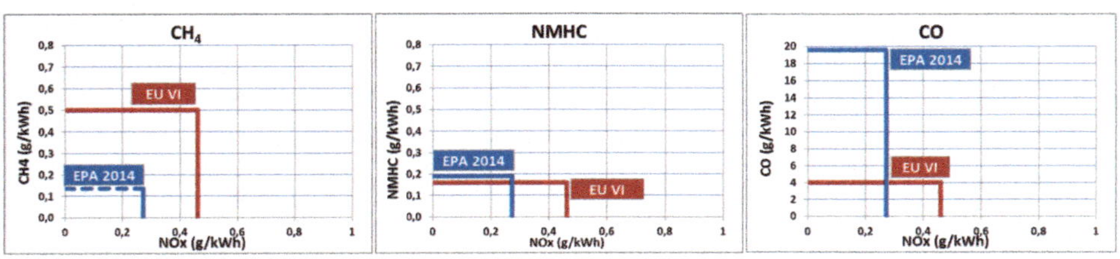

Diagram 5.122 EU and US emission legislation for commercial vehicles

Table 5.5 Comparison of fuel characteristics (Natural gas values from the NIST Standard Reference Database)

	Fuel state	Storage pressure (bar)	Storage temperature (°C)	Density (kg/m^3)	Lower calorific value (MJ/kg)
Diesel	Liquid	1	20	820–845	42–43
CNG	Gaseous	200–250	20	160–190	32–47
LNG	Liquid	7–10	<−125 °C	360–375	48–49

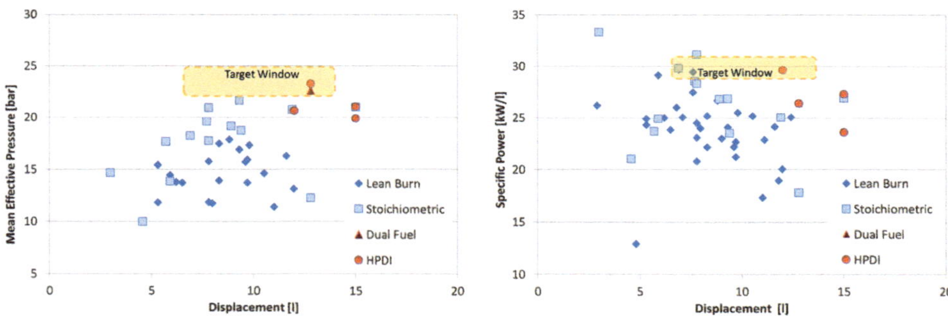

Diagram 5.123 Mean effective pressure and specific power of current gas engines for commercial vehicles (Source: AVL)

lies mainly on efficiency, power density and drivability. The challenge for gas engines consists of being measured against current EURO VI commercial vehicle diesel engines with a thermal efficiency between 42 and 44 %, despite all the limitations of an Otto-cycle combustion process. The mean effective pressure of modern natural gas engines is currently just over 20 bar and the corresponding power density lies between 25 and 28 kW/l. The target values for new developments will continue to orient themselves on the diesel engine. This means that future gas engines must achieve a specific power density of approximately 30 kW/l and BMEP levels of 23–25 bar, as do comparable diesel engines. Currently, only a handful of engines fulfil these requirements, as shown in Diagram 5.123.

Today, end-customers rarely compromise with regard to transport efficiency and driving performance in comparison to diesel engines. However, a global overview shows that various emission guidelines and market requirements are leading to a range of gas engine technologies. The following four gas engine concepts are currently predominantly used according to the requirements for commercial vehicle powertrains:

1. Lean burn concept (spark ignited)
2. Stoichiometric concept (spark ignited)
3. Dual-Fuel concept (compression ignited)
4. High pressure gas direct injection (compression ignited)

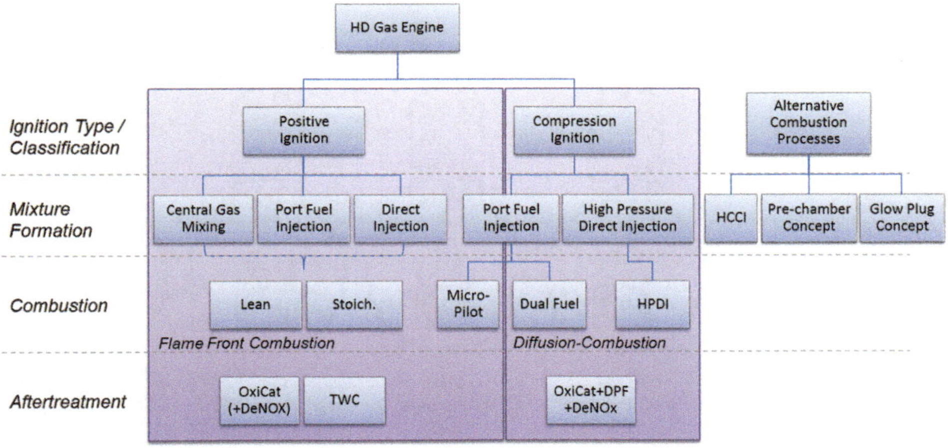

Diagram 5.124 Structuring of the processes for gas-powered commercial vehicles

Diagram 5.124 shows a structured view of the ignition method, mixture formation process, combustion process and exhaust gas aftertreatment systems, the details of which will be covered in subsequent sections.

5.3.2 Thermodynamic Basics

This section will present a few relevant thermodynamic interrelationships for the full-load behavior, efficiency and emissions of different natural gas engine concepts.

5.3.2.1 Full-Load Potential

Full-load, that is the achievable mean effective pressure, p_e is directly proportional to the mixture calorific value H_G, along with air-fuel ratio λ and effective efficiency η_e according to the equation $p_e = \lambda \cdot H_G \cdot \eta_e$. Diagram 5.125 shows the theoretical full-load potential compared to conventional diesel engines using the mixture calorific value of different natural gas and methane mixture formation processes. The diesel combustion process used an excess air ratio of $\lambda = 1.2$ and methane a value of $\lambda = 1$. The boundary conditions concerning efficiency, air requirement, speed and displacement are assumed to be constant. There are advantages for all natural gas processes compared to diesel operation due to the absolute lower mixture calorific value with diesel engine specific air-fuel ratios compared to stoichiometric methane operation. In particular, the injection of low temperature natural gas into the inlet manifold leads to a cooling of the cylinder charge mass. The gaseous injection of methane at approximately 120 K into the inlet manifold results in a decrease in temperature of approximately 20 K, which leads to a theoretical increase in power potential of 7 %. Additional positive effects of the charge cooling with

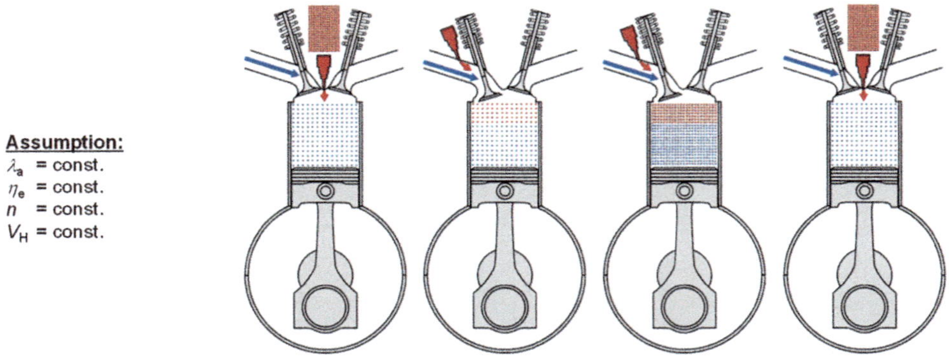

Fuel	Diesel	CNG	LNG	CNG
Mixture formation	Direct injection	Intake pipe	Intake pipe „cryo“	Direct injection
Mixture temperature [K]	293	293	273	293
Excess air ratio [–]	1.2	1	1	1
Mixture cal. value [MJ/m³]	2.93	3.17	3.40	3.50
Specific power [%]	100	108	116	119
	„State of the art“		„Advanced/Research “	

Diagram 5.125 Theoretical comparison of different methane (natural gas) mixture formation concepts ($\lambda = 1$) with diesel direct injection ($\lambda = 1.2$)

respect to combustion anomalies and the formation of NO_x will be explained in more detail in Sect. 5.3.5. The direct injection of CNG compared to port fuel injection of natural gas results in approximately 10 % higher BMEP and thus specific power. The actually achieved full-load depends heavily on the attainable efficiency (influenced by combustion, knock, emissions, etc.) and the attainable air delivery ratio (or volumetric efficiency), whereby the differences documented in Diagram 5.123 also occur.

5.3.2.2 Efficiency

In order to assess the achievable efficiency potential of combustion processes, the ideal process for a complete engine is used as a baseline [124]. The compression ratio and the properties of the gas are the relevant parameters. Assuming a real natural gas combustion process with stoichiometric air-fuel ratio in part-load and one observed (at first without considering the feasibility) a variation of the parameters ε, λ and the 50 % mass fraction burned (MFB50) point, then the effect of efficiency would be as shown in Diagram 5.126.

In reality, the limited boundary conditions knock-limit, acceptable component temperatures predominantly derived from diesel engines and emission limits lead to significant deviations from the theoretical thermodynamic consideration. The influence, as measured on a real engine, of the compression ratio on efficiency is displayed in Diagram 5.127 and shows the limited increase in efficiency with compression ratio limited by knock for part-load and full-load point.

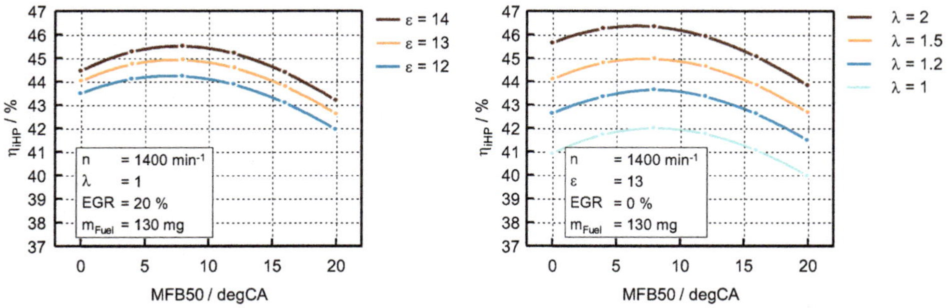

Diagram 5.126 Theoretical indicated high-pressure efficiency as a function of mass fraction burned 50 % for a ε-variation with EGR (*left*) and a λ-variation without EGR (*right*) (Source: IVT, TU Graz)

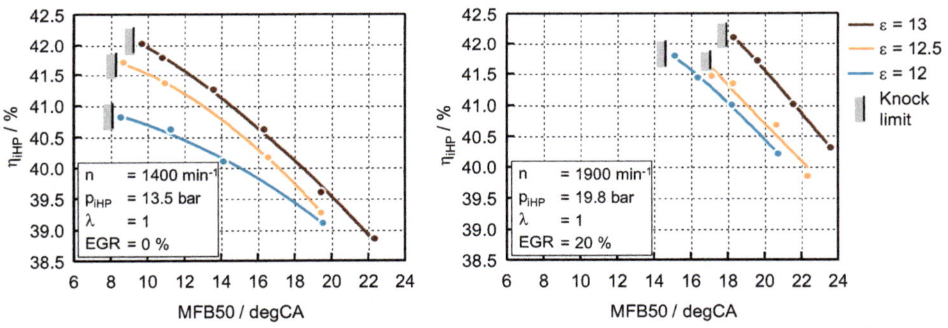

Diagram 5.127 Indicated high pressure efficiency as a function of the MFB50 % point for part-load (*left*) and full-load (*right*), measured on a single cylinder research engine (Source: IVT, TU Graz)

The influence of a lean burn combustion can only be investigated in reality within certain limits. These limits are primarily defined by the occurrence of too high cyclical variations, a safety margin to the misfire limit and too high emissions of unburned fuel. They are shown in Diagram 5.128 starting from stoichiometric operation.

The relationship between knock limit, ignition limit (misfire limit) and exhaust gas temperature is shown in Diagram 5.129.

Split of Loss Analysis

Diagram 5.130 shows split of loss analysis for a single cylinder research engine with a swept volume of approximately 2 l. The unit under test was used to analyse different natural gas combustion methods and compare them with pure diesel operation. In particular, the following combustion processes on the basis of EURO V were compared to each other at part load ($n \approx 1500$ rpm, $p_i \approx 12$ bar) and a full-load point ($n \approx 1900$ rpm):

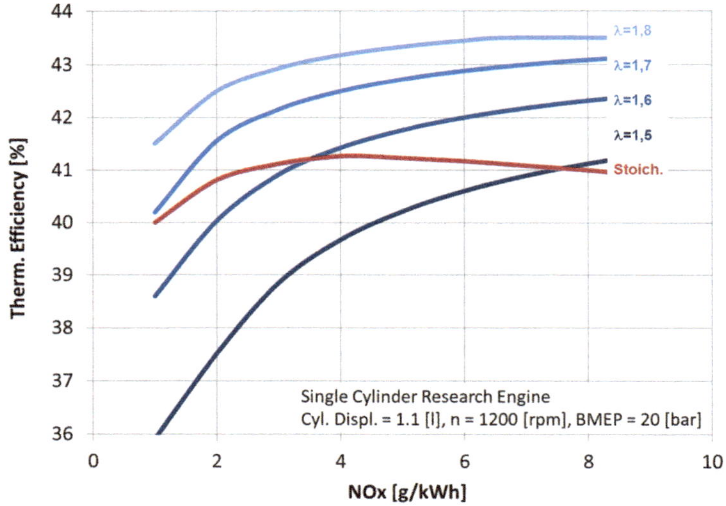

Diagram 5.128 Influence of excess air ratio on efficiency [125]

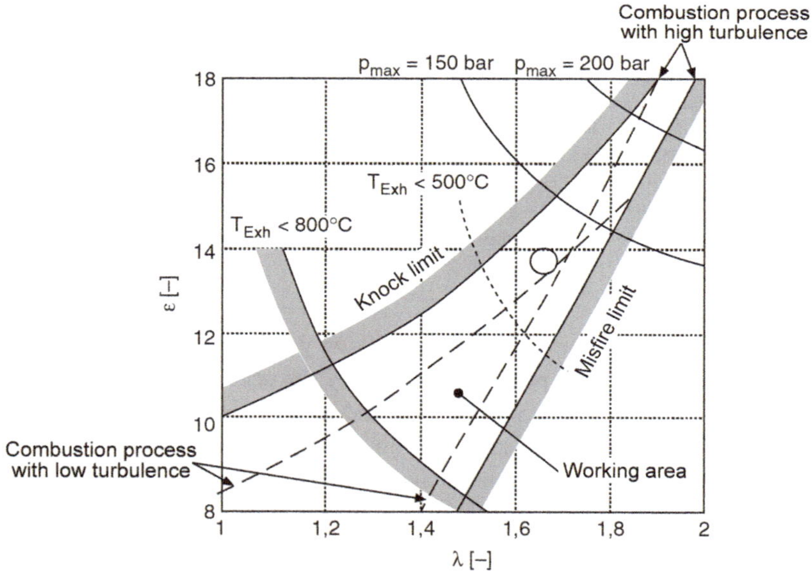

Diagram 5.129 Operating area of the natural gas combustion process [126]

- Diesel operation with EGR
- Natural gas (CNG) lean operation without EGR
- Natural gas (CNG) $\lambda = 1$ operation with EGR
- Natural gas/diesel Dual-Fuel operation with EGR

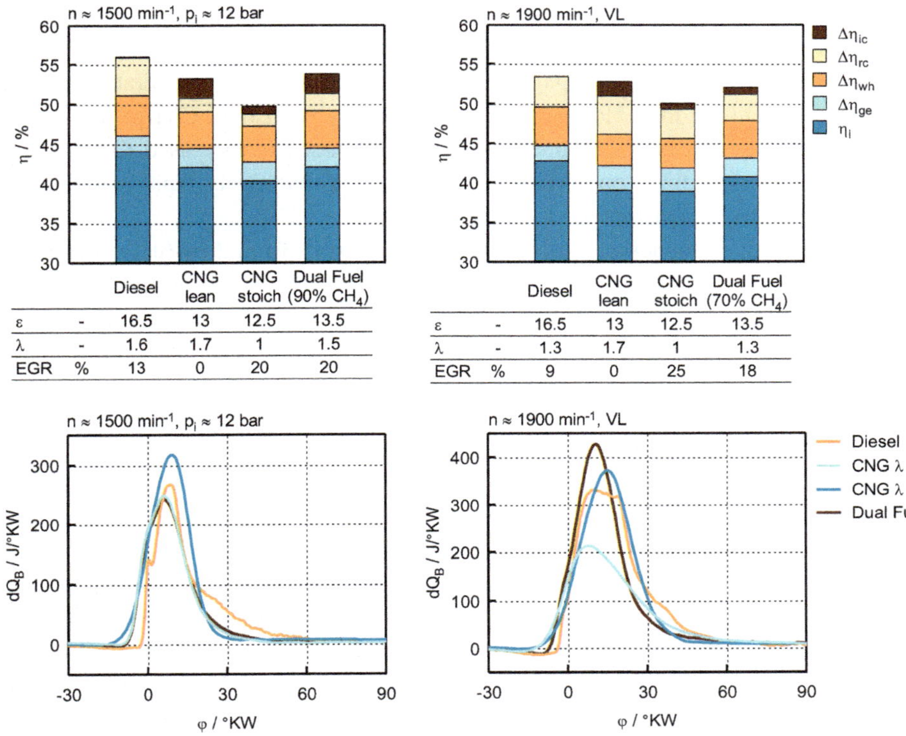

Diagram 5.130 Split of loss analysis and corresponding combustion traces of different natural gas combustion methods compared to pure diesel operation at part-load (*left*) and full-load (*right*) on a single cylinder research engine (Source: IVT, TU Graz)

The basis for the loss analysis is represented by the efficiency of the complete engine with real charge (including EGR). The diesel engine efficiency is the highest, since here the compression ratio is the highest (16.5) and the excess air ratio is considerably higher than 1. Both points with diesel combustion are characterized by almost complete combustion with low CO and HC emissions. The non-pre-mixed combustion results after a first combustion pulse to a delay in the final phase of fuel conversion. This results in relatively high losses in the real combustion. This effect is intensified through the addition of EGR, which generally slows down combustion.

The efficiency in lean CNG operation of the complete engine is still very high due to the very lean operation at $\lambda = 1.7$, despite a significantly lower compression ratio. At this relatively high air-fuel ratio for natural gas, the combustion is very close to the misfire limit. The flame is quenched next to cold walls before the fuel is completely burned. This causes increased CO/HC emissions and thus increased losses due to the incomplete combustion. The combustion is however very fast and compact, meaning that the losses in the real combustion process are low.

Stoichiometric combustion of natural gas in these comparison tests results in the lowest efficiency of the complete engine due to the lowest values for air-fuel ratio and compression ratio. Since flame front propagation is higher in stoichiometric operation than in lean operation, there are lower losses due to incomplete combustion. Charge dilution in the form of EGR also results in flame quenching and hence to increased CO and HC emissions compared to diesel combustion. Due to the comparatively rapid combustion, the losses in the real combustion are lowest in this case.

A relatively high efficiency of the complete engine results from dual-fuel combustion mode, where the gas injected into the inlet manifold is ignited by a diesel injection, caused by the slightly higher compression ratio compared to other natural gas combustion methods and the lean air-fuel ratio. The losses from incomplete combustion in this case are very dependent on air-fuel ratio and load. Depending on the amount of diesel substitution, different global air-fuel ratios occur in the combustion chamber. With a low share of natural gas (and also at low loads), the air-fuel ratio can lie outside the ignition limit and thus very negatively influence the diesel injection operation with regard to unburned fuel. Furthermore, the resulting slower combustion leads to higher losses in real combustion compared to the other CNG methods.

The wall heat losses are all on very similar levels for all part-load points despite considerable differences in compression and air-fuel ratios, since both parameters partially compensate each other. They differ to a certain extend depending on load and combustion duration at full-load measurements. It has to be noted that the gas exchange losses of the single cylinder research engine are only approximately comparable, since the boundary conditions of a real exhaust gas turbocharger cannot be accurately represented.

5.3.2.3 Emissions

Diagram 5.131 shows the raw emissions trade-off of NO_x and PM for the single cylinder diesel research engine for EGR variations, and the NO_x and PM raw emissions for the natural gas combustion under investigation at 1500 rpm and approximately 12 bar BMEP. The lean burn concept (CNG $\lambda > 1$) complies with the limits for emission level EURO V concerning NO_x and particulate emissions without additional exhaust aftertreatment. The

Diagram 5.131 Soot-NO_x trade-off (Source: IVT, TU Graz)

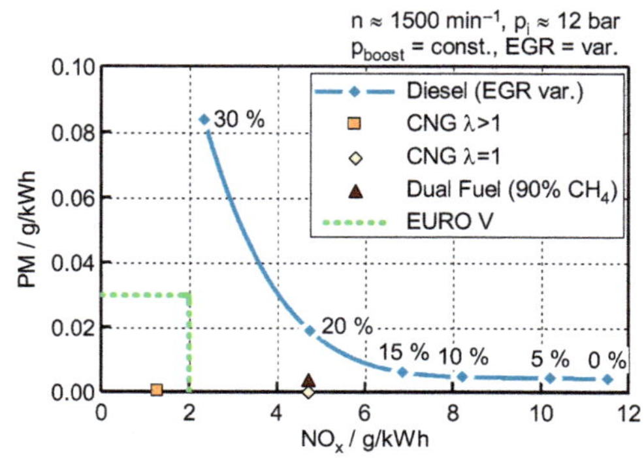

particulate emissions are extremely low due to the homogenous air-fuel mixture. The higher combustion temperatures occurring with the stoichiometric combustion concept (CNG $\lambda = 1$) lead to much higher NO_x raw emissions in comparison to the lean combustion concept. These are brought to the required emission levels through the use of a 3-way catalyst (TWC). The dual-fuel concept (in this case with a gas energy ratio (GER) of 90 %) also lies below diesel combustion with respect to NO_x and particulate levels.

The efficiency and raw emissions change of the investigated natural gas combustion concepts relating to the conventional diesel concept for a part-load point ($n \approx 1500$ rpm, $p_i \approx 12$ bar) and a full-load point ($n \approx 1900$ rpm) can be seen in Diagram 5.132. The natural gas combustion concepts achieve significant CO_2 savings over the diesel combustion concept despite the disadvantages due to efficiency [127]. The reason for this is the low C/H ratio in natural gas and in gas—diesel mixture, whereby the CO_2 saving for a dual-fuel concept with increasing diesel share is lower. The NO_x and particulate emissions are, as described earlier, at a lower level than for the diesel concept and can be reduced to the emissions limits with appropriate exhaust aftertreatment systems. A trend in the HC and CO raw emissions in natural gas operation towards pure diesel combustion can be recognized. This is due to quenching effects with increasing dilution and the partial extinguishing of the flame outside the ignition limit, as well as high charge dilution. The HC emissions consist mainly of methane, which due to its low reactivity requires high catalyst light-off temperatures [128]. A downstream 3-way catalyst is used for the exhaust aftertreatment for stoichiometric combustion concepts. A methane oxidation catalyst is used with lean and dual-fuel concepts for the reduction of the methane emissions. The high air fuel ratio and the resulting lower exhaust gas temperature significantly reduces the conversion rate of the catalyst compared to stoichiometric operation. Even though EURO V emission limits can be achieved it is still an open issue whether, and under which conditions, EURO VI limits can be met.

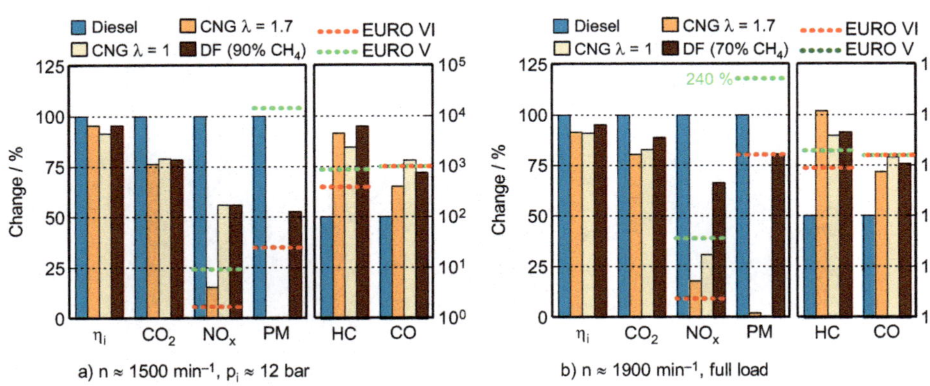

Diagram 5.132 Comparison of efficiency and emissions of different natural gas combustion concepts compared to pure diesel operation at (**a**) part-load, and (**b**) full-load on a single cylinder research engine (Source: IVT, TU Graz)

5.3.2.4 LNG Tank Technology

Pressurized gas tanks are predominantly used today to store natural gas in a vehicle at a pressure of between 200 and 250 bar. More details concerning the gaseous storage of natural gas are provided in Sect. 5.1. Due to the considerably higher storage density, the liquid storage of natural gas and methane specifically for commercial vehicles is an interesting solution. Current LNG tanks are largely designed for the North American market where space requirements are not as much in the foreground as they are in the European market. Typical storage pressures lie between 7 and 10 bar and this pressure level is directly applicable for use in engine operation with external mixture preparation. Today's LNG tanks are made of stainless steel and are designed for maximum pressures of up to 15 bar, Diagram 5.133 shows an example installation.

Storage at LNG filling stations is typically between −160 to −145 °C (cold LNG), which corresponds to vapour pressures of approximately 1–3 bar and densities of 422–399 kg/m^3. LNG is available with a high degree of purity due to the cryogenic production process. The different boiling points of the natural gas components at low temperatures leads to a practically perfect separation of impurities from methane and results in a high degree of natural gas purity.

However, the high purity of LNG also has a negative effect on the lubrication effectiveness of gas engine components (injector systems, valves, etc.) and must be considered in the design process.

Currently available LNG tank systems have largely been designed for the storage of saturated LNG at pressures between 7 and 10 bar (Diagram 5.134). The injection pressures

Diagram 5.133 LNG tank system (Source: Volvo Trucks)

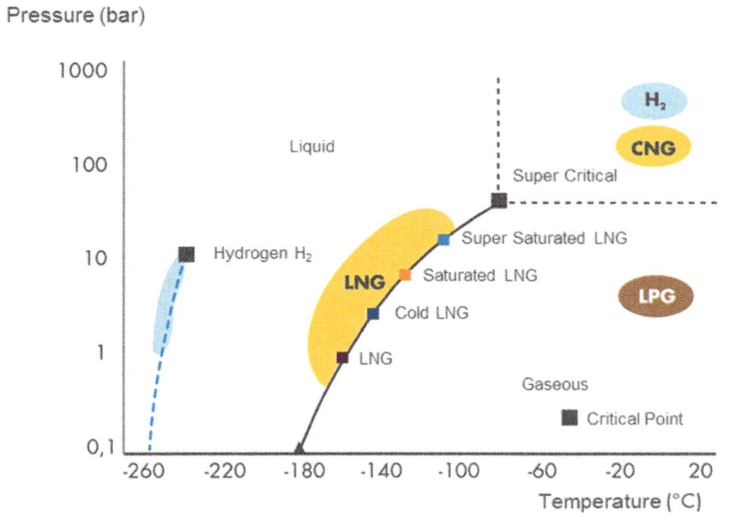

Diagram 5.134 Pressure and temperature ranges of LNG [129]

of port fuel injected SI engines and dual-fuel engines currently on the market lie in this pressure range.

The storage of saturated LNG leads to the following disadvantages compared to cold LNG concerning the functionality of LNG tank systems and the corresponding filling station infrastructure:

- The low density of saturated LNG reduces the range for the same tank size by up to 15 %.
- The filling pressure compared to cold LNG is higher. This results in increased levels of safety being required for the user when filling.
- The cold LNG available at the filling station must be heated to saturated LNG before the filling process can begin, which leads to effort for the filling station operator and time required to fill the tanks.

Injection pressures of up to a maximum of 20 bar (currently 10 bar) are expected for future spark ignited and dual-fuel natural gas engines in order to further improve the efficiency and to comply with future emission legislation. The storage of saturated LNG at these pressure levels no longer makes sense from a technical and economic point of view, since the wall thickness required for the inner tank skin would result in a much higher tank mass and correspondingly higher costs. Above all, current LNG filling stations are not equipped for the generation of saturated LNG up to 20 bar. Due to the disadvantages mentioned (range, time to fill, etc.) when using saturated LNG, tank manufacturers are concentrating their research activities on the storage of cold to very cold LNG. The main difficulties in storing cold LNG are the generation of the required system pressure for engine operation (cryogenic pump is required) and the increased requirements on tank insulation. The required injection pressure for high-pressure direct injection concepts (currently max. 300 bar) is produced by an LNG pump integrated into the tank (e.g. layout shown in Diagram 5.150).

Diagram 5.135 shows the schematic layout of a multi-layer vacuum insulated LNG tank without additional LNG pump for the storage of saturated LNG, as can be found today. The tank is filled with liquid gas from above via a filling receptacle at temperatures between −100 to −125 °C and a maximum pressure of up to 16 bar. The extracted gas is converted from liquid to gaseous form via a cooling water heat exchanger. The high pressure of the gas after the vaporization phase is then reduced to the operation pressure for the low-pressure common rail. The pressure in the storage tank increases over time due

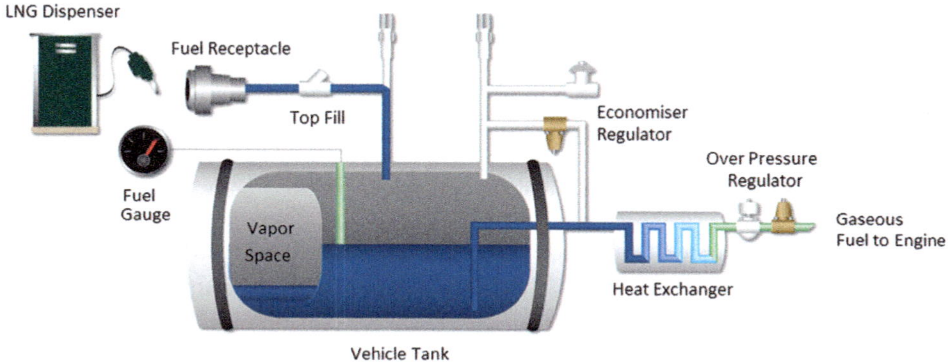

Diagram 5.135 LNG storage (Source: Chart Industries) [126]

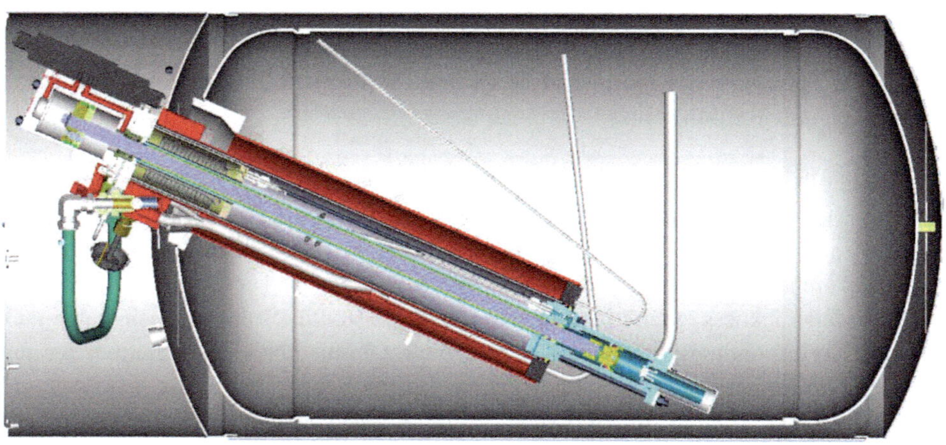

Diagram 5.136 LNG storage with integrated cryogen pump (Source: CWI/Taylor-Wharton) [131]

to the influence of external heat absorption which causes the liquid gas in the tank to evaporate. The tank pressure must be reduced during long periods of standstill via an overpressure valve, which vents the gas into the atmosphere. During engine operation, the natural gas is fed to the engine from the gaseous phase via a pressure-controlled valve [130].

5.3.3 Spark Ignited Natural Gas Engines

The majority of natural gas engines for commercial vehicles currently on the market are based on a spark ignited engine concept that is either run in lean-burn, mixed mode or stoichiometric mode. The example in Diagram 5.137 is of a typical commercial vehicle engine with a swept volume of 2 l/cylinder and shows the operating modes plotted against the specific NO_x emissions that vary according to the NO_x emissions requirements.

NO_x emissions rise rapidly with increasing air excess from the stoichiometric air-fuel ratio onwards and reach their maximum caused by the increase of the adiabatic flame temperature between $\lambda = 1.1–1.2$ [124]. Beyond this point, the specific NO_x emissions fall with increasing air dilution. However, the emissions of unburned hydrocarbons, predominantly CH_4, rise steeply due to the increasing flame quenching with rising air excess.

Since efficiency is higher with increasing excess air ratio (see Sect. 5.3.2.2), lean-burn concepts were mainly used in the past. A tightening of emission legislation has, however, led to the lean-burn operating window to be placed very close to the misfire limit in order to reduce the NO_x levels in accordance with EURO IV and EURO V emission

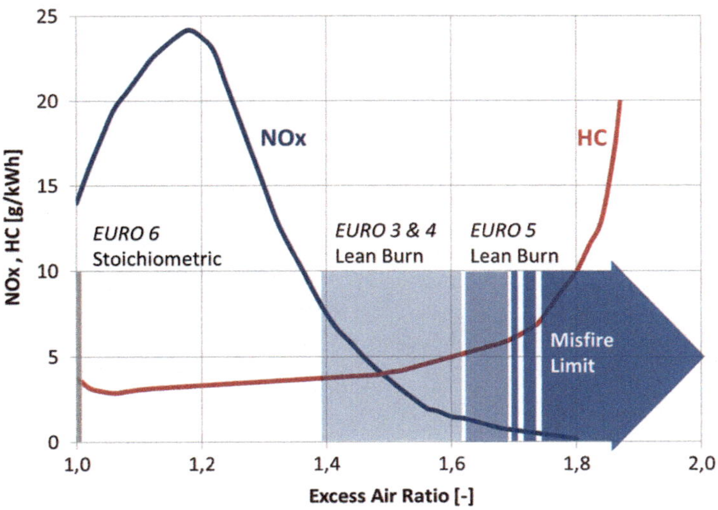

Diagram 5.137 Lean operating limits for increasing lean-burn

requirements. A further increase of excess air ratio to fulfil the current EURO VI and future emission legislation is no longer feasible. Instead, only stoichiometric combustion concepts are currently being used and combined with three-way catalyst.

5.3.3.1 Mixture Formation

Nowadays, mixture formation in natural gas commercial vehicles is performed via single point fuel injection, SPFI, or multi point fuel injection, MPFI, as shown in Diagram 5.138. Low pressure fuel injection (Diagram 5.141) that can occur early or late in the working cycle depending on the pressure level is still in the early stages of development for commercial vehicles.

In a conventional CNG system, the natural gas is stored in gaseous form in several high-pressure storage cylinders at approximately 200 bar. The gas pressure, depending on the system design, is limited to approximately 7 bar via shut-off valves and a pressure controller and feeds to the gas injectors. An intake throttle is used to control the air mass flow to the inlet manifold, into which the natural gas is injected, as shown in Diagram 5.139.

Since fuel injectors with a sufficient flow rate are not always available for commercial vehicle gas engines, several injectors are often grouped together into a module with a gas mixer to form a central gas mixer. This technology can be clearly seen in the example of the Daimler M936NGT shown in Diagram 5.140. A central gas mixer can also be achieved with just one gas injector and a gas mixer, for example the Cummins Westport natural gas engine ISX12G.

One of the advantages of central gas mixing is the very good mixture homogenization, which is normally done in a venturi mixer in combination with a long mixing path to the inlet valve. A further advantage is the simple integratability in an existing engine concept where the inlet manifold from a diesel engine can be used. The disadvantages of

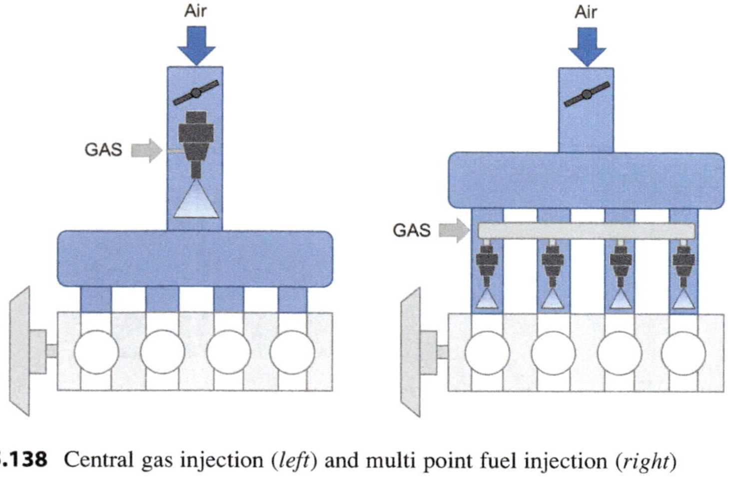

Diagram 5.138 Central gas injection (*left*) and multi point fuel injection (*right*)

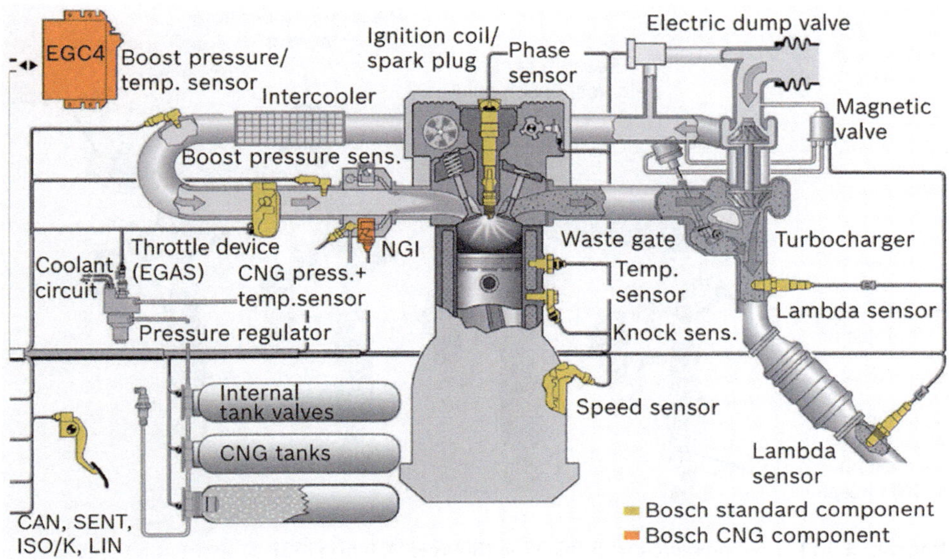

Diagram 5.139 Natural gas system for spark ignited commercial vehicles (Source: Bosch)

Diagram 5.140 Example application of a gas mixer (Daimler M936NGT)

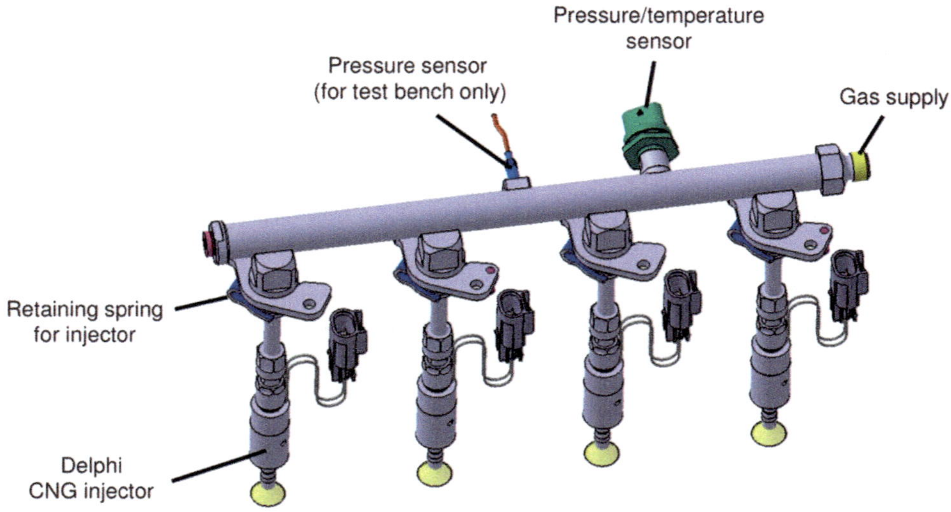

Diagram 5.141 Low pressure direct injection (Source: Delphi) [132]

centralized gas mixture formation are the variations in combustion gas/air mixture during transient phases, since a large amount of mixture is stored in the inlet manifold. Additionally, cylinder-specific measurement of fuel metering is not possible and cylinder-specific faults with respect to OBD cannot be easily diagnosed. The advantages of inlet manifold or port fuel injection with one or two injectors per cylinder are the drastically improved quality of the mixture during transient phases, the possibility of cylinder-specific injection quantity control and gas shut-off in motoring conditions. Mixture homogenization is more difficult to optimize due to the very short time and distance available for the mixing process. Injection duration is limited by the volume of the inlet manifold and the cylinder "crosstalk", which is why the flow rate of the injectors has to be exactly matched to engine performance. Currently, due to the relatively low production volumes of natural gas commercial engines, this does not seem to be possible.

Low-pressure direction injection with a pressure level of approximately 20 bar during the low pressure cycle or the early compression phase has all the advantages of port fuel injection and also increases the volumetric efficiency (see Sect. 5.3.2.1). Low-pressure direct injection also enables long injection times, even when the inlet valve is closed, charge motion and level of turbulence can also be positively influenced using a suitable injection strategy. Systems with low-pressure direct injection for passenger cars are already about to enter series production, whereas for commercial vehicles they are still only in the pre-development phase.

5.3.3.2 Ignition

Ignition systems on gas engines for commercial vehicle applications today exclusively use spark ignition via a spark plug. The ignition system is identical in principle to the ignition

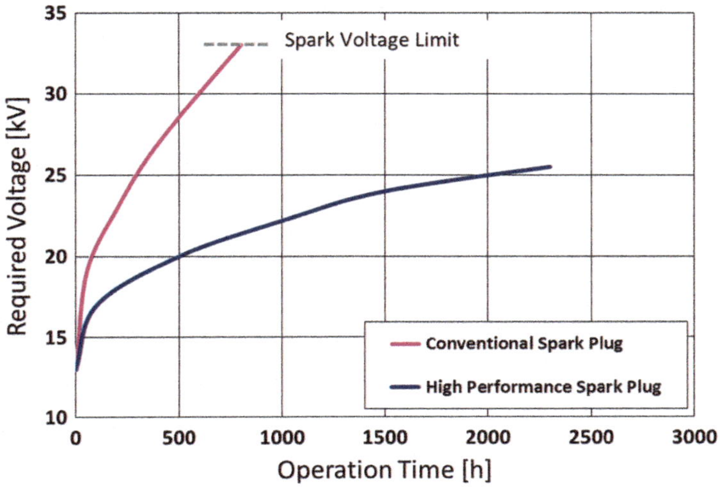

Diagram 5.142 Ignition voltage increase with advancing electrode wear

concepts used for passenger car Otto engines (compare with Sect. 5.2.4). As a rule of thumb, one ignition coil is employed per cylinder together with the most durable spark plug possible in order to achieve the long service life required.

Commercial vehicle engines of the typical cylinder bore diameter and speeds range exclusively use spark ignition. The flame speed in the combustion chamber is fast enough at stoichiometric combustion to achieve thermodynamically sufficient burn rates, even with substantial charge dilution by EGR (see Sect. 5.3.2.2). However, for lean burn operation, where the operating range for low NO_x emissions are very close to the misfire border, very high ignition voltages (>30 kV) are necessary, which in turn cause higher spark plug electrode wear.

The electrode wear of spark plugs is still one of the crucial issues, particularly in comparison with Diesel-ignited engines. As shown in Diagram 5.142, advancing electrode wear, depending on the electrode material, increases the ignition voltage until the ignition system can no longer sustain the secondary voltage and misfire results. To achieve the replacement intervals of 100,000 to 120,000 km for spark plugs, commercial vehicles normally use spark plugs with iridium, or iridium-rhodium electrodes.

5.3.3.3 Combustion and Exhaust Aftertreatment

Lean burn concepts are mainly used up to emission level EURO V, as already mentioned, since this concept is adaptable without major changes to the underlying diesel engine platform. The thermal loading is only slightly higher compared to the diesel engine, which enables the adoption of diesel components. However, the disadvantages are the limited power density and unsatisfactory transient behavior. Lower NO_x emissions are achievable through lean-burn concepts, but the ability to run lean is limited by the ignition border at higher air dilution rates, as shown in Diagram 5.137. As discussed earlier, it is no longer sufficient to use a lean-burn concept to reduce the NO_x to fulfil current EURO VI and future emission legislation. An SCR exhaust-aftertreatment system is therefore necessary to reduce NO_x (Diagram 5.143). The greatest challenge lies in the durability of a methane-selective oxidation catalyst, which must attain an efficiency of over 90 % over an emissions life period of 700,000 km. Due to the high sensibility to sulfur, the long-term stability of methane conversion is very difficult to maintain, particularly since catalyst ageing as a result of sulfur poisoning shifts the light-off temperature towards much higher values and this makes additional $DeSO_x$ strategies necessary to maintain the conversion rates of methane to an acceptable level. Currently available substrates do not work for a cost-effective lean-burn concepts for EURO VI, hence the stoichiometric combustion concept will, for the time being, represent the only engine concept for spark ignited engines.

This is the reason why the stoichiometric combustion concept is strongly favored for modern commercial engine concepts from emission level EURO VI onwards. With a three-way-catalyst, it represents by far the most cost-effective solution. The greatest challenges in this concept are the very high thermal loads on the piston, cylinder head, exhaust valves and turbine housing. Compared to a diesel engine, the heat rejection to the engine cooling system for a stoichiometric combustion concept can be up to 15 % higher, which must be considered during the vehicle design phase.

Cooled exhaust gas recirculation (EGR) has proven itself to be an effective means of reducing the high thermal loads. In the example shown in Diagram 5.144, EGR rates between 20 and 25 % are able to reduce the turbocharger inlet temperature by

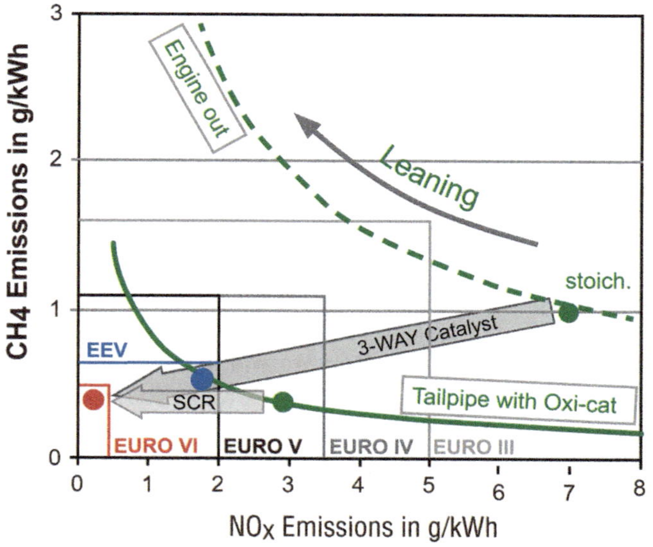

Diagram 5.143 Emissions trade-off for CV gas engines [128]

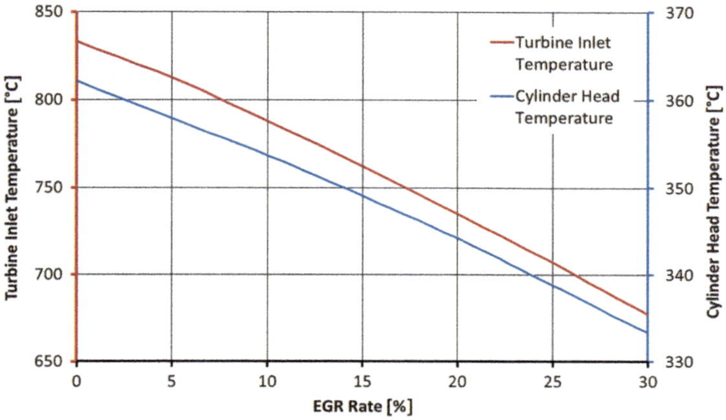

Diagram 5.144 Reduction of thermal load via exhaust gas recirculation [125]

approximately 100 °C. In addition, the recirculated, cooled exhaust gas also helps to reduce knock sensitivity at high loads and thus enables a higher geometric compression ratio in association with a higher thermal efficiency.

However, the positive effects of exhaust gas recirculation are limited by the pressure difference between the exhaust manifold and the inlet plenum, which limits the generation of EGR at low speeds. This typically leads to the use of relatively low turbine sizes with correspondingly high waste-gate mass flows. At part load, EGR contributes to a reduction in pumping losses, but is restricted by reduced combustion stability and associated misfire limits.

The application of cooled EGR and optimized coolant flow in critical areas of the fire deck enables the high heat rejection to be managed to a certain degree [125], but still is an engineering challenge with respect to the required durability. Over and above this, the extension of the knock limits towards higher loads and the trade-off between power density and efficiency require intensive optimization effort. The stoichiometric combustion concept has a considerable advantage over the lean burn concept with regards to the lower complexity of exhaust after-treatment, since it enables the use of a three-way catalyst. However, it should be noted that a TWC used for natural gas engines contain a 3–4 times higher precious metal loading than gasoline Otto engines, which is required for the more complex oxidation of methane. The operating window of stoichiometric natural gas engines compared to gasoline engines is shifted towards richer operation, since this is where the highest conversion of methane can be attained, as shown in Diagram 5.145.

The narrower window for natural gas engines with TWC results from the worse reactivity of CH_4 with O_2, which is the dominant reaction in lean-burn mode. Long-chain hydrocarbons in the lean-burn operating regime generally show a higher reactivity than methane and are responsible for the rapid fall in NO_x conversion efficiency above $\lambda > 0.99$, as shown in Diagram 5.145.

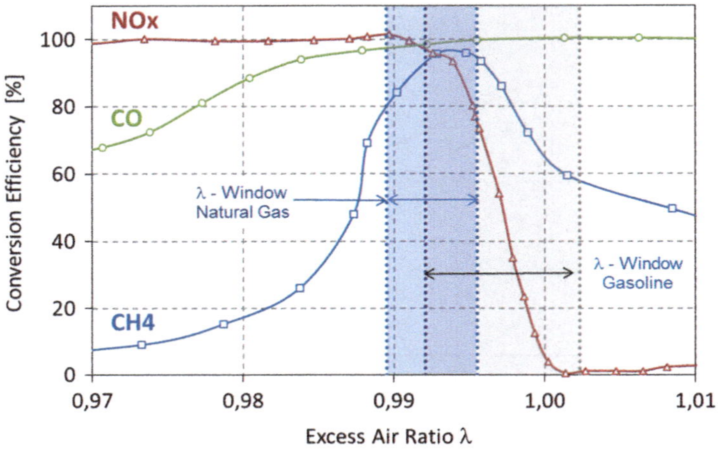

Diagram 5.145 Operating window for a gas engine with three-way catalyst

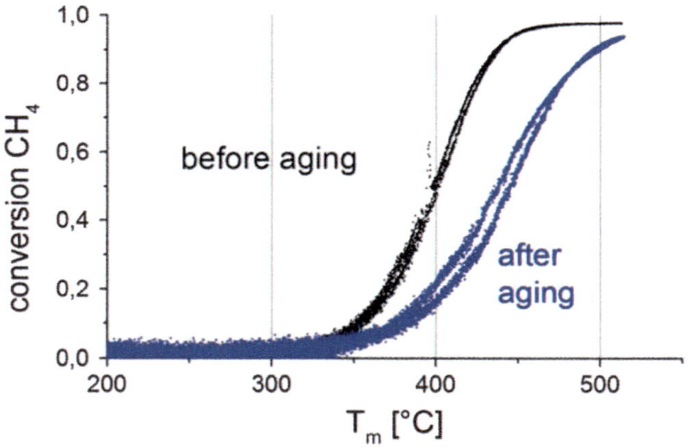

Diagram 5.146 Methane conversion, conversion efficiency measurements for and after aging (100 h at 450 °C) [133]

Achieving corresponding long-term stability of efficient methane conversion is a huge technical challenge, since as a result of thermal and chemical catalytic ageing processes the threshold is shifted towards higher temperatures. Thermal ageing can lead to a drastic shift in the conversion rate of methane, as shown in the example for a substrate of a three-way catalyst, which was thermally aged to simulate a life of 700,000 km.

Fresh substrates, as shown in Diagram 5.146, can achieve methane conversion rate of above 98 % at only 450 °C, whereby the conversion rate efficiency at 450 °C collapses to 65 % for an aged substrate. Even at temperatures above 550 °C, conversion efficiency rates above 90 % are hardly achievable. This characteristic must be taken into account in the development phase of natural gas engines and exhaust-aftertreatment systems. Ideally, locating the catalyst close to the engine together with the implementation of thermal management is necessary to achieve the corresponding minimum exhaust gas temperature for sufficient conversion of methane in the WHTC test cycle, which is relevant for commercial vehicles.

5.3.4 Natural Gas Engines with Compression Ignition

Alongside spark ignited engine concepts, different approaches of diesel-ignited engine concepts have established themselves for commercial vehicle natural gas engines, these are so-called dual-fuel engines and engines using high-pressure gas direct injection. Both engine concepts are based on the principle of compression ignition and use directly injected diesel as an ignition source for the combustion of natural gas.

The dual-fuel concept, where gas is injected into the inlet manifold (Diagram 5.147, left), can be differentiated into two approaches, a micro-pilot procedure and the conventional dual-fuel approach, where the majority of the diesel amount is replaced by natural gas. The micro-pilot concept, which employs a diesel pilot injection to ignite the gas/air mixture, is currently exclusively used on large engines. Both of the approaches used for today's commercial applications, the substitution-dual-fuel approach and the high-pressure gas direct injection (Diagram 5.147, right), are still fundamentally different in view of mixture formation, combustion concept and also with regards to the peripheral gas system components that will be described below.

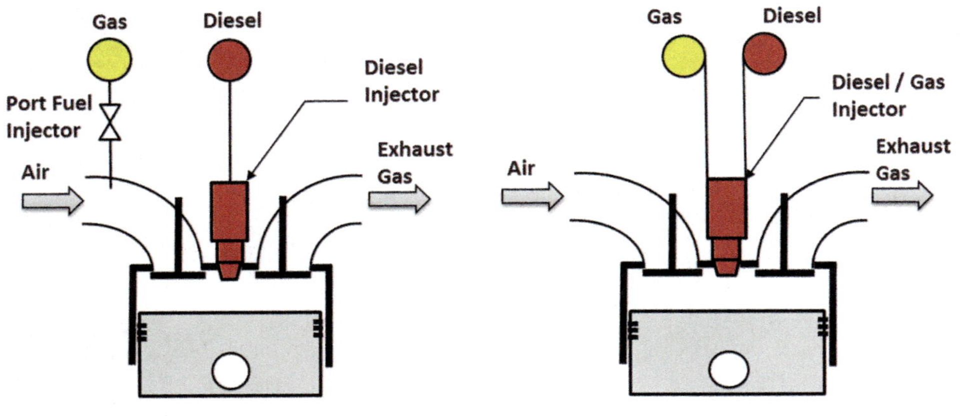

Diagram 5.147 Dual-fuel approach (*left*) and high-pressure gas direct injection (*right*)

5.3.4.1 Dual-Fuel Engine Concept

Dual-fuel engines for commercial vehicle applications are typically adapted from conventional diesel engines. Natural gas is injected into the inlet manifold and hence mixed with the diesel engine's combustion air, as shown in the schematic diagram in Diagram 5.148. Ignition and combustion events are basically no different to those in conventional diesel processes. The base engine is typically not modified with respect to injection equipment, combustion chamber and turbocharger, just supplemented with a gas tank system (CNG or LNG), gas control path and gas pressure controller. The adaptation to a dual-fuel engine can, depending on the application, substitute 60–80 % of the diesel fuel with natural gas. Cold start, lower part load and operating the engine at low speeds are normally run in pure diesel operation. The actually achievable substitution rate mainly depends on the architecture of the base engine, the average BMEP level and the complexity of the engine management system used, which is mainly run independently from the diesel engine management system.

One of the advantages of the dual-fuel concept is that the system can be changed to pure diesel operation even at full load and thus a long driving range can be achieved even with limited gas storage volume.

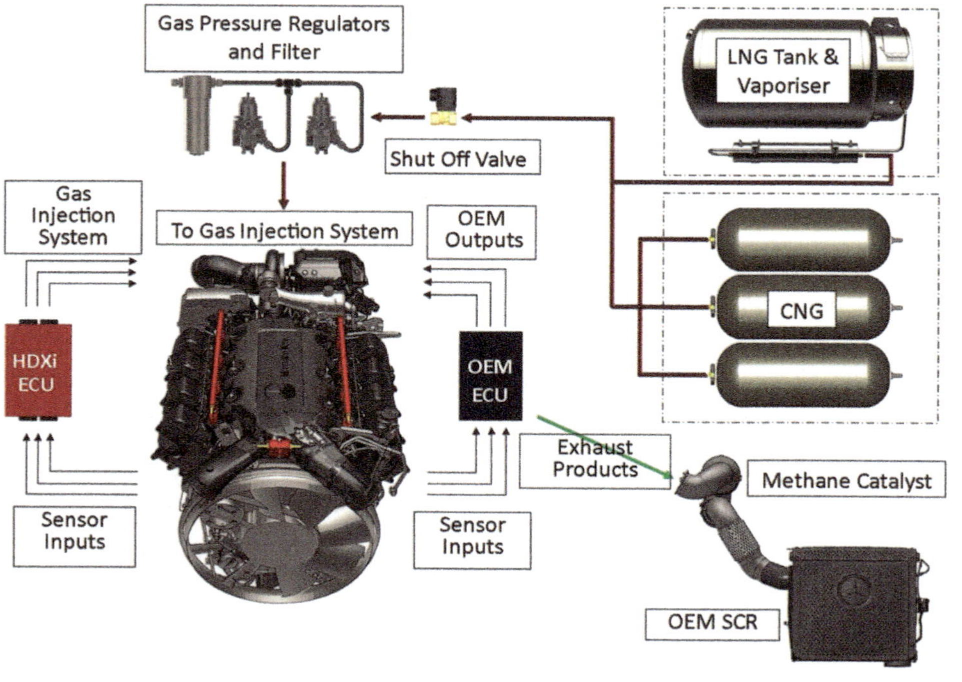

Diagram 5.148 Dual-fuel system for commercial vehicle engines (Source: Hardstaff Group)

The technical issues that must be solved regarding higher substitution rates are manifold. Poor flame propagation and incomplete combustion occur at lower part loads due to the lean-burn engine operation. In order to avoid the danger of misfire, this part of the engine control map is normally run in pure diesel mode. At high loads, the factors limiting the achievable substitution rates [134] are knocking combustion cycles, spontaneous self-ignition and high diesel injector nozzle tip temperatures. The major challenge however is compliance with current and future emissions limits. The dual-fuel combustion concept, which is marked by high methane slip, requires methane conversion rates for EURO VI that are not achievable with today's oxidation catalysts. In addition, dual-fuel engines require the same exhaust-aftertreatment components to achieve compliance with current NO_x and particulate emissions limits as modern diesel engines.

5.3.4.2 High Pressure Gas Direct Injection

An innovative concept for heavy-duty commercial vehicle engines is the high-pressure gas direct injection, where natural gas is injected directly into the combustion chamber at high pressure using a special diesel/gas injector (Diagram 5.149) [135]. A small pilot injection of diesel fuel prior to the injection of gas leads to diffusion combustion of the injected gas that is very similar to that of the diesel process. This high-pressure gas direct-injection approach enables substitution rates of over 90 %, whereby power density, BMEP and transient behavior are practically identical with the base diesel engine. There are also no knock problems due to the combustion approach and the injection of gas late in the combustion cycle. This enables the high compression ratio of the diesel engine to be retained and, due to the lean combustion and unthrottled part load operation, the fuel consumption over the cycle is comparable to the diesel engine.

The high gas pressure required for direct injection favors LNG gas storage over CNG, since a gas pressure up to 300 bar is required for the high-pressure injection. Energetically this is best achieved by compressing liquid LNG. The LNG pump required

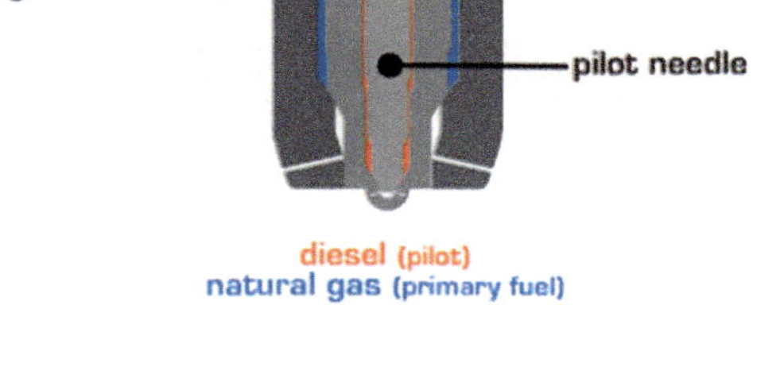

Diagram 5.149 Coaxial diesel/gas injector for high pressure gas direct injection (Source: Westport)

to do this is shown in Diagram 5.150 and is mounted directly in the LNG tank and driven hydraulically (Diagram 5.136).

The compressed gas is fed to vaporiser, filtered and stored in gaseous form at approximately 40 °C in an intermediate accumulator. The accumulator compensates for pressure variations and stores a small amount of gas for an engine start. The injection pressure for the diesel pilot injection is simultaneously produced by the diesel high-pressure pump and fed, together with the natural gas, to a conditioning module where the pressure of the pilot injection and the gas injection are regulated. Both fuels are then fed via distribution lines to a twin-needle injector. The adaptation effort for a base diesel engine is substantially higher than for an inlet manifold injection dual-fuel engine, since the wide-ranging design changes are necessary for the integration of the complete system

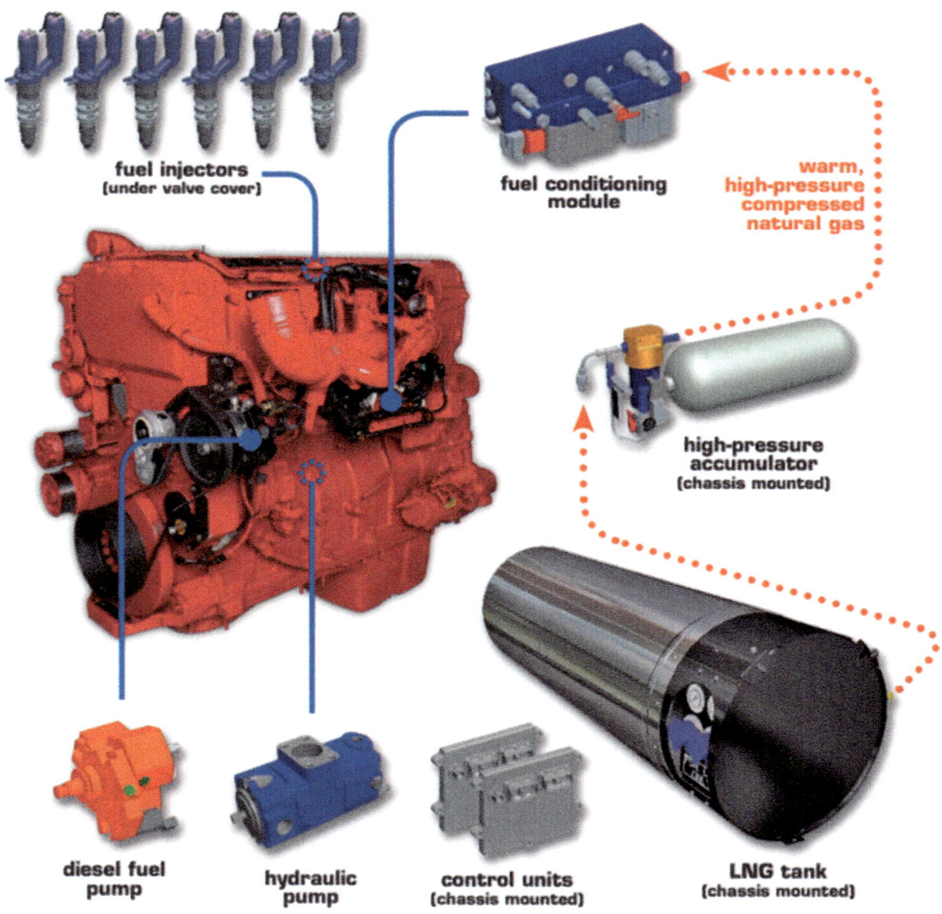

Diagram 5.150 High-pressure gas direct-injection—Westport HPDI (Source: Westport)

comprising injector and peripheral components. Regarding exhaust-aftertreatment, the required technology does not differ from EURO VI diesel engines. Oxidation catalyst, DPF and SCR are necessary to comply with the emissions limits for NO_x and particulates, which can be taken directly from the diesel application. Methane emissions, in contrast to inlet manifold dual-fuel engines, are manageable with conventional oxidation catalyst technology due to the diffusion combustion.

5.3.5 Alternative Ignition and Combustion Concepts

The reasons for dealing with alternative ignition and combustion concepts are multi-faceted. Drivers could be special emissions requirements, the local availability of special fuels or the desire to be independent from an energy carrier. A few of the following approaches are derived from large engines, where a diverse knowledge base and intensive experience exists due to a wider range of applications and the intensive occupation with alternatives, refer also to Chap. 6. So far however, none of the following approaches have been able to achieve (economical) significance in the passenger car and commercial vehicle domains.

Cryogenic Mixture Formation

Displacement of charge air by gaseous fuel leads to a reduction of the calorific value of the mixture and hence a lower power density in combustion processes with external gas-mixture formation. This can be partially compensated for both with (late) inner mixture formation and also with charge cooling (Diagram 5.125). In addition to this goal of cryogenic mixture formation, the reduction tendency for knock with falling temperatures and the reduced NO_x formation are further desirable effects. The boundary potential (using simplified assumptions) for the mixture temperature reduction with a stoichiometric mixture is approximately 20 K, assuming a gaseous addition of methane at 120 K (the boiling point is 112 K). An application of the vaporization heat would be tempting with a theoretical total temperature drop of 40 K, but is not insinuated here due to the still open issue of practicality. Although this method has been convincingly verified in experiments [136, 137], compared with hydrogen-air mixtures at −40 °C, i.e. a mixture calorific increase of 25 %, the effect is very low (Diagram 5.151). Due to the low potential and the difficulties of implementation, the incentive to use the application in mixture formation is significantly lower, hence the use of vaporization energy, for example for intercooling, appears recommendable.

Pre-chamber Concept

The results obtained in recent years are remarkable, particularly with regard to with pre-chamber approaches relating to efficiency in gas engines (see also Chap. 6) and suggest the necessity for investigative work into this approach also with commercial vehicle engines. In addition to the pre-chamber concepts, pre-chamber spark plugs are

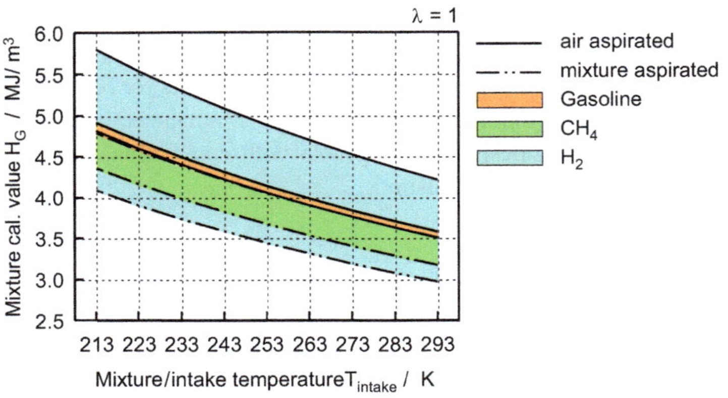

Diagram 5.151 Mixture calorific value as a function of temperature and fuel (Source: IVT, TU Graz)

also sometimes being employed, however, not yet for the application to commercial vehicles. Even if the main advantage of a pre-chamber combustion system, the rapid implementation of even very lean mixtures in association with large cylinder bores, is not relevant to the same degree, advantageous characteristics could still be of interest. However, no results from these investigations have been published.

HCCI Concept

The desire for an almost emissions-free (NO_x) combustion also leads to the consideration of this combustion concept also for gas engines for commercial vehicles. One look of the fuel characteristics of methane with a self-ignition temperature of almost 600 °C shows the challenges that prevent successful implementation with natural gas. Investigations with hydrogen, which has a practically identical self-ignition temperature, albeit with a factor of 10 lower ignition energy, give an impression of the difficulties. Even under these conditions, homogenous self-ignition cannot be represented meaningfully commercial vehicle applications and their mode of use [138]. Investigation on larger engines [139] show that no functioning HCCI concept for commercial vehicle powertrains is possible with natural gas without additional measures, for example an additional fuel.

Glow Plug Concept

Currently, engine concepts that use ignition via hot surfaces [140] are being investigated in order to guarantee ignition even under difficult conditions and the increase the lifetime of the ignition system. These approaches, however, have not been implemented in commercial vehicle engines, apart from in the early days of engine development, and are not foreseen.

Simultaneous Methane-Gasoline Combustion

The simultaneous operation of natural gas and gasoline, mentioned in [141] amongst other sources, whereby different ratios of the two fuels are used depending on the operating point, is based on the Otto principle for gasoline operation with the corresponding boundary conditions. Accordingly, the focus of applicability will be on passenger cars rather than commercial vehicles.

5.4 Hybridization of Passenger Cars

Florian Winke and Michael Bargende

The automotive industry is investigating the possibilities of different technologies to achieve a sustainable reduction in traffic-related CO_2 emissions. In the short term, an efficient lowering of these emissions can be achieved via the systematic optimization of conventional powertrain systems. Concepts utilizing alternative fuels or alternative powertrain concepts offer further potential in the medium term, whereby the intelligent hybridization of the powertrain can play a significant role. In the long term, there is the possibility of using purely electrically powered vehicles to achieve a substitution of the primary energy source for transportation by regenerative energy sources and thus to pave the way for completely emission-free mobility. Electric vehicles thus represent a complete replacement of existing powertrain systems, while the optimization of conventional systems, the further development of concepts utilizing alternative fuels and the improvement of alternative powertrain systems can complement each other to a large degree (Diagram 5.152).

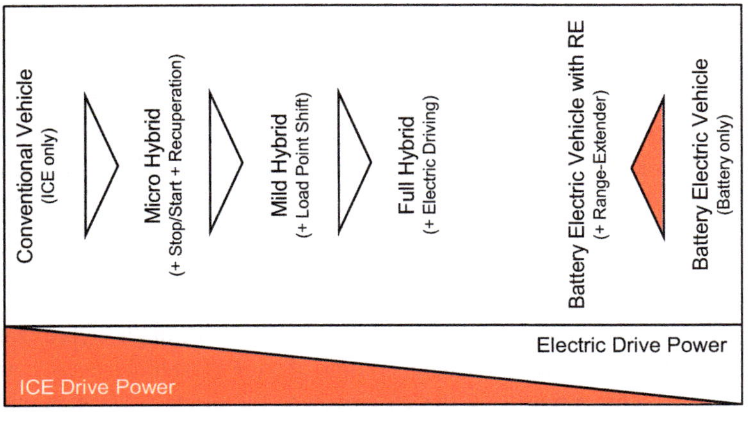

Diagram 5.152 Different variations of electrified powertrains

Hybrid powertrains offer the possibility of avoiding the existing problems of purely electrically driven vehicles to a large extent. Above all, the lower battery cycle depth enables the battery costs and weight to be lowered while simultaneously increasing the lifetime of the battery system. Since different electrical vehicle components are used in hybrid technology, these are often regarded as bridging technologies. A complete substitution by electric powertrains is, however, not assured and only conceivable in the long term [142].

The possible increase in efficiency due to the hybridization of combustion engine powered drivetrains is generally based on three pillars:

- Recuperation of kinetic energy (Regenerative Braking)
- Elimination of unfavorable combustion engine operation points by switching the engine off (Start/Stop, Electric Driving)
- Substitution of non-optimal combustion engine operating points with those with higher efficiency and lower emissions (Load Point Shifting)

Furthermore, the combination of a combustion engine with one or more electric drives normally permits a "downsizing" or a "rightsizing" of the conventional powertrain, since peak loads can be intercepted by the electric motor.

The implementation of alternative fuels and alternative powertrain concepts are mainly regarded as being in competition with one another. However, particularly the use of a natural gas engine concept within a hybrid powertrain shows that these approaches can complement one another. Some of the resulting advantages will be discussed in the following.

5.4.1 General CO_2-Potential

The use of natural gas as an alternative fuel within a hybrid powertrain concept offers an excellent possibility of achieving exceptionally low carbon dioxide emissions. Even without hybrid specific adaptations of the employed engine concept, significant reductions in greenhouse gas emissions can be realized due to the possible increases in efficiency through the hybridization and the advantages of natural gas described in the previous chapters.

The efficiency characteristics of turbocharged natural gas engines are particularly suitable for use in a hybrid system. A range of interconnections can be derived regarding the above-mentioned mechanisms to increase efficiency through hybridization. The achievable advantages through the recuperation of kinetic vehicle energy are mainly influenced by the vehicle mass and the efficiency of the electrical system and are thus independent to a large degree of the engine concept employed. However, the substitution

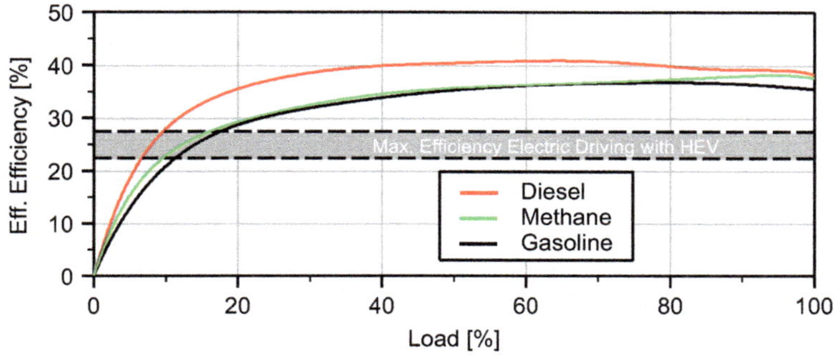

Diagram 5.153 A comparison of efficiency characteristics for different engine concepts (cf. [143])

of operating points in hybrid operation result in specific advantages when used with turbocharged natural gas engines, since the increase in efficiency over the relative load is steeper than for a gasoline engine (see Diagram 5.153). Large gains in efficiency can be achieved through small shifts in load points, whereby the losses in the electrical system can be kept very low.

If one considers the system efficiencies for electrical driving in a hybrid vehicle (without plug-in function), then there is always a restricted area, limited upwards by a range of factors. The reason for this is that the required electrical energy cannot be exclusively supplied from recuperation. Constant driving optimized for maximum efficiency at low load therefore requires intermittent operation of the drive aggregates with alternating electric and hybrid operation. Whereas the combustion engine is switched off during electrical driving and the electrical energy is supplied by the battery, in hybrid operation the load point of the combustion engine is raised to a more efficient area by using the electric motor in generator operation, whereby the battery is charged. In terms of system efficiency, this results in a linking of the individual component efficiencies of the combustion engine, electric motor, power electronics and battery.

Based upon the efficiency characteristics of various engine concepts, it can be seen that for the shown CNG concept the efficiency of pure combustion engine operation exceeds the maximum efficiency of the described pulse operation (electrical and hybrid driving) at lower loads compared to the gasoline concept. Moreover, due to the relatively steep gradient at very low loads, only a smaller shift in the load point is necessary to run the combustion engine in a favorable efficiency area. The conclusion is that CNG engine concepts are almost ideal for powertrains with a low degree of hybridization. In particular, mild hybrid systems with 48 Volt vehicle supply offered by many suppliers [144–146] are able to achieve large reductions in fuel consumption in such a system.

Highly turbocharged engine variants can be designed due to the high knock resistance of natural gas, which result in CO_2 advantages at high loads compared to gasoline or diesel aggregates. The low propensity of methane to knock means that when running in natural gas mode, operating points in close proximity to full-load can be run at the most efficient MFB50 point (compare also Sect. 5.2.6). However, the thermal load of components can prove to be problematic. While the early MFB50 points can contribute to limiting the exhaust gas temperature, the missing evaporation enthalpy compared to liquid fuels and stoichiometric operation up to full-load result in higher combustion chamber temperatures and correspondingly higher thermal load [147].

5.4.2 Hybrid Specific Engine Modifications

Operating a combustion engine in a hybrid system is significantly different to operation in a conventional powertrain. The most important differences result from the considerably more frequent operation at higher loads and the high number of starts/stops of the engine, whereby a true idle as it exists in a conventional powertrain is no longer mandatory. There is, above all, the possibility to make the engine more "phlegmatic", meaning the smoothing of peak engine loads. This results in lower load gradients during dynamic engine operation. This possibility results from the coupling of a highly dynamic electrical machine that can compensate for the dynamic behavior of the combustion engine during abrupt changes in performance demand [148, 149].

The reduced dynamic requirements on the combustion engine open up different possibilities to implement efficiency-increasing measures that would not be applicable in a conventional powertrain due to the deterioration in dynamic response. For example, by over-dimensioning the turbocharger employed, the turbine pressure ratio can be reduced and hence the work required for gas exchange can be reduced. As an alternative, an extension of the turbine neck diameter was investigated in [150]. This measure results in comparable effects without consequences for the engine packaging. Diagram 5.154 shows the influence of the diameter extension.

Since this effect is increased with higher mass flow, efficiency benefits can be achieved particularly in the nominal power range. Under certain circumstances, turbine efficiency can be degraded in this approach that is however more than compensated for with regard to the fuel consumption by the significant reduction in exhaust backpressure. For example, fuel consumption could be reduced by 1.5 % in the NEDC cycle in [150]. The reduction of the full-load torque at low speed can be compensated for by the torque capacity of the electric drive. The described dynamic behavior of the electric motor can also compensate for the deterioration in dynamic response.

Using an actively controlled waste gate enables an optimization of the gas exchange in the partial-load area through targeted controlling of the boost pressure. To achieve this, the waste gate is kept completely opened until naturally aspirated engine full-load is achieved by completing opening the throttle valve. This results in an engine operation in

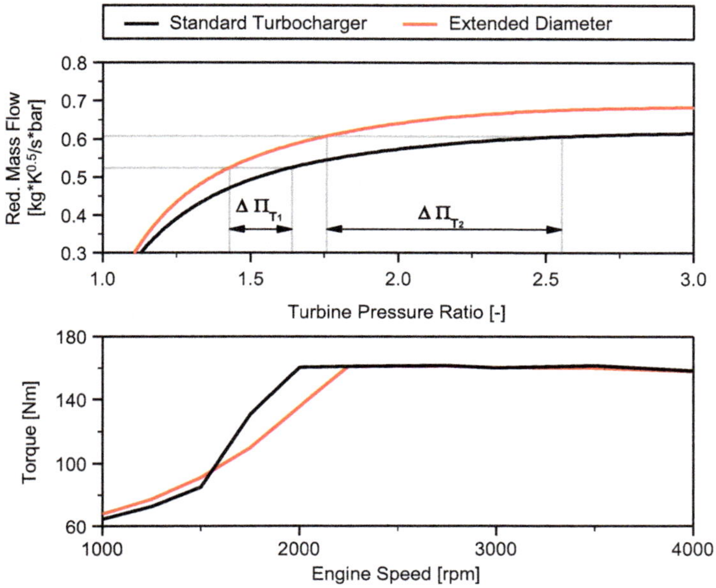

Diagram 5.154 Influence of an extended turbine neck diameter on TPR and full-load torque [150]

the lower load ranges that broadly corresponds to a naturally aspirated engine. For load demands that are above the naturally aspirated full-load, the waste gate is closed until the required torque has been reached. This procedure of closing the waste gate is limited by reaching full-load, resulting from limitations of either peak cylinder pressure or exhaust temperature. This approach is called boost-pressure-based load control.

In conventional turbocharged engines, the waste gate remains closed up to a boost pressure defined by a pre-loaded force and only opens continually with increasing boost pressure at higher load. Compared to an actively controlled waste gate, this results in higher exhaust turbine speeds in lower load areas that produce an increased charge pressure level. Therefore, when adjusting the desired load point, part of the charge pressure must be dissipated via the throttle valve, which is the reason for the worse gas exchange efficiency and hence lower indicated engine efficiency. This effect is accepted in engines for conventional powertrains because of the improved dynamic behavior. In a hybrid system however, this potential can be utilized. Diagram 5.155 portrays the effects described here in a graphical form.

The largest differences occur in medium load areas, where a significantly higher turbocharger speed and associated higher boost pressures and higher exhaust gas backpressure occur for conventional variants. This area also shows the greatest advantage in indicated engine efficiency.

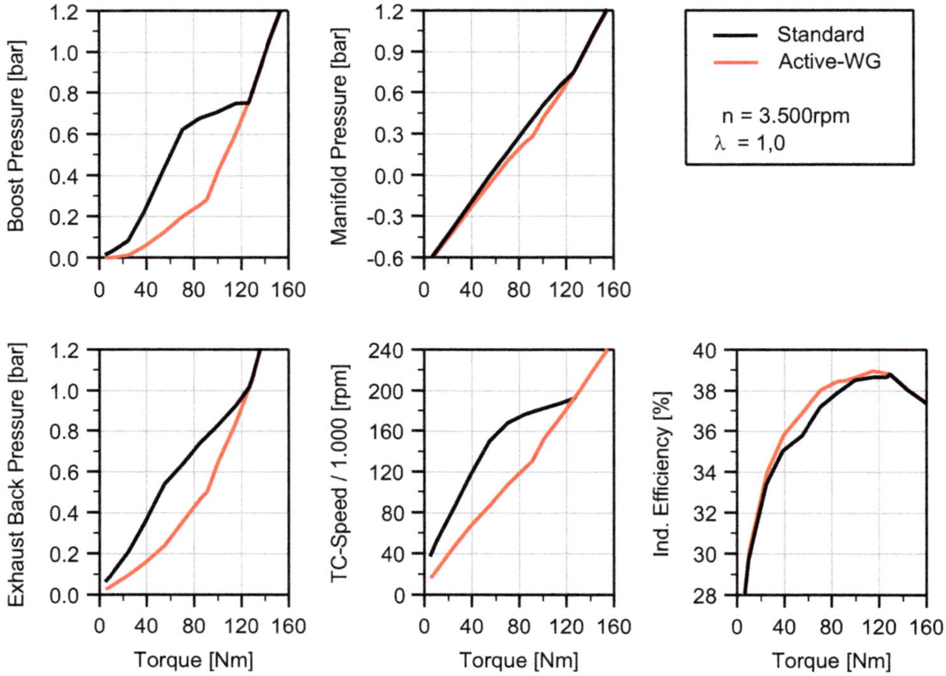

Diagram 5.155 Conventional waste gate application versus active waste gate [150]

5.4.3 Pollutant Emissions and Cold Start

The peculiarities regarding the emission of pollutants by natural gas hybrids are determined by the hybrid-specific engine operation. Characteristics of this are on the one hand the operation at higher loads and on the other the long phases of engine stop during electrical driving mode or in start/stop mode.

The largest differences to gasoline-powered hybrids can be found in the areas of high loads due to the lack of mixture enrichment at full-load. The deviation from stoichiometric operation results in the three-way catalytic converter no longer being able to oxidize the hydrocarbons and carbon monoxide in the exhaust gas due to the lack of oxygen and thus losing its effectiveness for these two emission types. Since natural gas engines can be run stoichiometrically up to full-load, the three-way catalytic converter maintains its full effectiveness over the whole operating range. This is particularly noticeable in hybrid mode in that at high loads a further shift in load points can occur by operating the e-motor as a generator, without having to lose the good conversion efficiency of the catalytic converter.

As for all combustion engine driven powertrains, the warm-up has a particular influence on the pollutant emissions of a natural gas hybrid vehicle. For example with conventional powertrains, between 60 and 80 % of the cumulated pollutant emissions are emitted during the phase leading to light-off (50 % conversion point) of the catalytic converter [151]. This problem is exacerbated when operating with natural gas due to the significantly higher converter light-off temperature compared to gasoline operation (compare also Sect. 5.2.7) and the higher possible degree of boosting with the exhaust turbocharger turbine as an additional heat sink in the exhaust system. In a hybrid system, the additional possibility of driving electrically with shut-off combustion engines can also lead to a cooling of the catalytic converter (light-out) after a successful warm-up.

These circumstances can be taken in to account via a range of measures. The previously described active waste gate strategy can contribute to increasing the catalytic converter input temperature up in to the upper areas of partial-load and thus guarantee a higher heat influx into the converter. The opened waste gate bypasses the turbine acting as a heat sink, which results in higher temperatures for the converter despite lower turbine input temperatures (Diagram 5.156). Additionally cooling can be prevented by more effective insulation. In this case, the possible elimination of engine overrun operation proves to be beneficial, since a flushing of the exhaust-aftertreatment system with fresh air can be avoided.

The hybrid start with delayed injection activation (high speed start) represents an approach to reduce unburned hydrocarbons during the engine start [152]. For this start procedure, the engine speed up is exclusively driven by the electric machine so that the first injection only occurs after idle speed has been reached and a significant reduction of inlet manifold pressure can be observed. Above all, this enables an improvement in start comfort. Additionally, the improved mixture formation of the first combustion events enables a higher combustion efficiency and thus lower HC emissions.

The larger dimensioned battery in terms of voltage and energy content in hybrid vehicles also permits the efficient use of an electrically heated catalytic element in order

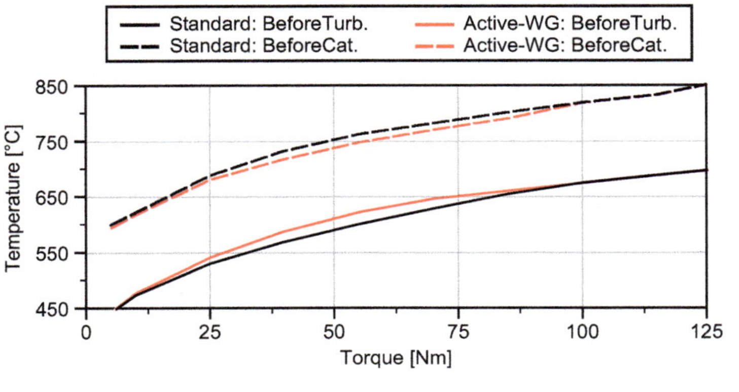

Diagram 5.156 Exhaust gas temperatures with conventional WG compared to Active-WG [150]

to achieve an additional acceleration of the heating up process. The additional heating of the exhaust gas stream in combination with conventional measures to heat the converter opens up the possibility to reach catalytic light-off point much faster. Alternatively, conventional heating activities can be reduced or completely eliminated in favor of lower fuel consumption. The selection of the appropriate heating strategy is always dependent on the relevant system and also from the energetic evaluation of the electrical energy. Depending on the hybrid concept, recuperative energy won during brake manoeuvres can be used to feed the electrically heated converter in order to prevent the catalytic converter cooling during engine overrun phases. This can be particularly useful if the battery has already reached maximum charge before or during the brake manoeuvre [153].

The application of an electric catalytic converter in [150] has been shown to achieve a significantly early light-off point for cold start and restart under load after the engine has been shut down. Such a procedure is typical when starting the combustion engine during electric driving. A practically immediate light-off, in particular for short engine shutdown periods, can be achieved with the use of very little electrical energy. Diagram 5.157 shows the effects of auxiliary electrical heating on the temperature of the catalytic converter and the HC conversion for a restart under load. The time between engine shutdown and restart in the example illustrated is 7 min. The ability to blow fresh air around the engine, as it would occur during electric driving mode was not available in the case investigated, hence faster cooling of the exhaust-aftertreatment system is to be expected.

The use of an electrical heating element as an alternative to the known engine-related measures is always subject to an energetic evaluation of the hybrid strategy. The greatest potential results from the use of a predictive operating strategy [154] with regard to the application after stationary phases. Such a strategy permits possible phases of engine shutdown to be identified early and evaluated appropriately.

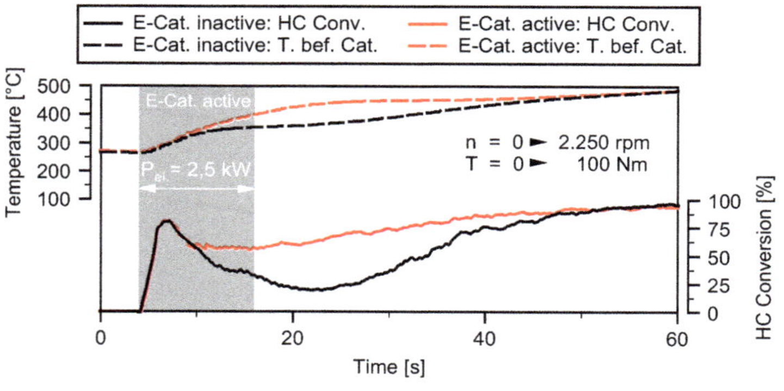

Diagram 5.157 Converter heat-up process and conversion rates for catalytic restart under load [150]

5.4.4 Practical Examples

The Federal Ministry of Economics and Technology sponsored a joint research project "Powertrain Concept with Natural Gas Hybrid" at the end of 2006. The cooperation partners in this project were Adam Opel AG, Robert Bosch GmbH, the Institute for Combustion Engines and Automotive Engineering (IVK) at the University of Stuttgart and the Research Institute of Automotive Engineering and Vehicle Engines Stuttgart (FKFS). The aim of the project was to create a prototype on the basis of an Opel Astra Caravan that achieved less than 90 g/km of carbon dioxide in the NEDC cycle (Diagram 5.158).

In order to reach the required emissions targets, the production engine of the base vehicle was replaced by a highly turbocharged, monovalent three-cylinder natural gas engine with 1.0 l displacement and a 25 kW permanent magnet synchronous motor with a separating clutch organized in a parallel hybrid layout. The electric motor is located between the combustion engine and the automated 5-gear transmission. The traction battery is located in the luggage compartment. The pressure tanks, located under the rear floor of the vehicle, are certified for a 20 kg of natural gas at a system pressure of 200 bar. The vehicle mass is 1626 kg. A unique feature of the prototype vehicle is the hybrid coordinator. The predictive and adaptive operating strategy evaluates the information received from a range of sensors during driving so that the operating strategy can be optimized using known trip parameters for when the vehicle travels the identical route. Parameters taken into account include road gradient, road bends and stationary times [149, 150, 154].

A further concept vehicle was created within the scope of the CLEVER project (Clean and Efficient Vehicle Research). The partners in this project were EMPA, ETH Zurich, Volkswagen AG and Robert Bosch GmbH. The aim of the project was to create a vehicle with 40 % lower CO_2 emissions than a conventional, gasoline-driven vehicle. A VW Touran was used as a base vehicle to create a parallel hybrid with a monovalent 1.4 l natural gas turbocharged engine and a 30 kW e-motor. The transmission used was a

Diagram 5.158 Prototype vehicle: Outside view and engine bay

manual 6-gear 4WD transmission. The layout was designed such that the cardan shaft was connected to the electrical drive instead of driving the rear wheels. The e-motor was located under the rear floor of the vehicle and was thus adjacent to the traction battery in the luggage compartment of the vehicle [155].

Other vehicles with natural gas hybrid drives have been demonstrated in the past by different institutions. In 2001, DaimlerChrysler presented the Dodge PowerBox, a rear-wheel drive turbocharged six-cylinder natural gas engine combined with an electrically driven front axle. This configuration, which results in additional tractive force, is called a "through the road" hybrid. In 2006, IFP in cooperation with Gaz de France were able to present a Toyota Prius modified to run with natural gas. In 2008, this was followed by a project with further partners to create a mild hybrid vehicle on the basis of a Smart fortwo, which was also driven by a natural gas engine. Toyota presented a study of a natural gas hybrid powertrain in a Toyota Camry at the end of 2008, which utilized the Japanese manufacturers well-known power split hybrid concept [149, 156].

5.4.5 Summary

Hybridization of the powertrain offers an excellent possibility to reduce the CO_2 and pollutant emissions. While the use of such alternative powertrains is normally considered to be in competition with the use of alternative fuels in conventional powertrains, the combination of both these measures can create synergy that increases the effect of the individual advantages.

The use of natural gas engines in hybrid powertrains can, apart from the immediate advantages of hybridization, achieve further reductions in emissions through the "phlegmatic" operation of the combustion engine. The descriptor "phlegmatic" can be understood to mean the smoothing of abrupt changes in load resulting in reduced demands on engine transient response. The highly dynamic electrical drive can be used to compensate for the reduced transient behavior. This results in various possibilities to use this concept on natural gas engines through technical modifications.

One possibility would be the use of an over-dimensioned exhaust gas turbocharger. This can achieve a reduction in the turbine pressure ratio and as a result reduced gas exchange work. Since this effect increases with higher mass flow, advantages in terms of efficiency can be achieved particularly in the nominal power range. A reduction in full-load torque at low speed is to be expected, which can however be compensated for by the electric drive with its distinctive torque advantages in this area.

An actively controlled bypass valve (active waste gate) can be used to optimize the gas exchange at low loads that permits targeted control of the boost pressure. In this way, the waste gate can be kept completely opened until the engine's naturally aspirated full-load point has been achieved. This results in an engine operation at low loads that closely resembles that of a naturally aspirated engine. Load requirements that exceed the naturally aspirated full-load result in the waste gate being partially closed until the desired boost pressure or the required torque has been achieved. The largest differences compared to conventional operation can be observed in the medium load area. An active waste gate strategy enables significantly lower turbine speeds and, associated with this, lower boost pressure and lower exhaust backpressure. In this area, the largest gains in efficiency can be observed.

The use of an electrically heated catalytic converter in conjunction with the already available high voltage battery offers the possibility to reduce pollutant emissions, since the higher voltage level permits the efficient operation of such systems. Electrical heating during the cold start phase results in the light-off point being reached significantly earlier and hence the conversion window. The active waste gate strategy mentioned above also results in an increased converter inlet temperature, since the turbine, which acts as a heat sink, is largely bypassed. The cold start problem that results from the higher light-off temperature compared to gasoline engines can thus be largely alleviated.

5.5 The Market for Natural Gas Fuel

Timm Kehler, Florian Feix, Claudia Petersen and Michael Schaarschmidt

5.5.1 Natural Gas as Fuel: Advantages and Market Development of the Refueling Station Network

Although electric vehicles currently receive more public interest, automakers have been increasing their efforts regarding offering vehicles with natural gas drivetrains for some years. This trend is motivated by political objectives that increasingly call for reduced greenhouse gas and pollutant emissions, but also intend on reducing oil imports. The German Federal Government's 2013 Mobility and Fuel Strategy calls to reduce the final energy consumption in transport compared the levels by around 10 % in 2020 and by 40 % in 2050. To achieve this goal, vehicles in road traffic will need to use a diversified array of fuels in connection with alternative drive technologies [157]. In addition, the EU has introduced a directive stating that fleet-wide CO_2 averages meet the limit of 130 g/km by 2015, and 95 g/km by 2020; similar efforts are expected for commercial vehicles. These policy objectives have not only intensified automakers' efforts to optimize gasoline and diesel motors and develop electric drives; they grant natural gas an important place in the considerations and strategies of the automobile industry.

5.5.1.1 Advantages of Natural Gas Fuel

Political Advantages

Natural gas offers a variety of advantages compared to gasoline and diesel fuel that support environmental and climate policies. If these fuels were to be substituted by natural gas, greenhouse gas emissions from road traffic would be significantly reduced without affecting mobility or performance. A vehicle with a natural gas drivetrain emits approximately 24 % less CO_2 than a gasoline-powered vehicle with the same power output. A vehicle fueled with bio-natural gas reduces CO_2 emissions by as much as 97 %, and is as climate-friendly as an electric vehicle powered by renewable electricity [158]. Further support arises from the stringent Euro-6 tailpipe emission standard that has been in place since January 2013 for commercial vehicles and busses over 3.5 tons and since September 2014 for passenger vehicles. Natural gas vehicles fulfill these norms without complex exhaust after-treatment technologies such as SCR catalysts or particle filters [159].

Natural gas vehicles further extend their lead with regard to vehicle noise emissions. Vehicles equipped with natural gas engines create only half as much noise as their diesel-engine counterparts [160].

Economic Advantages

Economic benefits further support the case for natural gas as vehicle fuel. In the first half of 2014, one kilogram of natural gas costs on average 1.10 Euro at German filling stations, and was 30 and 50 % more cost effective than diesel and gasoline. In 2012, the year that is accounted for the highest overall fuel prices in the latest German history diesel prices were as much as 43 % and gasoline prices were up to 65 % higher than natural gas [161]. There are similar price advantages in other countries, evidence that the price for natural gas as vehicle fuel is not subject to the large fluctuations of gasoline and diesel. The decoupling of the natural gas price from the crude oil price, encouraged by recent discoveries ensuring the availability of raw natural gas for several more decades, explains this tendency. Above all, the possibility of economically sustainable production and transport introduced by unconventional natural gas has fused the global natural gas markets and increased competition in the past decade. The growing number of natural gas-producing countries is more internationally dispersed than those that extract petroleum. Growing supply, long-term resource availability and stagnant demand due to efficiency improvements predict that only moderate price increases are to be expected.

Of all alternative fuels, natural gas vehicles are considered to be the most-effective solution to reduce CO_2 emissions compared to a conventional gasoline or diesel vehicle. This finding was confirmed by a study of the Energy Economics Institute of the University of Cologne (EWI, 2014) (Diagram 5.159). The greater the driven distance, lower the costs for natural vehicles. As beyond a certain mileage, there are no additional costs associated with CO_2 avoidance. Driving a natural gas car for an annual distance of 17,000 km saves customers money on a total cost of ownership basis versus a comparable conventionally powered vehicle. The EWI study also confirmed economic benefits for compact- and middle-class vehicles (when purchased) compared to conventional reference models because their higher acquisition costs are offset by lower vehicle excise duty. This scenario shows that the natural gas vehicles cover costs after an average of 2000 km per year—far lower than the average annual mileage of 16,000 km [162].

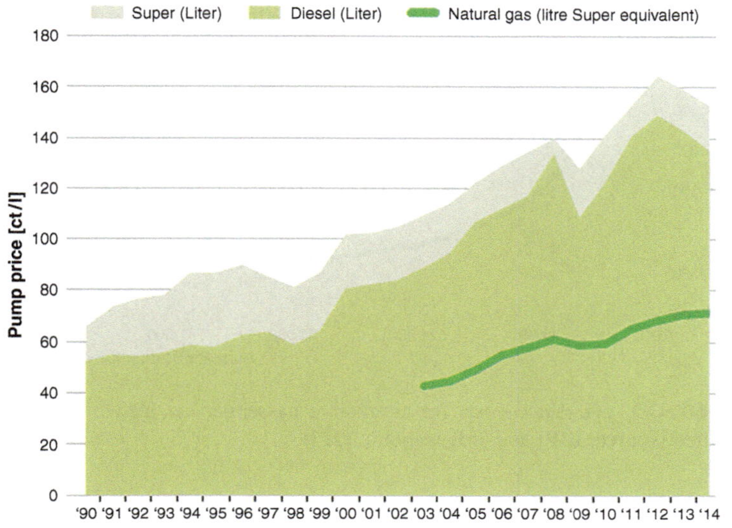

Diagram 5.159 Development of fuel costs in Germany 1990–2014 (Source: Statistical Bundesamt and erdgas.mobil from January 2015)

5.5.1.2 Market Development of Natural Gas Filling Stations

Following early activities in the 1990s, the German gas industry has systematically invested in the construction of a broad, comprehensive network of CNG filing stations since about 2000. Municipal transport companies who wanted to adapt their bus fleets to a drastic reduction, first applied the technology of natural gas drivetrains and initiated the expansion (Diagram 5.160). Towns and municipalities opted for natural gas buses because the Euro III buses of the early 2000s emitted significantly more pollutants.

Implementing CNG buses and their associated filling stations were environmentally ambitious (undertakings) and as such required funding by the Municipal Transport Financing Act. As a result, the first generation of natural gas fueling stations was often located, for example, at the corporate headquarters and transport depots of public utilities. Consequently, many of the fueling stations very difficult to locate, a special payment system was applied and refueling in remote areas was unattractive to the economy and the drivers.

Following the liberalization of the energy market in 2005, competition increased among the municipal utilities. Many providers wanted to sell gas at refueling stations as a supplementary service(,) and invested in new natural gas filling stations hoping for a market recovery and an increase in the variety of models. The aim was for natural gas vehicle drivers across Germany to have refueling points available within a reasonable distance of their work and home. Surveys revealed that around 1000 stations were required.

The installation rate for natural gas stations peaked around 2010. Meanwhile, with 923 filling stations—most of them funded by natural gas utilities—a dense nationwide

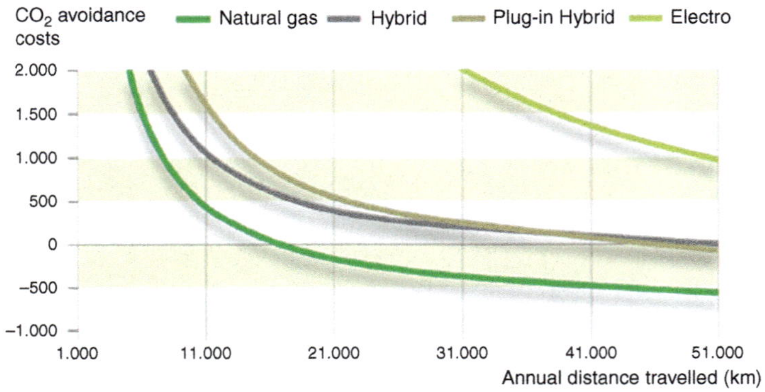

Diagram 5.160 CO_2 avoidance costs for alternative passenger car technologies versus annual distance travelled (source: EWI and erdgas.mobil 2013)

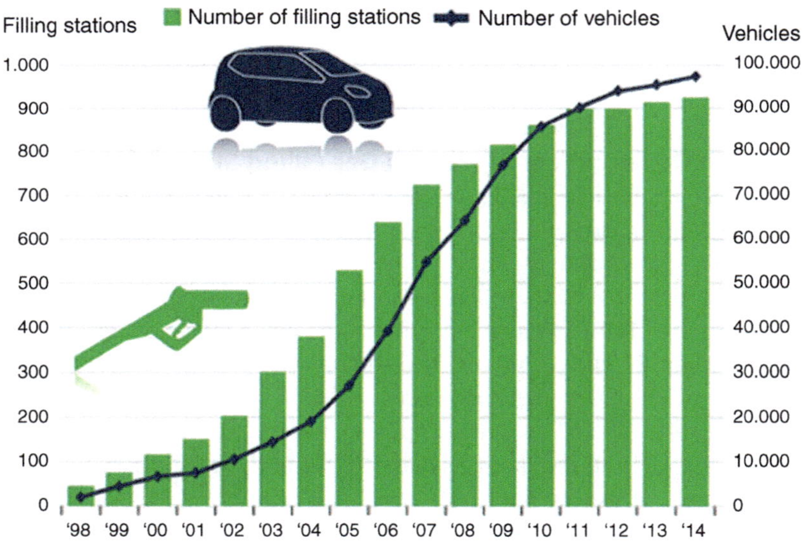

Diagram 5.161 Number of filling stations and vehicles 1998–2014 (source: KBA and erdgas. mobil 2014)

network is already in place. Of this number, erdgas mobil planned and implemented more than 500 stations.

The development of the natural gas vehicle fleets has not kept pace with the build-up of filling stations. So far, too few motorists have opted for a natural gas car, so that many (Diagram 5.161) filling stations are underutilized and cannot cover their costs [163]. With almost 100,000 natural gas vehicles reported in Germany in 2014, there are about

100 natural gas vehicles for every natural gas station. A profitable operation of the refueling stations is expected at a ratio of about 200 NGV's per station.

5.5.2 Business Case for Natural Gas Filling Stations: Market Obstacles and Opportunities

5.5.2.1 Structure and Function of Natural Gas Filling Stations

Natural gas used as fuel can be so-called L-gas, produced in Germany and the Netherlands, or H-gas, a higher energy content natural gas from Norway and Russia. The engine control systems of natural gas vehicles adjust the amount of fuel for the combustion and the ignition according to the different gas conditions and the driver recognizes performance penalty. However, the lower energy content of L-gas results in a shorter driving range.

In contrast to gasoline and diesel, which are petroleum derived and undergo a complex refining process, natural gas is delivered to the pump in a state virtually unchanged from its natural form. All 923 natural gas service stations currently operating in Germany are supplied directly via a supply line from the gas grid. At the gas station, the fuel is dried, compressed to 280 bar, stored in pressure-resistant steel containers and from there dispensed directly into the storage tanks of cars. Delivery via pipeline is the most efficient way to deliver gas to a natural gas filling station.

As an alternative to a grid connection, natural gas can also be transported to the filling station by truck. However, extra energy must be expended to first convert the gas into liquid form, in this case one speaks of *Liquefied Natural Gas* (LNG). Natural gas liquefies when cooled to $-162\ ^\circ C$. In order to remain in this liquid state, it must be transported and stored under pressure in highly insulated containers. LNG is traded at European import terminals such as Gate in Rotterdam or Zeebrugge in Belgium, where natural gas arrives as a liquid via ship from producing countries in Africa and the Middle East. From there, the LNG is loaded onto trucks for transport to the filling station.

LNG is stored in liquid form at the filling station, where it is warmed and re-gasified as necessary, compressed to 280 bar and finally dispensed via fueling nozzle as CNG. This transportation and storage medium is an option for gas stations that do not have a cost efficient access to the gas grid. In Spain, where the gas network is significantly less developed than in Germany, there are already twelve of these L-CNG stations [164].

LNG can also be dispensed at a filling station in its liquid state. The advantage of dispensing natural gas in this form is its high energy density. The energy equivalent of one liter of diesel requires 1.6 l of storage volume for LNG, whereas CNG stored at 200 bar of pressure requires a volume of 5 l. Thus, LNG vehicles offer a much longer driving range from an equivalent tank volume.

The downside of LNG is that when vehicles are parked for extended periods, the liquid fuel warms up and gasifies, and can escape from the tank as the pressure increases. Therefore, LNG is only suitable for trucks and buses since they are typically operated on a daily base and over long distances. These vehicles have a driving range similar to diesel when operating with LNG in the tank. The liquefied gas is re-gasified before it enters the engine, whereupon it performs identically to CNG [165].

CNG Filling Stations (Compressed Natural Gas)

A compressed natural gas filling station consists of a gas dryer, compressor, high-pressure storage accumulator, high-pressure pipeline system (through which the gas flows to the dispenser), and gas dispenser, usually equipped with two nozzles. These components should be optimized with regard to each other based on the typical volume of fuel delivered—i.e. the number of vehicles that refuel each day. A standard natural gas station in Germany is designed to serve 60–80 cars and light commercial vehicles per day.

Design, construction, equipment installation, testing, commissioning and operation of CNG filling stations that deliver fuel from the gas grid are subject to strict regulation. Stations in Germany must comply with the technical regulations of regulation G651 of the German Association of Gas and Water (DVGW), and their operation must comply with the Ordinance on Industrial Safety (Diagram 5.162) [166].

Natural Gas Dryer

The natural gas dryer removes moisture from the gas before it reaches the compressor. Gas is generally already dry, but it can absorb moisture while in the pipeline network. Water buildup can cause problems during refueling, for example by icing fittings on the gas dispenser.

Compressor

The natural gas for a gas station is usually taken from the gas distribution grid at pressures between a few millibars and up to 40 bars. The compressor compresses it to the required operating pressure of 280 bar and feeds it into the high-pressure accumulator. From the storage, the natural gas flows directly through pipelines to the dispenser. A standard CNG filling station is usually equipped with a compressor that has a capacity of 100–150 Nm^3/h.

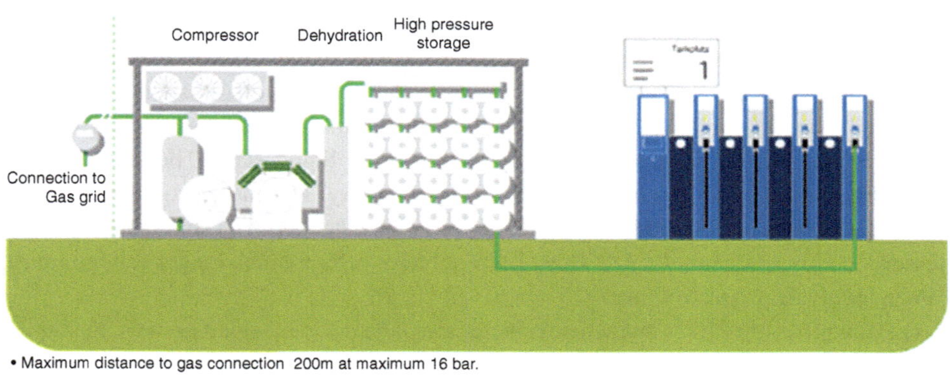

Diagram 5.162 Schematic of a CNG filling station (source: erdgas.mobil 2014)

Storage

The gas storage array acts as a buffer between the compressor and the dispenser. Vehicles are directly refueled from the gas storage via the large difference in pressure compared to the tank in the vehicle. The storage setup typically consists of 28 permanently fixed gas cylinders each with an 80-l geometric volume, which are made of seamless steel cylinders. As a result, the permitted number of load cycles is increased so that periodic inspections are required only every 10 years. The storage setup, compressor, dryer and gas control systems are usually located together in a compact concrete building.

Pipeline

The gas storage and fuel dispenser are linked by a so-called 'shell and tube' high-pressure pipe system, which is designed for an operating pressure of 280 bar. The bundled tubes are supplied by the manufacturer in drums and can be quickly installed above ground or in the soil. The pipes are pressure tested at the manufacturer's plant. The high operating pressure ensures that even during the summer, when the pressure of the gas delivered to the dispenser increases due to thermal expansion, vehicles can still be filled with the correct amount of fuel.

Gas Pump

A standard high-performance natural gas pump in Germany fills an average mid-size car with a CNG capacity of 20 kg in about 2–3 min; higher filling rates are possible for CNG-powered trucks and buses. The pump is equipped with two independent nozzles and measuring devices (Coriolis flow meters) so two cars can be fueled simultaneously. Since the components are able to withstand 280 bar of pressure, the full 280 bar of reservoir pressure can be delivered to the nozzle; only at the nozzle the fueling pressure is reduced to 260 bar. This is required for an optimal vehicle fueling process even at high ambient temperatures. The pump operates autonomously, i.e. independent of the gas compressor's controller.

Standardized natural gas dispensers are adapted to the individual design of a petroleum company, and blend seamlessly into any public gas station. Their connection to the service station's POS system is as straightforward as a gasoline or diesel fuel dispenser; therefore, paying for natural gas fuel is identical to paying for gasoline or diesel [167]. In addition, the gas station technology is modular, making it easy to expand capacity as demand rises.

LNG Filling Station (Liquefied Natural Gas)

An LNG filling station is supplied by tanker trucks. The storage tanks transport and store LNG. They are vacuum insulated in order to hold the liquefied gas at cryogenic temperatures for a long period. The pressure in the storage tank corresponds exactly to the working pressure in the tank systems of the trucks and buses that use LNG as fuel (Diagram 5.163). A centrifugal pump (cryogenic pump) delivers the liquefied gas from the storage tank into the insulated fuel tank on the vehicle.

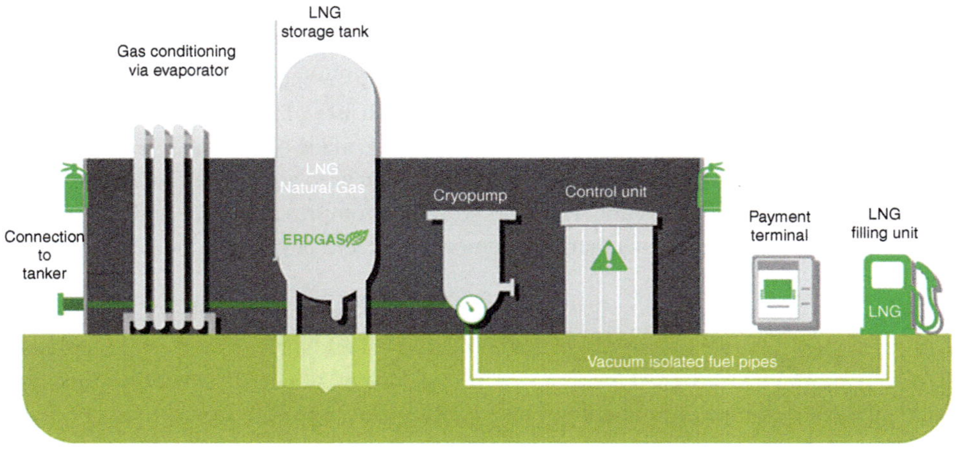

Diagram 5.163 Schematic of an LNG filling station (source: erdgas.mobil based on Cryostar SAS)

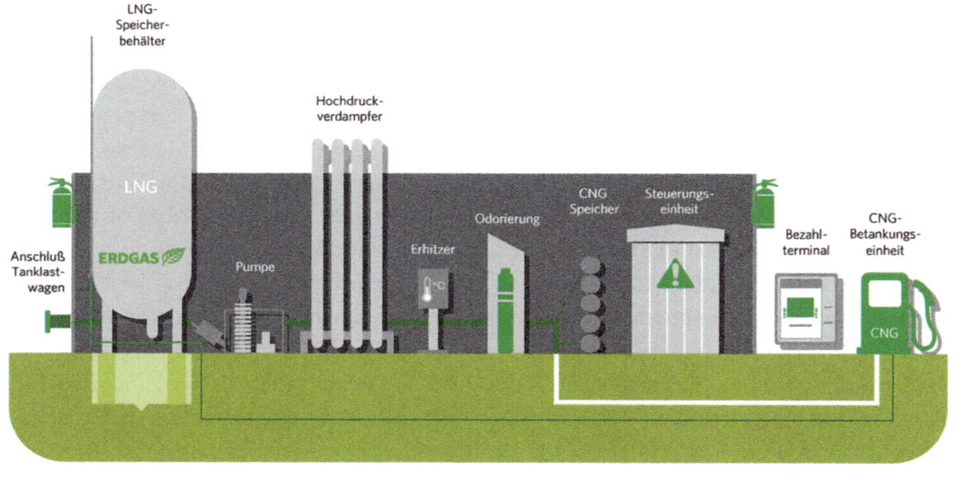

Diagram 5.164 Functional diagram of an L-CNG gas station (Source: erdgas mobil on the basis if Cryostar)

L-CNG Filling Station (Liquefied-to-Compressed Natural Gas)

Before refueling, LNG is first compressed and then converted into a gaseous state through an evaporator. From there, the refueling process is the same as that of a conventional CNG filling station. The gas flows via pressure differential from the CNG storage tanks into the vehicle tank, where it is stored without any gas loss (Diagram 5.164).

5.5.2.2 Business Model: CNG Filling Station Operation

The natural gas filling station business in Germany—unlike the gasoline and diesel business, which is dominated by a few mineral oil companies—is relatively fragmented. Municipal utilities are behind most of Germany's 923 CNG stations, and each typically operates no more than one or two filling stations in its network area. The gas industry has invested a total of ca. 300 million Euro in the CNG fueling infrastructure between the 1990s and 2014. Constructing one gas station costs approximately 250,000 €.

However, even after more than 10 years of development, the amount of gas sold by gas utilities as vehicle fuel is still small compared to sales to households, commerce and industry. Consequently, the utilities have worked to optimize their offerings. Many natural gas pumps at hard-to-find depots were taken out of service; today, more than 90 % of all natural gas stations are found at service stations of major fuel suppliers. Motorists using these facilities can not only fill up on gas, but also benefit from long opening hours and services such as shops and car washes.

In addition, the information available to drivers has improved. The introduction of the 'green leaf' symbol has been particularly helpful, as it has lent a figurative and symbolic brand identity to natural gas fuel. All of this has increased the attractiveness of natural gas as fuel; enhancing the visibility of the existing infrastructure has helped grow the number of natural gas vehicles to nearly 100,000 in early 2014. While this is a major milestone, more growth must be generated, as sufficient fuel sales and profitable operation is only possible with about 200 vehicles per station.

With higher gas sales and fuel turnover at the pump, gas suppliers could also reduce the investment levy added to fuel prices to cover the service station's construction cost. The operation and investment costs of the filling station represent about 1 % of the price for gasoline and diesel fuel, while operators of natural gas filling stations must add nearly 20 % to the price due to low utilization.

Higher utilization rates would also benefit the filling station operators. With each additional vehicle that fuels with natural gas, a tenant receives more commission—under contract with the gas utility that supplies the station—in exchange for settling transactions with the customers and performing daily maintenance on the dispensers and equipment. Drivers of natural gas cars are generally interesting customers for filling station tenants because they not only promise additional business in the gasoline station, they are also likely to return more frequently due to the relatively small number of CNG filling stations.

The cost of CNG at the pump is based on the following cost components: gas reference price, network usage charges, capital costs of the filling station, maintenance costs, electricity costs for the compressor, commission and energy tax. LNG avoids any system charges, but adds costs both for the loading of tank trailers at the LNG terminal as well as the cost of transport to the filling station. As a result of the fuel liquefaction process, LNG is expected to be slightly more expensive than CNG. The price inflation of CNG has risen an average of 2.6 % per year since 2009, below the 5.4 % value for diesel. Due to constantly growing global demand for petroleum products and expected scarcity of supply, in contrast to increasing natural gas supplies amidst moderate demand growth, the decoupled price development of both fuels is expected to continue.

5.5.2.3 Market Barriers in Germany

Selling natural gas as fuel remains a small business. Both the fleet of natural gas vehicles in Germany as well as the number of service stations have hardly grown in recent years. Although 8923 (38.2 %) more new natural gas cars were sold in 2013 than in the previous year, on balance the end of 2013 yielded an increase in vehicle population of only 2.8 %. This is because the number of de-registrations of vehicles that had reached the end of their useful lives greatly increased. However, the slight upward trend continued in 2014. In the first half of this year, 14 % more new registrations were recorded compared to the same period a year earlier [168].

The number of natural gas filling stations remained roughly constant in 2013. Dispensers were added in 19 new locations; however, 14 pumps were taken out of service due to poor profitability figures. This trend of replacing poorly located natural gas stations with better ones has been observed for about 3 years.

More vehicles are required for better utilization of existing stations. A vehicle growth rate of 2.8 %, as in 2013, is far too little for profitable operation of service stations in the foreseeable future. At current growth rates, the natural gas vehicle population of 120,000 that would be achieved in 2020 would not help station operators.

A look at the reasons for the relatively slow growth of natural gas mobility in Germany reveals market conditions that obstruct the spread of natural gas vehicles and the expansion of the retail network.

Temporary Energy Tax Reduction

The biggest obstacle to further development of natural gas mobility is the lack of clarity regarding the continuation of the tax reduction on natural gas as fuel. Today, the full energy tax is not levied on natural gas sold at fuel stations. However, this discount applies only until the end of 2018; after this, the tax would rise from 18 cents per kilogram of natural gas to about 41 cents per kilogram [169]. Not only would fuel costs increase by 25–29 %, the payback period for the additional upfront cost of the natural gas vehicle would increase considerably.

An extension of the tax reduction beyond 2018, as the CDU, CSU and SPD proposed in their coalition agreement of 2013, would create more investment protection for both station operators and vehicle owners. Drivers who assess the price development for natural gas over a long period of time would make the decision for a natural gas car. In addition, uncertainty about the resale value of natural gas cars would be alleviated.

Non-transparent Pricing

A further obstacle to the spread of natural gas vehicles is the lack of price transparency at gas stations. Price masts and dispensers display natural gas fuel prices in units of Euro per kilogram. One kilogram of natural gas, however, contains much more energy than a gallon

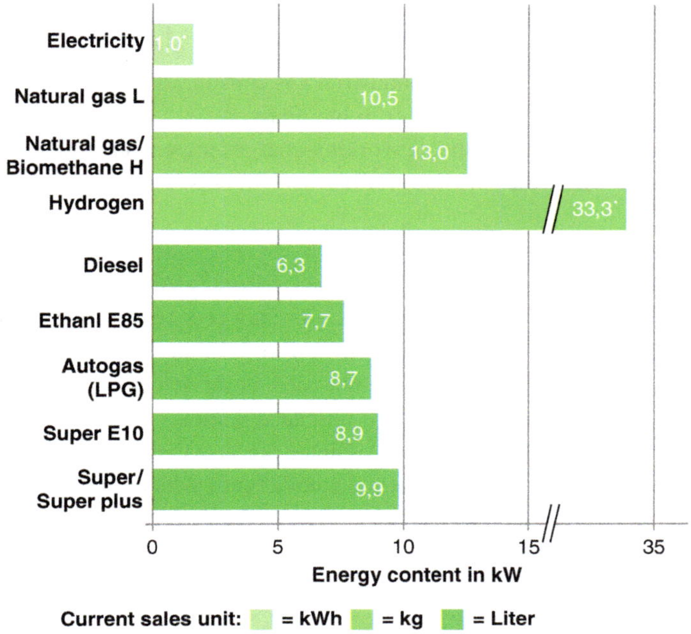

Diagram 5.165 Energy content of available fuel options pro sales unit (source: German Energy Agency, position paper "Transparent Price Information for a Changing Fuel Market", 2013)

of gasoline or diesel. This difference is not clear in the pricing. The result is that natural gas appears more expensive from a quick look at the price mast than it actually is.

The gas industry has criticized the lack of pricing parity between liquid fuels and natural gas for years. Gas companies demand that the attractiveness of alternative fuels must be quickly recognizable (Diagram 5.165) so that they can establish themselves in competition.

The Initiative for Natural Gas Mobility (Initiative Erdgasmobilität), backed by a group of gas suppliers, automakers and station operators with support from the German Energy Agency (Dena GmbH), has launched a proposal to increase transparency in the pricing displays. The advertizing effectiveness of large gasoline station signposts should not be underestimated, and prices should be indicated there as a 'gasoline liter equivalent,' as is done in Switzerland. Natural gas and gasoline prices would thus be easier to compare to each other, and motorists can thus base their decision on these grounds. In the first half of 2014, the natural gas sold nationwide was on average 54% less expensive than gasoline [170].

Classification of Network Charges

So-called network charges apply to the use of the pipeline infrastructure that supplies natural gas stations. However, the current pricing scheme makes it difficult to expand the refueling infrastructure. Network usage charges vary based on peak demand; a disadvantage of CNG filling stations is that most vehicles are filled during peak times, thus inducing higher fees. In contrast, industrial gas consumers with similar consumption rates pay less despite significantly higher usage hours.

According to the gas industry the networking charges impede an efficient operation of natural gas stations. Thus, they without further investments.

Load Restrictions in Commercial Vehicles

Due to their relatively heavy on-board fuel tanks, natural gas vehicles weigh more than equivalent gasoline or diesel powered vehicles. This reduces the permissible vehicle payload, which can negatively affect vehicles up to 3.5 tons in weight, especially light commercial vehicles. These vehicles are often used as service or package delivery vans. For such purposes, fleet operators usually purchase vehicles with the highest possible permissible vehicle payload. Natural gas vehicles are at a disadvantage from the outset because they are heavier than the same vehicle model with a diesel engine even before any cargo is loaded. A technology-specific calculation of the permissible total vehicle weight could help offset this disadvantage.

Company Car Taxation

The taxation scheme of company cars is another possible barrier to the further spread of natural gas vehicles. Company car tax is measured (simply) based on the list price of a new car. Eventhough in terms of climate change and environmental aspects natural gas vehicles are a good solution, they are outpaced in regard to price mechanisms as they more expensive due to lower (smaller) production numbers. The company car tax effectively cancels out the benefits offered by the reduced energy taxation.

Conservative Car Dealers

There is substantial competition in the new car market, where discounts are commonly offered to car buyers to encourage purchases. Every car dealer's performance is evaluated by high sales volumes, and it is easiest to achieve high volumes with gasoline and diesel models, since these are well known and easily explained. In contrast, customers require a great deal of education prior to the sale of a natural gas vehicle. A number of issues must be addressed: Is the refueling network sufficient? How much does CNG cost? Is refueling difficult? Is a natural gas vehicle safe, and where can it be maintained? This takes time, and uncertain or ill-advised customers will return to the proven gasoline or diesel technologies. According to the gas industry, there are too few incentives available to car dealers to encourage them to sell natural gas vehicles. Combined with a frequently encountered combination of ignorance and unfounded concerns about the safety of natural gas technology, progress for natural gas vehicles is difficult.

Safety Concerns with Natural Gas

Although natural gas vehicles must undergo the exact same regular safety tests and TÜV examinations as gasoline and diesel vehicles, the public frequently cites safety concerns. This also applies to natural gas filling stations, which are sometimes regarded with skepticism although they must meet the same extensive safety and security requirements as gasoline and diesel fuel stations.

5.5.2.4 Opportunities

Extension of the Energy Tax Reduction

There is strong evidence on the political side that natural gas—and in particular, its renewable counterpart biomethane—is a desirable and climate-friendly alternative to the dominant gasoline and diesel fuels. In June 2013, the former CDU/FDP coalition government proposed an extension of the temporary tax relief in their mobility and fuel strategy. At the end of 2013, the ruling CDU/SPD government announced its intention to extend the tax reduction beyond 2018 in its coalition agreement. The green party also announced its support of natural gas mobility during its governing period from 1998 to 2005.

A reduced tax rate of 1.39 cents per kilowatt-hour currently applies to natural gas as fuel, compared to the statutory tax rate of 3.18 cents per kilowatt-hour specified by the Energy Tax Act. The discount, which applies to both CNG and LNG, was introduced to incentivize the use of environmentally friendly natural gas as fuel for cars and trucks.
A brief study by the scientific institutes DLR, IFEU, LBST and DBFZ on behalf of the (then) Federal Ministry of Transport, Building and Urban Development (BMVBS) confirmed the value of extending the tax reduction. A reduced tax rate serves "to further promote the development of the CNG vehicle market and CNG infrastructure. This may pave the way for the integration of renewable energies, such as biomethane or methane from renewable electricity," according to the study [171].

The EU Wants to Build a Natural Gas Filling Station Network

On April 15, 2014, the European Parliament adopted a directive of the European Commission to build a Europe-wide network of service stations for alternative fuels. This includes natural gas and biomethane. The "Clean Power for Transport" policy calls for natural gas to be available for vehicles in cities and large population centers in all member states. The goal for 2025 is to establish an adequate number of CNG refueling stations along highways at a maximum interval of 150 km.

The directive also notes the need to display prices transparently. It thus proposes that the price of fuels at the filling station be based on their energy content equivalent to gasoline, especially for natural gas and biomethane.

The background of the directive is the political desire to significantly reduce Europe's oil dependency and to promote alternative fuel technologies through the expansion of fueling station infrastructure. In this way, both environmental and economic progress can be achieved.

According to the EU Council, expanding the retail network for alternative fuels could reduce expenditures on petroleum by 2.3 billion Euro per year by 2030. Another billion Euros per year could be saved by mitigating the impact of price fluctuations in the oil market through a diversified fuel supply and increased security of supply.

The EU member states now have 2 years to transpose the directives into national law. Germany has already exceeded the intended 2020 target regarding the number and distribution of CNG filling stations [172].

Development of the Natural Gas Vehicle Model Range

Several automakers have strongly supported the development of natural gas mobility. The Volkswagen Group in particular has introduced natural gas variants of several important models in recent years and has thus made alternative fuel vehicles available in almost every vehicle class. The following vehicles were recently introduced in sequence: VW eco up!, Seat Mii EcoFuel, Skoda Citigo G-Tec, Seat Leon TGI, Skoda Octavia G-Tec, VW Golf TGI, Audi A3 g-tron and Fiat 500 l Natural Power. Mercedes Benz also contributed to the range of available vehicles with its B-Class and E-Class vehicles. Market experts believe that this expansion has tripled the sales potential of natural gas vehicles in 2014 compared to 2012 (Diagram 5.166).

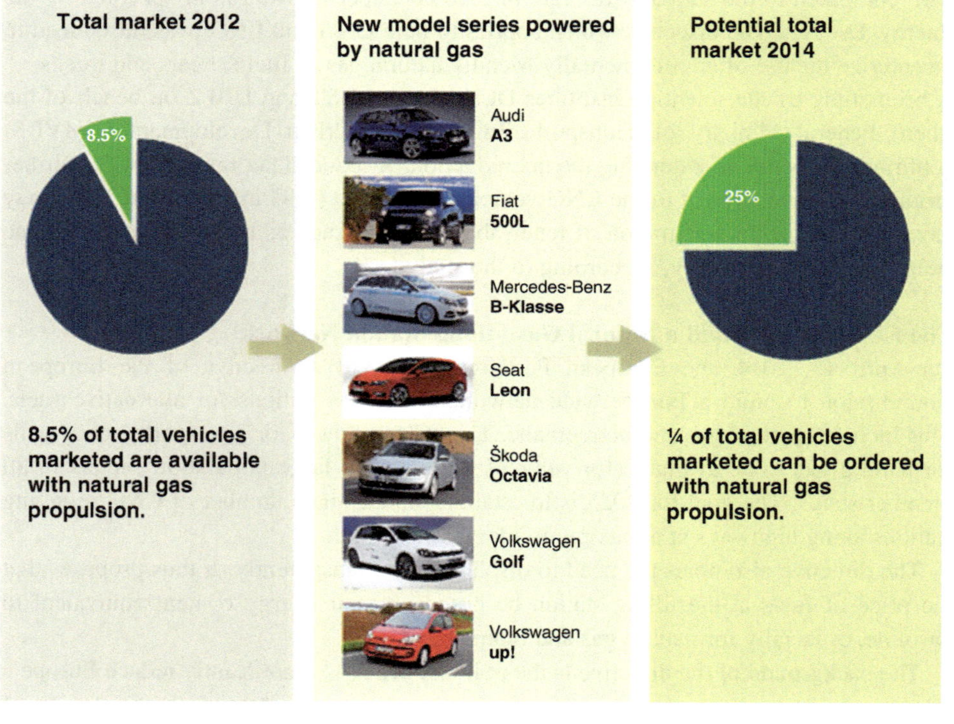

Diagram 5.166 Market revival through new models (source: erdgas.mobil, based on KBA registrations, 2012)

The Volkswagen Group especially wants to develop the small car segment with natural gas drives. At the 35th Vienna Motor Symposium in May 2014, the company introduced its concept for a new generation of small power units to be offered in natural gas models. This 1.0-l three-cylinder TSI motor, which will be used in the Polo and Golf models, will also include a CNG variant with turbocharging.

Even the purchase prices of natural gas vehicles, which were still significantly more expensive than diesel vehicles in 2012, have become more attractive. Experts believe that the price difference between diesel and natural gas versions of the same model is disappearing as diesel cars become more expensive due to the Euro VI emissions requirements effective as of September 1, 2014. CNG vehicles could even become cheaper than their diesel counterparts in the near future when lower road tax is accounted for.

Alternatives for Commercial Vehicles

While CNG-powered vans, garbage trucks and buses have already proven themselves as environmentally friendly vehicles for everyday use in urban traffic, LNG is the only viable alternative for long haul freight. In liquid form, natural gas can be transported aboard a truck in sufficient quantity to make typical long haul distances economically viable with natural gas.

LNG can also reduce fuel costs—and therefore operating costs—by almost 20 % versus comparable diesel models. This cost advantage can be crucial in the fiercely competitive, low-margin transport industry.

This advantage is expected to grow in the next few years as demand for low sulfur diesel from ship traffic on the North and Baltic Seas, which will switch from heavy fuel oil to lower sulfur fuel, increases [173].

In mid-2014 there were only about 300 trucks with LNG drivetrains in Europe and around 45 LNG stations [174]. They have been operated to date under national exemptions. Internationally accepted technical requirements for LNG vehicles (UN/ECE R110) entered into force in July 2014. Later this year, the commercial vehicle manufacturer Iveco is expected to launch the first approved LNG tractor-trailer as a production model on the German market.

References

References for Section 5.1

1. DIN 51624:2008–02 Kraftstoffe für Kraftfahrzeuge—Erdgas—Anforderungen und Prüfverfahren
2. PTEC Pressure Technology GmbH, D-51399 Burscheid, Germany. www.ptec.eu
3. HAM-LET® ADVANCED CONTROL TECHNOLOGY. www.ham-let.com
4. Rotarex Automotive S.A., L-7505 Lintgen, G.-D. du Luxembourg. www.rotarex.com
5. VTI Ventil Technik GmbH, D-58706 Menden, Germany. www.vti.de
6. Dynetek Europe GmbH, D-40885 Ratingen, Germany. www.luxfercylinders.com

7. Heep C, Jauss A (2004) Patentschrift EP1286103B1, Behälter zum Speichern von Gas unter hohem Druck. Machbarkeitsstudie Erdgas-Strukturtank. Benteler Automobiltechnik GmbH, D-33102 Paderborn/EON, D-45131 Essen/Adam Opel AG, D-65423 Rüsselsheim
8. Powertech Labs Inc (2005) CNG Cylinders 101. NG Transit Users Group Meeting
9. Hürter H, Jauss A Patentschrift DE102005046913 A1, Verbinder für Druckleitungen
10. Perthel K, Jauss A Patentschrift DE10361781, Elektromagnetisches Ventil
11. Zumbroich I, Jauss A Patentschrift EP 1884651B1, Ölabscheider für gasbetriebene Brennkraftmaschinen
12. Flügge W (1981) Statik und Dynamik der Schalen, 3. Auflage. Berlin
13. Girkmann K (1986) Flächentragwerke, 6. Auflage. Wien
14. Dankert J, Dankert H (2006) Technische Mechanik Statik, Festigkeitslehre, Kinematik/Kinetik. 4. Auflage. Wiesbaden
15. Toray Industries (America), Inc Dokument: T700SDataSheet.pdf. www.toraycfa.com (24.10.2014)
16. Hürter H, Jauss A Patentschrift DE 10308676B4, Schutzkappe für Flaschenventil einer Gasflasche
17. (2013) Abrasion protection for composite compressed gas tanks. In: Technik heute Nr. 9
18. Fell S, Schultheis V, Schürmann H, Bleier A (2011) The hierarchical building-block approach of testing and fe-analysis for strength prediction of high pressure wet wound composite vessels
19. Powertech Labs Inc (2007) Root cause analysis and report for CNG cylinder field failures
20. Seattle Fire Department Arson: Incident #26564 auto fire with compressed natural gas (CNG) fuel tank explosion
21. Wiedmann HK (2005) Der Saarbrücker Gasbus-Unfall, Informationstag: Der Betriebsleiter in Grenzsituationen, Erfahrungen nach schweren Unfällen. Versorgungs- und Verkehrsgesellschaft Saarbrücken mbH
22. Deutsche Edelstahlwerke "Stammbaum" der austenitischen rostfreien Stähle. www.dew-stahl.com
23. VDM Metals GmbH (2014) Hochlegierte Werkstoffe für besondere Beanspruchung. In: VDM Report Nr. 26. www.vdm-metals.com
24. Francis R (2001) Galvanic corrosion. "A practical guide for engineers". NACE International, Houston, TX. Siehe auch Informationsstelle Edelstahl Rostfrei, Postfach 10 2205, 40013 Düsseldorf. www.edelstahl-rostfrei.de, Merkblatt Edelstahl 829
25. Sarikaya A, Jauss A Patentschrift EP 2287458A2. Method of detecting at least one malfunctioning high-pressure gas tank

References for Section 5.2.1

26. http://de.wikipedia.org/wiki/Hybridantrieb#Bivalente_Antriebe (2.9.2014)
27. http://de.wikipedia.org/wiki/Flexible_Fuel_Vehicle (2.9.2014)
28. http://de.wikipedia.org/wiki/Autogas (5.9.2014)
29. Wollenhaupt H, Waltner A, Wunderlich K (2013) Die neue Erdgas-Motorengeneration von Mercedes-Benz aus der M270/M274-BlueDIRECT-Motorenfamilie. 8. Tagung Gasfahrzeuge, Stuttgart
30. http://www.opel.de/selbststaendige-grosskunden/motoren-umwelt.html (5.9.2014)
31. Bargende M, Berner H-J, Bohatsch S, Hofmann B (2008) Ein strahlgeführtes CNG-Brennverfahren mit minimalen CO_2-Emissionen. Der Antrieb von morgen: MTZ-Konferenz Motor, München
32. Eichhorn A (2014) Systematische Bewertung von fünf Brennverfahren an einem aufgeladenen 2-Zylinder-Ottomotor. Dissertation, Herzogenrath

References for Section 5.2.2

33. Bargende M (1991) Ein Gleichungsansatz zur Berechnung der instationären Wandwärmeverluste im Hochdruckteil von Ottomotoren. Dissertation, Technische Hochschule Darmstadt, Darmstadt
34. Hanel D (2004) Molekulare Gasdynamik—Einführung in die kinetische Theorie der Gase und Lattice-Boltzmann-Methoden. Heidelberg
35. National Institute of Standards and Technology (2011) Thermophysical properties of fluid systems. http://webbook.nist.gov/chemistry/fluid/
36. Berner H, Chiodi M, Bargende M (2003) Berücksichtigung der Kalorik des Kraftstoffes Erdgas in der Prozessrechnung. 9. Tagung "Der Arbeitsprozess des Verbrennungsmotors", Graz
37. Otto F (2009) Grundlagen Verbrennungsmotoren—Simulation der Gemischbildung, Verbrennung, Schadstoffbildung und Aufladung. Heidelberg
38. Grill M (2006) Objektorientierte Prozessrechnung von Verbrennungsmotoren. Dissertation, Universitat Stuttgart, Stuttgart
39. Pischinger R, Klell M, Sams T (2009) Thermodynamik der Verbrennungskraftmaschine. Heidelberg
40. Zeldovich Y (1946) The oxidation of nitrogen in combustion and explosions. In: Acta physicochimica, USSR, vol 21
41. Wollenhaupt H, Wunderlich K, Waltner A (2013) Die neue Erdgas-Motorengeneration von Mercedes-Benz aus der M270/M274-BlueDIRECT-Motorenfamilie. 8. Tagung "Gasfahrzeuge –Antrieb mit Zukunft", Stuttgart
42. Bargende M, Berner H, Bohatsch S, Ferrari A, Hofmann B (2008) Ein strahlgeführtes CNGBrennverfahren fur minimale CO_2-Emissionen. MTZ Motortechnische Konferenz, München
43. Bohatsch S (2011) Ein Injektorkonzept zur Darstellung eines ottomotorischen Brennverfahrens mit Erdgas-Direkteinblasung. Dissertation, Universität Stuttgart, Stuttgart
44. Grill M, Billinger T, Bargende M Quasi-dimensional modeling of spark ignition engine combustion with variable valve train. SAE-Paper 2006-01-1107
45. Wenig M (2013) Simulation der ottomotorischen Zyklenschwankungen. Dissertation, Universität Stuttgart, Stuttgart
46. Bossung C, Bargende M (2014) Turbulenzmodellierung fur quasidimensionale Prozessrechnung. Abschlussbericht zum FVV-Vorhaben Nr. 1067. Forschungsvereinigung Verbrennungskraftmaschinen
47. Heywood J (1988) Internal combustion engine fundamentals. McGraw-Hill Series in Mechanical Engineering
48. Gulder O (1984) Correlations of laminar combustion data for alternative S.I. Engine Fuels. SAE-Paper 841000
49. Brequigny P, Mounaim-Rousselle C, Halter F, Moreau B (2013) Impact of fuel properties and flame stretch on the turbulent flame speed in spark-ignition engines. SAE-Paper 2013-24-0054
50. Millo F, Costa S, Mollo D, Ceratto D (2013) Effects of natural gas composition on engine performance and combustion characteristics. 8. Tagung "Gasfahrzeuge—Antrieb mit Zukunft", Stuttgart

References for Section 5.2.3

51. Robert Bosch GmbH (Hrsg) (2005) Ottomotor-Management: Systeme und Komponenten 3. überarb. u. erg. Aufl
52. Binder A (2007) CNG mixture preparation. Internal report, Robert Bosch GmbH
53. Chiodi M, Berner H, Bargende M (2006) Investigation on different injection strategies in a direct-injected turbocharged CNG-engine. SAE Technical Paper 2006-01-3000
54. Yadollahi B, Boroomand M (2013) The effect of combustion chamber geometry on injection and mixture preparation in a CNG direct injection SI engine. Fuel 107:52–62
55. Eichhorn A, Lejsek D, Hettinger A (2014) Methodischer Vergleich verschiedener Brennverfahren für einen aufgeladenen 2-Zylinder-Ottomotor. 11. Internationales Symposium für Verbrennungsdiagnostik, Baden-Baden
56. Eichhorn A (2014) Systematische Bewertung von fünf Brennverfahren an einem aufgeladenen 2-Zylinder-Ottomotor. Dissertation, TU Darmstadt
57. ECE-R110: Regulation No. 110, Part I., Specific components of motor vehicles using compressed natural gas (CNG) in their propulsion system, United Nations, 30 May 2008

References for Section 5.2.4

58. Pischinger S (2007) Verbrennungskraftmaschinen I, Lecture Script. RWTH Aachen University, 26th edn
59. Dingel O (2004) Gasfahrzeuge, expert Verlag
60. (2003) Untersuchungen zur Senkung der Zündkerzentemperatur am Gasmotor mit Vorkammer, 3. Natural Gas Conference, Dessau
61. Maly R (1984) Spark ignition: its physics and effects on the internal combustion engine, fuel economy in road vehicles powered by spark ignition engine. Plenum, New York
62. Hohner P (1999) Adaptives Zündsystem mit integrierter Motorsensorik, Thesis, University of Stuttgart
63. Maly R (1998) Die Zukunft der Funkenzündung. MTZ 59:7/8
64. Auzas F (2008) Décharge radiofréquence produite dans les gaz à pression élevée pour le déclenchement de combustion, Thèse, Université Paris Sud–XI
65. Bohne S, Rixecker G, Brichzin V, Becker M (2014) Corona ignition for natural gas engines with charge dilution, 1. International Engine Congress, Baden-Baden

References for Section 5.2.5

66. Sauerstein R, Weiske S, Münz S (2011) Turbocharging the compressed natural gas (CNG)—fuelled SI engine: a series solution for lowest CO_2-Emissions. SiA Congres International, Strasbourg
67. Sauerstein R, Sens M, Dingel O (2005) Geregelte zweistufige Abgasturboaufladung und ein CNG-Brennverfahren - eine vielversprechende Kombination. HdT Conference "Engine Process Simulation and Supercharging", Berlin
68. Boland D (2011) Wirkungsgradoptimaler Betrieb eines aufgeladenen 1,0 l Dreizylinder CNG Ottomotors innerhalb einer parallelen Hybridarchitektur, Thesis, University of Stuttgart

69. Fladung O, Hagelstein D, Noodt F, Pott E (2006) Aufladekonzepte für Ottomotoren mit Erdgasbetrieb, 11. Supercharging Conference, Dresden
70. Getzlaff J, Dingel O, Avramopoulos I, Kahrstedt J (2003) Ansätze zur Optimierung des Brennverfahrens eines Erdgasmotors für zukünftige Pkw-Anforderungen, 6. International Conference "Engine Combustion"
71. Kuhlbach K, Brinkmann F, Werner J, Kiener T, Becker M (2013) Innovative two-stage turbocharging system with cooled regulating valve for gasoline engines. Aachen Colloquium Automobile and Engine Technology
72. Sauerstein R, Becker M, Bullmer W, Dabrowski R (2011) Die geregelte zweistufige Abgasturboaufladung am Ottomotor—Auslegung, Regelung und Betriebsverhalten. Vienna Engine Symposium

References for Section 5.2.6

73. Wollenhaupt H, Wunderlich K, Waltner A (2013) The new Mercedes-Benz CNG engine generation as part of the BlueDIRECT engine family M270/M274. In 8th conference gaseous fuel powered vehicles—propulsion systems towards a sustainable future, Stuttgart
74. Pischinger R, Graßnig G, Taucar G, Sams T (1997) Thermodynamik der Verbrennungskraftmaschine. Springer, Berlin
75. Kemmler R, Frommelt A, Kaiser T, Schaupp U, Schommers J, Waltner A (2002) Thermodynamischer Vergleich ottomotorischer Brennverfahren unter dem Fokus minimalen Kraftstoffverbrauchs. In 11. Aachen Colloquium Automobile and Engine Technology, Aachen
76. Herden W, Vogel M (2002) Visionen idealer strahlgeführter BDE-Brennverfahren. In: Diesel- und Benzin-Direkteinspritzung. Expert-Verlag, Essen

References for Section 5.2.7

77. Wunderlich K, Waltner A, Merdes N, Vent G, Kreitmann F, Weller R (2013) The new Mercedes-Benz CNG engine generation as part of the BlueDIRECT engine family M270/M274. In: 22nd Aachen Colloquium Automobile and Engine Technology, Aachen
78. Kneer R (2006) Lecture Script Feuerungstechnik, Lehrstuhl für Wärme- und Stoffübertragung. RWTH Aachen, Aachen
79. Pischinger F (1989) Verbrennungsmotoren Band II, Lecture Script. RWTH Aachen, Aachen
80. Woschni G (1988) Verbrennungsmotoren 2. Auflage, Lecture Script, TU Munich, Munich

References for Section 5.2.8

81. Mollenhauer K, Tschöke H (eds) (2007) Handbuch Dieselmotoren (3rd newly revised edn). Springer, Berlin
82. Geiger J (2013) Alternative Erdgas: Entwicklungsanforderungen an zukünftige Pkw-Erdgasmotoren. SPECTRUM, Sonderausgabe Gasantriebe, Oktober 2013 (brochure of FEV GmbH, D-52078 Aachen)

83. Wunderlich K, Merdes M, Waltner A, Vent G, Kreitmann F, Weller R (2013) The new Mercedes-Benz CNG engine generation as part of the BlueDirect engine family M270/ M274. 22nd Aachen Colloquium Automobile and Engine Technology, pp 71–88
84. Thien U (Magna Steyr AG & Co. KG), Schaffer K, Eichlseder H (TU Graz) (2006) 1,6 l Kompressormotor für CNG Motorisierung eines Sportfahrzeugs. 27. International Vienna Engine Symposium, pp 115–136
85. Trijselaar A (2012) Knock prediction in gas-fired reciprocating engines. Master thesis University of Twente, Enschede (NL), January 2012
86. Bey R, Ohrem C, Biermann JW, Bütterling P (2013) Downsizing concept with two-cylinder CNG engine. MTZ 09/2013, vol 74, pp 12–19
87. Bartz WJ, Möller UJ (eds) (2000) Expert Praxislexikon Tribologie Plus. Expert Verlag, Renningen, p 286
88. ADAC Autotest (2013) VW up! 1.0 EcoFuel BMT high up! 1
89. Eichler F, Szengel R, Helbing C, Worm J (2013) The new EA211 1,4 l TSI CNG. 22nd Aachen Colloquium Automobile and Engine Technology, pp 89–102
90. Boland D (2011) Wirkungsgradoptimaler Betrieb eines aufgeladenen 1,0 l Dreizylinder CNG Ottomotors innerhalb einer parallelen Hybridstruktur. Doctor's thesis, University of Stuttgart
91. Klan S (2004) Beitrag zur Evolution von Aluminium-Gusslegierungen für warmfeste Anwendungen. Doctors's thesis, Technische Universität Bergakademie Freiberg
92. Klimesch Ch, Bechtle S, Vogt O, Ziegler S (KSPG AG), Röpke S, Schüller A (VW AG) (2011) Herausforderungen bei hoch beanspruchbaren Zylinderköpfen. GIESSEREI 98 06/2011, pp 60–66
93. LITEKS, NANOFRIKS und KS 309—Innovative Kolbentechnologien für Ottomotoren. Brochure of KS Kolbenschmidt GmbH (a company of KSPG AG), D-74172 Neckarsulm
94. Der Erdgasantrieb EcoFuel mit dem 1,4 l-110 kW-TSI-Motor. Selbststudienprogramm 425. Volkswagen AG, After Sales Qualifizierung, Service Training, VSQ-1
95. Basshuysen van R, Schäfer F (2002) Handbuch Verbrennungsmotor—Grundlagen, Komponenten, Systeme, Perspektiven, 1st edn. Friedr. Vieweg & Sohn, Braunschweig/ Wiesbaden
96. (1983) Handbuch (user guide) TRW Thompson GmbH (aktuell TRW Automotive GmbH, Engine Components, D-30890 Barsinghausen), 6th edn
97. MAHLE GmbH (ed) (2013) Ventiltrieb—Systeme und Komponenten. Springer Vieweg/Springer Fachmedien, Wiesbaden
98. Lewis R, Dwyer-Joyce RS (Series ed, Dowson D) (2002) Automotive engine valve recession. Professional Engineering Publishing Limited, London, Bury St. Edmunds
99. Ventilsitzringe für Gasmotoren. Product Information PI 0028 of MS Motor Service International GmbH, D-74196 Neuenstadt (a company of KSPG AG) bzw. http://www.ms-motor-service.com/produkte-kataloge/ks-trw-produkte/produkte-details/prod/ventilsitzringe/materialien/ (called on 04-24-2014)
100. Kawata H, Maki K (2007) Recent trends in heat resistant/wear resistant sintered alloys. Hitachi Powdered Metals Report No. 6, pp 3–11
101. Biemelt A (2011) Der dynamische Erdgasantrieb. Schrift der MBtech Group GmbH & Co. KGaA, D-71063 Sindelfingen, Segment Powertrain Solutions
102. Simon V, Oberholz G, Mayer M. Abgastemperatur 1050 °C—Eine konstruktive Herausforderung. Academy BorgWarner Turbo Systems (bwts_library_105_106.pdf)
103. Spezialzündkerze für aufgeladene Erdgas-Motoren. amz: auto/motor/zubehör No. 10-2009
104. Press Information on the occasion of the 61th IAA (Truck Show, Hannover), September 2006, of Emitec Gesellschaft für Emissionstechnologie mbH, D-53797 Lohmar (D)

105. Schmierstoffe für stationäre Gasmotoren. Schrift der Fuchs Europe Schmierstoffe GmbH (930082657 05/2012 2.5), D-68169 Mannheim
106. Dominsky S (2009) Schmierstoffe/Spezial-Motoröle—Auch mit Gas gut geschmiert? Kfz-betrieb 5:36–39

References for Section 5.2.9

107. Hübelt J SILENA-Studie—Untersuchung des Standes der Lärmminderungstechnik bei Mikro-BHKW. Akustik Forschung Dresden
108. Göhringer V (2008) Beitrag zur experimentellen Bestimmung des Strukturübertragungsmaßes von Dieselmotoren. Universität Stuttgart, Dissertation. Schriftenreihe des Instituts für Verbrennungsmotoren und Kraftfahrwesen, Band 38, Bargende M, Hrsg. ISBN 978-3-8169-2875-1
109. Flotho A, Spessert B (1988) Geräuschminderung an direkteinspritzenden Dieselmotoren. Teil 1 und 2. In: Automobilindustrie 3 und 5
110. Abou-Arab TW, Mohamed HS Thermo-mechanical (vibration & noise) performance of CNG CONVERTED PETROL ENGINE. SAE 2009-01-1682
111. Wenzel W (2006) Akustikmaßnahmen für aufgeladene Motoren. In: MTZ 67, Nr. 5
112. Pischinger S, Mirlach R, Heuer S, Silies A (2009) Akustik neuer ottomotorischer Brennverfahren. In: MTZ 70, Nr. 10, S. 770–777
113. Freytag C, Maunder M, Perchanok M (2010) Improving the acoustic validity of turbochargers in 1D gas dynamics simulation. Ricardo Software 2010 European User Conference, Prag, 20 Apr 2010
114. Brohmer A, Kemmerling J, Nehl J, Heuer S, Schäfer E (2002) NVH-Entwicklungsaspekte bei Ottomotoren mit Direkteinspritzung. In: MTZ 63, Nr. 11, S. 906–913
115. Rütten O, Habermann K, Izanloo H, Kabasforousha M (2006) From prototype to production: development of a new natural gas bifuel engine family. In: Dingel O (Hrsg.) Gasfahrzeuge II. Expert-Verlag
116. Gesk M (2008) NGI2—Ein innovativer Injektor für Erdgasfahrzeuge. 6. Management Circle Fachkonferenz "Start frei für Erdgasfahrzeuge", Stuttgart
117. Patro TN, LaRue DA Alternate fueled powertrain—an insight into its combustion related NVH issues. SAE 1999-01-1758
118. Karim GA, Wierzba I (1983) Comparative studies of methane and propane fuels for spark ignition and compression ignition engines. SAE West Coast International Meeting, Vancouver, BC
119. Czimmek PR (2013) Dispersion type dampers for acoustic noise reduction of a fuel injector. Patent DE112005000440 B4
120. Sachdeva A, Mansuri S Reduction of pulsation noise in CNG vehicles. SAE 2013-26-0098
121. Fang L, Deng-feng W (2011) Additional fuel system optimization for CNG vehicle base on MDO method. International Conference on Transportation, Mechanical, and Electrical Engineering (TMEE)

References for Section 5.3

122. Official Journal of the European Union (2009) Directive 2009/30/EC of the European parliament and of the council of 23 April 2009

123. Navigant Research (2013) Natural gas trucks and buses—medium and heavy duty natural gas trucks and buses, natural gas supply, refuelling infrastructure, CNG, and LNG global market analysis and forecasts. Report 4Q, 2013
124. Pischinger R, Klell M, Sams T (2009) Thermodynamic of the internal combustion engine (German), 3. Auflage. Springer, Wien
125. Figer G, Graf G (2014) Potential and limitations of heavy duty commercial vehicle gas engines. JSAE Proceedings, JSAE 20145285
126. Chart Industries, Inc. http://www.chartlng.com/App_Themes/ChartLNG/images/lng-fuelingsystems-large.jpg
127. Pischinger S, Umierski M, Hüchtebrock B (2003) New CNG concepts for passenger cars. SAE Pap. 2003-01-2264
128. FEV GmbH (2013) Natural gas as alternative fuel—natural gas powertrain for commercial engines (German). Spectrum Special Edition Natural Gas Powertrains
129. Shell (2013) Natural gas—bridging technology for the future? www.shell.com
130. Kitterer H, Sturm A (2011) IVECO Cursor8+ CNG—powertrain for heavy duty trucks for compressed or liquefied natural gas (German); VDI-Reports Nr. 2128, VDI Verlag GmbH, ISSN: 0083-5560
131. Wegrzyn J (2003) LNG fuel systems technology on-board LNG pumps, storage tanks, and heat exchangers, natural gas vehicle technology forum, Dallas TX, January 29, 2003
132. Rößler K, Otto F, Preuhs J-F, Farah P (2012) Direct injection of natural gas for passenger cars —prospects and challenges. 7th IAV conference: gas-powered vehicles, Potsdam, 24–25 September 2012
133. Gremminger A, Deutschmann O, Munshi S, Grunwaldt J (2014) Investigation of mechanism of catalytic methane reduction (German). FVV Spring Convention
134. Königsson F, Stalhammar P, Angstrom H (2011) Characterization and potential of dual fuel combustion in a modern diesel engine. SAE Technical Paper 2011-01-2223
135. Goudie D, Dunn M, Munshi S, Lyford-Pike E, et al (2004) Development of a compression ignition heavy duty pilot-ignited natural gas fuelled engine for low NO_x emissions. SAE Technical Paper 2004-01-2954
136. Trapp C, Laiminger S, Chvatal D, Wimmer A, Schneßl E, Pirker G (2011) New jenbacher gas engine generation from GE (German). 32. Vienna Engine Symposium
137. Leitner D (2012) Potential of an hydrogen car engine with cryogenic port fuel injection (German). Doctoral Thesis, TU Graz
138. Eichlseder H, Spuller C, Heindl R, Gerbig F, Heller K (2010) Concepts for diesel-like hydrogen combustion. MTZ 71(1):60–66
139. Kogler G, Wimmer A (2007) Potential of HCCI for large natural gas fueled engines, Paper No. 162, CIMAC Congress 2007, Wien
140. Pachta-Reyhofen G (2007) Faszination large engines—forrunners for vehicle engines? 28th Vienna Engine Symposium
141. Flierl R, Temp A, Wegmann A, Barrois A, Schmitt S (2011) Simultaneous combustion of methane-gasoline and methane-diesel, MTZ 10:S. 806–814

References for Section 5.4

142. Wehlen T (2009) CO_2-Reduzierung durch Systemlösungen vom Mikro- bis zum Vollhybrid. 4. Tagung Gasfahrzeuge, Stuttgart
143. Bargende M (2012) Future powertrains for passenger cars—in the crossfire between desire and reality. AutoUni Volkswagen AG, Wolfsburg

144. Kampmann S, Busch R, Christ A (2014) Gasoline engine—fit for the future powertrain. 14th Stuttgart International Symposium, Stuttgart
145. Götte C, Graf F, Klein B (2014) 48-V-Elektrifizierung—Bezahlbare Hybridisierung mit hohem Kundennutzen. 9. MTZ-Fachtagung—Der Antrieb von morgen, Wolfsburg
146. Nalbach M, Körner A, Hoff C (2013) "Der 48-V-Mikro-Hybrid—Ein neues Leistungsbordnetz. In: Automobiltechnische Zeitschrift 4, S. 296–300
147. Wollenhaupt H, Waltner A, Wunderlich K (2013) Die neue Erdgas-Motorengeneration von Mercedes-Benz aus der M270/M274-BlueDIRECT-Motorenfamilie. 8. Tagung Gasfahrzeuge, Stuttgart
148. Auerbach M (2013) Phlegmatisierung des Dieselmotors im Hybridverbund. Dissertation, Stuttgart
149. Böhm M (2011) Steuerung und Regelung des Antriebsstrangs eines Erdgashybridfahrzeugs. Dissertation, Stuttgart
150. Boland D (2010) Wirkungsgradoptimaler Betrieb eines aufgeladenen 1,0 l Dreizylinder CNG Ottomotors innerhalb einer parallelen Hybridarchitektur. Dissertation, Stuttgart
151. Raab F (2007) Untersuchungen zur Reduzierung der Kohlenwasserstoff-Emissionen im Kaltstart von Ottomotoren mit Saugrohreinspritzung. Dissertation, Magdeburg
152. Müller S (2010) Der Startvorgang von hybridisierten Ottomotoren—Untersuchung, Bewertung, Optimierung. Dissertation, Darmstadt
153. Maus W, Brück R, Konieczny R, Scheeder A (2010) Der E-Kat als Thermomanagementlösung in modernen Fahrzeuganwendungen. Motortechnische Zeitschrift 5:340–346
154. Riemer T (2011) Vorausschauende Betriebsstrategie für ein Erdgashybridfahrzeug. Dissertation, Stuttgart
155. Soltic P (2012) Der CNG Parallel-Hybrid Versuchsträger "Clever". 7. IAV-Tagung: Gasfahrzeuge—der konsequente Weg zur CO_2-Reduzierung, Potsdam
156. Hofmann P (2010) Hybridfahrzeuge. Wien

References for Section 5.5

157. Rumpke A, Peters D, Kalinowska D (2013). im Auftrag BMVBS: Entwicklung einer Mobilitäts- und Kraftstoffstrategie für Deutschland—Voruntersuchung, Berlin.
158. Rumpke A, Peters D, Saar D et al (2010) Erdgas und Biomethan im künftigen Kraftstoffmix. Berlin
159. Knorr H Erdgastechnik von MAN—MAN Euro Vl-Erdgasmotoren. MAN Truck & Bus AG. Präsentation vom 08.11.2012 in Nürnberg
160. Willner K (2013) Testing of unregulated emissions from heavy duty natural gas vehicles. Malmö, Svenskt Gastekniskt Center AB
161. erdgas mobil, basierend auf eigenen Berechnungen und Daten vom ADAC und MWV, Berlin 2012/2014
162. Growitsch C, Hecking H, John C et al (2013) Potenziale für Erdgas im Straßenverkehr—eine ökonomische Analyse. Energiewirtschaftliches Institut an der Universität zu Köln, Köln
163. erdgas mobil: Statistik Erdgastankstellen und KBA-Zulassungsstatistik
164. NGVA Statistik (2013) http://www.ngvaeurope.eu/european-ngv-statistics
165. Iveco (2014) Infobroschüre zu Iveco Stralis LNG
166. DVGW (2009) Arbeitsblatt G 651/VdTÜV M 510. Erdgastankstellen, Bonn
167. E.ON Gas Mobil (2014) AG, erdgas mobil: Das Funktionsprinzip der Erdgastankstelle
168. erdgas mobil: Neuzulassungen und Bestand alternative Antriebe basierend auf KBA-Zulassungsstatistik

169. Energiesteuergesetz: § 2 Abs. 1 Nr. 7 und 8 sowie § 2 Abs. 2 Nr. 1
170. Dena (2013) Positionspapier. Transparente Preisinformation für einen Kraftstoffmarkt im Wandel, Berlin
171. Heidt C, Lambrecht U, Hardinghaus M, et al (2013) im Auftrag BMVBS: CNG und LPG—Potenziale dieser Energieträger auf dem Weg zu einer nachhaltigeren Energieversorgung des Straßenverkehrs. Heidelberg
172. NGVA News (2014) European Parliament approves directive to support roll-out of CNG and LNG infrastructure in Europe. Brüssel. 15.04.2014. http://www.ngvaeurope.eu/european-parliament-approves-directive-to-support-roll-out-of-cng-and-lng-infrastructure-in-europe
173. International Maritime Organization (2005) Marpol 73/78, Annex VI: Prevention of air pollution from ships
174. NGVA Statistik (2013) http://www.ngvaeurope.eu/european-ngv-statistics

Off-Highway Gas Engines

6

Robert Böwing, Georg Tinschmann, and Christian Trapp

6.1 Large Gas Engines for Industrial and Maritime Applications

6.1.1 Areas of Application

Gas engines currently have an extremely diverse range of applications. Off-highway gas engines—i.e. engines that are not intended for road vehicles—are increasingly also being used for applications that just a few years ago were the exclusive preserve of diesel engines.

Gas-fueled combustion engines with outputs above 500 kW are currently mainly used for on-site power (electricity or combined heat and power), for driving compressors in the transport and distribution of gas (mechanical drive) (www.dieselgasturbine.com/Market-Surveys/) [1] and for the propulsion of gas tankers (marine), Diagram 6.1.

A glance at Table 6.1. shows the following figures for the various applications of engines with outputs greater than 500 kW (www.dieselgasturbine.com/Market-Surveys/):

Diesel engines dominate and the proportion of gas engines in power generation units and mechanical drives is around 10–12 %.

Due to the low gas price (http://www.bp.com/en/global/corporate/about-bp/energy-economics/statistical-review-of-world-energy/statistical-review-downloads.html), particularly in the USA, but also as a result of exhaust legislation, it is expected that the gas engine will be able to increasingly gain shares in all the segments mentioned above [2] and will penetrate applications that are currently the exclusive preserve of Diesel engines.

Examples include other marine applications, use in locomotives (Rail) and also use in mining vehicles (Construction/Mining).

R. Böwing • G. Tinschmann • C. Trapp (✉)
GE Jenbacher GmbH & Co. KG, Achenseestraße 1-3, 6200 Jenbach, Austria
e-mail: Christian.Trapp@ge.com

R. van Basshuysen (ed.), *Natural Gas and Renewable Methane for Powertrains*, Powertrain, DOI 10.1007/978-3-319-23225-6_6

Diagram 6.1 Typical applications for gas engines with outputs above 500 kW. (**a**) Compressor and pump drives (mechanical). (**b**) Power. (**c**) Powering gas tankers (Marine)

Table 6.1 Number of off-highway engines supplied with outputs above 500 kW for various applications and fuels (as of 2012) (www.dieselgasturbine.com/Market-Surveys/)

	Total	Diesel		Heavy fuel oil		Natural gas and dual fuel	
Fuel and application	(Qty)	(Qty)	(%)	(Qty)	(%)	(Qty)	(%)
Power generation	31,696	27,980	88.3	236	0.7	3471	11.0
Mechanical drives	5642	5057	89.7	7	0.1	578	10.2
Marine applications	8840	6623	74.9	2194	24.8	23	0.3

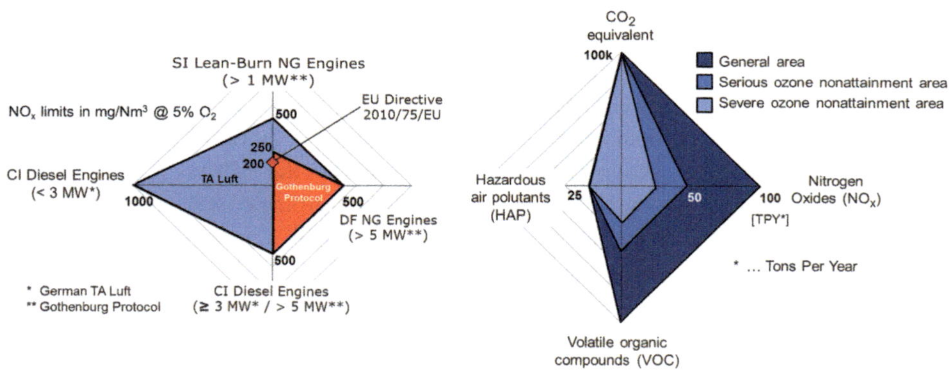

Diagram 6.2 NO_x limits for large gas engines in the EU [3–5] and limits in the USA [6, 7]

6.1.2 Emissions Legislation and Fuel Prices

Numerous sets of legislation apply to gas engines across the world. If we limit ourselves to natural gas as a fuel, many states in Europe have derived their limit values from the German "Technische Anleitung zur Reinhaltung der Luft" (Technical Instructions on Air Quality Control), abbreviated to TA Luft [3], Diagram 6.2. For lean-burn gas engines with input fuel powers above 1 MW, the applicable limit is 500 mg NO_x/m_N^3 (based upon an O_2 concentration in the exhaust gas of 5 %).

For lean-burn gas engines, the "Gothenburg Protocol" that has been ratified by some European states [4] means that values 50 % lower than those stipulated by the TA Luft must be achieved. Since the start of 2013, the EU Directive 2010/75/EU, which stipulates 200 mg NO_x/m_N^3, must be implemented for engines with a fuel power greater than 15 MW when used in power plants with a fuel power greater than 50 MW [5].

In the medium term, we can expect to see not only a tightening up of NO_x limits, but also a lowering of the limits for formaldehyde (HCHO), carbon monoxide (CO), methane (CH_4) and hydrocarbons (C_nH_m).

A glance at the emission limits in the USA reveals that a range of local directives has been issued by the US states. In some cases, NO_x limits below 100 mg NO_x/m_N^3 are in force, which lean-burn gas engines can only fulfil with exhaust aftertreatment. Some regions have moved to a system in which pollutant emissions (NO_x, CO, etc.), and also CO_2e emissions, are limited per year and per site.

CO_2e means that emissions of pollutants such as CH_4 are weighted and included in the calculation for CO_2 emissions (EPA Title V) [7]. If the emissions produced are limited by plant, engines with very low emissions have an advantage, because operators have greater freedom in the maximum installable power per site.

A look at global gas prices is also interesting (Diagram 6.3) (http://www.bp.com/en/global/corporate/about-bp/energy-economics/statistical-review-of-world-energy/statistical-review-downloads.html). Particularly in the USA, the gas price has become decoupled from the oil price and basically two price systems apply. Whereas in Europe and Asia the gas price is linked to the oil price, in the USA, and to some extent also in Great Britain, gas is freely traded. It is currently thought that the gas price will remain significantly below

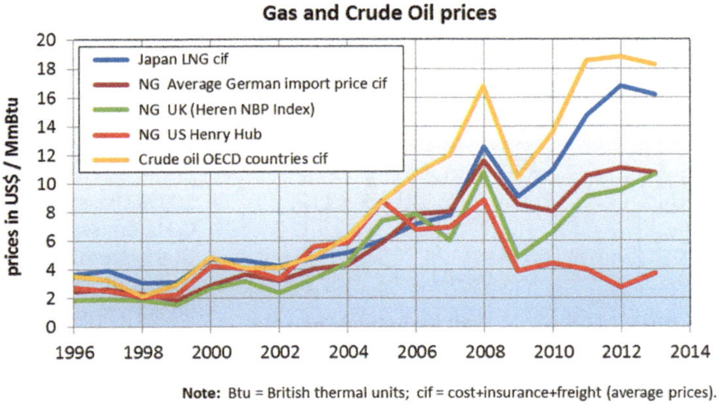

Diagram 6.3 Worldwide gas prices in comparison to prices for crude oil in US$/MmBtu (http://www.bp.com/en/global/corporate/about-bp/energy-economics/statistical-review-of-world-energy/statistical-review-downloads.html)

the oil price in the future and that increasing intercontinental transport of gas will lead to the convergence of global prices.

6.1.3 Gas Engines in Shipping

Driven by the *International Maritime Organization* (IMO), the emission limits for nitrogen oxide in shipping, particularly in coastal waters (Emission Control Areas, ECAs), will fall by around 70 % in comparison to 2011 in the near future [8]. At the same time, the use of low-sulfur fuels in coastal waters is stipulated. Investment costs and also fuel costs for drive systems with diesel engines are rising significantly. Dual fuel engines provide a very interesting alternative, as they fulfill IMO Tier III in coastal waters in gas mode and can still be operated with the low-priced heavy fuel oil on the high seas.

Dual fuel engines with outputs of up to 50 MW are currently available. In medium-speed engines, low-pressure gas injection into the inlet port is typically used (Diagram 6.4c) [9–11] and the diesel ignition oil represents around 1 % of the total

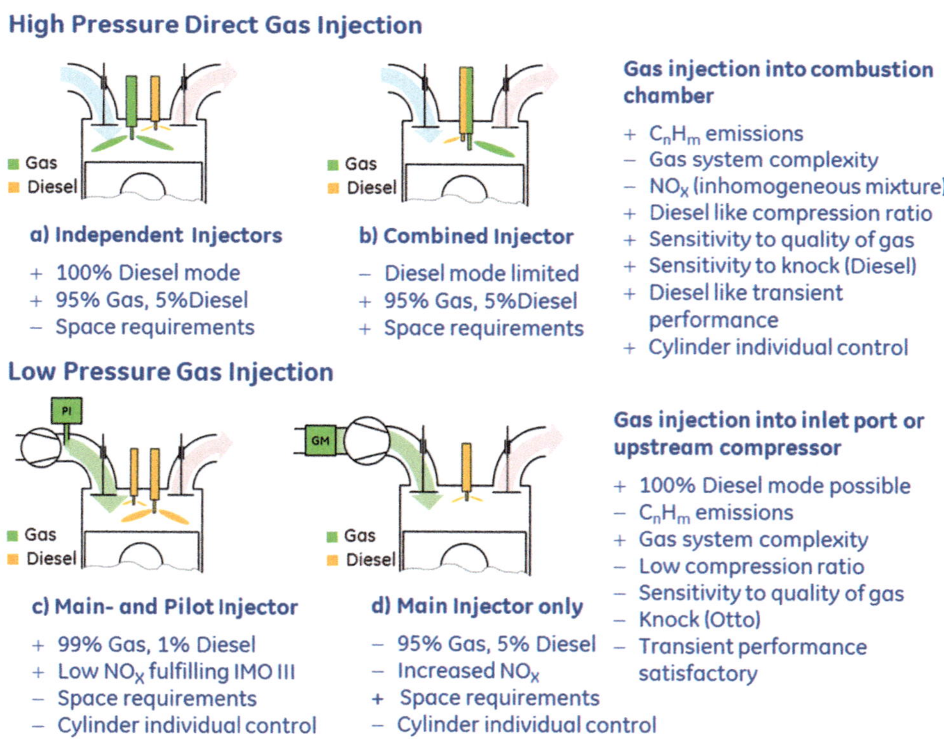

Diagram 6.4 The main Dual Fuel concepts in use today

input energy. Using this concept, IMO Tier III is fulfilled without exhaust aftertreatment.

Direct gas injection is also used in slow-speed two-stroke engines (Diagram 6.4a) [12]. The concept is characterized by insensitivity to knocking and very low C_nH_m emissions, but additional measures are needed for compliance with IMO Tier III, such as exhaust gas recirculation or an SCR catalyzer.

Pure gas engines are also used on ships in ports, on fixed routes and where there is a good gas infrastructure. A redundant operating mode with diesel fuel is not available. The combustion process and carburation system are generally taken from engines in on-site power systems (Sect. 6.2, Diagram 6.9), with factors such as power, turbocharging and safety equipment being adapted to the requirements of shipping and the application type in question. Tugboats require drive engines with a large characteristic map width and good transient behavior; gas engines are now also found on board [13].

6.1.4 Gas Engines for Locomotive Drives

The low gas price in the USA is giving rise to changes such as the development of large heavy-haul locomotives with gas engines [14], Diagram 6.5. They use a modified 3.5 MW diesel engine that burns around 80 % of its input energy in the form of gas in mid-cycle in gas mode. One important feature is redundant operation with diesel fuel. The dual fuel engine works with a diesel main injector and gas injection valves in the inlet port, Diagram 6.4c, d.

Challenges for development are:

- Safe operation of the system as a whole, including the gas tender,
- 100 % diesel operation must be possible at all times,
- Performance comparable with diesel engine,

Diagram 6.5 Heavy-haul locomotive with dual fuel gas engine [14]

- Highest gas percentages and reliable control of knocking,
- Low service costs,
- Option of retrofitting.

6.2 Mixture Preparation and Combustion in Large Gas Engines

6.2.1 Requirements for On-Site Power and Mechanical Drive Applications

Gas-fueled combustion engines with outputs above 500 kW are currently used mainly for on-site power (electricity or combined heat and power, Diagram 6.1) and for driving compressors or electrical drives within the extraction and distribution of natural gas (mechanical drives) (www.dieselgasturbine.com/Market-Surveys/).

Whereas for on-site power supply the life-cycle costs are significantly influenced by fuel costs in many market segments, if the ever-stricter emissions limits are adhered to, in compressor drives the investment costs, robustness and adherence to extremely strict emissions limits with widely fluctuating gas qualities are the main focus [1].

These different requirements are mirrored in the selection of the engine design for these different applications: currently, on-site power supply systems almost always use lean-burn combustion concepts without exhaust aftertreatment, whereas in compressor drives the number of engines with stoichiometric combustion processes and three-way catalytic converters is increasing more and more. Common to both applications is that they almost always use turbocharged engines to increase power density (Diagram 6.6).

Diagram 6.6 The J624 large gas engine with two-stage turbocharging and lean-burn combustion concept from GE [15]

6.2.2 Turbocharging and Gas Exchange Process

The turbocharging and gas exchange concept in today's large gas engines has a decisive influence upon the power and efficiency of the products:

To reduce the specific costs of large gas engines and keep the installation space needed the same as total power rises, a significant increase in specific power (and thus the effective mean pressure) at only slightly higher production costs is a high-priority development goal. Increasing specific power while holding the speed range constant also allows the friction element to be reduced, and thus efficiency increased.

Gas engines with lean-burn combustion concept, the most common type of engine used in on-site power systems, use a high excess air up to lambda > 2 (for natural gas) to minimize nitrogen oxide emissions by reducing the peak temperature and to improve efficiency by increasing the polytropic exponent.

The high amount of excess air can also be used to slightly reduce the tendency to knock. However, the early or late closing of the inlet valve as in a Miller or Atkinson process, and the associated reduction of the effective compression ratio and the process temperature in the combustion chamber, are significantly more effective to reduce knock tendency.

To realize all these points, high inlet pressure and with this high compressor pressure ratios are needed, along with excellent turbocharging efficiencies and very good intermediate cooling of the compressed air or compressed mixture. Large gas engines with single-stage turbocharging and compressor pressure ratios of up to five achieve mean effective pressures of up to 22 bar [16–23]. The J624 from General Electric with two-stage turbocharging achieves 24 bar at an electrical efficiency (measured up to the generator terminal) of 46.5 % at a compressor pressure ratio of up to 8 [15]. To achieve these efficiencies, the turbocharger efficiency (including any intermediate cooling or aftercooling of the charge) must be significantly above 70 %. With single-stage turbocharging, around 68 % is feasible, whereas with two-stage turbocharging, a figure of 75 % is currently being achieved in series production [24].

6.2.3 Mixture Preparation

Two main approaches to prepare the gas/air mixture are used in large gas engines: central mixing of the combustible gases with the air before the turbocharger (we call these mixture-charged engines) and mixing within the ports in the cylinder head just before the inlet valves for each individual cylinder. Less widespread is the injection of the gas at high pressure directly into the combustion chamber.

For all these approaches, the development goal is to obtain a homogeneous mixture of air and combustible gas with the lowest possible pressure drop.

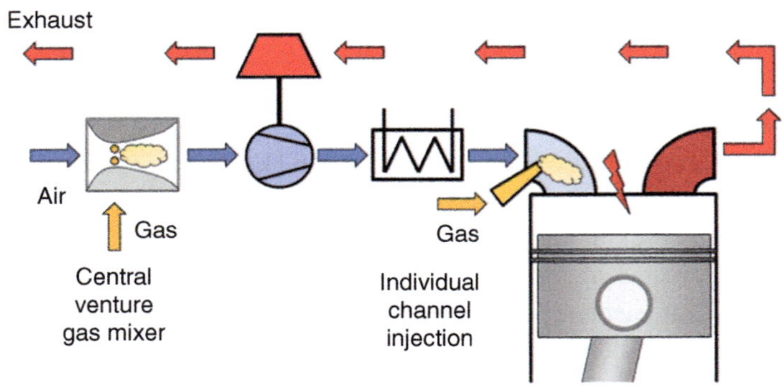

Diagram 6.7 Schematic diagram of the two most commonly used approaches to carburetion

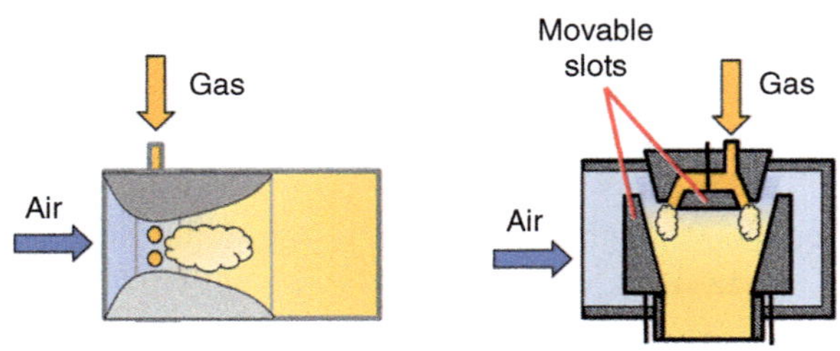

Diagram 6.8 Venturi and slot-type gas mixer with regulation inserts

Central Gas Mixer

The arrangement of a central gas mixer before the turbocharger is the simplest option for adding the combustible gas to the air. The mixers are designed either as slot-type or venturi mixers (Diagram 6.8), and the mixture quality is regulated either by moveable inserts or by an upstream gas-pressure regulation valve. A central gas mixer before the turbocharger allows the dosing of gases at low pressures of approx. 50–100 mbar, so biogas or landfill gas can be used without further compression. If a gas pressure regulation valve is used for increased dosing accuracy, the pressure must be at least 200 mbar (Diagrams 6.7 and 6.8).

Port Injection for Individual Cylinders

A significantly higher gas pressure is required when injection takes place into the ports before the inlet valves for each individual cylinder: the gas has to penetrate into the channel against the boosted inlet pressure, and therefore a pressure differential of at least 1 bar is needed in relation to the inlet pressure. Since the distance to the cylinder and with this the time to homogenize the mixture is significantly shorter than that for a central gas mixer, the injection or mixing unit has to distribute the gas as thoroughly as possible across the entire cross section. Lances that are aerodynamically optimized in terms of pressure loss protrude into the center of the channel flow. As well as significantly reducing the quantity of combustible mixture in the suction section, channel injection has the following advantages: individual adjustment of the fuel/air mixture for each cylinder, gas supply cut-off in the event of rapid load shedding and cylinder shutdown at low loads (Diagram 6.7).

High-Pressure Gas Injection into the Combustion Chamber

A few large gas engines are based upon injection of the combustible gas directly into the combustion chamber at high pressure. Depending upon the timing of injection, gas pressures between 10 bar (suction phase) and 300 bar (compression phase shortly before TDC) are required. One advantage of direct injection of the gas into the combustion chamber is improved transient behavior of the engine. When the gas is injected shortly before top dead center, there is also a significant reduction of the knocking of the final gas due to late mixing. This means that even gases with very low methane numbers (30–50, for example in oil and gas extraction) can be used directly. The critical point here is the very short mixture preparation time because in spite of this a homogeneous mixture with the desired fuel/air ratio should exist everywhere in the combustion chamber at the time of ignition.

When the combustible gas is injected directly, the losses when the fuel is compressed in a gaseous state should be taken into consideration.

6.2.4 Combustion Process

Three main combustion processes are currently mainly used in large gas engines for on-site power and for compressor drives (Diagram 6.9): Combustion in an open combustion chamber, similar to a standard four-stroke engine from the automotive sector,

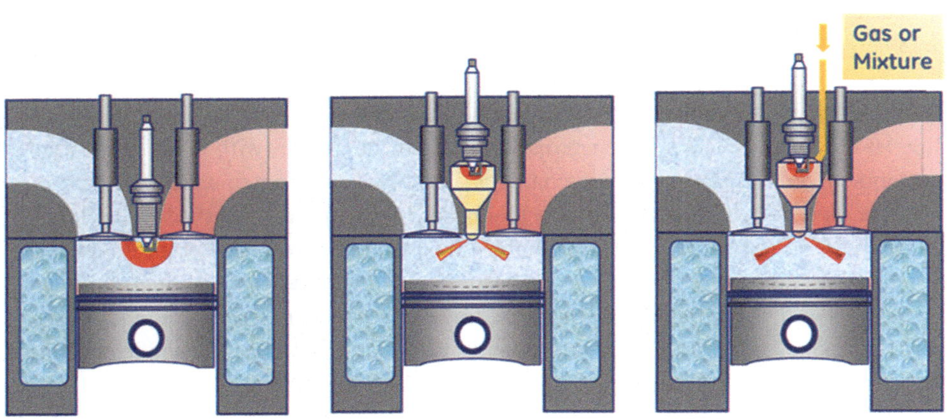

Diagram 6.9 Schematic diagram of a combustion process with an open combustion chamber (*left*), unflushed pre-chamber (*middle*) and flushed pre-chamber (*right*)

the use of an un-scavenged pre-chamber or a pre-chamber spark plug and finally the use of a scavenged pre-chamber.

So-called micropilot combustion processes, in which the gas is not ignited by a spark plug but by a directly injected, self-igniting quantity of diesel (1–5 % of the energy) are much rarer but not unheard of [25].

In the lean-burn process, the combustion concept is designed such that the emission limits, for example according to the German TA Luft, can be satisfied without exhaust aftertreatment. These limits currently relate primarily to nitrogen oxide, formaldehyde and in some countries also hydrocarbons and carbon monoxide. The focus on hydrocarbons, and here in particular methane (regarded as a greenhouse gas), as well as particulate matter, etc., may intensify and expand in the future.

Even today, there is an increasing conflict of aims between the nitrogen and hydrocarbon emissions to be achieved in lean-burn combustion processes (especially if low nitrogen oxide limits are prescribed, e.g. by the IED 2012). Hydrocarbon emissions also need to be significantly reduced to achieve a high level of efficiency and low greenhouse gas emissions.

Open Combustion Chamber

A combustion concept with an open combustion chamber is used both in lambda-1 and lean-burn engines. In the latter, a high level of turbulence permits the ignition and combustion of a mixture with a lambda value of up to 1.85. This microturbulent flow is generated by the decay of the in-cylinder charge movement, a high swirl or squish, which is often promoted by a special piston crown bowl. In such concepts, the flow speed close to the spark plug electrodes is of critical importance (danger of blowing out ignition spark or the initial flame kernel), which is why spark plugs with a flow shield are sometimes used here.

Concepts with an open combustion chamber are currently being used with success at mean effective pressures up to 22 bar in engines with bore diameters less than 170 mm in compliance with TA Luft (500 mg/Nm^3 NO_x with 5 % oxygen in the exhaust gas) [26].

Unscavenged Pre-chamber/Pre-chamber Spark Plug

If higher efficiencies, lower nitrogen oxide emissions or engines with larger bores are the development goal, unscavenged pre-chambers, usually in the form of a pre-chamber spark plug, are used. The chamber (or cap) around the spark plug is flushed with fresh air through bores connected to the main combustion chamber during the compression phase, with the flow being guided in such a way as to intentionally increase the microturbulence in the chamber. In addition, the mixture that flows into the chamber, or is already there, is heated by the chamber walls. These two measures lead to a shifting of the lean ignition limit, and the flares, which shoot out of the chamber at a pressure of 1–3 bar (pre-chamber spark plug with 1–2 cm^3 chamber volume) after the ignition, then ignite the main combustion chamber. Combustion processes with an unscanvenged pre-chamber frequently use a moderate swirl and/or squish to aid complete combustion [16, 18, 27, 28].

Scavenged Pre-chamber

If even lower emissions (for example the IED 2012) and/or higher outputs (>22 bar mean effective pressure) need to be achieved in engines with bore diameters greater than 170 mm, scavenged pre-chambers are used. In these, in addition to the lean mixture from the main combustion chamber, gas flows via an additional active or passive valve into the pre-chamber, so that the ignition and combustion in the pre-chamber takes place almost stoichiometrically. After a homogenization of the pre-chamber charge in the compression phase, flames at an overpressure of up to 80 bar shoot into the main combustion chamber and, if the chamber is correctly designed, achieve rapid combustion with optimized burn-out even with very lean mixtures (lambda > 2.2). The matching of the mixture and the generation of microturbulence in the pre-chamber, as well as the interaction of the flame flares with the main combustion chamber, are critical [14, 15, 20, 24].

In the case of scavenged pre-chambers, the main turbulence is achieved by the flame torches in the main combustion chamber, so that assistance by swirl or squish is not usually necessary (Diagram 6.10).

6.2.5 Control

Power Control

The output of large turbocharged gas engines is usually controlled by a combination of throttle valve (primarily in the lower load range) and compressor bypass valves; less commonly a waste-gate or a variable turbine geometry is used for this. In engines with

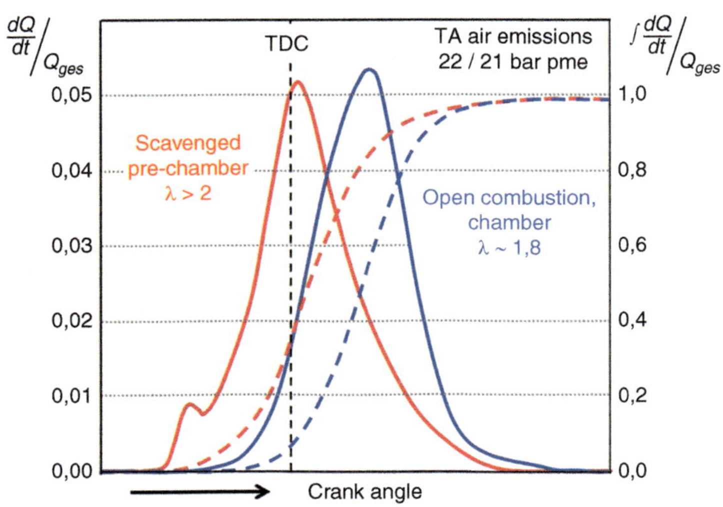

Diagram 6.10 Combustion sequences with different combustion processes

port injection, cylinder deactivation can be used in the lower load range to dispense the use of a throttle valve.

Emissions Control

Various concepts are available to control emissions (mainly nitrogen oxide) in the lean-burn combustion process: the determination of the calorific value of the mixture based upon the pressure and temperature in the suction manifold and generator power output, the measurement of exhaust temperature or combustion chamber temperature using thermocouples and the measurement of the flame transit time using ionization current probes are widespread in on-site power systems and, after suitable calibration, all permit the air/fuel ratio to be determined. In lambda 1 engines, and in a few lean-burn engines, lambda sensors or NO_x sensors are used in the exhaust tract.

6.3 Use of Special Gases in Power Plant Engines

6.3.1 Overview

Combustion engines that are used in power plants for on-site power can be operated not only with natural gas, but also with a wide range of other gases. This field of application is becoming ever more attractive due to dwindling resources, rising energy prices and increasing statutory climate obligations. This section provides an overview of the various fuel gases and their use in the stationary combustion engine (Diagram 6.11).

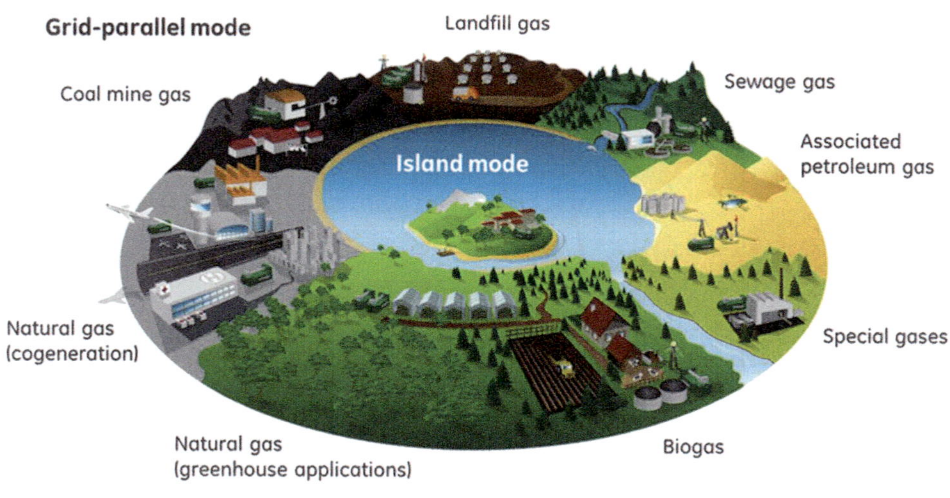

Diagram 6.11 Various combustion gases and their use in stationary combustion engines (GE Jenbacher)

6.3.2 Gas Types

The fuel gases used in the gas engine can be divided into fossil gas and biologically and technically produced gases. The gases differ in terms of their raw materials and their generation, and they therefore also have different compositions and combustion properties. Impurities and rapid changes in the gas composition should also be taken into consideration [29, 30].

Fossil Gases

Natural gas is a fossil gas that is found in underground deposits. It is primarily composed of methane and is an excellent, future-proof fuel for stationary engines thanks to its worldwide availability, estimated reserves, good combustion properties and composition (low C/H ratio for low carbon dioxide emissions). On the other hand, the greenhouse potential of methane should be taken into account and it is therefore important that complete combustion takes place. Its composition varies depending upon the extraction area and the addition of liquefied gas by gas suppliers.

Mine gas escapes during coal mining and comprises mainly natural gas, nitrogen and oxygen. We normally differentiate between coal-bed methane (90–95 % CH_4), gas from active mines with ventilation (25–60 % CH_4) and gas from abandoned mines (30–80 % CH_4). To increase safety in mines and prevent the escape of the greenhouse gas methane into the atmosphere, gas engine units are increasingly being used across the world. The gas must be largely dry and free of dust when used in combustion engines [29, 31].

Associated petroleum gas is the gas dissolved in petroleum under pressure (methane, ethane, propane, butane, etc.), which is released when oil is extracted and the pressure falls. The gas, which can vary greatly in composition, is often simply flared off on site. For ecological, economic and legal reasons, however, it is increasingly being used in gas engines. This provides a supply of power on site, without the associated transport costs.

Liquefied petroleum gas (propane, butane, etc.) known as LPG is easy to store and is therefore used in vehicles where it represents an alternative to natural gas, which is known as CNG (Compressed Natural Gas). Its use in stationary gas engines is rather limited due to the unfavorable combustion properties (knock resistance). From a safety point of view, it is important to bear in mind that propane and butane are heavier than air and sink to the ground, whereas methane and ethane are lighter than air and rise [29].

Biologically Produced Gases

Biologically produced gases arise due to the anaerobic decomposition of organic matter. Microorganisms convert carbohydrates, proteins and fats into methane and carbon dioxide. As a renewable energy source they are, in theory, a CO_2-neutral alternative to fossil fuels. The gases can contain trace elements such as ammonia, hydrogen sulfide or silicon compounds and are usually saturated with moisture. These contaminants must be removed or reduced, on a case-by-case basis, by gas treatment to protect the engine and other components (catalytic converters, heat exchangers, etc.) against deactivation, corrosion, deposits and wear [32–34].

Biogas arises due to the decomposition of biomass from renewable raw materials, organic wastes or residues from livestock farming. It forms in the fermenter of biogas plants.

Landfill gas arises due to the decomposition of wastes in landfill sites. The gas is drawn off from the waste dump via drainage pipes and gas wells and fed to the gas engine. This prevents greenhouse gas methane from escaping and stops unpleasant smells arising.

Sewage gas arises during the fermentation of sewage sludge in the digestion tower. The gas is used in the engine to cover the plant's electricity and heat demand and to eliminate odor problems.

Technically Produced Gases

Steel gases such as blast furnace gas and converter gas, or in a wider sense coke gas, are process gases that arise during the production of steel or coke. Unlike fossil and biologically produced gases that are mainly composed of methane, the main energy carriers in this case are hydrogen and carbon monoxide. Steel gases are characterized by their low calorific values and fluctuating gas compositions. They can contain sulfur compounds, dust and tar and therefore require cleaning. From a safety point of view, highly toxic carbon monoxide should be given particular consideration [30, 35–38].

Gases from gasification processes of biomass (e.g. wood) or waste (e.g. plastic) can also be used in gas engines if their purity and stability are sufficiently high. The energy carriers in the gas are methane, hydrogen and carbon monoxide. Sulfur compounds and tar are problematic and carbon monoxide should be taken into consideration for safety reasons [39].

Gases from the chemical industry can arise as a by-product of a manufacturing process. These gases can have a very low calorific value. For example, there are applications in the formalin industry in which the calorific value is only 5 % of that of natural gas (0.54 kWh/Nm^3 at 18 % H_2 and 82 % N_2).

Hydrogen as a pure gas has, in the past, only extremely rarely been used in large stationary engines. It has occasionally been added locally to natural gas operation. Its addition to the natural gas network is currently being discussed under the slogan of "Power-to-Gas".

Overview

Modern stationary engines can efficiently burn gas with extremely varied methane numbers (approx. 0–155), calorific values (approx. 0.5–34 kWh/Nm^3) and minimum air requirements (approx. 0.5–12 Nm^3/Nm^3). Table 6.2 provides an overview.

Table 6.2 Composition and properties of different fuel gases (guide values)

	CH_4 % Vol.	C_2H_6 % Vol.	C_xH_y % Vol.	H_2 % Vol.	CO_2 % Vol.	CO % Vol.	N_2 % Vol.	O_2 % Vol.	MZ –	H_u kWh/ Nm^3	L_{min} Nm^3/Nm^3
Natural gas	65–99	0–25	0–18	–	0–3	–	0–15	–	60–99	8.5–10.5	9.5–10.5
Coal mine gas[a]	25–50	0–1	–	–	0–3	–	30–65	9–12	100–120	2.5–5.0	1.7–4.5
Associated petroleum gas	35–90	2–20	5–25	0–0.5	15–45	–	0–45	0–0.5	40–80	5.0–13	5.0–12
Biogas	45–70	–	–	–	25–40	–	1–12	0–3	117–142	4.5–6.5	4.3–6.2
Landfill gas	40–60	–	–	–	30–40	–	5–20	0–4	123–154	4.0–6.0	3.8–5.7
Sewage Gas	55–70	–	–	–	30–40	–	–	0–2	130–140	5.5–7.0	5.2–6.7
Coke gas	20–30	–	1	50–65	1–5	5–8	4–6	1	40–50	4.0–5.0	3.4–4.4
Blast furnace gas[b]	0–1	–	–	1–5	17–25	20–25	50–55	1	120–130	0.8–1.1	0.55–0.8
Converter gas	0–2	–	–	0–33	0–20	55–70	0–15	0–2	50–100	2.6–3.1	1.8–2.3
Wood gas	0–12	0–4	0–6	12–45	0–25	12–45	0–60	0–3	40–90	1.4–4.8	1.1–4.1

[a]Active mine
[b]Blast furnace gas

6.3.3 Use

Challenges

The use of fossil gases, such as coal mine gas and associated gas, or biologically produced gases such as biogas, landfill gas and sewage gas, has for many years represented the state of the art. Mean effective pressures and efficiencies are usually similar to those in natural gas operation. Since the gases can vary significantly in their properties, corresponding adjustments must be made to the mixture preparation, combustion method and turbocharging of the gas-specific engine variants. Impurities in the gas must be removed or reduced to protect the engine and permit the use of catalytic converters and heat exchangers. Fluctuations in gas pressure and composition must be taken into account in the design and compensated for by the control system [31–34].

The use of technically produced gases such as steel gases, gases from gasification processes and gases from the chemical industry is still associated with major challenges. The volumetric flow rates of these fuel gases can be more than ten times as high as those for natural gas operation. Dust, tar or moisture in the gas must, as far as possible, be eliminated. Rapid and large fluctuations in the gas composition, such as a 25 % change in calorific value in 30 s, call for measures to be implemented in the plant and engine. Special gas characteristics encourage pre-ignition and backfire, which can also be triggered by deposits, particulate matter in the gas, oil in the combustion chamber or rapid changes in the gas composition, etc., and can limit the maximum feasible output. Different combinations of hydrogen and carbon monoxide (and methane) require different engine variants [30, 35–38, 40].

Engine Adaptations

In the case of low calorific gases, special gas mixers are used for mixture preparation that permit very large quantities of combustion gas to be added at very low admission pressures. Several gas trains are used per engine. Gas preparation systems and dehumidifying devices protect the engine and the components in the exhaust system. Fluctuations in gas composition can be reduced by plant-based measures such as gas storage tanks and gas flares or must be controlled by engine-based measures such as the rapid variation of the ignition timing, air/fuel ratio or engine performance [30, 35, 38].

Changes may be made to valve timings based upon gas exchange and turbocharging. The turbocharger design depends upon the combustible gas and installation conditions. Special hot-gas designs are used when exhaust temperatures are high. Flame arresters in the inlet system limit the effects of any back-firing. Their design is specific to the gas type and, ideally, they are positioned close to the inlet valves. The inlet system as a whole, including the air filter, is constructed as robustly as possible. In addition, pressure relief flaps are often provided in the exhaust system to alleviate exhaust-side deflagrations.

For optimal combustion, the combustion method, compression ratio and maximum mean effective pressure are adapted to the gas properties. Knocking tendency, auto-ignition tendency, laminar flame speed, calorific value and gas purity are taken into account as well [30, 37, 40]. In the case of biologically produced gases, combustion methods with an open combustion chamber or a pre-chamber are used. In the case of technically produced gases, open combustion chambers are generally used with a tuned charge motion. In the case of hardly flammable gases such as blast furnace gas, high-performance ignition systems with adjusted spark durations and special spark plugs are used.

Uncontrolled combustion such as pre-ignition and back-firing are suppressed or their effects limited by various measures. The focus here is on influencing parameters such as deposit formation, component temperatures, oil consumption, oil composition and residual gas content, as well as gas purity and fluctuations in gas composition. As an alternative, or in addition, to knocking sensors, sensors detecting cylinder pressure or combustion chamber temperature are used to register combustion. Robust engine operation without shutdowns is possible based upon ignition adjustment (in individual cylinders), rapid power reduction, mixture leaning or gas-flow interruption [30, 38].

Multi-gas and mixed-gas operation should be regarded as special cases. In applications using blast furnace gas, for example, if the gas composition deteriorates, a second gas (natural gas or coke gas) can be added to keep the combustion parameters within the desired range [38].

Since, when fueled with technically produced gases, engines are operated at comparatively low mean effective pressure (e.g. 60 % of that for the natural gas variant), the mechanical and thermal motor design is often modified. The engine oil and oil-changing interval are adapted to the gas in question to avoid acidification and sulfur corrosion [34].

Applications

Cogeneration is an ecologically and economically attractive technology for the use of non-fossil fuels (special gases). Stationary gas engines convert the energy in these special gases cleanly and efficiently into electrical energy and thermal energy with decentralized availability.

The use of biogas engines is strongly influenced by the statutory framework and has been booming in Germany, for example, since 2004. Amongst European gas engine manufacturers, for example, biogas variants currently enjoy a market share of around 50 % of the total number of engines sold in the power class up to approx. 1.5 MW.

Example: On-site power using 12 GE J620 converter gas engines in Aceralia, Spain. In 2004, one of the world's first CHP plants running on converter gas from steel production was commissioned. Twelve gas engine units run on converter gas, three of which can also run on natural gas as an alternative. The electrical output of the plant is 20.4 MW and its thermal output is 25.2 MW. The converter gas is made up of 60–75 % CO, 1 % H_2, 13 % N_2, 13 % CO_2 and 4 % H_2O. The calorific value is around 2.2 kWh/Nm^3 (Diagram 6.12).

The alternative natural gas operation is made possible because each engine has a separate gas train for natural gas in addition to the four parallel gas trains for converter gas. As part of a comprehensive safety concept, two different types of main gas valve are installed between the factory and the CHP system. This guarantees safe operation whether the engines are running or stationary and a quick response in the event of a gas alarm. For safe maintenance, nitrogen flushing is used to clear the toxic CO from gas-carrying components. Individual exhaust systems for each engine prevent the backflow of hot exhaust gas to the shut-down engines during maintenance work. Pressure relief valves can alleviate any deflagrations in the exhaust system. In the case of the J620 gas engines, in addition to engine management, the shape of the combustion chamber, the compression ratio and the turbocharger layout have all been optimized to permit robust operation at high overall efficiencies [30, 35].

Diagram 6.12 CHP system in Aceralia, Spain, with 12 Jenbacher J620 converter gas engines from GE

References

1. Sorge G (2013) GE Waukesha 275GL+ die nächste Generation in Gasverdichtermotoren [GE Waukesha 275GL + the next generation of gas compressor engines]. 8th Dessau Gas Engine Conference, Dessau
2. Fiveland SB (2014) Driving customer value with natural gas across large engine markets. AVL Large Engine TechDays, Graz
3. Bundesministerium für Umwelt, Naturschutz und Reaktorsicherheit [Federal Agency for Environment, Nature Protection and Reactor Safety] (2002) Erste Allgemeine Verwaltungsvorschrift zum Bundes-Immissions-schutzgesetz (Technische Anleitung zur Reinhaltung der Luft—TA Luft) [First General Administrative Regulation on the Federal Immission Control Act (Technical Instructions on Air Quality Control)], internet version, p 95 ff. http://www.taluft.com/taluft20020730.pdf
4. Gothenburg Protocol (1999) Protocol to the 1979 convention on long-range transboundary air pollution to abate acidification, eutrophication and ground-level ozone, Gothenburg. http://www.unece.org/env/lrtap/multi_h1.htm
5. DIRECTIVE 2010/75/EU. http://ec.europa.eu/environment/air/ pollutants/ stationry/index.htm
6. Clean Air Act of 1963. United States Public Law P.L. 88-206. United States Statutes at Large 77 Stat. 392
7. EPA Title V. http://epa.gov/apti/video/pdfs/tailoring.pdf
8. IMO (2009) revised MARPOL Annex VI: regulations for the prevention of air pollution from ships and NO_x technical code 2008.
9. Menage A, Gruand A, Berg P, Golloch R (2013) The new dual fuel engine 35/44 DF from MAN diesel & turbo SE. Paper no. 291. CIMAC Congress, Shanghai
10. Troberg M, Portin K, Jarvi A (2013) Update on Wärtsilä 4-stroke Gas Product Development. Paper no. 406. CIMAC Congress, Shanghai
11. N.N. Wärtsilä 34DF Engine technology. www.wartsila.com
12. Juliussen LR, Mayer S, Kryger M (2013) The MAN ME-GI engine: From initial system considerations to implementation and performance optimization. Paper no. 424. CIMAC Congress, Shanghai
13. Vlaskos I, Gagliardi D, Spiller M, Thuemmler K (2013) Analysis and evaluation of innovative hybrid powertrain architectures combining gas engines and electric propulsion for tugboats paper no. 235. CIMAC Congress, Shanghai
14. Hall E (2014) Dual fuel for medium speed engines in transportation. AVL Large Engine TechDays, Graz
15. Klausner J, Lang J, Trapp C J624—der weltweit erste Gasmotor mit zweistufiger Aufladung [J624—the world's first gas engine with two-stage turbocharging]. Motortechnische Zeitschrift 04/2011
16. Böwing R, Spreitzer K, Drehobl E, Vormstein H, Snuis H-B, Drexel C (2011) Thermodynamische und regelungstechnische Maßnahmen zur Steigerung von Effizienz und Robustheit der MWM Gasmotoren [Thermodynamic and control measures to increase the efficiency and robustness of MWM gas engines]. 13th conference "Der Arbeitsprozess des Verbrennungsmotors" [The Working Process of the Internal Combustion Engine], Graz
17. Auer M, Friedrich C, Waldenmaier U, KNAFL A, Stiesch G (2013) Vorgehensweise in der Brennverfahrensentwicklung für mittelschnellaufende Gasmotoren bei MAN Diesel & Turbo [Procedure for the development of the combustion process for medium-speed gas engines at MAN Diesel & Turbo]. 14th conference "Der Arbeitsprozess des Verbrennungsmotors" [The Working Process of the Internal Combustion Engine], Graz

18. Sander U (2013) The new MTU type L64 of series 4000 gas engines. CIMAC Congress, Shanghai
19. Juliussen L, Mayer S, Kryger M (2013) The MAN ME-GI Engine: From initial systems considerations to implementation & performance optimization. CIMAC Congress, Shanghai
20. Lee Y, Park H, Kim K, Son J, Jung C (2013) Newly updated combustion system for HIMSEN gas engine H35/40G. CIMAC Congress, Shanghai
21. Vlaskos I (2013) The power & efficiency upgrade approach for the development of the new Caterpillar 10 MW medium speed gas engine. CIMAC Congress, Shanghai
22. Solbakken H, Eide T, Nordrik R (2013) The new Bergen B35/40 lean burn marine gas engine series & practical experiences of SI lean burn gas engines for marine mechanical drive. CIMAC Congress, Shanghai
23. Troberg M, Portin K (2013) Update on Wärtsilä 4-stroke gas product development. CIMAC Congress, Shanghai
24. Trapp C, Birgel A, Spyra N, Kopecek H, Chvatal D (2013) GE's all new J920 gas engine—a smart accretion of two-stage turbocharging, ultra-lean combustion concept & intelligent controls. CIMAC Congress, Shanghai
25. Beran R, Baufeld T, Ludu A, Almer W (2005) Entwicklung eines Micro-Pilot Gasmotors zur Erreichung höchster Wirkungsgrade auch bei kleineren Bohrungsgrößen [Development of a micro-pilot gas engine to achieve extremely high efficiencies even with small bore sizes]. 4th Dessau Gas Engine Conference 2005
26. Fuchs J, Gebhardt AS, Leitner A, Thalhauser J, Tinschmann G, Trapp C (2012) Technologiebausteine für direkt gezündete Hochleistungsgasmotoren [Technology modules for directly ignited high-performance gas engines]. 7th MTZ Heavy Duty Conference 2012
27. Hampson GJ, Chiera D (2013) Die zweite Generation einer Vorkammerzündkerze für die Nachrüstung und Ausweitung der Magergrenze [The second generation of a pre-chamber spark plug for retrofitting and the expansion of the lean-burn limit]. 8th Dessau Gas Engine Conference, Dessau
28. Sotiropoulou E, Tozzi L (2011) Passive Vorkammerzündkerzen: Gestern und Heute [Passive pre-chamber spark plugs: then and now]. 7th Dessau Gas Engine Conference, Dessau
29. Zacharias F (2001) Gasmotoren [Gas engines]. Vogel Fachbuch, Würzburg. ISBN 3-8023-1796-3
30. Amplatz E, Schneider M, Trapp C (2011) Verwendung von Sondergasen in stationären Gasmotoren [Use of special gases in stationary gas engines]. Heavy-Duty, On- and Off-Highway Engines, MTZ Conference, 15th and 16th Nov 2011, Kiel, Germany
31. Mohr H, Baufeld T (2009) Grubengasnutzung in Stationärmotoren—Potentiale und Technik [The use of mine gas in stationary engines—potential and technology]. 6th Dessau Gas Engines Conference, 26th–27th Mar 2009, Dessau-Roßlau, Germany
32. Ernst A, Sander U, Kibele C, Odermatt M, Wasgindt A (2011) Die neue Baureihe 4000 für Biogasanwendungen [The new series 4000 for biogas applications]. 7th Dessau Gas Engines Conference, 24th–25th Mar 2011, Dessau-Roßlau, Germany
33. Pflaum H, Hofmann P, Geringer B (2009) Potenzial von unterschiedlichen Biogasqualitäten im ottomotorischen Berieb [Potential of different biogas qualities in four-stroke operation]. 12th Conference "Der Arbeitsprozess des Verbrennungsmotors" [The working process of the internal combustion engine], 24th–25th Sept 2009, Graz, Austria
34. Lehmann A, Stellwagen K, Plohberger D (2009) Einfluss der Schmierölwahl auf den Betrieb von Gasmotoren mit Biogas [The influence of the choice of lubricating oil on the operation of gas engines with biogas]. 6th Dessau Gas Engines Conference, 26th–27th Mar 2009, Germany
35. Amplatz E, Schneider M, Trapp C (2009) Sondergase aus Industrieprozessen—neue Ressourcen für Energieerzeugung mit Verbrennungsmotoren [Special gases from industrial processes—new

resources for power generation using combustion engines]. Gas Vehicles Conference 2009, Stuttgart, Germany
36. Schneßl E, Kogler G, Wimmer A (2009) Großgasmotorenkonzepte für Gase mit extrem niedrigem Heizwert [Large gas engine concepts for gases with an extremely low calorific value]. 6th Dessau Gas Engines Conference, 26th and 27th Mar 2009, Dessau, Germany
37. Schneßl E, Pirker G, Wimmer A (2009) Optimierung von Brennverfahren für Sondergasanwendungen auf Basis Simulation und Versuch am Einzylinder-Forschungsmotor [Optimization of combustion processes for special gas applications based upon simulation and experiment on the single-cylinder research engine]. Heavy-Duty, On- and Off-Highway Engines, MTZ Conference, 17th and 18th Nov 2009, Friedrichshafen, Germany
38. Trapp C, Böwing R, Zauner S, Amplatz E, Arnold G, Kopecek H, Wimmer A, Schneßl E (2014) Nutzung von Gichtgas im Großmotor mit Hilfe eines auf Zweigasbetrieb angepassten Regelungskonzeptes [Use of blast furnace gas in the large engine with the aid of a control concept adapted to dual gas operation]. 3rd Rostock Large Engines Conference, 18th–19th Sept 2014, Rostock, Germany
39. Burkhardt H (2013) Holzvergasung—Technologie der Zukunft oder Vergangenheit? [Wood gasification—technology of the future or of the past?]. 8th Dessau Gas Engines Conference, 21st–22nd Mar 2013, Dessau-Roßlau, Germany
40. Chmela F, Krenn M, Pirker G, Schlick H, Wimmer A (2013) Simulation des Brennratenverlaufs beim Betrieb mit Sondergasen [Simulation of the combustion rate trend during operation with special gases]. 8th Dessau Gas Engines Conference, 21st–22nd Mar 2013, Germany

7 Natural Gas Is Caviar for the Engine

Michael Bargende

We have known this statement for so long now that probably nobody knows anymore which inspired person formulated it first. The meaning the statement conveys is that methane, as the main constituent of natural gas, enables a practically completely particulate-free, i.e. "deposit-free" combustion and its high octane rating leads to high knock resistance, which enables high compression ratios and hence high degrees of efficiency.

In earlier times, when the carburettor was the standard for mixture preparation systems for gasoline engines, further advantages emerged through the use of a gas mixer. Hence relatively simple components could create a very homogeneous gas-air mixture and there was no danger of either carburettor icing in winter or vapour bubbles in summer.

Today, a lot has changed, thanks to the advent of electronically controlled injection and engine management systems in conjunction with gasoline direct injection and the availability of gasoline with octane ratings of up to 100: Compression ratios in gasoline engines—even with turbo charging—have already reached values where no significant gains can be expected through further increases. In actual fact, the increasing ignition voltage requirements and the cylinder peak pressure at the most efficient point of combustion are becoming limiting factors for a further increase in the compression ratio. The increasingly higher ignition voltage required for increasing compression ratios is, in the case of natural gas, of even higher importance, since methane already requires

M. Bargende (✉)
Forschungsinstitut für Kraftfahrwesen und Fahrzeugmotoren Stuttgart (FKFS), Pfaffenwaldring 12, 70569 Stuttgart, Germany
e-mail: Michael.bargende@fkfs.de

R. van Basshuysen (ed.), *Natural Gas and Renewable Methane for Powertrains*, Powertrain, DOI 10.1007/978-3-319-23225-6_7

20 % higher minimal ignition energy than gasoline. The highest ignition voltage requirement is reached when the retarded ignition timing must be set to achieve maximum boost pressure in order to generate more exhaust gas enthalpy, which however still lies in the compression phase.

Methane emissions during cold start up until the full conversion temperature of the catalytic converter together with the ever-stricter exhaust emissions limit values are becoming increasingly important, since the light-off temperature of methane is approximately 150 K higher than gasoline. This is a problem, particularly in Europe, since here the limits concern the emissions for total hydrocarbons (THC), whereas in the USA the limits concern organic gases without considering methane (NMOG *Non-Methane Organic Gases*). In the USA, methane emissions are often called "methane slip"—simply because their potential for global warming (GWP *Greenhouse Warming Potential*) is evaluated and penalized accordingly. However, this has a significantly lower effect on the design of exhaust aftertreatment systems.

Overall it can be noted that gasoline engine technologies are in the stage of series development, or are already in series production, whose application would only have been possible in monovalent natural gas engines just a few years ago due to the more favorable fuel characteristics.

In the future, what are the arguments in favor of natural gas fuel if gasoline engine technology is developing so quickly?

First of all, the very high potential for CO_2 reduction in relation to all other technological possibilities. The CO_2 emissions using natural gas are more than 20 % lower as compared to a gasoline engine at the same degree of efficiency (see Sect. 5.2.6).

Even when compared to battery-powered electric vehicles, there is significant potential for CO_2 reduction, if the CO_2 emissions produced during electricity generation are seriously taken into account (compare [1]). In Germany, these indirect CO_2 emissions by electric vehicle are calculated to be approximately 100 g CO_2/km. A comparable (with regards to transport capacity) natural gas driven vehicle emits—depending on the technology employed—between 70 and 80 g CO_2/km. The situation is similar if not even worse in most industrial countries. Exceptions are countries where electricity is predominantly generated via nuclear power stations.

Natural gas engines are therefore a very honest possibility to effectively reduce CO_2 emissions and should therefore be subsidized at least as intensively as electric vehicles.

If we look towards a future with a dwindling availability of the oil-based fuels gasoline and diesel, then synthetically manufactured fuels assume high relevance.

The production of synthetic methane via the methanation of hydrogen represents a very realistic path for a sustainable fuel supply. Audi has recently demonstrated a 6 MW pilot plant in northern Germany in cooperation with Stuttgart ZSW (*Zentrum für Sonnenenergie- und Wasserstoff-Forschung Baden-Württemberg*) that shows that this process is possible on a large scale with remarkably good efficiency. The particular advantages of this technology are manifold:

- The necessity for intermediate storage to provide grid stability is becoming increasingly more relevant due to the continued increase in regenerative electricity generation based on wind farms and photovoltaic. The production of synthetic methane out of currently unused wind power solves this problem.
- The gas grid in Germany (and in other industrial nations) is the largest energy store and has the particular advantage that the feed-in and gas extraction points do not have to be identical. There is hence no transportation required from producer to consumer.
- The production of synthetic methane permits a gradual transition from a fossil fuel—natural gas—to a synthetic fuel.
- The methanation process can use CO_2. Hence the process is CO_2 neutral with regard to emissions.
- By using the excess from fluctuating regenerative energies in Power-to-Gas plants with the efficiency achievable today of approximately 60 %, this technology represents one of the most efficient possibilities for a CO_2-neutral production of synthetic fuels. In contrast to battery-driven electric vehicles, gas-driven vehicles have no restrictions concerning either the range or the tank-filling/battery-charging method.
- Synthetically produced methane represents a realistic alternative to diesel fuel in the heavy-duty truck sector in a post-oil era. Already today, dual-fuel trucks are on the road with up to 85 % gas use and only 15 % diesel consumption with considerably extended ranges, which is made possible by using two different fuel types. Of course, there is a range of technical problems to be solved and the scope of optimization potential has not yet been exhausted. Despite this, a path seems to have been found that allows us to look optimistically into the future and see how fuel supplies could look in a post-oil era. In the medium term however, the development of gas engines for passenger cars will continue to benefit from the development of gasoline engines, upon which they will also continue to be based. Increases in efficiency due to downsizing, turbocharging, direct injection and even better engine control—all developed for gasoline engines—will lead to a similar increase in efficiency in gas engine derivatives. A truly independent gas engine development in passenger engine construction is only expected if, due to a future scarcity of gasoline, gas engines succeed in penetrating the market by a higher two digit percentage. Then we will perhaps see gas engines in series production with central direct injection and stratified combustion or very lean-run gas engines with diesel ignition jets. It is conceivable that particularly ignitable fuels used as ignition fuels could be synthetically produced, because specific advantages compared to a high voltage electrical ignition can be accessed. However, these developments could be notably accelerated if similar political measures of support were to be initiated for gas engines to those currently usual for electric drives. The question must be asked here, how much will the enormous increase in the relevance of natural gas in the USA through access to large, unconventional reserves force support for the research and

development of gas engines in Europe. In the commercial vehicle sector, future developments are mainly dependent on two influencing factors: the further tightening of emissions legislation and the price development of diesel fuel. It is relatively clear now that it will not be easy for dual-fuel engines with a variable substitution rate natural gas-diesel and homogeneous air-gas mixture to be compliant with future emission limits with regard to methane emissions. This is particularly regrettable since these engines come very close to base diesel engines with respect to efficiency and thus fuel consumption. The market penetration of pure gas engines with spark ignition and normally worse efficiency as compared to the diesel pendant will be heavily dependent on the price development of diesel fuel. Here too, politics has a high degree of influence.

All in all, in our view, the future of gas engines in a post-oil era is beyond dispute, perhaps even as the dominant drive source in the commercial vehicle market for intercity long-haul, i. e. when people and goods need to be transported over long stretches and no (electrified) train connection is available.

The question that cannot be answered today is simply: What will the path to this future scenario look like? This question is largely unanswerable because the influencing factors are hardly of a technical or ecological nature, but depend rather on economic and political developments. Examples of this are the oil price developments and nationally widely diverging policies.

In accordance with this, all scenarios that are based on possible and probable technical and ecological developments and yet ignore economic and political developments should be treated with caution.

Referring back to the first sentence: Natural gas—and in the future synthetic methane—is still caviar for the engine and will certainly become a very important if not even the sustainable fuel of the future.

Or to cite a one-time companion of Steve Jobs, Steve Wozniak: The goal is clear, but the road there remains exciting!

Reference

1. Icha P (2014) Entwicklung der spezifischen Kohlendioxid-Emissionen des deutschen Strommix in den Jahren 1990 bis 2013. Umweltbundesamt, Dessau

Printed by Printforce, the Netherlands

NSTERN

REITKUNST ALS THERAPIE

Müller Rüschlikon

Impressum

Einbandgestaltung: R2 I Ravenstein, Verden

Titelfoto: Radoyka Despodovska
Foto auf der Umschlagrückseite: Radoyka Despodovska
Bildnachweis:

Autorin und Verlag haben dieses Buch nach bestem Wissen und Gewissen mit viel Liebe zum Detail erarbeitet. Für einen eventuellen Missbrauch der Informationen in diesem Buch können weder die Autorin noch der Verlag oder die Vertreiber des Buches zur Verantwortung gezogen werden. Eine Haftung für Personen-, Sach- und Vermögensschäden ist ausgeschlossen.

Dieses Buch wurde nach den geltenden Vorschriften der General Product Safety Regulation (GPSR) entwickelt. Bei Fragen oder Problemen wenden Sie sich bitte an den Verlag unter folgender E-Mail-Adresse: gpsr@paul-pietsch-verlage.de

ISBN 978-3-275-02352-3

Postfach 103743, 70032 Stuttgart
Ein Unternehmen der Paul Pietsch Verlage GmbH & Co. KG

1. Auflage 2025

Sie finden uns im Internet unter
www.mueller-rueschlikon-verlag.de

Gesamtleitung: Claudia König
Innengestaltung: R2 I Ravenstein, Verden
Druck und Bindung: Graspo CZ, 76302 Zlin
Printed in Czech Republic

Im Rahmen eines sensiblen Sprachgebrauchs werden, wo immer es möglich war, genderneutrale Begriffe verwendet. Da die Meister der Alt-Klassik sowie die meisten Hauptakteure der heute bekannten Stall- und Reitmeister in der Regel männlich waren, wurde auf eine Anpassung dieser Begriffe verzichtet.

Folgende Pferde sind auf den Bildern in diesem Buch zu bewundern:

Dantano, 16-jähriger Hengst
Vancouver, 10-jähriger Hengst
Leo, 21-jähriger Wallach
Floro, 8-jähriger Wallach
Coronado, 9-jähriger Hengst

Mascha Schumann-Rudnitzki mit
Honey, 15-jährige Stute
Sissi, 6-jährige Stute
Szofi, 7-jährige Stute
Bella, 6-jährige Stute

Dr. Christina Ziegenhagen mit
Inca, 11-jähriger Wallach
Emma, 17-jährige Stute

Gina Uhlig mit
Davidoff, 18-jähriger Wallach

Sie finden die Autorin Sonja Leitenstern im Internet unter www.sonja-leitenstern.de

sowie auf Instagram unter @sonjaleitenstern

INHALT

DANKSAGUNG

Zuallererst gilt mein Dank meiner wunderbaren Community. Ihr seid an meiner Seite und das Tag für Tag. Wir haben miteinander gedacht, gefühlt, gelacht, diskutiert und parallel nebeneinander gelebt, aber doch miteinander verbunden. Ihr habt mich ermutigt, getragen, herausgefordert und inspiriert. Danke dafür.

Ein ganz besonderer Dank gilt meinen Onlinekursschülerinnen und -schülern. Ihr seid das stabile Fundament, auf dem ich meine Arbeit Tag für Tag aufbauen darf. Eure Lernfreude, euer Vertrauen und eure Neugier geben meinem Tun Tiefe und Richtung. Ohne euch hätte meine Vision von Reitkunst keinen Bestand, ihr seid das Rückgrat meiner beruflichen Entwicklung.

Danke an meine Reitschüler, dass ihr mir erlaubt habt, euch zu begleiten und dass ich dabei selbst so viel mitnehmen durfte.

Ein herzliches Danke gilt auch meinen Pferden. Ihr seid meine loyalen Mitarbeiter, Weggefährten, meine Sportkameraden. Ihr seid Künstler, Stars, echte Persönlichkeiten und vor allem: ihr seid meine Lehrer. Jeden Tag zeigt ihr mir, worauf es wirklich ankommt. Mit euch zu arbeiten ist ein Geschenk, das ich niemals für selbstverständlich halte.

Ein gigantisches, unerschütterliches Danke gilt meiner besten Freundin, Dr. Christina Ziegenhagen. Du bestärkst mich, du siehst mein Potenzial, selbst dann, wenn ich es selbst nicht mehr erkenne. Du glaubst an mich, auch wenn in selbst dazu nicht mehr fähig bin. Du bist mein seelischer Rückzugsort, und jedes Gespräch mit dir ist Substrat für meine Seele.

Und dann sind da noch zwei Menschen fest, verlässlich, unerschütterlich in ihrer Unterstützung. Nie im Rampenlicht, aber stets im Hintergrund tätig. Oftmals mein Wohlergehen selbstverständlich über ihr eigenes stellend. Danke Mama für deine Stärke, deinen Humor und für deinen Schutz. Danke Vati für dein gutes Herz, deine lösungsorientierten Strategien und für unseren regelmäßigen Gedankenaustausch zum Calcium-Phosphor-Verhältnis.

VORWORT

In jedem Stall stehen sie, die Pferde mit Sehnenschäden, Arthrose am Halswirbel, mit Ekzemen oder einem kaputten Huf, der einfach nicht besser werden will.

Im Rahmen einer konservativen Therapie werden diese Tiere meist schön langsam über Spaziergänge bewegt, gelegentlich kommt der Veterinär vorbei und spritzt das eine oder andere. Das oftmals innig geliebte Freizeitpferd verkümmert zusehends, aber was soll man schon machen bei einer schlechten körperlichen Verfassung und einem depressiven Tierchen, das sich oft nur schlecht motivieren lässt?

Ich sage: Dressur nach altklassischer Manier, Dressur im Sinne barocker Reitmeister wie einem la Guérinière und ihren preußischen Nachfolgern wie Otto von Monteton, aber auch einem Baucher oder Racinet. Sie alle verstanden viel davon, ein Pferd nachhaltig gesund und fit für die Reiterei zu halten.

Was sie mangels Wissenschaftsfortschritt noch nicht wussten, steht uns heute zur Verfügung: Dressur als Form des intensiven Kraftsports mit hoher mechanischer Spannung am Muskel führt zu voluminösen Muskelpaketen. Und die sind die Antwort auf nahezu alle der oben genannten Freizeitpferdprobleme.

Das vorliegende Werk *Reitkunst als Therapie* dient nicht dem Zweck, den Menschen zu therapieren. Sondern einzig und vorrangig dazu, sich mit einer gesunderhaltenen Reitweise auseinanderzusetzen, die die Pferdegesundheit schützt und erhält.

WIE MAN PFERDE UNWISSENTLICH KAPUTT REITET

KURZ UND KNAPP:

- Die heutige Freizeitreiterei basiert auf Kriegstraining.
- Die moderne Auffassung über Pferdetraining reitet Pferde kaputt.
- Spaziergänge sind nett, aber kein Training.
- Wer springen will, muss Dressur reiten.
- **Jedes** Pferd kann Dressur erlernen.

Die heutige Reiterei schafft es, Pferde gleichzeitig zu **unterfordern** und zu **überlasten**. Denn wenn das einzelne Tier nicht zufälligerweise für extrem teures Geld für die Dressur nach FN-Maßstäben gekauft und vorbereitet wurde, dann ist es angeblich gar nicht in der Lage dazu, körperlich und auch kognitiv anspruchsvolle Dressurarbeit zu leisten. Stattdessen wird es spazieren geführt wie ein überdimensionales Hündchen, nachdem es brav seine Runden an der Longe absolviert hat.

Das alles ist nicht sonderlich spannend für ein Tier, das zwar Routinen liebt, aber durchaus auch in der Lage dazu ist, komplexe Bewegungsabläufe zu verinnerlichen. Im Grunde sogar völlig unabhängig von der Rasse!

Parallel dazu werden moderne Reitpferde ohne eine angemessene Grundausbildung am Boden und an der Hand an Reitfans übergeben, die oft ebenso am Anfang einer Reitkarriere stehen, wie ihre frisch erworbenen Tiere. Das führt eher früher als später unwissentlich und in den meisten Fällen hoffentlich auch unwillentlich zur Fehlbelastung des equinen Trageapparats. Kein Wunder, dass das heutige moderne Reitpferd ständig mit großen und kleinen Zipperlein zu kämpfen hat. Arthrosen, Sehnenschäden und *Rücken* führen zu reichlich Besuchen bei allerlei medizinischen und osteopathischen Fachkräften. Eine Beobachtung von Baucher, die ich tagtäglich teile:

> *»Wenn es einmal so weit gekommen ist, dann ist es beinahe unabwendbar, dass die Situation noch schlimmer wird. Bald schon wird der Reiter der Machtlosigkeit seiner eigenen Anstrengungen überdrüssig sein und das Pferd für die Konsequenzen seiner eigenen Ignoranz verantwortlich machen. Das Pferd, das er nun als »Gaul« behandelt, hatte vielleicht ganz hervorragende Möglichkeiten und mit mehr Einsicht und Kenntnis hätte man daraus ein Reitpferd machen können, das durch seinen gelehrigen und braven Charakter ebenso wie durch seine anmutigen und angenehmen Gänge beeindruckt hätte. Ich habe oftmals beobachtet, dass Pferde, die angeblich unzähmbar sind, sich als die energievollsten und lebensfrohesten erweisen, sobald es einem gelungen ist, ihren körperlichen Schwächen, welche sie in der Verwendung ihrer Kräfte lähmten, Abhilfe zu schaffen* (Baucher).«

Sei ehrlich: Reitest du wirklich gesunderhaltend und pferdegerecht?

Der durchaus umstrittene Baucher ist hier recht deutlich in seiner Einschätzung, wie sich Anspruch und Fähigkeit zur Pferdeausbildung unterscheiden. Im Podcast *Hinter verschlossenen Stalltüren* fällt in einer Nachbesprechung zu einem Interview mit mir dort sinngemäß der Ausspruch: »Die FN-Richtlinien sind der Klappentext, die klassische Dressur ist das umfassende Werk (Hölz/Seiler 2025).« Das trifft es wirklich gut, denn die Richtlinien der FN sind im Kern eine massive und zum Nachteil des Pferdes geschehene Verkürzung der klassischen Dressurkunst.

Freizeitreiterei mit Kriegsvergangenheit

> *»Mißachtet man die Vergangenheit, so verleugnet man Tradition und Wesen der Reitkunst* (Loch 2010)*.«*

Wer heute auf Turniere geht und einmal genau hinsieht, der findet erstaunlich viele Hinweise auf die militärische Basis der heutigen Freizeitreiterei.

Auf Hochglanz polierte Pferde und Stiefel, Begrüßungs- und Aufstellungssequenzen, dazu klare Ranglisten, Richtlinien über Richtlinien und so weiter und so fort. Im Grunde ist das keine große Überraschung, denn die heutige Freizeitreiterei entstammt einer langen vorwiegend militärischen Tradition:

> *»Die Macht-, Herrschafts- und Eroberungspolitik der Weltgeschichte wohnt eine wesentliche Konstante inne: die Beteiligung von Pferden. Ohne die Verwendung von Pferden wäre die Austragung von Kriegen bis zum letzten Weltkrieg nicht möglich gewesen. Ohne die Verwendung von Pferden wäre auch die landwirtschaftliche Nutzung bis Mitte des 20. Jahrhunderts undenkbar* (Fahnenbruck 2025)*.«*

Bis zur Übernahme des Kriegsgeschehens durch Militärmaschinen waren Pferde das Go-to-Mittel zum Truppentransport und bis kurz vor Beginn des 1. Weltkriegs bewiesen sie sich vor allem sowohl als Kampfgefährten als auch als regelrechte Waffen. Ein in der Dressur ausgebildetes Kriegspferd wendete in der Regel auf kleinstem Kreis und reagierte auf feinste Hilfengebung, während der Reiter einhändig die Zügel führte und mit der zweiten Hand die Waffe schwang. Als fertig ausgebildetes Schulpferd im Sinne der barocken Reitmeister galten Pferde erst, wenn sie all das und noch etwas mehr beherrschten, was meist gute 10 Jahre Ausbildungsdauer bedeutete.

Heutzutage beruft sich die für die Ausbildung in Deutschland maßgeblich verantwortlich zeichnende Reiterliche Vereinigung (FN) ausgerechnet auf die kastrierte Fassung einer Heeresdienstvorschrift, die im Zuge des massiven Materialverschleißes im und nach dem ersten Weltkrieg die Militärpferdausbildung massiv zusammenstrich.

Aus hochspezialisierten und kompetenten Kriegspferden wurden Truppentransporter, die auch von wenig qualifizierten Soldaten hin- und herbewegt werden konnten.

Pferdehaltung als Freizeitvergnügen

Pferde halten, rein fürs Privatvergnügen? Das konnten sich in vergangenen Jahrhunderten bis hin zum Preußischen Kaiserreich nur die Eliten sprich Adlige leisten. Wer etwas auf sich hielt, der präsentierte seine edlen Pferde natürlich nur zu gern, musste aber auch finanziell gut dastehen. Das erforderliche Equipment samt hauseigenem Reitmeister und einer ausgiebigen Reitausbildung für den adligen Nachwuchs gehörte zum guten Ton.

In altklassischer Reitkunst ausgebildetes Warmblut. Sonja mit Dantano.

Die moderne Reiterei hat sich von beiden Wurzeln weit entfernt. Glücklicherweise sind Pferde hierzulande ein kaum noch relevanter Bestandteil des Militärs, wenn man von Paraden und Show-Einlagen einmal absieht. Die Bundeswehr nutzt Pferde höchstens noch in unwegsamen Gelände – als Transporttiere (Kümmel 2022). Die entsprechende hochspezialisierte Ausbildung zum Nahkampf mit dem Pferd ist wortwörtlich Geschichte.

Im Freizeitbereich dagegen sind wesentlich mehr Menschen und Pferde involviert. Die FN spricht von 800.000 Turnierregistrierungen und gut 3 Millionen Pferdefans ohne Turnierlizenz zählen sich zu den Hobbyreitern und -reiterinnen (FN 2025). Wer die Reiterei ernst nimmt, der reitet aber schon Turniere. Hier geht es immerhin um sportliche Repräsentation, korrekt?

Hirnloser Wettkampf ohne Sinn

Die Richtlinien der FN sind eine Konstante in der Beurteilung der sogenannten Reitfähigkeiten im Profisport hierzulande. Auch für die meisten Hobby- und Freizeitreitfans sind die FN-Richtlinien der Leitfaden schlechthin für eine pferdegerechte Reitweise, mit der sich auch dekorative Schleifchen verdienen lassen.

Dabei haben wir doch gerade eben erst gelesen, dass sich die Reitkunst aufgrund der fehlenden Relevanz für das Militär weiter und weiter reduziert hat. Was am Ende noch übriggeblieben ist, ist sogar noch weniger als ein Klappentext, sondern kaum mehr eine Fußnote in der Geschichte der meisterlichen Pferdeausbildung.

Trotzdem werden Turniere abgehalten, die Industrie hinter der Bewertung und Präsentation von Freizeitpferden ist groß und erfolgreich. Die Schäden am Bewegungsapparat der Pferde auch, denn anhand der FN-Richtlinien lässt sich eben kaum langfristig gesunderhaltend reiten, insbesondere weil versammelnde Arbeit als hehres Ziel ganz am Ende der Ausbildungsskala angesetzt wurde. Logisch, ein Truppentransporter oder ein

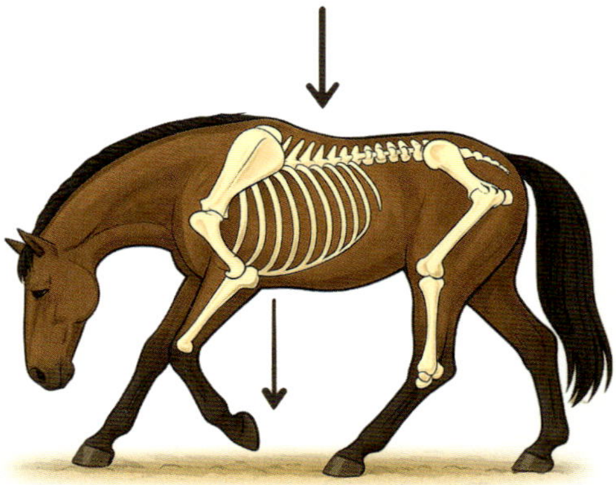

Da sich der Rumpf im Vorwärts-Abwärts schlechter gegen den Zug des Nackenband heben kann, kann das heute gängige Vorwärts-Abwärts nicht geeignet sein, um Rumpfhebung zu erzeugen.

hobbymäßiger über Stangen getriebenes Freizeitpferd lässt sich auch ohne Versammlung mehr oder weniger gut von oben lenken.

Ein lang einsatzbereites Reitpferd dagegen erhält erst mit der Versammlung die Fähigkeit, viele Jahre lang fit und gesund als Reittier zu dienen.

Wissen schafft Pferdegesundheit

Um das zu erreichen, diesen Stand eines langfristig gesunden und fitten Tieres ohne verkalktes Nackenband durch das hierzulande so typische tiefe Vorwärts-Abwärts, ohne Sehnenschaden und Arthrosen oder gar eine fulminante Trageerschöpfung, da muss im Grunde nur die Allklassik herangezogen werden.

> *»Wie man Musik mögen kann, ohne etwas von der Theorie zu wissen, wobei der Genuß wesentlich größer ist, wenn man die Theorie versteht, gilt das gleiche auch für die Reiterei und praktisch für jeden Sport, jede Kunst oder Wissenschaft. Es ist beinahe selbstverständlich, daß uns die Kunst umso mehr Freude und Nutzen bringt, je mehr wir über ihre Theorie und Technik wissen.* (de Souza 1927)«

Auch wenn die klassischen Reitmeister die heute verfügbaren biologischen Hintergründe kaum gekannt haben konnten, so wussten sie eben doch Pferde rein über das Training lang reitbar zu halten. Das sollte uns zu denken geben!

Bestes, aber an dieser Stelle kurz gehaltenes Beispiel: Zahlreiche Reitmeister von Pluvinel bis Baucher haben seitenlange Abhandlungen dazu verfasst, wie das Maul des Pferdes beweglich und seine **Halsbasis** ohne harte Hand ausgerichtet werden sollte.

Die extrem moderne Auffassung, nur das Vorwärts-Abwärts mit nahezu am Boden schleifender Nase sei die richtige Art und Weise, um Pferde losgelassen zu reiten, widerspricht diesen altklassischen Reitlehren.

Schaut man sich die aktuelle Forschung an, wird eins klar: Die Reitmeister lagen richtig, ohne je einen Röntgenapparat oder dergleichen gesehen zu haben.

Was allein der Zug am Nackenband anrichtet

So zeigen die Untersuchungen von Dr. Clayton recht klar, dass besonders bei tiefen und runden Haltungen erhöhte Spannungen im Halsbereich auftreten – und die zu Veränderungen insbesondere der Nackenplatte führen (Zsoldos/Licka 2015, Clayton/Hobbs 2017). Das Nackenband ist eigentlich ein elastisches Bänderkonstrukt, welches durch konstante Fehlhaltung **verkalkt** und demzufolge an Elastizität einbüßt.

Die aktuell in den Ställen viel zu häufig sichtbare sehr tiefe Vorwärts-Abwärts-Dehnungshaltung verändert zudem die Position des Schulterblattes. Das wiederum fördert ein rückständiges Vorderbein, das Schulterblatt selbst kann sich nicht komplett frei bewegen. Diese Einschränkung erfolgt logischerweise durch das extra Reitgewicht plus einer für das Tier insgesamt zu tiefen Kopfhaltung. Der vom Reitgewicht erhoffte Raumgriff kann nur durch extra Arbeit weiterer Gelenke erreicht werden.

Die Auswirkungen der tiefen Kopfhaltung:

Das **Ellbogengelenk** zeigt vermehrt Flexion und Extension. Der Vierbeiner in Dehnungshaltung arbeitet - wenn wir das wirklich so nennen wollen - verstärkt mit dem Ellbogen, statt freischwingend aus dem Schulterblatt heraus.

Im **Vorderfußwurzelgelenk** findet ebenfalls wesentlich mehr Bewegung statt, als rein physiologisch nötig wäre. Für die vom Menschen auf dem Pferderücken geforderten Schrittlängen beugt und streckt das Pferd das Karpalgelenk häufiger. Eine mögliche Folge ist die Überstreckung.

Eingerollt statt tief und vorwärts ist übrigens kaum besser: Untersuchungen am Pferd belegen, dass das Nackenband regelrecht verkalkt, wenn über Hyperflexion (stark eingerollter Hals, auch Rollkur genannt) zu viel Zug auf die Halslinie ausgeübt wird (Lazarczyk et al. 2020). Gerade diese stark eingerollte Kopfhaltung ist in Dressurturnieren oft gewollt und gilt als besonders ausdrucksstark.

Tatsächlich scheint es keine guten Daten darüber zu geben, wann genau denn nun eine Hyperflexion zur nackenbandschädlichen Variante wird und wann das berühmte Einrollen noch nicht so arg schlimm für den Pferdehals sei. Bis dahin scheint es mir klug zu sein, wenn überhaupt, dann zumindest sehr sparsam und umsichtig mit dem Thema Nackenband, Vorwärts-Abwärts und der Hyperflexion umzugehen.

Reha-Fälle überall

Die aktuelle und häufig zu beobachtende Form der Dehnungshaltung, ist übrigens nicht im Mindesten Teil der FN-Richtlinien. Da wird immerhin noch von einer Dehnung bis Höhe Buggelenk gesprochen, häufig werden die Pferdeköpfe aber

Nur in der Versammlung ist gewährleistet, dass sich der Rumpf aktiv hebt – während er im Arbeitstrab mit passiver Muskulatur im Halteapparat höchstens hoch- und runterschaukelt.

Wie viel sinkt der Rumpf im Trab …

Trab

6-8 cm Schwankung (Widerrist)

… und wie wenig in der Passage?

Passage

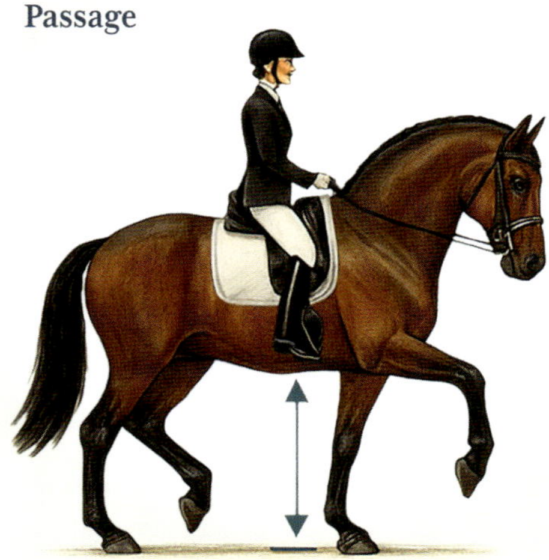

1,5-2,5 cm Schwankung

wesentlich tiefer fallen. Und mit ihnen fallen auch die biomechanischen Bewegungsmuster, die das Reitgewicht mit dem Schwerpunkt des Pferdes besser vereinen.

Eine echte Rumpfhebung ist mit einem derart tiefen Kopf und einer größtenteils mangelhaften Reitpferdemuskulatur eben nicht zu machen. Und so schleppen sich Pferd und Mensch trotz aller Liebe zueinander und für den sogenannten Reitsport mittels rudimentärer Anleitungen, die kaum für eine langanhaltende Pferdegesundheit gedacht waren, über die Reitplätze.

Schonender Spaziergang statt kräftigendem Bodybuilding

Nichts könnte unpassender und Pferdegesundheit vernichtender sein, als der Versuch, über entspannte Spaziergänge dem Pferd Gutes tun zu wollen. Das Gegenteil ist der Fall. Nette Spaziergänge durchs Gelände sind mit Sicherheit genau eins: nett. Sie entspannen - vor allem den Menschen - sie helfen dabei, Zeit tot zu schlagen und sie helfen ganz sicher dabei, dass sich das Gefühl einstellt, etwas *Gutes* getan zu haben.

Nur leider bringen Spaziergänge als alleinige Bewegungsform nichts, wenn es um die **Stärkung der Reitpferdemuskulatur** geht oder darum, vorhandene gesundheitliche Probleme auszukurieren. Sie bringen nichts, wenn ein zu dickes Pferd abnehmen oder ein zu dünnes Pferd Muskulatur aufbauen soll. Sie helfen nicht dabei, Bänder und Sehnen, den Rücken und die Gelenke fit für die Belastung durch das Reitgewicht gleich hinter den Schulterblättern zu machen.

All das kann nur Krafttraining, oder genauer: altklassische Dressurkunst nach Art der Alten Meister. Eine langsam und gezielt ausgeführte Hankenbeugung in Versammlung bringt dem einzelnen Pferd - egal ob es alt ist oder ein jüngerer Reha-Fall - viel mehr, als gemütliche Spaziergänge durch die Natur.

Krafttraining ist schonender Sport mit kontrollierten langsamen Bewegungen.

Was Krafttraining kann (und Spaziergänge nicht)

Krafttraining wirkt sich auf mehrere physiologische Mechanismen aus, die zusammengenommen Schmerzen reduzieren können.

Krafttraining = effektiver Schmerzkiller.

- **Verbesserung der Muskelkraft:** Schwache Muskeln führen zu einer höheren Belastung von Gelenken, Sehnen und Bändern. Krafttraining stärkt die Muskeln, was die Belastung auf diesen Strukturen reduziert.
- **Stabilisierung von Gelenken:** Besonders bei Arthrose oder Rückenschmerzen schützt ein besser stabilisiertes Gelenk vor unkontrollierten Bewegungen oder Belastungen, welche Schmerzen verursachen können.
  Das Besondere am Krafttraining ist dabei vor allem die bewusst langsam ausgeführte Bewegung, wodurch geringere kinetische Energie (Bewegungsenergie) auf die Gelenke einwirkt als im Vergleich zum zügigen Spaziergang oder dem so beliebten Cardio-Abschleudern an der Longe.
- **Freisetzung von Endorphinen:** Durch Krafttraining wird die Produktion von Endorphinen angeregt, die als körpereigene Schmerzmittel fungieren und damit eine analgetische, also schmerzlindernde Wirkung haben.
- **Desensibilisierung des Nervensystems:** Regelmäßiges Training kann die Empfindlichkeit des zentralen Nervensystems (ZNS) gegenüber Schmerzreizen verringern. Das wird oft als Verbesserung der *Schmerzschwelle* bezeichnet.
- **Ausschüttung entzündungshemmender Botenstoffe:** Beim Training können entzündungshemmende Zytokine wie IL-10 freigesetzt werden. Sie reduzieren entzündungsfördernde Prozesse im Körper. Besonders bei chronischen Entzündungen und damit zusammenhängenden Krankheiten wie Arthrose wirkt sich Bodybuilding entzündungshemmend und damit auch schmerzstillend aus.

- **Stärkung von Sehnen und Bändern:** Durch das Krafttraining wird die Kollagenproduktion verbessert. Das hilft bei der Regeneration von Bindegewebe, was wiederum die Strukturen (Bänder, Sehnen, Faszien) widerstandsfähiger gegen Belastungen macht. Langsam ausgeführtes und kontrolliertes Belastungstraining regt außerdem den Stoffwechsel der Knorpelzellen an, was deren Gesundheit fördert und vor allem Arthroseschmerzen lindern kann.

Krafttraining wirkt damit sowohl **direkt** auf Schmerzmechanismen als auch **indirekt** durch die Förderung von Heilungsprozessen. Bei intensiven Übungen mit hohem Kraftaufwand und/oder komplexen Bewegungen wie der Hankenbeugung ist das zentrale Nervensystem stark beteiligt. Je mehr Wiederholungen und Sätze absolviert werden, desto mehr Feuerkraft, also Energie zur Weiterleitung der Signale wird verbraucht. Mit der Zeit sinkt die Fähigkeit, motorische Einheiten im Muskel (engl. motor units) maximal zu rekrutieren. Bei der altklassischen Dressur ist es deshalb entscheidend, das Pferd an, aber nicht über seine Leistungsgrenzen hinaus zu treiben. Kurze Lektionseinheiten mit gerade als herausfordernd vom Pferd wahrgenommenen Wiederholungssätzen reichen bereits aus, um die positiven Effekte des Krafttrainings abzugreifen. Ein vollständiger Energieburnout wird aber bitte vermieden.

Dass sich das **Selbstbewusstsein** deines Pferds durch die Bewältigung der Dressurlektionen steigert und es zunehmend Vertrauen in seine körperliche Leistungsfähigkeit gewinnt, sollte ebenfalls klar sein.

Mit Bodybuilding wird die Lebensqualität verbessert, chronische Schmerzen werden effektiv und vor allem langanhaltend reduziert.

Kurz:

Versammlung reduziert Schmerzen. Auf der Vorhand mit Reitgewicht rumrutschen macht dagegen welche!

Springen oder (Dressur) tanzen?

Die Frage danach, ob Springreiten nicht die sinnvollere Betätigung für Reitpferde sei, kommt immer und immer wieder auf, insbesondere wenn es um vermeintlich natürliche Aspekte der Bewegung geht. Dabei wird gern und ausdauernd auf die Tatsache hingewiesen, dass über-Dinge-hinweg-springen ja durchaus Teil der normalen Bewegungsform eines Pferdes sei. Natürlich stimmt das - im Kern und für die Tatsache, dass zumindest von Natur aus eben kein Gewicht auf dem Rücken des Pferdes mittransportiert werden muss. In der Praxis wählen Pferde durchaus öfter den leichteren Weg um Hindernisse herum.

> *»Horses aren't natural jumpers, although in the wild they will jump obstacles that get in their way. This is what makes watching a horse jumping fences so fascinating as you see them using their bodies to an extreme (Higgins 2021).«*

»Pferde sind keine natürlichen Springer, obwohl sie in freier Wildbahn über Hindernisse springen, die sich ihnen in den Weg stellen. Das macht es so faszinierend, ein Pferd beim Springen über Hindernisse zu beobachten, denn man sieht, wie es seinen Körper bis zum Äußersten einsetzt.«

Die formale Dressurgrundlage für die Springreiterei ist so rudimentär, dass sie sich in zwei Sätzen zusammenfassen lässt: Zunächst sollte *der Mensch* mit Springwunsch einen leichten Sitz und Balance vorweisen. Außerdem seien das Reiten von Übergängen, Überstreichen, Handwechsel und Wendungen Teil der Basis. (FN 2018) Das wars. Wie sich über diese grundlegenden Techniken eine angemessene Reitpferdemuskulatur entwickeln soll, bleibt unklar.

Im nachfolgenden Trainingskapitel wird das Thema **kinetische Energie und Körperschwerpunkt** immer wieder eine große Rolle spielen. An dieser Stelle greife ich kurz vor und sage: Ohne entsprechenden muskulären Unterbau ist Springtraining für Reitpferde gesundheitlich mindestens bedenklich,

immerhin wird hier massiv kinetische Energie aufgewandt, die abrupt beim Sprung über das Hindernis im Bewegungsapparat festgehalten und beim Aufsetzen auf dem Boden auch wieder irgendwohin freigegeben werden muss. Die Belastung für die Gelenke und den gesamten Bewegungsapparat ist enorm! Das gilt aber generell für alle Reitformen, die sich für schneller-höher-weiter begeistern.

Ich zitiere Steinbrecht nicht häufig, an dieser Stelle aber hat er unwissentlich von heutiger wissenschaftlicher Forschung tatsächlich völlig richtig gelegen:

> *»Nachdem wir uns darüber klar geworden sind, dass die Dressur den Zweck hat, durch systematisch geordnete gymnastische Übungen die Muskeln des Pferdes zu befähigen, dem Knochengerüst die für den Reitdienst erforderlichen Richtungen zu geben ...* (Steinbrecht).«

Wir müssen die Muskulatur durch Übungen so weit stärken, dass sie befähigt ist, unser Pferdeknochengerüst in versammelter Haltung zu bewahren (= erhalten) und zu gleich entsprechend formen. Wenn das gelungen ist, dann und zwar erst dann, sind explizite Zusatzbelastungen wie Sprünge überhaupt denkbar! Die Frage bleibt natürlich, ob ich, sobald Seitengänge, Piaffen, Passagen bis hin zur Galopppirouette gemeistert sind, zum Spaß tatsächlich noch im Kreis über Hindernisse springen wollen würde.

Das Pferd ist einfach nicht für die Dressur gemacht ...

Ein typischer Trugschluss, der zwischen Spring- und Dressurfans häufig aufkommt, ist der, dass es vor allem auf die genetische Disposition und damit auf die Rasse des Pferdes ankäme, ob es nun eher zum Springen oder zur Dressur geeignet sei.

Wenn ein Pferd schulmäßig in Versammlung ausgebildet ist, kannst du im Parcours ganz anders taxieren. Sonja mit Leo.

Kaltblut, Warmblut, Vollblut: Über die Mär der Eignung zur Hohen Schule

Ja, einige Züchtungen sind quasi prädestiniert dafür, wahlweise besonders beeindruckend zu sprinten oder elegant tänzelnd durch die Halle zu schreiten.

Die Ursache dafür ist zumindest in Teilen auf die genetische Veranlagung zurückzuführen. Einige Varianten im Mystatin-Gen (MSTN) sorgen dafür, dass vorrangig Vollblüter und insbesondere Quarter Horses mit dieser genetischen Besonderheit schneller eine kräftige Muskulatur entwickeln (Hill et al. 2010).

Sie sind damit ideal für Bewegung im Bereich Kraft, Schnellkraft und Sprintleistung ausgestattet. Diese strukturellen Vorteile sorgen dafür, dass der Vierbeiner insgesamt profitiert durch

- die passive Stoßdämpfung aufgrund von einem *mehr* an Muskelmasse.
- eine biomechanische Reserve für das Training und in Belastungssituationen.
- einen ästhetisch und züchterisch gewünschten Körperbau.

Nun sind insbesondere Quarter Horses aber eher nicht bergauf konstruiert, was sich spätestens bei der Passage in einem höheren Trainingsbedarf bemerkbar macht.

Hohe Schule mit Ponys

Wie ist das nun mit Ponys? Auch hier gibt es solche, die eine ganz gute bauliche Grundlage mitbringen - und solche, die noch etwas mehr leisten müssen, um die Hohe Schule zu meistern. Auch wenn Ponys relativ selten in hoher Versammlung zu sehen sind, sie können das!

Wie genau du dein Pony arbeitest und fütterst entscheidet langfristig über seine Gesundheit und darüber, wie es Muskulatur aufbaut und erhält.

Bei Ponys mag das anfangs seltsam wirken, aber ja: Auch ein Mini-Pferdchen kann in Versammlung

Iberer werden häufiger klassisch angehaucht ausgebildet, alle anderen leider kaum. Sonja mit Coronado.

gearbeitet werden. Diese kleinen Energiebündel profitieren absolut davon, wenn sie Seitengänge, Piaffen oder am Ende sogar die Courbetten meistern.

Neben der körperlichen Betätigung ist Dressur nach Art der Alten Meister ja auch eine kognitiv herausfordernde Betätigung. Nichts gegen Möhrchensuchspiele oder Ähnliches, um dein Shetty beschäftigt zu halten. Nur liefert die Dressur neben körperlich herausforderndem Krafttraining für Pferde aller Größen und Rassen eben auch eine anspruchsvolle geistige Beschäftigung.

Sie vermittelt jedem Tierchen, ob Groß ob Klein, ein ausgezeichnetes Gefühl für die eigene Leistungsfähigkeit, die Körperwahrnehmung wird gestärkt und – das gilt vor allem für introvertierte oder wenig selbstbewusste Vierbeiner – das Vertrauen in das eigene Können nimmt dank der Hohen Schule deutlich zu.

Dressurkunst mit sozialen Pferden

Ich bezeichne sie gerne als wandelndes Häufchen Oxytocin, weil sie sich so sehr kümmern und bis zur absoluten Selbstaufgabe hinweg für alle und alles um sie herum Sorge tragen. Die Rede ist natürlich von Iberern, die sich häufig *so* sozial zeigen, dass sie ihre Menschen betüdeln und teilweise die totale Befehlsverweigerung hinlegen. Denn der schwache und der Situation nicht gewachsene Mensch weiß ja gar nicht, was gut für ihn ist und wird dann entsprechend von seinem Iberer vor sich hergetrieben.

Ein häufiger zu beobachtendes Beispiel aus der Stallpraxis: Mensch und Iberer sind auf der Koppel, es dämmert und für den Schutz der Herde wäre es so langsam an der Zeit, sich in eine sichere Umgebung zu begeben. Springst du als Mensch nicht zuverlässig als Herdenoberhaupt ein und übernimmst diese Aufgabe, fühlt sich der Iberer in der Pflicht, dich zunächst mit leichtem Stupsen und wenn du nicht richtig aufpasst, auch mit nachdrücklicherem Zwicken in Richtung Stall und Sicherheit zu treiben.

Wichtig zu verstehen ist hier, dass sich diese so sozialen Tiere mit der Führungsrolle oft nicht wohlfühlen, die Verantwortung für das Herdenglück belastet sie - und das Oxytocinbündel rauscht geradewegs in den Burnout.

Auch dagegen hilft die Hohe Dressurkunst, denn da musst du als Mensch den Ton angeben, du legst fest wer da wen bewegt, präsentierst Durchsetzungsvermögen und übernimmst die Führung. Das entlastet den sozialen Iberer von seiner selbstauferlegten Bürde, für dich sorgen zu müssen.

Zusätzlich stärkt es sein Vertrauen in sich selbst und auch das tut dem empathischen Tierchen gut. Statt sich selbst als Imposter (Hochstapler) anzusehen, kann der Iberber in der ihm zugewiesenen und passenden Rolle wachsen.

Tanzschule für schwere Brocken

Schwere Warmblüter wie Friesen, Kaltblüter und andere vermeintlich schwerfällige Großpferde sind genetisch muskulär meist gar nicht so schlecht ausgestattet, werden aber nur selten bewusst in Versammlung geritten. Die Größe, das damit einhergehende Gewicht und die daraus entstehende kinetische Energie stellen eine ziemliche Belastung für den gesamten Bewegungsapparat dar.

Mit ihrer tendenziellen weichen Mittelhand sind Kaltblüter auch nicht unbedingt ideal ausgestattet, um Menschen herumzutragen. Der ursprüngliche Züchtungszweck als Arbeits- und Zugpferde vorrangig in der Landwirtschaft spiegelt sich eben auch in der körperlichen Ausstattung wider, bei der die Tragkraft kaum eine Rolle spielte.

Wenn der Kaltblüter heutzutage dann untrainiert mit dem Reitgewicht im Kreis herumstolpern soll, ist eine ordentliche versammlungslastige Arbeit eigentlich unverzichtbar.

ALTKLASSISCHE DRESSUR FÜR REHA-PFERDE – UND ALLE ANDEREN

KURZ UND KNAPP:

- Muskelverlust verringert die Lebensqualität – bis zum Tod.
- Bodybuilding verändert das Gehirn.
- Mehr Muskelmasse ist das Ziel – und Dressur als Krafttraining mit hoher mechanischer Spannung im Muskel der Weg dahin.
- Du kannst lernen, die richtigen Reitpferdemuskeln zur richtigen Zeit einzusetzen.
- Bänder und Sehnen brauchen Muskeln, um optimal zu funktionieren!

Versammlung ist Vertrauen in deine Führungskompetenz

Dass Pferde für uns an die Versammlung herangehen, ist ein riesiger Vertrauensbeweis. Statt sich all seine Schubkraft für die Flucht nach vorn zu bewahren, lehnt es in der Versammlung seinen körperlichen Schwerpunkt nach hinten und beugt die Hinterhand nach unten. So entsteht das, was wir uns als Tragkraft wünschen und was zugleich das Pferd vorerst etwas ausbremst.

Bei hoher Versammlung und entsprechendem Trainingsgrad verfügt der Vierbeiner dann aber über eine enorme Schubkraft, die das Pferd im Falle des Fluchtfalles nach vorne katapultieren würde. Und nutzt diese (hoffentlich!) in deinem Sinne, weil es dir und deinem Führungsvorbild folgt.

Pferde wählen die tiefe Hankenbeugung eher *nicht* natürlicherweise von allein über lange Zeiträume hinweg. Menschen auf seinem Rücken herumzuschleppen, wäre aber logischerweise auch nicht Teil dessen, was nicht-domestizierte Pferde üblicherweise tun würden. Deshalb die Versammlung: Wir erhalten über die Versammlung die Gesundheit unseres Pferdes, indem wir die empfindliche Rückenmuskulatur stärken und für das menschliche Reitgewicht präparieren, welches den körperlichen Schwerpunkt des Pferdes leicht nach vorn verlagert. Über den Sitz und die Aufrichtung oder das Anheben der Halsbasis lässt sich der Schwerpunkt anpassen. Idealerweise natürlich hin zu einer biomechanisch optimalen Position, damit das Pferd unter seinem Menschenrucksack nicht ständig das Gefühl bekommt, gleich vornüber zu kippen!

Reiten im Sinne der Altklassik fordert eine Menge Vertrauen vom Pferd und eine Menge Verantwortungsgefühl vom Menschen für sein Tier.

Altklassische Dressurkunst verändert Pferdekörper

Mangelnde Ausbildung einerseits und mangelnde Kenntnisse der allermeisten durchschnittlichen Reitmenschen andererseits verlangt dem modernen

In der Jungpferdeausbildung ist es Gang und Gäbe, große schnelle Bewegungen zu forcieren. Diese schaden dem Jungpferd genauso wie dem älteren Tier!

heutigen Reitpferden viel ab. So viel, dass sie weit vor ihrer Zeit ernsthafte Verschleißerscheinungen zeigen.

Eine Beobachtung aus der Praxis ist die Folgende: Die meisten Menschen lieben ihre eigenen Vierbeiner wirklich heiß und innig. Gleichzeitig sehen sie ihren Freizeitreitspaß als genau das an, nämlich als Spaß. Reitpferde suchen sich die Betätigung als Hilfsmittel für spaßige Ausritte in die Natur aber nicht selbst aus. Sie werden in dieses Leben mit all seinen Konsequenzen hinein geboren.

Selbst beim gemütlichen Spazierenreiten sinkt der Widerrist = Stockmaß ein, die Biomechanik dahinter macht eben auch vor der Freizeitgestaltung nicht halt. Schon allein im Schultergürtel braucht es ausreichend Stützmuskulatur, damit nicht nur die harten Knochen möglichst alt werden, sondern alle Pferde gesund bis ins hohe Pferdealter werden.

Tipp:

Wenn du dich mit dem von mir durchaus kritisch betrachteten Vorwärts-Abwärts befassen möchtest, lies idealerweise in meinem ersten Buch das Geschichtskapitel zum Einstieg. Von der griechischen Reitkunst bis zur Gegenwart hat sich in der Reitlehre viel getan, das heute gängige Vorwärts-Abwärts ist da aus Sicht der Reitgeschichte relativ blutjung. An dem weit verbreiteten Grundgedanken, die FN wäre das Maß aller Dinge in der Ausbildung von Reitpferden, darf man ruhig ein bisschen zweifeln. Denn so entstehen beim durchschnittlich genutzten Freizeitpferd ernsthafte Verletzungen am Bewegungsapparat.

Die immer häufiger auftretende Trageerschöpfung ist kein Zufall, sondern direkte Folge mangelnder Muskulatur und einer anhaltenden Fehlbelastung. Die Pferde werden zwar immer rittiger, aber auch weicher im Bindegewebe, was den Weg zur Trageerschöpfung nur noch stärker ebnet. Zum Glück lässt sich das aber nicht nur beheben. Mittels Krafttraining nach Art der Alten Reitmeister lassen sich auch moderne Pferde gymnastizieren.

Dressurkunst ist Arbeit

»Das Geheimnis beim Reiten ist, weniges richtig gut zu tun.«

Dieser Ausspruch wird Nuno Oliveira zugesprochen und ist für den Start in die altklassische Dressur ein guter Einstieg.

Eine Remonte kommt mit einer natürlichen Schiefe in die Ausbildung, so wie sie jedes Wirbeltier besitzt. Mit der heute typischen Form des Trainings wird das kaum verbessert. Zugleich erlernt es aber auch Bewegungen, und hier meine ich ausdrücklich das heute gängige und überhaupt nicht in Frage gestellte Vorwärts-Abwärts, die seinen Bewegungsapparat nachhaltig schädigen.

Die von der Altklassik eingesetzten sinnvollen Techniken, um zunächst das Pferd in Balance (= gewisser Versammlungsgrad) zu bringen, bevor es überhaupt geritten wird, sind in Vergessenheit geraten. Das kannst du im Detail in meinem Buch *Klassische Reiterei neu denken* nachlesen. In diesem Abschnitt hier geht es mir vielmehr darum, vorhandene Fehlbelastungen zu erkennen und sie über geeignete Trainingseinheiten zu korrigieren. Dass das geht, sollte unbestritten sein.

Falls du dir an dieser Stelle sorgen darum machst, weil dein Vierbeiner kein klassisches Dressurpferd ist: Verfrachte diesen Gedanken besser gleich in die Ablage P (wie Papierkorb). Grundsätzlich profitiert jedes Pferd von altklassischen Lektionen, mit denen voluminöse Muskulatur aufgebaut wird.

Natürlich gibt es gewisse Limitationen, die sich aus der Grundstruktur und genetischen Veranlagung des individuellen Pferdes ergeben. Einen wesentlich größeren Einfluss auf die Perspektive in Richtung Schulreiterei aber hat die Art des bisherigen Trainings. Denn das entscheidet maßgeblich darüber, wie beispielsweise die Hankenbeugung gelingt, in welchem Winkel die Oberarme zum Schulterblatt stehen oder wie gut sich die Muskelgruppen an Hals, Rücken und Kruppe zur besseren Verteilung des Reitgewichts eignen etc. Die Pferdephysiologie ist veränderbar, sowohl im positiven wie im negativen Sinne.

Ein oftmals ursächlich kaputt gerittenes Reha-Pferd mit **Muskelatrophie** und **Sehnenschaden** kann mit optimierter Ernährung und angepasstem Training durchaus aufgemuskelt und lahmfrei durch sein weiteres Reitpferdeleben schreiten. Ein im Grunde gesundes und fittes Pferd kann durch ungünstiges

BODYBUILDING MACHT MUSKELN UND VERÄNDERT DAS GEHIRN

KURZ UND KNAPP:

- Bodybuilding verändert die Struktur einzelner Gehirnareale.
- Depressionen und erhöhte Kortisollevel sind miteinander verknüpft.
- Bodybuilding reduziert Kortisollevel – und sorgt für Glückshormone.
- Tanzen (und Musik) aktiviert das Glückshormon Dopamin – und Dressur ist wie Tanz.

Krafttraining für Pferde ist vor allem im Reha-Sport eminent wichtig, denn die lahmen, die alten und die kraftlosen Pferde haben mindestens eins gemeinsam: So richtig fröhlich sind sie nicht.

Schmerzen zermürben, die typische Ruhigstellung nach Verletzung schränkt den Bewegungsspielraum ein. Vor allem trageerschöpfte Pferde, sofern die Trageerschöpfung überhaupt bemerkt wird, wird man versuchen mit ewigem Longieren zu therapieren. Auch Spaziergänge am Strick sind jetzt nicht unbedingt herausfordernd. Trageerschöpfung ist so ein Thema, dass glücklicherweise präsenter wird, aber es zählt leider auch zu den Ursachen, die bei Reha-Fällen nicht oder nicht schnell genug erkannt werden. Da wird dann hier und dort herumgedoktert, ohne die eigentliche Ursache auch nur ansatzweise zu beheben.

Dabei wäre Kraftsport perfekt geeignet, weil bei den langsamen und kontrollierten Bewegungsabläufen kaum kinetische Energie auftritt, welche dann die Gelenke und Sehnen noch weiter belasten würde.

Die Reha-Fälle werden im Laufe der Zeit immer müder, fetter und trauriger.

Krafttraining baut das Gehirn um

Bewegung fordert und fördert die Entwicklung von Muskelstrukturen. Wie sich gezeigt hat, verändert Krafttraining aber auch die Signalgebung zwischen Muskeln und Knochen, Bändern und Sehnen. Da scheint es doch logisch, dass auch das Gehirn durch anabolikafreies Krafttraining profitiert.

In einer Meta-Studie wurde genau das untersucht. Siehe da: Gehirn-, Muskelstrukturen und Muskelfunktionalitäten stehen in einem signifikanten Verhältnis zueinander (Kilgour et al. 2014). Der Körper baut sich durch effizientes Training um, ebenso wie sich das Gehirn mit seinen Strukturen an seinen Körper anpasst. Vor allem der Hippocampus (Zentrale für Gedächtnis und Lernen) sowie der präfrontale Kortex (kognitive Entscheidungen) verändern sich auf positive Weise durch athletische Betätigung (Paruk et al. 2020).

Beim Menschen ist das hinreichend dokumentiert. Nehmen wir einmal an, dass das genauso auch für das Pferd gilt: Reichlich Bewegung an und für sich wäre also eine gute Möglichkeit, um positive Effekte im Gehirn zu erzielen. Nun sollen Reha-Pferde und natürlich auch alle anderen Vierbeiner nicht nur möglichst kognitiv fit sein, sondern auch möglichst munter. Im besten Falle sogar fröhlich!

Krafttraining gegen Depression

An dieser Stelle drehen wir eine kurze Schleife in die Welt der Depressionsforschung – keine Sorge, gleich gehts zurück dazu, warum Krafttraining genau das richtig für Pferde ist. Wohlgemerkt für alle, nicht nur das kaputtgerittene Reha-Pferd. Auch die heutigen modernen Schulpferde und Wallache im Allgemeinen sind aufgrund ihrer Lebensumstände häufig anfällig für depressive Schübe. Erstere kämpfen mit verdeckten Schmerzen durch Fehlhaltungen und wenig gesunderhaltendes Training bei ständig wechselnden, oft nicht besonders gut ausgebildeten Reitschülern. Letztere sind schon aufgrund des Wegfalls des Testosterons potenziell schwermütiger.

Gehirnveränderungen bei Depressionen durch Kortisol

Forschende aus Leipzig haben in einer noch relativ kleinen Studie beobachtet, dass Menschen mit diagnostizierten Depressionen häufig einen vergrößerten Hypothalamus aufweisen (Schindler et al. 2019). Das ist der Teil der im Gehirn, der alle vegetativen und endokrinen Vorgänge regelt.

Das heißt, der Hypothalamus steuert, wie sich unser Nervensystem reguliert und welche beziehungsweise wie viele Hormone den Blutkreislauf betreten. Genau dieser Part im Gehirn scheint beim Menschen mit Depression vergrößert. Damit einhergehend sind häufig erhöhte Kortisollevel nachweisbar. Auch der Zusammenhang zwischen Hippocampus-Volumen und der Art der Depression spielt eine Rolle, die sich auf die Ausschüttung des Stresshormons Kortisol auswirkt, das konnten Forschende der Uni Wien kürzlich belegen (Rabl et al. 2025).

Ob mehr Kortisol als Folge einer depressiven Erkrankung ausgeschüttet wird oder die höheren Kortisolwerte depressive Erkrankungen (mit-)verursachen, wird in der Forschung gerade erst diskutiert. Klar ist, dass diverse Studien einen Zusammenhang zwischen laufenden Entzündungen durch vermehrte Cytokinausschüttung, höheren Kortisolwerten und damit verbundenen Veränderungen im Gehirn nachweisen können (Amasi-Hartoonian et al. 2022). Klar ist auch, dass Kortisol zahlreiche biochemische Prozesse im Körper befeuert, die die Betroffenen ernsthaft krankmachen.

Wichtig: Kortisol ist aber kein *böses* Hormon, das es unbedingt zu minimieren gilt. Wir brauchen es ebenso wie Adrenalin für zahlreiche bedeutsame körpereigene Abläufe. Wer mit Depressionen zu kämpfen hat, der will aber sicherlich das Kortisollevel ganz generell auf ein eher durchschnittliches Niveau senken, statt ständig mit erhöhten Werten zu tun zu haben.

Gleich schließt sich der Kreis, denn wie lässt sich eine Depression und damit auch ein erhöhtes Kortisolniveau in den Griff bekommen? Über (unter anderem) Kraftsport!

In einem 218 Studien umfassenden Review wurde schließlich untersucht, wie sich verschiedene Sport- und Bewegungsformen auf Depressionen auswirkt. Neben Laufen und Yoga war vor allem Krafttraining eins der Mittel, die hervorragende Ergebnisse über alle Studien hinweg zeigte. Je intensiver, desto besser waren die beobachteten Effekte (Noetel et al. 2024) auf die Studienteilnehmenden!

Beim Reha-Fall ist klassisches Cardio so wenig geeignet wie beim Reitpferd, um Bänder und Gelenke zu schonen. Yoga fällt für Pferde in der Regel flach und wäre mit Blick auf den Einsatz als Reitpferd ja auch wenig hilfreich. Bleibt das Krafttraining, dass schon aufgrund der geringeren kinetischen Energie wesentlich schonender für den Bewegungsapparat ausfällt.

In einem Review wurden die Auswirkungen von Krafttraining mit sorgfältig dosierten Bewegungsabläufen untersucht, wobei speziell Menschen mit Arthrose im Fokus standen. Das Ergebnis nach Auswertung von insgesamt 27 Studien mit über

Ohne Schweiß kein Preis. Mascha mit Honey.

1.700 Studienteilnehmenden? Krafttraining reduziert nicht nur Schmerzen und verbessert die körperliche Stärke, es optimierte auch die Funktionalität von Knie- und Hüftgelenken bei den Arthrosegeplagten (Lim et al. 2024).

Plus: Langjähriges Training erzeugt eine dauerhafte und vor allem auch aktiv genutzte Muskulatur, die vermehrt Myokine ausschüttet und insgesamt dazu beiträgt, das Immunsystem effektiv und langfristig zu stärken (Zunner et al. 2022).

Die Botenstoffe des Glücks: Endorphine, Dopamin und Serotonin

Endorphine machen klassischerweise glücklich, sorgen aber auch für Entspannung und wirken als körpereigene Schmerzmittel zur Linderung bei kleineren Wehwehchen. Zugleich schafft der Ausstoß aus Glücksbotenstoffen wie Endorphinen und **Serotonin** es, die körpereigene Stressresistenz zu erhöhen. Beim Training steigt das Kortisol, aber danach scheint es bei regelmäßigem Krafttraining effektiver zu sinken – und niedrig zu bleiben. Das wirkt sich dann auch positiv aufs Gehirn aus (American Psychology Association 2020).

Achtung: Lauftraining mit höherer Intensität (mind. 30 Minuten Dauer) scheint dagegen eher zu einem dauerhaft höheren Kortisollevel zu führen (Guseman et al. 2008).

Dopamin wird vorrangig über Bewegung erzeugt und entsprechend in inaktiven Phasen wenig ausgeschüttet. Krafttraining trägt dazu bei, dass es verstärkt produziert wird, um im Blutkreislauf zu zirkulieren und dort diverse wichtige Prozesse anzuschieben (Marques et al. 2021).

Das ethische Dilemma, das keins ist

Pferde, vor allem Reha-Fälle zeigen oft Antriebsschwäche, sie wirken müde und regelrecht depressiv. Bewegung, vor allem aber körpermodulierendes und muskelaufbauendes Krafttraining kann nicht nur bei konkreten Bewegungseinschränkungen helfen, Entzündungen oder Schmerzen lindern; es kann auch gezielt den Antrieb steigern und wesentlich die Stimmung verbessern.

Nur was ist, wenn das Pferd anscheinend so gar keine Freude an der Bewegung hat? Wenn es sich durch jede Lektion schleppt und es - vor allem anfangs - einfach echt zäh wirkt? Ist dann Aufhören und Spazierengehen oder gar die Rentnerweide die vermeintlich tierschonendere Lösung?

Ganz klar: nein.

Kein Pferd sucht sich das Leben als Reitpferd aus, sondern wird in dieses Leben als wohlbehütetes, oft geschätztes und gehätscheltes Haus- und Sporttier hineingeboren. Als Menschen haben wir die Verantwortung für diese Tiere zu übernehmen. Wie bei einem Schutzbefohlenen auch, ist es unsere Pflicht, uns bestmöglich um die Gesundheit und das allgemeine Wohlbefinden zu kümmern.

Das depressive Pferd mit Arthroseschaden ist auf der Rentnerweide kaum besser aufgehoben, obwohl es dort zwar eine gewisse Bewegung haben wird. Das bremst den körperlichen Zerfall aber kaum, denn das Pferd wird auf der Weide wohl kaum von allein Kraftsport praktizieren.

Das hat aber nichts mit einer vermeintlichen Unnatürlichkeit von Krafttraining für Pferde zu tun, sondern die Lebensumstände unserer heutigen modernen Reitpferde sind schlicht so, dass unsere Schutzbefohlenen extra Training benötigen, um möglichst lange gesund zu bleiben. Wild lebende Mustangs erreichen im Durchschnitt ein Alter von 15 Jahren. Das durchschnittliche Freizeitpferd bringt es, abhängig von Einsatzzweck und Pflege, auf runde 20 bis 25 Jahre. Großpferde sterben etwas früher, kleinere Pferderassen wie Isländer erreichen oft die 30 Jahre.

Im Humanbereich akzeptieren wir schon längst, dass Krafttraining die reinste Medizin ist, die Schmerzen und Entzündungen lindert, Alterserscheinungen reduziert und sich stabilisierend auf die Psyche auswirkt (Westcott 2012). Warum übertragen wir das nicht auf Pferde?

Statt also beim ersten lustlosen Seitengang abzubrechen, kannst du stattdessen aktiv mit Stimm- und, wenn das für das individuelle Tier angebracht ist, auch mit Futterlob arbeiten, um es zu motivieren.

Der Mensch fungiert an dieser Stelle ganz klar als Gesundheitscoach, der motiviert, fördert und fordert. Auch dann, wenn das Pferd vielleicht mal nicht ganz so begeistert reagiert. Hast du selbst aber eigentlich keine Lust auf kognitiv anspruchsvolles Tanztraining aka Dressurarbeit am und mit dem Pferd, dann lässt sich auch dein Tier schlechter dafür begeistern. Da steckt also auch Arbeit für dich mit drin, nicht nur fürs Pferd!

Liefere deinem Vierbeiner kurzgefasst die besten Gründe, sich gerne kraftraubend im Training zu bewegen. Das reduziert mit der Zeit auch immer mehr Fehlhaltungen, die Schmerzen nehmen ab und die Freude an der Bewegung kommt zurück.

Für Reitpferde ist das Krafttraining generell ein Muss, denn kaum ein Tier ist tatsächlich ohne jedes Zipperlein.

Dank Krafttraining: Bewegung wird zum Tanz

Was ist noch schöner als Bewegung? Bewegung in Perfektion. Das gilt gleichermaßen für Pferde wie für Menschen: Gelungener Tanz ist Bewegung in Perfektion. Tanz ist also eine Kunstform und nicht umsonst wird die echte Dressurkunst nach Art der Alten Meister gern auch mit tänzerischen Bewegungsabläufen verglichen. Die bestens ausgebildeten Pferde gleiten, schweben und tänzeln über die Bahn, sie wirken federleicht und zeigen dabei kontrollierte Kraft und somit präzise Bewegungen.

Da ist es keine große Überraschung, dass sich seit den 1980ern eine eigene Wettkampfform dafür entwickelt hat. Dressurchoreografien entstehen, die zu passender Musik dargeboten werden. Was im hiesigen Schulsport mit Bändern und Kegeln zu eher fragwürdigen Bodenturnübungen gehörte, ist im Reitsportwettbewerb auf einem wesentlich höheren Level im Idealfall hübsch anzusehen.

Heutige Musikküren als Teil der Dressurwettbewerbswelt lassen leicht den Eindruck entstehen, dass *Pferde zur Musik tanzen*. Die Bewegung des Pferdes ist an die Musik angepasst, Takt (des Pferdes) und Schrittlänge orientieren sich an den musikalischen Vorgaben. Das ist grundfalsch, denn richtigerweise muss es natürlich andersherum sein. Die Spanische Reitschule, so beschreibt es die Dressurexpertin Sylvia Loch, handhabt das glücklicherweise etwas anders.

> *»In Wien ist die Musik als herrliche Untermalung für die Arbeit der Pferde gedacht; sie wird niemals zum Hauptthema. Es gibt keine plötzlichen Änderungen von Melodie oder Takt zur Vervollkommnung der Lektionen. Stattdessen geht die beinahe spirituelle Ausstrahlung von den Pferden selbst aus. Sie tanzen, weil sie leicht und von lebhaftem, jedoch kontrolliertem Schwung erfüllt sind und nicht, weil der Reiter dem Rhythmus entsprechend die Tritte verkürzt oder verlängert! (...) Ihre Bewegungen sind flüssig, weil sie ungezwungen sind, nicht weil dies die Musik so vorschreibt! Dies ist der eigentliche Sinn des Musikreitens nach Musik, was niemals in Musik nach Dressur umgekehrt werden darf* (Loch 2010).«

Dabei ist die Spanische Hofreitschule noch nicht einmal das, was ich unter altklassischer Reitkunst verstehe, denn die herausragende Arbeit eines La Guérinière wird auch in Wien nicht mehr gezeigt und praktiziert.

Inwieweit ein Friese, ein Isländer oder ein kompaktes Warmblut in eleganten Figuren und tänzerischer Anmut die Lektionen absolviert, ist natürlich abhängig davon, wie die Muskulatur per Krafttraining für ebendiese Lektionen vorbereitet ist.

Fehlt die Kraft für eine ausdrucksstarke Courbette, Piaffe oder den Spanischen Schritt, wird das Ergebnis nicht sehr anmutig ausfallen. Dafür ist schlichtweg ein monate- bis jahrelanger Kraftaufbau nötig!

Diese Lektionen sind ja auch tatsächlich eher *Abfallprodukte guter Arbeit* dieses Kraftaufbaus, mit denen sich das mit Bodybuilding gestählte Pferd in anmutigen Gesten wie Piaffe oder Courbette auszudrücken versteht.

Kleiner Hinweis am Rande:

Ich spreche immer von der klassischen Courbette. Die Wiener Courbette der Spanischen Hofreitschule meine ich ganz ausdrücklich nicht.

Das heißt nicht, dass Pony oder Kaltblut dazu generell nicht in der Lage wären. Die barocken Spanier haben da natürlich einen gewissen Züchtungsvorteil. Daneben spielt aber auch die generelle genetische Veranlagung mit hinein. Das Stichwort dazu lautet COMT und du wirst dazu gleich noch mehr im Abschnitt zum Pferdetemperament lesen.

Dein Gehirn liebt Musik (und Tanz)

Aus der Humanforschung ist bekannt, dass sich Gehirne durch Musik und durch Tanz verändern. Wer regelmäßig tanzt oder Musik hört/spielt, zeigt veränderte Gehirnwellen; die bildliche Vorstellungskraft und – das kennen wir aus dem Profisport – damit auch die Ausführung der Bewegungen verbessert sich.

> *Both short- and long-term dance training lead to significant changes in brain structure and neural activity, including, in particular, neural activity during dance performance, imagery, and dance observation (Foster et al. 2023).*

Auf deutsch: »Beides, das Kurz- und -langzeit-Tanztraining führten zu signifikanten Veränderungen in der Gehirnstruktur und den neuronalen Aktivitäten, insbesondere neuronale Aktivitäten während einer Tanzperformance, der bildlichen Vorstellung und bei der Beobachtung von Tanzdarbietungen.«

Denn: Vor allem bei tänzerischen Aktivitäten ist es hilfreich, die vertrauten Bewegungen einer Choreographie regelmäßig zu beobachten. Ob sich das so auf Pferde übertragen lässt, sei mal dahingestellt.

Aber was an dieser Erkenntnis spannend sein dürfte: Als Dressurfan dürfte sich das wiederholte Anschauen von Dressurlektionen, wahlweise auf der Bahn und/oder per Videomitschnitt durchaus *für* dich lohnen, um auf einer ganz elementaren Basis im Gehirn die Weichen für die nächsten Lektionen mit dem eigenen Vierbeiner zu stellen.

Such dir dabei gute Vorbilder aus, die mit einer feinen Hilfengebung und nach Vorbild der Altklassik Lektionen zeigen. In meinen Online-Kursen kannst du dir natürlich genau das ansehen: Sowohl Pferde, die gerade erst mit dem Krafttraining starten als auch Vorbilder dazu, wie Seitengänge, Piaffe und Co. Idealerweise aussehen sollten.

Neben der Veränderung an der Gehirnstruktur durch Beobachtung kommt ein wirklich hübscher **Belohnungsfaktor** dazu, der vor allem bei Musik gut erforscht ist. Denn beim Anhören von vertrauter und geliebter Musik schüttet das Gehirn bereits dann fröhlich stimmende Botenstoffe aus, wenn der Song oder das Musikstück noch gar nicht richtig angefangen hat. Das reine Wissen um das Kommende reicht schon aus, um die Botenstoffausschüttung im Gehirn in Schwung zu bringen (Salimpoor et al. 2011).

Merke: Das Gehirn merkt sich positiv besetzte Reize (hier: Musik) und beim kleinsten Anzeichen dieser positiv besetzten Reize schickt es die gute-Botenstoffe-Kavallerie los, um diese positiv besetzten Reize gleich nochmal so gut auskosten zu können.

Auch das können wir auf Pferde und die Dressurkunst übertragen:

Eine Lektion gut zu meistern, dafür belohnt zu werden und die Übung erneut zu starten kann dazu führen, dass das Gehirn die guten Botenstoffe losschickt, obwohl die eigentliche Bewegung noch gar nicht zu Ende ausgeführt ist.

Lies diesen Satz nochmal. Dieser Mechanismus im Gehirn deines Pferdes sorgt also dafür, dass die positive Verknüpfung aus absolvierter Lektion und Anerkennung durch seinen Menschen für waschechte Glückshormone sorgt. Immer und immer wieder, kaum dass die Lektion abgefragt wird und noch vor der nachfolgenden liebevollen Wertschätzung!

Altklassische Dressurlektionen fordern viel vom Pferd, sie sind kognitiv und körperlich anstrengend. Deshalb zeigen Pferde anfangs oft eine eher angestrengte Mimik, so wie jeder Sportler und jede Sportlerin das bei echten körperlichen Herausforderungen eben auch präsentiert. Das ändert aber rein gar nichts am empfundenen Glück infolge der sportlichen Betätigung.

Das lässt sich für den Umgang mit deinem Vierbeiner nutzen! Lobe, was das Zeug hält und zwar bitte auf die Art und Weise, die deinem Pferd am meisten behagt.

Ruppige Klopfer auf den Pferdehals sind nicht immer angebracht. Es sei denn, du hast das entsprechend konditioniert.

Meine Pferde kennen den festen Klopfer auf den Hals als Signal für die Pause. Sie wissen dann ganz genau, dass sie bis hierhin (meist) wunderbar mitgearbeitet haben und sich jetzt eine Runde entspannen dürfen.

Vor allem futtermotivierte Schleckermäuler freuen sich im Rahmen des körperlich anstrengenden Krafttraining über leckere Snacks.

Bonus fürs Gehirn: Dressur im Grünen

Aus der Humanforschung ist bekannt, dass schon kurze Spaziergänge in der freien Natur für sinkende Kortisolspiegel sorgen. Eine Meta-Studie konnte zusätzlich einen sich normalisierenden Blutdruck, eine höhere (= optimalere) Herzfrequenzvariabilität (HRV) sowie die Verringerung von Angst- und Stresssymptomen nachweisen (Wenfei et al. 2021).

Das erklärt sich vor allem aus unserer sehr langen Evolutionsgeschichte, die vorrangig draußen – und keinesfalls eingesperrt in kleinen Zimmern – stattfand. Wälder, Steppen, Wüsten, Berge und alles dazwischen gehörten für unsere Vorfahren zum tagtäglichen, vertrauten und damit auch sicheren Umfeld. Gilt übrigens gleichermaßen für Pferde und ihre Wildpferdahnen.

Die waren ja auch kaum tagtäglich viele Stunden in engen schmalen Boxen und schon gar nicht in willkürlich zusammengewürfelten Herden ohne Familienzusammenhalt zusammengezwängt, sondern haben sich im Familienverbund über ziemlich weitläufige Steppen bewegt.

An dieser Stelle nur kurz ein Stichwort: **Haltungsstress**. Das ist Stress, der durch eine nicht optimale Haltungsform für das individuelle Pferd entsteht. Weder seine Boxen- oder Offenstallnachbarn, noch die jeweilige Haltungsform kann sich das einzelne Pferd aussuchen.

Introvertierte Tiere leiden oft eher still unter stressverursachenden Situationen, über die sie keinerlei Kontrolle haben. Der Offenstall ist genauso wenig die optimale Lösung für alle Pferde wie die Boxenhaltung, es kommt viel mehr immer auf das individuelle einzelne Tier an.

Natur = gut. Natur + Bewegung = noch besser!

In der Natur sein tut gut, verstanden. Und wenn jetzt noch moderate und an die Bedürfnisse des jeweiligen Tiers angepasste Bewegung dazu kommt? Dann dürften sich die Effekte noch verstärken. Da wird nicht nur aktiv das allgemeine Stressniveau gesenkt, sondern auch gleich noch der muskel- und gehirnstärkende Effekt der Bewegung integriert.

In diesem Sinne: Dressurkunst ist nicht auf Hallen und Bahnen beschränkt. Sie kann – und darf! – im Grundsatz überall stattfinden. Gerade Reha-Pferde profitieren von dieser positiven Kombination aus Natur und einem an das individuelle Pferd angepasstem Krafttraining. Achte hier aber auf die passende Bodengestaltung, wenn es sich beispielsweise um Reha-Pferde mit auskuriertem Sehnenschaden handelt. Dann ist vorübergehend ein für das individuelle Pferd günstigerer Boden die bessere Lösung.

Dicke Muskelpakete sind das Ziel, kein Nebeneffekt der Dressur

Das ultimative Ziel für ein in klassischen Lektionen geschultes Reitkunstpferd ist die Bezeichnung als Reitkunstpanzer (Inside-Witz meiner Follower)!

Ein dicker, tragender Muskel ist nicht nur stark, er hält den Knorpel gesund. Er ist biologischer Schutz. Er ist die Gelenkluftmatratze.

Voluminöse Muskelpakete an den Hanken, am Rücken, an Widerrist und Kruppe dienen dem Schutz und der Stabilität des gesamten Bewegungsapparats. Gesunderhaltendes Reiten ist so für viele Jahre möglich, ohne das Reitpferde weit vor ihrer Zeit auf die Rentnerweide abgeschoben werden müssen, weil die Knochen schlapp machen und Arthrose überhandnimmt.

Um diesen Reitkunstpanzer ausreichend aufzupolstern, wird der Zustand der Muskel-Hypertrophie, also ein intensives Muskelwachstum angestrebt. Dafür ist Training mit starker Muskelbeteiligung, sogenanntes High Resistance Training (RT-Training) ideal. Dabei werden Muskelgruppen aktiviert und in langsamen, aber kontrollierten Bewegungen stark beansprucht.

Eine Übersichtsarbeit von Krzystofik aus dem Jahr 2019 hat gezeigt, dass RT-Training ideal ist, um Bodybuilding abwechslungsreich und muskelaufbauend zu betreiben (Krzystofik et al. 2019).

Keine Muskeln ohne Muskelfasern

Jeder Muskel besteht im Kern aus Muskelfasern, die sich in unterschiedliche Typen untergliedern lassen. Sie entscheiden, wie schnell, wie explosiv-kraftvoll oder wie ausdauernd eine Bewegungsfolge absolviert werden kann.

Jedes Pferd, ob Araber oder Quarter Horse, verfügt grundsätzlich über die Muskelfasern Typ-1 und Typ-2. Die für ihre Ausdauer berühmten Araber haben im Verhältnis mehr Typ-1-Fasern, während das Quarter Horse über eine größere Menge und eine optimale Verteilung von Typ-2-Fasern verfügen. Beide Pferderassen sind in der Lage, grundsätzlich sowohl lange Cardio-Einheiten oder auch die Lektionen nach Art der Alten Meister zu absolvieren. Je nach Veranlagung fallen die unterschiedlichen Bewegungsarten dem einen Tier leichter, dem anderen schwerer – und umgekehrt. Umso weniger dein Pferd genetisch auf Kraftleistung hin gezüchtet wurde, umso mehr musst du an Zeit und Arbeit in die Ausbildung stecken. Der Körper muss sich anpassen können und an seinen Aufgaben wachsen dürfen.

Damit ein Reitpferd für lange Zeit mit dem Reitgewicht knapp hinter den Schulterblättern zurechtkommt, ohne am Ende krumm und lahm zu sein, braucht es vor allem die Muskelfasern des zweiten Typs, weil sie nun mal die Kraftfasern sind. Außerdem ist es im Falle des Reitpferdes nun mal so, dass es ein Zusatzgewicht mit sich herumschleppen können soll – da ist mehr Kraft aufgrund der vermehrt gebildeten Typ-2-Muskelfasern extrem von Vorteil.

Veranlagung ist toll, aber nicht alles

Eine Mischung aus Typ-1- und Typ-2-Fasern finden sich vorrangig in Warmblütern wie Hannoveranern und Oldenburgern. Friesen, aber auch Quarter Horses und barocke Spanier bringen oft eine Mehrzahl an Typ-2B-Fasern mit sich, was sie besonders für kraftvolle Bewegungen prädestiniert.

Die spanischen/barocken Züchtungen wie PRE, Lusitanos und Lipizzaner sind natürlich speziell für die hohe Versammlung gezüchtet und entsprechend mit reichlich Fasern des Typs 2A ausgestattet. Das liegt primär an genetischen Faktoren, wie die Variation des Myostatin-Gens (MSTN), welches das Muskelwachstum fördert (Hill et al. 2010) und insbesondere bei Quarter Horses und der Rennlinie unter den Vollblütern auftritt, hier in Form einer MSTN-SINE Variation.

Spannenderweise lässt sich der Ursprung der ursprünglichen Mutation in Vollblütern sogar noch eingrenzen auf sagenhafte drei nach Großbritannien importierte Pferde, die der Ursprung für die heute wegen ihrer enormen Rennleistung bekannten britischen Rennpferdzüchtungen zu sein scheinen (Bower et all. 2012).

Genetik ist aber nicht alles, auch der Araber kann mit versammlungslastigem Training ordentlich Muskulatur aufbauen!

Die Masse und damit auch die Kraft der Typ-2-Fasern wächst mit jeder Lektion. Bereits vorhandene Typ-2-Fasern werden mit dem richtigen Training effektiver genutzt und vergrößern sich. Eine Umwandlung von Typ-1- auf Typ-2-Fasern ist eher begrenzt und in einem sehr kleinen Rahmen möglich – aber möglich ist sie!

Muskelfasern im Überblick

Typ-1-Muskelfasern sind perfekt für lange, gleichmäßige Belastungen. Ihre Energie beziehen Typ-1-Muskelfasern aus Sauerstoff, sie arbeiten also aerob. Werden diese Fasern durch lange und ruhige Ausdauereinheiten angesprochen, knabbert der Körper vorhandene Fettreserven weg. Der eigentliche Kalorienverbrauch pro Minute ist aber relativ gering. Die Muskelmasse selbst wird durch die Beanspruchung von Typ-1-Fasern kaum zur Vermehrung gereizt. Der Grundumsatz bleibt entsprechend auf einem konstanten Niveau, ohne sich zu erhöhen – und das trotz reichlich Cardio!

Einsatz der Typ-1-Fasern: Wanderritte und gemütlicher Trab, jeweils im langsamen und ruhigen Tempo, aber auch langanhaltende Haltearbeit und alles, wo Stabilität und Balance gefragt sind. Fasern des Typ-1 sind nicht stark genug, um eine kraftvolle Versammlung zu ermöglichen.

Typ-2A-Muskelfasern reagieren schnell und sind relativ ausdauernd. Sie lassen sich durch das Training modulieren und sind vor allem indirekte Kalorienkiller. Die bevorzugte Energiequelle für Typ-2A-Fasern ist weniger der Sauerstoff, sondern eher der Blutzucker, bei anhaltender Belastung werden auch Fettreserven angefordert (anaerob).

Besonders angesprochen fühlen sich Typ-2A-Fasern durch Kraft- und Intervalltraining, was für eine aktive Kohlenhydratverbrennung sorgt. Gleichzeitig steigt bei ausreichend Proteinzufuhr über die Nahrung auch die Proteinsynthese hin zu mehr Muskelmasse. Das erhöht den Grundumsatz, es wird mehr Energie verbrannt.

Apropos verbrannt: Der sogenannte Nachbrenneffekt (EPOC) ist bei Typ-2A-Fasern besonders wirkungsvoll, hier wird noch Stunden nach einem intensiven Training reichlich Energie, insbesondere auch Fett verwertet. Typ-2A-Fasern sind auch an Prozessen beteiligt, die die Hormonausschüttung (Wachstumshormone und Testosteron) sowie die Fettverbrennung anheizen. Hengste und Stuten profitieren insbesondere von der Testosteronausschüttung, die beide in unterschiedlichen Mengen produzieren. Nur beim Wallach fällt dieses Geschlechtshormon eher weg.

Einsatzformen der Typ-2A-Fasern: Perfekt für die Versammlung und für anhaltende, aber kraftvolle Bewegungsformen wie die Piaffe über mehr als zwei Tritte hinweg.

Typ-2X-Muskelfasern sind für kurze, explosive Bewegungen gedacht, reagieren sehr schnell und zeigen wenig Ausdauerleistung. Dafür sind sie aber wahre Kalorienkiller, verbrauchen also viel Energie und sind am besten für das Muskelwachstum (Hypertrophie) geeignet.

Gezieltes Training dieser speziellen Muskelfasern steigert die Kraft, den Ausdruck und die Dynamik des Pferdes in der Dressur. Die formschöne Regenrinne mit voluminöser Rückenmuskulatur gehört beim nach altklassischer Manier ausgebildetem elegant dahintänzelnden Reitkunstpanzer übrigens zur Grundausstattung dazu.

Hinweis für das praktische Training: Um Typ-2X-Fasern zu trainieren, sind Lektionen in hoher Versammlung wie Mezair oder Courbetten ideal. Durch die enorme Lastaufnahme mit der Hinterhand plus federnde, schnelle und kraftvolle Bewegungen werden genau diese Muskelfasern ideal angesprochen.

Einsatzarten der Typ-2-Fasern allgemein: Springen, sprinten oder schnelle Beschleunigungen.

Die Muskelfasern des Typ-2 ermüden aber auch schneller. Regenerationszeiten sind wichtig, damit sich die Muskelstrukturen erholen können. Fordere die Leistung in den Lektionen ein, aber überfordere dein Tier nicht durch fehlende Pausenzeiten im lockeren Stand!

Hypertrophie = Muskelwachstum

Hypertrophie ist das Wachstum von Muskelfasern, insbesondere des Typs 2. Sie werden bei Beanspruchung dicker und legen bei passendem Training ordentlich an Volumen zu. Und das bedeutet: mehr Kraft und mehr Power.

Für dieses mehr an Volumen reicht es anfangs schon aus, wenn nur ein Teil der Muskelfasern arbeitet, um das Gewicht zu bewegen. Das spart zunächst mal Energie. Im Laufe des konstanten Krafttrainings erhält das Gehirn aber zunehmend Signale, dass die einzelnen Muskelfasern Unterstützung benötigen. Deshalb werden mehr Muskelfasern rekrutiert, also zur Mitarbeit hinzugezogen.

Diese Muskelfaserrekrutierung sorgt zu Beginn des Trainings dafür, dass sich deine Muskelkraft erhöht, ohne das direkt sichtbare Muskelberge erscheinen. Gehirn und Körper aktivieren weitere Muskelfasern, um deinen Kraftbedarf zu erfüllen. Erst im Verlauf einer längeren Trainingszeit wachsen die Muskeln, insbesondere die Fasern des Typ 2 auch richtig sichtbar.

Um das Ganze besonders spannend zu machen, läuft gleichzeitig noch ein weiteres Programm im Körper. Denn während durch entsprechende Reize am Muskel und eine proteinreiche Ernährung Muskeln aufgebaut werden, wird im Rahmen der Energieoptimierung immer auch Muskelmasse abgebaut (Atrophie). Du erinnerst dich? Alles, was nicht ausgiebig genutzt wird, wird eingespart.

Damit das Team Hypertrophie gewinnt und der Reitkunstpanzer die gewünschte Gestalt annimmt, sind vor allem fünf Punkte entscheidend:

1. **Training**, insbesondere Krafttraining in moderater Intensität und mit ausreichenden Wiederholungen pro Einheit plus einer gewissen Regelmäßigkeit im Alltag. Einmal pro Woche Krafttraining mit deinem Vierbeiner ist besser als nichts, zwischen drei- bis viermal pro Woche wäre deutlich angemessener!

2. **Ernährung**, mit Fokus auf eine ausreichende Proteinzufuhr und einer ausgewogenen Menge an Mikronährstoffen im Futtermix.

3. **Schlaf und Regeneration**, die beide stark abhängig sind von der Haltungsform und dem jeweiligen Temperament des Pferdes. Der optimale Ort für erholsamen Schlaf kann für das eine Tier im Offenstall zu finden sein, beim anderen stellt sich die beste Erholung aber unter Umständen erst mit Boxenhaltung ein. Die individuell optimale Haltung ist aus Raum- und Geldgründen vielleicht nicht immer (sofort) umsetzbar, sollte aber immer mit bedacht werden, wenn es um die Pferdegesundheit geht.

4. **Hormone**, möglichst im Gleichgewicht, wobei sowohl Training als auch Ernährung eine große Rolle bei der Hormonbalance spielen.

5. **Alter und Genetik**, die die Geschwindigkeit und die Konstanz der Hypertrophie beeinflussen. Bei älteren Tieren ist meist mehr Geduld gefragt, bis sich erste voluminöse Muskelpakete zeigen. Erfahrungsgemäß nehmen Pferde ab circa 17 Jahren Proteine weniger gut aus dem Futter auf. Da kann es absolut sinnvoll sein, sowohl die Menge als auch die Qualität der Aminosäuren zu erhöhen, um dem älter werdenden Pferd den Muskelaufbau zu erleichtern.

Muskelspannung für Muskelzuwachs

Entscheidend im Training ist außerdem die umgesetzte Muskelspannung, beziehungsweise genauer die Zeit an sich, die ein Muskel unter Spannung steht (engl. time under tension, TuT). Beim Krafttraining entsteht durch Krafteinwirkung auf den Muskel mechanische Spannung, die die Muskelfasern zum Wachstum anregen. Die aktivierende Spannung kann sowohl über schnelle als auch über langsamere Bewegungen entstehen.

Im Humanbereich wird häufig noch mit der Annahme trainiert, dass viele schnelle Sätze mit hoher Belastung das beste Ergebnis erzielen. Das bedeutet oft acht bis zwölf Wiederholungen und mehrere Trainingssätze, durch die sich Bodybuilding-Enthusiasten durchhetzen, um den Trainingsplan zu absolvieren, bevor die Muskeln schlapp machen.

Dabei liegen schon seit Längerem Forschungsergebnisse vor, die eine höhere Belastung bei gewollt langsamer Ausführung der Bewegung als wesentlich erfolgreicher einstufen (Burd et al. 2010). Die Bewegungen finden unter bewusster Verlangsamung statt, wobei gleichzeitig mit hoher Muskelbelastung gearbeitet wird. Die Empfehlungen fallen unterschiedlich aus, liegen für den Humanbereich aber in der Regel bei 40 bis 60 Sekunden pro Satz, mit wenigstens sechs und maximal 15 Wiederholungen unter stetiger Muskelreizzunahme.

Das Ziel für eine optimale Muskelarbeit unter TuT-Prämissen (Time under tension, Muskelspannung) ist eine anschließende Muskelermüdung, die zu erfolgreicher Hypertrophie führt. Erfolgreich meint an dieser Stelle: Ein intensiverer Anstieg der Proteinsynthese im betreffenden Muskel, eine intensivierte Signalgebung, um die Mitochondrien aufzubauen und ein verbesserter allgemeiner Zellstoffwechsel.

Wenn du eine dieser schönen, aber superanstrengenden Raketen unter dem Sattel hast, startest du am besten mit Seitengängen in deutlicher Abstellung. Langsam, bewusst und kontrolliert ausgeführt, reduzieren sie praktischerweise die überschäumende Energie deines Vierbeiners und sorgen für nette Muskelpakete im gesamten Rumpfbereich.

Natürlich profitieren auch lethargische Pferde von dieser Lektion, denn das **mehr** an Mitochondrien durch **mehr** Muskelmasse verleiht ihnen auch etwas **mehr** an Energie, die sie gut gebrauchen können.

Bessere Ergebnisse mit Slow-Motion-Training

Prof. Dr. Theodor Stemper, Sportwissenschaftler an der Universität Wuppertal, fasst die Bedeutung der neueren Erkenntnisse zu TuT (Time under tension, Muskelspannung) mit Bezug auf Trainingspläne im Humanbereich so zusammen:

> *»Wenn die »time under tension« eine so große Rolle für den Trainingseffekt spielt, dann reicht es nicht aus, wie in den Plänen üblich, lediglich die Last (Gewicht) und die Wiederholungszahl vorzugeben. Die Art der Ausführung, hier vor allem das Bewegungstempo, ist dann offenbar auch wichtig, wenn nicht sogar noch wichtiger.* (Stemper 2017)«

Er kommt außerdem zu dem Schluss, dass vor allem Menschen mit Vorschädigungen und ältere Personen von diesem verlangsamten, dafür aber intensiverem Training profitieren können, um Muskelmasse sowohl zu erhalten als auch – im besten Falle – weiter aufzubauen.

Das lässt sich 1 zu 1 auf unsere Reha-Fälle und die Seniorenabteilung übertragen: Langsame, kontrollierte Bewegungen sind vor allem für diese Tiere ideal, um gelenkschonend gearbeitet zu werden und gleichzeitig bewusst schützende Muskelpakete aufzubauen.

In der altklassischen Reitkunst starten wir natürlich immer langsam. Jeder einzelne Seitengang und jede Piaffe werden zunächst im Slow-Motion-Rhythmus ausgeführt. Das allein ist anfangs anstrengend genug für dich - und für dein Pferd. Erst, wenn die entsprechende Muskelmasse aufgebaut und der Bewegungsablauf ausreichend verankert ist, kannst du über eine höhere Frequenz nachdenken.

Um insbesondere die Typ-2-Muskelfasern durch kraftvolle Bewegungen anzusprechen, eignet sich die Piaffe in Slow Motion und auch entsprechend anstrengend gestaltete Seitengänge mit passender Abstellung und Rumpfhebung. Bis dahin ist die echte Reitkunst eine, die mit hoher mechanischer Spannung im Muskel und damit starke Muskelbeteiligung arbeitet - und das Ganze im absoluten Schneckenkriechgang.

Drei gute Gründe für Hypertrophie

Ein wichtiger Leitsatz vorweg: Nur genutzte Muskelmasse schützt langfristig.

#1 Mehr Muskelmasse, mehr Grund zum Futtern

Ein volumiger Muskelaufbau führt zu einem höheren Grundumsatz, da Typ-2-Fasern größer werden und bereits im Ruhezustand mehr Energie verbrennen.

So verliert dein Pferd leichter Fettgewebe, kann die Versammlungsarbeit dank der vorhandenen Muskelberge effizienter umsetzen und dadurch noch mehr Muskeln aufbauen.

Natürlich nur, wenn der Futterplan keinen Energieüberschuss durch Massen an Kohlenhydraten und Stärke produziert. Eine ausgewogene und eher proteinlastige Ernährung ist nicht nur beim Menschen, sondern auch beim Pferd unverzichtbar für eine optimale Aufmuskelung.

#2 Mehr Muskelmasse, mehr Schutz

Eine aktive und voluminöse Muskelmasse schützt durch Stoßdämpfung. Sie wirkt außerdem viskoelastisch (Sarvazyan et al. 2014):

- puffert Stöße ab,
- reduziert Druck auf Gelenke,
- füttert den Knorpel durch bessere Versorgung der Gelenkflüssigkeit (Synovia),
- schützt vor degenerativen Veränderungen.

Synovia ist eine mit Proteinen, Glukose und unter anderem Hyaluronsäure versetzte Flüssigkeit, die von der Gelenkinnenhaut gebildet wird. Die wiederum wird über die umliegende Muskelstruktur mit Bausteinen für die optimale Synovia-Zusammensetzung versorgt. Ein dicker, tragender Muskel ist ein aktiver Gelenkschutz, besonders am Rücken, den Schultern und in der Hankenregion.

Muskelmasse ist Funktion, Schutz und Leistung.

Ein kräftiger, gut entwickelter Muskel ist kein Schönheitsideal, sondern ein biologisches Schutzsystem für Gelenke, Sehnen, die Funktionsfähigkeit des Rückens und damit für die Langlebigkeit deines Pferdes.

Beim Menschen lässt sich die Schutzfunktion einer höhere Muskelmasse direkt mit einem reduzierten Risiko für Sarkopenie und damit verbundenen Fallverletzungen in Verbindung bringen. Die Meta-Studie von Steffl et al. aus dem Jahr 2017 bestätigt das, nach einer Auswertung von 25 bereits existierenden Studien (Steffl et al. 2017).

Parallel dazu berichten vor allem ältere Menschen mit chronischen Schmerzen nach Beginn eines Krafttrainings über reduzierte Schmerzlevel. In einem Review von De la Corte-Rodriguez et al. 2024 kamen die Forschenden zum Schluss:

> *»There is solid evidence of the analgesic effect of physical exercise in many pathologies, such as chronic low back pain, neck pain, osteoarthritis, fibromyalgia, and rheumatoid arthritis.«*

Auf Deutsch: »Es gibt stichhaltige Beweise für die schmerzlindernde Wirkung von körperlicher Betätigung bei zahlreichen Erkrankungen, wie z. B. chronischen Kreuzschmerzen, Nackenschmerzen, Arthrose, Fibromyalgie und rheumatoider Arthritis.«

So kann ein Pferd mit Muskelatrophie aussehen!

#3 Mehr Muskelmasse, viel bessere Laune

Schon mehrfach habe ich das hier und in meinem ersten Buch beschrieben, trotzdem folgt jetzt eine Wiederholung. Denn eine aktive und leistungsfähige Muskulatur bedeutet in aller Regel, dass sich Entzündungsprozesse im Körper reduzieren.

Der Stoffwechsel läuft auf vollen Touren, körpereigene Schmerzstiller wie Endorphine werden ebenso wie körpereigene glücklich machende Hormone vermehrt ausgeschüttet.

Durch eine aktive Fettverbrennung, die erst durch den Energie-Mehrbedarf einer voluminösen Muskelpanzerung so richtig in Schwung kommen kann, reduzieren sich die stimmungsvermiesenden Signale aus dem viszeralen Fettgewebe.

In einer zehn Studien umfassenden Review-Analyse konnte außerdem durch Ke-Vin et al. ein gehäuftes Zusammentreffen von Sarkopenie und Depressionen nachgewiesen werden. Das bedeutet noch nicht, dass es sich um eine kausale Verknüpfung handelt, aber: Wer an Muskelschwund mit den damit verbundenen körperlichen Einschränkungen leidet, scheint auch wesentlich häufiger an Depressionen zu erkranken (Ke-Vin et al. 2017).

Übrigens hängt nicht nur die generelle Stimmung von der verfügbaren Muskelmasse im Körper ab, sondern es scheint auch eine gewisse Verknüpfung zwischen der Menge an aktiv genutzter Muskulatur und sinkenden kognitiven Funktionen zu geben. Das legt zumindest die Auswertung einer kanadischen Kohortenstudie nahe, die an über 8.000 älteren Menschen durchgeführt wurde. Diejenigen mit niedrigen Werten bezüglich der Muskelmasse zeigten im Rahmen der Studie nach drei Jahren einen erkennbar stärkeren kognitiven Verfall (Tessier et al. 2022).

Was bedeutet das für dein Pferd (und dich)? Ganz klar: Mehr Muskelmasse hebt nicht nur die Laune, es schützt auch vor Tüdeligkeit im Alter.

	Atrophie	Hypertrophie
Muskeln	bauen sich ab/verlieren Masse	bauen sich auf/gewinnen Masse
Ursachen	• Bewegungsmangel • Krankheit • Immobilisation/Ruhigstellung • wenig geeignetes, konventionelles Training	• Krafttraining unter hoher mechanischer Spannung im Muskel
Auswirkungen	• Leistungsabfall • Schwäche/Kraftverlust • erhöhte Verletzungsanfälligkeiten	• Verbesserte Leistungsfähigkeit • mehr Tragkraft insbesondere im Rücken • Geringeres Verletzungsrisiko

Abb. Tabelle Atrophie vs. Hypertrophie, eigene Zusammenstellung

Hypertrophie versus Atrophie

Use it or loose it (Nutze es oder verliere es) ist keine leere Phrase, sondern ein im Grunde wichtiges Prinzip für Muskelstrukturen.

Wenn Muskeln nicht genutzt werden, verschwinden sie. Sarkopenie ist die schwere Folge. Der Zwischenschritt nennt sich Atrophie (Muskelschwund) und ist im Grunde eine Ressourcenoptimierungsmaßnahme des Körpers. Nur keine Energie verschwenden für Gewebe, das nicht gebraucht wird!

Ein Team der Universität von Kentucky hat 2022 einen Muscle Atrophy Scoring System (MASS) entwickelt, bei dem insbesondere bei älteren Pferden verstärkt Atrophie nach festgelegten Kriterien entdeckt werden kann (Herbst et al. 2022).

Ein Review aus dem Jahr 2023 fasst zusammen, dass selbst normale Alltagsbewegungen wie Treppen steigen oder einfaches gehen für ältere Menschen mit beginnender oder fortgeschrittener Sarkopenie schwer zu bewältigen sind. Im Gegensatz dazu zeigt der Review aber auch klar, dass insbesondere exzentrisches Training (Teil des Krafttrainings, bei dem auf kontrollierte und langsame Bewegungsabläufe gesetzt wird) die Muskelkraft erhalten und damit die Alltagsbewältigung erleichtern kann (Cvečka et al. 2023).

Reitpferde sind oft schon dann in ihrer Leistungsfähigkeit merklich eingeschränkt, wenn sie noch längst nicht im sogenannten Rentenalter angekommen sind.

Muskelfibrose: Schleichender Muskelumbau ohne Chance auf Umkehr

Muskelfibrosen (fibrotische Myopathie) treten dann auf, wenn Muskulatur über lange Zeit hinweg nicht oder in Fehlbelastung genutzt wird.

> *»Skeletal muscle fibrosis impairs muscle function, negatively affects muscle regeneration after injury and increases muscle susceptibility to re-injury, therefore, it is considered a major cause of muscle weakness* (Mahdi 2018)*«.*

Auf deutsch: »Fibrosen an der Skelettmuskulatur beeinträchtigen die Muskelfunktion, wirken sich negativ auf die Muskelregeneration nach Verletzungen aus und erhöhen die Anfälligkeit der Muskeln für erneute Verletzungen. Deshalb gilt die Muskelfibrose auch als Hauptursache für Muskelschwäche.«

Im Sinne einer energieoptimierten Verwendung baut der Körper diese nicht ausreichend verwendeten Strukturen zu einer Form des Bindegewebes um. Was einst Muskulatur war, ist nun hart, unelastisch und in seiner Funktionsfähigkeit nicht mal ansatzweise mit dem Wunderwerk Muskelfaser vergleichbar.

Die Folgen sind weitreichend: Der einst weichere Gang wird stechschrittartig, Lahmheit und eine insgesamt weniger flüssige Bewegungsfolge folgen auf dem sprichwörtlichen Fuße. Dein Vierbeiner kann Schwierigkeiten beim Gangartwechsel zeigen oder bestimmte Bewegungsabläufe rundheraus verweigern. Nicht, weil es dich ärgern will oder faul ist, sondern weil die Bewegung schlichtweg schmerzt!

Damit Fibrosen entstehen können, brauchen sie

1. **Zeit.** Sie entstehen nicht über Nacht!
2. **Inaktivität.** Fibrosen profitieren durch einen Mangel an aktiver Muskelnutzung.
3. **Fehlbelastung.** Die fehlerhafte Belastung wird oft noch durch Inaktivität ergänzt.

Beim Reitpferd haben wir die Sondersituation, dass Irritationen bei oft papierdünnen Rückenmuskeln ein Hauptthema für Fibrosen sind. Das Reitgewicht sitzt auf diesem ungenügenden Muskelchen, denn Muskel können wir das an der Stelle eigentlich guten Gewissens gar nicht mehr nennen. Das Gewicht drückt die Blutzufuhr ab, die Nährstoffversorgung fällt bestenfalls ungenügend aus. Das begünstigt Fibrosen enorm.

Nicht ganz sicher, ob es bei deinem Vierbeiner schon zu Fibrosen gekommen ist?

Hier folgt eine Schritt-für-Schritt-Anleitung, um Fibrosen zu erkennen.

1. Schritt: Tasten!
Du findest feste, gummiartige Muskelpartien, insbesondere entlang des langen Rückenmuskels, der Kruppenmuskulatur oder entlang des Trapezmuskels. Der betreffende Muskel fühlt sich unter deinen Händen nicht mehr elastisch, sondern eher *verhärtet* und *knotig* an. Manchmal sind auch *sehnenstrangartige* oder fasrige Strukturen unter der Haut ertastbar.

2. Schritt: Angucken!
Schau dir dein geliebtes Pferd im Detail an. Fibrosen zeigen sich durch sichtbar abgeflachte oder eingefallene Muskelgruppen zum Beispiel am Rücken. Stichwort Regenauffangbecken, dass uns im Kapitel zum langen Rückenmuskel und dem Rumpfheber schon begegnet ist. Diese eingefallenen Muskelgruppen können aber auch an der Kruppe oder den Schultern auftreten. Starke Asymmetrien oder ein insgesamt irgendwie *hohl* wirkendes Erscheinungsbild sprechen ebenfalls für eine Fibrose.

3. Schritt: Arbeitsbeobachtung!
Dein Pferd zeigt Schmerzreaktionen bei leichtem Druck auf die bereits in Schritt 1 und Schritt 2 identifizierten Bereiche, hat einen steifen Rücken, nicht die geringste Lust auf Bewegung oder zeigt, wenn es sich schon bewegt, steifere oder schleppende Bewegungen. Es ist unwillig beim Putzen, Satteln oder Longieren.

Unter Umständen kommen auch noch Magenprobleme aufgrund der ständigen Schmerzen und des Unwohlseins dazu, kein Wunder bei diesen ganzen mit der Fibrose zusammenhängenden Problemen!

Tipp: Mach Fotos zur Dokumentation im Wochenabstand von deinem Tier. Möglichst seitlich, von oben und von hinten. Hast du deinen geliebten Vierbeiner tagtäglich vor der Nase, fallen dir die zunächst kleineren Veränderungen vielleicht nicht direkt auf. Hast du Bildmaterial im größeren zeitlichen Verlauf, lässt sich eine beginnende Fibrose leichter identifizieren.

Was der Veterinär tun kann

Ein mit Fibrosen vertrauter Veterinär kann den Verdacht differenzialdiagnostisch absichern. Wichtig ist die Abgrenzung zu Atrophie, zu einer tatsächlichen Schutzspannung (Verspannung) oder einer Nervenirritation. Schmerzmittel oder Entzündungshemmer schlagen nicht an, wenn die Beschwerden durch eine Fibrose bedingt sind.

Tierärztliche Diagnosehilfen im Detail:

- Tastbefund (Palpation), Ultraschall oder Thermografie zeigen untypische Strukturen (faserig, unregelmäßig)
- in Spezialfällen helfen Muskelbiopsien zur Identifikation

Blutbilder sind oft unauffällig, denn Fibrosen sind eben kein akuter Entzündungsvorgang, der in den Blutwerten ins Auge springt. Sie sind auch kein Muskelkater und keine Schutzspannung – sondern ein struktureller Umbau.

Einmal in Gang gesetzt, lässt sich dieser Umbau nur durch extrem harte Arbeit und viel Mühe aufhalten. Rückgängig machen ist dagegen kaum möglich. An dieser Stelle ist also unbedingt Prävention gefragt!

Was du gegen die Entwicklung von Fibrosen tun kannst

Das beste Mittel gegen Muskelfibrosen ist logischerweise: Kraftsport zur Muskelaktivierung und -stärkung.

Aktives **Longieren in Aufrichtung** hilft bei der grundlegenden Aktivierung der Muskulatur, bestenfalls mit dem Idealbild der Versammlung im Hinterkopf. Dabei bitte weder das Tier im Vollspeed durch die Halle scheuchen noch extra den Kopf hochbinden!

Seitengänge sind regelrechtes Fullbody-Workout, sprich die Muskulatur insbesondere des Rumpfs, die Vorder- und die Hinterhandmuskulatur werden aktiviert und – sofern sauber ausgeführt – auch symmetrisch belastet. Seitengänge erhöhen zudem sehr stark die mechanische Spannung in einzelnen Reitpferdemuskelgruppen, weil die Hintergliedmaßen- und Vordergliedmaßenvorführer im Seitwärts zum ersten Mal tatsächlich unter Körpergewicht arbeiten müssen.

Denn: Ohne Belastung einfach einen Fuß (oder Arm!) anheben zu müssen, ist zunächst ja kaum eine Anstrengung. Erst unter Gewichtsbelastung wird es zu einer gewissen Herausforderung. Diese Beanspruchung ist genau das, was die Hypertrophie einleitet. Darauf kommt es uns in der Altklassik an, um Pferde gesunderhaltend aufzumuskeln.
Je früher du erste Anzeichen für Fibrosen erkennst und aktiv dagegen antrainierst, desto besser.

Ausführlichere Hankenbeugung und Übergänge zwischen den Seitengängen und in Schulgangarten ergänzen die präventive Muskelarbeit zur Kompensation einer beginnenden Fibrose.

Medikamentöse oder chirurgische Therapien bringen in aller Regel keine Besserung, letztere werden aufgrund der zu erwartenden Komplikationen in der Regel nicht empfohlen (Hellige 2009).

Das reine Reiten auf der Vorhand solltest du bei einer bestehenden Fibrose besser vermeiden wie der Teufel das Weihwasser! Gleiches gilt für Arbeiten in dauerhafter Dehnung oder das stumpfe monotone Laufen ohne jede ordentliche Muskelbeteiligung. *Drübertrainieren* ist ebenfalls keine gute Idee, denn die Fibrose ist eben kein Fitness-, sondern ein Strukturproblem. Mehr von der fehlerhaften Bewegungsabfolge macht es nicht besser, sondern reitet dein Pferd und dich nur tiefer in die endlose Behandlungsspirale hinein.

Um beginnende Fibrosen aufzuhalten, musst du den Muskel kräftigen – nicht dehnen.

Muskelzuwachs ist nur was für die jungen – oder?

Alte Pferde können keine Muskulatur mehr aufbauen, was einmal verloren ist, bleibt verloren. Diese Mär geistert gerne noch durch die Stallflure, dicht gefolgt von Achselzucken und der Ansicht, dass so ein Tierchen mit 10 Jahren ja eh schon lange zum alten Eisen gehört. Das ältere Pferde aktiv Muskulatur durch mangelnde Aktivität verlieren können, wir nur zu oft einfach hingenommen.

Dabei wird gerne vergessen, dass domestizierte Pferde eine durchschnittliche Lebenserwartung von 20 bis 35 Jahre haben können, einzelne Exemplare werden auch deutlich älter! Nur der Verschleiß durch ausgiebiges Abschleudern an der Longe und ewig lange Spaziergänge im Schneckengang sorgen dafür, dass Pferde in der Lebensmitte längst nicht so fit und gelenkschonend aufgemuskelt sind, wie sie es sein könnten. Krafttraining mit einer erhöhten mechanischen Spannung im Muskel stärkt vor allem die älteren Tiere enorm aus gesundheitlicher Sicht.

Hier bietet sich wieder ein Blick in die wesentlich umfangreichere Humanforschung an. Glücklicherweise lässt sich vieles auch auf unsere Pferde übertragen. Ein Forschungsteam der Universität

Auch ein 21jähriger Warmblüter muss nicht von Atrophie oder Fibrose zerfressen sein! Sonja mit Leo.

von Kopenhagen hat sich 2016 der Frage gewidmet, ob ältere Personen mit und ohne vorhandene chronische Erkrankungen von Krafttraining profitieren können. Über 450 Personen zwischen 62 und 70 Jahren nahmen an der Studie »Live Active Sucessful Aging (LISA)« teil, bei der wahlweise leichtes, mittelschweres und schweres Krafttraining absolviert wurde. Am Ende konnte nur der Teil der Gruppe, der das schwere Krafttraining absolvierte, seine Muskelkraft auch beim Follow-up einige Jahre später erhalten. Die Studienteilnehmer in den Gruppen mit leichten oder mittelschweren Trainingseinheiten zeigten dagegen eine stagnierende oder sogar sinkende Muskelkraft in den Beinen und standen damit nach Studienende sogar schlechter dar! Offensichtlich ist nur Krafttraining mit hoher Intensität dazu geeignet, die Muskelkraft an sich auch im höheren Alter zu erhalten:

> *»The LIve active Successful Ageing (LISA) study, a large-scale randomised controlled trial (n=451), showed that strength can be maintained over 12 months following 1 year of heavy resistance training (HRT), but not after moderate training* (Bloch-Ibenfeldt et al. 2024).«

Auf deutsch: »Die Live active Successful Ageing (LISA)-Studie, eine groß angelegte, randomisierte, kontrollierte Studie (n=451), hat gezeigt, dass die Kraft nach einem Jahr schweren Widerstandstrainings (HRT) über 12 Monate erhalten werden kann, nicht aber nach einem moderaten Training.«

Die größte Erkenntnis aus dieser Studie lässt sich vom Menschen aufs Pferd übertragen: Je älter, desto essenzieller ist das bewusste Krafttraining, um Muskelmasse zu bewahren. Das sichert nicht nur allgemein die Mobilität im Alltag, es schützt auch aktiv die Gesundheit (Gylling et al. 2020).

Die Studie zeigt aber auch, dass es nie zu spät ist, mit Krafttraining anzufangen!

Für deinen Reha-Fall oder dein Seniorenpferd lohnt sich also auch ein sogenannter später Start noch – und zwar immer! Um es mit den Worten vom Muskelexperten Prof. Dr. Froböse zu sagen:

> *»Muskeln sind in ihrer Anpassungsfähigkeit so dynamisch, dass sie in jedem Alter schnell und effizient Verbesserungen der Leistungsfähigkeit zeigen und uns damit die Ressourcen geben, unsere Lebensqualität lange zu erhalten* (Froböse 2023).«

MINI-MUSKELKUNDE FÜR REITFANS

KURZ UND KNAPP:

- Muskulatur arbeitet in Muskelketten und *die* entscheiden über echte Versammlung.
- Kinetische Energie kommt bei Bewegung ins Spiel.
- Quadriceps und Rumpfheber stärken die Tragkraft.
- Bänder und Sehnen halten den Bewegungsapparat zusammen.
- Arthrose ist kein Grund, mit Dressur gar nicht erst zu beginnen.

Ein gut bemuskeltes Pferd ist weder dick noch dünn, sondern eher Typ Schwarzenegger. Diese Unterscheidung fällt vielen Reitfans, aber auch Fachpersonen, schwer. Erfahrungsgemäß ist es im Alltag hilfreich, die Funktion der Reitpferdemuskeln zu kennen, um sich bestmöglich für das eigene – hoffentlich ordentlich aufgemuskelte – Pferd gegenüber vermeintlich besser informierteren durchzusetzen.

Ventrale und dorsale Muskelketten für optimale Bewegungsmuster

Die meisten Reitpferdemuskeln sind *Hilfsabspreizer*, also Muskeln, die bei der sogenannten Abduktion unterstützen beziehungsweise diese sogar selbst durchführen.

Mit **Reitpferdemuskeln** sind ausdrücklich diejenigen gemeint, die für den Zweck des Reitens enorm wichtig sind.

Bei einer einfachen Bewegung wie dem Auffußen oder einer komplexeren Bewegungsfolge wie für eine Courbette sind in jedem Fall immer zahlreiche Muskelgruppen beteiligt, die als Agonisten (Hauptmuskeln) und Synergisten (Hilfsabspreizer) gemeinsam den Pferdekörper stabilisieren und in die gewünschte Position bringen.

Die **Muskeln** selbst arbeiten in der Regel in Paaren zusammen: einer zieht sich zusammen (Kontraktion), der andere dehnt sich (Flexion). Selten ist nur ein einzelnes Muskelpaar beteiligt, wenn eine Bewegung ausgeführt wird. Deshalb springen in der Regel gleich ganze **Muskelgruppen** an, um eine Kniebeuge (Mensch) oder die Hankenbeugung (Pferd) auszuführen und um Bewegungsfolgen wie Laufen, Rennen oder Springen zu ermöglichen.

Beispiel Abduktion: Um das Bein abzuspreizen, wird der *M. gluteus medius* (mittlerer Gesäßmuskel) aktiviert. Zur Unterstützung der Abspreizung sind auch der *M. gluteus minimus* (kleiner Gesäßmuskel) und der *Tensor fasciae latae* (Teil der äußeren Hüftmuskulatur) involviert.

Die gesamte Biomechanik hinter der Pferdebewegung wird über zwei wichtige Gruppen gesteuert, namentlich die **ventralen und die dorsalen Muskelketten**. Erst über die Zusammenarbeit von Muskelgruppen in Muskelketten entstehen fließende Bewegungen.

In der Dressur liegt der Fokus insbesondere auf der ventralen Muskelkette, denn diese stützt und stärkt den Vierbeiner für die menschliche Zusatzbelastung auf dem Rücken beim Reiten. Über dieses komplexe System aus stützenden und aktivierten Muskeln ist sichergestellt, dass der Hauptmuskel nicht die Hauptlast der Bewegung - und hier vor allem die dabei entstehenden Zug- und Scherkräfte - allein zu tragen hat.

Die **ventrale Muskelkette** verläuft entlang der Körperunterseite des Pferdes, sie verbindet dabei die Hinterhand (über Becken und Oberschenkel) mit dem Schädel und ist entscheidend für das Heranstellen der Hinterhand, das Aufwölben des Rückens und zusammen mit der dorsalen Kette für die Aufrichtung des Pferdes zuständig.

Alle unterhalb der Wirbelsäule verlaufenden Muskeln (= ventral) sind an der Beugung der Wirbelsäule beteiligt. Ihre Komponenten sind vor allem in Versammlung maximal aktiviert, um unter Zuhilfenahme der dorsalen Muskelkette den Rumpf zu stabilisieren und den Schwerpunkt leicht nach hinten auf die Hinterhand zu verlagern:

- **untere Halsmuskulatur** (z. B. *M. brachiocephalicus, M. Sternocephalicus*)
  In Versammlung: Involviert im Aufwölben und Runden des Halses, unterstützt die Halsneigung.
- **Bauchmuskulatur** *(M. rectus abdominis, M. obliquus externus/internus abdominis, M. transversus abdominis)*
  In Versammlung: Besonders stark aktiviert, beugt und stabilisiert den Rücken, verhindert dadurch das Verkrampfen der langen Rückenmuskeln unter dem Reitgewicht.
- **vordere Oberschenkelmuskulatur** *(M. quadriceps femoris, M. tensor fasciae latae M. sartorius,)*
  In Versammlung: Stark beansprucht, um das Becken zu kippen und die Hinterhand beugen.
- **tiefe Becken- und Hüftmuskulatur** *(M. iliopsoas, M. psoas minor)*
  In Versammlung: Beugung des Lumbosakralgelenks (ISG), welches man getrost als oberes Ende der Hanken betiteln darf, ermöglicht das Untertreten und die Versammlung der Hinterhand.
- **Zwerchfell** (Diaphragma)
  In Versammlung: Stark eingebunden, da es über die Atmung und Spannung die Rumpfstabilität unterstützt und mit den Bauchmuskeln interagiert.

Da Pferde über kein Schlüsselbein verfügen, wird der gesamte Rumpf über ein komplexes System aus Bindegewebe und Muskulatur getragen. Kommt das Reitgewicht knapp hinter den Schulterblättern dazu, sind die Teile der ventralen Muskelkette besonders gefragt, um ein Absacken nach unten idealerweise zu verhindern, oder doch wenigstens abzupuffern.

Die **dorsale Muskelkette** verbindet **Hinterhand, Kreuzbein, Rücken, Hals und Kopf miteinander**. In der versammelnden Arbeit ist sie für Selbsthaltung und Energieübertragung entscheidend, wobei insbesondere das **Nackenband** eine Schlüsselrolle als elastische Verbindung zwischen Kopf und Rücken spielt.

Über die dorsale Muskelkette sind folgende Muskelgruppen und Strukturen miteinander verbunden:

- **Rumpfträger** *(v. a. M. serratus ventralis)*
  In Versammlung: Übernimmt eine hoffentlich zentrale Rolle bei der Hebung und Stabilisierung des Rumpfes, verhindert das Absacken auf die Vorhand, tritt als enormer Stoßdämpfer auf.
- **Rückenmuskulatur** *(M. longissimus dorsi, M. erector spinae)*
  In Versammlung: Soll das Vorderteil des Pferdes heben, trägt zur Hebung der Vorhand bei, unterstützt dadurch die Aufrichtung.
- **Hinterhandmuskulatur** *(M. gluteus maximus, M. gluteus medius, M. soleus)*
  In Versammlung: Retraktion durch Wegdrücken des Bodens nach hinten-unten.
- **Hintere Oberschenkelmuskulatur** *(M. biceps femoris, M. Semitendinosus)*

In Versammlung: Kontrollierte Sprung- und Kniegelenksbewegung gemeinsam mit ventraler Muskelkette.

Der Rücken wölbt sich durch die ventrale Muskelkette auf, das Pferd trägt sich selbst und kann die Hinterhand beugen. Dabei muss vor allem das Nackenband vor zu dauerhafter Handeinwirkung unbehelligt sein, damit es uns nicht mit Nackenbandansatzverkalkungen und folgenden Entzündungen und Schmerzen reagiert.

Pferdekörper in Bewegung: es ist Physik!

Tauchen wir ganz kurz gedanklich in den Physikunterricht der siebten oder achten Klasse ein. In der Grundlagenvermittlung zur Mechanik spielte auch die kinetische Energie eine Rolle. Genau die brauchen wir, um uns eins klar zu machen: Insbesondere ohne schützende Muskelpuffer und Muskelmasse ist eine hohe Geschwindigkeit problematisch und schafft eine massive Belastung für die Gelenke und den gesamten Bewegungsapparat, denn hier wirken massive Kräfte!

Kinetische Energie ist die Bewegungsenergie eines Körpers (E_{kin}) und steigt mit der Masse (m) und dem Quadrat der Geschwindigkeit (v^2). Beim Pferd wird

Die größere Amplitude mitsamt dem passiven Auf- und Abbouncen im heutzutage beliebten Vorwärts-Abwärts wird nicht bei der Passage gemessen – was gut ist, angesichts der größeren kinetischen Energie, die höhere Amplituden mit sich bringen.

sie durch Muskelkraft erzeugt, um Bewegungen wie Galoppieren, Springen oder Versammeln auszuführen.

Kinetische Energie wird umso größer, je größer die Masse eines Körpers ist und je größer die Geschwindigkeit dieses Körpers ist.

Die **potenzielle Energie E_{pot}**, auch als Lageenergie bekannt, bezeichnet die Fähigkeit eines Körpers aufgrund seiner Höhe relativ zu einer tieferen Ebene, **Arbeit** zu verrichten. Die **potenzielle Energie** hängt ab von der Höhe h, in welcher sich der Körper befindet, und von seiner Masse m.

Kinetische und potenzielle Energie werden im Rahmen der Bewegung erzeugt, kurzfristig in flexiblen Strukturen wie Muskeln, Sehnen und Bändern gespeichert und dann wieder entladen, sobald der Huf auf dem Boden aufkommt.

Muskelkraft ist notwendig, um das Pferd zu beschleunigen, Sprünge abzufedern oder Bewegungen zu kontrollieren. Sie wirkt als *Stoßdämpfer*: Gut trainierte Muskulatur nimmt kinetische Energie auf, speichert sie kurzfristig und gibt sie dosiert und kontrolliert wieder ab.

Gelenkverschleiß entsteht, wenn die auf die Gelenke wirkenden Kräfte (besonders Stoß- und Scherkräfte) nicht ausreichend von Muskeln, Sehnen und Bändern abgefangen werden. Überlastung, schlechte Technik oder mangelnde Muskulatur beschleunigen den Abbau von Gelenkknorpel und führen zu Arthrose oder anderen Schäden. Je stärker kinetische Energie auf Gelenke und Co. einwirken kann, desto stärker können Verschleißerscheinungen auftreten.

Grundlegend sind Pferdefesseln, Sehnen und Bänder aber natürlich auf schwungvolle Bewegungen ausgelegt:

> *»When a joint is supported by elastic tissues the net joint power curve shows a typical pattern of energy absorption in early stance followed by a similar amount of energy generation in late stance. This pattern is seen across several of the horse's joints during trotting, notably the fetlock which hyperextends as it is loaded in early stance under control of the digital flexor tendons and the interosseous ligament on the palmar side of the joint. Subsequently, the supporting elastic tissues recoil as the fetlock flexes in late stance (Claytin/Hobbs 2019).«*

Auf Deutsch: »Wenn ein Gelenk durch elastisches Gewebe gestützt wird, zeigt die Nettokraftkurve des Gelenks ein typisches Muster der Energieabsorption in der frühen Phase, gefolgt von einer ähnlichen Menge an Energieerzeugung in der späten Phase. Dieses Muster ist bei mehreren Gelenken des Pferdes während des Trabens zu beobachten, insbesondere bei der Fessel, die sich unter der Kontrolle der digitalen Beugesehnen und des Ligamentum interosum auf der palmaren Seite des Gelenks überstreckt, wenn sie im frühen Stand belastet wird. Anschließend ziehen sich die stützenden elastischen Gewebe zurück, wenn sich die Fessel im späten Stand beugt.«

Das zusätzliche Reitgewicht verursacht bei fehlendem Muskelpuffer auch zusätzliche Verschleißschäden, wobei der Verschleiß bei ständigem Herumscheuchen in der Halle mit ungünstigem Bewegungsmuster durchaus auch ohne extra Reitgewicht auftreten kann.

Am Ende sind es ja oft Zahlen, die überzeugen. Deshalb rechnen wir mal kurz die durchschnittliche kinetische Energie als Zustandsgröße bei einem 500 Kilogramm Pferd in der Piaffe (langsame, kontrollierte Bewegungsausführung) aus.

Die Formel dafür lautet: $E_{kin} = 1/2 * m * v^2$
E_{kin} in Joule
m in Kilogramm = 500
v^2 in Meter/Sekunde = 0,1 (langsame Ausführung, nahezu Stand)

Bei unserem Beispielpferd ergibt das eine kinetische Energie von **0,0025 Kilojoule** (2,5 Joule $= 1/2 * 500 * 0{,}1^2$) für die Piaffe.

Beim durchschnittlichen **Galopp** bewegt sich ein Pferd zwischen 5,8 und 9,7 Meter pro Sekunde. Mit obiger Formel sind wir bei einer kinetischen Energie von **8,41 Kilojoule** (8.410 Joule $= 1/2 * 500 * 5{,}8^2$) oder sogar **23,511 Kilojoule** (23.522 Joule $= 1/2 * 500 * 9{,}7^2$).

Im Trab bewegen sich Pferde durchschnittlich mit 3,3 bis 5 Meter pro Sekunde – rechne einfach selbst aus, welche Kräfte da aufs Pferd einwirken. Die typische Folgefrage lautet dann: **Ja, aber wie ist das beim Springreiten?**

Physikprofis atmen an dieser Stelle bitte kurz durch. Es ist stark vereinfacht und zusammengefasst, aber für eine Vergleichbarkeit mit den einwirkenden Kräften beim Springreiten ist es sinnvoller, die *potenzielle Energie* zu berechnen. Denn hier wollen wir ja nicht nur die vorwärts gerichtete, sondern auch die Energie im Sprung beachten.

Hier lautet die Formel: $E_{pot} = m * g * h$
E_{pot} in Joule

m in Kilogramm

g steht für den Ortsfaktor, grob gesagt die Gravitationskraft und beträgt 9,81

h steht für Hubhöhe in Metern

Bei einem durchschnittlichen Hindernis mit einem Meter Höhe wirken allein **4,905 Kilojoule** (4.905 Joule = 500 * 9,81 * 1) aufs Pferd. Höhere Hindernisse bewirken deutlich höhere Energieentwicklungen, bei 1,5 Metern sind es bereits um die **7,357 Kilojoule** (7.357 Joule). Die Schwerkraft (Gravitationskraft) sorgt dafür, dass die gesamte Masse des Pferdes beim Aufprall abgefangen werden muss – sonst gibt es weit mehr als blaue Flecken für Mensch und Pferd zu beklagen.

Je nach Pferderasse fallen die Springhöhen und die Geschwindigkeiten unterschiedlich aus. Quarter Horses können auch wesentlich schneller auf kurzer Distanz galoppieren, Springpferde mit entsprechendem Training meistern durchaus auch höhere Hürden.

Kurz und knapp für Menschen, die Physik nicht unbedingt zu ihren Spezialinteressen zählen: Je mehr Energie in Joule, desto mehr und stärkere Kräfte wirken. Diese Kräfte müssen entsprechend auch vom Bewegungsapparat sinnvoll verarbeitet werden. Bänder und Sehnen werden bei starkem Energieeinsatz stärker beansprucht als bei Bewegungsformen, die eine geringere Energie in Joule ergeben.

Dicke Muskelpakete fungieren als Puffer, um diese Belastungen auf die sensibleren Gelenk- und Sehnenstrukturen abzufedern.

Die Faszien und das beliebte Dehnen

In meinem ersten Buch kam es schon ausführlichst zur Sprache, aber für den Fall, dass es dir auch mit Reha-Pferd gern ans Herz gelegt wird, in aller Deutlichkeit:

Sogenannte verklebte Faszien lassen sich durch Stretching nicht verändern, weder lockern noch lösen noch gleitfähiger machen, weil es sich bei Bewegungsbeeinträchtigung mit Faszienbeteiligung eher um eine mangelnde Flüssigkeitsaustauschstörung handelt.

Warum? Faszien sind bindegewebsartige kollagenhaltige Strukturen, die im Kern überall im Körper vorhanden sind. Sie sind in mehreren Schichten aufgebaut, inklusive einer gelartigen Substanz, die primär aus Wasser besteht. Entzündungen oder Inaktivität können den Flüssigkeitszufluss und -austausch reduzieren, wodurch die Schichten weniger optimal übereinander gleiten. Das fühlt sich vielleicht an wie eine *Verklebung*, aber genau genommen ist es eben **keine** im mechanischen Sinne.

Die interessante Frage an dieser Stelle wäre wohl, ob bei konstanter und optimaler Hydration die häufig gut bekannten Symptome einer verklebten Faszienregion weiterhin bestehen bleiben – oder ob es sich mit einem ordentlichen Eimer Wasser (für das Pferd) oder einem schönen Glas (für den Menschen) bereits im Kern beheben ließe.

In einem Review aus dem Jahr 2022 (Suda et al. 2022) wurde untersucht, inwieweit sich bestimmte kursierende Aussagen dazu, dass Faszien verkleben und entsprechend Schmerzen im Bewegungsapparat verursachen könnten, überhaupt belegen lassen.

Die medizinische Evidenz für diese speziellen Aussagen seitens einer mittlerweile sehr bekannten

Muskeltraining ist immer auch Faszientraining.
Sonja mit Coronado.

Therapiemarke ist dünn, in vielen Fällen sogar eher nicht existent. Sprich: Stretching löst keine sogenannten verklebten Faszien, zum einen, weil das Stretching selbst keinen Effekt auf die Muskelstrukturen hat und zum anderen, weil Faszien im Grunde nicht verkleben können. Stretching scheint auch rein gar nichts zur Beweglichkeit von Gelenken beizutragen, so haben es Forschende in einem Review ermittelt, der insgesamt 18 Studien und über 549 Teilnehmenden untersuchte (Harvey et al. 2017).

Passives langes Dehnen kann zusätzlich per Neuromodulation kurzzeitig dazu führen, dass die Schutzspannung aufgehoben wird. Nur hat auch das langfristig negative Folgen, so werden im Anschluss an lange Dehnungsphasen sowohl Kraftverlust als auch negative Effekte auf die neuronale Signalgebung vermutet (Trajano et al. 2017).

Stretching hilft also schon mal nicht. Wie sieht es dagegen mit den allseits beliebten Massagen aus?

Auf physiologischer Ebene wirken Massagen durchaus, allerdings mit extrem kurzer Dauer. Crommert et al. haben das schon 2015 mit knapp drei Minuten Wirkdauer beziffert. Danach ist der mechanische Massageeffekt direkt wieder verflogen. Was bleibt, sind physiologische Effekte, insbesondere die Erleichterung, gegen verkrampfte Muskeln und scheinbar verklebte Faszien etwas Handfestes getan zu haben.

Die Faszienstruktur scheint sich zumindest beim Menschen mit den Jahren zu verändern, vor allem im Alter werden dickere und weniger flexible Bindegewebsstrukturen beobachten (Wilke et al. 2018).

Was gesichert sowohl präventiv als auch im Reha-Fall gegen steife Strukturen und einer damit einhergehenden erhöhten Schutzspannung im Muskel hilft, das ist dann wohl die muskelstärkende Versammlungsarbeit.

Versammlung ist Sehnenschutz

Altklassische Dressur (= Versammlung) ist Sehnenschutz, denn: Durch die Versammlung verlängert sich die Standbeinphase. Das Pferd setzt das Bein auf, bleibt länger in Stützfunktion und hebt es dann kontrolliert wieder ab. Und das bedeutet zugleich, dass ein langsamer und kontrollierter Sehnenspannungsaufbau erfolgt. So werden geringere Spitzenbelastungen in den Strukturen erzeugt. Schließlich ist die Bewegung in der Versammlung kontrollierter, bewusster und kinetische Energie fällt geringer aus.

Die Sehnen freuen sich, denn sie lieben vor allem langsame und zyklische Belastungen. Aus der Humanforschung wissen wir, dass sich Sehnen besser aufbauen, wenn sie langsam, regelmäßig und mit zunehmender Spannung belastet werden.

Auch ein federnder Laufstil ist sehnenschonend, da die Sehnen als natürliche Federn fungieren und elastische Energie speichern. Nicht nur das, sie recyceln die Energie noch in der Bewegung, verringern die Belastung am Muskel und erhöhen die Effizienz (Lichwark/Wilson 2005).

Damit das Federn klappt, muss das Pferd aber kürzertreten. Weniger Raumgriff bedeutet auch, dass der Vorderhuf näher am Körperschwerpunkt aufsetzt. Das reduziert die Bremskräfte und den harten Aufprall, der Sehnen und Gelenke schneller schädigt.

In einem Meta-Review konnte das für Sportbegeisterte im Jogging nachgewiesen werden. Höhere Schrittzahlen bei ansonsten gleicher Belastung führten über zahlreiche Studien hinweg zu geringer empfundenen Schmerzen vor allem bei Knie- und Hüftproblemen, die ja in direktem Zusammenhang zu belasteten Sehnen und Bändern der Beine stehen (Anderson et al. 2022).

Schnell, schneller – Sehnenschaden

Was Sehnen so gar nicht lieben? Das wären dann wohl die kurzen und harten Belastungen, wie sie etwa auf der Vorhand im Trab mit hohem Tempo ausfallen. Das ist das grobe Gegenteil von dem, was eine Versammlung vom Bewegungsapparat des Pferdes fordert. In der Versammlung geht es um elastische Federung, kein hartes Abfangen.
Das heißt auch, dass eine gute Passage auf das Sehnenstabilitätskonto einzahlt, die häufig so geschätzte Trabverstärkung bucht tendenziell eher von diesem Konto ab. Pferde mit Sehnenschäden befinden sich da längst schon im Minus und werden trotzdem fleißig weiter mit unsinnigem Gerenne in die (Sehnen-)Schulden getrieben.

Das Kreuz mit dem Läuferknie

Wir starten wieder mit Erkenntnissen aus der Humanforschung. Alle die gern laufen, lesen jetzt mal ganz besonders aufmerksam mit.

Rund 40 Prozent aller Laufverletzungen betreffen die Knie. Das »Läuferknie« ist sogar eine eigene Diagnose (iliotibiales Bandsyndrom, kurz ITBS), mit dem landauf und landab Menschen zur orthopädischen Fachpraxis geschickt werden.

»Schon beim Gehen lastet das etwa eineinhalbfache unseres Körpergewichts auf ihnen, beim Treppensteigen das Vierfache und beim Springen das bis zu Zehnfache. Beim Joggen mit seiner beschleunigten Vorwärtsbewegung kann es bis zum Achtfachen kommen. Für einen 80-Kilo-Mann sind das bei jedem Schritt 640 Kilogramm pro Knie (Himmer/Lohmann 2025).«

Kleine Matheaufgabe zwischendurch: Bei einem durchschnittlich 550 Kilogramm schweren Pferde macht das dann wie viel Belastung pro Kniegelenk beim gemütlichen Abschleudern an der Longe? 550 x 8 = 4.400 Kilogramm. Pro. Knie.

Aber zurück zum menschlichen Läuferknie: Kommt es wiederholt zu stechenden Schmerzen an der Außenseite des Knies, liegt das wahrscheinlich daran, dass der Tractus iliotibialis wiederholt über den äußeren Oberschenkelknochen reibt, was in direkter Folge zu Entzündungen im umliegenden Gewebe führt. Der Tractus iliotibialis ist ein dicker sehniger Strang, der von der Hüfte abwärts in Richtung des äußeren Schienbeinkopfs hin verläuft. Er stabilisiert und verspannt das Knie zusätzlich zu den oft besser bekannten Kreuz- und Außenbändern.

Die konservative Therapie beim Läuferknie? Kühlen, schonen und Entzündungshemmer einnehmen, um die akute Reizung in den Griff zu bekommen. Und dann natürlich gezielter Muskelaufbau!

»Die meisten Menschen denken bei Krafttraining eher an Bizeps und dicke Oberschenkel, aber besonders die kleine Muskulatur um das Knie herum sollte trainiert werden. Das schützt, stützt und stabilisiert die Knie. Gezieltes Krafttraining ist der beste Freund eines jeden Läufers (Himmer/Lohmann 2025).«

Was bedeutet das jetzt übertragen auf das Pferd? Wenn die Muskulatur nicht den passenden Rahmen für den Tractus iliotibialis setzt, dann kann auch das Pferd ein sogenanntes Läuferknie erleben. Mit üblen Konsequenzen bis hin zur Lahmheit. Die wird aber nur zu oft mit mehr Schonung statt mit mehr Krafttraining sozusagen kuriert. Kaum scheint das Knie wieder in Ordnung, beginnt der Kreislauf aus Überlastung und pseudo-knieschonender, aber wenig hilfreichem Standbetrieb.

Langsame kontrollierte Bewegungen wie in der Versammlung als Form des Krafttrainings sind dagegen besser geeignet, um Knie- und auch Ellbogengelenke mit voluminösen Muskelpuffern zu umgeben und Schmerzen in diesen Regionen zu reduzieren. Ein 27 Studien und über 1.000 Personen umfassender Review konnte das (erneut) belegen (Lim et al. 2024). Selbst massive Gelenkschäden lassen sich auf diese Weise relativ schmerzarm überstehen.

Autsch, das Knie hängt!

Die Kniescheibe (Patella) ist beim Pferd über ein wirklich raffiniertes System von Bändern mit dem Ober- und dem Unterschenkel verbunden.

Eine Patellaluxation, auch als Verhakung der Kniescheibe bekannt, ist eine Erkrankung, bei der die Patella nicht korrekt über das Kniegelenk gleitet und sich dort, nun ja: verhakt. Häufig hängt die Kniescheibe im Stand oder beim Übergang zwischen Ruhe und Bewegung. Das kann dann zu einem charakteristischen zombiemäßigen Gangbild führen, bei dem das betroffene Bein gestreckt gehalten wird und sich erst nach einem Ruck wieder normal bewegt.

Ursachen und begünstigende Faktoren für die Patella-Verhakung:

Teils spielt die **Genetik** den Pferden einen Streich. Shetlandponys haben häufiger eine flache Führungsrinne, was das Herausspringen der Kniescheibe durchaus begünstigen kann. Vorbeugendes Training der Hanken und insbesondere des *M. quadriceps* femoris über stärkende Versammlungsübungen sind hier absolut angesagt.

Das gilt auch für Tiere, die eher steil gestellte Hinterbeine mitbringen und entsprechend dieser baulichen Voraussetzungen von Versammlungstraining profitieren werden, um die Knie insgesamt und besonders die Patella zu stabilisieren.

Nach Stürzen oder Verletzungen kann die Patella-Fixation eine Folge des erlebten Traumas sein. Stabilisierung des Gelenks sowie eine insgesamt proteinreiche Ernährung zur Regeneration nach der Verletzung reduzieren die Folgeschäden.

Schwach ausgeprägte Muskulatur, speziell im Bereich der *M. quadriceps femoris* sowie eine im Allgemeinen mangelnde Tragkraft führen ebenfalls dazu, dass sich die Patella verabschiedet.

Junge, unausbalancierte oder Reha-Pferde neigen dazu, in Ruhephasen das Bein zu überstrecken – und schon ist es um die Kniescheibe geschehen.

Guck genau hin: Symptome der Patella-Fixation

Wenn du dir nicht sicher bist, ob sich die Kniescheibe verhakt, dann achte auf die folgenden Details:

1. Das betroffene Bein bleibt gestreckt, es folgt ein kurzer Schritt mit anschließendem normalem Gang.
2. Das betroffene Bein wird manchmal hinterhergezogen oder schleift mit der Zehe über den Boden
3. Das Kniegelenk ist versteift.
4. Der Takt geht verloren, speziell bei Kurven mit dem betroffenen Bein auf der Innenseite.
5. Eine mögliche Folge: Vermeidung von engen Wendungen.

Präventive Muskelstärkung durch Versammlungsarbeit

Perfekte Prävention erfolgt durch eine vorbeugende Stärkung *M. quadriceps*, also dem hinteren Oberschenkelmuskel. So wird der mechanische Halt der Sehne erhöht, die Kniescheibe bleibt an Ort und Stelle und das Risiko der Fixation wird erheblich gesenkt.

Die Studienlage dazu ist (noch) überschaubar.
Aus veterinärmedizinischer Sicht kommen Ross/Dyson allerdings zu dem Schluss, dass intensiver Muskelaufbau die beste Option sei, um eine Patella-Fixation bestmöglich zu vermeiden (Ross/Dyson 2010). Eine Studie an Hunden mit Knieverletzungen von der tiermedizinischen Universität in Liverpool kam zum Schluss, dass ein Training mit bewusstem Einsatz des Körpergewichts inklusive Balanceübungen signifikant bessere Ergebnisse bei der Therapie von Knieverletzungen (insbesondere bei Patella-Beteiligung) erbrachte (Brantberg et al. 2024).

Über Seitengänge in deutlicher Abstellung aktivierst du die Abduktoren (abspreizende Muskeln) sowie den Kniegelenksstrecker. Langsame Sequenzen mit kontrolliert ausgeführten Bewegungen bieten sich für den Start an. Wenn das im Schneckenkriechgang gut gelingt, können die Bewegungen kraftvoller mit Seitwärts im Trab kombiniert werden.

Wichtig: Der Muskelaufbau rund um das betreffende Kniegelenk umfasst das gesamte Bein inklusive der Außenrotatoren! Es benötigt Zeit und regelmäßige Bewegungen, die kontrolliert und unter hoher mechanischer Spannung im Muskel erfolgen sollten.

Zusätzlich sollten die Ernährung und die Nährstoffversorgung stimmen. Es braucht ausreichend Aminosäuren im Futter, um die Muskulatur aufzubauen und um die Sehnen und Bänder zu stärken, die die Kniescheibe künftig besser an Ort und Stelle halten.

GESUNDHEITS-CHECK-IN

Einige Rassen zeigen eine gewisse Anfälligkeit für Herzerkrankungen, vor allem im Alter. Warm- und Vollblüter sind dabei deutlich eher von Herzkrankheiten betroffen als die robusteren Rassen wie Haflinger oder Isländer. Die Größe spielt hier auch eine gewisse Rolle: Kaltblüter und großrahmige Warmblüter verlangen ihrem Herz-Kreislaufsystem allein schon wegen der schieren Größe und Masse ihres Körpers deutlich mehr ab, als das bei einem Pony mit knapp einem Meter Stockmaß der Fall wäre.

Kreislaufschwäche bei Hitze, Koliken bei Wetterwechsel und frühzeitig bemerkte Herznebengeräusche sind erste Anzeichen dafür, eher etwas früher als später aufmerksam über die Herzgesundheit deines Pferdes zu wachen.

Check-in fürs Herz: Kleiner Check im Alltag für Seniorenpferde

1. **Puls und Atmung im Ruhezustand und nach dem Reiten beobachten.**
   Erholt sich dein Tier genauso schnell oder eher langsamer als früher?
2. **Zahnfleischfarbe kontrollieren.**
   Schön rosig bedeutet, dass alles ok ist.
   Blass bis bläulich deutet aber eher auf eine Kreislaufschwäche hin.
3. **Herztöne checken.**
   Beim Routinecheck in der Veterinärpraxis kannst du bei Pferden ab 15 Jahren ruhig mal das Herz belauschen lassen. Herzgeräusche sind in dem Alter gar nicht mal so selten.

Beobachtest du eine gewisse Mäkeligkeit beim Futter, Verdauungsprobleme und leichte Wassereinlagerungen (Ödeme) an Brust oder Bauchunterseite, kann das ebenfalls auf Herzprobleme hinweisen.

Wer sich nicht besonders wohl fühlt, isst meist weniger – das gilt auch fürs Pferd. Bleibt das entstehende Problem mit dem Herzen unerkannt, kann der Appetit aufgrund einer Minderdurchblutung des Verdauungstrakts noch weiter zurückgehen. Die Verdauung wird noch etwas anfälliger für Störungen und dann sind oft auch die Ödeme nicht mehr weit.

Gleiches gilt für eine plötzliche Wetterfühligkeit oder neu auftretende Kreislaufprobleme. Dein alterndes Pferd wirkt plötzlich schlapp? Lass sein Herz durchchecken!

Bei Pferden ab 15 Jahren achtest du besser sorgfältig auf weitere Symptome, falls dir dein Tier trotz versammlungsreichem Bodybuilding und einer passenden Ernährung zwischendurch einfach nicht so richtig fit vorkommt.

Zwischen 18 bis 20 Jahren kannst du die Herzgesundheit immer mit bedenken. Für ein gutes und vor allem gesundheitsförderndes Training gehört der regelmäßige Gesundheitscheck alle ein bis zwei Jahre samt Herz-Check-up da einfach immer dazu.

FÜNF LEKTIONEN FÜR GESUNDE PFERDE

KURZ UND KNAPP:

- Seitwärts (hilft auch bei Arthrose!)
- Piaffe
- Passage
- Spanischer Schritt
- Schulgalopp

Echte Versammlung ist ein Zeichen für Kraft, Tragfähigkeit und koordiniertem Muskeleinsatz. Pferde, die das beherrschen, verfügen über die dazu nötige Muskulatur.

Sie sind waschechte Reitkunstpanzer, die nicht trotz, sondern wegen ihrer voluminösen Muskelberge leichtfüßig durch die Halle tänzeln können. Viel Muskulatur schafft nämlich erst die Möglichkeit, sich stabil und leichtfüßig zu bewegen, wenn entsprechende Synergien aus Muskelkraft und einer optimalen neuronalen Verknüpfung geschaffen werden. Genau das ist das Ziel jeder Dressurarbeit.

Im besten Falle startet die Ausbildung schon mit der Remonte, um den Bewegungsapparat für das spätere Reiten mit Reitgewicht optimal zu stärken. Klassische Reitmeister haben die sogenannte *natürliche Richtung* eines nicht ausgebildeten Pferdes als Ausgangspunkt angesehen.

Sie liegt im Grunde dann vor, wenn du dein Tier am Halfter durch die Halle frei traben lässt. Häufig nehmen die Vierbeiner dabei auch den Kopf tiefer- das ganze sieht entspannt aus, ist aber für die Situation mit Reitgewicht auf dem Rücken ganz und gar nicht optimal. Denn mit den dabei anfallenden langen Schritten ist das keine Versammlung, es ist einfach nur laufen. Es kommt durchaus der Eindruck auf, genau dieses Laufen alleine wäre das Ziel. Zumindest schaut es in den Reithallen landauf, landab häufig so aus.

Dabei können Pferde, sofern ihnen jetzt nicht gerade ein Bein fehlt, ganz prima laufen. Dafür braucht es kein extra Training! Was wir über die Dressur im Sinne eines Bewegungstrainings erreichen wollen und unbedingt auch erreichen sollten, ist etwas ganz anderes. Nämlich, dass Pferde ihre Muskeln nicht nur zum Herumlatschen verwenden, sondern zusätzlich das Skelett und die Gelenke auf spezifische Art und Weise arbeiten lassen.

So wird in der Passage die Körperspannung rund um den Rumpf bewusst und ganz gezielt erhöht. Bei der Piaffe sehen wir eine durchgängige Hankenbeugung, die die Hinterhand formt und Muskulatur sowie Gelenke daraus folgend vor einwirkender Stoßbelastung schützt. Das Reitgewicht verursacht dann weniger Schaden, darum geht es im Kern bei der altklassischen Dressur!

Heutzutage ist Versammlung recht unmodern, deshalb sehen modern gearbeitete und versammlungsarm gerittene Pferde im Vergleich zu versammlungslastig gerittenen Pferden eher wie Antilopen aus.

Mini-Übung für dich:

Stell dir Pferd A und Pferd B vor. Welches willst - und kannst - du viele Jahre lang glücklich reiten?

Pferd A in natürlicher Richtung nutzt seine Muskulatur, um von A nach B zu gelangen. Nur eben mit einem Reitgewicht auf dem Rücken, das langfristig für eine negative Verformung des Pferdekörpers sorgt. Stichwort und Schreckgespenst an dieser Stelle: Trageerschöpfung.

Pferd B geht den Pfad der künstlichen Richtung, was im Sinne der Alten Meister ein versammeltes, sprich ausgebildetes, Pferd bedeutet. So erlernt es eine neue Art des Gleichgewichts und eine Form der Bewegung, die es quasi immun gegen die Belastungen durch das Reitgewicht auf dem Rücken machen.

In der heutigen modernen Reitwelt passt genau das aber natürlich nicht. Da wird nach einer viel zu kurzen Zeit am Boden mehr oder weniger direkt losgeritten.

Das heißt aber nicht, dass die mangelnde Gymnastizierung einfach so stehen bleiben darf. Zu absolut jedem Zeitpunkt im Pferdeleben darf und sollte die Dressurarbeit starten, um Muskulatur aufzubauen und die Gesundheit langfristig zu erhalten. Optimalerweise erlernen Mensch und Pferd die Dressurlektionen dabei anhand von Grundsätzen, die die Gesundheit und damit das Pferdewohl in den Vordergrund stellen. Bauchgefühl allein genügt hier nicht! La Guérinière hat das besonders schön auf den Punkt gebracht:

> *»Jede Wissenschaft und jede Kunst hat ihre Grundsätze und Regeln, die zu neuen Entdeckungen und zur Perfektion führen. Warum sollte die Reiterei die einzige Kunst sein, für die nur Übung erforderlich ist?* (La Guérinière 1733)«

Das Seitwärts züchtet uns die Muskulatur für höhere versammelte Aufgaben (siehe Bild!) heran.

Ob es sich bei dem betreffenden zu gymnastizierenden Tier um ein Reha-Pferd mit Arthrose handelt, um ein in die Jahre gekommenes Freizeitpferd oder eine noch völlig unausgebildete Remonte spielt dabei zunächst keine Rolle. Alters- und leistungsangepasstes Gymnastizieren ist zu jedem Zeitpunkt nicht nur möglich, sondern auch unbedingt zu empfehlen!

Meine Erfahrung nach sind die folgenden fünf Lektionen ideal, um angemessen voluminöse Muskeln aufzubauen und damit effektiv die Sehnen, Bänder und Gelenke für die weitere Beanspruchung zu stärken.

Lektionen, die jedes Pferd lernen kann **und** was es lernen sollte

Jedes Pferd kann im Grunde die nachfolgenden Lektionen erlernen – wir haben ja bereits gelesen, dass es nicht die eine genetische Disposition gibt, die ein Vierbeiner für Dressur als geeignet darstellt:

1. Seitwärts
2. Piaffe
3. Rumpfhebung
4. Passage
5. Spanischer Schritt
6. Galopp nach Art der Alten Meister (Schulgalopp!)

Mit Geduld und Wissen lässt sich aus jedem Pferdetyp eine angemessene Rumpfhebung, eine Piaffe oder Passage herauskitzeln.

Was ganz sicher jedes Reitpferd erlernen sollte, das sind Seitengänge inklusive einer sauber erarbeiteten Rumpfhebung.

Dabei arbeiten wir uns in jeder Lektion langsam vom Schritt bis in die nächsten Gangarten vor. Es geht dabei nicht darum, möglichst zügig in die nächste Gangart zu wechseln, sondern darum, jede Lektion mit angemessener Perfektion

Rumpfhebung konnte nur in der Versammlung gemessen werden. Studien zeigen, dass der Rumpf in versammelten Bewegungen wie Passage oder Piaffe aktiv angehoben wird. In schwungvollen Gängen bleibt er dagegen meist auf seinem unnatürlichen Tiefpunkt.

zu meistern. Eine wirklich hervorragend ausgeführte Traverse im Trab muss nicht durch eine eher mäßig gelungene Galopp-Traverse ersetzt werden.

Beim Gymnastizieren ist das Tempo der Killer, wie es so schön heißt – und das wussten auch die Alten Meister nur zu gut, wie dieses Zitat von La Guérinière beweist:

> *»Ein von allen Meistern aufgestellte Regel besagt: Man soll nie den Galopp fordern, bevor das Pferd im Trabe so geschmeidig gemacht ist, daß es den Galopp von selbst anbietet, ohne sich auf die Hand zu stützen. Man soll also abwarten, bis das Schulterherein gut ausgebildet, der Gehorsam auf beide Schenkel in der Passage mit der Kruppe an die Wand vorhanden ist und die Piaffe in den Pilaren leicht ausgeführt wird. Das ganze Pferd muß geschmeidig und gehorsam sein. Erst nach Erreichen dieses hohen Ausbildungsgrades wird es auf leichte Hilfen willig galoppieren (La Guérinière 1733).«*

Denn: Je höher das Tempo, desto größer die kinetische Energie und desto höher die Belastung auf die in der Bewegung eingebundenen Strukturen wie Muskeln und Sehnen. Wie im Abschnitt zur kinetischen Energie schon zu lesen war, wirkt sich die (steigende) Geschwindigkeit massiv auf die Energieentwicklung aus.

Gelenkschonende Arbeit am Pferd ist – zunächst – möglichst langsam und erfolgt besser als kontrollierte Bewegungsfolge mit maximaler mechanischer Spannung im Muskel, um die Muskulatur zum Wachstum anzuregen.

Apropos Wachstum: Der Botenstoff brain-derived neurotrophic factor (BDNF) aus der Gruppe der Myokine hilft dabei, neuromuskuläre Synapsen umzuformen, wenn regelmäßig Übungen aus dem Krafttraining angewandt werden. So werden die leistungsstarken Muskeln für explosive kraftvolle Bewegungen immer mehr, während Muskelfasern für den Ausdauersportbereich abgebaut beziehungsweise umgewandelt werden (Delezie et al. 2019).

Die Basis: Seitwärts

Auf den Punkt: Seitengänge trainieren die abduzierenden (abspreizenden) und außenrotierenden Muskeln der Hinterhand, insbesondere die mittlere Gesäßmuskulatur und die tiefliegende Beckenmuskulatur. Sind diese Muskelgruppen aufgrund von Seitengängen gut entwickelt, lässt sich ein nach innen rotierendes Standbein mit eingedrehter Zehe kurz vor dem Abfußen präventiv vermeiden.

Eine Stärkung dieser Muskulaturbereiche ist auch nötig, um eine tiefe Hankenbeugung zu erzeugen und zu erreichen, wodurch die Lastaufnahme verbessert und einer Trageerschöpfung vorgebeugt wird. Die erste kleine Hankenbeugung zeigen Pferde bereits beim Seitwärts, was im Rahmen der Altklassik logischerweise absolut erwünscht ist. Zusätzlich sind Kraftübungen wie Seitengänge perfekt, um das Muskelwachstum anzuregen. Das Training allein wirkt sich schon auf die Skelettmuskulatur aus und zwar selbst auf kleinster biochemischer Ebene:

> *»A single bout of exercise induces changes in gene expression in skeletal muscle. Regular exercise results in an adaptive response involving changes in muscle architecture and biochemistry, and is an effective way to manage and prevent common human diseases such as obesity, cardiovascular disorders and type II diabetes (Bryan et al. 2017).«*

Auf Deutsch: »Eine einzige sportliche Betätigung führt zu Veränderungen der Genexpression in der Skelettmuskulatur. Regelmäßiges Training führt zu einer Anpassungsreaktion, die Veränderungen in der Muskelarchitektur und der Biochemie mit sich bringt, und ist ein wirksames Mittel zur

Start ins Seitwärts mit beginnender Rumpfhebung.
Mascha mit Bella.

Bewältigung und Vorbeugung von Volkskrankheiten wie Fettleibigkeit, Herz-Kreislauf-Erkrankungen und Diabetes Typ II.«

Blickschulung für gelungene Seitengänge

- Gelungene Seitengänge zeichnen sich durch diese Punkte aus:
- Das Pferd bewegt sich vorwärts und seitwärts.
- Die Lastaufnahme erfolgt über die leicht gebeugten Hinterbeine.
- Die Dampfrinne und/oder herzchenförmige Kruppe werden sichtbar.
- Die Abstellung beträgt zwischen 30 Grad bis 80 Grad, gerne auch 85 Grad.

Bei für das individuelle Pferd angemessener Ausführung, sprich einer ordentlichen Beanspruchung, lässt sich die sogenannte Dampfrinne (Linie am Bauch) und/oder eine herzchenförmige Kruppe beobachten. Bei geringer ausgeprägter Kruppenmuskulatur fällt die Fläche zwischen den Hüfthöckern tendenziell eher flach aus, vergleichbar mit einer Tischplatte.

Wie genau das Seitwärtsgehen in seinen Varianten möglichst formvollendet dem Pferd nahegebracht werden kann, das haben viele großartige Reitmeister umfassend beschrieben.

Da es mir in diesem Buch vorrangig um die präventive und therapeutische Ausübung der Dressurkunst geht, lassen wir eine Schritt-für-Schritt-Anleitung hier aus. In meinen Online-Kursen bespreche ich das Seitwärts sowie alle noch folgenden Lektionen

Video

Hier siehst du in einer kurzen Video-Sequenz, wie ein optimales Seitwärts aussehen kann. Fallen dir die genannten Punkte auf?

ausgiebig – ein Blick in die Videoformate lohnt sich also. In diesem Buch geht es mir vor allem um konkrete Anwendungsfälle für die Arbeit mit Senioren, Reha-Fällen und wie auch noch-nicht-Reha-Fälle davon profitieren werden.

Warum Pferde seitwärts gehen sollten

In der modernen Dressurwelt sind Seitengänge durchaus umstritten. Die einen meinen, das wäre nur etwas Tierquälerei, für die anderen sind das Lektionen auf so hohem Niveau, dass es für normale Freizeitpferde schlicht nicht lohne, sich damit zu befassen. Wieder andere, und dazu zähle ich auch mich, schwören auf das Trainingspotenzial der Seitengänge in ihrer Vielfalt.

Das Seitwärts und die damit einhergehende diagonale Hilfengebung ist der rote Faden in der Ausbildung.

Das sah schon der Reitmeister La Guérinière so und drückte das in einem Satz aus, der auch heute noch aufgrund seiner grundlegenden Wahrheit gilt: »Schulterherein ist die erste und die letzte Schule, in der man ein Pferd unterrichten muss (La Guérinière).

Der Reitmeister Steinbrecht, der tendenziell eine eher etwas anglomane Einstellung zur Reitpferdeausbildung einnahm, hat zu dem Thema auch so seine Meinung:

> *»Wenn ich nunmehr dieses wichtige Kapitel schließe, so geschieht es mit der ernsten Mahnung an jeden Reiter, das Studium des richtigen Schulterherein unermüdlich zu betreiben und es als Hauptstütze der ganzen Dressur zu betrachten, durch die er aus dem Pferde nicht nur alles gewinnen und entwickeln kann, was die Natur ihm mitgegeben hat, sondern auch Fehler in Haltung und Gang, die sich durch falsche Arbeit eingeschlichen haben, gründlich zu beseitigen vermag. Reiter, die die wunderbare Macht dieser Lektion vollkommen erkannt haben, mühen sich nicht ab, ihre Pferde durch dauernden Wechsel der verschiedensten Stellungen biegsam und gehorsam zu machen, denn sie wissen, dass aus dem richtigen Schulterherein alle Schulen hervorgehen, deren sie zur Erreichung des vorgestreckten Ziels bedürfen, sei dies die wohlbegründete Richtung ins Gleichgewicht, sei es die auf die Hanke.* (Steinbrecht 1998).«

Beide sind sich da also recht einig in einer positiven Bewertung der Seitengänge, die die absolute Basis in der Reitpferdearbeit bilden. Das Pferd lernt hier immerhin, auf Kommando seine Beine zu verschränken und sich durch die Halle oder im Gelände seitwärts zu bewegen!

Racinet führt in seinem Werk *Auf dem falschen Fuß* das hübsche Beispiel einer in seinem Besitz befindlichen Quarter Horse Stute an, die ganz hervorragend im versammelten Galopp im 90 °-Grad-Winkel seitwärts neben einem Traktor herlief, sodass er sich dort bequem unterhalten konnte. Also wenn das keine Gewandtheit ist, was dann?

Seitwärts ist perfekt für eine optimale **Balance**. Dr. Hilary Clayton hat das, insbesondere in Bezug auf Versammlung ausführlich erforscht. Sie konnte anhand von Bewegungsanalysen belegen, dass Versammlung in der Tat den Schwerpunkt des Pferdes nach hinten verlagert, was sich positiv auf die Tragfähigkeit des versammelten Pferdes auswirkt (Clayton 2003).

Seitwärts schafft zudem die Grundlage, um ein besseres Gespür für das Gleichgewicht zwischen der Schenkelinnenseite (medial) und der Schenkelaußenseite (lateral) des Körpers zu entwickeln.

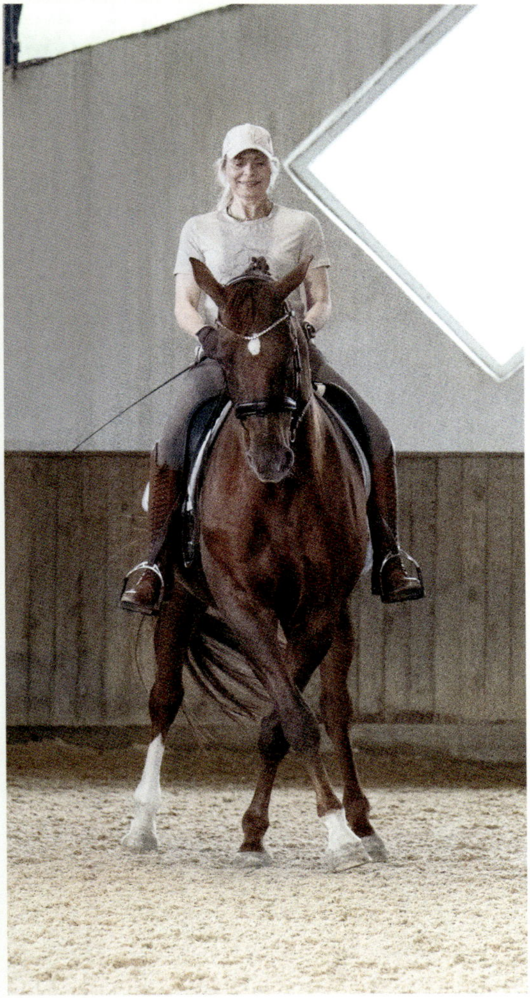

Seitwärts. Christina mit Emma

Steinbrecht sieht im Seitwärts ein optimales Mittel, um Schub- und Tragkraft über die Abstellungsgrade auszubalancieren:

> *»Miteinander in Harmonie gebracht, regeln sie dann das Verhältnis von Schub- und Tragkraft der Hinterhand, wozu der Grad der Seitenstellung die beste Handhabe ist. So oft das Übergewicht der Schubkraft der Belastung der Hinterhand entgegenwirkt, muss der Gang durch vorübergehend verstärkte Seitenstellung gemäßigt werden (Steinbrecht).«*

Mini-Übung für Seitwärts für dich:

Leg dich auf die Seite. Das oben liegende Bein wird lang ausgestreckt zur Seite und damit lateral angehoben. Die Abduktoren und die Außenrotatoren der Hüfte aktivieren sich. Dadurch wird die Frontalebene im Einbeinstand stabilisiert, wobei sich der Einbeinstand mit der Standbeinphase beim Pferd gleichsetzen lässt. Der Zweck der Übung ist zum einen, dass du deine Abduktoren und die Außenrotation am eigenen Körper spürst und die Biomechanik dahinter dann besser auf dein Pferd übertragen kannst. Zum anderen ist es einfach eine gute Übung für deine untere Körperhälfte.

Mini-Übung fürs Pferd:

Seitengänge in erkennbarer Abstellung, gern in mehrfacher Wiederholung am Stück mit anschließender ausreichender Pause. Kurze Sets mit mehrfachen Wiederholungen reichen anfangs aus. Achte auf eine langsame und kontrollierte Ausführung. Insbesondere im Seitwärts dürfen wir nicht eilen, weil dann mehr kinetische Energie entsteht. In der modernen Variante unter vollem Tempo geritten, fallen Seitengänge bei nicht ausreichend aufgebautem Muskelpuffer negativ auf. In Versammlung dagegen wird nur wenig kinetische Energie auftreten und die positiven Aspekte des Seitwärts können voll ausgeschöpft werden.

Eine optimale Balance und das sichere Körpergefühl kommen *nicht* zustande, wenn

- die Abduktoren (abspreizende Muskeln) und Außenrotatoren vergleichsweise schwach sind,
- während gleichzeitig die Adduktoren (heranziehende Muskeln) und die Innenrotatoren zu viel

Schutzspannung wegen einer bestehenden Kompensation aufbauen. Das Geritten-werden sorgt hier dafür, dass diese Kompensation wahrscheinlicher auftritt - wie gesagt, sofern die Muskulatur entsprechend weniger stark ausgeprägt ist.

Statt der erhofften Geraderichtung und Balance entsteht bei dieser Kombination aus einer muskulär schwachen und zugleich unter Kompensationsspannung stehenden Hinterschenkelmuskulatur eine Rotation des Pferdebeins gegen Ende der Standbeinphase (Hufsohle befindet sich am Boden). Das schafft einen leichten Zug zur Körpermitte hin und führt damit zu Instabilität und einem erhöhten muskulären Korrekturaufwand.

Simples im Mittelschritt gehen sieht bei diesem Pferd dann nicht so aus, wie es sollte. In den anstrengenderen Gangarten kann man das erst später erkennen.

Nicht sicher, ob eine unerwünschte Rotation in der Standbeinphase bei deinem Tier der Fall ist? Dann achte auf folgende Anzeichen:

- Die Hufzehe (vorderster Punkt des Hufs) dreht sich im Laufe der Standbeinphase nach innen, obwohl das Pferd geradeaus im Mittelschritt geht. Das zeigt sich besonders deutlich bei viel Raumgriff.
- Die Tragkraft ist verringert, insbesondere betroffen ist dabei die diagonale Tragkraft.

Es scheint kontra intuitiv zu sein, ausgerechnet eine Lektion mit leichter Beinrotation dazu zu nutzen, die nicht erwünschte Innenbeinrotation in den Griff zu bekommen, aber gerade auf Grund der Stärkung der Abduktoren und der Außenrotation ist diese Übung perfekt als Korrekturmaßnahme!

Hinten runter, vorn hoch – und auf

Seitengänge sind ideale Hilfsmittel dafür, eine bessere Genickbeugung zu erreichen. Das betrifft vor allem die Pferde, die wie Sternengucker durch die Halle laufen und kaum von sich aus in die Genickbeugung kommen. Und das, obwohl bei Seitengängen eine etwas höhere Aufrichtung, meist allerdings aufgrund von Handeinwirkung, der Fall ist.

Warum kommt es dann trotzdem dazu, dass sich Pferde über das Seitwärts stärker im Genick beugen? **Seitengänge aktivieren die ventrale Muskelkette.** Das wirkt sich auch auf das Zungenbein und das Genick aus.

Denn über die Beugung des *M. Iliopsoas* (Hüftbeuger) und seiner Einwirkung auf die Lendenwirbelsäule und die Hintergliedmaßen hinweg erstreckt sich die Aktivierung über den *M. Rectus abdominis* (gerader Bauchmuskel) bis hoch zum *M. Longus colli* und *M. Scalenus* (tiefe Halsmuskeln), die sich auf die Zungenbeinmuskulatur auswirken. Letztere verbindet Kehlkopf, Zunge, Genick und Brustbein miteinander. Die Wirbelsäule beugt sich (mehr), was die Ausrichtung des relativ schweren Kopfes unterstützt und die Position des Genicks in versammelter Bewegung erleichtert.

Das in Versammlung gut beobachtbare typische rhythmische Kauen im Takt der Hinterbeine ist ein Ausdruck dieser Tiefenaktivierung der ventralen Muskelkette.

Kauen und Schlecken sind zugleich das Zeichen für die Aktivierung des Vagusnervs und somit auch Nachweise für einen balancierten Zustand des Nervensystems.

Seitwärts in der Bodenarbeit. Anna mit Floro.

Jedes Pferd hat seine guten und schlechteren Tage. Manchmal gelingen die ersten Seitwärts-Sets nicht besonders gut. Manchmal gelingen die ersten Seitwärtsintervalle beinahe lehrbuchmäßig - oder aus meinen Online-Kursen - und sind damit schon nahezu perfekt dargestellt! Ich schreibe bewusst nahezu perfekt, denn in meinen Kursen zeige ich natürlich immer wieder auch Pferde, die die 100-prozentige Perfektion noch nicht erreicht haben.

Perfektion ist aber leider vergänglich, dranbleiben ist angesagt!

Seitwärts steigert auch die **Koordination**, denn über die Seitengänge wird das Kleinhirn aktiviert wird. Das ist sowohl für das motorische Lernen als auch für die Kontrolle aller Bewegungen zuständig. In Kombination mit Balance-Übungen lässt sich vor allem für ältere und gebrechliche Menschen (gilt auch für Pferde) ein Gewinn an Lebensqualität durch eine verbesserte Mobilität und eine höhere Bewegungssicherheit verzeichnen (Dunsky 2019).

Fehlersuche: Pferd streift im Seitwärts

Wenn das Pferd stärker im Seitwärts permanent mit den Karpalgelenken aneinander streift, dann lassen wir an dieser Stelle noch mal Steinbrecht zu Wort kommen. Er meint dazu nämlich:

> *»Anstatt Fehler dieser Art durch Verminderung der Seitenrichtung und Verstärkung der Versammlung abzustellen, werden sie oft noch vermehrt, weil der Reiter glaubt, die übertretenden Beine durch stärkeres Seitwärtstreiben*

anregen zu müssen. Wie sollen diese aber räumig genug treten, wenn die zu überschreitenden Beine nicht durch Belastung genügend verkürzt gehalten werden? Reiner Gang und reines Übertreten müssen daher stets bei dieser Arbeit als Leitfaden dienen, nämlich bestimmen, wie viel der Reiter die Vorhand vorzurichten und die zu überschreitenden Beine in ihrer Seitwärtsbewegung zu beschränken hat. Im Schulterherein, bei dem die inneren Beine über die äußere treten, der innere Schenkel die Rippen biegt und gleichzeitig seitwärts treibt, hängt die Richtigkeit der Biegung und des Ganges von der Beherrschung des äußeren Hinterbeins ab (Steinbrecht).«

Steinbrecht sagt also, dass die inneren Beine über die äußeren Beine kreuzen sollen, möglichst langsam und mit kürzeren Tritten. Auf diese Weise lässt sich das Aneinanderschlagen verringern. Im langsamen Tempo ist dieses Aneinanderstreifen aber nicht grundsätzlich problematisch – schwierig im Sinne von gelenkbelastend wird es erst im rasanten Tempo, wenn die kinetische Energie zunimmt. Je mehr das Pferd die passende Muskulatur erarbeitet, hier unter anderem der Trapezmuskel, desto weniger werden sich die Karpalgelenke berühren.

Die Trapezmuskeln sowie insbesondere der Serratus (Schultermuskel/Hauptrumpfmuskel) ermöglichen das Anheben und Vorführen der Vordergliedmaßen im Seitwärts, indem sie das überkreuzgehende Vorderbein unter Gewichtsbelastung abspreizen, also das Bein weg vom Körper führen.

Sind diese Muskeln nicht oder besser gesagt *noch* nicht gut ausgebildet, erfordert das saubere Anheben der Vorderbeinchen durchaus Kraft, vor allem bei schmalen Brustmuskelaufbau und eher zarten und langen Spinnenbeinen. Hier sind wenige in kontrollierter Bewegung und deshalb sauber ausgeführte Seitengänge in kurzen Reprisen sinnvoll, um zunächst die Trapezmuskeln zu stärken.

Zu Steinbrechts Zeiten waren die typischerweise verfügbaren Pferde tendenziell eher breiter gebaut, während heutige moderne Züchtungen schmalbrüstiger und zugleich langbeiniger erscheinen. Diese züchtungsbedingte Gebäudeveränderung sorgt natürlich auch dafür, dass sich heute Gangprobleme zeigen, die die Alten Meister inklusive eines Steinbrecht nicht in einer derartigen Häufigkeit beobachten mussten.

Schadet Seitwärts dem Arthrose-Pferd?

Auch wenn dein Vierbeiner Arthrose in den Gliedmaßen oder in der Halswirbelsäule hat, solltest du unbedingt Seitengänge und Versammlungstraining im Allgemeinen setzen. Denn egal welche Arthrose dein Pferd plagt, viel Muskulatur ist eben immer entzündungshemmend und schützt die Gelenke.

Selbst dann, wenn das Gelenk im Röntgenbild ausschaut wie ein Steinbruch, kann der Gelenksinhaber völlig schmerzfrei sein. Die Schmerzreaktion hängt vor allem davon ab, wie die umliegende Muskulatur aufgebaut ist und wie dick der Gelenkknorpel ausfällt. In einer Studie mit 600 Personen waren die Schmerzlevel immer dann höher, wenn auch der Gelenkknorpel im Knie geringer ausfiel (Bacon et al. 2020).

Krafttraining hält die Belastung der Gelenke niedrig, während gleichzeitig die Muskelkraft zur Stützung dieser besonders empfindlichen Region zunimmt. Aus der Humanforschung ist bekannt, dass kombinierte Trainingseinheiten aus Kraft-Ausdauer mit Überkreuz-Bewegungen insgesamt die Mobilität erhöhen, Schmerzen durch Muskelaufbau senken und die Beweglichkeit steigern (Tanaka et al. 2013).

Ist Seitwärts bei einem schiefen Pferd sinnvoll?

Die Schiefe allein ist im Grunde völlig natürlich. Sie gibt dir auch recht gute Hinweise darauf, welche Vordergliedmaße bei heutiger moderner Reitweise üblicherweise zu Problemen führen wird. Beim linksdominanten Pferd wäre die vordere linke Gliedmaße tendenziell eher gefährdet, Schäden durch Fehlbelastungen zu entwickeln.

Die natürliche Schiefe wird erst beim Reitpferd mit dem nicht so natürlichen menschlichen Gewicht auf dem Rücken problematisch. Über die Geraderichtung lässt sich die Rotation der Brustwirbelsäule korrigieren, um die Belastung der Vordergliedmaßen symmetrischer zu machen. Wie viel oder wenig Stellung nötig ist, hängt vom individuellen Pferd, seiner Schiefe und deinem reiterlichen Gefühl ab und lässt sich pauschal nicht sagen.

Wichtig ist, dass die Geraderichtung insbesondere bei wenig bemuskelten Pferden möglichst frühzeitig geschieht, um Verschleißschäden durch einseitige Belastungen am Bewegungsapparat zu vermeiden oder, wenn bereits aufgetreten, um sie möglichst zügig zu reduzieren.

Tipp:

Beim markant linksdominanten Pferd wäre es überaus sinnvoll, sich das Seitwärts zunächst von links nach rechts zu erarbeiten, denn dann bringe ich das linke Vorderbein in eine längere Hangbeinphase. Seitengänge generell sind dazu gedacht, die Hangbeinphase auf der gewählten Gliedmaße zu verlängern und auf diese Weise auch zunehmend mehr Kontrolle über die Hangbeinphase zu gewinnen.

Dadurch allein wird das Pferd symmetrischer und zwar immer, nicht nur in den Seitengängen. Das kannst du vom Sattel aus millimetergenau orchestrieren, indem du die Brustwirbelsäulenrotation beobachtest und den Widerrist exakt mittig einstellst. Dafür braucht es zwingend einen Sattel, der nicht verrutscht ist, sonst kannst du de facto tun was du willst, das Pferd bleibt schief und die einseitig stärker ausfallende Rotation seiner Brustwirbelsäule bleibt bestehen.

Vor allem bei Pferden mit kleinem Widerrist muss beachtet werden, dass der Sattel wirklich mittig aufliegt. Bei einem Pferd mit einem Rumpf wie ein Fass verrutscht der Sattel schon mal, weil der Widerrist in der Muskel- oder Fettmasse regelrecht verschwindet. Ein balancierter Sitz ist hier nötig, das kann auch zu einer recht großen Herausforderung werden.

Beim Reiten der Seitengänge ist das langfristige Ziel, dass

- der Sattel so mittig wie möglich liegt,
- der Widerrist so mittig wie möglich (senkrecht zur Hallendecke) bleibt,
- die Rotation nach links und nach rechts symmetrisch ausfällt.

Beim versammelten Pferd in Rumpfhebung ist die Rotation insgesamt verringert, das unversammelte Pferd zeigt eine wesentlich stärkere Rotation.

Verursacht Seitwärts ein engstehendes Hinterbein?

Ganz im Gegenteil! Das Seitwärtsgehen sorgt für eine korrekteren Stand. Die Beinachsen werden sogar durch das Seitwärts korrigiert oder besser, die vermehrte Muskelmasse sorgt dafür, dass die Beinachsen optimaler an der Senkrechten ausgerichtet sind.

Die Pferde stehen dann tendenziell etwas breiter da, weil der Rumpfheber in seinen Ausläufern zwischen Rippenkasten und Vordergliedmaßen gelagert ist. Gleiches gilt für die Muskulatur der Hintergliedmaßen, die sich bei Volumenzuwachs in einem breiteren Stand bemerkbar machen.

Wenn die Abduktoren zu schwach sind, werden die Pferde hinten tendenziell schmäler – fehlendes Training und fehlende Muskulatur sorgen also für ein engstehendes Hinterbein. Es sind eben nicht die Seitengänge, die ja meist erst dann in Betracht gezogen werden, wenn das Tierchen schon ein paar Jahre modernes sogenanntes Reitsporttraining mit oftmaliger Tendenz zu atrophierten Muskeln hinter sich hat.

Manche Pferde zeigen das Verhalten, dass sie unversammelt große lange Schritte (Mittelschritt) machen. Dabei lastet das Körpergewicht in der Standbeinphase noch voll auf dem Standbein, gleichzeitig rotiert es aber nach innen. Das heißt, die Trachten zeigt nach außen, die Zehe zeigt am Ende der Standbeinphase nach innen.

Verursacht wird das durch nicht ausreichend gekräftigte Muskeln im Hinterbein, insbesondere der *M. quadriceps* und der Muskel der Oberschenkelaußenseite (Tensor facscia latae, kurz TFL), unter Beteiligung von Hüft- und Kniegelenk.

Fehlt es an Beweglichkeit, verursacht durch Schutzspannungen oder sind beteiligte Muskelgruppen durch Verspannungen und Schmerzen beeinträchtigt, dann entsteht genau dieses Rotationsgeschehen:

> *»Eine unzureichende Hüftbeweglichkeit in der Rotation der Transversalebene führt bei einem Schritt mit dem rechten Fuß nach vorn dazu, den Femur des linken Beines zusammen mit dem Becken in eine Drehung nach links und außen zu führen. Der Mangel an Rotation wird nun von einem der nächstgelegenen Gelenkpartner kompensiert und resultiert in diesem Falle in einer starken Innenrotation im Kniegelenk, sofern der linke Fuß sich noch am Boden befindet* (Herzog 2022).«

Eine Stärkung der Abduktoren und der Außenrotatoren sorgt also genau genommen dafür, dass wir die Innenrotation reduzieren und auf diese Weise auch gleich noch die beteiligten Gelenke dauerhaft durch eine stärkere Muskelbeteiligung entlasten.

Arthrose beim Pferd – hilft da Seitwärts?

Die gängige Lehrmeinung ist ganz klar: Abgenutzter Gelenksknorpel verursacht Entzündungen, die wiederum zu Schmerzen führen. Am Ende wird dann schmerzhafte und körperlich die Leistungsfähigkeit einschränkende Arthrose diagnostiziert.

Tatsächlich ist der mechanische Ablauf andersherum: Die Entzündungen treten zuerst auf und beschädigen den Gelenksknorpel. Dieses Entzündungsgeschehen führt zum weiteren Abbau des Knorpels, der dann weitere Reizungen begünstigt. Denn was an dieser Stelle in der Regel fehlt, das ist schützende Muskulatur außen herum!

Klassische Therapie beinhalten vor allem Schonung in Kombination mit entzündungshemmenden Medikamenten und angepasster Bewegung. Letztere besteht aus vermeintlich gelenkschonendem Spazierengehen, gern auch kombiniert mit lockerem Abschleudern an der Longe.

Was dagegen wirklich hilft? Muskulaturstärkendes Krafttraining, idealerweise über Kreuz ausgeführt, um Adduktoren und Abduktoren gleichermaßen zu kräftigen und damit die Versorgung des Gelenkknorpels wieder anzuregen.

Denn über geeignete Trainingsmaßnahmen wird die Durchblutung in eben dieser umgebenden Muskulatur gestärkt, was wiederum in Form einer verbesserten Nährstoffversorgung des gesamten Muskel-Gelenk-Komplexes zu einer idealerweise optimal versorgten Gelenkschmiere (Synovialflüssigkeit) führt.

Damit das Arthrosegelenk möglichst zurück auf »null« gestellt wird, müssen spezialisierte Immunzellen (Makrophagen) zum Zuge kommen. Während M1-Makrophagen vor allem für entzündungsfördernde Zustände verantwortlich sind, sind M2-Makrophagen für Aufbau und Regeneration zuständig.

Seitwärts und Über-Kreuz-Bewegungen für gesunde Gelenke

Reha-Fälle profitieren insbesondere von langsamen, kontrolliert ausgeführten Bewegungen im seitwärts. Aus der Humanforschung ist bekannt, dass selbst bei bewusst einseitigem Krafttraining beide Beine davon profitieren (Bowen et al. 2022).

Die biomechanischen Vorgänge bei Seitengängen ähnelt einem Ganzkörperkreuztraining. In einer randomisierten kontrollierten Studie mit 137 älteren Arthrose-Patienten führte ein kombiniertes Cross-Training-Programm zu signifikanten Verbesserungen bei Schmerz, Gehfähigkeit, Flexibilität und Muskelkraft im Vergleich zur Kontrollgruppe (Péloquin et al.1999).

Für Arthrose-Pferde lässt sich das übertragen: kräftigende Muskelarbeit über Seitwärtsbewegungen in geschwindigkeitsarmen Einheiten mit kontrollierten bewussten Bewegungen stärkt langfristig sowohl das betroffene Gelenk als auch die umgebenden Muskelgruppen (Hodges/Daneels 2019).
Die biochemischen Abläufe hinter diesem Muskelzuwachs regulieren zugleich die Schmerzen (Knutzen et al. 2007), was die Kooperationsbereitschaft seitens des Pferdes deutlich erhöhen kann. Bis es soweit ist, bist du als Animateur gefragt, um dein Tierchen in die hilfreiche, anfangs aber vielleicht noch unangenehme Bewegung hinein zu lotsen.

Hankenbeugung als Zwischenschritt zur Rumpfhebung

Kniebeugen beim Menschen sind analog zur Hankenbeugung beim Pferd – und bei beiden kann das Knie- und Beinkrafttraining für hübsch ausgeformte *M. quadriceps* (Muskulatur der Oberschenkelvorderseite) sorgen. Der Quadriceps ist zunächst als Hintergliedmaßenvorführer und als einziger aktiver Kniegelenksstrecker beim Pferd tätig, er zählt zu den mit am stärksten ausgebildeten Muskelgruppen im Körper. Bei der Seitwärtsbewegung des Beins ist er indirekt in abspreizender Funktion beteiligt

Auch bei unseren Vierbeinern kann sich dieser Bereich zu einer voluminösen Form entwickeln, die es dem Pferd mittels Hinterhand vor allem bei versammelter Arbeit ermöglicht, präzise und kontrolliert eindrucksvolle Bewegungen zu zeigen.

Gleichzeitig schützen die Muskelgruppen rund ums Knie ebenjenes, die Patella (Kniescheibe) wird über den Quadriceps und den komplexen Bandapparat stabilisiert. Das geschieht bei der Beugung des Kniegelenks:

1. Der Quadriceps verkürzt sich zunächst,
2. er bleibt unter Spannung, damit die beugende Muskulatur die Beugung der Hanke durchführen kann (konzentrische Arbeit),
3. damit die Kniescheibe (Patella) nicht einfach ohne Halt wegknickt.

So verhindert der aktivierte *M. Quadriceps femoris* das unkontrollierte Einknicken des betreffenden Beins. Verlängert sich der Quadriceps unter Spannung (exzentrisch), kontrolliert er zugleich die Beugung des Kniegelenks.

Mini-Übung für dich:

Setz dich langsam in die Hocke. Beobachte dabei deine Oberschenkel. Der Muskel dehnt sich, aber er arbeitet auch spürbar, denn sonst würdest du einfach auf den Boden plumpsen! Kinetische Energie wird hier kurzfristig in den Strukturen gehalten und beim Aufrichten wieder abgegeben.

Über die Hankenbeugung, an der der Quadriceps, wie wir gesehen haben, extrem beteiligt ist, wird auch die ventrale Muskelkette aktiviert. Schon setzen daraufhin all die Nebenprozesse ein, die in der Versammlung gewollt sind: Ein im richtigen Maß gebeugter Hals mit dem knöchernen Genick als obersten Punkt, ein bewegliches und nachgiebiges Maul, ein aktivierter Schultergürtel und eine

ebenso stark beteiligte Rumpfmuskulatur basieren auf einer aktivierten ventralen Muskelkette.

Umso besser die Hankenbeugung gelingt, desto wahrscheinlicher ist es, dass sich eine Rumpfhebung anschließen lässt. In der Passage wird über die Hankenbeugung ausreichend Spannung in der Rumpfmuskulatur erzeugt, um den Rumpf insgesamt anzuheben.

Doppelte Progression (Hankenbeugung)

Die doppelte Progression ist ein Prinzip aus dem Krafttraining, das du wunderbar zum Training der Hinterhand nutzen kannst.

Dein Leitsatz dabei: Wir steigern erst die Wiederholungen, dann die Last.

So gehst du vor:

1. Phase: Qualität und Wiederholbarkeit
Erarbeite die Hankenbeugung vorzugsweise an der Hand in Entlastung. Nur dein Pferd und du in der Bahn oder draußen auf einem geeigneten Platz. Kein Sattel, kein Reitgewicht!

Euer Ziel besteht hier darin, dass dein Vierbeiner mehrere Wiederholungen kontrolliert und kraftvoll zeigen kann. Das gelingt am besten über mehrere Piaffe-Schritte, einige längere Seitwärtsreprisen oder im fortgeschrittenen Stadium über extreme Lastaufnahme mit der Hinterhand durch Schulgalopp oder Mezair. Viele Pferde meistern den Schulgalopp am besten, wenn sie dabei seitwärts gearbeitet werden. Ein Schulgalopp nach vorne folgt meist erst im Anschluss an die Variante im seitwärts.

2. Phase: Steigerung der Belastung
Die Bewegungen sind reproduzierbar, stabil und harmonisch? Wunderbar, dann kannst du die gleiche Qualität der Übung auch vom Sattel aus

Studien beweisen: Rumpfhebung ist in Versammlung messbar (und sieht schön aus!).

Kontrahierter Quadriceps im Sonnenschein.
Mascha mit Bella.

einfordern. Dein Reitgewicht wirkt als Zusatzgewicht, ganz wie eine (oder mehrere) extra Gewichte im Fitnessstudio eingesetzt würden. So erhöhst du über die zusätzliche Belastung die mechanische Spannung im Muskel.

Der Anforderungsgrad der Übung steigt. Parallel dazu verbesserst du die Muskulatur der Hanken, des Rückens und der Rumpfstabilisatoren. Insbesondere der lange Rückenmuskel muss beim gerittenen Schulgalopp auch dein Gewicht mit anheben.

Und wozu soll das Ganze gut sein? Ganz einfach: Wenn du deinen geliebten Vierbeiner zu früh unter Last, also mit Reitgewicht nimmst, dann muss es die Hankenbeugung ohne ausreichende Muskulatur und ohne korrekte Eigenbalance zeigen. Damit steigt das Risiko für Vorstadien eines Muskelkaters, Gegenwehr und fehlerhafte Bewegungsmuster. Gerade letztere sind dann immer und immer wieder der Grund dafür, dass du reichlich Termine mit der Pferdephysio, dem Veterinär und dem orthopädisch geschulten Hufschmied ausmachen darfst.

Fazit: Trainiere zuerst Wiederholbarkeit, dann extreme Tragkraft, sprich tiefe Hankenbeugung.

Die Hankenbeugung ist wie eine Kniebeuge fürs Pferd. Niemand sollte bei den ersten Kniebeugen

Bergziegenartiges Piaffieren ist total in Mode, bringt dem Pferd aber viel weniger gesundheitlich als die altklassische Variante.

mit Zusatzgewicht starten, wenn der mechanische Ablauf der Bewegung noch gar nicht sitzt!
Das Zusatzgewicht bist in diesem Falle du – als Reitgewicht mit einer für das Pferd eigentlich ungünstigen Positionierung direkt hinter den Schulterblättern.

Aus meiner Sicht hat sich die geduldige Arbeit mit der Hand in Vorbereitung extrem bewährt, um Pferde für die eigentlichen Kraftakte – also das geritten werden – vorzubereiten und dauerhaft gesund zu erhalten.

Die Piaffe

Auf den Punkt: Piaffen fungieren als Hilfsmittel zur Balancierung. Sie stärken insbesondere die Hinterhand und alle dort verorteten Muskelgruppen und legen die Basis für eine Verstärkung der Tragkraft. Piaffen eignen sich ebenso wie das Seitwärts dazu, energiegeladene Raketen unter dem Sattel in langsame, aber konzentrierte und kraftvolle Bewegungsroutinen zu lenken. Bei eher introvertierten oder leicht depressiven Tierchen ist die Piaffe das Mittel, um langsam und stetig das Selbstbewusstsein (wieder) aufzubauen.

Eine gelungene Piaffe zeichnet sich durch diese fünf im Grunde ganz einfachen Punkte aus:

- Die Nase des Pferdes ist an oder leicht vor der Senkrechten.
- Das knöcherne Genick ist der höchste Punkt, bei mächtigen Hälsen kann das optisch täuschen.
- Die Hanken (beide!) sind gebeugt.
- Der Hinterhuf ist unter dem Iliosakral-Gelenk (ISG).
- Die Vorderbeine sind möglichst an der Senkrechten.

Eine Piaffe, die ihren Sinn nicht verfehlt.
Sonja mit Coronado.

Dr. Clayton fasst die Merkmale einer Piaffe folgendermaßen zusammen:

> *»Piaffe: a highly collected, cadenced, elevated diagonal movement giving the impression of remaining in place. The hindquarters are lowered; the haunches with active hocks are well engaged, giving great freedom, lightness and mobility to the shoulders and forehand. Each diagonal pair of legs is raised and returned to the ground alternately, with spring and an even cadence. In principle, the height of the toe of the raised forefoot should be level with the middle of the cannon bone of the other supporting foreleg. The toe of the raised hind foot should reach just above the fetlock joint of the other supporting hind leg* (Clayton/Hobbs 2019).«

Auf Deutsch: »Piaffe: eine sehr gesammelte, kadenzierte, erhöhte diagonale Bewegung, die den Eindruck vermittelt, an Ort und Stelle zu bleiben. Die Hinterhand ist gesenkt, die Hinterbeine mit den aktiven Sprunggelenken sind gut eingesetzt und geben den Schultern und der Vorhand große Freiheit, Leichtigkeit und Beweglichkeit. Jedes diagonale Beinpaar wird abwechselnd angehoben und wieder auf den Boden gesetzt, mit Federung und gleichmäßiger Kadenz. Grundsätzlich sollte sich die Spitze des angehobenen Vorderfußes auf gleicher Höhe mit der Mitte des Röhrbeins des anderen tragenden Vorderbeins befinden. Die Zehe des angehobenen Hinterfußes sollte bis knapp über das Fesselgelenk des anderen tragenden Hinterbeins reichen.«

Denk daran: Die Beurteilung fällt bei einer Foto-/Videoaufnahme viel leichter als in der Bewegung. Außerdem schaut das rein physiologisch bei unterschiedlichen Pferdetypen ganz unterschiedlich aus, obwohl die Kriterien eins bis fünf der optimalen Piaffe ganz grundsätzlich erfüllt sind.

Warum sollten Pferde die Piaffe erlernen?

In der Piaffe zeigt sich, ob Pferde über ein gutes **Körpergefühl** verfügen und ob sie in der Lage sind, **maximales Tragevermögen** zu demonstrieren. Außerdem ist es eine echte Balancekunstvorführung, denn hier zeigt das Pferd im Grunde eine einbeinige Kniebeuge unter voller Belastung durch das eigene Körpergewicht.

Die Hinterhand ist aktiviert, mit schwingenden Hinterbeinen unter dem Schwerpunkt plus einer klaren Beugung im Hüft- und im Sprunggelenk. Gleichzeitig ist der Rücken aufgerichtet und stabil, die Stirn-Nasen-Linie liegt leicht vor der Senkrechten. Das heißt auch, dass sich das Pferd selbst trägt – da ist keine harte oder tiefe ziehende Hand seitens des Reitgewichts nötig.

Die Bergziegenvariante einer Piaffe wollen wir aber bitte nicht: Dann sind die Vorder- und Hinterbeine nahezu V-förmig mit stark rückständigen Vorderhufen und Hinterbeinen, die sehr weit unter dem Schwerpunkt stehen.

In einer Beobachtungsstudie von Clayton et al. von 2017 wurde die vertikale Schwankung des Pferderückens über Marker am Widerrist und am Brustkorb ermittelt. Für die Piaffe wurde eine beeindruckend niedrige Schwankung ermittelt:

1. 30 bis 50 Millimeter vertikale Schwankung im gesammelten Trab
2. 15 bis 25 Millimeter vertikale Schwankung bei der Passage
3. **Unter 10 Millimeter vertikale Schwankung bei der Piaffe**

Jüngeres Pferd an der Hand in der Piaffe. Mascha mit Bella.

Diese vertikale Schwankung, also die Abwärtsbewegung des Rumpfs nach Bodenkontakt ist der Anteil, der am deutlichsten reduziert ist. Das *Einsinken* findet aufgrund der Versammlung kaum noch statt.

Beim weit fortgeschrittenen Pferd machen die Piaffetritte einen schwungvollen Eindruck. Um noch einmal Dr. Clayton zu Wort kommen zu lassen:

> *»Some horses leap from one forelimb to the other or bounce from one hind limb to the other, which can fool the eye into thinking there is an aerial phase when, in fact, there is always at least one hoof on the ground. Limb support sequences include periods of tripedal and even quadrupedal support* (Clayton/Hobbs 2019).«

Auf Deutsch: »Manche Pferde springen von einem Vorderbein auf das andere oder hüpfen von einem Hinterbein auf das andere, was dem Auge eine Luftphase vorgaukelt, während in Wirklichkeit immer mindestens ein Huf auf dem Boden steht. Sequenzen, in denen die Gliedmaßen gestützt werden, umfassen Perioden der drei- und sogar vierbeinigen Unterstützung.«

Die Praxis zeigt: Wenn sich das Pferd durch vorangegangene Arbeit Versammlungsfähigkeit erwirbt, dann ist es immer leichter, die Tritte so weit zu verkürzen, dass das Pferd im Schritt auf der Stelle tritt. Dann entsteht die Piaffe auch ganz leicht und ohne, dass man vorn festhalten und hinten draufdreschen müsste, wie es heutzutage leider oft noch der Fall ist.

Die Piaffe entwickelt sich stattdessen im Grunde über Seitengänge, bei denen die Tritte bei fortschreitender Arbeit immer kürzer werden. Zugleich wird im Rahmen der Erarbeitung die Bewegung über die Seite mehr und mehr reduziert, bis am Ende eine Piaffe zu bewundern ist. Es muss nur sukzessive der Rhythmus energischer werden.

Schon klappt es ganz von alleine und das Pferd bietet die Piaffe von selbst an, ohne dabei sonderlich angestrengt zu wirken. In der modernen Piaffe bekommst du allerdings allein vom zuschauen schon Angst davor, dass das Pferd mit der Nase voran nach vorne kippt.

Festsitzendes Iliosakralgelenk (ISG) lockern

Das Iliosakralgelenk ist die Schnittstelle zwischen den Hinterbeinen und dem Rücken, es befindet sich direkt unter dem höchsten Punkt der Kruppe. Bei wenig bemuskelten Pferden lässt es sich auch mit wenig Druck unter Umständen sogar ertasten.

Sogenannte »Blockaden« oder Schutzspannungen sind meist auf muskuläre Schwäche oder Fehlbelastungen zurückzuführen.

Als Teil der konservativen Therapie werden tatsächlich sogar Seitengänge empfohlen (Baumann 2024), aber häufig wird auf niedrigere Abstellungen verwiesen. Dass Seitengänge in größerer Abstellung in der Regel sinnvoller sind, um Adduktoren und Abduktoren hinreichen anzusprechen, hast du schon in der Lektion zu Seitengängen gelesen.

Da das ISG als Kraftüberträger fungiert, sind langsame und kontrollierte Bewegungen sicherlich angebrachter als kleinere Abstellungen, weil sie die Muskeln der Kruppe kaum ausreichend ansprechen. Die Piaffe eignet sich als Lektion hervorragend dazu, die komplette Kruppenregion samt ISG zu konservieren, weil sie die Muskeln dort aufbaut.

Über die Piaffe erreichst du außerdem, dass sich alle beteiligten Gelenke und Muskelgruppen in fließenden, nahezu federnden Schritten bewegen. Ein vom Körper in Form von Schutzspannung angelegter Schmerz im Bereich des ISG lässt sich unter anderem auch durch die Piaffe langfristig gesehen ursächlich auflösen, immerhin wird der gesamte Bereich der Hinterhand bis nach vorn über die ventrale Muskelkette aktiviert.

Der Muskel, an dem alles (= Rumpf) hängt

Beim Menschen sichert das Schlüsselbein die knöcherne Verankerung zwischen Schulterblatt und Brustbein – Pferde müssen diese Absicherung aber über eine muskuläre Lösung erreichen, um nicht einfach nach vorne zusammenzubrechen.

Zum Glück gibt es dafür den ***M. serratus ventralis*** (gegliedert in pars cervicis und pars thoracis), der das Schulterblatt (Scapula) mit den Hals- und Brustwirbeln verbindet. Er ist wesentlicher Bestandteil der *Aufhängung* des Rumpfes zwischen den Vordergliedmaßen und übernimmt eine, wenn nicht sogar *die* zentrale Rolle im Rumpftrageapparat deines Pferdes.

Die Rumpfhebung durch Aktivierung der rumpftragenden Muskulatur sind essenziell für eine stabile Oberlinie und damit für ein möglichst verschleiß- und schmerzfreies Reiten.

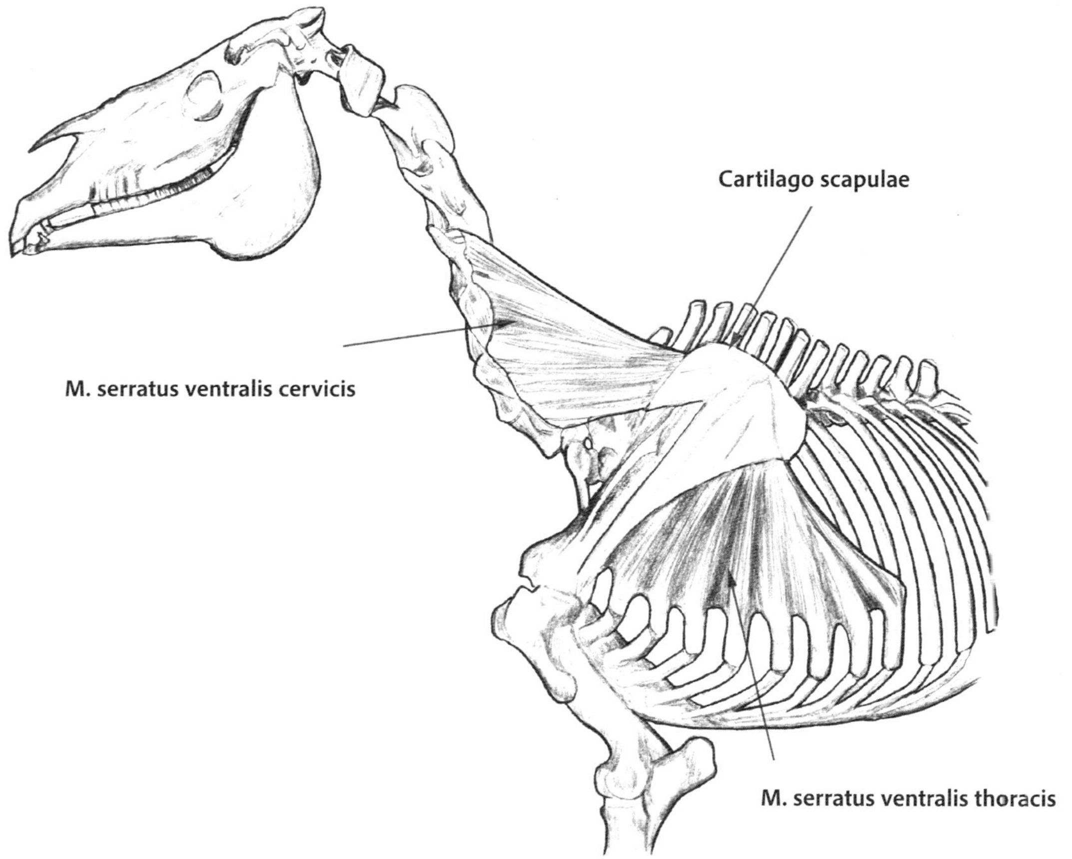

Nahezu unbekannt in der reiterlichen Welt ist die Funktion des *M. serratus* und wie er durch den Tausch von Punctum mobile und Punctum fixum die Rumpfhebung erst ermöglicht.

Rumpfhebung ist der beste Stoßdämpfer der Vorhand. Sonja mit Dantano.

Mini-Anatomiewissen: Muskelansätze sind austauschbar

Anatomisch betrachtet hat jeder Muskel einen Ansatz- und einen Muskelkraftverarbeitungs*punkt*. Passenderweise wird das Ganze auch als Punctum bezeichnet.

- **Punctum fixum:** Der Fixpunkt, auf den der Punctum mobile zu bewegt wird.
- **Punctum mobile:** Der bewegliche Punkt, an dem die Muskelkraft umgesetzt wird.

Die Zuordnung von Punctum fixum und Punctum mobile ist **lage- und funktionsabhängig** und kann sich je nach Bewegungsmuster und Lektion verändern. Sprich: Das steht nicht fest, sondern entscheidet sich erst im Rahmen der ausgeführten Bewegung!

Bei nicht versammelten Bewegungen insbesondere auch im Vorwärts-Abwärts oder insgesamt mit wenig Versammlung inklusive Aufrichtung kann das im Falle des Serratus bedeuten, dass der Punctum fixum den Rippenkasten darstellt, während der Punctum mobile das obere Ende des Schulterblattes ist. So kommt der Muskel seiner Aufgabe nach,

- das Schulterblatt nach hinten-unten zu ziehen,
- er unterstützt dabei die Vorführung des Vorderbeins,
- und federt (zumindest etwas auch insgesamt in seiner Rolle) das Rumpfgewicht ab.

Beim einem nicht versammelten Pferd kracht der Rumpf eher nach unten, weil der Rumpfheber seiner Tätigkeit als Heber des Rumpfs nicht hinreichend nachkommt.

In **versammelter Bewegung** wird das Schulterblatt zum Punctum fixum, während Brustkorb zum Punctum mobile wird. Nur so kann die Aktivierung des Serratus den Rumpf zwischen den Vorderbeinen vermehrt anheben und seinem vollen Potenzial als Rumpfträger nachkommen! Die Stoßdämpfung der gesamten Vorhand wird dadurch perfekt.

In der Versammlung übernimmt der Muskel eine Maximierung der stabilisierenden Rolle, während er bei freier Bewegung vor allem für Beweglichkeit und Stoßdämpfung sorgt. Über die Versammlung wird anfangs der Brustteil des Rumpfhebers aktiviert, während der Halsteil des Rumpfhebers davon unabhängig über konstante Arbeit weiter aktiviert werden muss.

Im Vorwärts-Abwärts wird insbesondere der Halsteil des Rumpfhebers gedehnt, was aber keine maximierte Stoßdämpfung ermöglicht. Damit Punctum fixum und Punctum mobile tauschen können, ist eine gewisse Aufrichtung nötig. Die Schnauze über den Hallenboden schnüffeln zu lassen, ermöglicht eben keine Reitpferdeaufrichtung.

Ein gezieltes Training der Versammlung fördert nicht nur die Tragkraft, sondern schützt über die maximierte Stoßdämpfung auch die Gelenke der Vorhand langfristig.

Das Zusammenspiel von Rumpfheber und Rücken

Erst in versammelter Arbeit wird folglich Reitpferderumpfhebung wirklich erreicht, die es wiederum möglich macht, den langen Rückenmuskel enorm aufzubauen.

Pferd hat »Rücken« = Papierdünne Rückenmuskeln trotz viel Bewegung?

Manche reichlich an der Longe gearbeiteten Pferde haben erstaunlich dünne Rückenmuskeln. Dabei werden sie doch so viel und reichlich bewegt! Und überhaupt, als Fluchttiere sollte Rennen doch das perfekte Training für fitte Reitpferde sein.

Ja nun, so klappt das aber nicht. Reitpferde, die nicht versammlungslastig gearbeitet werden, wirken im Rücken häufig *leer*. Sie sind oft druckempfindlich an der Sattellage oder weisen dort regelrechte Mulden auf, manchmal lassen sich sogar tiefe Kuhlen neben dem Widerrist beobachten. Der ganze lange Rückenmuskel *(M. longissimus dorsi)* – zuständig wie schon beschrieben für die Vorhandhebung und die Rumpfstabilität – ist dünn wie Papier und schwach.

Woran liegt das? An Atrophie! Der lange Rückenmuskel wird über Spaziergänge, Abschleudern und versammlungsfreies Reiten nicht ausreichend gefordert – und verkümmert. Damit ein Muskel, ganz egal wo im Körper, ausreichend gefordert, sprich aktiviert wird, muss er

- kontrahieren,
- gegen Widerstände arbeiten,
- gezielt durchblutet werden.

Für den Körper in seiner Gesamtheit sind diese drei Punkte das Signal, dass diese Struktur definitiv gebraucht wird. Aus Energiesparsamkeit wird jede Struktur, die nicht oder nicht wirklich oft benötigt wird, aber abgebaut.

Sprich: Muskelmasse wird abgebaut, wenn sie nicht benötigt wird. Ein bis zwei Wochen Dauer*nicht*nutzung reichen beim Menschen bereits aus. Beim Pferd sind die Zeiträume leider nicht viel länger, bis der Abbau startet. Die eingesparten Ressourcen kommen anderen Strukturen zugute, während der betreffende Muskel langsam verkümmert.

Beim langen Rückenmuskel ist das für das Pferd regelrecht dramatisch, vor allem wenn weiterhin mit Reitgewicht knapp hinter den Schulterblättern für konstante Irritationen oder sogar Mikroverletzungen gesorgt wird. Dann ersetzt der Körper die vorhandenen Muskelfasern stetig durch Kollagenfasern, also durch Bindegewebe (Muskelfibrose). Für den Muskel bedeutet das: Er wird unelastisch bis zäh, kontrahiert nicht mehr aktiv und lässt sich relativ schwer zurück in den ursprünglichen Zustand versetzen.

Die Muskelfibrose ist häufig als harter Strang unter der Haut tastbar und wirkt wie ein nicht lösbarer Muskelknoten oder die klassische *Verspannung*, die ja auch Menschen mit Rückenschmerzen nur zu gut kennen dürften.

Muskelfibrose: Regenauffangbecken im Pferderücken

Manchmal ist der Muskelschwund schon so weit fortgeschritten, dass sich regelrechte Kuhlen rechts und links der Wirbelsäule im Rücken des Pferdes bilden. Zur Erinnerung: Wir wollen Regen*rinnen*, keine Regen*auffangbecken* auf dem Pferderücken!

Was kannst du tun, um eine Muskelfibrose vor allem am Rücken zu vermeiden?

Früherkennung ist entscheidend. Denn Fibrose lässt sich nur schwer bis gar nicht zurückbilden, aber sehr wohl vermeiden. Dazu ist es aber notwendig, dass du qualitativ hochwertige Seitengänge trainierst, die die gesamte Rumpfmuskulatur entsprechend auch fordern. Und dazu zählt nicht Kopf-herein!

Im Bereich des Rückens ist die Fibrose besonders kritisch, denn um es ganz drastisch auszudrücken: Wir als Menschen sind in der Verantwortung, unsere Tiere gesunderhaltend zu bewegen, zu ernähren und zu halten. Allein durch die Druckbelastung durch das Reitgewicht sind die Rückenprobleme quasi schon vorprogrammiert!

Eine vorliegende Fibrose in der Sattellage bereitet dem Reitpferd erhebliches Unbehagen, denn die für das Reiten nötige schützende Muskulatur ist verkümmert. Und das wohl kaum, weil der Vierbeiner sich nicht oder nicht ausreichend bewegt hat, sondern weil die für die Bewegung verantwortlichen Menschen das Pferd im Grunde genommen falsch trainiert, eventuell auch unzureichend ernährt haben. Bewegung allein schafft keinen Deluxe-Rückenmuskel. Den benötigen wir aber dringend für das Reiten!

→ Weitere Informationen zur Fibrose findest du im Kapitel zu Muskelhypertrophie.

Nun kann Bindegewebe ja viel leisten, aber die Druckbelastung durch ein Reitgewicht kann es beileibe nicht gut kompensieren. Und der ungenutzte Rückenmuskel wird aus Energiegründen seiner Ressourcen beraubt, da er seiner eigentlichen Aufgabe, vornehmlich der Vorhandhebung nicht mehr im eigentlich benötigten Umfang nachkommt.

Das besagte Pferd trabt weiter brav an der Longe im Kreis, nur die funktionale Rumpfarbeit fehlt dauerhaft und der lange Rückenmuskel verkümmert einfach weiter und mit ihm auch zusätzliche Strukturen.

Firth/Rogers konnten belegen, dass die korrekte Entwicklung und das Training der Tragemuskulatur inklusive des langen Rückenmuskels essenziell sind, um über die biomechanischen Belastungsketten (dorsal/ventral) die distalen Gelenke zu entlasten (Firth/Rogers 2005). Wenn du ein dauerhaft lahmfreies Pferd haben willst, brauchst du eine starke Oberlinie.

Mit fliegenden Fahnen versammelt: die Passage

Auf den Punkt: Die Passage ist ein stark versammelter, erhabener Trab mit deutlich verzögerter Schwebephase und hoher Ausdruckskraft. Charakteristisch sind:

- Diagonal versetzte Schrittfolge mit Schwebephase,
- Halsansatz und Rumpf sind im Ideal gut angehoben, Rückenmuskeln hilft dabei,
- gesamte Körperspannung ist erhöht,
- Hankenbeugung existiert, mit wenig Beugung im Lumbosakralgelenk.

bis in die Waagerechte gestreckter Unterarm, während die Hinterbeine federnd arbeiten

Fliegen kann nicht schöner sein, denn die insbesondere nach altklassischen Grundsätzen erarbeitete Passage lässt das Pferd scheinbar schwerelos über den Boden schweben. Wobei diese Schwebephase doch eine eher rasseabhängige optische Täuschung ist, die aber natürlich trotzdem schön anzuschauen ist. Ihre Ausführung ist im Rahmen der FN eine Sache, die relativ nahtlos an den versammelten Trab anschließen kann, aber dies freilich nur für entsprechend hochbegabte Tiere. Dabei kann nahezu jedes Pferd mit entsprechender vorbereitender Arbeit die Passage erlernen und von ihren gesundheitsfördernden Aspekten für das Reitpferd profitieren.

Lockere Schultern, schöne Passage

> *»Die Passage entsteht weder aus einer Beugung des Nierenbereichs (gemeint ist das Lumbosakralgelenk, A. d. A.) und der Hanken noch aus dem Untertreten der Hinterbeine. Sie zeichnet sich durch eine verlängerte Schwebephase aus, die zum Teil vom Schub der Hinterbeine, vor allem jedoch von der erhabenen Bewegung der Schultern herrührt* (Karl 2006/2007).«

Wobei natürlich beim federnden Abstoßen vom Boden eben doch eine gewisse Beugung der Hanken nötig ist, auch wenn das Karl nicht ganz so sieht.

Karl sieht die größeren Chancen auf eine hervorragende und damit gesundheitlich optimal umgesetzte Passage darin, zunächst den Spanischen Schritt einzuführen, um erst im Anschluss daran die Passage zu erarbeiten.

Der Weg über den Spanischen Schritt ist durchaus ein guter Ansatz. Aus meiner Erfahrung heraus können begabte Pferde durchaus allein durch die Perfektion der Seitengänge im Schritt und im Trab oft schon von sich aus zur Passage trainiert werden.

Fliegen kann nicht schöner sein.
Sonja mit Leo.

Das Exterieur hat großen Einfluss darauf, wie die Passage von außen interpretiert wird. Sonja mit Dantano.

Bietet mir das Pferd das von sich aus an, ist es auch meine Aufgabe diese richtigen Momente unbedingt sofort lobend zu verstärken. Für andere Pferde ist es beispielsweise sinnvoller, die Passage eher aus der Piaffe heraus zu entwickeln.

Im professionellen Reitsport wird übrigens eher mit Gas und Bremse gearbeitet, um einen sogenannten negativen Schwebetritt zu erhalten. Das meistern nur die extrem talentierten Pferde. Für die große breite Masse der eher durchschnittlichen Tiere ist es sinnvoller, den altklassischen Weg zu absolvieren.

Retraktion trägt zu Rumpfhebung bei

In der Beobachtungsstudie von Clayton et al. von 2017 wurde die vertikale Schwankung über Marker am Widerrist und am Brustkorb ermittelt.

In Studien wurde die Rumpfhebung vom Boden ausgehend gemessen und ist nachweislich in der Passage gewährleistet.

Die beispielhaften Werte, die im Rahmen dieses Versuchs auftraten, sind für Altklassik-Fans nicht sonderlich überraschend:

- 30 bis 50 Millimeter vertikale Schwankung im gesammelten Trab
- **15 bis 25 Millimeter vertikale Schwankung bei der Passage**
- Unter 10 Millimeter vertikale Schwankung bei der Piaffe

Diese vertikale Schwankung, also die Abwärtsbewegung des Rumpfs nach Bodenkontakt ist der Anteil, der am deutlichsten reduziert ist. Hier können wir endlich von (messbarer) Rumpfhebung sprechen!

Das *Einsinken* findet aufgrund der Versammlung kaum noch statt. Das frühe Hinterhand-Aufsetzen in der Passage ermöglicht eine stabile Vorwärts-Aufwärts-Bewegung. Woher wissen wir das? Weil in dieser Studie ganz klar beobachtet wurde, dass der Hinterhuf vor dem diagonalen Vorderbein aufsetzt. Taktfans bekommen darüber bitte keinen Anfall, die Bewegung ist eben so. Da muss auch nichts sonderlich *natürlich* und im Takt ausfallen, denn: Reiten per se ist **nicht** natürlich.

Die Hinterhand ist deshalb noch weiter nach hinten zurückgezogen, besonders tragfähig und vermindert das Nach-Vorne-Kippen (Pitch down). Die längere Standphase in der Hinterhand führt dazu, dass der Vierbeiner mehr Zeit hat, den Impuls aufzubauen und vor allem, das ist der immens wichtige Part, zu *kontrollieren*.

Dieser Zeitgewinn erlaubt mehr aktive Muskelarbeit, die wiederum überhaupt erst die Rumpfhebung ermöglicht! In der Passage muss das Pferd eben nicht wie schon im versammelten Trab unter dem Reitgewicht wie verrückt nach vorne eilen, um nicht unversehens auf die Schnauze zu fallen. Wie viel schlimmer das im gängigeren Arbeitstrab sein dürfte, kann man sich ja ganz gut ausmalen.

Die Zeitunterschiede in der Schwebephase haben Clayton et al. übrigens ebenfalls erfasst und konnten belegen, dass in der Passage weniger Zeit (in Millisekunden, ms) im Freiflug ohne Bodenkontakt anfällt:

> *»The aerial phases have shorter absolute (52 ms) and relative (4.8%) durations than collected trot* (Clayton/Hobbes 2019).«

Auf Deutsch: »Die Schwebephasen haben eine kürzere absolute (52 ms) und relative (4,8 %) Dauer im Vergleich zum gesammelten Trab.«

Der gezielte und frühe Auftritt in der Passage erlaubt das dynamische Gleichgewicht zugunsten einer kontrollierten Bewegung und Tragkraft.

Über die Retraktion der Hintergliedmaße wird in der Passage außerdem ein enorm wichtiger Schritt hin zur Rumpfhebung gemeistert. Denn wenn sich das Hinterbein hinter dem Körperschwerpunkt bewegt und das Tier aktiv vom Boden abdrückt, dann

- hebt sich der Rumpf des Pferdes aufwärts,
- richtet sich der Brustkorb gegen die Schwerkraft auf
- und das ganze Tier wird insgesamt viel tragfähiger.

Clayton et al. konnten zudem gezielt nachweisen, dass die Hinterhand in der Passage deutlich höhere vertikale Kräfte entwickelt als im Trab und so der Rumpf entgegen die Schwerkraft nach oben angehoben wird.

> *»The larger vertical impulses in the hindlimbs contribute significantly to the elevation of the horse's forequarters and center of mass* (Clayton et al. 2017).«

Auf Deutsch: »Die größeren vertikalen Impulse in den Hintergliedmaßen tragen wesentlich zur Anhebung der Vorderhand und des Schwerpunkts des Pferdes bei (Clayton et al. 2017).«

Die Kombination aus einer verlängerten Retraktion bei geringerer vertikaler Schwankung und einer relativ kurzen Schwebephase verringert das Risiko, dass das Pferd den Rumpf vertikal kippt (pitching) (Clayton/Hobbs 2017).

Stell dir dazu möglichst eine Wippe vor, mit dem Schwerpunkt des Pferdes als Balancepunkt in der Mitte. Starke vertikale Schwankungen und/oder hohe Geschwindigkeiten führen dazu, dass sich der Kipppunkt der Wippe nach vorn neigt – die Wippe wird instabiler, das Kippen nach vorn wahrscheinlicher. Bei einer verlängerten Retraktion ist das Gewicht konstanter auf der Wippe verteilt und durch die erreichte Rumpfhebung bleibt der Kipppunkt optimal für eine angemessene Balance.

Eine kontrollierte und vor allem stabile Oberlinie ist die logische Konsequenz, was wiederum wirklich hilfreich ist, wenn das Pferd Zeit seines Lebens einen Menschen auf seinem Rücken herumtragen können soll.

Da sieht man mal wieder, dass weder Versammlung noch Lektionsarbeit irgendwelche Zirkustricks sind, die keiner braucht. Stattdessen wird noch klarer, dass wir hier in ernstzunehmende Ausbildung zugunsten der Gesunderhaltung unserer Pferde investieren sollten. Da darf auch gerne die Passage dazugehören!

Hoch das Bein! Der Spanische Schritt

Auf den Punkt: Im Spanischen Schritt schreitet das Pferd majestätisch dahin, jeweils eine Vordergliedmaße ist waagerecht angehoben und wird nach vorn gestreckt. Imponiergehabe vom Feinsten, noch schöner und stolzer wird das Tier sich höchstens in der Passage präsentieren. Die gleiche Biomechanik lässt sich auch über den Spanischen Trab erreichen, der auch gestreckte Passage genannt wird.

Warmblüter im Spanischen Schritt.
Sonja mit Dantano.

Charakteristisch sind:

- Die Vordergliedmaße wird angehoben und vollständig nach vorn gestreckt,
- das Schulterblatt nähert sich der Waagerechten,
- der Oberarm des in der Luft befindlichen Beins nähert sich der Senkrechten,
- der Unterarm des in der Luft befindlichen Beins nähert sich der Waagerechten,
- bei abwechselndem Anheben und Aufsetzen der Vorderbeine.

Was imposant aussieht, dient ganz klar der Mechanik hinter der Schulterbewegung und dem Kraftgewinn über die Hinterhand:

> *»Das Pferd übt das Ausbalancieren durch das abwechselnd verstärkte Anheben der Schultern, was bei korrekter Ausführung mit einer vermehrten Laufaufnahme hinten verbunden ist.* (Diacont/Weber 2022)«

Zusätzlich gewinnt das Pferd dank der opulenten Geste nach vorn Zeit, um sich aus dem Standbein aufwärtszudrücken. Das ist, siehe die Studien von Clayton zur Passage, Part der Rumpfhebung. Zusätzlich nimmt die Schultergürtelspannung über die Aktivierung des *M. serratus* und den Tausch von Punctum fixum und Punctum mobile zu.

Eine fehlende Bemuskelung der gesamten Schulterregion kann in Verlust von Knorpel resultieren, Entzündungsprozesse wie Arthrosen nehmen zu – der Bewegungsradius schrumpft aufgrund der wachsenden Instabilität des Bewegungsapparats, sodass eine vom Nervensystem angelegte Schutzspannung (= Schmerzen) großräumige Bewegungen mit potenziellem Verletzungsrisiko verhindern helfen soll. Mehr Muskelmasse füttert dagegen die gesamte Region inklusive des Knorpelgewebes im Buggelenk, sodass Schmerzen reduziert und Entzündungsprozesse effektiver bekämpft werden können.

Letztlich sind Schmerzen und Fehlhaltungen die Folge der ursprünglichen, anfangs vielleicht noch eher leichten unterkomplexen Bemuskelung. Unser bestes Mittel dagegen? Natürlich Krafttraining, und zwar mit möglichst großem Bewegungsradius (range of motion). Dazu zählen auch die Seitengänge, die aber möglichst nicht in voller Geschwindigkeit ausgeführt werden sollten.

Die maximale mechanische Spannung im Muskel und kontrolliert ausgeführte Bewegung sorgt für eine bessere Nährstoffversorgung der Knorpel sowie des umliegenden Gewebes. Es werden mehr Myokine ausgeschüttet, die wiederum unter anderem entzündungshemmend wirken.

Der Spanische Schritt ist aufgrund der federnden Schrittfolge auch die ideale Sehnenpflege, wovon insbesondere ältere Pferde profitieren (Lichtwark/Wilson 2005). Bei jüngeren Tieren sind anfangs langsame Sequenzen der reinste Sehnenschutz, um die noch schwachen Sehnen vor Überlastung zu bewahren. Auch hier sind kontrollierte Bewegungen in Versammlung sinnvoller als überstürztes durch-die-Halle treiben!

Im Falle der Schultern ergeben sich so gleich mehrere Lösungen für steife, unbewegliche Schultern: Entzündungsprozesse insbesondere rund um das Schultergelenk reduzieren sich, die Muskelneubildung wird in allen beteiligten Muskelgruppen angeregt, die Bänder und Sehnen besser mit Nährstoffen versorgt und das Tier trainiert Balance und Koordination über den kadenzierten Wechsel zwischen dem Anheben und Ausstrecken des Beins sowie dem Schritt nach vorn über die Hinterbeine.

Spanische Schritte für alle!

Kann jedes Pferd den Spanischen Schritt erlernen? Karl sieht das eindeutig so:

»Die einzige Übung, mit der man den Schultern des Pferdes die für die Passage notwendige Freiheit und Erhabenheit der Bewegung verleihen kann, ist der Spanische Schritt. Und

Ein gut ausgeführter spanischer Schritt vermittelt dem Pferd eine Idee der optimalen Rumpfhebung. Sonja mit Coronado.

diesen kann ein kompetenter Reiter jedem Pferd beibringen, auch einem sehr gewöhnlichen (Karl 2006/2007).«

Die langsame, kontrollierte Vorwärtsbewegung fördert die **Balance** des Vierbeiners, der zumindest für eine Schrittdauer stabil und sicher auf drei Beinen stehen muss.

Der Spanische Schritt und die Maultätigkeit

Und nur zur Erinnerung: Manche Pferderassen sind extrem gesprächig und arbeiten in herausfordernden Lektionen wie dem Spanischen Schritt permanent mit dem Maul mit. Das heißt, sie demonstrieren das, was die Alten Meister wohl ein durchlässiges und nachgiebiges Maul nennen würden. Sie zeigen eine

Hankenbeugung mit deutlicher Vorhandhebung, das stärkt den langen Rückenmuskel!

sich bewegende mobile Zunge. Das ist **kein** direktes Signal für Stress oder ein Ausdruck für negative Anspannung. Es ist schlicht normales Pferdeverhalten und in der altklassischen Arbeit mit Pferden auch ausdrücklich erwünscht, denn über das mobile Maul zeigt sich auch immer wieder die Aktivierung des Parasympathikus, der für Stressabbau und Entspannung verantwortlich ist.

Galopp nach Art der Alten Meister (Schulgalopp)

Auf den Punkt: Der Schulgalopp in seinen unterschiedlichen Ausprägungen ist im Grunde eine Art von gesprungenen Kniebeugen, die mit beiden Beinen ausgeführt werden. Sie stehen damit im Gegensatz zur einbeinig belastenden Piaffe, die einer ebenso einbeinig ausgeführten Kniebeuge entspricht. Im Schulgalopp findet keine Schwebephase statt.

Zu den Kunstgangarten der Hohen Schule zählen neben dem Schulgalopp noch Tummeln, Mezaire, Courbetten.

Warum sollten Pferde den Schulgalopp erlernen?

Moderner Renngalopp ist der Standard, der noch am ehesten im Freizeitsportbereich zu sehen ist. Das Pferd bewegt sich mit möglichst viel Tempo nach vorne, aber kaum noch nach oben und bringt dabei auch noch reichlich *Schwung*, sprich eine Schwebephase mit.

Beim versammelten Schulgalopp in seinen verschiedenen Varianten ist das doch deutlich anders. Der Fokus liegt weniger auf dem Schwung im Sinne von reichlich Geschwindigkeit, sondern vielmehr darauf, eindrucksvolle kraftvolle Bewegungen mit höchstem Ausdruck zu präsentieren und zugleich jederzeit Balance und Kontrolle zu erhalten.

Schulgalopp und damit einhergehende Schulsprünge stammen ursprünglich aus Zeiten der Kriegsreiterei. Hier war maximale Versammlung und Kontrolle gefragt, um im Reiterkampf mit feinster Hilfengebung – meist einhändig! – zu bestehen und notfalls auf der Stelle zu wenden.

Heutzutage sind Kämpfe hoch zu Ross glücklicherweise aus der Mode geraten. Stattdessen lohnt die Erarbeitung des Schulgalopps, um die Balance und die **Schwerpunktverlagerung** nach hinten zu trainieren.

Auch hier hat Clayton, dieses Mal im Team mit Burns, untersucht, wie sich dieses Bewegungsmuster auf die Biomechanik des Pferdes auswirkt.

Per Videoanalyse wurde der versammelte Galopp mit der Galopppirouette anhand von elf hochklassigen Dressurpferden verglichen, die Aufzeichnung stammt von den Olympischen Spielen 1992 in Barcelona.

Im Ergebnis zeigte sich beim versammelten Galopp eine Schwebephase von circa 0,013 Sekunden. Bei der Galopppirouette dagegen ließ sich keine echte Schwebephase identifizieren. Auch zeigte sich hier ein klarer 4-Takt-Rhythmus (Burns/Clayton 1997).

Das **veränderte Timing** basiert darauf, dass das Hinterbein vor dem diagonalen Vorderbein aufsetzt. Das beeinflusst die Hebung und Balance – und zwar deutlich. Denn durch das frühe Absetzen der Hinterhand wird eine Brustkorbanhebung ermöglicht und es findet eine **leichte Schwerpunktverlagerung nach hinten** statt. Umkippen nicht länger möglich, trotz der doch recht erheblichen Geschwindigkeit und der Beeinträchtigung durch das Reitgewicht auf dem Rücken.

Im Gegensatz zum heute üblichen Galopp ist die Galopppirouette kein *Galopp in der Luft*, sondern ganz klar eine kontrollierte Hebebewegung. Durch den konstanten Bodenkontakt, immerhin ist jederzeit wenigstens ein Huf am Boden, ist das Pferd auch insgesamt trittsicherer.

Weiche Mittelhand? Schulgalopp!

Es gibt einfach nichts, was die Mittelhand so gut formt wie der Schulgalopp. Dabei werden

Nahkampfgadget aus alten Zeiten: Rückwärtsgalopp. Auch heutzutage als Krafttraining sinnvoll. Sonja mit Leo.

insbesondere die Muskeln und Muskelgruppen des Rumpfs einschließlich des Brustkorbs, der Flanken und des Rückens angesprochen. Natürlich kommen hier sowohl die ventrale als auch die

Schulgalopp baut die Mittelhand und die Oberschenkelmuskeln auf. Sonja mit Vancover.

Der versammelte Galopp ist für die meisten Pferdetypen eine Herausforderung. Christina und Emma.

dorsale Muskelkette als jeweils ganzes zum Tragen. Schenkelweichen und Seitengänge dienen der Vorbereitung. Die Passage und der Spanische Schritt ergänzen die Hinführung zum Schulgalopp, weil sie die wunderschöne Gestik und das starke Anheben der Front opulenter machen.

Die meisten Pferde werden den Schulgalopp sowieso nur erlernen, wenn sie wiederholt im Seitwärts angaloppiert werden.

Ch12 Emma mit Christina.

Merke:

Aus dem Arbeitsgalopp heraus entwickelt sich kein Schulgalopp. Ein ausdauernder Sprinter wird halt auch keine Kniebeugen in Perfektion zeigen, wenn er tagein tagaus das Geradeausrennen trainiert. Von daher: Idealerweise startest du im Seitwärts und galoppierst hier an und stärkst zugleich den Schultergürtel und das Anheben der Front über Passage und Spanischen Schritt. Voilà, Schulgalopp!

Wie gut oder schlecht der Schulgalopp umgesetzt wird, ist tatsächlich sowohl vom Gebäude des Pferdes als auch vom Arbeitsaufwand abhängig, der in diese Lektion investiert werden kann. Grundsätzlich ist ein heutiges modernes Reitpferd auch mit Seitwärts, Piaffe und Passage bereits gut vorbereitet und lässt sich über Variationen hervorragend gymnastizieren. Der Schulgalopp ist die darauffolgende nächste Herausforderung, die bei manchen Pferden sehr gut, bei vielen aber nicht unbedingt sofort gelingt. Hier ist Geduld gefragt.

Versammlung im Galopp – ohne Hände!
Sonja und Daniano.

Ziel ist es, mit immer geringerer Hilfendichte auszukommen. Sonja und Dantano.

Hilfen? Hilfe!

Lassen wir zuerst einen der gar nicht so alten Meister zu Wort kommen:

> *»In neuerer Zeit, wo theils durch den vermehrten Gebrauch orientalischer und englischer, oder nach englischen Grundsätzen gezüchteter Pferde, teils durch die grössere Verbreitung der Wettrennen und des Jagdreitens und die hierdurch gesteigerten Anforderungen an Schnelligkeit, theils durch andere Umstände (A. D. A.: gemeint ist beispielsweise die Mode) das Reiten der Pferde in schnellen Gängen meist als das höchste Ziel der Reitkunst angesehen wird, ist das Schulreiten, leider, vernachlässigt worden. Indessen haben nicht diese eben angeführten Umstände allein, sondern auch andere zu dieser beklagenswerten Vernachlässigung beigetragen, so z. B. Der fühlbare Mangel an tüchtigen Schulreitern als Lehrer.* (Seeger 1884)«

Schon zu Seegers Lebzeiten war die Schulreiterei selbst und mit ihr eine feine Hilfengebung rückläufig. Statt sich über beeindruckende Reitkünste zu profilieren, rasten die adligen Herren über Stock und Stein. Die Pferde brachen sich dabei nicht selten die Beine und die Reiter ihr Genick.

Die Kunst der feinen Hilfengebung geriet in Vergessenheit, nicht zuletzt dank der weit verbreiteten Annahme, dass sich das Pferd anzulehnen habe. Was ein barocker Reitmeister darunter verstanden haben könnte und was heutzutage darunter verstanden wird, ist grundsätzlich gegenteilig und könnte viele weitere Bücher füllen.

Das Warum hinter den leichten Hilfen

Pferde suchen sich das Leben als Reitpferd nicht aus. Das müssen wir uns als reitende Menschen immer und immer wieder vor Augen führen. Ob als hochgeschätzte Haustiere, als teures Mittel zur sportlichen Betätigung oder Part der eigenen Persönlichkeitsentfaltung: Pferde sind dem Willen und insbesondere auch dem Können ihrer Menschen ausgeliefert.

Harte, fordernde Hände, grober Sporeneinsatz oder ständiges Peitschen mit der Gerte oder welchem Gerät auch immer wird dein Pferd nicht zu einer besseren Zusammenarbeit motivieren. Es kann aber durchaus ernsthafte Folgen für seinen Bewegungsapparat haben, wenn Hilfen grob, hart oder konstant zum falschen Zeitpunkt eingesetzt werden. Denn dann gerät es in Vermeidungs- und Fehlhaltungen, die biomechanisch ernsthafte Probleme verursachen.

Was Hilfengebung sein kann

Hilfengebung ist teils Konditionierung, teils Kommunikationsmittel.

Endprodukt: körperlich maximale Anstrengung beim Pferd und nahezu null Hilfengebung seitens des Reiters. Sonja und Dantano.

Die **Konditionierung** entsteht dann, wenn dein Vierbeiner eine Anweisung (Hilfe) mit einer hoffentlich gewünschten und klar definierten Handlung verknüpft.

Ein Beispiel aus meiner Stallpraxis: Kopfscheue Pferde mit dem Schwamm abzukühlen kann ganz schön aufwendig sein, wenn sie ständig den Kopf wegreißen. Bei meinen Vierbeinern gehe ich so vor, dass ich den Schwamm am Pferdegesicht anlege und ruhig, aber bestimmt die Anweisung gebe, den Kopf zu senken. Sobald der Kopf runterkommt, entferne ich den Schwamm auch schon. Wird der Kopf wieder hochgerissen, bleibt der Schwamm - oder kommt wieder! Diese Verknüpfung von

- der Anweisung (=Sprachbefehl) mit
- einer erhofften Handlung (= Schwamm weg vom Kopf!)

ist nichts anderes als Konditionierung.

Sie funktioniert ohne Strafen, ohne starken körperlichen Zwang und ist auch über kleine, zunächst eher unscheinbare Gesten umsetzbar. Ein schöner Nebeneffekt entsteht hier dadurch, dass du dabei immer ganz klar Führung und Sicherheit zeigen kannst. Du stellst eine vom Pferd durchaus leistbare Anforderung und bist geduldig, aber klar bis zur Ausführung.

Diese Art der **Kommunikation** zwischen deinem Pferd und dir wiederum stellt sicher, dass

a) das Tier aufnahmebereit ist, also deine Anweisung wahrnimmt und

b) du sichergestellt hast, dass dein Tier diese Anweisung auch versteht.

Trifft a) auf b), dann scheint die Hilfengebung fast schon wie Telepathie zu funktionieren. Ruppiges am Zügel zerren oder touchieren sind nicht nur nicht *nötig*, sie würden die feine Balance aus Kommunikation und Reaktion auch einfach nur stören.

Eine feine und damit sensible Kommunikation über die Hilfen ist anstrengend, denn du musst zu einhundert Prozent beim Vierbeiner bleiben, um nur ja nichts zu verpassen.

Bei einem optimalen Einsatz der Hilfen ist die Prämisse, dass die treibende Hilfe nur genauso lange bleibt, bis die Reaktion vom Pferd kommt, dann wird sie **sofort** reduziert.

Beispiel: Einzelne Wade anlegen → Pferde geht seitwärts von der Wade weg → Anlegen der Wade reduzieren

Einen desensibilisierenden Hilfeneinsatz bekommt man wesentlich häufiger zu sehen, dann bleibt die treibende Hilfe nämlich bestehen, obwohl die Reaktion bereits erfolgte.

Beispiel: Einzelne Wade anlegen → Pferd geht seitwärts von der Wade weg → Wade bleibt angelegt → Pferd registriert, dass sich sofortige Reaktion nicht auszahlt.

Hilfengebung über den Sitz

Im Sattel stehen ändert nichts am Reitgewicht - 70/80 Kilogramm bleiben 70/80 Kilogramm, egal ob da jemand im leichten Sitz reitet - oder nicht. Was sich aber ändert ist, dass das Reitgewicht nahezu vollständig auf dem vorderen Schultergurt lastet.

Gerade die Pferde, die noch kaum in eine echte Reitpferdebalance hineingefunden haben, erleben das als enorm stressig. Warum? Na, weil sie bisher schon das Gefühl haben, jederzeit vornüberfallen zu können. Und nun, mit dem übermäßig eingesetzten Zügeln und dem Reitgewicht im wortwörtlichen Nacken verstärkt sich das Gefühl nur noch mehr!

Hilfengebung über den Sporn

Der Sporn berührt im Einsatzfalle die Bauchmuskulatur des Pferdes. Das kann unter Umständen dazu führen, dass Bauchmuskeln angespannt und damit auch der Rücken gebeugt werden. Braucht es den Einsatz von Sporen, um eine Beugung des Rückens zu erreichen? Natürlich nicht. In meinem Training kommt der Sporn nur sehr selten zum Einsatz.

Solche Spielereien werden auch gerne an der Longe ausprobiert. Dann wird mit irgendwelchen

Gerätschaften an der Bauchunterseite am Reflexpunkt der Linea Alba herum manipuliert, um eine Rumpfhebung zu *trainieren*. Wobei hier natürlich genau nichts trainiert wird. Dabei ist es völlig egal, ob für dieses Kratzen am Reflexpunkt Stöcke, Stifte, komische Bratpfannen oder sonst irgendwelche Küchengerätschaften mit marketingoptimierten neuen Namen genutzt werden.

Viel wichtiger als seltsame Formen der sogenannten Hilfengebung (die keine sind), sind pferdegerechte Pausen zwischen den Lektionen.

Pausengestaltung für Reitkunstpanzer

Wie gestalte ich die Pause beim Versammlungstraining? Für eine effiziente und damit wirklich erholsame Pause ist es sinnvoller, wenn dein Pferd lernt, sich im Stand mit hingegebenen Zügeln zu entspannen. Das kann anfangs noch nicht jedes Pferd, schon gar nicht mit Reitgewicht knapp hinter dem Widerrist.

Ohne die Einrahmung durch den Zügel kann sich das für den Vierbeiner anfangs so anfühlen, als würde er jeden Moment vornüberkippen. Die Pause im Stand und mit hingegebenen Zügeln ist allerdings aus neurologischer und aus trainingstechnischer Sicht ideal. Natürlich gibt es manchmal trotzdem gute Gründe, aktuell oder dauerhaft trotz Pause weiter im Schritt zu gehen.

Häufig wird aber völlig grundlos zur Pause zwischen zwei Einheiten ohne hingegebenen Zügel und im flotten Schritt weiter im Kreis herumgeritten. Das ist genau genommen *keine* Pause, sondern nur eine andere Form der Bewegung. Gerade bei kurzen Bewegungsabläufen mit bewusstem Krafteinsatz sind Pausen zwischen den Sets angebracht, um die Leistung des Muskels bei der folgenden Übung zu erhalten und eine allgemeine durch freigesetzte Milchsäure verursachte Muskelerschöpfung zu reduzieren (Staniszewski et al. 2024).

Ich würde das für meine Pferde jedenfalls präferieren, um ihre Leistungsfähigkeit insgesamt zu erhalten. Ruhiges Stehen erlaubt dem zentralen Nervensystem (ZNS) und damit dem Kreislauf herunterzufahren, ohne zusätzlich Energie für an dieser Stelle sinnloses im Kreis herumschlappen bereitstellen zu müssen. Die Bewegungsmuster müssen nämlich zunächst *sortiert* werden.

Das zentrale Nervensystem lernt nicht während der eigentlichen Aktion, sondern erst danach. Eben dann, wenn es zur Ruhe kommt. Erst dann, *in Ruhe* konsolidiert das motorische Gedächtnis neue Bewegungen, die Koordination wird angepasst und die Bewegungseffizienz verbessert. Ohne Pause gibt es also keine motorische Entwicklung. Schön dumm, denn genau das wollen wir mit den Lektionen doch erreichen!

Wie lange ist eine gute Pause?

Je schwerer und neurologisch fordernder die Versammlungsphase ist, desto eher ist Ruhe im Stand in der Pause die bessere Option.

Im Grundsatz sind 90 Sekunden bis drei Minuten eine ideale Basis für eine wirksame Pause. Bei starken Versammlungsintervallen sind auch gerne mal fünf Minuten als ruhige Standpause angebracht.

Eine zu kurze Pause macht sich im Rahmen einer ZNS-Fatigue (zentrale Ermüdung) bemerkbar. Dann zeigt dein Vierbeiner:

- Konzentrationsabfall,
- ungenaue Bewegungen,
- vermehrte Koordinationsfehler,
- sinkende Muskelansteuerung.

Letzteres zeigt sich durch geringere Kraft trotz leerem Muskelgefühl.

Das ZNS braucht Pausen!

Die neuromuskuläre Steuerung ist unglaublich energieaufwendig, vor allem bei neu zu erlernenden Bewegungsabläufen. Denk nur daran, wie du dich beim Erlernen einer neuen Sportart fühlst. So ein

In der Handarbeit hat man eine andere Perspektive aufs Pferd. Mascha mit Honey.

Set an mehreren, sauber ausgeführten Kniebeugen sieht vielleicht erst einmal einfach aus, aber die Umsetzung kostet dann doch erstaunlich viel Muskelkraft – und eben auch kognitive Energie!

Denn das ZNS ist bei jeder Bewegung, vor allem aber bei den präzisen und kontrollierten Bewegungen voll aktiv. Es sendet dabei elektrische Signale (Aktionspotenziale) über die Nervenbahnen an die Muskelfasern. Bei intensiven Trainingsreizen sinkt die Erregbarkeit der Nervenzellen. Die Signalgeschwindigkeit und die Fähigkeit zur Rekrutierung großer motorischer Einheiten verringert sich.

Der Körper benötigt seine kleinen Pausen zwischendurch, um diese Systeme wieder aufzuladen.

Die Materialbeschaffung braucht Zeit!

Über ruhige Standpausen verschaffst du dem Körper deines Pferdes wertvolle Zeit, um neues Baumaterial in Form von Proteinbausteinen für die Muskeln auf den Weg zu bringen. Die ganzen Reparatur- und Aufbauarbeiten starten schon während des Trainings, dauern aber darüber hinaus noch einige Tage an. Feinste Muskelrisse müssen ausgebessert, Glukosespeicher aufgefüllt und Mikronährstoffe an ihren Bestimmungsort gebracht werden. Das aber gelingt nur bei einem zentralen Nervensystem in Ruhe.

Außerdem ermöglichen wir über die Pause im Stand die Chance auf eine hohe muskuläre Spannung. Wenn ich das Pferd stattdessen im Mittelschritt durch die Halle rasen lasse, verschenke ich dieses Potenzial für die Bildung einer optimalen, das Volumen vergrößernde mechanische Spannung im Muskel.

Keine Leistungsbeurteilung ohne Blickschulung

Welche Haltung oder Bewegung bekomme ich vom Pferd angeboten – und ist das für das Pferd dauerhaft gesund? Wenn ja – schön! Aber falls das nicht der Fall ist, dann ist das der Punkt, um in der Lektionsarbeit genauer hinzusehen. Genau das ist das Stichwort in diesem Abschnitt: Hinsehen! Es geht darum, deinen Blick für die feinen Details deiner Arbeit am und mit dem Pferd zu lenken.

Achte nicht nur aufs Maul ..

... oder auf die Ohren. Denn einzelne Gesichtsbestandteile wie Maul oder Ohren beschreiben nur einen kleinen Ausschnitt der Mimik deines Pferdes. Sicher, vom Rücken aus kannst du oft nur das Spiel der Ohren in seiner ganzen Ausdrucksstärke beobachten.

Aber spätestens bei der Bodenarbeit oder bei der Arbeit mit Foto-/Videoaufnahmen kannst du dir das gesamte Pferdegesicht anschauen – und wirst staunen, wie fein die Mimik eines Pferdes eigentlich ist, wenn man sich als Mensch mal die Zeit nimmt und genauer hinsieht.

In einer gerade frisch veröffentlichten chilenischen Studie filmte man 36 Pferde in unterschiedlichen Alltagssituationen. Die mimischen Ausdrücke wurden im Anschluss daran kategorisiert und zusammengefasst. Über 800 Ausdrücke kamen am Ende zusammen. Nur ein Teil ist auf die typischen und offensichtlichen Marker wie Ohren, Maul und Zunge zurückzuführen. Stattdessen verwenden Pferde eine viel größere Kommunikationsbandbreite über Nutzung der seitlichen Gesichtsmuskeln, um ihre Stimmung und ihre Bedürfnisse gegenüber Artgenossen auszudrücken.

Dabei zeigte sich, dass unter Reitfans typischerweise als aggressiv gelabeltes Verhalten wie zurückweisende, flach angelegte Ohren und erweiterte Nüstern durchaus auch im Spiel untereinander auftreten, ohne sonstige aggressivere Untertöne.

Bemerkenswert, aber wenig überraschend ist es die erste Studie überhaupt, die die Gesichtsausdrücke unter Pferden im sozialen Kontext untersuchte. Die Forschenden halten am Ende fest, dass eine ganzheitliche Betrachtung aus Körperhaltung und Gesichtsausdruck optimaler ist, um den Zustand eines Pferdes besser einschätzen zu können:

> *»The within-context variation in facial behaviour observed, both within and between individuals, highlights the importance of utilising whole-body behaviours, alongside facial behaviour, when evaluating the perceived intent of social signals (Lewis et al. 2025).«*

Auf Deutsch: »Die kontextabhängige Variation der Mimik unterstreicht sowohl bei einem einzelnen Pferd als auch als Ergebnis der Interaktion zwischen verschiedenen Pferden, wie wichtig es ist, neben der Mimik auch das Verhalten des ganzen Körpers zu berücksichtigen. Nur so lässt sich die wahrgenommene Absicht von sozialen Signalen bewerten.«

In der Altklassik sind das im Grunde keine neuen Erkenntnisse, denn hier ist es nie nur das Spiel oder Ohren oder nur das bewegliche Maul allein, das uns viel über den Zustand des Pferdes verrät.

Lektionsarbeit selbst ist natürlich anstrengend, weder Pferde noch Menschen können dabei total entspannt gucken. Die gesamte Haltung während, vor allem aber davor und danach verrät uns enorm viel über den Zustand des Pferdes.

Bei der Blickschulung geht immer um eine möglichst umfassende Gesamtbetrachtung des individuellen Pferdes. Und es wäre schön, wenn das auch in der nicht von der Dressurkunst überzeugten Reitwelt ankäme.

Kriterien zur Blickschulung

Behalte bei der Beurteilung einer gezeigten Lektion, sei es bei deinem eigenen Reitpferd oder bei anderen Pferden auf der Bahn immer im Hinterkopf:

Es handelt sich hier um **Momentaufnahmen**. Pferde und auch Menschen haben mal gute und mal schlechte Tage. Das zeigt sich vielleicht in Form dieser scheinbaren Fehler:

- Das Vorderbein kann minimal rückständig in der Versammlung ausfallen.
- Die Nase kann auch mal minimal hinter die Senkrechte gleiten.

Der verlängerte vertikale Standbeinphase ist Bedingung für eine optimale Rumpfhebung (grüne Pfeile).

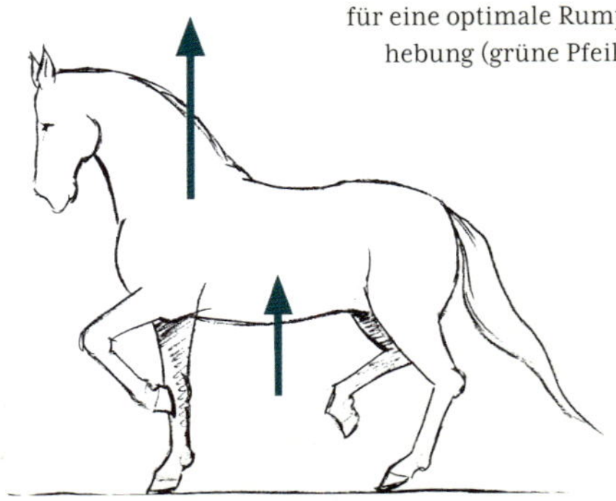

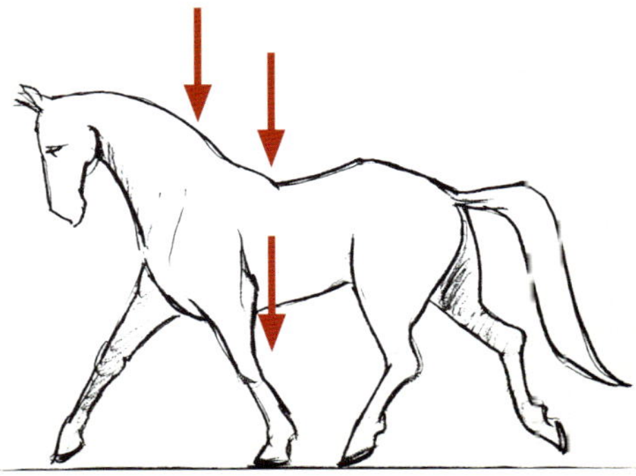

Trotzdem kann die Darstellung beispielsweise der Piaffe an diesem einen schlechteren Tag mit geringeren Leistungsreserven für dein Pferd eine herausragende Leistung darstellen, bei der es sich gerade wirklich angestrengt hat. Gerade die ersten vom Pferd eigenständig angebotenen piaffartigen Tritte sind es wert, dass du ausgiebig und sehr reichlich lobst.

Das generelle Ziel der Arbeit in der Bahn sollte stets die **optimale Muskelbeteiligung je Bewegungsablauf** beim Seitwärts, bei Piaffen, beim Spanischen Schritt usw. sein, ohne dabei sinnlos Meter um Meter in der Bahn zu schruppen, wenn diese Bewegungen für die Muskulatur, Bänder und Sehnen deines Vierbeiners ohne gesunderhaltenen Nutzen sind. Die Pausenzeiten sind davon natürlich ausgenommen.

Ganz im Gegenteil geht es ja darum, dass genau diese Bewegungen das Tier kräftigen und schützen sollen. Dafür braucht es natürlich vor allem voluminöse Muskelpakete, aber auch die korrekte Ansprache der einzelnen Areale. Auf dem Pferd sitzend ist das relativ schwierig einzuschätzen - aber möglich, wenn du weißt, worauf du achten musst.

Tipps für die eigene Blickschulung

So manche unschöne Details fallen erst mit etwas Abstand auf, aber auch Fortschritte in der täglichen Arbeit mit dem Pferd gehen gerne mal unter. Betriebsblindheit trifft auf Perfektionismus, beides bringt dich aber in der Dressurkunst nicht weiter. Es braucht ein gesundes Maß an Detailverliebtheit, wobei du immer wieder auch möglichst externes Feedback brauchst, um dein Pferd gesunderhaltend zu arbeiten.

Schritt 1: Fotos, Fotos, Fotos

Dokumentiere deine Arbeit am Boden und unter dem Sattel mit Fotos, noch besser mit Videos. Nutze Begleitpersonen oder ein Stativ, um möglichst Ganzkörperaufnahmen vom Pferd zu erhalten. Mach entweder Fotosequenzen oder verwende Videos, die du dann in Zeitlupe abspielen kannst.

Schau dir im Anschluss an die Lektion das Bildmaterial im Detail an und überprüfe, wie sich Genick, Kruppe, Hanken und Bauch-/Rückenlinie verhalten.

Tipp:

Wenn du mit gedachten Linien zur Beurteilung von Vorder-/Rückständigkeit nicht gut zurechtkommst, druck dir das Bildmaterial aus. Dann nimm ein Blatt Papier und decke Teile des Pferdes ab, wahlweise Oben-Unten oder Vorne-Hinten. Betrachte dann erst einmal nur diesen Teil.

Vergiss nicht, den Fortschritt zu dokumentieren! Bei Reha-Pferden oder bei solchen, die lange durch übermäßiges Touchieren in Fehlhaltungen gezwungen wurden, werden die Fortschritte erst über längere Zeitverläufe sichtbar.

Schritt 2: externes Feedback
Externes Feedback durch in der Altklassik geschulte Personen kann dabei helfen, bislang nicht behobene Technikfehler zu erkennen. Vor allem aber hilft es dabei, noch feiner in der Hilfengebung zu werden! Leider fällt die Auswahl an geeigneten Profis mit echtem Wissen zur altklassischen Dressur unglaublich dünn aus.

Es gibt natürlich sehr viele Personen, die nach FN-Richtlinien arbeiten und vermeintlich Ahnung von Dressur haben. Die helfen dir an dieser Stelle erfahrungsgemäß null weiter, weil es doch häufig genug kaum über teure Sozialkontakte hinaus geht und damit die reinste Geldverschwendung ist. Bevor du da unnütz Geld verbrennst, kannst du notfalls das 1:1-Trainingsfeedback auslassen und dich lieber im Sinne von *Do-it-yourself* über meine Online-Kurse selbst weiterschulen.

Externes Feedback durch Personen, die sich mit Pferdegesundheit auskennen, ist natürlich auch sinnvoll, aber die sind leider ebenfalls sehr rar. Idealerweise kennt die Person dabei sowohl eure Ausgangssituation als auch euer Ziel und versteht mehr von Pferdegesundheit als der durchschnittliche Veterinär das oft tut.

Kurz nochmal etwas Eigenwerbung: In meinen Storys und Beiträgen auf Instagram veröffentliche ich regelmäßig aktuelle Studienergebnisse und Erkenntnisse aus der Forschung, die praktische Auswirkungen auf die Gesundheit und das Wohlbefinden unserer Pferde haben können.

Schritt 3: Gehirntraining am besten Beispiel
Im Abschnitt zur Gehirnveränderung durch Bodybuilding kam es bereits zur Sprache. Dein Gehirn lernt - auch - durch gute Beispiele. Statt moderner Dressurwettkämpfe suchst du dir aber vielleicht lieber altklassische Reitdarbietungen aus, die genau die tänzerische Anmut zeigen, die auf den Prinzipien der barocken bis preußischen Reitmeister beruht.

Unbewusst speichert dein Gehirn ab, wie sich Mensch und Pferd als Einheit bewegen.

Ganz bewusst kannst du auch hier - Sequenz für Sequenz oder Bild für Bild - die Position der Fixpunkte analysieren:

- Wo ist das Standbein?
- An welchem Punkt der Bewegungsfolge sinkt die Kruppe nach hinten ab?

Blickschulung gefällig?

Da habe ich was für dich vorbereitet. Einfach den QR-Code scannen – und gucken.

Video

TRAINING: DOPING FÜR DIE HIRNCHEMIE

KURZ UND KNAPP:

- Gutes Doping mit körpereigenen Hormonen ist die schönste Belohnung.
- Depressive Wallache gewinnen Selbstbewusstsein durch Dressurkunst.
- Testo-Stuten lassen sich im Training etwas ausbremsen.
- Kortisol ist nicht der Teufel. Vermeiden ist gar nicht nötig!

Kennst du den Ausspruch: »In einem gesunden Körper wohnt ein gesunder Geist?« An dieser Weisheit ist tatsächlich was dran, sagt die Forschung. Denn der Aufbau der Gehirnstruktur hängt tatsächlich mit der Muskulatur und ihrer Funktionsfähigkeit zusammen (Kilgour et al. 2014).

Neuere Forschung zeigt außerdem, dass psychische Erkrankungen wie Depressionen mit einer Veränderung der Gehirnstruktur einhergehen – und dass diese Veränderungen bei den untersuchten Frauen auch abhängig vom Gewicht sind (Łapińska et al. 2024). In der Studie von Łapińska zeigten sich diese Gehirnumbauten insbesondere bei Frauen mit Mehrgewicht, eine Unterteilung der Studienteilnehmerinnen nach Muskelmasse wurde leider nicht unternommen, hätte aber sicherlich ebenfalls interessante Ergebnisse gezeigt.

In einer aktuellen Studienveröffentlichung, die auf Daten von über 7.000 Menschen im mittleren Alter basiert, wurde zwar nicht das Gehirn auf eventuelle Umstrukturierungen untersucht. Dafür konnte der komplexe Zusammenhang zwischen niedriger Muskelmasse und erhöhtem Risiko für Depressionen belegt werden. Die Forschenden empfehlen am Ende ganz klar, sowohl die Veränderungen an der Muskelmasse ab der Lebensmitte zu tracken, als auch die Ernährung entsprechend anzupassen, um dem Muskelabbau vorzubeugen (Gan et al. 2025).

Nein, nicht alle Übergewichtigen sind depressiv. Aber, und das ist ein großes Aber: Veränderungen in der Gehirnstruktur, Depression und Übergewicht beziehungsweise eine **sehr geringe Muskelmasse** gehen häufig Hand-in-Hand.

Der Grund dafür ist unter anderem, dass Botenstoffe aus entzündungsfördernden Fettdepots, wie oben belegt die Gehirnchemie verändern. Das Empfinden von Freude wird schwieriger, wenn die Bildung von Endorphinen (Glückshormone) auf molekularer Ebene beeinträchtigt ist. Und wer keine Freude empfindet, der bewegt sich von sich aus viel seltener. Ein Kreislauf, der auch bei Pferden zu beobachten ist.

Zum Glück gibt es einen kleinen Lichtblick, wenn dein Tier eher durchschnittlich gebaut ist. Bei einem Pferd mit mindestens durchschnittlicher Muskelmasse ist allein das Vorhandensein selbiger ja bereits ein Anzeichen dafür, dass über die Nahrung ein Mindestmaß an Proteinen aufgenommen wird, die dann über die Proteinsynthese zu Bausteinen für den gesamten Organismus verarbeitet werden. Allein die Tatsache, dass ein Pferd so viel Muskulatur aufbaut, zeigt das eine gewisse Aminosäurenbasis da ist. Beruhigend, nicht wahr? Bei einem sehr dicken Pferd oder einem, dass zwar äußerlich schlank erscheint, aber schlapp und lethargisch wirkt, stimmt die Aminosäurenversorgung aber unter Umständen nicht.

Auch optisch eher schlanke Pferde können lethargisch wirken und ihr Interesse an Herausforderungen verlieren. Die Motivation für alle möglichen Interaktionen mit dem Menschen ist dahin, das ewige im Kreis herumschleudern an der Longe führt höchstens noch zu schlurfenden Schritten.

Der typische, dezent ängstlich-vorsichtige Reitfan findet das aber oft ganz nett und interpretiert das als entspanntes, ja sogar als glückliches und zufriedenes Pferd, das sich einfach handeln lässt und keine großen Kraftanstrengungen erfordert. Ein ständig wie tot herumstehendes teilnahmsloses Pferd sollte bei Pferdemenschen ganze Kirchtürme voller Alarmglocken läuten lassen. Freude ist an dieser Stelle wirklich nicht angebracht!

Wir können davon ausgehen, dass sich auch hier die Gehirnstruktur verändert. Kommen Schwierigkeiten mit dem Bewegungsapparat dazu, wird das Tierchen zudem meist schlicht abgeschoben. Dann landet es untherapiert und weit vor seiner Zeit auf der Rentnerweide, wo sich sämtliche Symptome nur weiter verschlechtern.

Mit echter Reitkunst schaffen wir Abhilfe bei Bewegungsproblematiken **und** wir legen die Basis für eine Gehirnchemie, die Reitpferden ein gesundes, ausgeglichenes und langes Leben ermöglicht. Dressur im Sinne der Alten Meister stärkt das Selbstbewusstsein deines Pferdes. Ja, auch dann, wenn ihr am Anfang noch völlig zurecht über die ersten Schritte stolpert, wenn sich alles holprig und wenig

Freundlicher Wallach auf Kuschelkurs. Gina mit Davidoff.

elegant anfühlt. Die hohe Kunst der Versammlung besteht im Grunde aus imposanten Gestiken (= Imponiergehabe). Diese kraftvollen Gesten in den Lektionen zu erarbeiten - und dafür stetig gelobt zu werden - macht auch was mit der Psyche des Pferdes. Und zwar etwas Gutes!

Mit jeder Lektion festigt sich das Wissen um die richtigen Bewegungsabläufe. Davon profitieren am Ende alle: In der klassischen Schulreiterei ausgebildete Pferde haben in der Regel keine Schäden am Bewegungsapparat. Sie werden gezielt gymnastiziert und durch diese Gymnastik wird Muskelmasse optisch sichtbar aufgebaut, um genau diese Schäden zu vermeiden. So entsteht ein schützendes Muskelpolster, welches Gelenke und Sehnen zusätzlich abpuffert. Das ist nicht nur aktuell wichtig, sondern auch im Alter!

Tänzerisch anmutende Dressurkunst entsteht erst dann, wenn die Grundbedingungen stimmen. Deshalb ist es neben den Grundlagen in der Schulreiterei auch wichtig zu verstehen, wie sich aufgrund des anabolikafreien Bodybuildings in der Bahn die Chemie des Pferdehirns verändert.

Pferdetemperament: hitzig vs. phlegmatisch

Die einen kommen kaum aus dem Quark, die anderen sind nur schwer zu bremsen: Pferde haben recht unterschiedliche Temperamente. Im Kern zeigt sich die charakterliche Neigung aber, wie beim Menschen auch, in zwei Polen auf dem Spektrum. Auf der einen Seite sind die Schnellstarter, die hitzigen Turboraketen mit Hummeln im Hintern zu finden. Auf der anderen Seite sind die gemütlichen, langsamen, kaum aus der Ruhe zu bringenden Pferde angesiedelt. Zwischen diesen beiden Extrempolen pendelt sich im Grunde jedes Tier ein, mit mal stärkerer, mal schwächerer Ausprägung für eine Seite. Für die individuell pferdegerechte Versammlungsarbeit ist es sinnvoll, wenn du beim Krafttraining auch das Temperament deines Pferdes beachtest und dich nicht ausschließlich auf seine körperlichen Baustellen konzentrierst.

Die Schnellstarter: Hitzige Pferde ticken gern mal mehr und mal weniger dezent aus, hören Ameisen am Stall vorbei trippeln und neigen dazu, hier und dort die erwünschten Lektionen auf ihre ganz eigene Art umzusetzen. Wenn du ein Tier im Vollgalopp durch die Halle sprinten siehst, nachdem die Hallentür etwas zu laut zugefallen ist, könnte ein hitziges Pferdetemperament dahinterstecken. Das liegt an ihrem empfindlicheren Nervensystem, das auf Dopamin und Androgene sensibler reagiert. Dadurch zeigen sie mehr Bewegungsdrang und mehr Reaktionsbereitschaft. Typische Vertreter dieser Spezies sind alle Rassen mit einem höheren Vollblutanteil. Insbesondere auch die Iberer sind teilweise extrem auf Leistung gezüchtet und manche Vertreter gelten als extrem unentspannt an der Hand.

Hitzige Pferde

Hitzige Pferde setzen oft blitzschnell Stresshormone wie Kortisol und Adrenalin über ihre Nebennieren frei. Die Herzfrequenz steigt, die Aufmerksamkeit ist sichtlich auf bestimmte Reize ausgerichtet. Das ganze Tier gleicht einem gespannten Bogen.

Das liegt dann am Noradrenalin, auch das wird von den Schnellstartern schneller und in höherer Konzentration ausgeschüttet. Dazu kommt dann noch Dopamin, das regelt sowohl die Motivation als auch die Reizempfindsamkeit. Unter diesem Einfluss sind die hitzigen Vierbeiner dann wesentlich neugieriger, gerne aber auch mal ängstlicher als gleichgültigere Artgenossen.

Explosive Bewegungen und die unterm Sattel spürbare Muskelanspannung geben dir das Gefühl, auf einer ~600 Kilogramm-Bombe mit Kampf-oder-Flucht-Tendenzen zu sitzen.

Allerdings erholen sich die Raketengeschosse von diesem Stress-Boost auch schneller, ihr Nervensystem ist häufig flexibler und kehrt trotz der Stressspitze zügig wieder in die ausgeglichene Ruhelage zurück. Insgesamt sind die hitzigen Vierbeiner anfälliger für chronischen Stress, denn wenn das

Nervensystem erst mal aus dem Takt gerät, bleibt es auch gerne mal so.

Laufbegierige Energiebündel – ob Mensch oder Pferd – erleben bei intensivem Training das sogenannte **Runners High**. Ein Zustand, bei dem durch die körperliche Anstrengung schließlich Endorphine (Glückshormone) ausgeschüttet werden. Dieses körpereigene Doping sorgt für Euphorie und noch mehr Leistungsbereitschaft.

Statt dein Pferd über lange intensive Strecken hinweg müde zu reiten, wird es nur noch aufgedrehter und energiegeladener. Bis es schließlich beinahe umfällt, weil seine Energiereserven gnadenlos erschöpft sind. Die ausgeschütteten Endorphine plus das allgegenwärtige Adrenalin lassen das Tier seine körperlichen Laufgrenzen gnadenlos überschreiten. Es spürt die Erschöpfung einfach nicht. Gerade bei einem Pferd mit hitzigem Temperament ist stundenlanges Highspeed-Gerenne durch das Gelände oder in der Halle also exakt eins nicht: hilfreich, denn das Torpedo-Pferd spürt seine grenzwertige Erschöpfung keinesfalls.

Stattdessen erlebst du die negativen Effekte solcher fehlgeschlagenen Ermüdungsversuche hautnah:

- **Gegenteilige Effekte.**
  Müde wird das Tier nicht. Höchstens *noch* fitter. Und damit auch schwerer zu handeln. Außerdem leidet der Bewegungsapparat, weil es sich selbst und körpereigene Belastungsgrenzen einfach nicht wahrnimmt!
- **Steigender Stress.**
  Statt geistiger Klarheit erlebt das Pferd Stress pur. Sie sind mental überfordert, die erhoffte Ruhe und Gelassenheit stellen sich bei konstanter Trainingsüberforderung eben nicht ein.
- **Erhöhter Energieverbrauch.**
  All das Gerenne und Gespringe hält die Level von Kortisol, Adrenalin und Co. stetig oben. Das Nervensystem erhält keine Chance auf Ruhephasen. Das ganze Tier wirkt stets überdreht.

Was hilft wirklich? Um deinem Torpedo-Tier Ruhe und Gelassenheit zu bieten, kannst du:

- **gymnastizierende Arbeit einsetzen.**
  Langsame kontrollierte Übungen wie die Seitengänge bremsen dein Energiebündel nicht nur sicher aus, sie fördern auch die Konzentration und helfen dabei, eventuell vorhandene Verspannungen effektiv zu lösen.
- **progressive Muskelentspannung nutzen.**
  Menschen mit hohen Kortisolwerten neigen beispielsweise zu Nackenschmerzen. Auf große Anspannung sollte möglichst große Entspannung folgen – beim Pferd lässt sich das über versammelnde Arbeit erreichen, die nach der körperlichen Muskel-Anspannung dann eine stärkere Muskel-Entspannung ermöglicht.
- **Übungen zur Versammlung einbauen.**
  Die erfordern den Einsatz von echter Körperkraft und das beansprucht deinen Vierbeiner entsprechend. Ideal geeignet sind für den Anfang Übergänge zwischen den Seitengängen und Piaffe-Ansätze. Die Anstrengung der Bewegung im Seitwärts bremst das Torpedopferdchen zunächst und stärkt seine Koordinations- und Konzentrationsfähigkeit. Zugleich ist Krafttraining, und nichts anderes ist Versammlung an dieser Stelle, bewusstes und kontrolliertes Ausführen von Bewegungen und dadurch relativ verletzungsarm. Zum Vergleich: Bodybuilding wird mit 0,24–7,5 Verletzungen/1.000 Stunden bewertet (Ritsch 2020), während beispielsweise im Profi-Fußball 40,8 Verletzungen auf 1.000 Stunden Training und Wettkampf anfallen (VBG Report 2024).
- **Rituale und Routinen einführen.**
  Hitzige Pferde profitieren in der Regel von vertrauen Abläufen, von Routinen und einer ruhigen Reiterhand, die möglichst genau weiß, was sie tut. Wenn du dein Pferd sauber und gesunderhaltend gymnastizierst, ist ständige Abwechslung gar nicht nötig. Wenn natürlich eher kritische Bewegungsabläufe auf dem Plan stehen, sollten sich die schon eher abwechseln. Pferde sind aber im Grundsatz ganz zufrieden mit vertrauten Routinen und vorhersehbaren Abläufen. Hitzige Pferde profitieren davon ganz besonders, denn so fühlen sie sich sicherer und

können sich unproblematischer auf die Arbeit mit ihrem Menschen einlassen.

Der Bewegungsapparat deines hitzigen Pferdes wird dir für diese Form des Trainings durch dauerhafte Leistung und vor allem durch langfristige Gesundheit danken. Hast du dein hitziges Pferd erst einmal ins Gleichgewicht gebracht, hast du garantiert viel Freude an all dieser Energie, die dir dann für die schönste Reitkunst im Sinne der Alten Meister zur Verfügung steht.

Phlegmatische Pferde

Die Gemütlichen: Phlegmatische Pferde sind ausgeglichen, geraten nicht so schnell in Wallung und überhaupt, sind sie oft diejenigen, um die man gerne beneidet wird. So ein nettes, ruhiges Tierchen ist doch ganz easy unterm Sattel. Richtig?

Nun ja, oft drücken sie den tatsächlich empfundenen Stress nicht so offensichtlich aus, während sie innerlich wiehernd im Kreis galoppieren. Diese äußere Gemütlichkeit wird vor allem unter Belastung und beim Training als Vorteil gesehen. Dadurch erhalten die entspannter wirkenden Pferde aber nicht oder zumindest nicht immer die nötige Beruhigung und Bestätigung ihres Menschen – trotzdem sind auch sie durch innere Stressreaktionen belastet.

Der Hauptgrund für die Gelassenheit entspannter Pferde ist, dass die Nebennieren etwas gemütlicher arbeiten, als die der hitzigen Vierbeiner. Der Kortisolspiegel bleibt also eher niedrig. Adrenalin und Kortisol werden im geringeren Verhältnis und auch teils in geringerer Menge ausgeschüttet.

In Stresssituationen bleiben sie länger ruhig und können sie insgesamt auch länger augenscheinlich, also äußerlich gelassen bleibend, tolerieren. Ist das Kortisollevel aufgrund von Stress aber erst mal erhöht, bleibt das im Vergleich zu den Torpedopferden wesentlich länger so. Die Erholungszeit von Stressmomenten fällt also deutlich länger aus und das Nervensystem wird durch die anhaltende Stressspitze auch stärker belastet.

Die Rakete im Stall benötigt (mehr) Aminosäuretreibstoff

Die Turboraketen mit einem nervösen oder schlicht temperamentvollen Charakter verbrauchen unter Umständen deutlich höhere Mengen an Neurotransmittern als ihre phlegmatischen Gegenstücke, wenn sie sowohl turbogetriebene Raketen als auch schnelle Verwerter und schnell im Abbau der Stoffwechselprodukte sind.

Ein Beispiel aus der Praxis: Das Tier, das bei jedem ungewohnten Geräusch am Stall und auf der Bahn zusammenzuckt, den Schweif schmeißt und nervös unter dem Sattel tänzelt, benötigt abhängig von seiner genetischen Veranlagung höchstwahrscheinlich mehr Tryptophan und Tyrosin für die Neurotransmitterproduktion oder hat einen exorbitant schnellen Stoffwechselabbau.

Denn all diese Reaktionen auf die Umweltreize müssen ja über die Sinne aufgenommen, verarbeitet und vom Nervenleitsystem entsprechend bewertet und mit passenden Reaktionen quittiert werden.

Das Thema Verbrauch und Abbau betrifft phlegmatische und hitzige Pferde gleichermaßen. Diejenigen mit schnellem Abbau neigen auch gerne mal dazu, völlig kopflos rückwärts in Gräben zu krachen, weil sie in Stresssituationen so gar keinen kühlen Kopf bewahren können.

Ein hoher Verbrauch an Neurotransmittern kann dazu führen, dass auch der Bedarf an bestimmten essenziellen Aminosäuren wie Tryptophan, Tyrosin oder Methionin steigt. Essenzielle Aminosäuren kann der Körper nicht selbst herstellen, sondern muss sie über das – hoffentlich geeignete – Futter aufnehmen.

Ist das Futter nicht auf den höheren Verbrauch angepasst und der Bedarf damit nicht gedeckt, dann führt das relativ schnell zu Symptomen wie Stressanzeichen, zu Reizbarkeit oder zu Müdigkeit. Der Pferdekörper kann nun mal nicht unendlich Neurotransmitter produzieren, wenn die Bausteine für diese Produktion fehlen!

Die genetische Komponente kommt noch dazu. Tierchen mit bestimmten Stoffwechselvarianten und züchtungsbedingten Vorbedingungen reagieren empfindlicher auf entstehende Neurotransmittermängel. Da ist die Vorwarnstufe zu akzeptablen Bausteinengpässen schneller erreicht, als bei einer eher robusten Züchtung.

Apropos Züchtung: Auch die Zeugungsfähigkeit (oder ihr Mangel) macht etwas mit der Leistungsfähigkeit deines Pferdes - und mit seiner Gesundheit.

Hoden weg, Leistung runter

Eine Kastration hat für Hengste grundlegende Folgen. Einige sind erwünscht, andere nicht unbedingt. Natürlich freut sich jeder Pferdemensch, wenn das geliebte Tierchen nicht mehr völlig ferngesteuert hinter jedem hübschen Stutenhintern hinterherrennt. Nur mit den Nebenwirkungen sind nicht alle unbedingt vertraut.

Die eigentliche Kastration verursacht zunächst eine kurze Stressspitze. Sprich: Es wird vermehrt Kortisol ausgeschüttet. Langfristig pendelt sich das aber ein. Ist die eigentliche Operation vergessen, gehts dem Tierchen auch schon wieder gut.

Aber: Mangels Hoden verändert sich das hormonelle Gefüge der Wallache und damit unterscheidet es sich deutlich von der Hormonlage ihrer voll ausgestatteten Hengstkollegen. Die Ursache dafür findet sich im Testosteron, das vorrangig in den Leidig'schen Zwischenzellen der Hoden und den Nebennieren produziert wird.

Hengste sind aufgrund des im Vergleich zu Stuten und Wallachen durchschnittlich höheren Testosterons meist früher ausgewachsen, besser aufgemuskelt und weisen eine stabilere Knochenstärke auf. Sie sind im Schnitt auch wesentlicher hitziger und bringen ordentlich Selbstbewusstsein mit, was aber oft zur Kastrationsentscheidung führt. Schließlich sind Wallache sanftmütiger und einfacher zu handeln, richtig?

Wallach ungleich Hengst

Grundlage für die **Testosteronproduktion** ist Dehydroepiandrosteron, kurz DHEA. Es wird vorrangig in den Nebennieren synthetisiert, ist von der Kastration selbst also nicht primär betroffen. Für die DHA-Werte scheint es aktuell keinen festen Referenzwert zu geben, in Studien wird meist eine Bandbreite von **0,02 - 0,04 mg/ml** für Wallache, Hengste und Stuten vor Beginn der Supplementierung beschrieben (Woodward et al. 2007). Der DHEA-Spiegel bleibt also in der Regel gleich, Schwankungen sind aber nicht ausgeschlossen. Immerhin verändert die kleine Operation mit großer Wirkung die gesamte Androgenproduktion im Pferdekörper.

Bei Wallachen springen, sobald die Hoden zwecks Kastration ausfallen, die Nebennieren zur Produktion von Testosteron ein. Die Konzentration ist aber *deutlich* niedriger als bei Hengsten. Im Labor (Beispieldaten von Laboklin, Stand 2023) finden sich in der Regel diese Referenzwerte für Testosteron:

- Hengste 1,0 bis 5,0 ng/ml (saisonal schwankend)
- Wallache < 0,04 ng/ml
- Stuten < 0,04 ng/ml

Der Unterschied ist allein bei diesem Hormon schon deutlich, nicht wahr? Das macht sich entsprechend auch im Alltag bemerkbar. Die Studienlage zu Testosteronwerten bei Stuten ist nicht besonders umfassend und sicherlich noch ausbaufähig.

Wallache sind in der Regel

- ängstlicher und depressiver (in den Augen vieler Reiterinnen und Reiter aber auch glücklicher),
- haben potenziell weniger dichte Knochen,
- leiden eher an (unbeabsichtigter) Gewichtszunahme,
- bauen potenziell weniger (leicht) Muskulatur auf.

Sie zeigen auch weniger Energie und Antrieb, was sich vor allem beim Training bemerkbar macht. Neben dem fehlenden Schwung bei den Lektionen verstärkt das niedrige Testosteronlevel noch den Abbau der Muskelmasse. Stattdessen wird eher Fett angelagert. In dem wird – wie auch beim Menschen – Testosteron in Östrogene umgewandelt. Dieser Prozess nennt sich Aromatase und findet üblicherweise auch bei Hengsten statt.

Nun ist bei Wallachen die Ausgangslage aufgrund des niedrigeren Testosteronspiegels allerdings so, dass weniger Östradiol aus dem wenigen vorhandenen Testosteron produziert wird. Bei älteren Wallachen kann das, ähnlich wie beim Menschen auch, zu einer Abnahme der Knochendichte führen.

Jede zweite bis dritte Frau nach der Menopause weist ein erhöhtes Risiko für Osteoporose aufgrund der mangelhaften Versorgung mit Hormonen vor, während und nach der Menopause auf (Bundesselbsthilfeverband für Osteoporose e. V., 2024). Bei Männern mit niedrigeren Testosteronwerten macht sich dieser Hormonmangel oft ebenfalls mit brüchigen Knochen, depressiven Episoden, einem höheren Risiko für Verletzungen usw. bemerkbar. Also im Grunde wie bei Frauen auch, nur oft noch ein bisschen schlimmer.

Testosteron ist zudem die Grundlage für Dihydrotestosteron (DHT). Da es am Ausgangshormon mangelt, sinkt der Spiegel beim Wallach erheblich ab. DHT ist unter anderem für die Ausbildung sekundärer Geschlechtsmerkmale zuständig. Bei Hengsten wäre das etwa eine kräftige Halsmuskulatur oder eine ausgeprägte Behaarung.

Dieser gesamte körperliche Unterbau führt generell zu weniger Lust auf alles und speziell zu weniger Lust auf anspruchsvolle Ertüchtigungen. Dressurlektionen, die dann ja viel schneller an den Reserven zehren, machen da besonders wenig Laune – und siehe da, schon wird der Wallach höchstens noch mit mittelmäßigem Anspruch durch die Halle geschickt. Schließlich zeigt er sich auch einfach so unmotiviert und scheint mit ein bisschen im Kreis herumtraben auch ausgelastet genug zu sein.

Die fehlende Bewegung und Förderung kann aber die bei Wallachen häufiger auftretende Ängstlichkeit und Anfälligkeit zu Depressionen noch verstärken.

Um es salopp zu formulieren, hier eine kurze Weisheit zum Umgang mit Wallachen: **Wenn ich dem Hengst die Eier schon nehme, dann muss ich selbst sprichwörtlich Eier zeigen, um dem Tier Sicherheit und Führung zu geben.**

Natürlich gibt es auch die hitzigeren Wallache, die sich weiterhin spritzig und aufgeweckt zeigen. Die grundlegende körperliche Veränderung, das niedrigere Testosteron und die Auswirkungen auf seine gesamte mentale und körperliche Verfassung finden aber nach der Kastration trotzdem statt. Nur das Niveau der offensichtlichen Ausprägung ist ein etwas anderes.

Als verantwortungsvoller Pferdemensch behältst du das im Fokus und schaffst auch dann für optimale Haltungsbedingungen, wenn dein Wallach auf den ersten Blick (noch) nicht total depressiv erscheint, damit das auch schön so bleibt.

Das kannst du tun, damit es deinem Wallach nicht nur gut, sondern richtig umwerfend geht:

1. **Dressurlektionen einbauen.**
   Das fördert den Muskulatur- und Knochendichteerhalt, hebt die Stimmung und aktiviert den Stoffkreislauf. Über den Muskelaufbau aktivierst du zudem das Belohnungssystem des Gehirns. Die muskuläre Belastung fördert die Ausschüttung von Myokinen. Der Insulinkreislauf bleibt stabil, es kommt seltener zu einer Insulinresistenz. Die Insulinresistenz habe ich ganz ausführlich in meinem ersten Buch beschrieben.

Um es an dieser Stelle kurz zu halten: Körperzellen können weniger sensitiv auf den Botenstoff Insulin reagieren. Der Botenstoff ist hauptverantwortlich dafür, dass Energie in Form von Glucose in ausreichender Menge in die Zellen gelangt.

Bei einer verminderten Sensitivität verbleibt die Glucose im Blutkreislauf, der Körper lagert diese Energie in Fettdepots an, während zugleich wichtige Reparatur- und Aufbauarbeiten in den Zellen mangels Glucose liegen bleiben.

21jähriger Wallach ohne Depressionstendenzen.
Sonja mit Leo.

Eine Insulinresistenz führt häufig zu Gewichtsproblemen, der Muskelaufbau ist erschwert, Stoffwechselprozesse kommen ins Stocken.

2. **Kleine Lektionserfolge ausgiebig belohnen.**
   Lobe laut und lobe so, dass sich dein Pferd wirklich **wertgeschätzt fühlt**. Vorsichtig bei zuckerhaltigen Snacks, denn einige Wallache haben da teils ein Thema mit dem Gewicht.

Versammlungsarbeit ist Imponierhaltung! Arbeitet ein Pferd viel in dieser Haltung, reagiert es auch entsprechend psychisch darauf. Zusätzlich stärkst du seine körperlichen Fähigkeiten, was sich oft auch im Rang innerhalb der Stallgemeinschaft bemerkbar macht.

So mancher Vierbeiner, und hier sind sowohl Wallache als auch Stuten und Hengste gemeint, steigt mit stetiger Versammlungsarbeit sogar im Rang auf. Pferde – ob mit oder ohne Hoden – wollen in der Regel mitarbeiten und sich von ihrer besten Seite präsentieren. Mach es deinem Wallach einfach, dir all seine schönsten Tricks zu zeigen.

3. **Ernährung anpassen.**
   Weniger kohlenhydratreiches Getreide, stattdessen eine gute Basis an proteinreichem Futter und – falls dein Tier bereits mit überschüssigen Fettreserven ausgestattet ist – eine eingeschränkte Weidezeit.

Stuten sind zu krasser körperlicher Leistung fähig, Zyklus hin oder her. Mascha mit Szofi.

Einfache, schnell verdauliche Kohlenhydrate sind schlicht Muskelbenzin, schnell verbrannt und wenig nachhaltig. Für das Pferd langfristig besser geeignet sind Ballaststoffe, die vorrangig aus unverdaulichen Kohlenhydraten bestehen - und natürlich eine individuell richtige Mischung aus Aminosäuren. Dazu findest du weiterführende Informationen im Ernährungskapitel.

Überschüssiges Fett, vor allem rund um die Organe, stört den Stoffkreislauf massiv. Um die Gesundheit eines Pferdes nachhaltig zu stärken, ist eine gesunde und auf seinen individuellen Bedarf abgestimmte Fütterung absolut unverzichtbar.

Ignorier den Zyklus nicht!

Ja, Stuten haben einen Zyklus, dank der hiesigen sehr guten Nährstoffversorgung auch durchaus vom Frühjahr bis in den Winter hinein. Bei Wildpferden ist der saisonale Zyklus aber weiterhin der Standard. Tageslichtlänge, Temperatur, Ernährung sowie die Art der Haltung beeinflussen den Eisprung und wann der Zyklus aus saisonalen Gründen aus- und wieder einsetzt.

Auf der Nordhalbkugel ist das in der Regel von Januar bis Mitte März der Fall. Einige Stuten ovulieren trotz saisonaler Pause, zeigen aber keine nennenswerten Symptome an.

Für die Lektionserarbeitung ist der Zyklus gleich in mehrfacher Hinsicht relevant. Stuten sind schlicht zyklische Wesen, ihr Energiehaushalt schwankt stärker innerhalb des hormonellen Rahmens. Die Stoffwechselhormone (auch: Geschlechtshormone) spielen schließlich nicht nur bei der eigentlichen Rosse eine Rolle, sondern beeinflussen auch die Leistungsfähigkeit sowie im langfristigen Verlauf die Fähigkeit, gesunderhaltene Muskulatur aufzubauen!

Geschlechtshormone bei der Stute

Progesteron, Testosteron und Östrogen spielen bei der Stute eine ganz ähnliche Rolle wie bei Frauen mit Menstruationszyklus, allerdings mit spezifischen Unterschieden bei der Ovulationslänge und der Lutealphase (Aurich 2011).

- **Zykluslänge:** 21-22 Tage bei Pferden/~ 28 Tage beim Menschen
- **Zeitfenster für Empfängnis:** 3-7 Tage bei Pferden/5-7 Tage beim Menschen
- **Lutealphase:** 15-19 Tage bei Pferden/14 Tage beim Menschen

Diese Werte sind Durchschnittswerte, sodass die individuelle Stute durchaus auch kürzere oder längere Zyklen mit entsprechend angepassten Ovulationszeitfenstern haben kann (Popova 2021).

Zu Beginn des Zyklus wird bei nicht eingetretener Trächtigkeit die Gebärmutterschleimhaut resorbiert. LH- und Östradiollevel sinken ab, Progesteron und FSH steigen an. Der Follikel reift heran. Zum Zeitpunkt der Rosse beginnen Progesteron und FSH abzusinken, während das LH-Level zusammen mit Östradiol deutlich steigt.

Kleine Hormonkunde

- **Progesteron** ist für das Zustandekommen und den Erhalt der Trächtigkeit nötig. Es wird im Gelbkörper gebildet. (Zu) Niedrige Progesteronlevel sind häufig in Verbindung mit einem Zinkmangel und hohen Kortisol-Niveaus zu beobachten. Progesteron ist maßgeblich für einen gesunden und ausgeglichenen Schlaf (Nolan et al. 2021), für einen optimalen Fett- und Kohlenhydratstoffwechsel und sorgt gemeinsam mit Östrogen für gesunde Knochen (Seifert-Klauss/Prior 2010).
- **Testosteron** wird im eigentlichen Zyklus der Stute kaum Beachtung geschenkt. Für die Lektionserarbeitung und im Knochenstoffwechsel spielt das Hormon natürlich auch bei Stuten eine erhebliche Rolle! Schließlich wird dieses Hormon nicht umsonst gern als das *Hormon des inneren Antriebs* bezeichnet. Fällt der individuelle Hormonwert noch unter die üblichen Referenznormwerte, kann Frau sich einfach zu nichts aufraffen. Es ist auch das Hormon, das im Wesentlichen die Gewichtung bei moralischen Fragen, was sich verkürzt als Gerechtigkeitssinn bezeichnen lässt, beeinflusst (Brannon et al. 2019).
- **Luteinisierungshormon (LH)** entsteht im Hypophysen-Vorderlappen und sorgt sowohl für die Endreifung des Follikels sowie für die Ovulation (Eisprung).
- **Follikel-stimulierendes Hormon (FSH)** dient der Östradiol-Synthese und fördert die Follikel-Reifung.
- **Östrogen**, genauer **Östradiol** wird zur Steigerung von LH und zur Hemmung von FSH benötigt. Es wird aus Testosteron synthetisiert. Östrogen ist ein Kernbaustein für die Knochenbildung (Serota/D'Erasmo 2024), wirkt regulierend auf den Blutdruck, beugt Gefäßverkalkungen vor und unterstützt in Kombination mit Krafttraining auch den Muskelerhalt sowie -aufbau (Lowe et al. 2011).

Zumindest für Menschen sind Veränderungen unter anderem des Ruheherzschlags sowie Anpassungen im Hypothalamus in Abhängigkeit von der jeweiligen Zyklusphase belegt (Prinsen et al. 2025). Die Forschung dazu steht allerdings noch völlig am Anfang. Da kommt also unter Umständen noch mehr zyklusbasiertes Wissen auf uns zu!

Dieser Zusammenhang lässt wenigstens die Vermutung zu, dass sich auch das unter vergleichbaren hormonellen Zyklen befindliche Stutenhirn an die jeweilige Zyklusphase anpasst. Das bedeutet aber nicht, dass sie womöglich deutlich zickiger auftreten, sondern:

- **Stoffwechselhormone (auch: Geschlechtshormone) beeinflussen die Produktion und Ausschüttung zahlreicher weiterer Botenstoffe.** Das kann zu kurzfristigen Wesensveränderungen führen und geht in alle Richtungen: anhänglicher, dominanter, weniger stressresistent, anfälliger für irritierende Geräusche, ausgeglichener etc.
- **Diese unterschiedlichen Hormonlevel sorgen für Veränderungen im gesamten biologischen System:** Körpertemperatur, Herzschlag und Verdauung passen sich entsprechend an, denn das zentrale Nervensystem spricht auf diese hormonell gesteuerten Phasen an. Im Kern bedeutet das, dass sich je nach Zyklus der Körper deiner Stute in mehr oder weniger stressigen Phasen befindet. Und das ohne dass sich an den äußeren Begebenheiten irgendetwas ändern würde! Der gleiche Stall, die gleichen Stallnachbarn, die gleichen Tagesroutinen, aber je nach Zyklusphase kann dein Vierbeiner damit mal mehr, mal weniger gelassen umgehen.
- **Die Kombination aus körpereigenem Stresslevel und Zyklusphase kann zu einer höheren Verletzungsanfälligkeit führen.** Bänder sind unter Einwirkung der Hormonkonzentration in der ersten Zyklusphase weicher, was zu weniger sauber ausgeführten Lektionen bis hin zu Verletzungen bei Überlastung führen kann. Um Verletzungen zu vermeiden, sollten bewusstere und kontrollierte Bewegungsmuster eingesetzt werden. Eine hohe mechanische Spannung im Muskel schützt den sensibleren Bewegungsapparat zu diesem Zeitpunkt im Zyklus! Bei Profi-Sportlerinnen ist dieser Zusammenhang erst seit relativ kurzer Zeit bekannt (Hutterer 2021).

Achtung Klischee:

Sind Stuten in der Rosse zickiger und schlechter zu handeln? In einer polnischen Studie zeigte sich, das rossige Stuten **toleranter** und **freundlicher** gegenüber Menschen – und anderen Pferden – waren, als ihre gerade nicht rossigen Studienteilnehmerinnen (Stachurska et al. 2023). Sie suchten auch eher nach Körperkontakt und waren offener für Zuwendungen in Form von Streicheleinheiten. Eigentlich ganz logisch, schließlich sollte der Hengst der rossigen Stute in der Ovulationsphase auch näherkommen können.

Für unsere therapeutische Reitkunstarbeit mit Stuten ist mit Blick auf den hormonell ähnlichen Ablauf des Zyklus wichtig:

1. Behalte bei der Lektionserarbeitung den Zyklus deiner Stute im Hinterkopf. Nein, du musst keinen Zykluskalender führen, aber doch zumindest grob wissen, wann die nächste Rosse ansteht.
2. Befass dich bei umfangreichen Ernährungsumstellungen auch damit, welche Nährstoffe für die Progesteron-, Östrogen- und Testosteronhormone dringend benötigt werden.
3. Beobachte genauer, welche Maßnahmen aus therapeutischer Sicht teils wahllos besser oder schlechter anschlagen – und vergleiche das mit dem ablaufenden Zyklus.

Wer eine Stute besitzt, sollte durchaus einen stabilen gesunden und optimalen Zyklus bei seinem Pferd anstreben.

Warum der Zyklus eine Rolle spielt: Krebstherapie-Edition

In der Forschung waren Frauen, vor allem diejenigen mit Zyklus, bis in die 90er Jahre hinein häufig aus der Medikamentenforschung ausgeschlossen. Bei Pferden ist das mit Blick auf die umfangreiche Datenlage zwecks optimierter Züchtung schon ein bisschen besser. Optimal ist es aber auch hier nicht, denn: Die Zyklusphase wird bei Therapieoutcome-Vergleichen längst noch nicht standardmäßig erfasst.

Warum das aber wichtig sein könnte, zeigt eine Studie aus Belgien. Hier wurde an Labormäusen zur Chemotherapie bei Brustkrebs geforscht. Einige reagierten hervorragend auf die Therapie, bei anderen zeigte sich keinerlei Verbesserung. Bloß warum?

Ganz einfach: Startete die Therapie an den Labormäusen zu einem spezifischen Zeitpunkt des Mäuse-Menstruationszyklus, griff die Therapie. Zu anderen Zykluszeitpunkten dagegen waren die Brustkrebstumore weit weniger anfällig für die Behandlungsmethode. Erste Datenvergleiche mit menschlichen Brustkrebspatientinnen liefern zusätzliche Daten für diesen Zusammenhang (Bornes et al 2025).

Obwohl Krebszellen selbst keine Rezeptoren für Hormone haben, scheint der Zyklus und damit das Level an Östrogen, Progesteron oder Testosteron doch entscheidend dabei zu sein, ob sich die Krebszellen in einem grundsätzlich therapieoffenen Umfeld befinden – oder nicht.

Startet die Therapie zu Beginn eines Zyklus, war die Krebsbehandlung effizienter und führte eher zu einer Reduktion der Tumore. Grund dafür scheint nach ersten Forschungsergebnissen das Progesteronlevel zu sein, das in der ersten Zyklushälfte signifikant niedriger ausfällt, als in der zweiten Hälfte nach der Ovulation.

Startet die Therapie dagegen in der zweiten Hälfte, wuchsen die Tumore trotz Chemotherapie weiter.

Die Forschung darf hier gern noch weiter an den Gründen für diese Zusammenhänge feilen.

(Stress-)Hormonkunde

Hormone sind chemische Botenstoffe, die eine Vielzahl an Funktionen im Körper regulieren. Im Rahmen des gesunderhaltenen Pferdetrainings spielen vor allem die Hormone eine größere Rolle, die Aufmerksamkeit, Freude – aber auch die Regulation von Stress betreffen.

Die Produktion dieser Botenstoffe findet nicht ausschließlich im Gehirn statt. Stattdessen ist die Produktionsstätte nahezu global und verteilt sich beim Pferd auf den Hypothalamus und die Hypophyse im Gehirn, Eierstöcke und Hoden, die Schilddrüse und die Nebennieren. All diese Organe spielen zusammen, um den Hormonhaushalt und den Stoffwechsel optimal zu regulieren.

Und wenn alles tut, was es soll, dann sind die Pferde fit, im Mittel gut gelaunt und relativ resistent bei Stress. Geraten die Hormone durcheinander, kann das unter anderem zum Equinen Metabolischen Syndrom (EMS) führen (Menzies-Gow 2023).

Wie groß die Rolle einzelner Hormone für den Muskelauf- und den Fettabbau ist, wird bei Pferden gerne mal vergessen. Dabei entscheidet ein optimaler Hormonkreislauf durchaus darüber, ob dein Tierchen als Fässchen oder Muskelpanzer auftritt!

Ausgewählte Hormone in der Übersicht

Das endokrine, also das hormonelle System im Pferdekörper ist ganz schön komplex. Alle aufzuzählen, führt an dieser Stelle zu weit. Also folgt jetzt eine kurze und bewusst auf trainingsbeeinflussende Hormone reduzierte Liste, die du bei deinem Pferd im Hinterkopf behalten solltest.

Adrenalin (Epinephrine) und Noradrenalin (Norepinephrine)
Diese Catecholamine werden bei Stress rasch über das sympathische Nervensystem ausgeschüttet. Sie steigern Herzfrequenz, Blutdruck, Muskelanspannung und setzen Energie frei, um eine schnelle Flucht zu ermöglichen. Außerdem sind beide enorm wichtig für die Dressurarbeit, denn:

> »*Norepinephrine and epinephrine stimulate the mobilization of lipid and glucose stores. In adipose tissue, the catecholamines increase the activity of hormone sensitive lipase thereby stimulating lipolysis* (Geor et al. 2013).«

Auf deutsch: »Noradrenalin und Epinephrin stimulieren die Mobilisierung von Lipid- und Glukosespeichern. Im Fettgewebe erhöhen die Katecholamine die Aktivität der hormonsensitiven Lipase und stimulieren so die Lipolyse (Fettaufspaltung, A. d. A.).«

Der Gegenspieler beider Hormone ist der Parasympathikus, der nach Abklingen der Gefahr (oder der komplexen Bewegungsabfolge) für Entspannung sorgt. Limitierende Faktoren sind chronischer Stress und fehlende Erholungsphasen, die zu Überreaktionen führen können. Pausen sind nach Anstrengung wichtig! Wie gut oder schlecht ein Körper Adrenalin verarbeitet, hängt unter anderem von einer Genvariation (COMT) ab.

Kortisol
Kortisol ist das wichtigste Nebennierenrinden-Hormon beim Pferd und wird über die Hypothalamus-Hypophysen-Nebennierenrinden-Achse (HPA-Achse) ausgeschüttet. Es wirkt langsamer als Adrenalin, sorgt für anhaltende Energiebereitstellung und beeinflusst die Immunabwehr sowie den Stoffwechsel (Thau et al 2023).

Normale Kortisolwerte liegen bei Pferden bei rund 42 ng/ml im 24-Stunden-Verlauf, wobei die Werte sowohl am Tag als auch in der Saison schwanken können (Larrson et al. 1979).

Bitte aufpassen: Kortisol ist nicht nur DAS Stresshormon und böse-böse, sondern auch absolut notwendig, um jenseits von schlappem auf dem Boden herum oxidieren überhaupt funktionieren zu können! Es hat eben auch seine hilfreichen Aspekte, siehe:

> »*While cortisol is crucial for adaptation and survival, chronically elevated concentration can have detrimental effects, making it a critical indicator of specific diseases* (Sikorska et al. 2023).«

Auf deutsch: »Kortisol ist zwar für die Anpassung und das Überleben von entscheidender Bedeutung, doch kann eine chronisch erhöhte Konzentration schädliche Auswirkungen haben, was es zu einem wichtigen Indikator für bestimmte Krankheiten macht.«

Dauerhaft erhöhte Kortisolspiegel beeinflussen das Immunsystem allerdings auf negative Art und Weise, befeuern Entzündungen (Sikorska et al. 2023) und scheinen in Verbindung mit Muskelabbau bis hin zur Sarkopenie zu stehen (Katsuhara et al. 2022). Was jetzt ursächlich ist, wahlweise Muskelabbau oder der konstant erhöhte Kortisolspiegel, das muss noch weiter zuverlässig erforscht werden.

Als Gegenspieler fungiert am besten die Erholung durch Parasympathikus-Aktivierung: Herzschlag runter, Kortisol raus aus dem Organismus.

Dopamin
Dopamin beeinflusst Herzschlag, Blutdruck und sorgt dafür, dass weitere Neurotransmitter wie Adrenalin oder Noradrenalin ausgeschüttet werden. Es ist außerdem enorm wichtig für den Antrieb und die Konzentrationsfähigkeit, für die Entscheidungsfreudigkeit und als sogenanntes **Belohnungshormon/ Glückshormon**.

Sein Gegenspieler ist der Parasympathikus, also Entspannung und lockere Ruhe.

Dopamin hat eine bidirektionale Wirkung auf körperliche Bewegung: Es wird durch die Bewegung aufgrund der Aktivierung des Sympathikus vermehrt ausgeschüttet und aufgrund seiner Wirkung als Glückshormon führt es auch dazu, dass sportliche Betätigungen häufiger ausgeübt werden (Marques et al. 2021). Kurz: Um durch Sport glücklich zu werden, muss also erst mal Sport ausgeübt werden.

Serotonin
Serotonin wirkt stimmungsstabilisierend und fördert Wohlbefinden, Schlaf, Appetit sowie die Verdauung. Es hilft, Stressreaktionen abzumildern und unterstützt die Rückkehr in die Ruhephase. Ein Mangel kann zu erhöhter Stressanfälligkeit führen. Häufig wird ein Serotoninmangel auch mit depressiven Schüben bis hin zur ausgewachsenen Depression assoziiert, die wissenschaftlichen Belege dazu sind aber nicht ganz eindeutig. Klassische Anti-Depressiva basieren aber häufig auf einer Zufuhr von Serotonin, um die Stimmung zu stabilisieren (Almulla/Maes 2023).

Klassische Gegenspieler sind Stresshormone wie Adrenalin und Kortisol; limitierend wirken Nährstoffmangel und fehlende positive Interaktionen.

Letztere wurde in einer relativ kleinen, aber aufschlussreichen Studie an Pferden untersucht. Das Fazit der Forschenden am Ende lautete:

> *»The lower levels of 5-HT (zirkulierendes Serotonin, A. d. A.) measured at 12:00 and at 16:00 in Group S could indicate that factors as absence of exercise and isolation could influence 5-HT levels (Alberghina et al. 2015).«*

Auf deutsch: »Die niedrigeren 5-HT-Werte, die um 12:00 Uhr und um 16:00 Uhr in Gruppe S gemessen wurden, könnten darauf hinweisen, dass Faktoren wie fehlende Bewegung und Isolation den 5-HT-Spiegel beeinflussen könnten.«

Bei einer bestehenden entzündungsbedingten Histaminose kann es zu einem Absinken der Serotonin-Level kommen:

> *»During acute inflammation, extracellular serotonin levels are decreased as a function of increased brain histamine (Hersey et al. 2021).«*

Auf deutsch: »Während einer akuten Entzündung sinkt der extrazelluläre Serotoninspiegel als Folge des erhöhten Histaminspiegels im Gehirn.«

Für die grundlegende Gesundheit und das Wohlbefinden ist es deshalb wichtig, Entzündungen möglichst zu reduzieren. Krafttraining sorgt über den vermehrten Ausstoß von Myokinen dafür, dass insbesondere stille Entzündungen weniger Schaden anrichten (de Sá Souza et al. 2022).

Das vegetative Nervensystem (Sympathikus/Parasympathikus)

Der **Sympathikus** ist für die Aktivierung in Gefahrensituationen zuständig (Stress/Flucht).

Der **Parasympathikus** sorgt für Entspannung, Verdauung und Erholung.

Ein Ungleichgewicht, etwa durch chronischen Stress oder fehlende Erholungsphasen, stört die optimale Stressregulation und kann zu ernsthaften Herz-Kreislauf-Störungen (beim Menschen) führen (Abboud 2010).

Die Genvariation COMT

Das COMT-Gen (Catechol-O-Methyl-transferase) steuert ein Enzym, das dabei hilft, bestimmte Neurotransmitter (Katecholamine) abzubauen (Tenhunen et al. 1994). Das betrifft insbesondere die Botenstoffe Dopamin, Noradrenalin, Adrenalin und wie schnell – oder langsam – der Körper sie verstoffwechselt.

Das COMT-Gen und das dazugehörige Enzym beeinflussen damit auf entscheidende Weise das Stresslevel, die Motivation und die mentale Leistungsfähigkeit. Seine Auswirkungen wurden sowohl beim Menschen als auch bei Nagetieren und Pferden hinreichend untersucht (Dall'Olio et al. 2009).

Kleiner Funfact am Rande: Für Rennsportzwecke wurde bereits untersucht, inwieweit sich COMT-Enzym-Hemmstoffe zur Dopingbekämpfung eignet. Das Ergebnis war recht eindeutig, Katecholamin-basierte leistungssteigernde Substanzen lassen sich mit entsprechenden Hemmstoffen ausschalten (Stanley et al. 2022).

Ein Wallach mit sehr langsamer COMT. Sonja mit Leo.

Schnelle COMT (Val/Val)

... bauen Katecholamine (Hormone) schnell ab. Das bedeutet meist eine höhere Stressresistenz, dafür aber eine geringere Motivation und Begeisterungsfähigkeit.

- **Die Vorteile:** ruhiger in Stresssituationen/weniger anfällig für Überforderung, Angststörung oder emotionale Schwankungen.
- **Die Nachteile:** wirkt schnell emotionslos und motivationslos, zeigt wenig Begeisterung für Neues.

Pferde mit dieser Mutationsausprägung benötigen mehr Aminosäuren, um permanent Nachschub an Neurotransmittern wie insbesondere Dopamin, Noradrenalin und Adrenalin bereitstellen zu können. Die werden sehr schnell abgebaut, der Körper muss also ebenso zügig nachproduzieren. Damit er das kann, sind vor allem Tyrosin und Phenylalanin limitierende Aminosäuren, die in ausreichender Menge vorliegen müssen.

L-Tyrosin ist dabei die Vorstufe für Dopamin, Tryptophan unterstützt die Produktion von Serotonin und Phenylalanin wird zu Tyrosin umgewandelt.

Typische Anzeichen für Aminosäurenmangel bei schneller COMT sind:

- Antriebslosigkeit oder »müdes Verhalten«
- geringe Motivation beim Training
- langsames Lernen
- wenig Reaktion auf Reize

Langsame COMT (Met/Met)

... zeichnen sich durch einen langsamen Neurotransmitterabbau aus, insbesondere Dopamin verbleibt länger im Organismus. Das zeigt sich durch mehr Kreativität und Motivation sowie einer stärker ausgeprägten Anfälligkeit für Stress.

- **Die Vorteile:** Intensivere Emotionen, stärkere Motivation, mehr Leidenschaft und mehr Engagement.
- **Die Nachteile:** Höheres Risiko für Stress und alle damit einhergehenden Negativfolgen, höheres Risiko für Angstzustände oder Überforderung. Vor allem Iberer fallen häufig mit einem Verhalten auf, das zu genauer dieser Genvariation passt!

Bei dieser Ausprägung werden natürlich auch Aminosäuren benötigt, aber die richtigen im richtigen Maß. Denn hier fällt der Dopaminspiegel tendenziell höher aus, der Abbau dieses Hormons ist langsam. Dabei kann ein Ungleichgewicht zwischen dem Abbau von Dopamin und Noradrenalin entstehen, der Körper kommt nicht oder nur sehr langsam zur Ruhe, obwohl äußerlich alles zu passen scheint.

Für Pferde mit langsamer COMT sind diese Aminosäuren besonders wichtig:

- L-Tryptophan, das fördert die Produktion von Serotonin und hilft dabei, das Dopaminlevel auszubalancieren.
- Methionin, um die Entgiftung anzukurbeln und überschüssiges Dopamin abzubauen.
- Glycin und Taurin, die beide beruhigend und stabilisierend auf das Nervensystem wirken.

COMT in der Praxis: Lernen

Pferde mit schneller COMT-Variante zeigen oft ein besseres Durchhaltevermögen bei langfristigem Lernen. Strukturierte Aufgaben liegen ihnen. Dafür glänzen sie eher nicht bei Anforderungen, die eher kreative Ansätze erfordern.

Pferde mit langsamer COMT-Variante ziehen relativ schnell Verknüpfungen zwischen verschiedenen Aufgaben, weil sie kreativer und weniger strukturiert an Herausforderungen herangehen. Die Gedächtnisleistung ist enorm, vor allem Details bleiben gut im Kopf. Dafür lässt sich das Tier aber leichter ablenken und verliert gerade bei langen und monotonen Aufgaben schnell den Faden.

Das Stresshormon schlechthin: Kortisol

Du hast gerade gelesen, dass Kortisol an und für sich nichts wirklich *Schlechtes* ist, sondern absolut notwendig für den Organismus, um überhaupt irgendetwas auf die Kette zu bekommen. Aber was ist, wenn da tatsächlich zu viel Kortisol im Organismus unterwegs ist?

Dauerstress kann dazu führen, dass der Körper permanent Gluconeogenese (Chourpiliadis/Mohiuddin 2023) betreibt, dadurch ist der Kortisolpegel chronisch erhöht. Vor allem die Muskulatur, der Stoffwechsel und die Leber leiden unter dem Dauerstress und seinen Begleiterscheinungen.

Dieser erhöhte Kortisolspiegel animiert die Leber dazu, aus Eiweißen statt aus Kohlenhydraten neue Glucose (=Muskelbenzin) herzustellen, um den Blutzuckerspiegel zu erhöhen (= Gluconeogenese). Selbst wenn das Tier ausreichend Futter in Form von Faserstoffen und Co. zur Verfügung hat, behilft sich der Körper mit den Protein-Vorräten aus der Muskulatur, denn das geht viel schneller. Die Folgen sind vor allem für unsere Reitpferde fatal:

- **Muskelabbau**
  Die Eiweiße aus der Muskulatur dienen als Rohstoff für die Gluconeogenese. Bei Pferden sind insbesondere Rücken und Kruppe von diesem Raubbau an ihrem Muskellager betroffen. (Zu hohes) Kortisol ist ein Muskelkiller!

- **Leberbelastung**
  Die Leber ist mit der Verstoffwechselung von Proteinen zu Glukose beschäftigt (Morgan et al. 2024). Das kann zu Überlastungen bis hin zur sogenannten nicht-alkoholischen Fettleber führen (Onyango 2022).

Übergewichtige Pferde profitieren durch Krafttraining, um den muskelabbauenden Effekt zu dämpfen. Anna mit Floro.

- **Insulinresistenz und Stoffwechselprobleme**
  Der ständige hohe Blutzuckerspiegel kann dazu führen, dass die Körperzellen weniger gut auf Insulin reagieren. Die Reaktion ähnelt dem Equinen Metabolischen Syndrom (EMS). Da gleichzeitig die Muskulatur schrumpft, verstärkt sich das Problem mit der Zeit noch. Siehe oben: Die Leber ist bei anhaltendem Stress mit der Verstoffwechselung der Proteine beschäftigt, die Kapazitäten für weitere Entgiftungsprozesse fehlen. Über Krafttraining lässt sich eine (beginnende) Insulinresistenz aber gut in den Griff bekommen (Pereira et al. 2020).

Energieraub und Erschöpfung

Obwohl dein Pferd ausreichend oder sogar eigentlich zu viel frisst, bleibt es müde und schlapp. Seine Muskelmasse schrumpft, die schlechte Laune gibts dadurch noch oben drauf. Das beste Mittel dagegen: Stressreduktion, so weit wie möglich plus stärkende Versammlungsarbeit, die dem Muskelraubbau entgegenwirkt.

Absetzen und Anreiten als Stressbooster

Erhöhte Kortisolwerte zeigen sich nicht nur generell bei älteren Tieren, sondern schon beim Absetzen des Fohlens. Vor allem die gängige Methode des abrupten Absetzens durch räumliche Trennung von Fohlen und Stute führt zur Verdopplung der Kortisolwerte für mehrere Wochen (Delank et al. 2023). Diese unter Stress stehenden Jungtiere zeigen Verhaltensänderungen wie

- deutlich sichtbare Ängstlichkeit, Aggressivität oder Passivität,
- kein/kaum Spieltrieb,
- ängstliche/klagende Lautäußerungen,
- passives Herumstehen.

Warum ist das wichtig? Weil dein ausgewachsenes Pferd heute unter Umständen in seinen erste Lebensmonaten verstärkt Stress ausgesetzt war – und das unter bestimmten Bedingungen möglicherweise seine HPA-Achse beeinflusst. Beim Menschen werden diese Zusammenhänge seit einiger Zeit untersucht, wobei sowohl HPA-Achsen-Überaktivität als auch zu niedrigen Kortisollevel sowie zu niedrige Kortisollevel Ursachen für eine gedämpfte HPA-Reaktivität zu sein scheinen (Seal et al. 2021).

Neben seinem Pferdetemperament und einer gewissen genetischen Veranlagung spielen auch frühe Haltungsformen und sogar das Stresslevel der Mutterstute während der Schwangerschaft eine Rolle dabei, wie gut dein geliebtes Pferd mit Aufregung und Stress zurechtkommt.

Anreiten kann stressig sein – aber muss das so?

Das Anreiten ist immer ein extremer Stressmoment. Jedes Pferd, auch die vermeintlich ruhigen, haben hier mit enormen Kortisolanstiegen zu kämpfen. Die hitzigen Vierbeiner tragen das unverkennbar nach außen, die phlegmatischeren knabbern dagegen nur innerlich an dieser völlig neuen Herausforderung.

Die vorbereitende Handarbeit sollte deshalb besser nicht übersprungen werden. Sie ist deine sprichwörtliche Sicherheitsleine beim Anreiten. Deine Knochen werden es dir ganz sicher danken!

Der schöne Bonus durch die ausgiebige Arbeit am Boden: Solcherart vorgebildete Remonten halten leichter die Balance. Zunächst ohne und später eben auch mit dem Reitgewicht auf dem Rücken, geraten sie weniger stark in gelenkschädigende Fehl- und Ausgleichshaltungen. Die Tiere, denen eine ausgiebige Gewöhnungsphase im Sinne der Alten Meister vorenthalten wurde, tun sich da erfahrungsgemäß eher schwer mit dem Ausbalancieren des Menschengewichts.

Routinen schaffen Sicherheit

Ich kann es gar nicht oft genug sagen: Pferde lieben Routinen.

Routinen bis zum Tod

Reichlich makaber ist die Studie von Dai et al. aus dem Jahr 2025. Hier wurde untersucht, ob sich die Etablierung von Routinen zur Verladung von Pferden auf die Stresslevel besagter Tiere auswirken könnte. Tatsächlich waren diejenigen Tiere mit einer sechswöchigen Vorbereitungsphase, in der das Verladen ausgiebig trainiert wurde, am Ende mit niedrigeren Kortisolleveln ausgestattet. Das hatte nur leider keinen weiteren gesundheitlichen Effekt, denn trainiert wurde die Verladung zugunsten einer Fahrt ins Schlachthaus.

Aber: Routinen im Sinne von vertrauten, immer gleichen Abläufen über längere Zeiträume hinweg sind eine gute Basis, um die Kortisolwerte möglichst niedrig zu halten.

Im Rahmen der Etablierung von Routinen für entspanntere Pferde passt es ganz gut, dass sich das Anreiten in viele kleine Schritte zerlegen lässt.

Je vertrauter dein Vierbeiner mit jedem dieser Schritte ist, desto mehr Sicherheit gewinnt er – und damit auch du. Nachvollziehbare und verlässliche Abläufe wirken sich außerdem kaum noch auf den Stresspegel aus. Der Kortisolspiegel erhöht sich zwar kurzfristig bei Bewegung, kann danach aber wieder auf ein Ruheniveau absinken.

Schritt 1:
Zunächst teste ich, ob ich die Remonte im Stand den Hals *abbiegen* lassen kann. Dieses Abbiegen fungiert im Falle des Falles als *Bremse*, wenn es dem Pferd dann doch zu unheimlich wird. Dieser Schritt ist Teil meiner Arbeit am Boden.

Die Nährstoffversorgung beeinflusst stark die Stimmung – bei Mensch und beim Pferd. Sonja mit Coronado.

Schritt 2:
Danach folgen das **Anreiten und Anhalten**. Auch hier: Dank Bodenarbeit sind die Stimmsignale bereits bekannt.

Schritt 3:
Kurze Einheiten, ausreichende Pausen. Das in heller Aufregung befindliche Nervenkostüm der Remonte darf nicht überstrapaziert werden. Im vorigen Kapitel kam die Pausengestaltung schon zur Sprache. Für Jungpferde beim Anreiten sind diese Pausen Pflicht, keine Kür.

Happy Hengst: Ausschachten

Da bietet der Hengst die schönste Piaffe an – und mittendrin baumelt der Pferdepenis voll ausgefahren herum. Was gerne mal für Gelächter sorgt, führt aber auch zur Irritation und Sorge: Was, wenn das jetzt ein Zeichen für Überanstrengung, Dominanzverhalten oder Aggression ist? Wallache sind an dieser Stelle mit gemeint, zeigen das Verhalten in der Regel aber in einem deutlich reduzierten Verhältnis.

Das Ausschachten selbst tritt übrigens relativ häufig auf, nämlich im Durchschnitt alle ein bis eineinhalb Stunden – und zwar täglich. Es ist Teil des normalen Sexualverhaltens und betrifft sowohl Hengste

Der Hengst in der Bodenarbeit, hier allerdings nicht erregt. Sonja mit Coronado.

Imponierender Wallach unterm Sattel.
Christina mit Inca.

in Junggesellenverbänden als auch Hengste, die gemeinsam mit paarungsbereiten Stuten gehalten werden.

Im Gegensatz zu früheren Annahmen sind Erektionen und Masturbation, bei der der Hengst den Penis gegen seinen Bauch schlägt oder presst, kein Anzeichen für Aggression oder gar für irgendwelche Dominanzspielchen gegenüber dem Pferdemenschen. Sie sind schlicht Teil der normalen Verhaltensweisen eines gesunden Pferdes (Schweiger 2022).

Wir unterscheiden allerdings zwischen sexuell motiviertem Ausschachten und dem reflektorischen Ausschachten. Ersteres ist Teil des normalen Sexualverhaltens.

Zweiteres ist Part einer Reaktionskette, die oft mit Stress oder Anspannung assoziiert ist. Reflektorisches Ausschachten hat aber viele Gründe und ist bei einem ansonsten gesunden Tier ohne erkennbare Schmerzen kein Grund zur Sorge. Aus altklassischer Sicht heraus ist das **reflektorische**, also aus einem Reflex heraus passierende Ausfahren des Pferdepenis durchaus eine sehr gute Sache.

Ausschachten als Anzeichen für korrekt arbeitendes Pferd

Tatsächlich kann das Ausschachten bei einem ansonsten gesunden Hengst *das* Signal sein, dass die Versammlung samt Hankenbeugung korrekt gearbeitet wird und all die wichtigen Muskelgruppen richtig angesprochen werden. Diese drei Gründe sprechen dafür, dass das Ausschachten bei der Lektionsarbeit positiv bewertet werden sollte:

1. Eine aktivierte Bauchmuskulatur hilft beim Heben des Brustkorbs und mobilisiert den Beckenbereich, was reflektorisch zum Ausschachten führen kann.

2. Beim Reiten werden insbesondere die tiefen Beckenbodenmuskeln (*M. bulbospongiosus* und *M. ischiocarvenosus*), der *M. musculus gluteus* und die Hinterhandmuskulatur (*M. semimembranosus* und *M. semitendinosus*) aktiviert. Diese Muskelgruppen sind auch mit der Kontrolle der Genitalien und der Beckenstabilität verbunden.

3. Wenn ein Pferd korrekt mit seinem Rücken arbeitet und vermehrt Last über die Hinterhand aufnimmt, kann dies ebenfalls eine Mobilisierung des Beckens und der umgebenden Muskelgruppen fördern. Auch so kann das Ausschachten begünstigt werden.

Bonuspunkte gibt es für entspannte und zugleich konzentrierte Hengste. Denn dann wird der Parasympathikus, zuständig für Ruhen und Verdauen, aktiviert. Ähnlich wie beim Dösen oder bei einer entspannenden Massage schachtet der Penis reflektorisch aus. Im *Fight-or-flight-Modus* mit aktiviertem Sympathikus ist das undenkbar!

Das reflektorische Ausschachten ist also eher kein Zeichen für eine massive Kortisol-Überdosis, die das arme, hormongesteuerte Pferd beutelt. Es ist vielmehr dein Signal, dass dein Tier sich auf die Arbeit mit dir einlässt im Sinne, dass es sich unter entspannter Konzentration darum bemüht, die gestellten Aufgaben zu meistern und zusätzlich dabei genau die richtigen Muskelgruppen aktiviert, die für eine aktive Lastaufnahme und damit für die Versammlung nach altklassischer Manier gefordert sind.

Mehr Selbstbewusstsein? Reitkunst macht das möglich!

Mit steigender Kraft und besserer Körperkontrolle entwickelt dein Pferd durch die Lektionen mehr Selbstbewusstsein. Es merkt, dass es die geforderten Aufgaben leichter bewältigen kann, was sich oft einer freudigeren, energievolleren Mitarbeit zeigt. Neben der psychologischen Komponente findet diese Veränderung auch auf zellulärer Ebene statt.

Der Spanische Schritt ist typisches Imponiergehabe und verbessert das Selbstbewusstsein auch beim Wallach. Gina mit Davidoff.

Gezieltes Versammlungstraining stimuliert das zentrale Nervensystem (ZNS) – und damit auch das Gehirn. Denn alle Signale aus der Peripherie landen final im Gehirn! Durch bewusst eingesetztes und forderndes Versammlungstraining werden im Körper deines Pferdes Prozesse angestoßen, die die Leistungsfähigkeit und die Energieproduktion nachhaltig steigern. Und zwar so:

- **Die aktivierte Muskulatur sendet (viele) Signale ans Gehirn.** Das passiert bei einem trampeligen ersten Seitengang genauso, wie bei einer in Vollendung ausgeführten Piaffe oder einem energetisch vorgetragenen Spanischen Schritt. Die intramuskuläre Koordination bessert sich ebenso wie die allgemeine Koordination. Die Aktivierung der Muskulatur hin zum voluminösen Reitkunstpanzer ist der entscheidende Punkt. Je zielgerichteter und bewusster jede Aktivierung geschieht, desto eindeutiger ist die Rückmeldung ans Gehirn.

- Diese Signale zeigen dem Gehirn an, dass sich **Wachsamkeit und Reaktionsbereitschaft** verstärken sollten, schließlich setzt sich das ganze zarte ~ 650 Kilogramm Pferd in kontrollierte und zunehmend tänzerische Bewegung! Das zentrale Nervensystem ist wachsam, aktiv und bereit für die Lektionen. Dopamin und Adrenalin machen sich auf den Weg, um alle beteiligten Signalketten und die erforderlichen Energiereserven für die Muskelarbeit zu aktivieren. Denk dran: Positiver Stress ist gut fürs System!
- **Die aktivierte Muskulatur sorgt unter anderem für die vermehrte Ausschüttung von Myokinen.** Das Tier fühlt sich wohler. Zugleich wird das Belohnungssystem des Gehirns getriggert, Endorphine und Dopamin verstärkt auszuschütten. Das Signal an den restlichen Körper lautet: »Los, mehr davon!« Das dämpft parallel auch Schmerzen, zudem sorgen die aktivierten und insgesamt immer dicker werdenden Muskeln dafür, dass Bänder und Sehnen weit besser geschützt sind.

Sportbegeisterte rennen, springen oder heben Gewichte für genau diesen Effekt. Ihre Anstrengungen sorgen für ein nettes dopamingesteuertes High im Hirn. Ohne harte Drogen, nur dank Bewegung und einer auf zielgerichtete Trainingseinheiten ausgerichteten Signalgebung.

Physiologische Grundlagen für mehr Selbstbewusstsein durch Training

Das intensive Training im Sinne der Alten Meister aktiviert die Kraftwerke der Zellen, die **Mitochondrien**. Sie sind zuständig für die ATP-Produktion, die universelle Energiewährung im Körper. Je mehr gesunde und aktiv arbeitende Mitochondrien in den Zellen vorhanden sind, desto mehr ATP kann bereitgestellt, verarbeitet und in Stoffwechselprozessen sinnvoll eingesetzt werden. Dein in der Versammlung geschultes Pferd zeigt sich seltener erschöpft und bringt deutlich mehr Bereitschaft zu dynamischen Bewegungen mit sich.

Zugleich schütten die Muskeln **Myokine** aus, die wiederum tiefgreifende positive Stoffwechselanpassungen anstoßen: Das Muskelwachstum wird angeregt, Entzündungen reduziert und diverse Stoffwechselprozesse verbessert. Dank regelmäßiger Lektionsarbeit wird die **Kapillarisierung** der Muskulatur gefördert.

Es entstehen also neue feine Blutgefäße, die die Muskelfasern mit Energie in Form von Sauerstoff und Nährstoffen versorgen. Das verkürzt die Erholungszeiten und erhöht die Bereitschaft deines Vierbeiners, sich auch bei der nächsten Lektion von seiner allerbesten Seite zu zeigen und ordentlich mitzumachen.

In Studien wurde die Wirkung von anhaltendem Krafttraining auf die Kapillarisierung bei älteren Männern beobachtet. Die Ergebnisse per Muskelbiopsie zeigen unter dem Mikroskop ein verbessertes Gefäßnetz sowie positive Effekte auf die Muskelfaserstruktur (Leuchtmann et al. 2020).

Einige Myokine sind an der Optimierung des Fett- und Zuckerstoffwechsels beteiligt. Andere Myokinarten wirken sich direkt auf das Nervensystem aus, um die Wachsamkeit und Motivation deines Pferdes in direkter Folge zu steigern.

Noch ein positiver Effekt durch eine vermehrte Myokinausschüttung, der sich vor allem bei hormonell aus dem Gleichgewicht geratenen Pferden bemerkbar macht: Sie beeinflussen den Hormonhaushalt und balancieren den negativen Einfluss von (zu viel) Stress aus. Das sorgt für mehr Energie, Motivation und Leistungsbereitschaft.

Dein Vierbeiner wird nicht nur stärker, sondern fühlt sich einfach insgesamt viel vitaler. In einer 2023 veröffentlichten Meta-Analyse kamen Augustin et al. zu dem Schluss:

> *»Resistance training seems to have an antidepressant effect.«*

Auf deutsch: »Widerstandstraining scheint eine antidepressive Wirkung zu haben.«

Der Hengst im vollem Imponiergehabe
Sonja mit Coronado.

Die Effekte der Versammlungsarbeit spürst du beim hitzigen und beim phlegmatischen Tier auf ganz unterschiedliche Art und Weise. Das Energiebündel wird dich durch eine konzentrierte und etwas gelassenere Arbeitsweise beeindrucken, wenn du sein Erregungsniveau erst einmal durch intensive Versammlungsarbeit in den Griff bekommen hast.

Das gutmütige Tierchen dagegen bekommt ein bisschen mehr Schwung und zeigt sich interessierter an neuen Dressurkunstaufgaben. Das liegt weniger am insgesamt verbesserten Muskelzusammenspiel und der Übung von Bewegungsabläufen, sondern ist tief in den beschriebenen physiologischen Prozessen verankert.

Insbesondere die Arbeit an der Hand lässt sich gut mit Krafttraining mit eigenem Körpergewicht vergleichen. Nehmen wir das Reitgewicht dazu, wirkt das wie eine durchschnittliche 80 Kilogramm-Hantel obendrauf.

Die simple Formel zu energiegeladenen Pferden in der Dressurkunst lautet also: Mehr Mitochondrien = mehr Energie = mehr Spaß bei der Lektion.

DU BIST, WAS DU ISST

KURZ UND KNAPP:

- Ein gesunder Stoffwechsel legt die Basis für langfristig fitte Pferde.
- Dickes oder dünnes Pferd? Genetik, Stoffwechsel und Futterqualität entscheiden, wohin die Waage ausschlägt.
- Hormonelle Ungleichgewichte reduzieren die Trainingserfolge.
- Keine Muskeln ohne ausreichend Protein.
- Fell- und Hufgesundheit eignen sich prima als Indikatoren für eine aufs individuelle Pferd ausgerichtete optimale Ernährung.

Ein auf Versammlung ausgerichtetes Training ist die ideale Basis, um muskelbepackte Reitkunstpferde zu erhalten. Die körperliche Schulung allein reichen aber nicht aus. Auch die Ernährung muss ihren Teil dazu beitragen.

Das bedeutet im Einzelnen, dass eine an das jeweilige Pferd angepasste Fütterung mit Proteinen (Eiweiße), gesunden und körperlich notwendigen Fetten für das Gehirnwachstum, und Fasern (Ballaststoffe) absolut notwendig ist. Wenn es für das jeweilige Tier angebracht ist, sollten auch in geeignetem Maße Kohlenhydrate (Zucker) enthalten sein. Mikronährstoffe, Vitamine und Spurenelemente kommen obendrauf. Voilà, schon ist die gesunde Ernährung deines Vierbeiners gesichert. Richtig?

Leider ist es nicht ganz so einfach. Denn jedes Pferd hat seinen ganz eigenen Bedarf, der sich zudem noch im Laufe der Zeit ändern kann. Was vermeintliche Ernährungsberater gern als optimale Lösung für ein Pferd abhängig von seinem Gewicht und seinem Stockmaß verkaufen, sind selten aufs individuelle Tier zugeschnittene Ernährungsvorgaben. Da braucht es schon einen regelmäßigen Blick auf Blutwerte plus ein immer besseres Gefühl für die genetische Veranlagung eben diesen Tiers, bevor gute Fütterungsentscheidungen getroffen werden können.

Dazu kommt noch, dass es mit unserem modernen Gras und dem daraus entstehenden Heu komplex sein kann, den Proteinbedarf eines Pferdes korrekt und gut zu decken, insbesondere mit einem optimalen Aminosäurenverhältnis. Das gilt erst recht für **Schulpferde** (Pferde, die die hohe Schule gehen).

Rehafälle und die bereits komplett aufgegebenen und ausgemusterten Pferd auf der Rentnerweide haben oftmals gleich ganz verloren, wenn sie – oft mit den besten Absichten – nur noch raus auf die grüne Weide geschickt werden.

Dabei ist gerade die schöne grüne Weide das Problem: Wenig muskelaufbauende Bewegung, dafür reichlich zuckerreiches Gras. Die volle Dosis Zucker kann beim Reha-Fall Entzündungsprozesse zusätzlich beschleunigen, die Menge an Freien Radikalen steigt und am Ende schreiten die Alterungsprozesse noch weiter voran.

Denn gerade das hiesige zuckerreiche Weidegras ist es, das für zahlreiche suboptimale Stoffwechselprozesse verantwortlich ist. Bei deinem Rehapferd, bei Tieren mit Arthrosen oder bei denen, die wiederkehrende Verdauungsprobleme zeigen, wird das besonders deutlich. Über die Ernährung lässt sich nämlich durchaus steuern, wie aggressiv entzündliche Prozesse im Körper verlaufen.

Der heutzutage gängige **Haltungsstress und aus dieser Kohlenhydratlast resultierende Mikronährstoffminderversorgungs-Stoffwechselstress** kommen dann noch obendrauf.

Haltungsstress entsteht dann, wenn dein Vierbeiner keine passende Umgebung vorfindet und sich eben mit dem arrangieren muss, was ihm angeboten wird. Introvertierte, teilweise rangniedrigere Tiere können insbesondere in einem Offenstall still leiden und mit extremen Stresshormonwerten konfrontiert sein. Ranghohe, teilweise expressive, extrovertierte Tiere erleben unter Umständen ebenfalls reichlich Stress, weil sie permanent die anderen Herdenmitglieder managen, einschüchtern oder bewachen.

Lärm, wechselnde Belegung und für das jeweilige Tier nicht passende Boxennachbarn verursachen schlichtweg ebenfalls Stress. Und der schlägt nicht nur aufs Gemüt, sondern auch auf die Figur.

Stoffwechselstress kommt noch dazu: Stress ist ein wahrer Vernichter für Mikronährstoffe, Vitamine und teils auch für Aminosäuren. So wird die empfindliche Balance aus Makro- und Mikronährstoffen gestört und das bedeutet dann eben auch, dass nie ausreichend Material da ist für die zahlreichen alltäglichen Aufgaben im Körper deines Reitpferdes. Der ganze Körper steht unter Anspannung und knappst am Ende hier und da wertvolle Aminosäuren aus der Muskulatur ab oder löst Mineralien aus dem Knochen, nur um irgendwie durch den Tag zu kommen.

Gesunderhaltenes Krafttraining und eine auf das Pferd abgestimmte Ernährung plus eine möglichst optimale und auf das individuelle Pferd zugeschnittene Haltungsform gehen also Hand in Hand. Eins optimieren und den Rest links liegen lassen? Das kannst du vergessen, wenn du in glückliches und bis ins hohe Alter gesundes und glückliches Pferd haben möchtest. Und da du dieses Buch hier in Händen hältst, gehen wir einfach mal davon aus, dass genau das dein Ziel ist!

Aus dem Takt geraten

Bei manchen Pferden lassen sich verschiedene Faktoren beobachten, die auf eine unzureichende Deckung des individuellen Mikro- und Makronährstoffbedarfs hindeuten. Das ist oftmals Part einer ungünstigen Kombination aus genetischer Veranlagung und/oder eines - mitunter auch dadurch - gestörten Entgiftungsprozesses in der Leber. Die Grundlagen dazu folgen im nächsten Abschnitt.

Zunächst solltest du dir die Frage stellen, ob eine oder mehrere der folgenden Faktoren bei deinem Vierbeiner zutreffen. Erfahrungsgemäß treten häufig mehrere parallel auf, allerdings nicht zwangsläufig alle zusammen.

Faktor 1: Unverträglichkeiten gegenüber Medikamenten

Normale Medikamentendosen für gängige Mittel wie Entzündungshemmer (Phenylbutazon), Wurmkuren oder Impfungen werden nicht gut vertragen. Die Leber ist dann damit beschäftigt, das gegebene Medikament zu entgiften.

Wichtig:

Ja, Medikamente sind oft nötig und wichtig. Aber noch wichtiger wäre eben, von Anfang gar keine Medikamente zu benötigen. Insgesamt sind die Erholungszeiten länger und Nebenwirkungen sind auffälliger beziehungsweise fallen schwerer aus als bei anderen Pferden.

Faktor 2:
Neigung zu Leber- und Nierenproblemen

Im Blutbild zeigen sich chronisch oder wiederkehrend leicht erhöhte Leberwerte. Vor allem GGT, AST, ALP und Bilirubin sind auffällig. Ein dunkler Urin oder eine Neigung zu Harnwegsproblemen zeigt sich ebenso wie diffuse Symptome wie Mattigkeit, schlechte Regeneration und reduzierte Leistungsfähigkeit.

Faktor 3:
Hautprobleme

Wiederkehrende Ekzeme, Juckreiz und Hautausschläge verstärken sich nach Wurmkuren, Impfungen oder Medikamentengaben. Hufrehe und angelaufene Beine treten auf, ohne dass es dafür eindeutig identifizierbare Ursachen gäbe. Schimmel im Stall oder Pestizide vom Feld in der Nähe der Weide, sowie Zusatzstoffe im Futter machen sich sofort in einer Verschlechterung der Haut bemerkbar.

Faktor 4:
Fütterungsprobleme und Stoffwechselstörungen

Selbst langsame Futterumstellungen ziehen starke Reaktionen nach sich. Getreide, Melasse oder Zusatzstoffe werden schlecht vertragen. Kotwasser, Durchfall und Blähungen treten häufiger auf. Dein Pferd riecht stark – wahlweise nach Pferd oder verströmt einen üblen Schweißgeruch.

Faktor 5:
Sensible Reaktion auf Umweltgifte und Schwermetalle

Jeder Mücken- oder Insektenstich sorgt für auffällige Reaktionen. Abhängig von dem Grad der Umgebungsbelastung (Ammoniak/Schimmel im Stall) verschlechtert sich der Allgemeinzustand deines Tiers – und zwar rapide. Jede Veränderung, sei es eine neue Einstreu, ein neues Futtermittel oder sogar eine neue Wasserquelle, sorgt für Irritationen.

Faktor 6:
Hormonelle Ungleichgewichte und Verhaltensauffälligkeiten

Es lassen sich Zyklusprobleme bei Stuten oder Hormonungleichgewichte bei Stuten, Hengsten und Wallachen beobachten. Jede Intervention in Form von Impfungen, Wurmkuren oder Futterzusätzen bedeuten unerklärliche Verhaltensänderungen. Dein Vierbeiner ist müde und wesentlich stressanfälliger als vergleichbare Tiere, obwohl du ihm reichlich Ruhe gönnst.

Diese sechs Faktoren können auf eine unzureichende Ernährung hinweisen. Sie können aber auch Hinweise auf bestimmte, bisher unerkannte Grunderkrankungen oder genetische Vorbelastungen sein. Deshalb ist es wichtig, dass du deinem geliebten Vierbeiner in seiner Gesamtheit betrachtest – nicht nur vereinzelt das Fell, die Leistung im Training oder sein Futterverhalten.

Detox fürs Pferd

Momentchen mal, was heißt hier Detox?

Lass mich erst definieren, was Detox fürs Pferd **nicht** ist: Keine seltsamen grünen Smoothies, keine Heilsteine über der Stalltüre, kein vermeintlich heilsames (gerne auch virtuelles) Handauflegen zur energetischen Neuausrichtung oder dergleichen.

Wissenschaftliche Belege zur Wirksamkeit der verschiedenen Detox-Kuren aus Kräutern, Säften oder Smoothies fehlen sowieso (Wipplinger 2019). Denn der Säugetierkörper verfügt über ganz eigene und wirklich hervorragende integrierte Entgiftungsorgane. Detox pur, kostet auch (fast) nichts extra, abgesehen von einer entsprechend gesunderhaltenen Ernährung **plus ein gesundes Maß an soffkreislaufanschiebendem Kraftsport.**

Als Entgiftungsorgane übernehmen die Leber, aber auch die Nieren und die Haut ihren Anteil daran, Giftstoffe abzubauen und aus dem Stoffkreislauf auszuleiten.

Die Detox-Kapazität ist allerdings unterschiedlich und unter anderem abhängig von der Genetik sowie von bestehenden Krankheiten. Ein fittes Pferd mit hervorragender genetischer Grundausstattung wird weniger anfällig für Schadstoffe aus der Umwelt sein, resilienter mit Stress umgehen und sich schneller von Verletzungen erholen.

Wenn dein Pferd genetisch bedingt aber eine schwächere Entgiftungskapazität hat, zeigt es oft unspezifische wiederkehrende Symptome. Die gehen mit einer Überlastung der Entgiftungsorgane und hier insbesondere der Leber einher. Logischerweise tritt dann auch oft ein spezifischer Nährstoffmangel auf.

Mit normalem 08/15- Futter fällt es dem genetisch weniger gut ausgestattetem Vierbeiner auch eher schwer, diesen Mangel irgendwie auszugleichen. Eine gezielte Ergänzung ist also logischerweise notwendig, um die Entgiftungskapazität zu optimieren und langfristige Gesundheitsprobleme zu vermeiden.

Detoxorgan #1: die Leber!

Die Leber baut Schadstoffe über zwei Entgiftungsstufen ab. Sie laufen nicht immer synchron, bedeuten aber immer einen erhöhten Verbrauch von Glycin und Glutathion. Beide Aminosäuren sind wichtige Faktoren, um eine reibungslose Entgiftung sicherzustellen.

Aminosäuren und Co-Faktoren für die Entgiftung in der Leber

Glycin

Glycin gehört zur Gruppe der nicht-essentiellen Aminosäuren und erfüllt zahlreiche, enorm wichtige Aufgaben im Säugetierkörper:

> »*Collectively, glycine has crucial function in cytoprotection, immune response, growth, development, metabolism, and survival of humans and many other mammals.* (Razak et al. 2017).«

Auf Deutsch: »Glycin hat eine entscheidende Funktion beim Zellschutz, für Immunreaktionen, Entwicklung, Verdauung und das Überleben des Menschen und vieler anderer Säugetiere.«

Glycin ist vor allem in Kollagen reichlich vorhanden.

Glutathion

Glutathion ist ein Tripeptid, also ein Molekül, welches aus drei Aminosäuren besteht: Glutaminsäure, Cystein und Glycin. Alle drei Aminosäuren sind in einem optimalen Spiegel nötig, damit Glutathion gebildet werden kann.

Als Haupt-Antioxidans spielt das Tripeptid nicht nur, aber auch in der Regulation von zahlreichen Stoffwechselprozessen eine große Rolle. Plus: Ein ausreichend hoher Glutathion-Spiegel scheint vor Herz-Kreislauf-Erkrankungen zu schützen (Matuz-Mares et al. 2021)!

Cystein

Cystein ist ein wichtiger Bestandteil im Rahmen der Entgiftungsprozesse in der Leber. Aus der Humanforschung ist bekannt, dass sich die Supplementierung von N-Acetylcysteine positiv auf den Verlauf der Behandlung einer nicht-alkoholischen Fettleber auswirkt (Khoshbaten et al. 2010).

Cystein kommt in der klassischen Pferdefütterung kaum in relevanten Mengen vor. Primär findest du diesen Stoff in proteinreichen tierischen Lebensmitteln wie Eier, Fleisch oder Milchprodukte – nichts davon ist für Pflanzenfresser üblicherweise Teil des Speiseplans.

Arginin

Arginin wird im Harnstoffzyklus für die Entgiftung von Ammoniak benötigt. Arginin kann unter Belastungssituation insbesondere von Fohlen oder tragenden Stuten nicht von Grund auf selbst per de-novo-Synthetisierung hergestellt werden (Martinez et al. 2022).

Das heißt, dass Arginin im besten Falle ausreichend über die Nahrung aufgenommen wird oder supplementiert werden sollte, um neben den Entgiftungsprozessen in der Leber insbesondere auch den Muskelaufbau und die Gewebereparatur optimal zu unterstützen. Arginin ist zudem ein Hilfsstoff, um die Absorption weiterer Aminosäuren wie Proline, Threonine oder Glutaminsäure sicherzustellen (Kelley et al. 2014).

Die Entgiftungsphasen in der Leber

In der **Phase I** wandeln Enzyme wie beispielsweise Cytochrom P450 Giftstoffe in reaktive Zwischenprodukte um. In dieser Phase werden sowohl Methionin als auch Cystein benötigt, um die Basis für die zweite Phase zu schaffen. Co-Faktoren sind an dieser Stelle insbesondere Vitamin B6, Zink und Magnesium.

In der **Phase II** erhalten diese Zwischenprodukte ein radikales Umstyling und werden mittels Glycins, Glutathions, Taurins und Arginins neutralisiert. Damit sind die ursprünglichen Giftstoffe reif für die Ausscheidung, wasserlösliche Stoffe werden über den Darm oder über die Nieren abtransportiert. Ammoniak als Abbauprodukt von Aminosäuren wird zu Harnstoff und verlässt den Pferdekörper ebenfalls über die Nieren, die Blase und letztlich den Urin.

Sind diese Prozesse aufgrund von Aminosäuremangel gestört oder ist die Leber mit zu hohen Kortisolleveln überlastet, finden die Entgiftungsprozesse nicht oder nicht in ausreichendem Umfang statt.

Die Entgiftungsprodukte, insbesondere Ammoniak verbleiben im Körper – mit weitreichenden Folgen (Leberzirrhose!) (Katayama/Kakita 2024).

Antientzündliches Pferdefutter

Antientzündlich, also gegen Entzündungen – als Label für Pferdefutter? Klingt nach schlechtem und teurem Scherz, meint aber eine möglichst gesunderhaltende Fütterungsform, bei der Zucker, Zusatzstoffe und sonstige für Pferde unnötige Komponenten strikt aus dem Stall verbannt werden.

Stattdessen kommen für die ursprünglichen Steppengras gewohnten Tiere möglichst reichlich ballaststoffreiche und kohlenhydratarme Futterarten in Frage, die ihren Darm nicht belasten und trotzdem alles beinhalten, was das muskelbepackte Reitkunstpferd für seine tänzerischen Einlagen unterm Sattel so benötigt. Und nein, das meint eben nicht nur Heu mit Premiumsiegel, sondern eine ausgewogene Mischung aus verschiedenen Komponenten.

Denn im Pferdebereich versteht man unter einer anti-entzündlichen Fütterung eine gezielte Ernährung, die das systemische Entzündungsgeschehen im Körper senkt – also stille, chronische Entzündungen reduziert. Solche Entzündungen können z. B. im Bindegewebe, in Muskeln, Faszien, Gelenken oder sogar in der Darmwand auftreten – ohne dass sie sofort klinisch auffallen.

Hier sind die wichtigsten Komponenten einer antientzündlichen Fütterung beim Pferd:

- **Omega-3-Fettsäuren (statt Omega-6-lastiger Fette)**
  Omega-3-Fettsäuren wirken entzündungshemmend, während Omega-6-Fettsäuren (z. B. aus Maisöl oder Sonnenblumenöl) entzündungsfördernd sein können. Empfehlenswerte Quellen für gesunde und damit entzündungshemmende Fette sind unter anderem Leinöl (hochwertig, kaltgepresst), Leinsamen (geschrotet oder gequollen) oder Chia-Samen, Algenöl (wenn Fischöl ethisch oder geschmacklich nicht in Frage kommt).
- **Zucker- und stärkearme Fütterung**
  Ein hoher Zucker- und Stärkegehalt des Futters fördert Dysbiosen (Ungleichgewichte) im Darm und Low-grade-Inflammation im Gewebe. Gute Quellen für Fasern, sprich Ballaststoffe sind: Heu mit niedrigem Fruktangehalt; möglichst kein oder nur sehr wenig Getreide (v. a. kein Mais, kein Melasse-Müsli). Bei sehr schlanken Pferden kann die Fütterung von strukturreichem Heu, Heucobs oder Heulage aber nötig und wichtig sein. Achte da unbedingt auf den individuellen Bedarf deines Tiers!
- **Antioxidantien & sekundäre Pflanzenstoffe**
  Sie hemmen oxidativen Stress und stabilisieren die Zellmembranen. Sinnvolle Futterzusätze, um oxidativen Stress zu reduzieren sind unter anderem Hagebutten (liefern reichlich Vitamin C, Flavonoide), Mariendistel (dient der Leberentlastung, liefert Antioxidantien), Kurkuma (wirkt entzündungshemmend durch Curcumin, evtl. mit Pfefferextrakt kombinieren), Ingwer (ähnlich wie Kurkuma), Spirulina/Chlorella (moderat dosiert). Und nicht vergessen: Passende Aminosäuren zur aktiven Unterstützung der Entgiftungsprozesse in der Leber!
- **Mineralstoffbalance**
  Das meint hier vor allem Zink, Selen und Magnesium. Ein Mangel an Zink oder Selen kann entzündungsfördernd wirken. Zink ist wichtig für die Haut, die Schleimhäute und zur Regeneration. Selen ist ein eminent wichtiger Co-Faktor für antioxidative Enzyme. Und Magnesium wirkt sowohl muskelentspannend als auch entzündungsmodulierend.
  Im Kapitel zu Nahrungsergänzungsmitteln (NEM) finden sich zusätzliche Informationen zu diesen und zu weiteren Mineralstoffen, die für eine funktionierende Entgiftung beziehungsweise für einen optimalen Muskelaufbau deines Vierbeiners nötig sind.
- **Darmgesundheit fördern = Entzündungen senken**
  Ein chronisch gereizter Darm bedeutet eine daueraktive Immunantwort. Für mehr Ruhe im Darm sind Flohsamenschalen oder die Aminosäure L-Glutamin hilfreich, um die guten Darmbakterien mit zu füttern und die bereits gereizte Darmschleimhaut zu beruhigen.

Was gehört also nicht in die Nähe von Pferdemägen? Vor allem bitte kein Mais, keine Melasse, keine Weizenkleie und keine Pflanzenöle mit hohem Omega-6-Gehalt wie Sonnenblumen-, Distel- oder Maisöl. Das permanente Knabbernlassen an Heulage ohne eine individuell passende Raufutterkontrolle kann entzündliche Prozesse befeuern. Eine sehr starke Überversorgung mit Eiweiß führt unter Umständen zu einer Belastung der Leber und erhöhter Ammoniakproduktion. Wenn du Sojaschrot oder Hafer fütterst, solltest du unbedingt Leinsamen für eine optimale Verdaulichkeit ergänzen.

Eine anti-entzündliche Pferdefütterung ist nicht »besonders« - sondern eigentlich das, was artgerechtes und gesundes Füttern auszeichnet. Dann wirkt die Fütterung nämlich:

- stoffwechselentlastend,
- regenerationsfördernd,
- unterstützend für jedes Training, das auf Muskelaufbau und Bindegewebsgesundheit zielt.

Gute Entzündung – schlechte Entzündung

Gute Entzündungen treten im Rahmen von Gewebeverletzungen, durch Keimbelastung und zur

Bekämpfung von (Virus-)Infekten auf. Sie sind lokal und/oder zeitlich begrenzt. Diese Begrenzung besteht aus gutem Grund, denn Entzündungen belasten das Gewebe und versetzen den gesamten Organismus in Stress.

Entzündungen selbst sind aber nötig, um die Heilungsprozesse vor Ort anzuschieben: Das betroffene Gewebe wird stärker durchblutet. Erhöhte Temperaturen helfen bei der Keimvernichtung. Über erweiterte Blutgefäße werden Hilfsstoffe zur Reparatur schneller an den Ort der Verletzung gebracht und Schadstoffe zügiger abtransportiert.

Schlechte Entzündungen verlaufen niederschwellig und still, d. h. sie treten nicht offensichtlich zutage. Die Belastung für das gesamte Energiesystem des Körpers findet trotzdem statt und äußert sich langfristig durch Schäden am gesamten Bewegungsapparat und vor allem auch durch einen gestörten Insulin- und Stoffkreislauf.

Im Kern geht es bei einer antientzündlichen Ernährung darum,

- stille Entzündungen im Körper (Low grade inflammation, LGI) zu reduzieren.
- das Mikrobiom des Darms und damit die Qualität der Immunantwort auf äußere Einflussfaktoren wie Stress oder Krankheiten deutlich zu verbessern.
- LGI so weit zu senken, dass beginnende Alterungsprozesse aufgehalten oder – sofern möglich – bis zu einem gewissen Grad umgekehrt werden.

Low grade Inflammation ist Bestandteil des normalen Alterungsprozesses. Körpereigene Systeme, die Keime durch Infektionen bekämpfen, greifen langsam aber stetig auch eigentlich gesunde Abläufe an. Körperzellen werden im Laufe der Zeit weniger effizient, die Mitochondrien liefern eine geringere Leistung, die körpereigene Energiewährung ATP kommt seltener zum Einsatz.

> »*Low-grade inflammation induces a metabolic switch from energy-efficient oxidative phosphorylation to fast-acting, but less efficient, aerobic glycolytic energy production; increases reactive oxygen species; and reduces insulin sensitivity. These effects result in reduced glucose availability and, thereby, reduced cellular energy* (Lacourt et al. 2018).«

Auf Deutsch: »Geringgradige Entzündungen führen zu einer Umstellung des Stoffwechsels von energieeffizienter oxidativer Phosphorylierung auf schnell wirkende, aber weniger effiziente aerobe glykolytische Energieproduktion, erhöhen reaktive Sauerstoffspezies (Sauerstoffradikale, A. d. A.) und verringern die Insulinempfindlichkeit. Diese Effekte führen zu einer geringeren Verfügbarkeit von Glukose und damit zu einer geringeren Zellenergie.«

Zellschäden verursachende Freie Radikale, die für oxidativen Stress sorgen. Diese werden nicht oder nicht mehr in vollem Umfang aus den zellulären Prozessen entfernt.

> »*Increased oxidative stress can adversely affects human health, including molecular damage to proteins, lipids, and nucleic acids, and can have adverse health effects. Cell death (apoptosis or necrosis) is the result of damage to biological systems or the formation of numerous reactive chemical species by oxidative stress. Oxidative stress is associated with more than fifty diseases such as diabetes, hypertension, cancer, cardiovascular disease, and neurodegenerative diseases* (Tumilaar et al. 2024).«

Auf Deutsch: »Erhöhter oxidativer Stress kann sich nachteilig auf die menschliche Gesundheit auswirken, u. a. durch molekulare Schäden an Proteinen, Lipiden und Nukleinsäuren, und kann gesundheitsschädliche Folgen haben. Der Zelltod (Apoptose oder Nekrose) ist das Ergebnis von Schäden an biologischen Systemen oder der Bildung zahlreicher reaktiver chemischer Spezies durch oxidativen Stress. Oxidativer Stress wird mit mehr als fünfzig Krankheiten wie Diabetes, Bluthochdruck, Krebs,

Herz-Kreislauf-Erkrankungen und neurodegenerativen Erkrankungen in Verbindung gebracht.«

Infolgedessen nimmt die Zahl der beschädigten Zellen zu. Der Körper altert!

Über Kraftsport und eine gesunde Ernährung lässt sich dieser Prozess bis zu einem gewissen Grad aufhalten. Warum Kraftsport das reinste Gold für müde Muskeln ist, hast du bereits im Kapitel zu den Lektionen gelesen. An dieser Stelle konzentrieren wir uns also auf die Effekte einer idealen Ernährung, um Reitkunstpferde deutlich über das typische Maß hinaus fit und gesund erhaltend reiten zu können.

Der Ernährungsmediziner Dr. Riedl spricht in einer Folge seines Podcasts *So geht gesunde Ernährung* sinngemäß davon, dass sich eine inflammatorische, also eine entzündungshemmende Ernährung zwar kaum per Studienlage beweisen lässt (Riedl 2024).

Aber: Wissenschaftlich betrachtet lässt sich der Einfluss von Brokkoli oder Rohkost ja auch relativ schlecht nachweisen, wenn stille Entzündungen in der Regel multifaktoriell bedingt sind. Inflammatorische Ernährung bringt allerdings positive Effekte mit sich. Logisch, denn eine gesunde, ausgewogene Kost mit reichlich Ballaststoffen und einem optimal bedarfsdeckenden Proteingehalt mit einem sehr guten Aminosäurenverhältnis ist natürlich vorteilhaft - für Mensch und Pferd. Das insbesondere die Aminosäureversorgung mittels tierischer Basis erfolgen muss, hat Dr. Riedl leider so nicht auf dem Schirm.

Die Prävention in Form einer gesundheitsfördernden Ernährung ist zwar deutlich günstiger, aber eben auch weniger leicht zu vermarkten als Behandlungspläne und Medikamente. So fließen Forschungsgelder eher in die Symptombehandlung von Arthrose und Co., statt in die präventive Erforschung zum Zusammenspiel aus Stoffwechsel, Ernährung und Sport. Trotzdem macht die Forschung an der Stelle, wenn auch vergleichsweise langsam, Fortschritte, sagt Riedl.

Bis sich allerdings neue Erkenntnisse aus der Forschung als Allgemeinwissen verbreiten, dauert es noch etwas länger. Erste Studien zeigen aber recht deutlich, dass sich eine antientzündliche Ernährungsweise positiv auf die Reduzierung von laufenden Entzündungsherden im Körper sowie auf darauf beruhende vorzeitige Alterungsprozesse auswirken kann (Calder 2017).

Entzündungsbooster Sonne

Schimmel, Palominos, Falben oder Cremellos sind prädestiniert dafür, im Hochsommer oder bei allgemein starker Sonneneinstrahlung einen Sonnenbrand zu bekommen. Ein Sonnenbrand ist dabei nicht nur kosmetisch problematisch, sondern Anzeichen einer ernsthaften Schädigung der Haut, die mit klassischen Entzündungsmerkmalen wie Schwellungen und Rötungen bis hin zu Blasenbildung mit offenen Wunden einhergeht.

Bisher hieß es: Zu viel Sonne schädigt die DNA, Zellen altern und daher stammen dann die Falten in den Gesichtern von denjenigen, die zu viel Zeit auf der Sonnenbank oder am Strand ohne Sonnenschutz verbringen.

Das scheint jetzt kein Thema für Pferde zu sein, sofern es sich nicht gerade um die hellfarbigen eher empfindlicheren Pferdchen handelt. Tatsächlich steckt da doch etwas dahinter, was im Rahmen einer möglichst entzündungsfreien Lebensweise wichtig ist.

Forschende der Universität von Kopenhagen sind der Signalgebung in der Haut nachgegangen. Dabei stellten sie fest: Jeder Sonnenbrand verändert die mRNA (mitrochrondiale DNA), aber nicht direkt die DNA. Innerhalb der mRNA springt das ZAK-alpha-Protein auf eine starke UV-Bestrahlung an. Dieses Protein fungiert als Überwachungssystem für den Status der Zellen. UV-geschädigte Zellen sterben schneller ab, Entzündungszeichen treten vermehrt auf. ZAK-alpha erkennt diesen Status und meldet das im Rahmen der normalen Zellteilung weiter.

Je häufiger die Sonne ungebremst auf die Haut herunterknallen kann, desto sensitiver reagiert die RNA und das Frühwarnsystem von ZAK-alpha auf UV-Strahlung.

Alle nachfolgenden UV-strahlungsbedingten Hautreizungen fallen dann stärker aus. Das ZAK-alpha-Protein hat ja schließlich schnell gelernt, dass das zu Zelltod und Verderben in Form von Entzündungen führt. Da reagiert es entsprechend sensitiver und schaltet sich noch schneller ein, um die Haut vor weiterer Beschädigung zu schützen! So erklärt sich auch der Zusammenhang zwischen chronischen Hauterkrankungen und einer damit einhergehenden stärkeren Empfindsamkeit für Sonnenlicht (Vind et al. 2024).

Für den Menschen ist dieser ganze Ablauf hinreichend erforscht. Ähnliche Prozesse dürften sich auch bei Pferden abspielen. Das heißt natürlich nicht, dass Pferde nicht mehr auf die Weide dürften, immerhin sind sie ja im Großen und Ganzen für den Aufenthalt unter freiem Himmel bestens ausgerüstet. Aber: Zu viel und zu starke Sonneneinstrahlung kann bei den modernen Züchtungen unter Umständen eben doch Schäden verursachen – das solltest du einfach auf dem Schirm haben.

Nun wissen wir ja, dass Entzündungen zwar normal und wichtig sind, aber den Stoffkreislauf stark belasten. Hast du nun aber ein Pferd mit entsprechender Sensitivität in Form von heller Haut, hellem oder dünner werdendem Fell, dann schenkst du besser auch dem Sonnenschutz noch etwas mehr Beachtung. Nur um sicherzugehen, dass sich vorhandene Haut- und Fellproblematiken nicht noch aufgrund der UV-Strahlung verstärken und das Entzündungspotenzial im Pferdekörper so niedrig wie möglich bleibt.

UV-Decken können eine Option sein, um die Strahlungsintensität etwas zu begrenzen, vor allem wenn die Koppel nur begrenzte Schattenplätze zu bieten hat.

Der Stoff, aus dem Entzündungsreaktionen sind: Histamin

Ein Neurotransmitter (Botenstoff), von dem oft genug mit gefährlichem Halbwissen die Rede ist: Histamin. Wenn wir über inflammatorische Reaktionen sprechen, spielt es allerdings eine wichtige und leider oft noch unterschätzte Rolle.

Wichtig: Histamin wird nicht primär bei Allergien ausgeschüttet. Nur weil du Heuschnupfen hast, hast du nicht automatisch einen fiesen Histaminüberschuss im Körper. Eine sogenannte Histamin-Intoleranz (HIT) wird oft auch als Allergie wahrgenommen, ist im Kern aber eine Stoffwechselstörung (NDR 2023) und kann ganz verschiedene Ursachen haben. Einige genetische Dispositionen führen dann dazu, dass sich eine sogenannte Histamin-Intoleranz als Lebensmittelunverträglichkeit zeigt, die aber eigentlich gar keine ist (IMD Labor Berlin).

Die von mir sehr geschätzte Dr. Koch spricht primär von einer Histaminose anstelle der sogenannten Histamin-Intoleranz. Ich übernehme diesen Sprachgebrauch auch für mich, auch wenn der Begriff Histamin-Interolanz üblicherweise verwendet wird.

Histaminose

Es existieren Schätzungen von Allergieverbänden, das rund ein bis drei Prozent der Menschen in Europa eine sogenannte Histaminose aufweisen. Das sind zwischen **80 und 240 Millionen Menschen** von aktuell (Stand: 2025) 800 Millionen Menschen.

Parallel dazu ist auch bekannt, dass **circa 150 Millionen Menschen** in Europa von Allergien betroffen sind – allein 20 bis 30 Millionen davon sind in Deutschland zu finden, die allgemeine Tendenz ist steigend.

Das Robert-Koch-Institut geht davon aus, dass rund einem Viertel aller Kinder und

Erwachsenen im Laufe des Lebens eine Allergie diagnostiziert wird (Fehr et al. 2017).

Überschneidungen zwischen Allergikern und Menschen mit Histaminose werden übrigens nicht systematisch erfasst. Die Schnittmenge ist aber sicherlich nicht gerade klein.

Histamin selbst ist ein grundlegender Bestandteil des körpereigenen Signalgebung. Es wird über die Nahrung aufgenommen, kann aber auch im Körper produziert werden. Damit die Histaminproduktion im erforderlichen Rahmen funktioniert, müssen aber zunächst mal das Immunsystem, das Mikrobiom des Darms und das zentrale Nervensystem optimal ineinandergreifen. Das gelingt dummerweise nicht immer. Und die Gründe dafür sind vielfältig.

Das körpereigene **Verdauungsenzym Diaminoxididase (DAO)** beispielsweise wird benötigt, um Histamin über die Darmschleimhaut abzubauen. Fehlt es an Vitamin B6 im Stoffkreislauf, wird das Enzym nicht oder nicht in ausreichender Menge hergestellt. Gleiches gilt für Vitamin C, das sowieso eine umfassende Rolle in sämtlichen Immunabwehrprozessen sowie bei der Bewältigung von Atemwegserkrankungen spielt (Ghalibaf et al. 2023). Das zweite zum Abbau von Histamin unverzichtbare Enzym nennt sich **Histamin-N-Methyltransferase (HNMT)** und kommt seiner Tätigkeit vorrangig in Leber, Niere und Bronchialschleimhaut nach.

Histamin im Überschuss kann für wirklich unschöne Kettenreaktionen sorgen. Bestes Beispiel dafür:

- Die Verbindung aus erhöhten Histaminwerten und niedrigeren Serotoninleveln. Denn der Botenstoff Histamin **kann** die Freigabe von Serotonin blockieren, wenn eine Histaminose vorliegt. Das kann unter anderem zu Schlaflosigkeit und Überstimulation führen, was wiederum den allgemeinen Stresslevel erhöht. Das zentrale Nervensystem ist dysreguliert – und damit wird auch fein weiter Histamin ausgeschüttet. Stress killt Entspannung – und macht auch noch schlechte Laune.

Histamin als Teil der Immunabwehr

Mastzellen (Abwehrzellen) sind ein enorm wichtiger Bestandteil des Immunsystems und besonders an den Grenzen aktiv. Dazu zählen Haut, Schleimhäute in Darm, Nase oder Lunge, Blutgefäße sowie die Nerven. Einmal aktiviert, schütten die Mastzellen Botenstoffe aus, die den Körper fit für die (vermeintliche) Bedrohung machen. Mit dabei ist auch Histamin. Selbiges ist großartig, wenn es um die schnelle Reaktion auf mögliche Gefahren geht, denn es sorgt für erweiterte und durchlässigere Blutgefäße und eine schnellere Reizweiterleitung.

Leider sorgt es auch dann dafür, wenn es eigentlich gar keine reelle Gefahr gibt. Für uns ist es ein abendliches Glas Rotwein, für deine Mastzellen ist es womöglich das Äquivalent zu einem Säbelzahntigerangriff in freier Wildbahn.

Während du noch dein Glas Wein genießt, springt dein Immunsystem schon an und liefert dir so hübsche Begleiterscheinungen wie Juckreiz, Husten oder eine überaus aktive Verdauung. Rotwein enthält nämlich sowohl Histamin als auch Alkohol, der als Histamin-Liberator die Bildung von körpereigenem Histamin beschleunigen kann. Die beim Menschen ebenfalls recht typischen Kater-Kopfschmerzen können wahlweise vom Alkohol oder vom Histamin (oder beidem!) stammen.

Ob und wenn ja, wie häufig Kopfschmerzen beim Pferd auftreten, lässt sich jetzt natürlich nicht so genau sagen. Aber sicherlich ist ein histamingeplagtes Pferd in seiner Leistung beeinträchtigt und du als Reitmensch tust gut daran, das in die tägliche Arbeit mit einzukalkulieren, sofern du eine Histaminose vermutest.

Histamininduzierte Veränderungen	Symptome
geweitete Blutgefäße	Rötungen, Wärmegefühl
durchlässigere Gefäße	Schwellungen, Juckreiz
Nervenreizungen	Juckreiz, Kribbeln, Unruhe
verstärkte Schleimproduktion	laufende Nase, Hustenreiz, Darmaktivierung
verstärkte Magensäureproduktion	Verdauungsprobleme, Sodbrennen
Aus dem Takt geratener Schlaf-Wach-Rhythmus	Müdigkeit, Konzentrationsprobleme, häufigeres nächtliches Aufwachen, Einschlafprobleme
Nervensystem im Alarmzustand	Herzklopfen, Unruhe, erhöhter Ruhepuls

Tab. Reaktionen des Körpers auf Histaminausschüttung (eigene Zusammenstellung)

Alle diese Reaktionen sind natürlich im Kern wichtig: Sie bereiten den Körper auf eine mögliche Gefahrensituation vor. Kritisch wird es nur, wenn da gar keine echte Gefahrensituation vorliegt!

Ein Tipp aus der Praxis: Lauwarmes Wasser sorgt im Sommer für etwas Erleichterung, vor allem nach dem Training. Das kühlt das Pferd etwas herunter. Zusätzlich ist Vitamin C in mehreren Dosen über den Tag hinweg zu füttern ebenfalls eine gute Idee.

Bei Pferden mit allergischem Husten oder bei denen, die zum Schubbern neigen, wird oft auch Antihistaminika aus dem Humanbereich verabreicht. Eine Zulassung für Pferde hat das Menschenmedikament zwar logischerweise nicht, aber im Alltag hat es sich doch oft bewährt.

Gesunde Pferde und Histamin

Nun sind gesunde Pferde im Grunde mit einem ziemlich robusten Histaminsystem ausgestattet. Immerhin sind sie ja Fluchttiere! Ihr ganzer Organismus ist an Belastung gewöhnt, systemisch bedingt ist ihre Grenze zur vermehrten Histaminausschüttung also deutlich höher als beim Menschen.

Nur trifft das natürlich nicht auf alle Pferde gleichermaßen zu. Die Tiere mit Allergien, mit Ekzemen, mit Hautjucken und Atemwegserkrankungen sind prädestiniert dafür, ein deutlich empfindlicheres Histaminsystem vorzuweisen.

Gesunde, aber sensible Pferde und Histamin

Beim Pferd mit entsprechend sensiblerem System ist anstelle des Gläschen Weins für den Menschen dann eher Ausdauertraining oder Sprint eine *dumme* Idee, sobald eine gewisse Neigung zur vermehrten Mastzellenaktivierung vorliegt. Der hohe Pulsbereich und das längere Schwitzen aktivieren nämlich selbige. Histamin, das vorrangig in der Haut und in den Atemwegen freigesetzt wird, kann dann für Hautrötungen und Juckreiz nach dem Training sorgen.

Pferde mit Ekzem-Symptomatik erleben eine signifikante Verschlechterung ihrer Haut, wenn sie regelmäßig mit hohem Puls abgeschleudert oder über Stangen gescheucht werden. Sie sind mitnichten einfach nur ein *bisschen sensibel* oder bringen eine schlechte Genetik mit.

Das ihnen angebotene Training ist schlicht dazu geeignet, sie im vollen Galopp in den Histaminkollaps zu treiben, während das ahnungslose Reitgewicht im Sattel sich über das sportliche Tempo begeistert. Einige Studien am Menschen belegen die signifikante Verschlechterung der Haut durch eine vorliegende Histaminose bei anhaltendem pulsbeschleunigendem Training (Kulinski et al. 2019).

Wichtig:

Histamin ist selten die einzige Ursache für solche Phänomene. Prüfe immer die Kombination aus Verdauungs- und Immunsystem sowie genetische Komponenten, bevor du dich auf Histamin als Ursache von Hautproblemen festlegst.

Allergiepferde und Histamin

Bei Allergiepferden könnte man jetzt davon ausgehen, dass die Mastzellen hochsensibel auf etwaige *Bedrohungen* des Immunsystems reagieren könnten, die Histamin-Liberatoren also entsprechend heftige systemische Reaktionen auslösen könnten. Die Histaminausschüttung aufgrund von Histamin-Liberatoren fällt allerdings etwas geringer, milder und langsamer aus. Bei chronisch krankem Pferd ist das natürlich trotzdem problematisch. Jedes zu viel an Histamin ist genau das: zu viel!

Einmal Histaminose, immer Histaminose?

Zum Glück stimmt das so nicht unbedingt. Neuere Forschung zeigt, dass sich die Sensitivität des Körpers auf Histamin zumindest in Teilen regulieren lässt. Wie? Vorrangig über die **Nahrungszufuhr**. Je weniger Histamin extern aufgenommen wird, desto weniger muss der Körper mühsam wieder aufspalten. Beim Pferdefutter achte auf unverdorbenes Futter. Denn das ist leider die Hauptquelle für externes Histamin, die ganz einfach vermieden werden kann durch:

- frisches, hygienisches Heu,
- schimmelfreie Frisch-/Trockenfutterkomponenten,
- das Vermeiden von zweifelhafter Heulage
- und verdorbenem Mash.

Allein schon zu langes Einweichen kann zu einem steigenden Histamingehalt führen. Futterhersteller empfehlen deshalb in der Regel eine Einweichzeit von maximal 15 Minuten bei heißem und 30 Minuten bei kaltem Wasser, um eine mögliche verstärkte Histaminbildung zu begrenzen.

Bei **Allergiepferden** wird oft von Luzerne oder Soja abgeraten, das würde die Histaminausschüttung nur fördern. Sie enthalten Histamin-Liberatoren, also Hilfsstoffe, die die Produktion von körpereigenem Histamin auslösen können. Grundsätzlich lässt sich Sojaschrot aber gut durchgängig im gesamten Jahr verfüttern. Simples ausprobieren kann sich hier bezahlt machen, denn viele Pferde kommen mit Luzerne und Soja - beides ja hervorragende Proteinquellen - wirklich gut zurecht. Bei verdorbenem Futter mit entsprechend hohem Histamingehalt sieht das aber meist ganz anders aus!

Tipp:

Bei richtig üblen Allergieschüben kann es zwecks Proteinversorgung angebracht sein, auf das wesentlich teurere Kartoffelprotein umzusteigen. Es wird pur als Pulver angeboten und muss nass verfüttert werden. Es ist umständlich in der Fütterung und es schmeckt anscheinend ziemlich furchtbar, sodass es beim Verfüttern relativ viel Aufwand produziert, weil es im nassen Kraftfutter untergemengt und nahezu *ver-*

steckt werden muss. Pferde mit Sinn für guten Geschmack neigen dazu, das extra angerührte Spezialfutter einfach stehen zu lassen, sobald das Kartoffelprotein zu stark heraussticht.

Eine übermäßige Histaminreaktion kann auch durch die **Einnahme geeigneter Co-Faktoren** zumindest reduziert werden. Forschende konnten nach intravenöser Gabe von Vitamin C signifikant niedrigere Histaminlevel (Serum) bei Menschen mit Allergien und bei Infektionen der oberen Atemwege nachweisen (Ghalibaf et al. 2023).

Das ist dann nicht nur für das allergiegeplagte Pferd spannend, sondern vielleicht auch für den schniefenden Menschen, der das Pferd liebevoll versorgt!

Tipp:

Hagebutten sind eine günstige Vitamin-C-Quelle, die nicht als Histamin-Liberatoren fungieren, aber durchaus etwas Histamin enthalten. Der Vitamin-C-Gehalt ist hier aber der ausschlaggebende Punkt. Sie lassen sich getrocknet recht gut über den Tag hinweg verfüttern, sofern das eine Option darstellt. Alternativ stehen auch Vitamin-C-Infusionen zur Verfügung, die aber die wenigsten Pferdemenschen einfach mal eben so dem Pferd verabreichen wollen – oder können. Magnesium darf gerne auch noch mit dazu, denn es ist ein wichtiger Co-Faktor für das Enzym Diaminoxidase (DAO), welches wiederum zum Histaminabbau benötigt wird.

Proteine als Basis für fitte Körper

Warum ist die Proteinsynthese wichtig – und was hat eigentlich Testosteron damit zu tun? Hervorragende Fragen, die in der Pferdeernährung im Rahmen eines gesunderhaltenden Trainings viel öfter gestellt werden sollten.

Proteine sind bekanntermaßen die Quelle fetter Muskelpakete, viel mehr noch als das Testosteron, das den Hengsten unter den Reitkunstpferden einen gewissen, aber nicht uneinholbaren Vorteil verschafft. Was menschliche Bodybuildingfans gern in Form von aromatisierten Eiweißpülverchen in sich hineinschütten, muss bei Pferden ebenfalls ergänzt werden.

Zur Erinnerung: Die einst eher kohlenhydratarmen, dafür leicht proteinreicheren Steppengrasfutterquellen der Vorfahren unserer heutigen Züchtungen gehören längst der Vergangenheit an. Heimische saftig-süße Weidegründe sind für den noch immer eher ursprünglichen Makronährstoffbedarf heutiger Züchtungen weniger geeignet, wenn die modernen Pferde auch noch hier und da eine ordentliche Runde piaffieren sollen.

So steht die Proteinsynthese oft nicht mit ausreichend Rohstoffen aus besagten fetten Weidegründen da, um ihre Arbeit vollumfänglich zu bewältigen. Dabei ist sie die Grundlage für eine Vielzahl enorm wichtiger Prozesse im Körper.

Fitte Proteinsynthese = fitter Stoffkreislauf

Versammelnde Arbeit unter dem Reiter ist ideal, um die Proteinsynthese anzuschieben. Besonders gut geeignet sind Piaffe, Passage oder langsames, konzentriertes Traben plus Schulgalopp, Schulschritt und absolut alles im Seitwärts. Versammlung fördert die maximale Muskelaktivierung ohne eine gleichzeitige hohe Gelenkbelastung. Bloß braucht es dafür eben vor allem eins: passende Proteine.

Proteine aus der Nahrung werden im Verdauungstrakt in Aminosäuren zerlegt. Über den Blutkreislauf gelangen diese Aminosäuren an ihre jeweiligen Bestimmungsorte im Körper – Zellen in der Leber, im Darm oder in den einzelnen Muskelfasern.

Hauptbestandteile dieser Schönheit: Proteine und Wasser. Sonja mit Coronado.

An diesen Bestimmungsorten startet dann die eigentliche Proteinsynthese – und aus Bausteinen wie Leucin oder Lysin werden ganz am Ende der Nährstoffverwertung schließlich unter anderem (mehr) Muskelfasern für Reitkunstpferde.

Proteinsynthese

Die Proteinsynthese ist der Prozess, bei dem der Körper neue Proteine herstellt. Diese werden für den Aufbau, die Reparatur und die Funktion der Zellen benötigt. Die Verfügbarkeit bestimmter Aminosäuren in der Nahrung ist eine notwendige Grundlage für den Start der Proteinsynthese.

Bei der Proteinsynthese werden also die genetischen Informationen der Zielzelle in spezifische Proteine übersetzt. Diese spezifischen Proteine übernehmen verschiedenste Aufgaben:

- Muskelaufbau und -reparatur, insbesondere nach körperlicher Aktivität und/oder Krafttraining.
- Enzymproduktion, um chemische Reaktionen im Körper zu katalysieren.
- Hormonherstellung, denn einige Hormone wie das Insulin beruhen auf Proteinbausteinen.
- Strukturarbeiten, um Zellmembranen, Haut, Fell und Hufhorn aufzubauen.
- Immunabwehr, in dem Antikörper zur Abwehr von Infektionen hergestellt werden.

Proteinsynthese ist die Basis für schöne Reitkunstpferde. Christina mit Emma.

Erstlimitierende essenzielle Aminosäuren und ihre Rolle in der Proteinsynthese

Die drei essenziellen Aminosäuren **Lysin, Methionin und Threonin** sind in typischen Pferdefuttermitteln (Heu, Gras, Getreide) oft in zu geringen Mengen vorhanden und bestimmen gleichzeitig, wie effektiv der Pferdekörper Proteine aufbauen kann. Sie werden als **erstlimitierende** Aminosäuren (Mok/Urschel 2020) bezeichnet, weil ihr Mangel als erstes den gesamten Eiweißaufbau und damit viele Stoffwechselprozesse begrenzt:

> *»If any amino acid is temporarily limited in supply then this could also temporarily limit protein synthesis. If free amino acids generated by exogenous turnover cannot be immediately used for protein synthesis pathways then they become subject to oxidation or conversion to fats. This could become a significant issue when additional protein is provided in the diet in an attempt to match demand of the specific amino acids, since amino acids per se, cannot be stored. Thus, if protein synthesis during recovery and repair from exercise was limited by the short supply of key amino acids, the rates of oxidation and/or fat conversion of the »surplus« amino acids would escalate* (Dunstan 2020).«

Auf Deutsch: »Wenn die Versorgung mit einer Aminosäure vorübergehend eingeschränkt ist, kann dies auch die Proteinsynthese vorübergehend einschränken. Wenn freie Aminosäuren, die durch den exogenen Umsatz entstehen, nicht sofort für die Proteinsynthesewege verwendet werden können, werden sie oxidiert oder in Fette umgewandelt. Dies könnte zu einem erheblichen Problem werden, wenn mit der Nahrung zusätzliches Protein zugeführt wird, um den Bedarf an bestimmten Aminosäuren zu decken, da Aminosäuren an sich nicht gespeichert werden können. Wenn also die Proteinsynthese während der Erholung und Wiederherstellung nach dem Training durch das knappe Angebot an wichtigen Aminosäuren eingeschränkt wird, würde die Oxidation und/oder Fettumwandlung der »überschüssigen« Aminosäuren eskalieren.«

In manchen Quellen wird auch **Leucin** als potenziell limitierend genannt, insbesondere bei jungen Pferden und in bestimmten Fütterungssituationen. Für die meisten erwachsenen Pferde gelten jedoch **Lysin, Methionin und Threonin** als die wichtigsten limitierenden Aminosäuren.

Lysin

Erstlimitierende Aminosäure und notwendig für Bildung und Reparatur von Muskelzellen und Teil der Kollagensynthese zur Bildung und Reparatur von Knochen, Sehnen und Bindegewebe. Ohne Lysin als Basis kann die Proteinsynthese an und für sich nicht starten. Ein Lysin-Mangel macht sich oft über eine schwache Immunabwehr in Kombination mit reduziertem Wachstum und insgesamt wenig belastungsfähiger Muskulatur bemerkbar (Liu et al. 2025).

Threonin

Zweitlimitierende Aminosäure. Threonin nimmt eine tragende Rolle für die Darmgesundheit ein, ist Bestandteil der Schleimhautproteine und von Immunglobulinen. Es unterstützt vor allem die Synthese von Enzymen und Antikörpern und ist entsprechend entscheidend für den Stoffwechsel und für eine hilfreiche Immunabwehr. Ein Threonin-Mangel führt primär zu einer erhöhten Infektanfälligkeit, außerdem zeigen sich Wachstumsstörungen, weil die Futterverwertung beeinträchtigt ist.

Methionin

Stellt das sogenannte Startmolekül für die Proteinsynthese dar, jede neue Proteinkette startet mit dieser schwefelhaltigen Aminosäure. Es ist an zahlreichen Stoffwechselprozessen beteiligt, unter anderem bei der Entgiftung in der Leber. Es liefert die Vorstufe für Cystein und Taurin, die wiederum für Glutathion als Antioxidans benötigt werden. Methionin ist unverzichtbar für den SAM-E-Stoffwechsel. Ein Mangel zeigt sich in grundlegend schlechter Hornqualität, einem schwachen Immunsystem und ausbleibendem Trainingserfolg trotz hinreichendem Training mit hoher mechanischer Spannung im Muskel. Methionin-Mangel bremst die Proteinsynthese aus!

Proteine sind die Basis für rundum gesunde Pferde, die Leistung zeigen können. Mascha mit Schimmelstute.

Der perfekte Proteinmix im Futter...

... existiert nicht. So viel vorneweg, bevor es tiefer in die Materie geht. Die Hoffnung auf eine perfekte Versorgung geht schon deshalb nicht auf, weil jedes Individuum - Mensch und Pferd gleichermaßen - natürlich einen individuell perfekten Bedarf an spezifischen Mikro- und Makronährstoffen hat.

Futtermittelempfehlungen wie die Gruber Tabelle (Bayerische Landesanstalt für Landwirtschaft 2024) existieren natürlich trotzdem, mit deren Hilfe zumindest grob passende Mengen für das eigene Tierchen zusammengestellt werden können. Nur: Typische Futterrationen aus Gras, Heu und Getreide enthalten in aller Regel nur wenig Protein und anteilig noch geringere Mengen an Lysin oder Methionin. In Zahlen kann das dann so aussehen, wie in der Tabelle unten.

Und wie viel sollten Pferde nun zu sich nehmen? Die Literatur dazu ist sich uneins und natürlich schwankt der individuelle Bedarf sowohl von Pferd zu Pferd als auch beim einzelnen Tier in Abhängigkeit von durchgemachten Infekten und Verletzungen, dem Schweregrad seiner Tätigkeit oder ob es beispielsweise gerade tragend oder säugend ist.

Ob die Zusammensetzung der Aminosäuren optimal ist, zeigt sich im Grunde über ein idealerweise fittes und gesundes Pferd.

Bei Entgiftungsproblemen lohnt sich ein Blick ins Futterprofil hinsichtlich Cystein und Methionin. Ist der Muskelaufbau trotz anspruchsvollem Krafttraining mit hoher mechanischer Spannung im Muskel eher unbefriedigend, sollten Methionin und Lysin im Fokus stehen.

Übersicht wichtige Aminosäuren

Eine Auswahl wichtiger Aminosäuren, ihre Co-Faktoren und mögliche Folgen bei mehr oder minder starkem Mangel. Wie schnell oder langsam ein Pferd auf eine nicht ausreichende Aminosäurenversorgung im Einzelnen reagiert, ist höchst individuell!

Tab. Gehalt an Lysin in Standard-Pferdefutter (eigene Zusammenstellung, Nährwertrechner)

* in Gramm/100 Gramm Futtermittel

Futtermittel	Lysin *	Methionin *	Threonin *
Sojaschrot	3,5	0,7	1,49
Rapskuchen	2,25	0,55	0,60
Sonnenblumenkuchen	1,25	0,45	0,93
Leinsamen	0,5	0,19	0,87
Luzerneheu	0,55	0,16	0,30
Hafer	0,48	0,19	0,42

Aminosäure	Funktion im Pferdeorganismus	Co-Faktoren	Folgen bei Mangel
Histidin	Wichtige Rolle bei Gewebereparatur und Wachstum, antioxidative Eigenschaften, Schutz vor freien Radikalen	Vitamin B6, Zink (allg. Protein)	Verzögerte Gewebereparatur, geschwächte Haut, erhöhte Anfälligkeit für oxidativen Stress
Isoleucin	Beteiligt am Muskelaufbau, Energieversorgung, Proteinbiosynthese.	Vitamin B6, Biotin	Schlechter Muskelaufbau, Muskelschwäche, reduzierte Leistungsfähigkeit
Leucin	Fördert Muskelaufbau, unterstützt Proteinbiosynthese, beteiligt an Hautgesundheit	Vitamin B6, Biotin	Muskelabbau, schlechte Regeneration, verminderte Hautgesundheit
Lysin	Wichtig für Zellteilung, Wachstum, Muskelaufbau, Knochenwachstum, Kollagensynthese, Immunabwehr	Vitamin C (für Verstoffwechselung), Eisen	Verzögertes Wachstum, schlechte Wundheilung, Muskelabbau, auffällige Leberwerte, schwaches Immunsystem
Methionin	Essenziell für Eiweiß-, Fell-, Haut- und Hufstoffwechsel, Kreatinbildung, Entgiftung	Vitamin B6, Folsäure, Selen	Schlechte Fell- und Hufqualität, Müdigkeit, Abgeschlagenheit, verlangsamtes Wachstum, erhöhte Krankheitsanfälligkeit
Phenylalanin	Vorstufe für Tyrosin, wichtig für Hormon- und Neurotransmittersynthese, Proteinaufbau	Eisen, Kupfer, Vitamin B6	Wachstumsstörungen, Haut- und Fellprobleme, neurologische Auffälligkeiten
Threonin	Bildung von Enzymen, Hormonen, Antikörpern; wichtig für Muskel- und Hautgesundheit	Vitamin B6, Magnesium, Zink	Schwacher Muskelaufbau, geschwächtes Immunsystem, schlechte Haut- und Fellqualität
Tryptophan	Vorstufe für Serotonin, wichtig für Nervenfunktion und Stimmung, Proteinaufbau	Vitamin B6, Magnesium	Nervosität, Stressanfälligkeit, verminderte Leistungsbereitschaft
Valin	Beteiligt an Muskelaufbau und -erhalt, Energieversorgung, Immunsystem	Vitamin B6, Biotin	Muskelabbau, Schwäche, schlechte Leistungsfähigkeit

Abb. Zusammenstellung wichtiger Aminosäuren für Pferdestoffwechsel (eigene Darstellung), die limitierenden Aminosäuren sind markiert

Sojaschrot – ständig in der Kritik

Das Thema der angeblichen Unverdaulichkeit von Sojaschrot hält sich hartnäckig. Im Rahmen einer Dissertation an der LMU München hat man sich das glücklicherweise mit etwas mehr wissenschaftlicher Expertise angeguckt. Das Ziel hinter dem Versuch war eigentlich, die optimale Proteinmenge für Warmblutfohlen zu ermitteln. Dabei kamen allerdings auch – wenig überraschend für mich – die positiven Aspekte der Fütterung mit Sojaschrot zutage.

Füttert man also drei Gruppen von Warmblutfohlen Eiweißrationen wahlweise aus

- Heu und Hafer (Gruppe 1),
- Heu, Hafer und essenzielle Aminosäuren als Ergänzungsfutter (Gruppe 2),
- Sojaschrot (Gruppe 3),

dann passiert da etwas, das nur für Sojaschrot-Gegner überraschend kommen kann.

Denn das beste Wachstum im Sinne von Gewichts- und Skelettstrukturzunahme zeigten die Fohlen der Gruppe 3. Selbst im direkten Vergleich mit Gruppe 2, also bei einer traditionelleren Fütterung aus Heu, Hafer und Aminosäuren, schnitt Sojaschrot gleich gut ab. Es gab keine Unverträglichkeiten, die untersuchten Gesundheitswerte im Blut sowie Knochenmarker zeigten keine negativen Veränderungen. Die rein mit Hafer und Heu gefütterte Gruppe zeigte allerdings sehr wohl problematische Marker, die darauf hinweisen, »dass die Gruppe HAF eventuell einen subklinischen Mangel aufweisen könnte. Die Fohlen der Gruppe AS zeigten bei ähnlichen Proteinwerten wie die Gruppe HAF keinerlei Anzeichen für eine Mangelernährung (Krumbiegel 2011).«

Die Gruppe 1 mit der praxistypischen Proteinversorgung rein aus Heu und Hafer wies im Gegensatz zu den beiden anderen Gruppen auch ein signifikant geringeres Wachstum (Körpergewicht, Muskelzuwachs) und einen niedrigeren Body-Condition-Score auf. Das Hufhornwachstum sowie der Muskelzuwachs war bei den Hengstfohlen der Gruppen 2 und 3 darüber hinaus am besten (Krumbiegel 2011).

Ganz klar: Die im Sojaschrot enthaltenen Aminosäuren kommen dem optimalen Proteinprofil in der Fütterung einfach sehr nahe. Die wachstumsreiche Entwicklungsphase von Fohlen ist an dieser Stelle besonders aussagekräftig, denn wenn hier eine nicht ausreichende Versorgung an körpereigenem Baumaterial (Protein) vorliegt, dann zeigt sich das in bereits relativ kurzen Zeitspannen.

Ähnliche Ergebnisse lassen sich übrigens auch in der Schweinezucht finden (Bayerische Landesanstalt für Landwirtschaft, 2010). Hier zeigen sich ebenfalls keinerlei Probleme mit der Verdaulichkeit, sondern ganz im Gegenteil hervorragende Ergebnisse.

Fragen & Antworten zu Soja(schrot)

Kaum steht das Sojaschrot im Stall, schon kommen die ersten besorgten Nachfragen dazu. Jede und jeder hat da diverses dazu gehört und du stehst da und weißt nicht so genau, wie du reagieren sollst?

Die mit Abstand häufigsten Fragen samt fundierter Antworten findest du hier. Am Austausch ernsthaft interessierte werden sich dann bestimmt weiter mit der Thematik befassen.

Erhöht Soja nicht das Krebsrisiko?

Eine davon wird unweigerlich den möglichen Zusammenhang zwischen den im Soja reichlich vorhandenen pflanzlichen Östrogenen und hormonabhängigem Krebs behandeln. Sollten Stuten da nicht besser kein Soja bekommen, sicher ist sicher?

Zunächst einmal: In Soja sind Isoflavone enthalten. Diese Stoffe wirken zwar ähnlich **wie**, sind aber nicht gleichzusetzen **mit** künstlichen Östrogenen.

Tatsächlich konnte lediglich eine Studie einen Zusammenhang zwischen der Einnahme von künstlichen

Östrogenen und einer Zunahme an Brustkrebserkrankungen herstellen. Eine (Rossouw et al. 2002)!

Seitdem ist beim Menschen die Behandlung von Frauen kurz vor und während der Menopause aus (berechtigter) Sorge vor Krebserkrankungen extrem erschwert. Da fallen nämlich insbesondere Progesterol und Östrogen ab – mit massiven Folgen auch für die Knochengesundheit, Stichwort Osteoporose.

Nur ist die optimale Behandlung in diesem Falle neben Kraftsport und dem damit zusammenhängenden Muskelerhalt auch eine Hormontherapie, optimalerweise mit bioidentischen Hormonen (BIH). Künstliche Hormone werden aufgrund der möglichen Risiken, die diese eine Studie benennt, nur ungern verschrieben und bioidentische Hormone landen zu Unrecht in der gleichen Schublade.

Zurück zu unseren Pferden, bei denen das Thema Östrogen und Krebsrisiko ja gern dazu führt, dass sie eben kein Sojaschrot erhalten, weil sich dieser Unsinn so hartnäckig hält und weder Veterinäre noch sogenannte Futtercoaches zwischen pflanzlichen Isoflavonen und künstlichem Östrogen der ersten Stunde unterscheiden können.

Isoflavone binden sich an spezielle Östrogenrezeptoren, was am Ende sogar vor bestimmten Krebsarten zu schützen scheint. Ein Blick über den großen Teich belegt das: In Ländern mit reichlich Soja auf der Speisekarte waren die Brustkrebsraten bei Frauen signifikant niedriger als in westlich orientierten Ländern mit entsprechend anderem Speiseplan (Stanford et al. 1995), erst mit Angleichung an typischere westliche Ernährungs- und Lebensformen stiegen die Brustkrebsraten auch bei den Frauen dieser Länder an (Gomez et al. 2019).

Sprich: Etwas mehr Soja, idealerweise in fermentierter Form wie Miso, Natto oder Tempeh von klein auf täte uns allen im Rahmen der (Brust-)Krebsvorsorge eigentlich ganz gut!

Wird für Soja nicht der ganze Regenwald abgeholzt?

Nicht ganz falsch, aber: Da hast du ja die Wahl. Es gibt längst Futtermittelhersteller, die ausschließlich Sojaschrot aus zumindest europäischem Anbau verwenden. Auch in Deutschland lässt sich Soja mit guter Qualität anpflanzen. Vielleicht hat auch der Bauer um die Ecke längst auf Soja umgestellt, um den Futtermittelbedarf heimischer Rinder und Co. zu decken. Nachfragen könnte sich da lohnen.

Auf den Verpackungen achtest du am besten auf Einträge, die das verwendete Soja möglichst GMO-frei und/oder aus europäischem Anbau kennzeichnen.

Der Großteil der Sojaimporte aus mittel- und südamerikanischen Gebieten findet tatsächlich statt, um vorrangig nordamerikanische Rinderherden zu füttern. Der hierzulande für Tofu, Sojamilch und Co. verwendete Soja-Rohstoff stammt überwiegend aus Europa.

Verklebt das ganze Protein aus Sojaschrot nicht die Muskeln?

Ungünstige Aminosäure-Profile können für Reaktionen sorgen, die manchmal als starre oder verklebte Muskelstrukturen interpretiert werden. Dann reagiert das Pferd hitziger. Liegt das nun am im Soja enthaltenen Protein?

Eher am ungünstigen Aminosäureprofil, das im Zuge der Aufspaltung des Stickstoffs für eine vermehrte Ammoniakbildung sorgt. Unangenehm fürs Tier und wenig hilfreich, um die proteinhungrigen Körperstrukturen effizient mit allem zu versorgen, was gerade benötigt wird.

Pferde mit Verspannungen und dem, was so gerne als verklebte Muskeln bezeichnet wird, haben in der Regel ganz andere Probleme. Allem voran Magnesiummangel, der vor allem bei Sportpferden und denen mit einem hitzigen Temperament zu finden ist. Dazu kannst du, wenn du genauer hinguckst, wahrscheinlich auch einen Elektrolyt- oder Mineralstoffmängel wie Kalium, Natrium und Calcium nachweisen. So bildet sich relativ schnell ein Säure-Basen-Ungleichgewicht und daraus entstehen weitere Stoffwechselprobleme. Über nette Spaziergänge und gelegentliches Longieren lassen sich die leider nicht in den Griff bekommen.

Proteine aus Sojaschrot

Dank thermischer Verarbeitung ist die Verdaulichkeit des Proteins in Sojaschrot zufriedenstellend. Die in Soja natürlicherweise vorkommenden Antinährstoffe wie Trypsin-Inhibatoren gelten in der Regel als verdauungshinderlich. Immerhin hemmen sie gerade das Enzym Trypsin, das für die Verdauung von Proteinen im Dünndarm essenziell ist. Nur: Deshalb wird Sojaschrot ja eben thermisch bearbeitet. Dadurch werden die Antinährstoffe inaktiviert und die Verdaulichkeit sichergestellt. Ein erneutes Erhitzen ist deshalb nicht nötig!

Hat dein Pferd aufgrund von Wachstumsphasen, Reitkunsttraining, Trächtigkeit, Laktation oder nach Verletzungen einen höheren Proteinbedarf, dann deckt die Fütterung mit Sojaschrot diesen Mehrbedarf zuverlässig ab.

Sojaschrot enthält im Mittel:

- **Rohprotein:** 44–48 g
- **Lysin:** ca. 3 g
- **Methionin:** ca. 0,6 g
- **Threonin:** ca. 1,5–1,8 g

Wichtig: Pferde mit Allergien oder die, die sich als wirklich schlechte Entgifter zeigen, sind unter Umständen mit **Kartoffelprotein** besser beraten. Sojaschrot als Histamin-Liberator kann teilweise gut vertragen werden, teils kann es in Kombination mit anderen Faktoren wie Genetik oder weiteren Erkrankungen besser sein, darauf zu verzichten.

Kartoffelprotein geht leider enorm ins Geld und ist ausschließlich als Pulver verfügbar, sodass es wenig alltagstauglich ist. Bei Kartoffelprotein ist die Verteilung (abhängig vom Produzenten, hier Agrana) tatsächlich noch optimaler, mit einem mittleren Anteil von circa 5 Prozent Lysin, circa 1,5 Prozent Methionin und circa 2 Prozent Cystin und Cystein. Leider schmeckt Kartoffelprotein furchtbar, sodass es häufig auch mit Weizen (Achtung zusätzliche Einfachzucker!) gestreckt wird, um es den Pferden schmackhafter zu machen.

Fittes Pferd, schönes Fell

Die schönsten Reitkunstkoffer haben das schönste Fell. Ist dir das auch schon aufgefallen? Das ist kein Zufall, sondern entsteht durch die Kombination aus fit trainiertem Pferd plus einer darauf angepassten Ernährung. Letztere entscheidet größtenteils mit darüber, wie seidig-glänzend das Pferdehaar – je nach genetischer Veranlagung und rassetypischen Möglichkeiten – ausfallen kann.

Der Klassiker in der Pferdefellpflege schlechthin: **Zinkmangel** führt zu struppigem Fell. Ist der Zinkmangel ausgeglichen, verbessert sich der Fellzustand oft schlagartig.

Vitamine fürs Pferdefell

Vitamin C gegen brüchiges Haar, um die Kollagenproduktion für eiweißreiche Strukturen wie Fell und Horn zu unterstützen.

Vitamin A und E gegen trockene Haut und kaputte Haarstrukturen, um oxidativen Stress zu reduzieren und die Hautbarriere zu stärken.
Eisen gegen Haarausfall, um die Sauerstoffversorgung der Haarfollikel über die roten Blutkörperchen zu unterstützen.

B7 gegen Haarausfall und Haarbruch, um die Haarstruktur und insbesondere die dafür benötigte Keratin-Produktion zu stärken.

D3 ist streng genommen ein Hormon, kein Vitamin. Ein Mangel kann zu dünnerem Haar führen, das Haarwachstum gerät ins Stocken. Steigt der D3-Spiegel durch Supplementierung, erhöht sich in aller Regel auch die Haardichte wieder.

Hauttumore – Sarkoide

Hauttumore, die wie Warzen aussehen? Das können nur equine Sarkoide sein. Die Hautveränderungen

Schönes Haar dank hervorragender Ernährung.
Sonja mit Coronado.

werden durch das Bovine Papillomavirus (BPV) verursacht, welches ursprünglich bei Rindern auftritt. Allerdings hat dieses Virus den Sprung zu anderen Tierarten geschafft und befällt nun eben auch Pferde.

Die durch das Virus verursachten Papillome (Warzen) sind bei Rindern harmlos, führen beim Pferd allerdings zu den Sarkoiden. Das sind tumorartige Hautveränderungen, die du unbedingt möglichst frühzeitig behandeln solltest. Die Infektion mit BPV tritt häufig in Verbindung mit einem beeinträchtigten Immunsystem auf.

Ob das körpereigene übereifrige Verteidigungssystem nun zuerst da war oder das BPV für die Überreaktion sorgt, sei mal dahingestellt. Es scheint aber durchaus auch eine genetische Komponente zu geben, bei der das Immunsystem tipptopp in Ordnung sein kann, es aber trotzdem zu Sarkoiden kommt.

Bei Züchtungen sollte deshalb auch darauf geachtet werden, dass möglichst nicht zwei zu Sarkoiden neigende Tiere Nachwuchs hervorbringen, der dann wiederum ebenfalls anfälliger für das BP-Virus sein wird. Wichtig ist für dich: Dein Vierbeiner ist nach Infektion und sobald die ersten Sarkoide auftreten, grundsätzlich angeschlagen und damit empfänglicher für weitere Erkrankungen. Das will aber niemand.

Um equine Sarkoide möglichst zügig zu behandeln, stehen verschiedene Therapieformen zur Auswahl. Als vielversprechend hat sich die Behandlung mit Mistelpräparaten gezeigt. Über subkutan, also unter die Haut gespritzte Dosen konnte die Ausprägung zumindest in Teilen reduziert werden (Christen-Clottu et al. 2010).

Fellgesundheit = Hautgesundheit = Sonnenschutz

Achtung Sonne: Jaja, da ist das Thema schon wieder. Aber ein gesunderhaltend trainiertes Reitkunstpferd mit reichlich Muskulatur ist nur deshalb so gesund, weil es rundum gut versorgt wird und Entzündungen nach Möglichkeit vermieden werden. Da Sonnenbrände per se Entzündungsreaktionen des Körpers auf die UV-A- und UV-B-Strahlung darstellen, ist eine Vermeidung ja also nur logisch. Plus: Entzündungen schädigen Haut und Fell.

Je heller die Haut (nicht das Fell!), desto eher kann sie durch UV-Licht geschädigt werden. Rosa Hautstellen, insbesondere dekorative Blässen im Gesicht sind prädestiniert für Sonnenbrände.

Pferde in hübschen Modefarben sind etwas gefährdeter als die, die mit schwarzem Fell und melatoninreicher Haut jede Menge natürlichen Sonnenschutz mit sich herumtragen.

Beim Pferd zeigt sich Sonnenbrand ganz ähnlich wie beim Menschen durch Hautrötungen und Schmerzempfindlichkeit der betroffenen Stellen (Grad 1). Das betrifft vor allem nicht-pigmentierte, rosa Hautareale sowie die Region um Augen, Nase und Maul. Wahrgenommen werden Sonnenbrände bei Pferden viel seltener und wenn, dann meist bei stärkeren Rötungen oder wenn sich bereits die Haut abschält.

Das auch Pferde unter klimatischen Herausforderungen wie starker Hitze leiden können, hat FEI nach den Olympischen Spielen in Barcelona 1992 festgestellt. Erst seitdem sind konkrete Regelungen erarbeitet worden, wie Pferde in, wie FEI es in einem Paper ausdrückt »thermally challenging environments« (deutsch: »thermisch anspruchsvolle Umgebungen«) bestmöglich performen, also Leistung zeigen können (FEI, Marlin et al. 2018).

Bei starker Rötung und Blasenbildung (Grad 2) kann es zu Entzündungen durch Bakterien kommen. Großflächiger starker Sonnenbrand ist extrem schmerzhaft, führt zur Ablösung größerer Hautflächen (Grad 3) und kann neben bakteriellen Sekundärinfektionen auch mit Fieber einhergehen.

Für eine gute Hautgesundheit achte also auf passenden Sonnenschutz. Auf der Koppel eignen sich Bäume und Sträucher sowie passende Unterstände. Schutzmasken, leichte Sommerdecken mit UV-Schutz oder Hufglocken schützen zusätzlich die sonnensensiblen Bereiche rund um Nase und Augen, Rücken, Kruppe und Fesseln.

Sonnenstich & Hitzschlag

Dunkles Fell, auf das ungehindert die Sonne knallt: Ja, auch Pferde können einen Sonnenstich bekommen! Begünstigt wird das noch durch zu geringe Trinkmengen und falls zu den hohen Temperaturen in der Sonne eine relativ hohe Luftfeuchtigkeit hinzukommt.

Blanke Koppeln ohne natürlichen Baumschatten oder einen offenen, möglichst luftigen Unterstand sind deshalb an feuchtschwülen Tagen keine gute Idee. Der schlecht durchlüftete vollgepackte Stall mit geschlossenen Türen natürlich auch nicht. Gleiches gilt für Hänger oder Turnierzelte, die keine Abkühlung bieten.

Sonnenstich: Das Pferd zeigt sich apathisch, schwitzt vermehrt oder taumelt sogar unter dem Sonneneinfluss. Manchmal bricht der Kreislauf auch direkt zusammen und das Pferd klatscht auf der Weide um. Sofortige Abkühlung durch Schatten und reichlich trinken lassen hilft meist gut gegen den Sonnenstich.

Hitzschlag: Zu den akuten Symptomen zählt unter anderem eine Körpertemperatur von 39 °C oder höher, Dehydrationsanzeichen wie stehenbleibende Hautfalten, dunkle oder bläulich verfärbte Schleimhäute. Das körpereigene Kühlsystem funktioniert nicht mehr, die Tiere schwitzen kein Tröpfchen mehr aus. Bei Druck auf das Zahnfleisch bleibt die Druckstelle länger als drei Sekunden weiß – ab vier Sekunden besteht durchaus Lebensgefahr, der Kreislauf kann kollabieren.

Veterinäre können zusätzlich mit mobilen Geräten vor Ort den Hämatokritwert bestimmen, ein erhöhter Wert spricht für eine Dehydrierung (Krüger 2023).

Der Hitzschlag ist potenziell tödlich, denn die Hitze führt am Ende zu Sauerstoffmangel im Körper sowie zu einer stoffwechselbedingten Übersäuerung.

Welche Pferde sind besonders gefährdet? Vor allem untrainierte Tiere, die schneller schwitzen und somit schneller dehydrieren. Robustrassen mit dichtem Fell vertragen die Hitze auch schlechter als der typische Araber, der schon per se mit einer entsprechenden genetischen Disposition für heißere Regionen ausgestattet ist.

Und natürlich sind unsere Reha-Fälle und Senioren-Pferde schlechter für die Hitze gerüstet, denn ihr Immunsystem ist anfälliger, sie haben eher mit Kreislaufproblemen und Erschöpfung zu tun als gesunde und fitte Tierchen in der Lebensmitte.

Auch an dieser Stelle zeigt sich also: Pferde gesunderhaltend zu trainieren zahlt sich aus!

Empfindliche Pferde mit Neigung zu Sonnenbrand reagieren unter Umständen auf Pflanzen, die die Lichtempfindlichkeit (Photosensibilität) der Haut erhöhen (Gehle et al 2005). Das sind beispielsweise Luzerne, Buchweizen oder Johanniskraut. Normalerweise führen diese Futterbestandteile aber kaum zu Problemen, was die Verträglichkeit der Sonne und die Hautgesundheit angeht.

Wirkt dein Pferd trotz geeigneter Schutzmaßnahmen viel zu schnell wie von der Sonnenbank, dann kann das auf eine grundlegende Stoffwechselstörung zurückzuführen sein. Die sekundäre Photodermatitis (hepatogene/metabolische Photodermatitis) kennzeichnet sich durch eine Verdauungsstörung, bei der Abbauprodukte des Chlorophyll (grüner Farbstoff in Pflanzen) nicht ordentlich in der Leber abgebaut werden.

Über die Leberenzymwerte (AST, O-GT, GLDH) lässt sich bestimmen, inwieweit die Stoffwechselerkrankung vorliegt. Bei unspezifischen Hautveränderungen und Haarausfall kann zusätzlich der Plasma-Selen- und Zink-Gehalt überprüft werden (Niedermaier/Gehlen 2009). Aus dem Takt geratene Leberenzymwerte und niedrige Selen- sowie Zinkwerte sollten unbedingt behandelt werden, um die Gesundheit deines Vierbeiners zu stärken.

Kein Dressurballett ohne gepflegte Hufe

Eine großangelegte Studie aus den Niederlanden vor einigen Jahren zeigte: Von 942 untersuchten Pferden hatten 85 Prozent mit Hufproblemen zu tun. 85 Prozent! Neben Freizeitpferden waren auch Dressur-, Fahr- und Zuchtpferde sowie Springpferde unter den untersuchten Tieren.

Strahlfäule, Hornspalten, ausgeprägte Wachstumsringe, Sohlenverletzungen, White Line Disease, Hornklüfte und chronische Lahmheit fanden sich durchweg bei allen Betätigungsfeldern der Vierbeiner. Es sind also mitnichten die eher an sachte Spaziergänge gewöhnten Freizeitpferde mit kaputten Hufen geschlagen – sondern alle (Holzhauer et al. 2017)!

Die Frage »Welcher Huf ist der Idealhuf?« ist wahrscheinlich überflüssig, da der beste Huf für das betroffene Pferd davon abhängt, in welcher Umgebung er funktionieren soll und wie viel Leistung er erbringen muss. (Hampson 2011)

Bewegung allein sorgt nicht für stabiles Hufhorn. Da muss auch die Genetik sowie die Ernährung mitspielen. Schauen wir uns zunächst die genetische Komponente an, die durch Züchtung oft noch verstärkt wird.

Hufgesundheit ist genetisch

Vor allem die auf Leistung gezüchteten **Vollblüter** neigen zu vermehrten Schwitzen, selbst wenn sie sich im Vergleich zum ausgeglicheneren Warmblüter kaum anstrengen. Ihr auf Speed gezüchtetes Herz-Kreislauf-System arbeitet einfach immer am oberen Rand seiner Kapazitäten. Höhere Temperaturen plus Feuchtigkeit durch das vermehrte Schwitzen = ideale Bedingungen für wachstumsbegeisterte Bakterien am ganzen Körper inklusive Huf. Zudem ist das Bindegewebe der Vollblüter feiner, an den Hufen zeigt sich das durch eine empfindliche Lederhaut und ein dünneres, sensibles Hufhorn.

Die Schwitzerei sorgt zugleich für einen Verlust an Elektrolyten wie Natrium, Kalium und Magnesium. Wird das nicht ausgeglichen, betreibt der Körper Raubbau an allen Strukturen und das betrifft dann auch die Hufe. Kein Wunder, dass du hier öfter mit Infektionen zu tun hast.

Die hufeigenen Schweißdrüsen der Strahllederhaut sorgen zwar eigentlich für ein leicht feuchtes und zugleich saures Klima, was für die natürliche Wundreinigung hilfreich ist. Äußere Umstände wie unpassender Beschlag oder fehlende Hufhygiene können aber durchaus dazu führen, dass diese Schweißdrüsen verstopfen. Das betrifft grundsätzlich alle Pferde, nicht nur die Vollblüter. Fehlt das saure Milieu im Strahl, vermehren sich krankmachende Keime rasant – Infektionen sind die Folge.

Quarter Horses sind für ihre dünnen Sohlen und ihr empfindliches Hufhorn wohl bekannt. In der Zucht wird auf schnelles Wachstum und reichlich Leistung geachtet, da gerät die Hufgesundheit schon mal aus dem Blick. Stattdessen kommt das spektakuläre Quarter-Horse-Showpaket mit einer gewissen Neigung zu Stoffwechselproblemen wie EMS oder Insulinresistenz daher sowie einer gewissen Neigung zu Hufabszessen.

Besonders die feingliedrig gezüchteten Linien und die Tiere, die extrem früh auf Leistung getrimmt werden, haben Zeit ihres Lebens mit infektanfälligen Hufen zu tun. Quasi als Ausgleich dazu haben sie in der genetischen Lotterie aber auch echt gewonnen, denn Quarter Horses verfügen in der Regel über eine bessere muskuläre Ausstattung, die sich andere Pferderassen erst lang und aufwendig erarbeiten müssen.

Andalusier bringen häufig ein weicheres Horn mit, insbesondere die modernen, feingliedrigen Zuchtlinien, die für die Dressur gedacht sind. Das weiche Hufhorn zeigt sich vor allem, wenn sie in feuchten Regionen leben, oder wenn sie in einer eher feuchten und unhygienischen Boxenhaltung ausharren müssen.

Die **Friesen** sind körperlich ganz anders gestaltet als die Andalusier, zeigen aber ebenfalls eine Tendenz zum weichen Horn. Bei schlechter Hufpflege neigen sie zudem zu Strahlfäule, was dann Entzündungen begünstigt.

Besseres Futter, besserer Huf

Spröde Hufe, wiederkehrende Entzündungen im Huf oder ein vermindertes Wachstum sollten dich aufhorchen lassen. Denn dann kannst du sichergehen, dass zumindest die Versorgung mit verschiedenen Nährstoffen nicht unbedingt ideal ist. Was auch bedeutet, dass dein Vierbeiner dir nicht die Leistung im Training zeigen wird, obwohl er das höchstwahrscheinlich gerne würde.

Hufe bestehen zu einem überwiegenden Teil aus dem Protein Keratin. Fehlt es an den passenden Aminosäuren in der Pferdeernährung, um ebendieses Protein herzustellen, dann wird der Pferdehuf zur Problembaustelle. Ergänzt du bereits Zink und Biotin, um deinem Tierchen Gutes für seine Hufe zu tun, reicht das bei auffällig anfälligen Hufen wohl nicht aus. Zumindest dann nicht, wenn du überwiegend eiweißärmeres Futter wie Heu in die Box schüttest. Da stellt sich vergleichsweise schnell ein Mangel an den Aminosäuren Lysin und Methionin ein, Kreatin wird nicht ausreichend schnell gebildet und das Hufhorn bleibt instabil.

Mangelt es dazu noch an ausreichender Bewegung auf abwechslungsreichen, möglichst auch harten Böden? Dann werden weder die Durchblutung noch die Abnutzung ausreichend angeregt.

Ein Forschungsteam aus Brasilien verglich die Hufgesundheit von Pferden aus Stallhaltung mit solchen, die halbwild auf den kargen brasilianischen Böden herumzogen und keinerlei Berührungspunkte mit einem Schmied hatten. Das Ergebnis war recht eindeutig: Die halbwilden Tiere mit nacktem Huf wiesen signifikant bessere Huflamellen auf. Die in der Studie untersuchten domestizierten Reitpferde dagegen zeigten – auch aufgrund der oftmals schlechten Futterbedingungen – bereits erste Anzeichen für Hufrehe (Malacarne et al. 2023).

Nährstoffkunde für gesunde Hufe

Für einen gesunden Huf sind verschiedene Nährstoffe unverzichtbar. Dazu zählen unter anderem Zink, Kupfer, Selen und Biotin plus Omega-3-Fettsäuren, Methionin und Lysin.

Grundsätzlich sind alle essenziellen Aminosäuren wichtig für eine optimale Proteinsynthese. Leucin ist der Hauptanschieber für den generellen Prozess, Lysin ist die Basis für viele Stoffwechselprozesse, die konkret die Hufgesundheit betreffen. Beides sind aber limitierende Aminosäuren, die in der Ernährung nicht zu kurz kommen dürfen (Dunstan et al. 2020)!

Zink: Das Spurenelement ist nötig für zig Stoffwechselprozesse sowie die Bildung des Proteins Kreatin. Ohne ausreichend Zink im Blutserum ist die Zellteilung und -reifung erschwert. Zinkmangel führt schnell zu brüchigen schwachen Hufen und Fellproblemen inklusive Ekzema, weil sich die Tiere blutig schubbern, im weiteren Verlauf ist das Immunsystem geschwächt und kommt seiner Aufgabe nur noch in Teilen nach. Wundheilungsstörungen sind typischerweise mit einem Mangel an Zink verknüpft.

Kupfer: Ohne Kupfer bleiben Enzyme inaktiv, die für die Verbindung der Keratinstränge im Huf dringend benötigt werden. Auch die Kollagensynthese sowie die Pigmentierung (Farbgebung) sind auf ausreichend Kupfer angewiesen. Vor allem bei Füchsen werden Haut und/oder Fell rund um die Augen heller, wenn es an Kupfer mangelt. Ein Kupferdefizit führt in der Regel zu einer minderen Hufqualität und langfristig auch zu einem schwachen Bindegewebe.

Selen: Ein enorm wichtiges Antioxidant. Im Falle der Hufe sorgt es dafür, dass die für die Elastizität nötigen Fette ihrer Aufgabe nachkommen können, ohne durch Freie Radikale beeinträchtigt zu werden. Ein Selenmangel führt zu spröden Hufen und im weiteren Verlauf zu häufigeren Infektionen bis hin zu Abszessen. Eine Selenüberschuss beeinträchtigt allerdings die Kreatinsynthese. Hier

Die Ernährung beeinflusst die Hufgesundheit.
Christina und Inca.

ist also ein gutes Händchen für eine optimale, aber nicht übermäßige Versorgung gefragt. Es ist hier schon sinnvoll, gelegentlich eine Blutuntersuchung machen zu lassen, um eine adäquate, also wirklich gute Selenversorgung sicherzustellen, ohne dabei in einen Mangel zu geraten! In Anbetracht der Tatsache, dass Selen aber auch einer der kritischen Stoffe ist, um den Glutathion-Stoffwechsel anzuschieben sowie unverzichtbar für die Funktionsweise der Schilddrüse ist, lohnt sich der Aufwand aber.

Biotin: Ausreichend Biotin stärkt die Hufqualität und die Widerstandsfähigkeit bei stärkeren Belastungen. Biotin-Mangel führt entsprechend zu brüchigen Hufen und einem verlangsamten Wachstum.

Omega-3-Fettsäuren: Als eine der wichtigsten Fettsäuren im Stoffkreislauf, sind Omega-3-Fettsäuren unverzichtbar für die Elastizität des Bindegewebes. Das betrifft dann auch die Strukturen des Hufhorns. Grast dein Pferd überwiegend, wäre der

Bedarf an Omega-3 und Omega-6 wahrscheinlich gut gedeckt. Heu und Getreide sind in der Regel nicht ausreichend, um den Bedarf vollständig zu decken. Ein Mangel an Omega-3-Fettsäuren führt zu sprödem Huf. Leinsamen sind hervorragend dazu geeignet, um den Omega-3-Fettsäurenbedarf zu decken, aktuell werden zwischen 50 bis 100 Gramm pro 100 Kilogramm Pferdegewicht zur optimalen Deckung des Bedarfs empfohlen. Alternativ dazu kannst du Omega-3-Öl aus Algen/Fischöl ergänzen.

Methionin: Diese essenzielle schwefelhaltige Aminosäure wird dringend benötigt, um über den Zwischenschritt Cystein das für den Huf benötigte Kreatin zu produzieren. Fehlt es im Nährstoffsystem deines Tieres, kommt es zu Störungen im Huf- und Hornstoffwechsel – Fell und Hufe werden spröde und gehen kaputt. Der in Methionin enthaltene Schwefel sorgt für eine ordentliche Quervernetzung innerhalb des Horns, was die Widerstandsfähigkeit erhöht. Ein Mangel an Methionin führt zu spröden, brüchigen Hufen und zu Ekzemen. Hufabszesse sind die häufige Folge davon!

Cystein: Diese schwefelhaltige Aminosäureverbindung wird im Körper aus Methionin gebildet und ist ebenfalls für die Kreatinproduktion nötig. Fehlt es in der Nahrung an Methionin, kann es zu Engpässen bei der körpereigenen Synthetisierung von Cystein kommen. Ein Mangel an Cystein führt logischerweise damit ebenfalls zu brüchigen und instabileren Hufstrukturen.

Lysin: Die essenzielle Aminosäure wird im Rahmen der Eiweißsynthese benötigt. Sprich: Sie ist absolut grundlegend nötig, um eine vernünftige Verstoffwechselung von Proteinbausteinen zu ermöglichen. Die essenzielle Aminosäure Leucin ist zwar der absolute Champion, was den Anschub der Proteinsynthese angeht, aber Lysin folgt dem direkt auf dem sprichwörtlichen Fuße. Lysin ist hilfreich bei der Aufnahme von Calcium, das für die Festigkeit des Hufhorns entscheidend ist. Ein Mangel an Lysin macht sich durch eine schlechtere Hufqualität und Wachstumsverzögerungen bemerkbar.

Threonin: Diese Aminosäure spielt eine große Rolle bei der Kollagenbildung. Die Schichten (»weiße Linien«) aus Kollagen verbinden die verschiedenen Hufstrukturen miteinander. Ein Mangel an Threonin führt zu einem weniger starken Immunsystem, chronische Abszesse sind wahrscheinlicher. Eine durchschnittlich gute Versorgung mit Threonin über das typische Standardfutter ist meist nicht gegeben. Es kann absolut sinnvoll sein, auch diese Aminosäure bewusst zusätzlich zu ergänzen, um die Proteinsynthese zu unterstützen.

Hilfe, ein Hufabszess!

Hufabszesse treten häufiger bei Tieren auf, die

- genetisch bedingt eine weiche, spröde oder rissige Hornstruktur mitbringen. Risse und kleine Spalten bilden sich selbst bei normaler Belastung schneller und lästige Bakterien gelangen ins Hufinnere, wo sie Abszesse auslösen.
- vergleichsweise dünne oder empfindliche Hufsohlen haben. Diese Empfindsamkeit sorgt für die Entstehung von Prellungen in Form von Sohlenhämatomen. Die wiederum sind die willkommene Grundlage für wiederkehrende Hufabszesse und Entzündungen. Übrigens sind auch in der Halle herumliegende spitze Steinchen problematisch bei dünnen Hufsohlen. Dann genügt schon der Tritt aufs Steinchen, um einen Kanal zu eröffnen und der Weg für krankmachende Bakterien ist weit offen.
- Stoffwechselprobleme wie das Equine Metabolische Syndrom (EMS), Insulinresistenz oder Cushing (PPID) aufweisen. Infektionen am Huf sind dann zusätzliches Symptom für eine zugrundeliegende Erkrankung, die primär behandelt werden sollte. Dass du auch den Hufen durch eine angepasste Ernährung und Supplementierung Gutes tun solltest, steht aber bitte außer Frage!
- im Matsch stehen. Feuchte Haltungsbedingungen können das Hufhorn aufweichen, das ebnet Bakterien den Weg ins Hufinnere.

Eine getreidelastige, proteinarme Ernährung und wenig Bewegung befeuern Entzündungsprozesse im Körper. Ein Abszess ist nichts anderes als eine massive Entzündung in einem bestimmten Körperteil. Entsprechend wird die Entzündung am besten großflächig behandelt.

Bei genetisch bedingter Neigung zu Entzündungen und Abszessen kannst du vorbeugend passende Nährstoffe geben, die Futterration proteinreich gestalten und für ausreichend Bewegung auf geeigneten Böden sorgen.

Dressurkunst ist besonders geeignet, um deinem Tierchen und dir die schmerzhafte und langwierige wiederkehrende Hufentzündung weitgehend zu ersparen. Denn dann schütten die Muskelprotzpakete deines Vierbeiners entsprechend entzündungshemmende Botenstoffe aus, das Immunsystem agiert zügig und kraftvoll und das Hufhorn wird durch die wiederkehrenden positiven Belastungen zu einem kontinuierlichen Wachstum samt Trimm angeregt.

Irgendwelche fiesen Bakterien haben dann, Steinchen auf dem Hallenboden hin oder her, keine große Chance zur Vermehrung. Der gesamte Pferdeorganismus ist so fit durch das Bodybuilding-Programm der Dressurkunst, dass das Immunsystem pathogene Keime schneller erkennt und effektiver bekämpft.

Hufgesundheit in 3, 2, 1 ... und los!

Jedes Pferd hat seinen eigenen Bedarf. Was dem tierischen Stallnachbarn zu wunderschön glänzendem Fell und festem Horn verhilft, kann bei deinem Pferd noch längst nicht genug sein. Aber: Eine grundlegend optimierte Ernährung hilft dir schon mal dabei, dich an die optimalen Werte für deinen geliebten Vierbeiner heranzutasten.

Starte mit einer eiweißhaltigen Ergänzung in Form von Luzerne, Esparsette oder die königlichen Aminosäurenlieferanten Sojaschrot und Kartoffelprotein. Alle vier bringen in der Regel ein gutes Aminosäureprofil mit sich, sodass der Pferdekörper daraus die nötigen Proteinbausteine für gesunde Hufe gewinnen kann.

Als Quelle für Omega-3-Fettsäuren schüttest du zusätzlich noch Leinsamen mit in deinen Futtermix. Alternativ kannst du auch Leinöl - aber bitte in guter Qualität (kaltgepresst) und kühl sowie dunkel gelagert - verwenden.

In Summe sieht das dann so aus:

- Hochwertiges Mineralfutter mit Zink, Kupfer, Selen, Biotin als Treibstoff für die Kreatin-Produktion, möglichst als organisch gebundene Mineralstoffe (Chelate) für eine bessere Bioverfügbarkeit. Anorganische Mineralstoffe sind zwar wesentlich günstiger, wirken sich aber praktisch kaum auf die Serumwerte aus. Der erhoffte Effekt, nämlich die Gesundheit deines Pferdes zu stärken, tritt mit anorganischen Mineralstoffen gar nicht ein.
- Zusätzliche Aminosäuren aus Sojaschrot für die Eiweißsynthese.
- Essenzielle Fettsäuren in Form von Leinsamen für Elastizität und für das Bindegewebe.
- Weniger Melasse, Zucker oder stärkehaltige Futtermittel.

Bei Äpfeln musst du je nach Tier vorsichtiger vorgehen - der enthaltene Fruchtzucker kann die Verdauung durcheinanderbringen. Vor allem, wenn es nie bei dem einen geplanten Apfel pro Stallbesuch bleibt.

Achte besonders bei Pferden nach Verletzungen und bei Tieren mit hohem Belastungslevel durch Wettkampfteilnahme darauf, dass die Hufe keine Anzeichen von Nährstoffmangel aufweisen.

Beschlagen oder nackt?

Eins vorweg: Barhufhaltung ist keinesfalls für jedes Pferd geeignet. Hufgesundheit, Nutzung und Haltung spielen enorm wichtige Rollen bei der Entscheidung für oder gegen das Beschlagen.

Hufe ohne Hufeisen können stärker und widerstandsfähiger werden. Durch den direkten Kontakt mit verschiedenen Untergründen passt sich der Huf besser an und härtet sich auf natürliche Weise ab. Zusätzlich baut sich das Horn stärker auf und eine dickere, widerstandsfähigere Sohle entsteht. Auch die Blutzirkulation wird bei der Barhufhaltung verstärkt angeregt, was sowohl den Hufen als auch dem gesamten Beinsystem zugutekommt.

Schöner federn

Der nackte Huf kann Stöße besser abfedern als ein beschlagener Huf. Der Grund dafür? Barhufe dehnen und ziehen sich deutlich optimaler wieder zusammen, was eine bessere Stoßdämpfung erzeugt – kein Vergleich mit beschlagenen Hufen an dieser Stelle. Im Grunde ist das ja auch keine Überraschung, die Wildpferdahnen mussten ja auch mit dem nackten Huf über Stock und Stein schreiten.

Diese bessere Stoßdämpfung reduziert dabei natürlich die Belastung auf Gelenke, Sehnen und Bänder. Wenn du langfristig Lahmheit vorbeugen willst, ist der nackte Huf unter Umständen die bessere Wahl.

Richtig beschlagen

Ein zusätzlicher Gedanke noch: Wie gut ausgebildet ist denn die Person, die die Hufe deines Tieres beschlägt? Im besten Falle ist sie auf einem hervorragenden Wissensstand und versteht es, die Hufe möglichst optimal für die Haltungs- und Nutzungsform vorzubereiten.

Ab dem Tag des Beschlages mit einem Hufeisen wächst der Huf kontinuierlich, die Hufmorphologie ändert sich schleichend und damit auch die biomechanische Interaktion des Hufs mit dem Untergrund.

Der Huf mag am Beschlagtag ideal bearbeitet werden, aber vier Wochen nach dem Beschlag kann die Situation erheblich verändert sein. Im Naturhufmodell hat der Barhuf die Möglichkeit mit der Umgebung zu interagieren, d. h. mit jedem Schritt Abrieb (»Trimm«) zu erfahren und eine konstante Hufform als biomechanisches Modell beizubehalten (Hampson 2011).

Es kommt aber auch vor, dass Pferde erst aufgrund der bearbeiteten Hufe zu Lahmheit neigen. Insgesamt zeigen Studien aber, dass ein möglichst abwechslungsreicher Boden und reichlich Bewegung plus eine angepasste Ernährung entscheidend sind.

Brasilianische Forschende hatten genau das untersucht und dabei die Hufgesundheit von sowohl halbwilden Iberern als auch von solchen in Intensivhaltung verglichen. Erstere waren, wenig überraschend, mit wesentlich belastbareren und gesünderen Hufen ausgestattet. Und das, obwohl sie aufgrund der klimatischen Gegebenheiten das halbe Jahr im Schlamm herumzogen und das restliche halbe Jahr über ausgetrocknete harte Böden liefen (Malacarne et al. 2023).

Was bedeutet das jetzt für dein Pferd? Unter Umständen sind Hufeisen das einzig Richtige für dein Tier. Es ist aber durchaus möglich, dass die Barhufhaltung auf lange Sicht die gesündere Entscheidung sein könnte.

VÖLLIG UNTERSCHÄTZTE NÄHRSTOFFLÜCKEN IN DER PFERDEFÜTTERUNG

KURZ UND KNAPP:

- Jedes Pferd hat einen genetisch individuellen Bedarf.
- Nahrungsergänzungsmittel sind heutzutage nahezu unverzichtbar, um Pferde langfristig gesund halten.
- B-Vitamine sind nicht nur fürs Gehirn gut.
- Die Leber will gepflegt sein.
- SAM-e ist die eierlegende Wollmilchsau für besseres Wohlbefinden.

Jedes Tierchen hat seinen eigenen idealen Nährstoffspiegel. Und der ist längst nicht in Stein gemeißelt, sondern auch noch abhängig davon, ob es

- genetisch komplexe Eigenheiten mitbringt.
- eher entspannt oder mit hohem Stresspegel durchs Leben geht
- gerade frisch von der Jungtierweide kommt.
- mitten im anspruchsvollen Reitkunsttraining steckt.
- als Reha-Fall vermeintlich kurz vor der Rentnerweide steht, mit dem richtigen Training aber durchaus noch viele schöne Jahre unter dem Sattel vor sich haben kann.

Also merke: Individueller Bedarf plus auf die Lebensphase angepasster Bedarf = realer Nährstoffbedarf.

Keine Spielerei, sondern essenziell: individuelle Nährstoffberechnungen

Jungtiere stecken mitten im Wachstum. Fohlen kommen zwar mit rund 60 Prozent ihrer endgültigen Größe auf die Welt, trotzdem schießen sie allein im ersten Jahr um rund 40 Zentimeter in die Höhe. Knochen, Muskeln, Sehnen und Bänder sind enormen Veränderungen unterworfen, um da mitzuhalten. Da brauchst du als stolzer Pferdebesitzer nicht nur viel Geduld, sondern auch das passende Futter. Das muss immerhin alles enthalten, was dein Jungspund nach dem Absetzen von der Mutterstute benötigt, um kraftvoll aufzuwachsen.

Reitpferde, die die heute übliche FN-Ausbildung durchlaufen haben, bringen oft weniger Muskelmasse auf die Waage als solche, die eine eher klassische Reitkunstausbildung genießen durften. Entsprechend unterschiedlich können die Bedürfnisse für Nährstoffe, vor allem aber für Proteine und gesunde Fette, in der Futterzusammenstellung ausfallen. Ja, selbst beim gleich schweren, gleich großen Tier aus der gleichen Zucht!

Gesunde, fitte Pferde benötigen für gelungenes Bodybuilding einfach ein bisschen mehr von allem – im Gegensatz zu den Tierchen, die standardmäßig auf langen Bahnen in der Halle herumschlurfen und gelegentlich mal einen Spaziergang über Stock und Stein auf der grünen Wiese verrichten.

Versteh mich richtig: Jedes Pferd verdient eine angemessene Versorgung mit Nährstoffen, um gesund und fit bis ins hohe Alter zu bleiben. Aber selbst bei gewöhnlich gearbeiteten Reitpferden mangelt es oft an Grundlegendem, wie den passenden Aminosäuren, an Antioxidantien, Zink und Co. In der Reitkunst ausgebildete Tiere benötigen zusätzlich noch etwas mehr an Mikro- und Makronährstoffen, um ihren hohen Leistungsstand vollkommen genießen zu können.

Die klassische Trageerschöpfung lässt sich immerhin mit geeignetem Training wahlweise verhindern - oder eben, wenn die heutige moderne Reiterei schon ihre Spuren hinterlassen hat, durch anabolikafreies Bodybuilding zumindest wieder beheben. Das erfordert aber eine angemessene Fütterung mit allem, was das Pferd so benötigt, um ballettgleiche Tanzschritte bis ins hohe Alter vorzuzeigen.

Was uns zum Spezialthema **Rehapferd** bringt: Denn Rehapferde mussten oft schon einiges erdulden, um letztlich viel zu schnell auf wenig kraftförderndes, sondern missverständlich für schonend gehaltenes Abschleudern in der Halle umgestellt zu werden. Damit das verletzte Tier nicht zu schnell zu fett wird, werden ihm dann gerne die proteinreicheren Futterbestandteile gekürzt und mit ordentlich Heu gestreckt. Dass das nicht gesund ist, kannst du auch im Futterkapitel meines ersten Buchs nachlesen.

Deshalb an dieser Stelle zunächst in aller Kürze: Gerade verletzte Tiere benötigen mehr Protein, um zügig zu heilen. Sie brauchen passende Mikronährstoffe, um die laufenden Stoffwechselprozesse im Pferdekörper optimal zu unterstützen. Vor allem aber bedarf es einer an ihren gesundheitlichen Zustand orientierten Trainingspraxis, die sie bei der Heilung unterstützt.

Worauf genau du achten musst und wie sich die Nährstoffversorgung deines Tieres in den verschiedenen Lebensphasen unterscheidet, stelle ich dir in den folgenden Kapiteln genauer vor.

Stressreduziertes Absetzen könnte besser sein

Apropos Nährstoffversorgung: Ob dein Tier extrem stressempfindlich ist oder nicht, entscheidet sich schon ziemlich früh. Und wie es von seiner Mutterstute getrennt wird, hat einen ziemlich großen Anteil daran.

Die natürliche Säugedauer von Pferden beträgt zwischen sieben bis 18 Monaten und ist abhängig von der Qualität der vorhandenen Nahrung (Reiter/Reed 2023). Je schlechter die Nahrungssituation, desto länger wird das Fohlen weiter gesäugt, auch wenn es längst selbst Steppengras zupfen könnte.

Heutige Pferderassen säugen im Schnitt um die 9 Monate lang. Der eigentliche Absetzzeitpunkt schwankt etwas und scheint eine Kombination aus geeigneten Futterquellen und der folgenden Trächtigkeit der Stute zu sein. Das heißt aber längst nicht, dass sich das quietschfidele Fohlen einfach dem nächstbesten Herdenmitglied zuwendet. Es bleibt in enger Bindung zur Mutterstute und das wenigstens für 18 bis 30 Monate.

Den meisten Zuchtställen dauert das aber zu lange. Da ist oft bereits nach einem halben Jahr Schluss mit dem uneingeschränkten Zugang zur Stutenmilch. Häufig werden die Tiere relativ abrupt getrennt und das Fohlen an einen anderen Standort zur Aufzucht in einer Jungtierherde gekarrt.

Dabei ergänzt die Milch der Stute den Speiseplan des Fohlens oft entscheidend um verschiedene Makro- und Mikronährstoffe. Stutenmilch enthält unter anderem durchschnittlich

- 1 Prozent Fett
- 6 Prozent Milchzucker (Laktose)
- 2 Prozent Eiweiß (davon 55 Prozent Casein)

Die einzelnen Komponenten schwanken allerdings und sind von verschiedenen Faktoren (Alter, Rasse, Futtermittel) der säugenden Stute abhängig (Sonntag 1996).

Zurück zum Fohlen: Per Stutenmilch wird es also im Grunde optimal mit Nährstoffen grundversorgt, ab dem dritten Lebensmonat noch zu gut einem Drittel seines täglichen Energiebedarfs. Fällt die Stutenmilch beim Absetzen weg, ist das also leicht zu kompensieren. Richtig? Das kommt auf die Methode des Absetzens an.

Immerhin ist die Trennung von der Mutterstute für alle Beteiligten ziemlich stressig. In einer Studie am Veterinärmedizinischen Fakultät in München ließ sich nachweisen, dass sich der Kortisolspiegel der Fohlen binnen einem Tag nach abrupter Trennung von der Mutter verdoppelt. Drei Wochen dauerte es im Rahmen der Studie, bis das Kortisol-Niveau der Jungtiere auf ihr zuvor ermitteltes Basisniveau absank. (Delank et al. 2023)

Wie zuvor im Kapitel zu Kortisol besprochen, ist das Stresshormon Kortisol ein echter Nährstoffräuber. Fohlen, die abrupt um den üblicherweise sechsten Monat herum abgesetzt und von ihrer Mutter getrennt wurden, sind tendenziell leichter und zeigen eher stressbedingte Verhaltensweisen.

In einer französischen Studie der Universität Rennes zeigt sich, dass der Outcome für ein durch das Fohlen und seine Mutter bestimmtes Absetzen deutlicher besser ist: geringere Stressanzeichen, eine bessere Gewichtsentwicklung und insgesamt weniger stressbedingte Erkrankungen. (Henry et al., 2020)

Positiver Stress kann motivierend wirken und müde Pferde in Bewegung setzen. Negativer Stress wie der, den Fohlen durch ein abruptes Absetzen erleben, verursacht unter Umständen eher Probleme. Dazu kommt noch, dass die heutige Haltungsform »Modell Kleinstfamilie« aus Stute und Fohlen nicht sonderlich natürlich ist. Normalerweise ist der zeugende Hengst Teil der Herde und übernimmt gemeinsam mit den Stuten erzieherische Aufgaben. Das Fohlen aus moderner Haltung startet also mit einer ganz anderen Sozialisation in sein Pferdeleben, als es seine wildlebenden Steppenpferdvorfahren getan haben.

Erfahrungsgemäß kann es deshalb ganz und gar nicht schaden, den Absetzer mit reichlich Magnesium zu versorgen. Das reguliert das Nervensystem etwas und wirkt beruhigend. Glycin oder gleich ein Magnesiumglycinat könnten sich ebenfalls eignen, um dem Jungpferd die kommenden Umstellungen nach dem Absetzen etwas zu erleichtern.

Aus therapeutischer Sicht ist die bei Neukauf eines Pferdes immer auch die Frage relevant: Aus welchem Stall stammt das Tierchen denn – und wie steht es zu seiner Mutter?

Aktive Leberpflege mit Cholin

Schon mal von Cholin gehört? Falls nicht, bist du nicht allein damit. Cholin zählt zu den eher unbekannteren Nährstoffen in der Pferdeversorgung. Schade eigentlich, denn Cholin trägt zur Lebergesundheit bei. Ganz nebenbei scheint sich die Gabe von Cholin sich auch positiv auf die kognitiven Fähigkeiten auszuwirken (Kansakar et al. 2023).

Um die Gesundheit der Leber unserer heutigen modernen Reitpferde könnte es durchaus öfter gehen. Gefühlt hat fast jedes Pferd **Leber**, was im Grunde genommen kein Wunder ist: Die Nährstoffversorgung samt Aminosäurengehalt ist selten optimal auf das individuelle Pferd ausgelegt. Das Weidegras und das ergänzende Futter enthalten weit häufiger als uns lieb ist Pestizide in Mengen, die die Leber belasten können.

Dummerweise machen sich Leberprobleme eher durch viele kleine Symptome mit nachhaltiger Wirkung bemerkbar. Einen einzigen glasklaren Marker im Großen Blutbild gibt es leider nicht. Ein

Gesunde Leber, aufgewecktes Pferd.
Sonja mit Leo.

Zusammenspiel aus verschiedenen Symptomen wie Mattigkeit, Muskelabbau, Lustlosigkeit und Leistungsabfall plus erhöhte Leberwerte wie GGT, AST und Bilirubin lassen aber auf vorhandene Leberprobleme schließen.

Eine Leberproblematik entsteht unter anderem durch

- die Aufnahme von Giftpflanzen oder Giftstoffen über das Futter,
- den Befall mit Parasiten,
- Infekte (viral oder bakteriell),
- Überfütterung,
- Ungleichgewichte im Darm,
- fehlende Aminosäuren im Futter,
- Mangel an weiteren Nährstoffen – unter anderem Cholin.

Eine geschädigte Leber macht sich schließlich durch Begleiterscheinungen wie Leistungsabfall, Neigung zu Ekzemen, Muskelabbau, Apathie und einer insgesamt sinkenden Kondition bemerkbar. Das wird dann oft unter spontaner Lustlosigkeit verbucht, die sich aber schleichend immer weiter verschlechtert.

Am Ende hat das Pferd dann ständig mit Juckreiz zu kämpfen, erlebt wiederkehrende Koliken bis hin zu Kotwasser, Verstopfung und Magengeschwüren als Folge anhaltender Leberprobleme (Gehlen et al. 2010). Im denkbar schlechtesten Falle sind es Hufrehe als Anzeichen für eine mangelnde Entgiftungsleistung der Leber, die final zu einem genaueren Blick auf die Blutwerte und damit zu einer Behandlung der Leberproblematik führen. Häufig wird auch das Equine Metabolische Syndrom (EMS) hinter dieser Symptomatik vermutet (Stefaniuk-Szmukier et al. 2023).

Leberwerte verstehen

In der Regel werden die Marker für die Lebergesundheit über ein großes Blutbild ermittelt. Achte bei der Blutentnahme einfach darauf, dass der Veterinär deines Vertrauens sich besonders auf die für die Lebergesundheit wichtigen Werte konzentriert. Die Laborauswertung sollte dann folgende Werte enthalten:

1. **Gamma-Glutamyltransferasen (Gamma-GT, kurz GGT)** ist ein Marker für Leber-, Nieren- oder Ausscheidungsprobleme. Dieser Wert ist sehr sensibel, denn die Gamma-Glutamyltransferasen sind Aminosäuren, die nur bei akuter Leberüberlastung freigesetzt werden.

   - Leicht erhöhte GGT treten bei Pferden mit EMS auf und bei solchen mit leichter Fettleibigkeit.
   - Erhöhte GGT lassen sich bei einer Virus bedingten Hepatitis, einer Kolik (Gallenstau) oder einer Bauchspeicheldrüsenentzündung (Pankreatitis) beobachten (Gehlen et al. 2010).
   - Stark erhöhte GGT zeigen eine Überlastung des Leber- und Gallensystems an.

   Wichtig: Chronische Leberprobleme zeigen sich oft nur in erhöhten GGTs, während die restlichen Werte im Normbereich liegen. Schau hier also – im Zeitverlauf – genau hin!

2. **Aspartat-Aminotransferasen (kurz AST)** ist ein Marker für Muskel-/Leberprobleme. Der Wert alleine hat keine ausreichende Aussagekraft zum Zustand der Leber, also bitte immer im Zusammenhang mit weiteren Werten betrachten.

   AST sind an den Ab- und Umbauprozessen von Proteinbausteinen beteiligt, d. h. sie sind nicht nur in der Leber, sondern auch der Muskulatur allgemein, den Nieren oder der Lunge zu finden.

   AST können nach einem intensiven Training erhöht ausfallen, achte deshalb auf eine ausreichend lange Ruhepausen zwischen Lektionstraining und Blutentnahme!

   Erhöhte AST können – in Verbindung mit weiteren Markern – für geschädigte Leberzellen stehen oder Marker für eine Vergiftung oder eine Entzündung der Leber sowie der Gallenwege sein.

3. **Bilirubin** ist ein Marker für Leberschädigungen. Bilirubin tritt als Abbauprodukt des roten Blutfarbstoffs Hämoglobin auf. In der Regel wird der Gesamt-Bilirubin-Wert ermittelt, seine Aussagekraft für die Bestimmung der Lebergesundheit ist aber etwas begrenzt.

   Beim Menschen zeigen sich stark erhöhte Bilirubin-Werte in einer sichtbaren Gelbfärbung der Haut sowie des Augapfels. Beim Pferd treten erhöhte Bilirubin-Werte auf, wenn die Lebergesundheit beeinträchtigt ist, wenn Giftpflanzen aufgenommen wurden, bei bestimmten Medikamenten oder bei einer Störung des Gallenflusses.

   Wichtig: Ein erhöhter Bilirubin-Wert kann auch nach einer starken Verletzung mit Blutverlust auftreten oder wenn das Pferd hungert.

Eine gezielte Supplementierung zur Stärkung der Lebergesundheit wird im heutigen modernen Reitsport relativ selten in Betracht gezogen. Schade eigentlich! Wenn du das anders handhaben willst, dann kannst du Cholin in Form von **Cholinchlorid oder Phospatidycholin** im Speiseplan deines Tieres ergänzen.

Cholinchlorid kann kurzfristig die Leberfunktion verbessern, vor allem, wenn sich bereits Symptome einer Leberverfettung zeigen. Idealerweise wird Cholinchlorid dann aber mit Methionin und Betain kombiniert, um die Verstoffwechselung zu optimieren. Methionin ist eine essenzielle Aminosäure, die die Cholinbiosynthese unterstützt (Case et al. 1976).

Phospatidylcholin ist die ideale Lösung zur Behebung von Leberschäden, denn: Es ist direkt am Aufbau der Zellmembran beteiligt und schützt so die Leberzellen vor Schädigungen. Phospatidylcholin fördert die Synthese von very low density lipoproteins (VLDL). VLDL sind dringend nötig, um Fette aus der Leber abzutransportieren. Was passiert, wenn Fette unverhältnismäßig zahlreich in der Leber verbleiben? Ganz klar: Fettleber!

Einige Vorteile von Phosphatidycholin:

- direkte Unterstützung der Leberzellmembranen
- effizienterer Abtransport der Fette aus der Leber
- kann Leberfunktion insgesamt verbessern

Die Nachteile von Phosphatidycholin:

- Bioverfügbarkeit nicht sonderlich hoch, dadurch höhere Dosierung, entsprechend höhere Futterkosten

Niacin – nötig für Darm, Fell und Leistung

Niacin, besser bekannt als B3, ist essenziell für die Pferdegesundheit. B3 trägt zur Funktion des Verdauungstrakts bei und hilft dem Körper dabei, Nährstoffbestandteile aus der Nahrung gezielt aufzunehmen.

Besteht ein Mangel, verarbeitet dein Tierchen trotz sorgfältig ausgewählter Futterkomponenten womöglich nicht alles, was es so in sich hinein mampft.

Denn Niacin wird nicht nur im Dickdarm synthetisiert, es liefert auch die Grundlage für weitere Bakterienstämme und wiederum deren Synthetisierungsprozesse, um Nährstoffe aufzuspalten und in verwertbare Bausteine für den gesamten Organismus bereitstellen:

Haut und Fell profitieren von einer guten Niacin-Versorgung. Mascha mit Honey.

»The bacterially synthesized vitamin B3 contributes to local colonocyte nutrition and maintains the morphology of intestinal stem cells. A study from the FoCus cohort identified links between vitamin B3 and intestinal microbial composition. They found a significant association between vitamin B3 deficiency and low α-diversity and the abundance of Bacteroidetes in obese individuals (Hossain et al. 2022).«

Auf Deutsch: »Das bakteriell synthetisierte Vitamin B3 trägt zur lokalen Ernährung der Kolonozyten bei und erhält die Morphologie der Darmstammzellen aufrecht. In einer Studie aus der FoCus-Kohorte wurden Zusammenhänge zwischen Vitamin B3 und der Zusammensetzung des Darmmikrobioms festgestellt. Es wurde ein signifikanter Zusammenhang zwischen Vitamin B3-Mangel und geringer α-Diversität und der Häufigkeit von Bacteroidetes bei fettleibigen Personen festgestellt.«

Fehlt Niacin als Grundlage für diese Bakterienstämme, können diese ihre Arbeit nicht oder nicht in ausreichender Qualität ausführen.

Einige Erkrankungen verstärken diesen Effekt der ungenügenden Nährstoffaufnahme aufgrund von B3-Mangel noch. Vorrangig alles, was die Darmgesundheit und damit die Aufnahme von Nährstoffen über den Verdauungstrakt beeinträchtigt. Denk dabei an chronische Darmentzündungen und Parasiten. Also alles, was deinem Tierchen und damit auch dir nachhaltig die Laune aufgrund der teuren Veterinärkosten verhageln könnte.

Ein Mangel an B3 macht sich in der Regel durch

- Leistungsabfall und Trägheit,
- allgemeiner Schwäche,
- Verdauungsbeschwerden (Durchfall, Koliken),
- trockenes Fell,
- Schuppenbildung
- oder Hautentzündungen

bemerkbar.

Häufig tritt **Appetitlosigkeit** auf. Wenn dein Pferd also eher mäßig begeistert in seinem Futtereimer herumwühlt und sich nicht mal über Leckerchen als Anerkennung bei der Lektionserarbeitung freut, dann solltest du schleunigst die Futterzusammensetzung prüfen. Unter Umständen ist allein ein höherer B3-Spiegel schon alles, was deinem Tierchen zu deutlich mehr Appetit verhelfen könnte.

In schweren Fällen von B3-Mangel kommt es zu Pellagra, einer Erkrankung mit Hautläsionen und Veränderungen im Fell.

Pellagra erkennen und behandeln

Pellagra ist eine relativ seltene, aber umso schwerwiegende Erkrankung der Haut. Im Falle unserer Vierbeiner zeigt sich das in einem schlechten Fellzustand: Glanzlos, trocken, schuppig und teils auch einfach ausfallend. Die Haut darunter ist trocken und rissig.

Pellagra basiert auf einem massiven Mangel an Niacin – meist aufgrund einer entsprechenden Mangelernährung. Die Hauterkrankung tritt häufig in Kombination mit einer unzureichenden Eiweißzufuhr auf. Notfalls kann sich der Körper zwar aus der Aminosäure Tryptophan das fehlende Vitamin B3 selbst synthetisieren.

Enthält die Nahrung aber weder das eine noch das andere in ausreichender Menge, wird es allerdings problematisch. Denn wenn Tryptophan für die B3-Produktion verheizt wird, dann sinkt zwangsläufig die Stimmung. Tryptophan ist eine Vorstufe für das stimmungsaufhellende Serotonin, ein geringer Spiegel dieser speziellen Aminosäure führt deshalb in der Regel auch zu einem niedrigen Serotoninlevel. Da ist die schlechte Laune quasi vorprogrammiert, beim Menschen und beim Pferd.

Erfolgt keine geeignete Therapie in Form von ausreichend dosierten Niacin-Präparaten und einer hochwertigen eiweißreicheren Ernährung, führt Pellagra sogar zum Tod. B-Vitamine sind wasserlöslich. Bei Symptomen, die für einen B-Vitamin-Mangel sprechen, kannst du dich also ohne Gefahr für Überdosierungen an eine geeignete Supplementierung herantasten.

Auf der Bahn zeigen sich Pferde mit B3-Mangel gern durch Nervosität, Zuckungen und Verhaltensänderungen, die sich nicht (nur) durch gelegentliche Launen erklären lassen. Eskaliert der Mangel bis zur Krankheit Pellagra, entwickelt sich beim Menschen sogar Demenz. Wird dein Pferd also scheinbar aus dem Nichts tattrig, wirds Zeit für eine ordentliche Supplementierung! Gerne auch in höheren Dosen.

Im absoluten Notfall kann der Pferdekörper B3 selbst synthetisieren. Dafür braucht es aber einen Überschuss an B2, B6, Eisen und Magnesium im Stoffkreislauf. Wenn dein Tierchen schon einen B3-Mangel hat, ist das der reinste Raubbau an seinem Nährstoffhaushalt! Fehlt es aufgrund der Futterzusammensetzung auch noch an Aminosäuren wie Tryptophan, kann die Backuplösung zur Synthese von Niacin aus ebendieser Aminosäure erheblich beeinträchtigt sein.

Aber denk dran: Eigentlich sollte B3 gleich über die Nahrung oder passende Supplemente im Verdauungstrakt deines Pferdes landen, damit Tryptophan in ausreichender Menge für alle anderen Prozesse inklusive der Produktion Serotonin zur Verfügung steht.

Niacin ist auch wichtig für die Zucht, denn: In einer schweizerischen Studie konnten über die Supplementierung von Niacin positive Effekte bei Stuten erreicht werden. Nach Supplementierung war die Follikelreifung stabiler und die Chancen auf eine gesunde Trächtigkeit erhöhten sich (Pollard et al. 2022).

Lies genau: Zutatenliste Kraftfutter

Wichtig zu wissen: **Niacin** ist nicht in jedem Mineralfutter enthalten! Du prüfst also besser deine Futterbestände und suchst nach **Nicotinsäure, Nikotinamid oder Inositol-Hexanicotinat.**

Nicotinsäure wird häufig eingesetzt, weil es eine gute Bioverfügbarkeit aufweist und direkt verstoffwechselt wird. Die Wirksamkeit von Nicotinsäure auf das Haarwachstum ist seit längerem wissenschaftlich untersucht, die Ergebnisse sind vor allem für Menschen, die von kreisrundem Haarausfall betroffen sind, recht vielversprechend (Draelos 2005).

Wichtig: In hohen Dosen treten gelegentlich Nebenwirkungen wie Gefäßerweiterungen (Flush-Effekt) auf. Bei Pferden mit ordentlichem Fellwuchs ist das aber eher selten relevant.

Übrigens ist Nicotinsäure nicht zu verwechseln mit dem Nikotin aus Tabak. Letzteres ist ein starkes Nervengift, das wir auf keinen Fall an unsere Pferde verfüttern!

Nikotinamid kommt ebenfalls oft zum Einsatz im Mineralfutter. Flush-Effekte sind keine bekannt. Es ist ähnlich effizient wie Nicotinsäure und dient zur Synthese von Nicotinamidadenindinukleotid (kurz NAD+), ist aber gleichzeitig milder und deshalb verträglicher. Nikotinamid wählst du dann aus, wenn du ein echtes Sensibelchen im Stall stehen hast!

Inositol-Hexaniconiat wird häufig als Flush-freies Niacin vermarktet. Das Niacin aus dieser Verbindung wird langsamer verstoffwechselt und ist damit

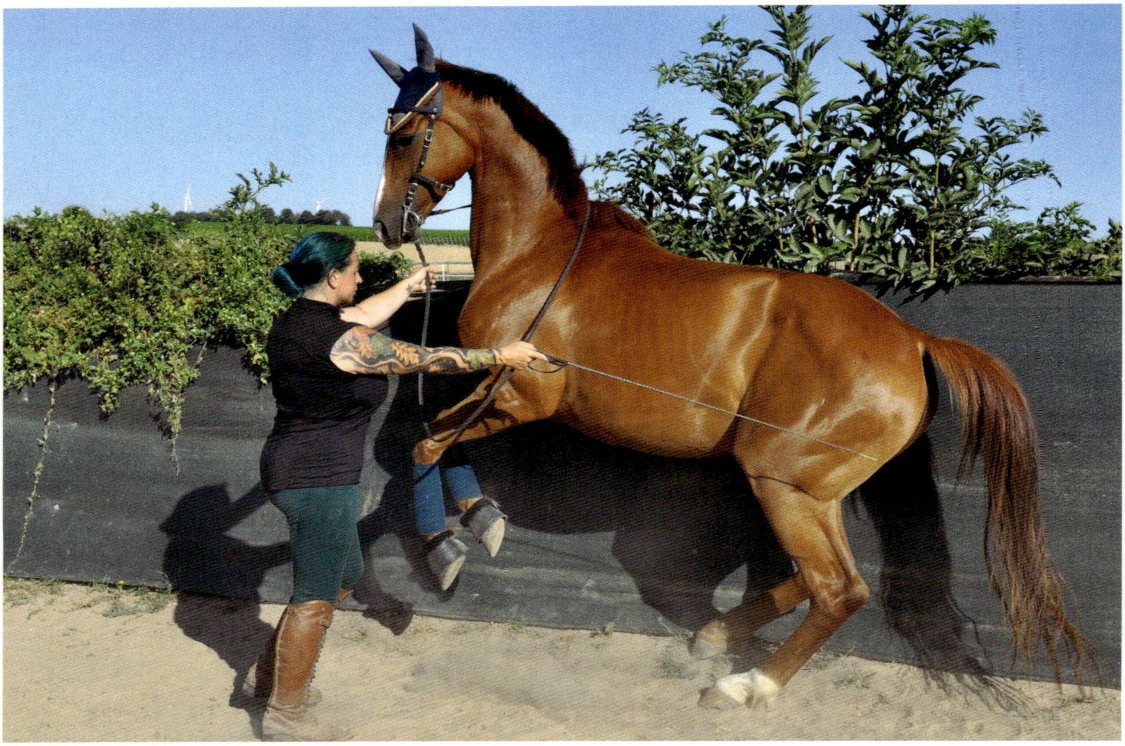

Stärkerer Leistungszuwachs durch Kreatin erlaubt auch jüngeren Pferden zeitnah kraftvolle Posen. Mascha mit Bella.

besonders schonend. In deinem Standard-Pferdefutter wirst du diese Form aber seltener vorfinden. Um an aktivierte B-Vitamin-Ergänzungsmittel zu kommen, wirst du unter Umständen eher auf den Humanmarkt ausweichen müssen.

Kreatin für mehr Leistung

Kreatin (auch: Kreatinmonohydrat) wird im Bodybuilding häufig zur Leistungssteigerung eingesetzt. Tägliche Dosen dieses Eiweißmoleküls können die Leistung kontinuierlich verbessern (Antonio et al. 2021). Sogenannte Loadingphasen, mit denen vor allem im Bodybuilding gern gearbeitet wird, sind für diese Effekte aber nicht notwendig.

Im Reitsport profitieren insbesondere Sportpferde in Sprint- oder Kraftdisziplinen von einer Supplementierung. Seniorenpferde und Rehafälle können das zusätzliche Kreatin gut im Rahmen des Muskelerhalts gebrauchen und um sich etwas schneller von Verletzungen zu erholen (Hall et al., 2021).

Insgesamt scheint sich Kreatin auch hervorragend zur präventiven Stärkung von Muskelkraft, Knochengesundheit und zum Erhalt der kognitiven Fähigkeiten zu eignen. Das legen jedenfalls diverse Studien der letzten Jahre nahe (bspw. Wallimann et al. 2011)

An die passende Dosierung darfst du dich allerdings herantasten. Eine offizielle Dosierungsempfehlung für Pferde existiert nämlich nicht.

Basierend auf der für Menschen gedachten Minimalempfehlung von 0,1 Gramm Kreatin pro Kilogramm Körpergewicht, sieht das für ein Reitpferd dann ganz grob folgendermaßen aus: Bei einem Lebendgewicht von 580 Kilogramm liegt die Erhaltungsdosis mit 0,05 bis 0,1 Gramm pro Kilogramm Körpergewicht bei 29 bis 58 Gramm Kreatinmonohydrat.

In einer Studie mit zwanzig Vollblütern testeten Forschende unterschiedliche Dosierungen. Die besten Ergebnisse im Sinne von geringeren Lactatwerten im Blutserum nach Training stellten sie bei Dosierungen von 20 und 40 Gramm pro Tag und Tier fest (Fagundes et al. 2011).

Die International Society of Sports Nutrition hat in einem Positionspapier diverse Gründe zusammengetragen, warum Kreatin insbesondere auch bei neurologischen Erkrankungen, bei Diabetes Typ 2 sowie gegen Alterungserscheinungen in Kombination mit einer höheren Proteinzufuhr und angepasster körperlicher Betätigung regelmäßig und in deutlich höheren Dosen eingenommen werden sollte (Kreider et al. 2022). Da lohnt es sich doch, die Minimaldosierung unter Umständen noch etwas nach oben zu verlegen.

Insbesondere ältere Tiere wie der 21jährige Wallach profitieren von einer optimalen Entgiftungsleistung. Sonja mit Leo.

Vitamin E und Selen schützen die Muskulatur

Die Kombination aus Vitamin E (Tocopherole) und Selen schützen die Muskeln vor oxidativem Stress. Der ist zwar in einem begrenzten Rahmen normal, kann aber aufgrund von Erkrankungen oder bei zunehmenden Alter Probleme bereiten. Dann werden Zellen durch schädliche Abbauprodukte beschädigt und ihre Funktionsfähigkeit eingeschränkt. Betrifft das die Muskulatur, bedeutet das Kraftverlust und eine verminderte Leistungsfähigkeit.

Oxidativer Stress ist da zu erwarten, wo entzündliche Prozesse eine größere Rolle spielen: bei Arthritis und Arthrose, entzündlichen Störungen im Verdauungstrakt und natürlich, wenn dein geliebter Vierbeiner deutlich zu schwer ist und entsprechend mit Fettdepots zu kämpfen hat, die Entzündungen begünstigen (Wallert et al., 2020).

Vitamin E ist fettlöslich. Kommt es seiner Aufgabe nach, verbleibt es danach selbst oxidiert im Körper. Vitamin C ist deshalb eine gute Ergänzung, um das oxidierte Vitamin E zu recyclen (Orfanos-Boeckel, 2022:139ff).

Denk dran: Pferde können zwar Vitamin C selbst herstellen, in vielen Fällen reicht die produzierte Menge aber nicht aus. Mit diesem Produktionsmangel sind nicht alle Säugetiere geschlagen. Ratten können beispielsweise unfassbar große Mengen selbst produzieren, wobei die gewonnen Zahlen je nach individuellem Studiendesign bei Werten zwischen 20 mg bis 200 mg pro Kilogramm Körpergewicht und Tag schwanken können (Longenecker et al. 1938). Achte bei der Auswahl des Supplements auf den Zusatz »natürliches Vitamin E« für eine bessere Bioverfügbarkeit.

Selen ist relevant für die Proteinsynthese (Schmidt/Simonović 2012). Ist der Spiegel zu niedrig, werden manche Stoffwechselprozesse nicht ordentlich ausgeführt. Das gilt speziell für das Glutathion-Entgiftungssystem, für die Aktivierung von Schilddrüsenhormonen und eine stabile Immunabwehr. Bei einem Mangel kannst du mit Selenmethionin arbeiten. Natriumselenit ist eine anorganische Verbindung, die teilweise ebenfalls zum Einsatz kommt und bei der eine Überdosierung ziemlich schwerfällt. Bei einem ausdrücklichen Mangel an Selen eignet sich aber organisches Selen wesentlich besser, um bei deinem Pferd die Proteinsynthese zu optimieren.

Eindrückliches Beispiel für alle, die mit Schilddrüsenproblemen zu tun haben: Ohne ausreichend Selen arbeitet das Enzym Deiodinase nicht korrekt und die Konversion von T4 in T3 gelingt nicht in ausreichendem Maße. Die Laborwerte für TSH, T3 und T4 sind oft trotzdem ok - aber nicht gut (Orfanos-Boeckel, 2022:63)!

Ohne Selen für eine optimale Proteinsynthese nützt dir die Supplementierung mit Vitamin E nicht viel, weil die bereits entstandenen Schäden durch oxidativen Stress nicht ordentlich behoben werden können.

Achte bei der Dosierung auf Folgendes: Kombiniere Vitamin E mit Selen und gegebenenfalls Vitamin C, um eine optimale Bioverfügbarkeit zu erreichen. Ich arbeite häufig mit Dosierungen in diesem Bereich:

- 1.000 bis 2.000 IU Vitamin E täglich plus
- 1 bis 2 mg Selen täglich
- ergänzend Vitamin C in angepasste Dosis

Denke daran, dass sich therapeutische Kontexte auch auf die Höhe der Nährstoffsupplementierung auswirken. Taste dich im Zweifel also an die für deinen Vierbeiner mit dem jeweiligen Gewicht, dem Trainingsstand und eventueller genetischer Vorbedingungen passenden Dosierungen heran.

Wichtig: Das ist die Erhaltungsdosis. Hat dein Vierbeiner einen bestätigten Mangel, besprich die angepasste höhere Dosierung entsprechend mit dem Veterinärmediziner deines Vertrauens, sofern er dir seine ausgewiesene Expertise bereits unter Beweis gestellt hat.

S-Adenosyl-methionin – Wundermittel für die Entgiftung

S-Adenosyl-methionin (SAM-e) ist eine körpereigene Aminosäureverbindung und wird aus Methionin und Adenosintriphospat (ATP) gebildet. **ATP** kennst du wahrscheinlich, wenn du dich bereits mit Stoffwechselprozessen beschäftigt hast. SAM-e spielt eine entscheidende Rolle bei verschiedenen biochemischen Prozessen, allen voran die Synthese von Neurotransmittern, Proteinen und Nukleinsäuren. Spätestens jetzt solltest du aufmerken, denn: Proteine und Neurotransmitter brauchen wir für fitte Pferde, die langfristig gesunderhaltend trainiert werden wollen.

SAM-e = Methionin plus ATP. Wichtige Co-Faktoren in diesem Prozess sind unter anderem Magnesium, Vitamin C und verschiedene B-Vitamine, ohne die SAM-e nicht oder nicht in ausreichender Menge produziert wird. Sobald es an SAM-e mangelt, können weitere Prozesse im Körper nicht ordentlich ausgeführt werden. Das betrifft unter anderem:

- die Entgiftung in der Leber,
- die Regulation des Nervensystems,
- den Serotonin- und Dopaminstoffwechsel,
- die Bildung von Antioxidantien sowie den Zellschutz.

SAM-e ist nämlich ein zentraler Player in der Methylierung. Das ist ein Prozess, bei dem Methylgruppen ($-CH_3$) auf andere Moleküle übertragen werden. Für alle ohne Chemie-Master: Ohne diese Methylierung können viele Stoffwechselprozesse im Körper schlichtweg nicht stattfinden. Wir werden fett, krank und lethargisch und das ohne eigenes Verschulden. Wie beim Menschen ist auch beim Pferd SAM-e essenziell und darf deshalb keinesfalls aus dem Stoffkreislauf herausfallen.

Dynamisch statt apathisch.
Anna mit Floro.

Energiewährung Adenosin-triphosphat (ATP)

ATP ist die grundlegende Energiewährung, die unsere Körper für die Ausführung von Stoffwechselprozessen und jegliche Form von Bewegung benötigen. Nichts, aber auch gar nichts läuft ohne eine ausreichende Portion ATP.

Auf chemischer Ebene findet die Energiespeicherung über die Bindung von Phosphatgruppen statt. Für die Zwecke hier musst du eigentlich nur wissen: ATP ist die Energiebank des Körpers und versorgt alle Zellen mit der Währung Energie, die ganz nach Bedarf gespeichert und wieder freigegeben werden kann. Entsprechend ist ATP auch einfach an allem beteiligt: Es transportiert Moleküle durch Zellmembranen, synthetisiert neue Moleküle oder stellt sicher, dass Muskelfasern kontrahieren. Ohne ATP ist die überlebensnotwendige Aufspaltung von Makromolekülen wie Proteine oder Nukleinsäure nicht möglich.

Die Produktion von Adenosintriphosphat erfolgt über

- Glykolyse
- Zitronensäurezyklus
- Oxidative Phosphorylierung

Um die ATP-Produktion zu optimieren, muss der Stoffwechsel rund laufen: Ausdauer- und Kraftsport steigern die Fähigkeit des Körpers, Energie effizient aufzunehmen und freizusetzen. Müde Pferde müssen also erst mal in die Bewegung kommen, um sich dann dank Lektionsarbeit fitter zu fühlen – und es am Ende auch wirklich zu sein.

SAM-e wird als Behandlungsoption für verschiedene Symptome und vor allem mit teils schwerwiegenden Erkrankungen in Verbindung gebracht. So wirkt SAM-e unter anderem

- bei Depressionen als Antidepressiva,
- zur Linderung von Arthrose,
- zur Symptomverbesserung bei Fibromyalgie,
- zur Unterstützung der Leberfunktion (Mato 2007),
- zur Verbesserung von Asthma-Symptomen.

SAM-e ist außerdem direkt an der Glutathion-S-Transferase beteiligt, kann sich also positiv auf den **Glutathionspiegel** auswirken. Glutathion ist besonders relevant zur aktiven Bekämpfung von Freien Radikalen, um körpereigene Abfallprodukte in der Leber zu entgiften und um die Zellteilung sowie den Zellstoffwechsel zu steuern. Ein niedriger Glutathionspiegel tritt insbesondere bei neuronalen Erkrankungen wie Alzheimer auf. Erste Studien konnten bereits nachweisen, dass sich ein SAM-e-Spiegel in gesunderhaltener Höhe positiv auf die Glutathion-S-Transferase auswirkt (Tchantchou et al. 2008), mit guten Aussichten für auf die kognitive Gesundheit!

Für die Pferdehaltung ist vor allem die positive Wirkung von SAM-e auf die Lebergesundheit relevant. Denn da scheint der Wirkstoff geeignet, um Heilungsprozesse zu fördern. Er wirkt außerdem schmerzlindernd und entzündungshemmend – alles Dinge, die Pferdemenschen zu schätzen wissen sollten!

In zwei Studien konnte die Wirksamkeit von SAM bei funktionellen Einschränkungen bei Patienten mit Arthrose belegt werden. Dabei wurde SAM-e als Alternative für die nebenwirkungsreicheren, aber häufiger verordneten nichtsteroidale Antirheumatika (NSRA) eingesetzt. SAM-e dämpfte in diesen Fällen nicht die Schmerzen, erleichterte den Patienten aber die eigentlichen Bewegungsabläufe (Soeken 2002).

SAM-e supplementieren

Wie lässt sich SAM-e nun in die Futterrationen einbringen? Die Verbindung ist leider ziemlich empfindlich und recht teuer. Eine Darreichungsform als Pulver scheint es nicht zu geben auf dem Markt. Die verfügbare Kapselversion ist leider sowohl unpraktisch als auch relativ teuer. Das macht es für Pferdebesitzer leider etwas kompliziert.

Wenn dein Tier im Kern gesund ist und keinen gravierenden gesundheitlich relevanten Mangel hat, kannst du über eine angemessene Versorgung mit Methionin und Co-Faktoren wie Magnesium, B-Vitamine und Vitamin C die SAM-e-Produktion anschieben. Methionin ist gerade in der Pferdefütterung absolute Mangelware.

Du hast im vorangegangenen Kapitel zu Sojaschrot ja schon gelesen, dass sich diese Futterpflanze schon aufgrund ihres Methioningehalts hervorragend zur Gesunderhaltung deines Pferdes eignet. Hast du den Verdacht, dass der SAM-e-Spiegel durchaus noch höher ausfallen könnte - und das gilt insbesondere für alle mit Arthrosepferden und solchen, die allgemein zu Gelenksverletzungen neigen - dann ist Sojaschrot eine hervorragende Ergänzung auf dem Futterplan.

Was nicht heißt, dass ich komplett von der Supplementierung von SAM-e abraten würde. Manchmal ist es weitaus sinnvoller, dieses Präparat gezielt in messbaren Dosen ins Pferd zu kippen, statt auf die natürliche Verstoffwechselung aus Futterpflanzen zu hoffen.

Tipp:

Dein Pferd sollte bei Supplementierung von SAM-e möglichst nüchtern sein, die Futteraufnahme beeinflusst die Resorption sonst leider negativ.

Zwischen 10 bis 20 Milligramm pro Kilogramm Körpergewicht sollte eine Dosis SAM-e enthalten. Arbeite dich langsam an die passende Dosierung für dein Tier heran.

Kollagen – Fleisch für Pferde?

Kollagen zählt zu den Nährstoffen, die unsere geliebten Vierbeiner relativ problemlos selbst aus den Aminosäuren ihrer pflanzlichen Nahrung synthetisieren. Und relativ heißt hier, dass das natürlich abhängig davon ist, wo und mit welcher Auswahl dein Tierchen sich den Bauch vollschlagen kann. Je nach Zusammensetzung des Weidegrases und seines sonstigen Futterangebots kann es sonst gut sein, dass die für die Kollagensynthese notwendigen Aminosäuren nicht oder nicht in ausreichender Menge vorhanden sind - Kollagenmangel wäre dann gar nicht so unwahrscheinlich.

Die Hauptbezugsquelle für Kollagen sind Gräser und Kräuter, zusätzlich auch Luzerne und eventuell, falls verfügbar Hülsenfrüchte. Hier ist vor allem das von mir hochgeschätzte Sojaschrot zu nennen. Aber auch Ackerbohnen, Erbsen oder Süßlupinen kommen gelegentlich im Pferdefutter zum Einsatz und liefern die notwendigen Bausteine für die Kollagenproduktion. Im Gras, weit stärker aber in Kräutern und vor allem im Sojaschrot sind Methionin, Lysin, Glycin, Prolin sowie weitere Aminosäuren enthalten, die wiederum für die Kollagensynthetisierung unverzichtbar sind.

Aktive Sporttreibende und Beautyinteressierte kennen den Stoff vielleicht, er wird ja gern als Wundermittel für pralle Haut (Pu et al. 2023), gesunde Gelenke und eine bessere Beweglichkeit (Hye et al 2023) beworben. Übrigens schiebt Krafttraining die Kollagensynthese an (Nulty et al. 2024) und kann so auch dazu beitragen, die Haut jugendlicher aussehen zu lassen (Nishikori et al. 2023).

Außerdem ist Kollagen hervorragend dazu geeignet, um das Haarwachstum anzuregen beziehungsweise die Haarstruktur zu stärken (Lee et al. 2024).

Was hat das mit Pferden zu tun? Kollagen (Kollagenhydroysat) als Nahrungsergänzungsmittel wird in der Regel aus tierischen Quellen hergestellt, ist streng genommen aber kein Fleisch. Das wird nämlich in so kleine Peptide zerlegt, dass das Endprodukt schlicht zum reinen Proteinbaustein wird.

Kollagen für eine optimale Sehnengesundheit.
Christina mit Inca.

Ähnlich wie andere Eiweißlieferanten aus Soja, Luzerne oder Erbse kann Kollagen deshalb ganz einwandfrei vom Verdauungstrakt deines Pferdes verarbeitet werden. Davon abgesehen zupfen unsere Pferde beim Grasen auf der Weide durchaus das eine oder andere größere Insekt mit ab. Verdauungsprobleme bekommt es von dem bisschen tierischen Eiweiß jedenfalls nicht.

Inzwischen gibt es auch vegane Kollagenprodukte, die aber zumindest für die Mengen, die bei einem durchschnittlich 500 bis 600 Kilogramm schweren Pferd nötig sind, noch relativ unbezahlbar sein dürften. Jedenfalls auf Dauer.

Kollagenhydrolysat auf tierischer Basis enthält

- Glycin,
- Prolin,
- Hydroxyprolin,
- Arginin,

jeweils in leicht verdaulicher Form, kann also schnell in Sehnen und Knorpel eingebaut werden.

Die Vierbeiner mit dem Potenzial für Bänder- und Sehnenschäden aus dem Spring- und Rennsport können diesen zusätzlichen Proteinbaustein jedenfalls gut gebrauchen. Alle anderen natürlich auch, denn von Kollagengaben profitieren natürlich alle Pferde, ob sie jetzt wie irre durch die Bahnen rennen oder nicht.

Die auf den Bewegungsapparat einwirkende kinetische Energie erhöht sich jedenfalls massiv bei höheren Geschwindigkeiten. Welche verschleißfördernden Kräfte auf das klassische Springpferd einwirken, kannst du dir wahrscheinlich ganz gut denken.

Der radikale Knorpelverschleiß aufgrund der massiven Belastung in diesem Reitsportsektor macht sich mit noch mehr Bänder- und noch mehr Sehnenproblemen bemerkbar, wenn Kollagen fehlt. Zur Regeneration des Knorpels kannst du Kollagenhydralysolat supplementieren und damit die Gelenkgesundheit samt der Beweglichkeit verbessern (Appelt 2005).

Bei Allergiepferden und solchen, die mit Ekzemen zu tun haben, ist Kollagen besonders empfehlenswert. Denn glücklicherweise bringt Kollagen kein Histamin in den sowieso schon belasteten Stoffkreislauf mit ein. Stattdessen ist es bekanntermaßen das Hilfsmittel für eine gesündere Haut – nicht nur beim Menschen, auch beim Pferd.

Starte mit einer Dosierung von circa 15 Gramm auf 100 Kilogramm Körpergewicht. Entscheidest du dich für eine Dauergabe, kannst du meist auf circa 8 Gramm pro 100 Kilogramm Körpergewicht reduzieren, um die Gelenke dauerhaft zu stärken. Aber beachte dabei: Kollagen alleine nützt gar nichts, wenn keine passenden Aminosäuren im Futter verfügbar sind. Für die Kollagensynthese sind zwingend weitere Aminosäuren nötig, damit alles restlos zugunsten des Stoffkreislaufs verwertet wird.

B12-Mangel: zwischen Lethargie und Aggressivität

Das Vitamin B12 (Cobalamin) ist in der Gruppe der B-Vitamine extrem komplex in seiner chemischen Struktur. Es hat einen starken Einfluss auf die Psyche deines Pferdes, ist essenziell für die Synthese verschiedener Neurotransmitter und nötig, um den Stresspegel zu regulieren.

Heftiges Training + Kollagenhydrolysat = stärkere Sehnen und Bänder

In einer aktuellen Doppel-Blind-Studie (Nulty et al. 2025) am Menschen wurde getestet, wie sich die Einnahme von Kollagen im Rahmen eines 12-Wochen-Krafttrainingsprogramms auswirkt. Das Ergebnis: Bei der im Fokus stehenden Patellasehne (Band, welches Kniescheibe mit Schienbein verbindet) wurde in der Gruppe, die täglich 30 Gramm Kollagenhydrolysat zu sich nahm, ein signifikanter Zuwachs des Sehnenquerschnitts gemessen. In der Gruppe mit dem Placebo war das – trotz gleichermaßen herausforderndem Krafttrainings – nicht zu beobachten.

Wichtig an dieser Studie: Bisherige Untersuchungen über die Wirkung von Kollagen auf Bänder und Sehnen beschränkten sich auf moderate Übungen. In der Studie von Nulty wurde mit Fokus auf starke Belastung und über langsam ausgeführte Wiederholungen mit steigender Intensität trainiert.

Falls dir das bekannt vorkommt: Richtig, genau so trainieren wir in der klassischen Reitkunst unsere Pferde! Zunehmende Herausforderungen mit stetiger Steigerung der Belastung von Schritt zu Galopp bei gleichzeitig gezielt langsamen und damit intensivem Bewegungsablauf.

Die Forschenden ziehen am Ende den Schluss, dass die nachgewiesene Vergrößerung des Sehnenquerschnitts und damit den Kraftzuwachs vor allem in der besseren Verfügbarkeit kollagenbildender Aminosäuren wie Glycin, Prolin und Hydroxyprolin zurückzuführen sei

Über pflanzliche Nahrung kann es nicht aufgenommen werden. Aber wie kommen Pferde in freier Wildbahn dann zu ihrem lebenswichtigen B12? Ganz einfach, sie fressen Dreck, der an Gräsern und Kräutern anhaftet. Die Aufnahmen von Verunreinigungen am Futter genügen in natürlicher Haltung meist, um den B12-Spiegel aufrecht zu halten.

Aber nur unter spezifischen Bedingungen: Das Futter deines Pferdes oder die Erde auf der Weide sind mit ausreichend Ausgangsmaterial in Form von Kobalt angereichert **und** der Magen-Darm-Trakt sowie das Immunsystem sind im Grunde intakt.

Kobalt ist dabei ein entscheidender Baustein, denn dieses essenzielle Spurenelement ist Teil des Vitamin-B12 Moleküls. Ohne Kobalt kann kein B12 produziert werden!

Die bakterielle Fermentation des Futters - oder des Drecks - hin zu unter anderem Vitamin-B12 findet im hinteren Teil des Verdauungstrakts über spezielle Mikroben statt. Konkret sind der Blind- und Dickdarm die Orte, die für eine optimale B12-Synthetisierung tip top in Ordnung sein müssen.

Um dein modernes Stall- und Wiesenpferd also mit ausreichend Cobalamin zu versorgen, muss schon so einiges stimmen. Erst bei einem Kobalt-Mangel oder einer gestörten Darmflora kann ein B12-Mangel auftreten.

B12 fällt nicht vom Himmel

Wichtig: Eine bestehende Gastritis, vermehrte Koliken und insgesamt ein aus welchen Gründen auch immer beeinträchtigtes Verdauungssystem sind sehr gute Anlässe, um die B12-Werte regelmäßig zu prüfen. Da das quasi nie über klassische Blutbilder beim Veterinär möglich ist, kannst du alternativ die im großen Blutbild erfassten Werte zu den roten Blutkörperchen beobachten. Vergrößerte rote Blutkörperchen sind ein guter Anhaltspunkt dafür, dass ein Vitamin B12-Mangel vorliegen kann.

Bei einer nicht intakten Darmschleimwand tritt ein B12-Mangel relativ häufig auf. Das allein wäre schon Grund genug, dieses essentielle B-Vitamin zügig in ausreichender Dosis zu supplementieren.

Denn bei Vitamin B12 gibt es leider ein Problem, welches durch die sehr spezifische Verdauungssituation auftritt. Extern zugeführtes Kobalt sowie B12 wird vor allem im **Dünndarm**, speziell im Krummdarm (Ileum) in den Blutkreislauf überführt. Spezielle Transportmechanismen und ein sogenannter intrinsic factor, der beim Pferd in der Bauchspeicheldrüse gebildet wird, sind dafür zwingend nötig.

Die mikrobielle Fermentation zur körpereigenen Bildung von B12 findet aber vorrangig im **Dickdarm** inklusive Blinddarm statt, der dummerweise erst hinter dem Dünndarm liegt. Wildpferde scheinen damit gut durchgekommen zu sein, immerhin haben sie es über zig tausend Jahre mit Steppengras geschafft, ohne massiven B12-Mangel zu überleben. Da scheinen auch Prozesse wie mikrobielles Recycling, eine optimierte Schleimhautaufnahme im hinteren Darm sowie eventuell **Koprophagie** (Kot fressen) bei Fohlen eine relevante Rolle zu spielen.

Die optimaleren Futterbedingungen aus Wildpferdezeiten sind heute aber so nicht mehr gegeben, eine ausreichende Eigenversorgung über mikrobielle Fermentation von Vitamin B12 im Dickdarm ist zumindest bei gesundheitlich angeschlagenen Pferden kein Selbstläufer.

Was eben auch bedeutet: Kobalt sollte in ausreichender Menge im Futter enthalten sein, damit Vitamin B12 überhaupt gebildet werden kann und bei bestehenden Erkrankungen oder einer gestörten Verdauung steckt B12 besser in den individuell angepassten Futterrationen mit drin.

Vitamin B12-Mangel erkennen

1. Ein Mangel am wichtigen Vitamin B12 macht sich bei deinem geliebten Vierbeiner gern mal mittels **Verhaltensänderungen** bemerkbar. Dazu zählt zunächst das sehr offensichtliche Kotfressen (Koprophagie), das bei Fohlen häufig beobachtet wird. Ausgewachsene Pferde nehmen normalerweise keinen Kot auf - tun sie es doch, steckt

B12 hat weitereichende Auswirkungen auf die Pferdegesundheit und auch auf die Psyche.

sehr häufig ein akuter Vitamin B12-Mangel dahinter. Sieh es als das letzte zur Verfügung stehende Mittel an, das deinem Tier noch zur Verfügung steht, um seinen B12-Bedarf noch irgendwie wenigstens am Rande zu decken.

2. Schwerwiegend und teils auch irreversibel sind Nervenschäden durch einen starken B12-Mangels. Viele B-Vitamine, insbesondere auch Vitamin B12 sind unverzichtbar für die Myelinisierung als Teil der schnellen und exakten Signalweiterleitung über die Nervenzellen. Fehlt es an B12, wird die Reizweiterleitung erschwert, Signale kommen gar nicht oder nur noch verzögert an. Das Gehirn funkt, aber die Nachricht bleibt irgendwo zwischen Hirn und Huf stecken.

Ein 2024 veröffentlichtes Konsens-Papier fasst typische Begleiterscheinungen und Folgen von Vitamin B12-Mangel zusammen und kommt zum Schluss, dass sich vor allem die teils unspezifischen Symptome wie Konzentrationsschwäche und Lethargie schwer erfassen lassen. Gut belegt und wesentlich häufiger sind dagegen Beeinträchtigungen der sensorischen und motorischen Fähigkeiten (Obeid 2024).

Dein Vierbeiner wird nicht einfach aus dem Nichts tollpatschiger, sondern ihm brechen unter Umständen aufgrund eines akuten B12-Mangels schlichtweg Nervenbahnen zur Signalweiterleitung weg!

3. Zusätzlich zeigt sich ein B12-Mangel auch gerne mal in einer Veränderung des Pferdetemperaments. Um Stresshormone und damit den Output des Nervensystems zu regulieren, wird B12 benötigt. Cobalamin reguliert wichtige Neurotransmitter wie
   - Dopamin,
   - Serotonin,
   - Noradrenalin.

Dopamin und Serotonin sorgen im Grunde für Entspannung und emotionale Stabilität. Zu geringe Mengen an Dopamin und Serotonin resultieren in Nervosität und Reizbarkeit bis hin zu Aggressivität. In direkter Folge wirkt dein stressanfälliges Pferd impulsiv und unausgeglichen. Das Training wird dadurch für beide Beteiligten keine große Freude bringen, Lektionen lassen sich nicht oder nur mit übermäßiger Mühe abarbeiten. Spaß macht das keinem, weder dir noch deinem Tier.

Fehlt es dagegen an Noradrenalin, schlägt das aufs Gemüt. Antriebslosigkeit, Lethargie und Müdigkeit sind die Folge. Energieproduktion und Nervenleitfähigkeit sind am Boden, zusammen mit der Motivation deines Tieres. Da ist auch nichts mit schönster tänzerischer Dressurkunst, wenn sich dein Vierbeiner kaum zum Verlassen seiner Box aufraffen kann.

Der Normalzustand deines Tieres bestimmt weitestgehend, wie es auf einen akuten B12-Mangel reagiert:

- **Junge oder temperamentvolle Pferde** werden nervöser und aggressiver.
- **Ältere und weniger energiereiche Pferde** tendieren eher zu passiver Lethargie.

Das hast du im Kapitel zum Pferdetemperament schon mehrfach gelesen. Die Grundpersönlichkeit deines Tieres bestimmt, wie es sich im Rahmen eines Nährstoffmangels verhält.

Sensible Tiere werden schreckhafter, insgesamt aber lethargischer. Die kleinen Energiebomben aus dem Stall dagegen zeigen sich dann zunehmend von ihrer dominant-aggressiven Seite.

Grund für die zunehmende Lethargie ist schlichtweg an Mangel an Energie. Denn ohne ausreichend B12 im Blut fehlt der Energietreibstoff fürs Gehirn, die Muskeln ermüden schneller. Dein Pferd zeigt sich matt und unmotiviert.

Gehört dein Tier zu den Tieren mit einem eher sensiblen Nervensystem, folgt manchmal auch eine paradoxe, also ins Gegenteil verkehrte Reaktion. Dann flippt dir das Tier ständig unter dem Sattel aus, wirkt impulsiv und scheint unkontrollierbar. Der bloße (starke) Mangel an B12 beeinträchtigt außerdem die Signalweiterleitung ans Gehirn in entscheidender Weise.

Ohne ausreichend Vitamin B12 ist zudem die Bildung von roten Blutkörperchen beeinträchtigt, sie sind dann per großem Blutbild nachweisbar vergrößert! Diese Blutkörperchen sind aber für den Transport von Sauerstoff, sprich Gehirntreibstoff, bitter nötig.

Kein Treibstoff in Form von Sauerstoff, kein energiegeladenes Pferdegehirn

Das gilt auch für die Fettsäurengewinnung - ohne B12 ist die Umwandlung von Fetten in nutzbare Energie beeinträchtigt. Und zu guter Letzt spielt Vitamin B12 auch noch eine enorm wichtige Rolle in der Myelinisierung der Nervenfasern, die wiederum die Grundlage für eine möglichst ungestörte Signalweiterleitung im Gehirn und im gesamten Nervensystem legen, weshalb ein Mangel eben auch gerne zu sensorischen und motorischen Fehlfunktionen führt.

B12-Mangel beheben

Die Bezeichnung **Cyanocobalamin** wirst du besonders häufig in Form von Nahrungsergänzungsmitteln bei B12-Mangel entdecken. Es wird auch gern dazu genutzt, um Lebens- sowie Futtermittel mit B12 anzureichern.

Cyanocobalamin ist teils schlechter bioverfügbar, das heißt: Es muss im Körper zunächst in die biologisch aktiven Formen Methylcobalamin oder Adenosylcobalamin umgewandelt werden, damit es für Stoffwechselprozesse nutzbar ist. Diese Umwandlung kann für Pferde - und Menschen - mit bestimmten Vorerkrankungen ineffizient sein. Zu diesen Vorerkrankungen zählt beispielsweise die MTHFR-Mutation oder auch verschiedene Entgiftungsstörungen.

Wenn Cyanocobalamin nicht oder nicht gut vertragen wird oder es anderweitige gesundheitliche Probleme gibt, kannst du alternative B12-Formen nutzen, wie

- **Methylcobalamin**
  für Methylierungsprozesse und Nervensystem,
- **Adenosylcobalamin**
  für Energiestoffwechsel und Mitochondrien,
- **Hydroxocobalamin**
  für eine gute Aufnahme dank stabiler Depotform.

Ist das noch Nahrungsergänzungsmittel (NEM) oder schon Doping?

Das essenzielle Spurenelement Kobalt wird gelegentlich anstelle vom bereits verwertbaren Vitamin B12 supplementiert. Dahinter steht die Hoffnung auf die vermeintlich natürlichere körpereigene mikrobielle Fermentation im Verdauungstrakt des Tieres.

Nur: Es gibt gar keine Möglichkeit vorherzusagen, wie viel Kobalt tatsächlich an den relevanten Stellen im Darmtrakt ankommt, wie effektiv die vorhandenen Mikroben im Dickdarm arbeiten und ob alle weiteren Bausteine zur Synthese von ausreichend Vitamin B12 vorhanden sind. Schon an dieser Stelle wäre es sinnvoller, schlichtweg zu Vitamin B12-Supplementen zu greifen, um eine stabile Grundversorgung sicherzustellen.

Dann kommt da noch etwas hinzu, dass gerade Rennsportbegeisterte nur zu genau wissen. Denn Kobalt gilt als Dopingmittel im Rennsport, immerhin könnte es auf relevante Weise die Leistung des Pferdes erhöhen. Aber natürlich nur in entsprechend enormen Mengen, die über die für das Pferd normale Synthetisierungsleistung hinausgeht. Belegt ist die leistungssteigernde Wirkung, bei der Kobalt die Sauerstoffaufnahme stark verbessern soll, allerdings nicht (Mobasherie et al. 2015).

Solltest du deshalb vorsichtshalber auf Kobalt verzichten, sobald du beispielsweise mit deinem Tier im Rennsport aktiv bist? Vitamin B12 ist essenziell für Pferde. Das gilt rundherum für sämtliche B-Vitamine. Sie bewusst wegzulassen scheint mir eine extrem unkluge Entscheidung zu sein.

Über Mineralfuttergaben dopingverdächtige Urinwerte für Kobalt zu erreichen, dürfte dir relativ schwerfallen. Auf Injektionen solltest du allerdings, sofern du mit deinem Tier an FN-regulierten Wettbewerben teilnimmst, definitiv verzichten!

Die Richtlinien der FN im Rahmen der Anti-Doping- und Medikamentenkontroll-Regeln (ADMR) im Detail:

> *»Die Verabreichung von Kobalt per Injektion sowie die orale Verabreichung großer Mengen Kobalt ist nicht ADMR-konform. Für Kobalt gibt es Grenzwerte für Urin (100ng/ml) und Blut (25ng/ml). Als Voraussetzung zur Einhaltung der Grenzwerte empfiehlt die International Federation of Horseracing Authorities (IFHA, ww.ifha-online.org), dem Pferd insgesamt nicht mehr als 5 mg Kobalt pro Tag zu füttern. Pferde sind bereits über die Grundfütterung in der Regel ausreichend mit Kobalt versorgt, so dass eine Supplementation nicht notwendig ist. Ein Kobaltmangel beim Pferd ist bisher nicht beschrieben worden. Kobalt ist Bestandteil des Vitamin B12. Das Verabreichen von verschiedenen hochdosierten Kobalt-haltigen Supplementen, die sowohl Kobalt als auch Vitamin B12 enthalten, kann somit ein Risiko darstellen (FN).«*

Aus meiner Erfahrung heraus sind Mineralfuttergaben mit B12-Supplementen spätestens dann nötig, wenn das Tier bereits über einen angeschlagenen Verdauungstrakt verfügt oder häufig starkem Stress mit all seinen nährstoffraubenden Begleitprozessen ausgesetzt ist.

KEINE REITKUNST OHNE **DICH**!

KURZ UND KNAPP:

- Reiten macht glücklich – aber warum?
- Du kannst nicht **nicht** mit deinem Pferd kommunizieren.
- Pferde orientieren sich an Führungspersönlichkeiten – bist du eine?
- Fehlende Konzentration kostet dich den Spaß an der Reitkunst.

Reiten = Hobby?

Reiten ist heutzutage für den Großteil derjenigen, die sich ein Pferd oder eine Reitbeteiligung leisten können, im Grunde ein schönes, aber teures Hobby. Von der Notwendigkeit, mit Pferd von A nach B zu kommen, schwere Holzstämme aus Wäldern zu zerren oder Felder zu pflügen sind wir ja glücklicherweise befreit.

Konzentriertes angestrengtes Pferd in der Rumpfhebung. Mascha mit Honey.

Sichtbarer Spaß am Reiten.
Christina mit Inca.

Trotzdem kann sich das *richtige* Reiten mehr nach Verpflichtung mit höchsten Anforderungen anfühlen statt nach Spaß und sportlicher Betätigung.

Warum eigentlich? Die Anzahl der Perfektionisten im Reitsport ist - wen wundert's - ziemlich hoch. Es wird geritten, bewertet und gemessen. Wer nicht zügig in die nächstbessere Klasse aufsteigt, hat versagt. Dazu wird auch gerne nur und ausschließlich nach oben verglichen - irgendwer ist halt immer schon weiter, schneller oder besser in der nächsthöheren Klasse unterwegs. Der Blick auf die eigene Leistung und auf individuelle Erfolge geht da schnell verloren.

Die Einteilung in Klassen deckt sich nun auch nicht unbedingt mit dem, was Pferde im Hobbyturnierbereich so von sich aus zeigen. Bestes Beispiel ist die Piaffe, denn die wird oft vom Pferd selbst angeboten, noch bevor es den versammelten Galopp beherrscht.

Nun gilt die Piaffe aber heutzutage als extrem komplex und wird deshalb erst auf Inter-2-Level verlangt. Der versammelte Galopp taucht dagegen schon in Klasse M oder L auf und sollte da auch besser sitzen, um am Ende mit Schleifchen nach Hause fahren zu können. Diese nicht am Pferd und seinem Lernfortschritt orientierten Anforderungen sind frustrierend für Pferd und Mensch.

Aus der Altklassik wissen wir außerdem: Mit Fokus auf versammlungslastige Lektionen wäre die Reihenfolge aus (1) Piaffe und (2) versammelter Galopp von Anfang an korrekt und vergleichsweise einfach umzusetzen. Stattdessen mühen sich alle an den Anforderungen der Hobbyreitklassen ab.

Im Freizeitbereich müssen Wettkämpfe mit militärischer Anmutung absolviert werden, um das sogenannte Hobby durch Auszeichnungen und Treppchenbesteigungen auch vor allen anderen zu rechtfertigen. Die Liebe zum Tier gerät da leicht mal aus dem Fokus. Der Spaß am Reiten oft auch.

Reiten = Beruf

Berufsreiterinnen und -reiter fangen ebenfalls oft aus Liebe zum Tier mit der Reiterei an. Aus Spaß wird aber schnell Ernst und dann ein Geschäft, das laufen muss. Damit der Lebensunterhalt aus Mangel an Erfolgen nicht wegbricht, kommen dann oft, um nicht zu sagen meistens, weniger tierliebende Methoden zur Anwendung.

In der Springreiterei ist das verbotene Barren Gang und Gäbe, schmerzende Salben oder Abhärtungsschläge gelten unter der Hand als akzeptable Mittel, um die Tiere auf die stetig steigenden Wettkampfanforderungen einzustimmen.

Der positive Blick aufs Pferd? Der geht hier richtig schnell verloren. Der Vierbeiner wird zum Hilfsmittel, um Stationen auf dem beruflichen Weg zu absolvieren und das Konto zu füllen.

Reitkunst als Balance aus Verantwortung und Freude

Nun ist es so: Heutige Reitpferde sind prädestiniert dafür, früh zu verschleißen und ihre schmerzenden Knochen auf der Rentnerweide herumzuschleppen. Im Eingangskapitel dürfte deutlich geworden sein, dass die typische moderne Reitpferdeausbildung nicht dazu geeignet ist, Pferde auf lange Sicht gesund und reitbar zu erhalten. Das hat sich der Vierbeiner selbst nun nicht so ausgesucht. Der Mensch ist in der Verpflichtung, sein Tier bestmöglich auszubilden, für eine möglichst angemessene Haltung zu sorgen und natürlich - auch ordentlich zu bewegen.

Hier sind mir zwei Dinge enorm wichtig.

1. Pferde sind im Grunde nicht dafür gemacht, ein Menschengewicht knapp hinter den Schulterblättern durch die Gegend zu tragen. Mit angemessener Reitkunstausbildung und lebenslangem anabolikafreien Krafttraining ist es *akzeptabel.*
2. Pferde sind aber auch nicht dafür gemacht, tagtäglich nur hübsch geputzt, spazieren geführt oder im Offenstall geparkt zu werden. Nichts davon entspricht seiner ursprünglichen Lebensform.

Heutige moderne Reitpferde sind mit einer geringeren Lebensqualität konfrontiert, leiden im Alter unter Sarkopenie (Abbau von Muskelmasse und Muskelkraftverlust), weil die Tiere weder angemessen ausgebildet noch angemessen lektioniert werden.

Über klassische Reitkunst im Sinne der Alten Meister lässt sich zumindest der Gesundheitsaspekt einer verantwortungsvollen Pferdehaltung relativ sicher umsetzen. Gleichzeitig kann das frustrierend sein, denn:

1. Die Allgemeinheit im Stall guckt höchstens kritisch, aber kaum jemand arbeitet sein Tier ähnlich gesunderhaltend. Menschen sind soziale Wesen und Anerkennung ist etwas, das wir uns alle wünschen. Ausgrenzung und Ablehnung ist deshalb das, was sich nicht nur einfach fies anfühlt, es kann auch dazu führen, dass eigentlich richtige Dinge wie eine gesunderhaltene Arbeit mit dem Reitpferd nicht stattfinden.
   An der Reitkunst interessierte Menschen verstehen deinen Wunsch nach einer gesunderhaltenden Pferdehaltung. Anerkennung, Tipps und Hilfen für echte klassische Dressur nach Art der Alten Meister gibt es in der Regel noch dazu. Wenn du vor Ort niemanden findest, schau online - unter meinen Followern finden sich viele, denen es auch so geht!
2. Die Lektionserarbeitung ist hart und dauert - die Erfolge stellen sich je nach Pferd manchmal recht zäh ein. Ein auf Versammlungsfähigkeit gezüchteter regulärer Lipizzaner wird sich bei den Lektionen von Anfang an geschickter anstellen als ein durchschnittlich talentiertes Islandpony. Beide sind in der Lage, am Ende in höhere Reitkunstsphären aufzusteigen. Aber bis das der Isländer den Seitwärtstrab oder eine Piaffe meistert, ist einfach aufgrund der körperlichen Rahmenbedingungen wesentlich schwieriger. Es ist keineswegs unmöglich, aber es fordert eben

Sonja mit Coronado.

sowohl vom Pferd als auch vom Menschen noch etwas mehr Geduld, ein große Hingabe – und harte Arbeit.

Lohnt es sich trotzdem, dein Pferd – ob jetzt Isländer, Lipizzaner, Iberer etc. ist völlig egal – mit geeignetem Krafttraining aufzumuskeln? Natürlich! Wie im Kapitel zum Training und zum Krafttraining angeschnitten, ist allein die verstärkte Ausschüttung von Myokinen im Rahmen des Krafttrainings ein enormer gesundheitlicher Bonus. Bänder und Sehnen werden gestärkt, der gesamte Bewegungsapparat vor den schädlichen Folgen von Muskelverlust und Knochenabbau geschützt.

Damit dir zwischendurch nicht die Freude am Reiten vergeht, mach dir bewusst, warum du tust, was du tust.

Aufs Gute konzentrieren

Der Beziehungsforscher John Gottmann hat einst die 5:1-Regel aufgestellt. Sie besagt, dass für jedes negative Erlebnis im Rahmen einer Beziehung mindestens fünf positive Erlebnisse folgen müssen, um diese eine negative Situation auszugleichen und zurück zu einer neutralen Basis zu gelangen. In Paarbeziehungen, die sich an dieser Regel orientierten, lief es meist wesentlich besser als bei denen, die die schlechteren Momente nicht aktiv ausgeglichen haben.

Der Grund dafür? Das menschliche Gehirn neigt dazu, sich auf Negatives zu fokussieren. Der unkonzentrierte Verkehrsteilnehmer auf dem Weg zum Stall bleibt präsent im Kopf.

Die rücksichtsvollen anderen Begegnungen an diesem Tag ... wahrscheinlich eher nicht.

In der Pferde-Mensch-Beziehung kannst du diese Regel bewusst anwenden, um dich auch durch schwierige Ausbildungsphasen zu bringen. Die Versammlungserarbeitung läuft einfach nicht so, wie du es dir vorstellst, dein Pferd ist heute eher mimosenhaft oder das Putzen einfach mies gelaunt? Dann gleich das aus - und zwar bewusst.

Es gibt immer etwas Gutes zu entdecken, auch an richtig dunklen Tagen. Und ganz wichtig: Es geht auch nicht ums Schönreden und toxische Positivität, bei der kein Raum für negative Beobachtungen und Gefühle bleibt. Aber unser Gehirn betrügt uns, wie oben angesprochen, indem es uns permanent die negativen Dinge ins Bewusstsein spült.

Achtsamkeitsübungen können dabei helfen, den Blick auf Gutes zu schärfen.

Übung 1:
Bei jedem Ritt überlegst du dir fünf Dinge, für die du bei diesem Pferd an diesem Tag dankbar bist. Du kannst es aufschreiben oder nur im Geiste notieren, ganz egal. Wichtig ist, dass du es regelmäßig machst. Wofür genau du dankbar sein kannst oder willst, ist ganz unterschiedlich:

- das Aufzäumen ging heute ganz einfach,
- der Vierbeiner ging nicht durch, als vorhin der Trecker eine Fehlzündung hatte,
- die eine richtig schöne Seitwärtsbewegung kam aus dem Nichts und wurde einfach so angeboten.

Übung 2:
Bei jedem Ritt machst du eine Glimmer-Liste. Glimmer sind als angenehm empfundene Sinneseindrücke, die sich als Gegensatz zu den allseits bekannten Triggern verstehen. Beispiele für Glimmer sind:

- der Geruch nach frisch aufgeschüttetem Heu,
- ein Lichtstrahl, der sich durch das Stallfenster auf dem Boden in interessanten Formen auffächert
- das Gefühl, wie sich der Zügel in deiner Hand bewegt.

So bleibt dir der Tag, der Moment oder auch diese eine bestimmte Lektion mindestens neutral oder sogar richtig positiv im Gedächtnis.

Das Glück liegt auf dem Rücken der Pferde – wirklich?

Krafttraining nach Art der Alten Meister sorgt für fitte Reitkunstpanzer, die elegant tänzelnd durch die Halle schweben. Schaut schön aus und hält das Tierchen fit - und was ist mit dir? Du bist wahrscheinlich wesentlich glücklicher mit Pferd, statt ohne.

Das Ding mit dem Glück

Anders als oft gedacht, ist es nicht allein Geld und Ruhm, die glücklich machen. Vielmehr sind Beziehungen zu geliebten Menschen und eine individuell gut verbrachte Zeit entscheidend. Im Falle von Reitbegeisterten sind damit gleich beide Punkte abgedeckt. Denn ein wohlumsorgtes Pferd baut zwangsläufig eine intensive Bindung zu seinem Menschen auf - und andersherum geschieht genau das gleiche.

Bonuspunkte gibt es dafür, dass du dich auch noch bewegst und reichlich Zeit an der frischen Luft bist. Reiten macht glücklich - und das nachhaltig.

Ein Harvard-Professor sucht das Glück

Dr. Waldinger ist Professor in Harvard und begleitet die längste Studie zum Thema Glück, die die Wissenschaft bisher gesehen hat. Sie ist 1938 mit zwei Gruppen junger Männer gestartet, die sich im Laufe der Zeit sukzessive auf Ehepartner und Kinder ausgeweitet haben. Mittlerweile läuft die Studie über 86 Jahre (Stand 2025) und liefert gar nicht so erstaunliche Erkenntnisse:

1. Geld allein macht nicht glücklich (beruhigt aber schon etwas, wenn es nicht völlig knapp ist).
2. Beziehungen und Sich-in-Verbindung-mit-anderen-Personen-Setzen beruhigt aber schon sehr.
3. Intensive soziale Interaktionen erhalten die kognitive und physische Gesundheit.

Die ursprünglichen Studienteilnehmer bedauern am Ende zwei Dinge:

1. Die Männer hätten gern weniger gearbeitet und mehr Zeit mit ihren Liebsten verbracht.
2. Die Frauen dagegen hätten sich gern weniger damit beschäftigt, was andere Menschen von ihnen denken.

Über alle Studienteilnehmer hinweg war aber deutlich: Glücklich und gesund waren diejenigen, die sich sinnstiftend mit geliebten Menschen und sozialen Interaktionen betätigt hatten (Fudala 2023).

Glücklich ist also, wer sich bindet und sich um andere kümmert, ohne sich viel um die Meinung anderer zu scheren.

Zumindest in Reitkreisen ist das oft so eine Sache. Leistung und Ästhetik erhalten einen enorm hohen Stellenwert – und das eben nicht nur beim Reitpferd, sondern auch beim reitenden Menschen obendrauf. Beim Pferd sprechen viele gute Gründe dafür, sich explizit der Leistung zu widmen, um das Tier möglichst lange rittig und schlicht gesund zu halten. Wer aber als Mensch aus verschiedenen Gründen aus der ästhetischen Rolle fällt, hat Pech. Die Mode, von der Reithose bis zum Reitstiefel ist in der Regel für schlanke Menschen ausgelegt. Wer außerhalb der gängigen Reitästhetik angesiedelt ist, hat oft mit bewertenden Blicken und Getuschel seitens der Stallkolleginnen zu tun.

Glücklich macht das nicht. Ganz im Gegenteil.

Zu dick für klassische Reitkunst?

In Wettkämpfen sind in der Regel schlanke, mindestens aber normalgewichtige Personen sichtbar. Wer mit Mehrgewicht zu kämpfen hat, kriegt oft nur die volle Breitseite an hämischen Blicken und Getuschel ab.

Nun hat Übergewicht nicht nur beim Pferd viele verschiedene Ursachen, beim Menschen verhält sich das ähnlich. Von diversen Stoffwechselerkrankungen über Spätfolgen von Krebserkrankungen, lange Kortisoneinnahmen und genetische Dispositionen bis zur immer noch relativ unbekannten Fettgewebsverteilungsstörung Lipödem könnte alles Mögliche hinter dem Gewicht stecken. Hormonelle Einflüsse sind ebenfalls ein enorm großes Thema, wenn es um Gewichtszu- und -abnahmen geht.

Lipödem betrifft überwiegend Frauen – die auch die größte Gruppe in der Reitgemeinschaft stellen. Wenn dir also jemand mit kräftigen Oberschenkeln, Waden oder Oberarmen begegnet, könnte es durchaus an einer krankhaften Veränderung des Fettgewebes liegen, statt an einer Vorliebe für kalorienhaltigen Snacks.

In einem Podcast (Schwarz/Kähler 2025) kommt das übrigens zur Sprache. Reiterin Steffi hat sich jahrelang nicht aufs eigene Pferd gesetzt, um es mit dem durch das Lipödem verursachte eigene Mehrgewicht nicht zu belasten. Das kann, je nach Pferd und Trainingsstand, auch die einzige pferdeschonende Lösung sein.

Getuschel gab's natürlich trotzdem. Das lässt sich während des Umgangs mit dem Pferd zwar ausblenden, aber ein gewisser übler Nachgeschmack bleibt. Was schade ist, denn klassische Reitkunst lädt ja geradezu dazu ein, sich über die Feinheiten der Dressur und der Lektionen auszutauschen, statt sich gegenseitig an Kleidergrößen zu messen.

Selbst zarte 50 Kilogramm Lebendgewicht können für ein gering bemuskeltes, nicht ausreichend in klassischer Reitkunst ausgebildetes Pferd komplett belastend sein. Aus meiner Praxis als Ausbilderin: Wird dazu noch auf heute moderne Weise mit reichlich Anlehnung und hartem Treiben in die Hand geritten, dann übersteht selbst dieses scheinbar zarte Reitgewicht kaum ein Pferd ohne Schaden.

Dann muss ein großer schwerer Warmblüter mit Cortison und Muskelrelaxanzien behandelt werden, weil er mangels schützender Muskelschichten und aufgrund der fehlenden Versammlung dem für uns so zart anmutenden Gewicht knapp hinter seinen Schulterblättern schlicht nichts entgegenzusetzen hat.

Ein hervorragend geschultes Reitpferd im Sinne der Alten Meister wird dagegen auch mit einem etwas höheren Gewicht zurechtkommen, wenn Sitz und Hilfengebung entsprechend feinfühlig ausgeführt sind.

Und für dich als Reiter oder Reiterin gilt natürlich ebenso wie für die Pferde, um die es uns hier geht, ein Grundsatz. Und der lautet: Muskulatur ist immer gut und (viszerales) Fettgewebe immer ein Grund, sich genauer mit dem eigenen Bewegungs- und Ernährungsverhalten auseinanderzusetzen. Eine proteinreiche Ernährung samt Krafttraining bringt am Ende schon viel für die Gesundheit.

Wenn wir insgesamt häufiger über Muskelaufbau und -erhalt und weniger über Kilogrammzahlen auf der Waage sprechen würden, wäre das schon ein guter Anfang. Du musst definitiv keine Fee sein zum Reiten, erst recht nicht, wenn du dich der altklassischen Reitkunst zuwendest. Denn die Tragfähigkeit dieser echten klassisch ausgebildeten Schulpferde ist deutlich höher als die von heutigen modernen Reitpferden.

Handle also besser ganz im Sinne für dich und dein Glück: Konzentriere dich auf dein Pferd, seine und deine Gesundheit, eure Lektionserarbeitung und euer Zusammenspiel als Team.

Mindfulness – mit Achtsamkeit durch den Stall

Dieses Sich-auf-sich-selbst-Konzentrieren ist leicht gesagt, aber manchmal schwer durchzuziehen. Wenn du zudem sensibel auf Ablehnung und Kritik reagierst, fällt das in Gegenwart von potenziell bewertenden Personen gleich noch schwerer.

Gerade im Reitsport sind Bewertungen von allem und jedem ja ständig an der Tagesordnung.

Diese Angst vor Ablehnung (engl. fear of rejection) kann massive Ausprägungen annehmen, die dich sogar komplett blockieren können. Sprich: Da bist du bei deinem Pferd und willst endlich Seitengänge üben, aber eine Handvoll Leute steht herum und scheint ständig zu tuscheln. Also lässt du es sein oder überbrückst den Moment mit anderem, in der Hoffnung, die unliebsamen Zaungäste verschwinden bald.

Dahinter steckt dein zentrales Nervensystem, genauer der Fight-, Flight- or Freeze-Response. Unterbewusst bist du im kompletten Flucht- oder Kampfmodus, der Herzschlag steigt, die Pupillen weiten sich. Je stärker die Angst vor Ablehnung ausgeprägt ist, desto stärker blockiert es dich.

Mit der Technik **Affect Labeling** aus der Psychologie kann diese Blockade aufgelöst werden und sie ist zum Glück auch ganz einfach und unauffällig anzuwenden. Dabei werden die wahrgenommenen Emotionen (bei dir selbst) und die gezeigten Reaktionen (bei anderen) benannt. Das kann lautlos und in Gedanken erfolgen, du kannst es aber auch laut aussprechen, wenn du genug Abstand zu allen anderen hast.

Diese Methode legt den Schalter zur sachlichen Situationsanalyse im Gehirn um. Der Einfluss der Amygdala (Stammhirn) wird kurzfristig abgeschwächt, die emotionsgeladene unbewusste Stressreaktion aufgrund der wahrgenommenen Ausgrenzung abgemildert. Der rationalere Part kann übernehmen und die Situation wortwörtlich retten. Das wurde von Liebermann/Eisenberger schon 2007 per MRT dokumentiert (Liebermann et al. 2007).

Name it to tame it (dt. Benenne es, zähme es)

Statt unbewusst in Panik vor dem Ausschluss aus der Gruppe zu verharren, benennst du also schlicht jedes Augenrollen, jedes Tuscheln, jedes bewertende Handzeichen so, wie du es wahrnimmst.

Zusätzlich beschreibst du deine eigenen Emotionen: Rast dein Herz, fühlst du dich unsicher, wird der Mund trocken oder die Hände zittrig?

Mit etwas Übung entspringt aus dem Affect Labeling heraus ein besseres Gespür dafür, wo dich scheinbar abwertende Blicke oder Getuschel von dem abhält, was du eigentlich tun willst – und was dich glücklich macht.

Sobald dir dein Pferd die allerersten Seitengänge im Schneckenkriechgang präsentiert oder tänzelnd zur ersten Piaffe ansetzt, wirst du definitiv glücklich sein. Unabhängig vom Stallgetuschel.

Eigen- und Außenwirkung für dich, fürs Pferd, für alle

Sobald sich die Stalltüre öffnet, beginnt die Kommunikation. Zwischen dir und deinem Pferd, den Pferden in den Nachbarboxen und jedem anderen, der da so gerade im Stall unterwegs ist. Dafür muss kein einziges Wort fallen. Körpersprache und Mimik reichen völlig aus, um allen Anwesenden genau zu vermitteln, wie es dir heute geht: Traurig, müde, wütend, im Kopf ganz woanders oder voll bei der Sache – alles wird unbewusst ausgewertet.

Selbst dein Geruch sagt unglaublich viel darüber aus, wie es um deinen emotionalen Zustand geht. Übrigens mögen Pferde am liebsten fröhliche Menschen und sie sind sogar in der Lage, zwischen Freude und Traurigkeit anhand eines Gesichtsausdrucks zwischen beiden Zuständen zu unterscheiden! Maskieren, also Emotionen vortäuschen oder verbergen, funktioniert mit Pferden allerdings nicht.

Sie sind tatsächlich in der Lage, Diskrepanzen in der Körpersprache, von Gesichtsausdrücken und beispielsweise der Stimmlage zu erkennen. Wenn du dein Tier anlächelst, innerlich aber völlig aufgelöst bist, wird dein Pferd das nicht nur erkennen – es wird auch entsprechend auf diese Unstimmigkeit reagieren (Jardat et al. 2023).

> *»Man kann nicht nicht kommunizieren, denn jede Kommunikation (nicht nur mit Worten) ist Verhalten und genauso wie man sich nicht nicht verhalten kann, kann man nicht nicht kommunizieren (Watzlawik 1969).«*

Pferde nehmen den Herzschlag und den Geruch ihres Menschen nicht nur wahr, sie analysieren diesen auch ganz genau. Körperliche und mentale Beeinträchtigungen hat das Tier sofort auf dem Radar.

Stabiles Innere, stabile Pferdepersönlichkeit

Eine ausgeglichene Führungspersönlichkeit wird mit einem Pferd leichter zurechtkommen, weil die Außenwirkung einfach stimmt. Das Fluchttier Pferd sucht nach einer verlässlichen Führung - und nimmt sie je nach Temperament und Charakter auch mehr oder weniger gerne an.

Treibt dich dein Pferd ständig durch die Halle, stimmt die respekteinflößende Außenwirkung nicht so ganz. Denn eins ist klar, Pferde testen immer, ob sich eine verlässliche Führungspersönlichkeit vor Ort befindet. Und diese Tests geschehen über eben dieses Herumschubsen. Lässt du zu, dass dich dein Vierbeiner durch die Gegend treibt oder behältst du auch in brenzligen Situationen immer die Kontrolle? Falls nicht, wird sich das nächstbeste tapfere Pferd notfalls der Aufgabe widmen, die Herde inklusive dir vor den großen und kleinen Gefahren des Alltags zu schützen.

Nun sind moderne Pferdezüchtungen bewusst auf Schönheit, auf Temperament oder Geschwindigkeit gezüchtet. Mit roten Ampeln, klappernden Traktoren oder heranstürmenden Hunden kommen sie aber immer noch nicht zurecht. Der Instinkt der Fluchttiere schlägt prinzipiell bei jeder möglichen Gefahr an. Dein Pferd kann auch nicht abschätzen, was gesund oder geeignet für es ist. Pferde sind unsere Schutzbefohlenen - ja, fast wie Kinder! - und wir sind in der Verantwortung, sie zu führen und ihnen einen gesunden Lebensstil zu ermöglichen. Das bedeutet auch, dass gerade beim Krafttraining manchmal Situationen aufkommen, die unangenehm sind. Kraftsport ist anstrengend, die muskelaktivierenden Bewegungen fallen schwer und lösen oft auch eine Abwehrhaltung aus. Das Ergebnis - Myokinausschüttung, Muskelzuwachs und damit Schutz vor Schäden am Bewegungsapparat und ein aktivierter Stoffwechsel - kann sich aber sehen lassen.

In der wilden Pferdeherde übernehmen die erfahrenen älteren Stuten und der ältere Hengst die Aufgabe, die Familie zu führen (Stuten), rechtzeitig zu warnen - und notfalls zu schützen (Hengst).

In der modernen Stallhaltung sind echte Familienverbände nahezu unmöglich zu reproduzieren. Stattdessen übernimmt der Mensch - im besten Falle also Du! - die Rolle des Beschützers und der Führungspersönlichkeit.

Sicherer auftreten in drei Schritten

Nun ist nicht jede Reiterin oder jeder Reiter ein echtes Führungspersönlichkeitsnaturtalent. Manchmal braucht das erst etwas Training. Und Fachwissen, was gute Führung eigentlich umfasst.

1. **Gute Führungspersönlichkeiten sind in der Lage, die Stärken ihres Gegenübers einzuschätzen - und entsprechend zu fordern und zu fördern.**

Dafür musst du aber zunächst dein Gehirn austricksen, das sich gerne nur die negativen Aspekte abspeichert und die besseren Momente einfach im Synapsen-Nirvana verschwinden lässt. Denn wahrscheinlich kannst du spontan und ohne nachzudenken wenigstens fünf Macken deines Vierbeiners benennen. Aber wie sieht es denn mit den guten Eigenschaften aus? Sobald du übrigens mit vielen Tieren arbeitest und entsprechend sehr viele unterschiedliche Pferdenaturen erlebst, bessert sich das automatisch etwas. Dann fallen auch positive Dinge schneller auf und bleiben länger im Gedächtnis. Bis dahin oder wenn euer dynamisches Duo aus Pferd und Mensch im Stall ganz auf die Zweisamkeit fokussiert bleibt, ist es an dir dich intensiv mit den Stärken und Schwächen deines Tiers zu beschäftigen und sie bei Bedarf zu fördern oder ausgleichend einzugreifen.

Dein Pferd lässt sich leicht ablenken? Dann ist es dein Job, unwillkommene Ablenkungen entweder vorausschauend zu umgehen oder dafür zu sorgen, dass dein Pferd deinen Anweisungen

Gina mit Davidoff.

und Hilfestellungen durch einen etablierten wertschätzenden, aber grundsätzlich konsequenten Umgang folgt.

Für den Anfang lässt sich ein leicht ablenkbares Pferd am besten in einer möglichst reizarmen Umgebung arbeiten. Dann dürfen nach und nach kleinere Reize dazukommen. Stellst du deinem Tier zusätzlich noch kleinere Aufgaben, die es trotz ablenkender Reize von außen meistern kann (Stichwort fordern!), dann wächst es an und mit seinen Aufgaben und lernt gleichzeitig, dass es Reizen gelassener entgegentreten kann (Stichwort fördern!).

Ein wertschätzender und konsequenter Umgang besteht vor allem aus deiner klaren Haltung. Das klassische Wer-bewegt-wen-Spiel lässt du besser nicht dein Pferd gewinnen. Stattdessen zeigt sich hier deine Qualität als Führungsperson: Du bleibst ganz und gar standhaft, egal was der Vierbeiner so abzieht.

Wenn dein Vierbeiner zu den eher phlegmatischen und introvertierten Pferden zählt, dann förderst und lobst du es bei jeder sich bietenden Gelegenheit, um es zu stärken und aufzubauen. Für hochsoziale Lebewesen wie Pferde ist das quasi die Goldwährung schlechthin. Also loben, loben, loben! Ja, das gilt für alle, also auch für die Raketenpferdchen. Den Introvertierten fallen aber sowohl die Piaffe als auch der Spanische Schritt relativ schwer - die sind immerhin das Imponiergehabe schlechthin.

Auch Stuten tun sich damit schwerer als Hengste, da spielt einfach die Evolution und das soziale Gefüge einer Herde mit hinein. Wer nicht ganz oben in der Hierarchie steht, rennt nicht imponierend herum, zumindest gilt das für Wildpferdeherden.

Heutzutage dagegen profitieren introvertierte und phlegmatische Pferde durch altklassische Lektionen wie Spanischer Schritt oder Piaffe, weil dieses konzentrierte Bodybuilding für eine erhöhte Myokinausschüttung sorgt. Die machen glücklich und halten gesund. Dazu stärken diese Übungen das Selbstvertrauen, was gerade eher in sich gekehrte Tiere positiv fördert. Sie trauen sich selbst mehr zu und genießen logischerweise auch das Lob, sobald sie eine Übung gemeistert haben.

Fazit aus diesem Abschnitt: Sobald ein Pferd etwas *gut* gemacht hat, lobst du. Sobald ein Pferd deine Grenzen austestet, bleibst du klar in der Führungsrolle.

2. Führungspersönlichkeiten erkennen Probleme und handeln entsprechend - und zwar rechtzeitig.

Das bedeutet für dich, dass du aufmerksam bist und das Verhalten deines Pferdes wahrnimmst, bewertest und es nicht aus Harmoniesucht ignorierst, nur weil das eine oder andere Verhalten sonst ein klares Eingreifen erfordern würde.

Über körpersprachliche Signale vermittelst du deinem Tier zusätzlich die Sicherheit, dass es dir sein Leben anvertrauen kann.

Atemübung für Gestresste, *bevor* das Stallgelände betreten wird: Vier bis fünf Sekunden lang ein- und acht bis zehn Sekunden lang ausatmen. Dazwischen kurz den Atem halten und aufmerksam aufs Innere horchen, wie sich Herzschlag und Atmung verändern. Mindestens fünf- bis zehnmal wiederholen.

Diese Methode senkt nicht nur aktiv den Herzschlag, es beruhigt auch die Atmung und das vegetative Nervensystem läutet eine Entspannungsphase ein. Auf diese Weise wirkst du nicht nur entspannter und in dir selbst ruhend, du wirst es auch. Und das vermittelt dem Fluchttier Pferd die nötige Ruhe und Zuversicht, dass du schon auf euch beide aufpasst.

Auch hier wieder der Hinweis: Wenn das wer-bewegt-wen-Spiel vom Pferd initiiert wird und das Tier dasjenige ist, das sich Raum nimmt und dich bewegt, dann vermittelt das keine Sicherheit und lieferst auch kein ausreichendes Signal für eine klare Führungspersönlichkeit. Gib dem Tier also erst gar keine Chance dazu, sich diesen Raum nehmen zu wollen!

3. Führungspersönlichkeiten sind beständig - und durchsetzungsfähig.

Ja, vielleicht versucht dein 600 Kilogramm schweres Haustierchen, dir den Weg zu versperren oder dich abzudrängen. Als Respektsperson gegenüber deinem Pferd lässt du das natürlich nicht zu. Sondern du machst klar, dass du ganz gelassen bestimmst, was passiert. Dafür braucht es eine entsprechende Körper-, vor allem aber die passende Geisteshaltung.

Kannst du dir vorstellen, dass sich ein François Robichon de la Guérinière von seinem Pferd durch die Halle schubsen ließ? Wohl kaum.

Führungspersönlichkeiten weichen nicht aus. Sie treten nicht zurück, sondern nehmen sich den

Raum, den sie beanspruchen wollen. Klappt das mit dem Pferd nicht so richtig? Dann ist das eine Chance, selbst zu reflektieren. Denn was Pferde wirklich gut können, ist das hier: Rückspiegeln, wie ihr Pferdemensch wirklich tickt.

Unsichere, nervöse, aufgeregte, abgelenkte und anderweitig nicht ganz konzentrierte Menschen stehen oft vor einem Tier, dass unsicher, nervös, abgelenkt, leicht reizbar oder unkonzentriert ist. Pferde nehmen Herzschlag und Atemfrequenz ihrer Menschen wahr. Ein gestresster Mensch ist also zugleich der beste Weg, um das Pferd aus der Ruhe zu bringen. Ein abgelenkter Mensch öffnet übrigens Tür und Tor dafür, dass das Pferd nicht nur testet, wer hier wem ausweicht - sondern das ganze Spiel haushoch gewinnt.

Für Menschen und hier insbesondere für Frauen ist Ausweichen im sozialen Gefüge eine Selbstverständlichkeit - und das gilt für Büroflure genauso wie im Bus, auf dem Gehsteig oder im Supermarkt. Sich höflich Platz lassen und Platz machen ist normal und wer gegen einen anderen rempelt, entschuldigt sich (hoffentlich!) und achtet künftig ein bisschen besser auf seine Umgebung.

Pferde ticken aber anders. Und dieses Andersticken verändert damit auch, wie sich Menschen in der Gegenwart von Pferden verhalten müssen, um als Führungspersönlichkeit wahrgenommen zu werden.

Sammeln - durchatmen - Stalltüre auf

In diesem Sinne ist Konzentration entscheidend. Fällt dir das schwer? Dann sammle dich kurz, bevor du den Stall oder die Koppel betrittst:

1. Wo sind die Gedanken?

Beim Pferd - oder bei Problem A bis Z im Job und der Familie? Nimm dir einen Moment, um dich auf die vor dir liegende Aufgabe einzustimmen. Gleich geht es nur ums Pferd, um eine oder zwei Lektionen in der Bahn oder eine Runde Füttern, Fellpflege etc.

Vor allem in den Lektionen ist deine volle Konzentration gefordert, denn für herausragende Reitkunstergebnisse ist es logischerweise nötig, dass die Hilfengebung und damit die Kommunikation mit dem Reitpferd zeitnah und klar erfolgt.

Das etwas abgedroschene Stichwort Achtsamkeit ist an dieser Stelle sowohl passend als auch nötig. Wer nicht achtsam in genau *diesem* Moment an und mit dem Pferd arbeitet, wer nicht einmal mitbekommt, wann und wie das Pferd drängelt und den Menschen zum Weichen bringt, der wird nur mittelmäßige Ergebnisse bei der Erarbeitung von altklassischen Reitkunstlektionen erhalten.

2. Wie steht es um deine körperlichen Stresssignale?

Pferde sind in der Lage, menschliche Herzschläge selbst aus größerer Entfernung wahrzunehmen. Das Fluchttier ist auf seine feinen Antennen angewiesen, um eventuelle Bedrohungen möglichst schnell zu erfassen. Der menschliche Stresspegel mit erhöhtem Puls und angespannter Atmung könnte ein Hinweis auf eine mögliche Bedrohung sein, obwohl der Mensch nur zwischen Terminen zum Stall hetzt, um mal eben eine Runde zu reiten.

Um das Zentrale Nervensystem zu beruhigen, eignet sich erneut die oben schon beschriebene Atemübung:

Drei Sekunden einatmen - kurz halten - sechs Sekunden ausatmen.

Das verlängerte Ausatmen aktiviert den Parasympathikus, das vegetative Nervensystem fährt Herzschlag und Atmung herunter. Typische Stresssignale entfallen - du wirkst nicht nur ruhiger, du wirst es auch! Diese Entspannungstechnik reguliert das Stressniveau bei Pferd und Mensch.

LITERATURVERZEICHNIS

A

Abboud, F. M. (2010). In search of autonomic balance: The good, the bad, and the ugly. ***American Journal of Physiology-Regulatory, Integrative and Comparative Physiology.*** https://doi.org/10.1152/ajpregu.00130.2010

ADMR-Suchmaschine. Stichwort Kobalt. Aufgerufen unter https://www.pferd-aktuell.de/turniersport/anti-doping-und-medikation/anti-doping-und-medikation-im-pferdesport (März 2025)

Agrana Stärke. (2009, 14. Oktober). ***Produktdatenblatt Agenaprot Kartoffeleiweiß.*** https://www.agrana.com/fileadmin/inhalte/austria/products/kartoffeleiweiss_datenblaetter.pdf

Alberghina, D., Giannetto, C., Rizzo, M., Panzera, M., & Piccione, G. (2015). Plasma serotonin in horses: Comparison between two different management conditions. ***Journal of Biological Research***, 88, 125–126. https://www.researchgate.net/publication/280547908_Plasma_serotonin_in_horses_comparison_between_two_different_management_conditions

Ali Mobasheri, C. J. Proudman. (2015). Cobalt chloride doping in racehorses: Concerns over a potentially lethal practice. ***The Veterinary Journal***, 205(3), 335–338. https://doi.org/10.1016/j.tvjl.2015.04.005

Almulla, A. F., & Maes, M. (2023). Although serotonin is not a major player in depression, its precursor is. ***Molecular Psychiatry***, 28, 3155–3156. https://doi.org/10.1038/s41380-023-02092-1

American Psychological Association. (2020, März 4). Working out boosts brain health. https://www.apa.org/topics/exercise-fitness/stress

Anderson, L. M., Martin, J. F., Barton, C. J., & Bonanno, D. R. (2022). What is the effect of changing running step rate on injury, performance and biomechanics? A systematic review and meta-analysis. ***Sports Medicine – Open***, 8(1), 112. https://doi.org/10.1186/s40798-022-00504-0

Angela Fehr, C. Lange, J. Fuchs, H. Neuhauser, & R. Schmitz. (2017). ***Journal of Health Monitoring***, 2(1), [Artikel]. https://edoc.rki.de/handle/176904/2578.2

Antonio, J., Candow, D. G., Forbes, S. C., Gualano, B., Jagim, A. R., Kreider, R. B., ... & Ziegenfuss, T. N. (2021). Common questions and misconceptions about creatine supplementation: What does the scientific evidence really show? ***Journal of the International Society of Sports Nutrition***, 18(1). https://doi.org/10.1186/s12970-021-00412-w

Appelt, K. (2005). Effekte der Gelatinesupplementierung auf den Knochen- und Knorpelstoffwechsel im Verlauf eines standardisierten Trainings beim Pferd. [Dissertation]. https://elib.tiho-hannover.de/receive/etd_mods_00002246

Aurich, C. (2011). Reproductive cycles of horses. ***Animal Reproduction Science***, 124(3–4), 220–228. https://doi.org/10.1016/j.anireprosci.2011.02.005

Augustin, N., Bendau, A., Heuer, S., Kaminski, J., & Ströhle, A. (2023). Resistance training in depression. ***Deutsches Ärzteblatt International***, 120, 757–762. https://doi.org/10.3238/arztebl.m2023.0196

B

Bacon, K., LaValley, M., Jafarzadeh, S., & Felson, D. (2024). Does cartilage loss cause pain in osteoarthritis and if so, how much? Annals of the Rheumatic Diseases, 79(8), 1105–1110. https://ard.eular.org/article/S0003-4967(24)01295-0/abstract

Baumann, J. (2024, 26. Juni). Wenn das Iliosakralgelenk des Pferds blockiert. ***Cavallo Magazin***. https://www.cavallo.de/medizin/wenn-das-iliosakralgelenk-des-pferds-blockiert/

Bayerische Landesanstalt für Landwirtschaft (LfL). (2024, Januar). Gruber Tabelle zur Pferdefütterung (9. Auflage). https://www.lfl.bayern.de/mam/cms07/publikationen/daten/informationen/gruber-tabelle-pferdefuetterung_lfl-information.pdf

Bayerische Landsanstalt für Landwirtschaft (LfL). (2010, Mai). Versuchsbericht S12/1. Verdauungsversuche mit Eiweißfutter – Sojaextraktionsschrot 43/48GVO/48NonGVO. https://www.lfl.bayern.de/mam/cms07/ite/dateien/26325_sojaschrot.pdf

Beule, J. (2023, August). Aminosäuren in der Pferdefütterung. https://www.agrobs.de/de/wissen-beratung/themenwelt/aminosaeuren-in-der-pferdefuetterung-919/

Bornes, L., van Winden, L. J., Geurts, V. C. M., et al. (2025). The oestrous cycle stage affects mammary tumour sensitivity to chemotherapy. ***Nature***, 637, 195–204. https://doi.org/10.1038/s41586-024-08276-1

Bower, M., McGivney, B., Campana, M., et al. (2012). The genetic origin and history of speed in the Thoroughbred racehorse. ***Nature Communications***, 3, 643. https://doi.org/10.1038/ncomms1644

Bowen, W., Frazer, A. K., Tallent, J., Pearce, A. J., & Kidgell, D. J. (2022). Unilateral strength training imparts a cross-education effect in unilateral knee osteoarthritis patients. ***Journal of Functional Morphology and Kinesiology***, 7(4), 77. https://doi.org/10.3390/jfmk7040077

Brannon, S. M., Carr, S., Jin, E. S., et al. (2019). Exogenous testosterone increases sensitivity to moral norms in moral dilemma judgements. ***Nature Human Behaviour***, 3, 856–866. https://doi.org/10.1038/s41562-019-0641-3

Brantberg, I., Grooten, W. J. A., & Essner, A. (2024). The effect of therapeutic exercise on body weight distribution, balance, and stifle function in dogs following stifle injury. ***Animals***, 14(1), 92. https://doi.org/10.3390/ani14010092

Bryan, K., McGivney, B. A., Farries, G., McGettigan, P. A., McGivney, C. L., Gough, K. F., MacHugh, D. E., Katz, L. M., & Hill, E. W. (2017). Equine skeletal muscle adaptations to exercise and training: Evidence of differential regulation of autophagosomal and mitochondrial components. ***BMC Genomics***, 18(1), 595. https://doi.org/10.1186/s12864-017-4007-9

Bundesselbsthilfeverband für Osteoporose e.V. (BfO). (2024). Osteoporose und Wechseljahre. Düsseldorf. https://www.osteoporose-deutschland.de/wp-content/uploads/2024/WechseljahreOsteo2024.pdf

Burd, N. A., West, D. W., Staples, A. W., Atherton, P. J., Baker, J. M., Moore, D. R., Holwerda, A. M., Parise, G., Rennie, M. J., Baker, S. K., & Phillips, S. M. (2010). Low-load high volume resistance exercise stimulates muscle protein synthesis more than high-load low volume resistance exercise in young men. ***PLOS ONE***, 5(8), e12033. https://doi.org/10.1371/journal.pone.0012033

Burns, T. E., & Clayton, H. M. (1997). Comparison of the temporal kinematics of the canter pirouette and collected canter. ***Equine Veterinary Journal Supplement***, (23), 58–61. https://doi.org/10.1111/j.2042-3306.1997.tb05055.x

E

Eriksen, C. S., Garde, E., Reislev, N. L., Wimmelmann, C. L., Bieler, T., Ziegler, A. K., Gylling, A. T., Dideriksen, K. J., Siebner, H. R., Mortensen, E. L., & Kjaer, M. (2016). Physical activity as intervention for age-related loss of muscle mass and function: protocol for a randomised controlled trial (the LISA study). ***BMJ Open***, 6(12), e012951. https://pmc.ncbi.nlm.nih.gov/articles/PMC5168596/

Eriksson Crommert, M., Lacourpaille, L., Heales, L. J., Tucker, K., & Hug, F. (2015). Massage induces an immediate, albeit short-term, reduction in muscle stiffness. ***Scandinavian Journal of Medicine & Science in Sports***, 25(5), e490–e496. https://pubmed.ncbi.nlm.nih.gov/25487283/

F

Fagundes, A., Almeida, F., Godoi, F., Migon, E., Santos, T., & Laranjeira, P. (2011). Creatine and maltodextrine dietetic supplementation in eventing horses at training.

Fahnenbruck, N. (2025). »... reitet für Deutschland«: ***Pferdesport und Politik im Nationalsozialismus*** (2. Auflage, S. 35). Rastede Edition Einwurf.

Firth, E., & Rogers, C. (2005). Musculoskeletal responses of 2-year-old Thoroughbred horses to early training. 7. Bone and articular cartilage response in the carpus. ***New Zealand Veterinary Journal***, 53(2), 113–122. https://doi.org/10.1080/00480169.2005.36488

Firth, E., & Rogers, C. (2005). Musculoskeletal responses of 2-year-old Thoroughbred horses to early training. 7. Bone and articular cartilage response in the carpus. ***New Zealand Veterinary Journal***, 53(2), 113–122. https://doi.org/10.1080/00480169.2005.36488

Foster Vander Elst, O., Foster, N., Vuust, P., Keller, P., & Kringelbach, M. (2023). The Neuroscience of Dance: A Conceptual Framework and Systematic Review. ***Neuroscience & Biobehavioral Reviews***, 150, 105197. https://doi.org/10.1016/j.neubiorev.2023.105197

Froböse, I. (2023). ***Muskeln. Die Gesundmacher. So bleiben wir fit, schlank und mental in Balance*** (S. 139). Ullstein Buchverlag Berlin.

Fudala, A. (2023, Februar 27). The good life: A discussion with Dr. Robert Waldinger. Harvard T. H. Chan School of Public Health. https://hsph.harvard.edu/health-happiness/news/the-good-life-a-discussion-with-dr-robert-waldinger/

G

Gan, D., Wu, X., Zhang, Y., et al. (2025). Dietary influence on the link between depression and muscle mass and muscle strength: exploring interaction and mediation effects. ***BMC Public Health***, 25, 764. https://doi.org/10.1186/s12889-025-21975-2

Gehlen, H., Deegen, B., & Ohnesorge, B. (2005). Photodermatis bei Pferden. ***Pferde Spiegel***, 8(2), 68–70. https://www.thieme-connect.com/products/ejournals/abstract/10.1055/s-0029-1237586

Gehlen, H., May, A., & Venner, M. (2010). Lebererkrankungen beim Pferd. ***Pferdeheilkunde***, 26(5). https://pferdeheilkunde.de/files/20100501.pdf

Geor, R., Harris, P., & Cohen, M. (2013). Equine applied and clinical nutrition: Health, welfare and performance (S. 40f). ***Elsevier Health Science***.

Ghalibaf, M. H. E., Kianian, F., Beigoli, S., Behrouz, S., Marefati, N., Boskabady, M., & Boskabady, M. H. (2023). The effects of vitamin C on respiratory, allergic and immunological diseases: an experimental and clinical-based review. ***Inflammopharmacology***, 31(2), 653–672. https://doi.org/10.1007/s10787-023-01169-1

Gomez, S. L., Yao, S., Kushi, L. H., & Kurian, A. W. (2019). Is Breast Cancer in Asian and Asian American Women a Different Disease? ***Journal of the National Cancer Institute***, 111(12), 1243–1244. https://doi.org/10.1093/jnci/djz091

Guseman, E., Zack, E., Battaglini, C., Viru, M., Viru, A., & Hackney, A. (2008). Exercise and circulating cortisol levels: The intensity threshold effect. ***Journal of Endocrinological Investigation***, 31, 587–591. https://doi.org/10.1007/BF03345606

Gylling, A., Bloch-Ibenfeldt, M., Eriksen, Ch., Ziegler, A., Wimmelmann, Ch., Baekgaard, M., Boraxbekk, C.-J., Siebner, H., Mortensen, E., Kjaer, M. (2020). Maintenance of muscle strength following a one-year resistance training program in older adults. ***Experimental Gerontology***, 139, 111049. https://doi.org/10.1016/j.exger.2020.111049

H

Herbst, A., Johnson, M., Gammons,G., Reedy, S., Urschel, K., Harris, P., Adams, A. (2022). Development and evaluation of a muscle atrophy scoring system (MASS) for horses. ***Journal of Equine Veterinary Science***, 110, 103771. https://doi.org/10.1016/j.jevs.2021.103771

L

La Guériniès, S. F. R. de la. (1733). École de cavalerie. Paris.

Lacourt, T. E., Vichaya, E. G., Chiu, G. S., Dantzer, R., & Heijnen, C. J. (2018). The high costs of low-grade inflammation: Persistent fatigue as a consequence of reduced cellular-energy availability and non-adaptive energy expenditure. ***Frontiers in Behavioral Neuroscience***, 12, Article 78. https://doi.org/10.3389/fnbeh.2018.00078

Larsson, M., Edqvist, L. E., Ekman, L., & Persson, S. (1979). Plasma cortisol in the horse, diurnal rhythm and effects of exogenous ACTH. ***Acta Veterinaria Scandinavica***, 20(1), 16–24. https://doi.org/10.1186/BF03546625

Lazarczyk, M., Michlik-Polczynska, K., & Skarzynski, D. (2020). Insertional desmopathies of the cranial attachment of the nuchal ligament in horses – A review. ***Veterinární medicína***, 65(8), 327–335. https://doi.org/10.17221/12/2020-VETMED

Lee, J. O., Kim, Y., Lee, J. M., Suk, J. M., Jung, I., Choi, S. Y., Yoo, K. H., Seok, J., & Kim, B. J. (2024). AP collagen peptides (APCPs) promote hair growth by activating the GSK-3β/β-catenin pathway and improve hair condition. ***Experimental Dermatology***, 33(7), e15137. https://doi.org/10.1111/exd.15137

Leuchtmann, A. B., Mueller, S. M., Aguayo, D., et al. (2020). Resistance training preserves high-intensity interval training induced improvements in skeletal muscle capillarization of healthy old men: A randomized controlled trial. ***Scientific Reports***, 10, 6578. https://doi.org/10.1038/s41598-020-63490-x

Lewis, K., McBride, S. D., Micheletta, J., Parker, M. O., Rincon, A. V., Wathan, J., & Proops, L. (2025). An ethogram of facial behaviour in domestic horses: Evolutionary perspectives on form and function. ***PeerJ***, 13, e19309. https://doi.org/10.7717/peerj.19309

Lichtwark, G. A., & Wilson, A. M. (2005). Effects of series elasticity and activation conditions on muscle power output and efficiency. ***The Journal of Experimental Biology***, 208(15), 2845–2853. https://doi.org/10.1242/jeb.01710

Lieberman, M. D., Eisenberger, N. I., Crockett, M. J., Tom, S. M., Pfeifer, J. H., & Way, B. M. (2007). Putting feelings into words. Psychological Science, 18(5), 421–428. https://doi.org/10.1111/j.1467-9280.2007.01916.x

Lim J, Choi A, & Kim B. (2024). The effects of resistance training on pain, strength, and function in osteoarthritis: Systematic review and meta-analysis. ***Journal of Personalized Medicine***, 14(12), 1130. https://doi.org/10.3390/jpm14121130

Liu, Y., Liu, Y., Bai, D., Dugarjaviin, M., & Zhang, X. (2025). Transcriptomic profiling reveals lysine-mediated proliferative mechanisms in Mongolian horse myogenic satellite cells. ***Animals***, 15(12), 1711. https://doi.org/10.3390/ani15121711

Loch, S. (2010). Dressur. Die Kunst der klassischen Reitweise (2. Aufl., S. 207, 213, 215). ***Kosmos Verlag***.

Longenecker, H., Musulin, R., Tully, R., & King, C. (1939). An acceleration of vitamin C synthesis and excretion by feeding known organic compounds to rats. ***The Journal of Biological Chemistry***, 129(2), [Pittsburgh]. https://digirepo.nlm.nih.gov/ext/mm/9803036X125/PDF/9803036X125.pdf

Lowe, D. A., Baltgalvis, K. A., & Greising, S. M. (2010). Mechanisms behind estrogen's beneficial effect on muscle strength in females. ***Exercise and Sport Sciences Reviews***, 38(2), 61–67. https://doi.org/10.1097/JES.0b013e3181d496bc

Łapińska, L., Szum-Jakubowska, A., Krentowska, A., Pawlak, M., Hładuński, M., Waszkiewicz, N., Karczewska-Kupczewska, M., Kamiński, K., & Kowalska, I. (2024). The relationship between brain structure volumes, depressive symptoms and body composition in obese/overweight and normal-/underweight women. ***Scientific Reports***, 14(1), 21021. https://doi.org/10.1038/s41598-024-71924-z

M

Martinez, R. E., Leatherwood, J. L., Bradbery, A. N., Silvers, B. L., Fridley, J., Arnold, C. E., ... & Wu, G. (2022). Equine enterocytes actively oxidize l-glutamine but do not synthesize l-citrulline or l-arginine from l-glutamine or l-proline in vitro. ***Journal of Animal Science***, *100*(4), skac077. https://doi.org/10.1093/jas/skac077

P

Pereira, R. M., Rodrigues, K. C. C., et al. (2020). Strength training reduces hepatic pyruvate carboxylase and gluconeogenesis in DIO mice. ***Journal of Endocrinology***, *247*(2), 127–138. https://joe.bioscientifica.com/view/journals/joe/247/2/JOE-20-0193.pdf

R

Rabl, U., Bartova, L., Sezen, P., Keller, J., Schatzberg, A., & Pezawas, L. (2025). HPA axis in psychotic and non-psychotic major depression: Cortisol plasma levels and hippocampal volume. ***Journal of Affective Disorders**, 377*, 14–22. https://doi.org/10.1016/j.jad.2025.02.014

Razak, M. A., Begum, P. S., Viswanath, B., & Rajagopal, S. (2017). Multifarious beneficial effect of nonessential amino acid, glycine: A review. ***Oxidative Medicine and Cellular Longevity**, 2017*, 1716701. https://doi.org/10.1155/2017/1716701

Razak, M. A., Begum, P. S., Viswanath, B., & Rajagopal, S. (2017, Erratum 2022). Multifarious beneficial effect of nonessential amino acid, glycine: A review. ***Oxidative Medicine and Cellular Longevity**, 2017*, 1716701. https://doi.org/10.1155/2017/1716701

Reiter, A. S., & Reed, S. A. (2023). Lactation in horses. ***Animal Frontiers**, 13*(3), 96–100. https://doi.org/10.1093/af/vfad003

Revista Brasileira de Zootecnia, 40, 1933–1940. https://www.researchgate.net/publication/262586179_Creatine_and_maltodextrine_dietetic_supplementation_in_eventing_horses_at_training

Riedl, M. (2025, März). Folge 42 Autoimmunerkrankungen – warum sich der Körper selbst bekämpft. In ***Dr. Matthias Riedl – So geht gesunde Ernährung [Podcast]***. Funke Verlag. https://podcasts.apple.com/de/podcast/dr-matthias-riedl-so-geht-gesunde-ern%C3%A4hrung/id1645482785?i=1000652366745

Ritsch, M. (2020). Verletzungen und Überlastungsschäden im Kraftsport. ***Sports Orthopaedics and Traumatology**, 36*(3), 260–270. https://doi.org/10.1016/j.orthtr.2020.07.004

Ross, M. W., & Dyson, S. J. (2010). ***Diagnosis and Management of Lameness in the Horse*** (2nd ed.). Elsevier.

Rossouw, J. E., Anderson, G. L., Prentice, R. L., LaCroix, A. Z., Kooperberg, C., Stefanick, M. L., ... & Writing Group for the Women's Health Initiative Investigators. (2002). Risks and benefits of estrogen plus progestin in healthy postmenopausal women: Principal results from the Women's Health Initiative randomized controlled trial. ***JAMA**, 288*(3), 321–333. https://doi.org/10.1001/jama.288.3.321

S

Salimpoor, V., Benovoy, M., Larcher, K., ... et al. (2011). Anatomically distinct dopamine release during anticipation and experience of peak emotion to music. ***Nature Neuroscience**, 14*, 257–262. https://doi.org/10.1038/nn.2726

Sarvazyan, A., Rudenko, O., Aglyamov, S., & Emelianov, S. (2014). Muscle as a molecular machine for protecting joints and bones by absorbing mechanical impacts. ***Medical Hypotheses**, 83*(1), 6–10. https://doi.org/10.1016/j.mehy.2014.04.020

Schindler, S., Schmidt, L., Stroske, M., ... et al. (2019). Hypothalamus enlargement in mood disorders. ***Acta Psychiatrica Scandinavica**, 139*(1), 56–65. https://doi.org/10.1111/acps.12958

Schmidt, R., & Simonović, M. (2012). Synthesis and decoding of selenocysteine and human health. ***Croatian Medical Journal**, 53*, 535-550. https://pdfs.semanticscholar.org/844d/626faff97bac2b-4c32b604fd5b30c735edc4.pdf

Schwarz, M., & Kähler, S. (2025). #23 Lipödem im Reitsport – für mehr Taktgefühl untereinander. ***Podcast Anreiz Pferd.*** https://www.podcast.de/episode/667555333/23-lipoedem-im-reitsport-fuer-mehr-taktgefuehl-untereinander-im-gespraech-mit-steffi-von-littletouchdown

Schweiger, E. (2022, April 7). Das erregte Pferd: Ein Grund zur Sorge? ***PferdeRevue.*** https://www.pferderevue.at/aktuelles/haltung/2022/das-erregte-pferd--ein-grund-zur-sorge-.html

Seeger, L. (1884). ***Das System der Reitkunst.***

Seifert-Klauss, V., & Prior, J. C. (2010). Progesterone and bone: Actions promoting bone health in women. ***Journal of Osteoporosis**, 2010*, 845180. https://doi.org/10.4061/2010/845180

Serota, A., & D'Erasmo, G. (2024). Estrogen exposure and skeletal health: Special populations and considerations. ***Journal of the Pediatric Orthopaedic Society of North America**, 7*, 100061. https://doi.org/10.1016/j.jposna.2024.100061

Sikorska, U., Maśko, M., Ciesielska, A., Zdrojkowski, Ł., & Domino, M. (2023). Role of cortisol in horse's welfare and health. ***Agriculture***, *13*(12), 2219. https://doi.org/10.3390/agriculture13122219

Soeken, K. L., Lee, W. L., Bausell, R. B., Agelli, M., & Berman, B. M. (2002). Safety and efficacy of S-adenosylmethionine (SAMe) for osteoarthritis. ***Journal of Family Practice***, *51*(5), 425–430. https://pubmed.ncbi.nlm.nih.gov/12019049/

Souza, G. B. (1922). ***Elementary Equitation***. E. P. Dutton & Co.

Srikanthan, P., & Karlamangla, A. S. (2014). Muscle mass index as a predictor of longevity in older adults. ***American Journal of Medicine***, *127*(6), 547–553. https://doi.org/10.1016/j.amjmed.2014.02.007

Stachurska, A., Kędzierski, W., Kaczmarek, B., Wiśniewska, A., Żylińska, B., & Janczarek, I. (2023). Variation of physiological and behavioural parameters during the oestrous cycle in mares. ***Animals***, *13*(2), 211. https://doi.org/10.3390/ani13020211

Stanford, J. L., Herrinton, L. J., Schwartz, S. M., & Weiss, N. S. (1995). Breast cancer incidence in Asian migrants to the United States and their descendants. ***Epidemiology***, *6*(2), 181–183. https://doi.org/10.1097/00001648-199503000-00017

Staniszewski, M., Tkaczyk, J., & Kęska, A., et al. (2024). Effect of rest duration between sets on fatigue and recovery after short intense plyometric exercise. ***Scientific Reports***, *14*, 15080. https://doi.org/10.1038/s41598-024-66146-2

Stefaniuk-Szmukier, M., Piórkowska, K., & Ropka-Molik, K. (2023). Equine metabolic syndrome: A complex disease influenced by multifactorial genetic factors. ***Genes***, *14*(8), 1544. https://doi.org/10.3390/genes14081544

Steffl, M., Bohannon, R. W., Sontakova, L., Tufano, J. J., Shiells, K., & Holmerova, I. (2017). Relationship between sarcopenia and physical activity in older people: A systematic review and meta-analysis. ***Clinical Interventions in Aging***, *12*, 835–845. https://doi.org/10.2147/CIA.S132940

Stemper, T. (2017). Time under tension: Die Bedeutung der Dauer der Muskelspannung für das Muskelaufbau-Training. ***Fitness & Gesundheit***, *2*, 80–82. https://www.fitness-gesundheit.uni-wuppertal.de/fileadmin/fitness-gesundheit/pdf-Dokumente/Publikationen/2017/FG_2-17-Stemper-neu.pdf

Suda, A. J., Kientopf, D., Leithner, A., Streeck, J., Colshorn, T., Dorotka, R., Schneider, M., & Höppchen, I. (2022). Keine Evidenz für die biomechanischen und pathophysiologischen Erklärungsmodelle muskuloskelettaler Erkrankungen nach Liebscher & Bracht. ***Zeitschrift für Orthopädie und Unfallchirurgie***, *160*(1), 13–18. https://doi.org/10.1055/a-1716-2274

Salimpoor, V., Benovoy, M., Larcher, K. et al. (2011). Anatomically distinct dopamine release during anticipation and experience of peak emotion to music. ***Nature Neuroscience***, *14*, 257–262. https://doi.org/10.1038/nn.2726

T

Tanaka, R., Ozawa, J., Kito, N., Yamasaki, T., & Moriyama, H. (2013). Evidence of improvement in various impairments by exercise interventions in patients with knee osteoarthritis: A systematic review and meta-analysis of randomized clinical trials. ***Journal of the Japanese Physical Therapy Association***, *16*(1), 7–21. https://doi.org/10.1298/jjpta.Vol16_003

Tchantchou, F., Graves, M., Falcone, D., & Shea, T. B. (2008). S-Adenosylmethionine mediates glutathione efficacy by increasing glutathione S-transferase activity: Implications for S-adenosylmethionine as a neuroprotective dietary supplement. ***Journal of Alzheimer's Disease***, *14*(3), 323–328. https://doi.org/10.3233/JAD-2008-14306

Thau, L., Gandhi, J., & Sharma, S. (2023). Physiology, cortisol. In ***StatPearls [Internet]***. Treasure Island (FL): StatPearls Publishing. https://www.ncbi.nlm.nih.gov/books/NBK538239/

Trajano, G. S., Nosaka, K., & Blazevich, A. J. (2017). Neurophysiological mechanisms underpinning stretch-induced force loss. ***Sports Medicine***, *47*(8), 1531–1541. https://doi.org/10.1007/s40279-017-0682-6

Tenhunen, J., Salminen, M., Lundström, K., Kiviluoto, T., Savolainen, R., & Ulmanen, I. (1994). Genomic organization of the human catechol O-methyltransferase gene and its expression from two distinct promoters. ***European Journal of Biochemistry**, 223*, 1049–1059. https://doi.org/10.1111/j.1432-1033.1994.tb19083.x

Tessier, A., Wing, S. S., Rahme, E., Morais, J. A., & Chevalier, S. (2022). Association of low muscle mass with cognitive function during a 3-year follow-up among adults aged 65 to 86 years in the Canadian longitudinal study on aging. ***JAMA Network Open**, 5*(7), e2219926. https://doi.org/10.1001/jamanetworkopen.2022.19926

Tumilaar, S., Geiner, H., Ari, D., & Kurnia, D. (2024). A comprehensive review of free radicals, oxidative stress, and antioxidants: Overview, clinical applications, global perspectives, future directions, and mechanisms of antioxidant activity of flavonoid compounds. ***Journal of Chemistry**, 2024*, Article 5594386. https://doi.org/10.1155/2024/5594386

V

Verwaltungs-Berufsgenossenschaft (VBG). (2024). *Sportreport* (S. 10 f.). Hamburg.

Vind, A. C., et al. (2024). The ribotoxic stress response drives acute inflammation, cell death, and epidermal thickening in UV-irradiated skin in vivo. ***Molecular Cell***. https://doi.org/10.1016/j.molcel.2024.10.044

W

Wallert, M., Börmel, L., & Lorkowski, S. (2020). Inflammatory diseases and vitamin E—What do we know and where do we go? ***Molecular Nutrition & Food Research**, 65*(1). https://doi.org/10.1002/mnfr.202000097

Wallimann, T., Tokarska-Schlattner, M., & Schlattner, U. (2011). The creatine kinase system and pleiotropic effects of creatine. ***Amino Acids**, 40*(5), 1271–1296. https://doi.org/10.1007/s00726-011-0877-3

Wenfei, Y., Xiaofeng, Z., & Qi, G. (2021). The effect of exposure to the natural environment on stress reduction: A meta-analysis. ***Urban Forestry & Urban Greening**, 57*, 126932. https://doi.org/10.1016/j.ufug.2020.126932

Wilke, J., Macchi, V., de Caro, R., & Strecco, C. (2019). Fascia thickness, aging and flexibility: Is there an association? ***Journal of Anatomy**, 234*, 43–49. https://doi.org/10.1111/joa.12916

Wipplinger, J. (2019, März 7). Detox: Der Mythos vom Entgiften. https://medizin-transparent.at/detox-der-mythos-vom-entgiften/

Westcott, W. L. (2012). Resistance training is medicine: Effects of strength training on health. ***Current Sports Medicine Reports**, 11*(4), 209–216. https://doi.org/10.1249/JSR.0b013e31825dabb8

Woodward, A. D., Nielsen, B. D., O'Connor, C. I., Skelly, C. D., Webel, S. K., & Orth, M. W. (2007). Supplementation of dietary long-chain polyunsaturated omega-3 fatty acids high in docosahexaenoic acid (DHA) increases plasma DHA concentration and may increase trot stride lengths in horses. ***Equine and Comparative Exercise Physiology**, 4*(2), 71–78. https://doi.org/10.1017/S1478061507811443

Z

Zehnder, C. (2009). Feldstudie zu Risikofaktoren für den Absatz von freiem Kotwasser beim Freizeitpferd [Unveröffentlichte Dissertation]. Ludwig-Maximilians-Universität München. https://edoc.ub.uni-muenchen.de/10321/1/Zehnder_Carolin.pdf

Zsoldos, R. R., & Licka, T. F. (2015). The equine neck and its function during movement and locomotion. ***Zoology (Jena, Germany)**, 118*(5), 364–376. https://doi.org/10.1016/j.zool.2015.03.005

Zunner, B. E. M., Wachsmuth, N. B., Eckstein, M. L., Scherl, L., Schierbauer, J. R., Haupt, S., Stumpf, C., Reusch, L., & Moser, O. (2022). Myokines and resistance training: A narrative review. ***International Journal of Molecular Sciences**, 23*(7), 3501. https://doi.org/10.3390/ijms23073501

ABBILDUNGSVERZEICHNIS

Radoyka Despodovska: S. 6, 9, 11, 12, 18, 20, 43, 54, 65, 73, 77, 80, 81, 84, 85, 86 o., 87 u., 89, 90, 91, 105, 112, 114, 116, 117, 121, 135, 143, 155, 161, 163, 169, 175

Herbert Kiebler: S. 63, 118, 136, 148, 166, 173

Jordan Klapper: S. 99, 119, 181

Fotografin Inga Langer: S. 29, 61, 71, 74, 94, 106, 138, 157, 160, 172

Veronika Weidlich: S. 50, 88

Grafiken von Jennifer Koban, Feeldesign: S.72, 76, 96